# PILE FOUNDATIONS IN ENGINEERING PRACTICE

# PILE FOUNDATIONS IN ENGINEERING PRACTICE


**Shamsher Prakash**

*Professor of Civil Engineering,*
*University of Missouri–Rolla,*
*Rolla, Missouri*

**Hari D. Sharma**

*Chief Geotechnical Engineer*
*EMCON Associates,*
*San Jose, California*

A WILEY-INTERSCIENCE PUBLICATION

John Wiley & Sons, Inc.

NEW YORK / CHICHESTER / BRISBANE / TORONTO / SINGAPORE

***Library of Congress Cataloging in Publication Data:***

Prakash, Shamsher.
Pile foundations in engineering practice/Shamsher Prakash, Hari D. Sharma.
p. cm.
"A Wiley-Interscience publication."
Includes bibliographies.
1. Piling (Civil engineering) I. Sharma, Hari D. II. Title.

TA780.P72 1989 — 89-31977
624.1'54—dc 20 — CIP
ISBN 0-471-61653-2

Printed in the United States of America

10 9 8 7 6 5 4 3 2 1

*To our spouses, Sally and Jaya, whose support was very valuable in completing this task*

# CONTENTS

# PREFACE

Pile foundations have been used since prehistoric time to transfer building loads to appropriate depths. In an effort to develop reasonable design methods, analytical and experimental studies on piles and pile groups have been performed extensively in the past four decades. Analytical studies have been directed toward prediction of bearing capacity under vertical loads, pile deflections under lateral loads, response of piles under dynamic loads, and the behavior of piles in permafrost. Numerical methods including finite difference and finite element techniques have also been applied. Also, a large amount of model and full-scale test data have been collected.

All the foregoing information has led to the development of design procedures of piles in different soil types, loading conditions, and environments. The purpose of this text is to present a concise, systematic, and complete treatment of the subject leading to rational design procedures for the practicing civil, geotechnical, and structural engineers. The book will be of equal benefit to graduate students specializing in foundation engineering.

This book contains eleven chapters. In Chapter 1, basic concepts of pile behavior under different types of loading are developed. More importantly, the changes in the soil properties particularly in clays under static and dynamic loading and on a long-term basis have been explained. In Chapters 2 and 3, details of different types of piles and their installation methods, respectively, are discussed.

Determination and selection of appropriate soil parameters for design of piles under different loading conditions and environment are presented in Chapter 4. Adequate attention is most often not paid by the design engineers to the factors affecting the selection of design parameters. This and other questions are explained in detail in this chapter.

In the subsequent four chapters, detailed information on behavior and design of piles have been included for (1) vertical loading and pullout in Chapter 5, (2) lateral, inclined, and eccentric loads in Chapter 6, (3) dynamic loads in Chapter 7, and (4) piles in permafrost in Chapter 8. A special feature in all these four chapters is that step-by-step design procedures are developed. Numerous solved problems are also included in each chapter.

Load test procedures and their interpretation are discussed in Chapter 9. The question of buckling of long, slender piles with and without unsupported length is the subject of Chapter 10.

A match of prediction and performance of piles and pile groups is of great importance in practice. This subject is discussed with the help of several case histories in Chapter 11.

Several parts of this text have been used in short courses for practicing engineers offered by the University of Missouri-Rolla. Input from several participants of these short courses resulted in many improvements.

Thanks go to the Civil Engineering Department, University of Missouri-Rolla, for the facilities offered and to the Interlibrary Loan of the Curtis Laws Wilson Library for procuring some difficult-to-find references.

Thanks also go to the American Society of Civil Engineers for permitting the use of material from their publications. Acknowledgment to other copyrighted material is given in other appropriate places in the text, figures, and tables.

In the preparation of this text, several of our colleagues and students helped in a variety of ways. Useful comments were offered by W. D. Liam Finn, M. T. Davisson, Norbert O. Schmidt, M. R. Madhav, and Swami Saran for improving the text. Solutions to some problems were prepared by George M. Manyando and Shamshad Hussain. Charlena Ousley, Janet Pearson, Allison Holdaway, Anna Hubbard, and Ida Lucero typed the text with painstaking effort. Anna Hubbard also prepared the subject and author indexes and the notations very patiently. (A special thanks is due to John Wiley's editorial and professional staff. Thanks go to all of them.)

SHAMSHER PRAKASH
HARI D. SHARMA

*Rolla, Missouri*
*San Jose, California*
*January, 1990*

# LIST OF SYMBOLS

| | |
|---|---|
| $A$ | area of cross-section of H-pile section |
| $A_1, B_1$ | coefficient (Table 6.5) |
| $A_m, B_m$ | moment coefficients for free head pile |
| $A_{mc}, B_{mc}$ | moment coefficient when subgrade modulus is constant with depth |
| $A_{me}, B_{me}$ | bending moment coefficients for dynamic loading |
| $A_p$ | area of pile tip |
| $A_p, B_p$ | soil reaction coefficient for free head pile |
| $A_s$ | area of pile shaft |
| $A_s, B_s$ | slope coefficients for free head pile |
| $A_v, B_v$ | shear coefficients for free head pile |
| $A_{x1}$ | horizontal displacement in sliding |
| $A_y, B_y$ | deflection coefficients for free head pile |
| $A_{yc}, B_{yc}$ | deflection coefficient when subgrade modulus is constant with depth |
| $A_z$ | maximum amplitude in vertical vibrations |
| $A_{\phi 0}$ | maximum amplitude in rocking |
| $A_\psi$ | maximum amplitude of vibrations in yawing (torsional vibrations) |
| $a$ | length of foundation |
| $a_\psi$ | $A_\psi M_{mz}/(m_e e_m r_c)$ = dimensionless amplitude of torsional vibration with quadratic excitation |
| $a_0$ | $r_0(\omega/V_s) = r_0(\omega/V_b) = r_0\omega\sqrt{\rho/G}$ = dimensionless frequency factor |
| $a_x$ | dimension along $x$ axis |
| $a_y$ | dimension along $y$ axis |

| | |
|---|---|
| $a_z$ | dimension along $z$ axis |
| $B$ | creep parameter, pile width, width of loaded area |
| $B_1$ | coefficient (Table 6.5) |
| $B_b$ | pile base or bell diameter |
| $B_{me}$ | moment coefficient |
| $B_x$ | modified mass ratio in sliding |
| $B_{yc}$ | deflection coefficient of pile in clay |
| $B_z$ | modified mass ratio in vertical vibrations |
| $B_\phi$ | inertia ratio in rocking vibrations |
| $B_{mc}$ | bending moment coefficient of pile in clay |
| $B_\psi$ | inertia ratio in torsional vibrations |
| $b$ | pile cap width; width of foundation; mass ratio |
| $b, c, w, k$ | experimental parameter (Table 8.3) |
| $C$ | clay, constant to represent penetration due to energy loss; volumetric heat capacity of permafrost, $J/m^3C$ |
| $C_1, C_2$ | integration constants; frequency-dependent parameters of vertical vibrations; soil adhesion forces |
| $\bar{C}_1, \bar{C}_2$ | frequency-independent parameters for vertical vibrations |
| $C_{bs}$ | allowable bond strength between concrete and rock |
| $C_c$ | compression index |
| $C_l$ | ratio of $K, K_T, K_b$ |
| $C_m$ | moment coefficient for fixed head, spring compression of element m in time interval, t |
| $C_N$ | correction factor for $N$ to account for overburden pressure |
| $C_p$ | empirical coefficient (equation 5.36) |
| $C_s$ | empirical coefficient (equation 5.37) |
| $C_{th}$ | thaw degradation constant |
| $C_u$ | coefficient of elastic uniform compression |
| $C_y$ | deflection coefficient for fixed head |
| $C_{yc}, C_{mc}$ | nondimensional factors in cohesive soils for fixed head pile |
| $C_{\omega 1}, C_{\omega 2}$ | dimensionless parameters of half space |
| $C_v$ | coefficient of consolidation |
| $C_{x1}, C_{x2}$ | frequency-dependent parameters for horizontal translation |
| $\bar{C}_{x1}, \bar{C}_{x2}$ | frequency-independent parameters for horizontal translation |
| $C_\delta$ | coefficient of elastic resistance of pile |
| $C_{\delta n}$ | pile stiffness at resonance |
| $C_\tau$ | coefficient of elastic uniform shear |
| $C_\phi$ | coefficient of elastic nonuniform compression |
| $C_{\phi 1}, C_{\phi 2}$ | frequency-dependent functions of the elastic half space for rocking vibrations |
| $C_\psi$ | coefficient of elastic nonuniform shear |
| $c$ | coefficient of internal damping; cohesion parameter of soil; experimental parameter in equation 8.1 |

| | |
|---|---|
| $c_a$ | adhesion, soil-pile adhesion; unit adhesion |
| $c_c$ | critical damping |
| $c_{lt}$ | long-term cohesion of permafrost |
| $c_1$ | long-term shear strength for ice-rich soil |
| $c_r$ | recompression index |
| $c_u = S_u$ | undrained shear strength of clay; cohesion parameter under undrained conditions when $\phi = 0$ |
| $c_{us}$ | average undrained shear strength of clay along pile shaft |
| $c_w^1$ | constant of equivalent viscous damping of one pile in vertical vibrations |
| $c_w^f$ | constant of equivalent viscous damping of pile cap in vertical vibrations |
| $c_w^g$ | damping coefficient of pile group |
| $c_x$ | damping coefficient in horizontal sliding |
| $c_x^l$ | damping constant of single pile in horizontal translation |
| $c_x^f$ | constant of equivalent viscous damping of pile cap in translation |
| $c_x^g$ | damping constant of pile group in horizontal translations |
| $c_{x\phi}$ | cross-coupled damping factor for coupled rocking and sliding |
| $c_{x\phi}^1$ | cross coupled damping constant of a single pile |
| $c_z$ | damping coefficient in vertical vibrations |
| $c_z^g$ | equivalent damping for a pile group in vertical vibrations |
| $c_\phi$ | damping coefficient in rocking mode |
| $c_\phi^1$ | damping coefficient of single pile in rocking |
| $c_\phi^f$ | damping coefficient of pile cap in rocking |
| $c_{\phi c}$ | critical damping in rocking |
| $c_\psi$ | damping constant of piles or footing in torsion |
| $c_\psi'$ | constant of equivalent viscous damping of a single pile in torsional vibrations |
| $D$ | diameter, downward drag force |
| $D_f$ | depth of pile tip below ground |
| $D_m'$ | soil plastic displacement around element $m$ in time interval $t$ |
| $D_r$ | relative density |
| $D_w^1$ | geometric damping ratio for a single pile |
| $d$ | depth factor |
| $d_m^*$ | displacement value of element $m$ in time interval, $t-2$ |
| $d_m$ | displacement of element $m$ in time interval, $t-1$ |
| $E$ | modulus of elasticity of pile material; actual energy delivered by hammer per blow in foot-pounds; Young's modulus |
| $E_b$ | bulk modulus |
| $E_c$ | constrained modulus |

| | |
|---|---|
| $E_d$ | dilatometer modulus |
| $E_h$ | average horizontal soil modulus along pile $= k_h$ |
| $EI$ | flexural rigidity of the pile; pile material flexibility |
| $E_p$ | modulus of elasticity of pile material; Young's modulus of pile |
| $E_s = E$ | modulus of elasticity of soil |
| $e$ | base of natural logarithms, coefficient of elastic restitution, voids ratio; eccentricity |
| $e_0$ | initial void ratio |
| $e_1$ | coefficient of elastic restitution |
| $F$ | side shear force; total upward adfreeze force or frost heave force |
| $F_{\mathrm{CL}1}$ | nondimensional frequency factor for piles embedded in soils in which soil modulus remains constant with depth |
| $F_{\mathrm{SL}1}$ | nondimensional frequency factor for piles embedded in soils in which soil modulus increases linearly with depth |
| $F_m$ | force exerted by spring in time interval, $t$ |
| $Fy$ | force in horizontal ($y$) direction |
| $F(t)$ | stress wave induced force at a point along the pile at time $t$ |
| $F'$ | yield displacement factor |
| $f_\rho$ | frequency of vibration |
| $f_{c'}$ | specified compressive strength of concrete |
| $f_c, f_\phi$ | resistance factors |
| $f_c$ | unit resistance of local friction sleeve of static penetrometer |
| $f_d, f_l, f_u$ | load factors |
| $f_n$ | natural frequency |
| $f_{nx}$ | natural frequency in horizontal sliding |
| $f_{nz}$ | natural frequency in vertical vibrations |
| $f_{n\phi}$ | natural frequency in pure rocking |
| $f_{n\psi}$ | natural frequency in yawing |
| $f_p$ | performance factor |
| $f_{pc} = p_e$ | effective prestress on the section |
| $f_q$ | load modification factor |
| $f_r$ | resistance modification factor |
| $f_s$ | side friction measured in cone penetration test; ultimate unit shaft (skin) friction |
| $f_{T1,T2}$ | torsional stiffness and damping parameters, respectively of a single pile |
| $f_{w1}, f_{w2}$ | vertical stiffness and damping parameters, respectively of a single pile |
| $f_{x1}, f_{x2}$ | horizontal (sliding) stiffness and damping parameters respectively of a free head pile |
| $f^p_{x1}, f^p_{x2}$ | horizontal (sliding) stiffness and damping parameters for a pinned head pile |
| $f_{x\phi1}, f_{x\phi2}$ | cross stiffness and cross damping parameters |

| | |
|---|---|
| $f_y$ | specified yield strength of reinforcement |
| $f_{\phi 1}, f_{\phi 2}$ | rocking stiffness and damping parameters of a pile |
| $G$ | shear modulus of soil |
| $G_b$ | shear modulus of soil beneath the pile tip |
| $G_e$ | group efficiency |
| $G_{max}$ | maximum value of shear modulus |
| $G_p$ | shear modulus of pile |
| $G_s$ | shear modulus of the soil on the sides of the pile |
| $G^* = G_1 + iG_2$ | complex shear modulus of soil |
| $G_1, G_2$ | real and imaginary parts of complex shear modulus of soil |
| $G_{1000}$ | shear modulus measured after 1000 minutes of constant confining pressure (after completion of primary consolidation) |
| $g$ | acceleration due to gravity |
| $H$ | height of fall of ram or hammer |
| $h$ | depth of embedment; length of pile above ground |
| $I$ | influence factor; moment of inertia of the pile |
| $I_d$ | material index |
| $I_G$ | coefficient of shear modulus increase with time |
| $I_r$ | rigidity factor |
| $I_{\rho F}$ | empirical coefficient for fixed-head pile in cohesive soils |
| $I_{\rho F}$ and $F_{\rho F}$ | empirical coefficient for fixed-head pile in cohesionless soils |
| $I_{\rho H}, I_{\rho m}, I_{\theta H}, I_{\theta M}$ | empirical coefficients for free-head piles in clays |
| $I_p$ | moment of inertia of pile; polar moment of inertia of the area |
| $I_x$ | moment of inertia of the area about the $x$ axis |
| $I_{xx}, I_{yy}$ | moment of inertia of pile group about $xx$ and $yy$ axes, respectively |
| $I_y$ | moment of inertia of the area about the $y$ axis |
| $J$ | an empirical factor; damping constant applicable to resistance at pile joint ($R_{12}$ in Fig. 5.7) |
| $J'$ | damping constant applicable to resistance at side of pile ($R_3$ to $R_{11}$ of Fig. 5.7) |
| $J_0$ | mass polar moment of inertia |
| $J_0, J_1$ | Bessel functions of first kind of order 0 and 1, respectively |
| $J_z$ | polar moment of inertia of the base contact area |
| $j_c$ | case method damping constant |
| $K$ | constant; coefficient of horizontal earth pressure; a dimensionless constant factor in equation 7.27 |
| $K_b$ | soil modulus for bottom layer; lateral earth pressure coefficient |
| $K_c, K_q$ | factors which are functions of $\phi$ and $s/B$ |
| $K_d$ | horizontal stress index |
| $K'_m$ | soil spring constant along element $m$ |
| $K_m$ | spring constant of element $m$ |

| | |
|---|---|
| $K_0$ | coefficient of earth pressure at rest |
| $K_p$ | Rankine's passive earth pressure coefficient |
| $K_q$ | bearing capacity factor based on pressuremeter test data |
| $K_R$ | flexibility factor |
| $K_r$ | relative stiffness |
| $K_{sp}$ | an empirical factor |
| $K_s$ | average coefficient of earth pressure on pile shaft, earth pressure coefficient |
| $K_T$ | soil modulus for top layer |
| $k$ | spring constant |
| $k_{h=k}$ | modulus of horizontal subgrade reaction |
| $k_s$ | coefficient of horizontal subgrade reaction in force per unit volume |
| $k_t$ | ratio of lateral load and lateral deflection |
| $k_v$ | ratio of axial load and axial settlement |
| $k_w$ | stiffness of pile in vertical direction |
| $k_w^l$ | stiffness constant of one pile in vertical direction |
| $k_w^f$ | stiffness constant of pile cap in vertical direction |
| $k_w^g$ | stiffness constant of pile group in vertical direction |
| $k_x$ | stiffness constant for translation along $x$ axis, equivalent spring constant of the soil in horizontal $x$ direction |
| $k_x^1$ | spring constant of single pile in translation |
| $k_z^f$ | spring constant of pile cap in translation |
| $k_x^g$ | stiffness constant of pile group in translation |
| $k_{x\phi}$ | cross coupled stiffness for coupled rocking and sliding |
| $k_{x\phi}^1$ | cross spring stiffness of single pile |
| $k_z$ | spring constant in vertical vibrations, equivalent spring constant of the soil in vertical direction |
| $k_\phi$ | spring constant in rocking vibrations |
| $k_\phi^1$ | spring constant of single pile in rocking |
| $k_x^f$ | spring constant of pile cap in rocking |
| $k_\phi^g$ | spring constant of pile group in rocking |
| $k_\psi$ | spring constant in torsion |
| $k_\psi^1$ | torsional stiffness of a single pile |
| $L$ | latent heat of water; low plasticity; pile embedment length; pile length |
| $L_A$ | length of pile in the active zone |
| $L_e$ | effective pile embedment; effective pile length |
| $LL$ | liquid limit |
| $L_r$ | embedded length of pile |
| $L_s$ | pile length that is socketed into the rock |
| $L_{\text{slurry}}$ | latent heat of slurry |
| $l$ | length of pile, any distance |
| $M$ | bending moment; model; moment; moment at pile head; $M_0 \cos \omega t$ excitation moment; silt |

| | |
|---|---|
| $M_G$ | applied moment on a pile group |
| $M_g$ | moment applied at pile head at ground level |
| $M_{max}$ | maximum bending moment |
| $M_0$ | ultimate moment for a pile under pure moment without any axial load; $m_c e_m r_e \omega^2$: amplitude of moment $M$ for quadratic excitation |
| $M_u$ | ultimate pile moment, ultimate moment capacity of pile shaft |
| $M_{uv}$ | moment caused by $Q_{uv}$ applied at eccentricity $e$ |
| $M_{hu}$ | moment caused by $Q_{hu}$ applied at height $h$ above ground |
| $M_x$ | moment at depth $x$ |
| $m$ | experimental constant; a factor $= M_0/(P_u L)$ |
| $m_0$ | rotating mass |
| $m_v$ | volume compressibility |
| $N$ | observed Standard Penetration Test Value |
| $\bar{N}$ | corrected Standard Penetration Test Value |
| $N_b$ | number of blows of $WXH$ energy needed to ram a unit volume of concrete into the base for Franki piles |
| $N_c, N_q, N_\gamma$ | nondimensional bearing capacity parameters |
| $N_G$ | normalized shear modulus increase with time |
| $N_h$ | rate of increase of $E_s$ |
| $N(Z)$ | axial force in the pile |
| $n$ | creep test constant (parameter); degrees of freedom of a multidegree system; number of cycles; number of piles in the group: scale ratio (Table 7.7) |
| $n_h$ | constant of horizontal subgrade reaction |
| $O$ | organic soil |
| $OCR$ | over consolidation ratio |
| $P$ | axial downward load; horizontal shear load; prototype |
| $P_{all}$ | allowable pullout capacity of a single pile |
| $P_f$ | pressure corresponding to $V_f$ in pressuremeter test |
| $P_G$ | applied axial pullout load on a pile group |
| $(P_G)_{all}$ | allowable pullout capacity of a pile group |
| $PI$ | plasticity index |
| $P_L$ | maximum limit pressure in pressuremeter test; pressure corresponding to $V_L$ in pressuremeter test |
| $PL$ | plastic limit |
| $P_0$ | pressure corresponding to initial volume in pressuremeter test; pressure corresponding to $V_0$ in pressuremeter test; pressure in dilatometer test corresponding to reading A |
| $P_{pul}$ | axial pullout (upward) load |
| $P_u$ | ultimate axial vertical load of pile; ultimate pullout capacity |
| $P_z$ | maximum unbalanced force in vertical direction, vertical component of resultant inertia force |

| | |
|---|---|
| $P_{z(t)}$ | time-dependent vertical force |
| $P_1$ | pressure in dilatometer test corresponding to reading B |
| $p$ | pile perimeter; soil reaction at a point on the pile per unit length along the pile |
| $p_a$ | atmospheric pressure |
| $p_{cd}$ | soil resistance below critical depth $x_r$ |
| $p_{cr}$ | soil resistance from ground surface to a critical depth $x_r$ |
| $\bar{p}_c$ | preconsolidation pressure |
| $p_k, p_m, p_u$ | points on $p$–$y$ curve corresponding to $y_k$, $y_m$, and $y_u$, respectively |
| $p_u$ | ultimate soil resistance |
| $p_x$ | soil reaction at depth $x$ |
| $Q$ | allowable lateral load; latent heat of slurry per meter of pile; lateral load; quake or maximum elastic ground deformation |
| $Q_{\alpha u}$ | ultimate central inclined load capacity |
| $Q_\alpha$ | inclined load on a pile |
| $Q_{\text{all}}$ | allowable lateral load |
| $Q_c$ | cone penetration resistance |
| $Q_{\text{dyn}}$ | dynamic resistance of soil to pile driving |
| $Q_{e\alpha}$ | eccentric and inclined load on a pile |
| $Q_{e\alpha,u}$ | ultimate pile load at an inclination $\alpha$ and eccentricity $e$ with the axis of the pile |
| $Q_{eu}$ | ultimate eccentric vertical load capacity |
| $Q_e$ | eccentric vertical load on a pile |
| $(Q_e)_G$ | total eccentric vertical load on pile group |
| $Q_f$ | frictional capacity along the pile perimeter or ultimate shaft friction |
| $Q_{fa}$ | actual shaft friction load transmitted by the pile in the working stress range |
| $Q_{fp}$ | ultimate shaft friction in pullout |
| $(Q_f)_{\text{all}}$ | allowable frictional capacity of the pile |
| $(Q_f)_G$ | ultimate friction capacity of a pile group |
| $(Q_f)_{\text{neg}}$ | negative skin friction |
| $Q_g$ | lateral load applied at pile head at ground level |
| $Q_{hu}$ | ultimate pile capacity under horizontal load |
| $Q_p$ | end-bearing capacity or ultimate tip resistance |
| $Q_{pa}$ | actual base load transmitted by the pile in the working stress range |
| $(Q_p)_{\text{all}}$ | allowable load at the pile base |
| $(Q_p)_G$ | ultimate point load of a pile group |
| $Q_u$ | ultimate lateral resistance |
| $(Q_u)_G$ | ultimate lateral load capacity of a group |
| $Q_{up}$ | magnitude of uplift forces in swelling and shrinking clays |
| $Q_{va}$ | applied axial compression pile load |

| | |
|---|---|
| $Q_{va}$ | axial downward load on pile |
| $(Q_v)_{all}$ | allowable bearing capacity of pile |
| $(Q_{vG})_{all}$ | allowable capacity of a pile group |
| $(Q_v)_{ult}$ | ultimate bearing capacity of pile |
| $(Q_{vG})_{ult}$ | ultimate capacity of a pile group |
| $Q_{vu}$ | ultimate pile capacity under vertical load |
| $Q_1, Q_2$ | lateral forces inclined at angles $+\delta_1$ and $-\delta_2$ with the horizontal |
| $q_a$ | allowable contact pressure on jointed rock |
| $q_c$ | cone penetration resistance; end resistance measured in cone penetration test |
| $q_0$ | horizontal at rest stress in soil at the elevation of pile tip |
| $q_p$ | ultimate unit point or end-bearing capacity |
| $q_u = 2S_u = 2c_u$ | unconfined compressive strength |
| $(q_u)_{core}$ | unconfined compressive strength of rock core |
| $R$ | pile radius; radius of plate; relative stiffness factor when modulus is constant with depth |
| $R_A, R_B, R_C$ | Axial forces on pile groups A, B, and C, respectively; reduction factor to account for scale effects in stiff fissured clays |
| $R_m$ | soil resistance along element $m$ in time interval $t$ |
| $R_n$ | load or reaction on any pile |
| $R_p$ | soil resistance at pile point $= R_u$ |
| $RQD$ | Rock Quality Designation |
| $R_s = (Q_v)_{ult}$ | static axial ultimate capacity |
| $R_s(t_m)$ | static soil resistance at time $t_m$ |
| $R_{um}$ | portion of $R_u$ applicable to weight $W_m$ |
| $R_u$ | ultimate soil resistance to driving |
| $r$ | adhesion factor: frequency ratio $\omega/\omega_n$, $f/f_n$; radial distance from pile, center to center spacing of piles |
| $r_0$ | effective radius of one pile, equivalent radius; radius of the pile |
| $r^1$ | radius of circular pile section |
| $r^2$ | radius of drilled hole |
| $S$ | center to center distance between piles, pile spacing; distance between geophones; pile point penetration per blow or permanent set of pile per blow |
| $S_b$ | shape factor |
| $S_{bu}$ | overall shape factor |
| $S_e$ | elastic compression of various parts |
| $S_d$ | spectral displacement |
| $S_G$ | pile group settlement |
| $S_n$ | clear distance between adjacent piles |
| $S_p$ | settlement of pile base or point caused by load transmitted at the base |

| | |
|---|---|
| $S_{ps}$ | settlement of pile point caused by load transmitted along the pile shaft |
| $S_s$ | settlement due to axial deformation of a pile shaft |
| $S_t$ | pile top settlement for a single pile |
| $S_T$ | equivalent length of embedded portion of the pile |
| $S_u$ | undrained shear strength |
| $S_{w1,w2}$ | frequency dependent dimensionless parameters of vertical resistance of soil along a vertical pile |
| $S_x$ | slope at depth $x$ |
| $S_{x1}, S_{x2}$ | frequency-dependent parameters of side layer for horizontal sliding |
| $\bar{S}_{x1}, \bar{S}_{x2}$ | frequency independent values of $S_{x1}$ and $S_{x2}$ |
| $S_1, S_2$ | frequency-dependent parameters of the side layer for vertical vibration |
| $\bar{S}_1, \bar{S}_2$ | frequency-independent parameters of side layer for vertical vibration |
| $\bar{S}_{\phi 1}, \bar{S}_{\phi 2}$ | frequency-independent values of $S_{\phi 1}$ and $S_{\phi 2}$ |
| $S_{\psi 1}, S_{\psi 2}$ | frequency-dependent side layer parameters for torsional vibrations |
| $\bar{S}_{\psi 1}, \bar{S}_{\psi 2}$ | frequency-independent values of $S_{\psi 1}$, $\bar{S}_{\psi 2}$ for torsional vibrations |
| $s$ | pile spacing |
| $s_d$ | spacing of discontinuities in the rocks |
| $s_e$ | elastic settlement |
| $s(z, t)$ | time-dependent soil reaction per unit length on vertical side of the footing |
| $T$ | relative stiffness factor when modulus increases with depth; time period—torque; torque applied in the vane shear test |
| $T_m$ | minimum soil temperature in freezing zone |
| $T_n$ | natural period |
| $T_{n1}$ | natural period in first mode of vibrations |
| $t$ | freezeback time; ratio of moment and lateral load for fixed head; time after application of load |
| $t_d$ | thickness of discontinuities in the rocks |
| $t_f$ | thickness of frozen soil |
| $t_m$ | time of first relative maximum in force and velocity measurement |
| $t^*$ | time used for starting computation of total driving resistance |
| $t_1$ | time after primary consolidation |
| $U$ | creep rate; displacement amplitude of pile displacement function |
| $u$ | assumed insitu hydrostatic pressure; displacement at any radius $r$; displacement in $x$ direction |

| | |
|---|---|
| $\dot{u}$ | velocity in $x$ direction |
| $V$ | velocity |
| $V_b$ | $\sqrt{G_b/\rho_b}$ = shear wave velocity of soil beneath pile tip |
| $V_c$ | longitudinal or compression wave velocity in infinite medium; $\sqrt{E_p/\rho_p}$ = longitudinal wave velocity in pile |
| $V_f$ | final volume in pressuremeter test |
| $V_L$ | upper limit of volume in pressuremeter test |
| $V_m$ | mean volume in pressuremeter test; velocity of element $m$ in time interval $t$ |
| $V_0$ | initial volume in pressuremeter test |
| $V_p$ | shear wave velocity of pile |
| $V_R$ | velocity of Rayleigh waves |
| $V_r$ | longitudinal wave propagation velocity in rod |
| $V_s$ | shear wave velocity |
| $V_x$ | shear at depth $x$ |
| $v$ | displacement in $y$ direction |
| $v_c$ | longitudinal wave velocity in pile |
| $v_m$ | velocity of element $m$ in time interval, $t-1$ |
| $v_r$ | velocity of propagation of stress wave |
| $v(t)$ | stress wave particle velocity |
| $v_1$ | velocity of pile cap at the instant of ram impact |
| $W$ | weight of ram or hammer |
| $W_m$ | weight of element m |
| $W_p$ | weight of the pile |
| $w$ | vertical displacement, weight per unit length; water content in percent of dry weight |
| $w_n$ | natural moisture content |
| $w_0$ | amplitude of vertical vibration of footing |
| $w_w$ | displacement in $Z$ direction |
| $w_{(x,t)}$ | complex pile displacement function at depth $z$ |
| $w(z)$ | complex amplitude of pile vibration at depth $z$ |
| $w_{1,2}$ | real and imaginary parts of displacement |
| $X$ | axis of $X$; depth of permafrost degradation |
| $X_r$ | depth of point of rotation |
| $x$ | axis of $x$; depth along pile; depth below ground |
| $x_n, y_n$ | distances from center of gravity of pile group for each pile in $x$ and $y$ directions, respectively |
| $x_0$ | depth below ground where maximum bending moment occurs |
| $x_r$ | coordinate of pile; critical depth below ground level |
| $\bar{x}, \bar{y}$ | eccentricities in $xx$ and $yy$ directions |
| $Y$ | axis of $Y$ |
| $Y_0, Y_1$ | Bessel functions of the second kind of order 0 and 1, respectively |
| $y$ | deflection; displacement; horizontal distance away from the pile, lateral pile deflection |

| | |
|---|---|
| $y_k, y_m, y_u$ | points on *p-y* curve |
| $y_0$ | maximum value of $y$ |
| $y_r$ | horizontal coordinates of pile |
| $Z$ | axis of $Z$; $x/T$ |
| $Z_c$ | height of center of gravity of pile cap above its base |
| $Z_m$ | accelerating force in element $m$ in time interval $t$ |
| $Z_{max}$ | $L/T$ |
| $z$ | displacement in vertical direction |
| $\dot{z}$ | velocity in vertical direction |
| $\ddot{z}$ | acceleration in vertical direction |
| $\alpha$ | inclination of load on vertical pile; thermal diffusivity of permafrost |
| $\alpha_A$ | axil displacement interaction factor for a typical reference pile in a group |
| $\alpha_f$ | a factor relating to ultimate moment $(M_u) = (A_s d/M_u)$ and the distance $(d)$ of extreme compression end to the center of tension bar of area $A_s$ |
| $\alpha_h$ | horizontal seismic coefficient |
| $\sigma'_{h1}$ | effective horizontal pressure (stress) at a point along pile length |
| $\alpha_L$ | lateral displacement interaction factor for a typical reference pile in a group |
| $\alpha_s$ | a number that depends on skin friction distribution |
| $\beta$ | inclination of batter pile; depth coefficient $= x/L$ |
| $\gamma$ | weight density or unit weight; unit weight of soil; shear strain |
| $\gamma'$ | effective unit weight of the soil |
| $\dot{\gamma}$ | shear strain rate induced in soil around pile due to shear stress $\tau$ |
| $\gamma_c$ | unit weight of concrete |
| $\gamma_d$ | dry density |
| $\gamma_s$ | unit weight of soil |
| $\gamma_{xy}$ | shear strain in the $xy$ plane |
| $\gamma_{xz}$ | shear strain in the $xz$ plane |
| $\gamma_{yz}$ | shear strain in $yz$ plane |
| $\gamma_\theta$ | shear distortion; shear strain |
| $\delta$ | angle of friction between soil and pile; angle of skin friction; loss angle see equation 7.61 |
| $\Delta T$ | initial temperature of permafrost °C below freezing |
| $\Delta E$ | energy loss |
| $\Delta L$ | a small pile element length |
| $\Delta t$ | a small time interval in seconds |
| $\Delta G$ | change in low-amplitude shear modulus from time $t_1$ to $t_2$ |
| $\Delta \sigma'_v$ | change or increase in effective vertical strain |
| $\varepsilon$ | longitudinal strain |

| | |
|---|---|
| $\bar{\varepsilon}$ | $\varepsilon_x + \varepsilon_y + \varepsilon_z$ |
| $\dot{\varepsilon}$ | uniaxial creep rate |
| $\varepsilon_c$ | strain at maximum stress |
| $\varepsilon_{50}$ | strain at one-half the maximum principal stress |
| $\varepsilon_x$ | longitudinal strain in $x$ direction; lateral strain in $x$ direction |
| $\varepsilon_y$ | longitudinal strain in $y$ direction; lateral strain in $y$ direction |
| $\varepsilon_z$ | longitudinal strain in $z$ direction |
| $\zeta$ | damping factor |
| $\zeta_x$ | damping factor in horizontal sliding |
| $\zeta_z$ | damping factor in vertical vibrations |
| $\zeta_\phi$ | damping factor in rocking |
| $\zeta_\psi$ | damping factor in torsional vibrations |
| $\theta$ | angular rotation; tilting; temperature below freezing point of water, °C |
| $\Lambda$ | complex frequency parameter of a pile |
| $\Lambda_1, \Lambda_2$ | real and imaginary parts of $\Lambda$, respectively |
| $\Lambda_0$ | real frequency parameter of pile |
| $\Lambda_r$ | dimensionless parameter |
| $\lambda$ | Lammes' constant; wavelength; ratio of $k_t$ and $k_v$ |
| $\lambda_R$ | Rayleigh's wave length |
| $\mu$ | coefficient of friction |
| $\dot{\mu}_0$ | lateral ground surface displacement rate |
| $\nu$ | Poisson's ratio |
| $\nu_s$ | Poisson's ratio for soil |
| $\rho$ | mass density of pile material; mass density of soil |
| $\rho_b$ | mass density of soil beneath pile tip |
| $\rho_p$ | mass density of pile material |
| $\rho_s$ | mass density of the soil on the sides of the embedded footing |
| $\Sigma$ | sum |
| $\sigma$ | principal stress |
| $\sigma_e$ | applied constant stress |
| $\sigma'_h$ | horizontal effective stress |
| $\sigma_m$ | mean normal pressure |
| $\sigma'_v$ | effective overburden vertical pressure |
| $\sigma_{v0}$ | vertical effective stress |
| $\bar{\sigma}_{vx}$ | vertical overburden pressure at depth $x$ |
| $\sigma'_{v1}$ | effective vertical pressure (stress) at a point along pile length |
| $\sigma_x$ | normal stress in $x$ direction |
| $\sigma_y$ | normal stress in $y$ direction |
| $\sigma_z$ | normal stress in $z$ direction |
| $\sigma_0$ | effective all-around stress |
| $\bar{\sigma}_0$ | mean effective confining pressure |
| $\sigma_1$ | major principal stress |

| | |
|---|---|
| $\sigma_2$ | intermediate principal stress |
| $\sigma_3$ | minor principal stress |
| $\tau$ | shear stress; induced shear stress in soil due to applied load $(Q_v)_a$ |
| $\tau_a$ | shear stress |
| $(\tau_a)_z$ | adfreeze bond strength |
| $\tau_f$ | adfreeze stress along the pile perimeter |
| $\tau_{th}$ | downward pressures due to thaw (permafrost degradation) |
| $\tau_{xy}, \tau_{yz}, \tau_{zx}$ | shear stresses |
| $\phi$ | friction parameter, angle of internal friction |
| $\phi'$ | friction parameter (effective) |
| $\phi_{lt}$ | long-term internal friction of permafrost |
| $\psi$ | torsional rotation |
| $\psi_0$ | maximum torsional amplitude |
| $\psi_r$ | resonant amplitude of pile rotation |
| $\lvert\psi_{(z)}\rvert$ | real torsional amplitude of pile at elevation $z$ |
| $\psi_1, \psi_2$ | real and imaginary parts of $\psi(z)$ |
| $\omega$ | angular velocity, circular frequency, operating frequency |
| $\omega_n$ | circular natural frequency |
| $\omega_{n1}, \omega_{n2}$ | first and second natural circular frequencies |
| $\omega_{nl1}, \omega_{nl2}$ | limiting natural circular frequencies |
| $\omega_{nx}$ | natural circular frequency in horizontal sliding |
| $\omega_{nz}$ | natural circular frequency in vertical vibrations |
| $\omega_{n\phi}$ | natural circular frequency in pure rocking |
| $\omega_{n\psi}$ | natural circular frequency in torsional vibration |

# 1

# INTRODUCTION

Piles and pile foundations have been in use since prehistoric times. The Neolithic inhabitants of Switzerland drove wooden poles in the soft bottoms of shallow lakes 12,000 years ago and erected their homes on them (Sowers 1979). Venice was built on timber piles in the marshy delta of the Po River to protect early Italians from the invaders of Eastern Europe and at the same time enable them to be close to the sea and their source of livelihood. In Venezuela, the Indians lived in pile-supported huts in lagoons around the shores of Lake Maracaibo. Today, pile foundations serve the same purpose: to make it possible to build in areas where the soil conditions are unfavorable for shallow foundations.

The commonest function of piles is to transfer a load that cannot be adequately supported at shallow depths to a depth where adequate support becomes available. When a pile passes through poor material and its tip penetrates a small distance into a stratum of good bearing capacity, it is called a *bearing pile* (Figure 1.1a). When piles are installed in a deep stratum of limited supporting ability and these piles develop their carrying capacity by friction on the sides of the pile, they are called *friction piles* (Figure 1.1b). Many times, the load-carrying capacity of piles results from a combination of point resistance and skin friction.

The load taken by a single pile can be determined by a static load test. The allowable load is obtained by applying a factor of safety to the failure load. Although it is expensive, a static load test is the only reliable means of determining allowable load on a friction pile.

*Tension piles* are used to resist moments in tall structures and upward forces (Figure 1.1c), and in structures subject to uplift, such as buildings with basements below the groundwater level, or buried tanks.

*Laterally loaded piles* support loads applied on an angle with the axis of the

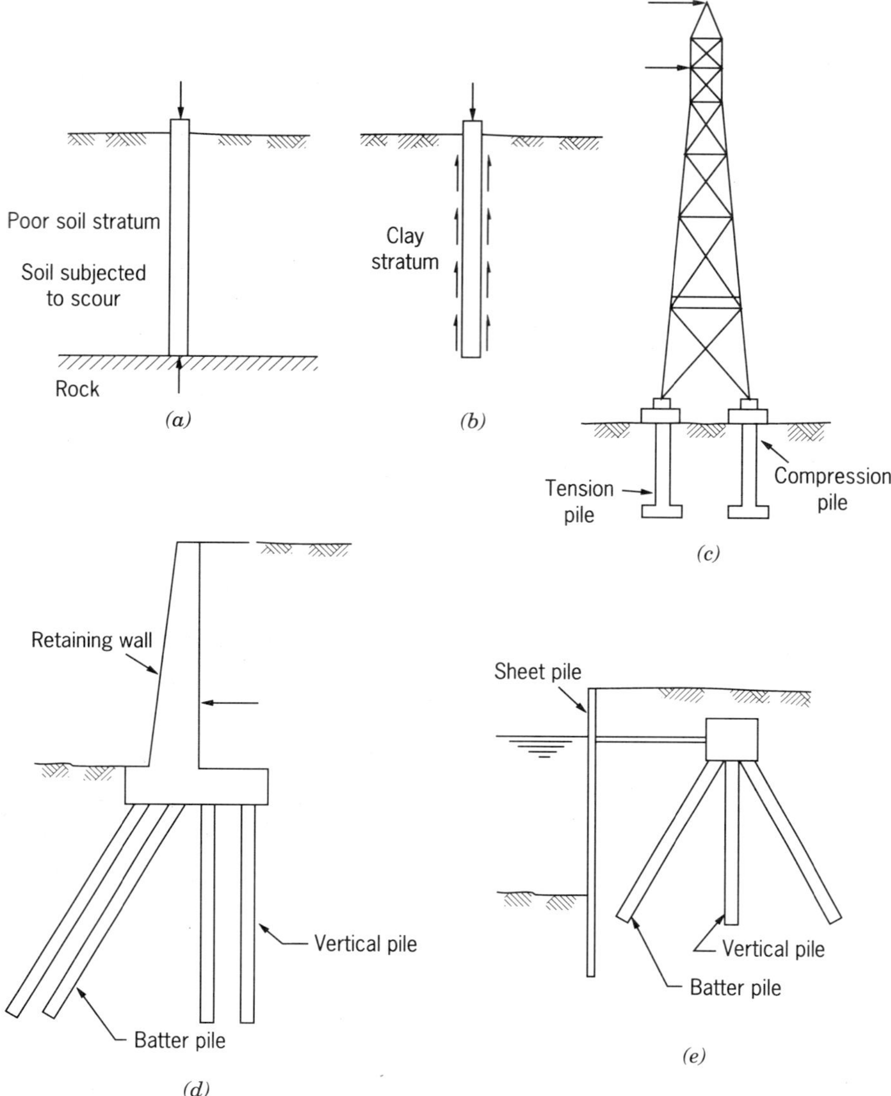


**Figure 1.1** Different uses of piles: (a) Bearing pile, (b) friction pile, (c) piles under uplift, (d) piles under lateral loads, (e) batter piles under lateral loads.

pile in foundations subject to horizontal forces such as retaining structures (Figure 1.1d and e).

If the piles are installed at an angle with the vertical, these are called *batter piles* (Figure 1.1d).

Dynamic loads may act on piles during earthquakes and under machine foundations.

Different types of piles based on their material are steel, concrete, timber, and composite piles (see Chapter 2).

Piles may be installed by any one of the following methods:

1. Driven precast
2. Driven cast-in-situ
3. Bored cast-in-situ
4. Screw
5. Jetting
6. Spudding
7. Jacking

The method of installation of a pile may have profound effects on its behavior under load and, therefore, its load carrying capacity. The method of installation may also determine the effect on nearby structures, for example, (a) undesirable movements and (2) vibrations, and/or structural damage. Much of the available data on installation effects are for driven piles in soft and loose soils, since driving of piles generally creates more disturbance than do other methods.

In this chapter, we first describe the mechanics of pile driving and its effects on pore pressures, and then we describe consolidation of clays based on field measurements.

During pile driving, the resistance to penetration is a dynamic resistance. When a pile foundation is loaded by a building, the resistance to penetration is a static resistance. Both the dynamic resistance and the static resistance are generally composed of point resistance and skin friction. However, in some soils, the magnitudes of the dynamic and static resistances may not be quite similar. In spite of this difference, frequent use is made of estimates of dynamic resistance by dynamic pile formulas and the wave equation (Chapter 5) for the static load capacity of the pile. Therefore, we also describe an understanding of the soil action during loading.

The concepts described in this chapter may not be directly used by a practicing engineer during the design. However, an understanding of these basic ideas will be helpful in explaining the pile behavior.

## 1.1 ACTION OF SOILS AROUND A DRIVEN PILE

The effect of pile driving is reflected in remolding the soil around the pile. Sands and clays respond to pile driving differently. First, we describe the behavior of clays and then the behavior of sands.

### Clays

The effects of pile driving in clays are listed in four major categories, De Mello (1969), as follows:

1. Remolding or disturbance to structure of the soil surrounding the pile
2. Changes of the state of stress in the soil in the vicinity of the pile
3. Dissipation of the excess pore pressures developed around the pile
4. Long-term phenomena of strength regain in the soil

The essential difference between the actions of piles under dynamic and static loadings is the fact that clays show pronounced time effects, and hence they show the greatest difference between dynamic and static action. These effects may be mechanistically described as follows.

Let us consider piles driven into a deep deposit of a soft impervious saturated clay. Since a pile has a volume of many cubic feet, an equal volume of clay must be displaced when the pile is driven. The pile-driving operation may cause the following changes in the clay:

1. The soil may be pushed laterally from its original position *BCDE* to *B′C′D′E′* (Figure 1.2) or from *FGHJ* to *F′G′H′J′*. If the clay has strength which is lost on disturbance, then relatively small amount of skin friction exists during driving.
2. Since the pile is being driven into a saturated impervious clay, the ground surface may heave considerably because of the displaced volume of clay.

In Figure 1.3, a pile of radius *oa* is shown embedded in a clay stratum. The changes in shear strength along the pile length and away from it are represented on figure *obcd* with *o* as the origin.

Curve *A* represents the shearing strength before the pile is driven and

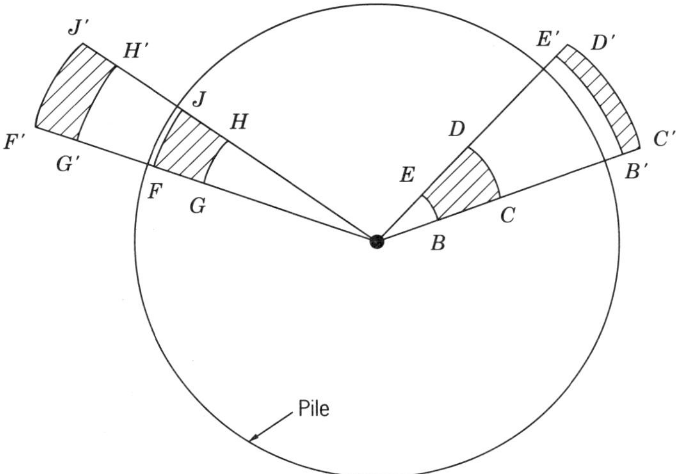


**Figure 1.2** The displacement and distortion of soil caused by a pile during driving.

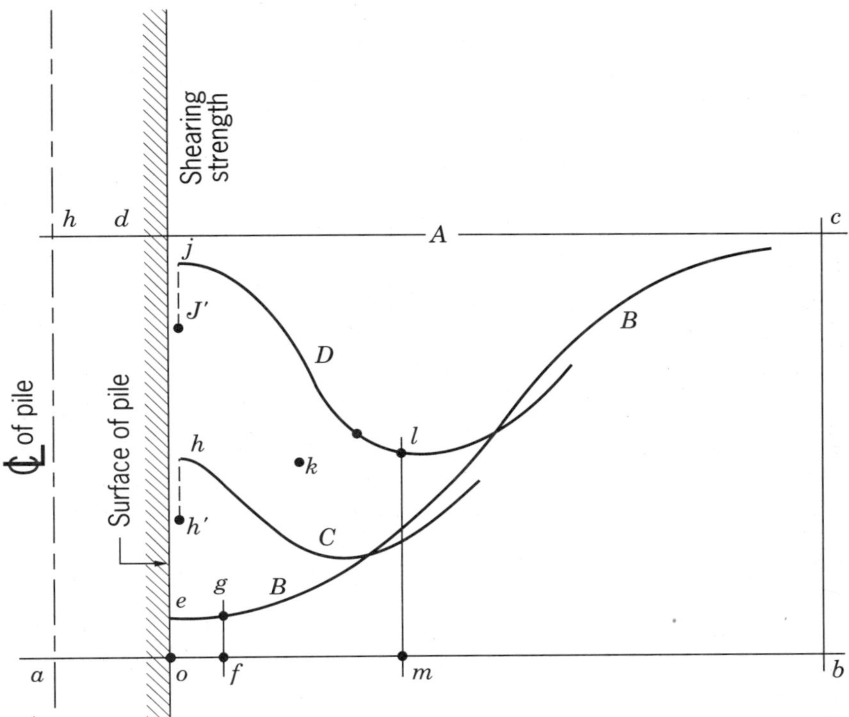


**Figure 1.3** Shearing strengths in saturated clay before and after pile-driving operations.

represents the undisturbed strength of the clay (quick strength). The strength at any point *b* at some distance away from *o* is *bc*.

Immediately after driving the pile, the shearing strength is represented by curve *B*. The clay that was at point *a* before driving has moved to point *o*; that originally at point *o* has moved to point *f*. The skin friction now is *oe*, which is the reduced shearing strength and is a small fraction of the original strength *od*.

The clay at point *o* has been remolded, and, therefore, the greater part of its intergranular pressure has disappeared. The total overburden pressure, consisting of intergranular pressure plus pore–water pressure, is essentially unchanged. Therefore, the lost intergranular pressure has been transferred to the pore water in the form of hydrostatic excess pressure. Thus, there is a large hydrostatic excess pressure in the clay adjacent to the pile immediately after pile driving. Since the disturbance to clay is less at a distance from the pile, therefore, the pore pressure increase is less. In addition, the lateral pressures adjacent to the pile increase considerably by the outward displacement of soil during driving. The gradients resulting from these excess pressures immediately set up seepage and start a process of consolidation. Since flow always takes place from points of high excess pressure to points of lower pressure, the direction of flow, therefore, is radially away from the pile. However, there may be some upward flow as well. During consolidation, clay particles move radially toward the pile because the water is flowing outward. The clay thus decreases in void ratio adjacent to the pile surface and expands a small amount at distances farther from the pile. Hence, after pile driving, soil builds up skin friction at a fairly fast rate. This is evidenced in a redriving test, which consists simply by allowing the pile to stand for a while and then driving it again (Taylor 1948). In Figure 1.3, *oh* represents the skin friction in redriving, and curve *C* represents the strength as a function of distance from the pile. If curve *C* represents strengths occurring a day or so after driving, curve *D* may represent strengths after a few weeks after driving. Since the soil at a distance from the pile expands slightly during consolidation, strength curves *C* and *D* may be a small distance below curve *B* in this region. If the pile is smooth, the resistance to shear at the surface may be less than the shearing strength in the clay a small distance from the pile surface. In this case, skin frictions are represented by points *h'* and *J'* instead of *h* and *j*.

If a loading test is run on this pile a few weeks after driving, the skin friction is represented roughly by distance *oj*. If a pile is pulled a few weeks after driving, a large mass of soil may stick to the pile and come up with it. The relative strength values at points explain this; for a nonuniform condition, the failure surface would not pass through *od* where the circumference is minimum, nor through *lm* where the strength is minimum, but would take place nearer to the radius where the product of strength and circumference is a minimum, perhaps at point *k* (Taylor, 1948).

The point resistance is generally large during driving because it equals the force required to cause all the remolding described above. Also, the soil that may

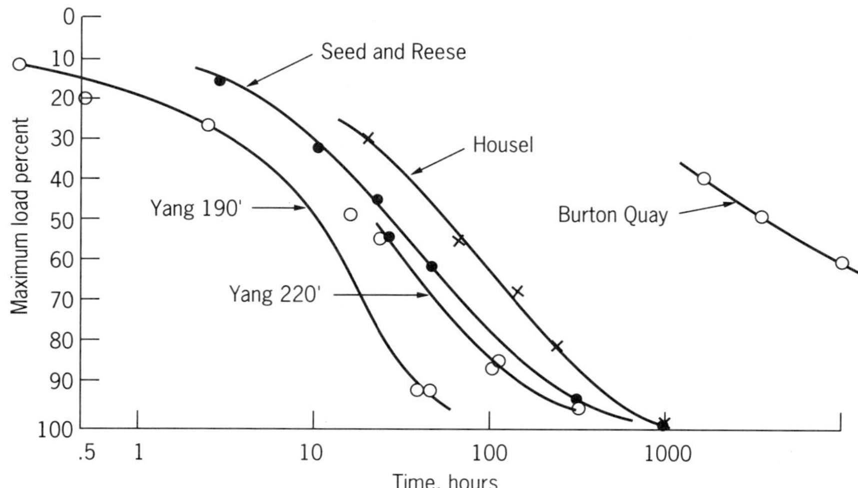


**Figure 1.4** Increase of load capacity with time (after Soderberg 1962).

have a high undisturbed strength has to be pushed out of the way. It cannot be compressed, because saturated soils are incompressible under quick loading conditions (e.g., as during pile driving). Moreover, there is no convenient place for the soil to go. Therefore, a column of soil, extending all the way to ground surface, must be heaved up to allow the pile to penetrate the soil below its tip. Practically all the resistance in many clays is point resistance during pile driving. De Mello (1969) suggested that immediately after driving, the amount of remolding decreased from about 100 percent at the pile–soil interface to virtually zero at about 1.5 to 2.0 diameters from the pile surface. Orrje and Broms (1967) showed that for concrete piles in a sensitive clay, the undrained strength had almost returned to its original value after nine months.

In addition to the dissipation of excess pore pressure, the rate of increase of soil strength after pile driving also takes place due to thixotropy in soils. Soderberg (1962) showed that the increase in ultimate load capacity of a pile (and hence, shear strength of the soil) was very similar in character to the rate of dissipation of excess pore pressure with time (Figure 1.4).

### Pore Pressures Developed during Driving

A number of measurements of the excess pore pressure developed in a soil because of pile driving have shown that the excess pore pressures at the pile face may become equal to or even greater than the effective overburden pressure.

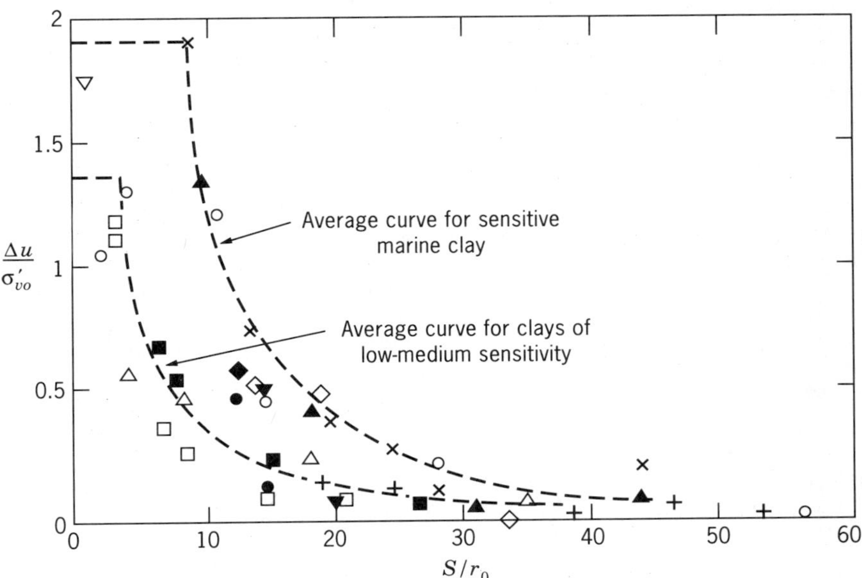


**Figure 1.5** Summary of measured pore pressures (after Poulos and Davis 1979).

(Lambe and Horn 1965, Orrje and Broms 1967, Poulos and Davis 1979, D'Appolonia and Lambe 1971).

In the vicinity of the pile, very high excess pore pressures are developed, in some cases approaching 1.5 to 2.0 times the in-situ vertical effective stress and even amounting 3 to 4 times the in-situ vertical effective stress near the pile tip. However, the induced excess pore pressures decrease rapidly with distance from the pile and generally dissipate very rapidly. In Figure 1.5, the excess pore pressure $\Delta u$ is expressed as $\Delta u/\sigma'_{vo}$, where $\sigma'_{vo}$ is the vertical effective stress in-situ prior to driving a single pile, and the radial distance $s$ from the pile is expressed as $s/r_0$ where $r_0$ is the pile radius. There is a considerable scatter in the points in this figure resulting largely from differences in soil type, the larger pore pressures being associated with the more sensitive soils (Poulos and Davis, 1979).

Beyond distance $s/ro$ of about 4 for normal clays, and about 8 for sensitive clays, a rapid decrease in pore pressure occurs with distance. In Figure 1.5, the excess pore pressures are virtually negligible beyond a distance of $s/r_0 = 30$.

## Sands

A pile in sand is usually installed by driving. The vibrations from driving a pile in sand have two effects:

1. Densify the sand, and
2. Increase the value of lateral pressure around the pile

Penetration tests results in a sand prior to pile driving and after pile driving indicate significant densification of the sand for distances as large as eight diameters away from the center of the pile. Increasing the density results in an increase in the friction angle. Driving of a pile displaces soil laterally and thus increases the horizontal stress acting on the pile. Horn (1966) summarized the results of studies of the horizontal effective stress ($\sigma'_h$) acting on piles in sand.

**TABLE 1.1 Horizontal Stress on Pile Driven in Sand***

| Reference | Relationship | Basis of Relationship |
|---|---|---|
| Brinch, Hansen, and Lundgren (1960) | (a) $\sigma'_h = \cos^2\phi' \cdot \sigma'_v = 0.43\bar{\sigma}_v$ if $\phi' = 30°$<br>(b) $\sigma'_h = 0.8\sigma'_v$ | (a) Theory<br>(b) Pile test |
| Henry (1956) | $\sigma'_h = K_p \cdot \sigma'_v = 3\sigma'_v$ | Theory |
| Ireland (1957) | $\sigma'_h = K\sigma'_v = (1.75 \text{ to } 3)\ \sigma'_v$ | Pulling tests |
| Meyerhof (1951) | $\sigma'_h = 0.5\sigma'_v$; loose sand<br>$\sigma'_h = 1.0\sigma'_v$; dense sand | Analysis of field data |
| Mansur and Kaufman (1958) | $\sigma'_h = K\sigma'_v$; $K = 0.3$ (compression)<br>$K = 0.6$ (tension) | Analysis of field data |

*After Horn (1966).

Table 1.1 shows a wide range in the value of the horizontal effective stress. It would seem logical that $K$ must exceed 1 and a value of 2 would seem to be reasonable (Lambe and Whitman, 1969).

## 1.2 DISPLACEMENTS OF GROUND AND BUILDINGS CAUSED BY PILE DRIVING

Pile driving generally causes a heave of the clay surrounding the pile and excess pore pressures followed by consolidation of the clay and dissipation of pore pressures. This movement may have a significant effect on adjacent structures. The piles driven earlier in a multiple-pile installation may heave during the driving of the later piles. If heave of adjacent structures and/or of the piles already installed is to be avoided, bored piles are sometimes used. The ratio of the total volume of initial heave to the total volume of driven piles within a foundation has

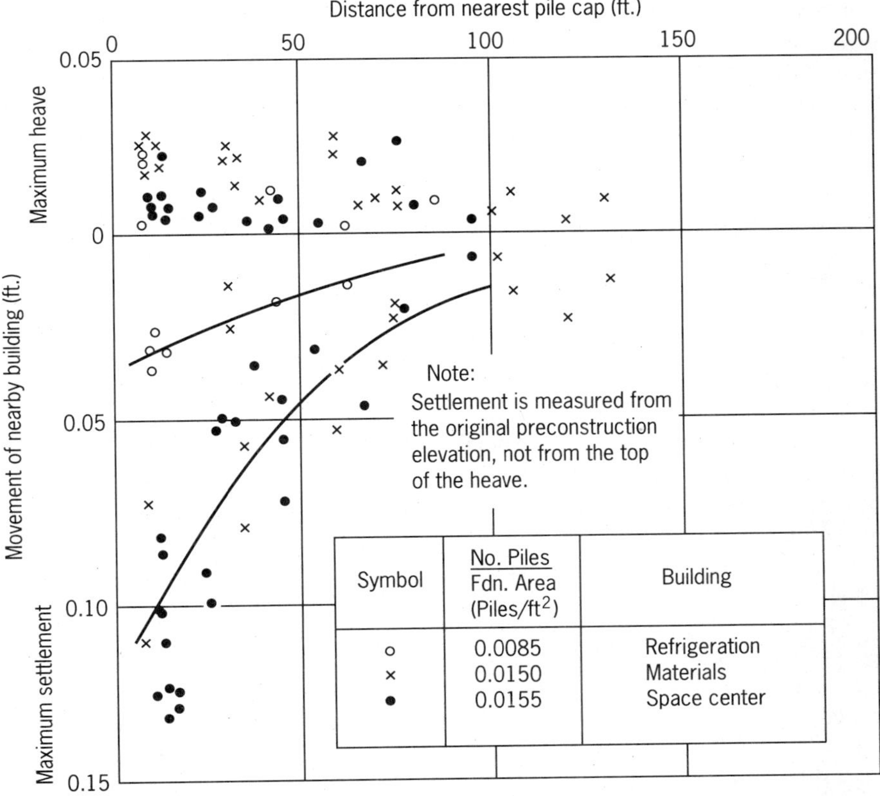


**Figure 1.6** Movements of nearby buildings caused by pile-driving operations (after D'Appolonia and Lambe 1971).

been found to be about 100 percent by Adams and Hanna (1970) for steel H-piles in a firm till, 50 percent for piles in clay by Hagerty and Peck (1971), 60 percent by Avery and Wilson (1950), and 30 percent by Orrje and Broms (1967) for precast concrete piles in a soft, sensitive, silty clay (Poulos and Davis, 1979). Orrje and Broms (1967) found that the heave near the edge of the foundation was about 40 percent of the value at the center. Adams and Hanna (1970) found that the maximum radial movement was about 1.5 in., and the maximum tangential displacement about 0.4 in. while the average vertical heave was about 4.5 in. As with vertical heave, very small lateral movements occurred beyond the edge of the group. Lambe and Horn (1965) reported the movement of an existing building due to driving of piles for the new building. It was found that, at the near corners of the existing building, a heave of about 0.3 in. occurred during driving. At the end of construction, a net settlement of about 0.35 in. had occurred. Despite the fact that the piles were preaugered to within about 30 ft of their final elevation, excess pore pressures of about 40 ft of water were measured near the corner of the existing building, even before a substantial building load was carried by the piles.

Figure 1.6 shows measurements of heave and settlement of buildings caused by pile driving, (D'Appolonia and Lambe, 1971). The settlement data plotted are for net settlement one to three years after the end of construction. Larger movements than those measured by Lambe and Horn (1965) were found, although the piles were again preaugered to within 20 to 30 ft of the final tip elevation.

Hagerty and Peck (1971) found that if the piles are first driven along the perimeter of the foundation, the heave of the soil surface in the central area of the foundation is increased and that of the surrounding area correspondingly decreased. Measurements of lateral movement showed that piles already driven tended to be displaced away when more piles were driven, and movements continue for a considerable length of time after completion of driving.

## 1.3 GROUP ACTION IN PILES

Piles are driven in groups at a spacing ranging from 3 to $4B$ where $B$ is the diameter or side of a pile. The behavior of piles in a group may be quite different than that of a single pile if the piles are friction piles. This difference may not be so marked in bearing piles.

Figure 1.7 shows assumed failure patterns under pile foundations (Vesic, 1967). The effect of load will be felt to a small distance below the tip of the pile.

A typical bearing pile usually penetrates a short distance into a soil stratum of good bearing capacity, and the pile transfers its load to the soil in a small pressure bulb below the pile tip (Figure 1.8a). If the stratum in which the piles are embedded and all strata below it have ample bearing capacity, each pile of the pile group is capable of carrying essentially the same load as that carried by single piles. If compressible soils exist below the pile tips, the settlement of the pile group may be much greater than the settlement observed in the single pile tests, although the bearing pressure may be smaller than the allowable value. This is

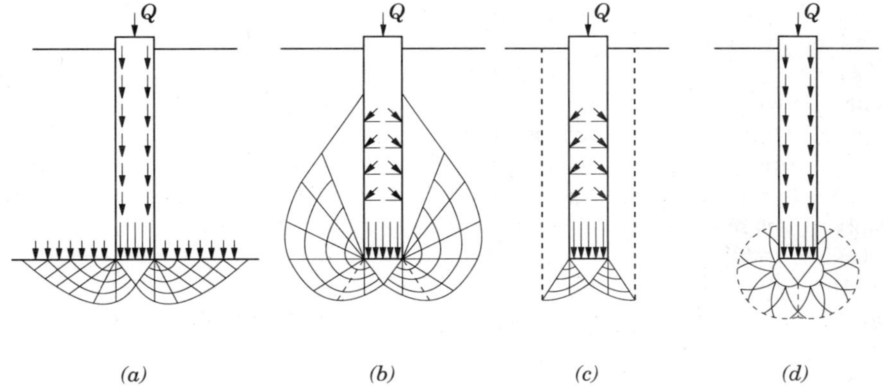


**Figure 1.7** Assumed failure patterns under deep foundations (Vesic 1967): (a) After Prandtl, Reissner, Caquot, Buisman, Terzaghi (b) After DeBeer, Jaky, Meyerhof (c) After Berezantsev and Yaroshenko, Vesic (d) After Bishop, Hill and Mott, Skemption, Yassin, and Gibson.

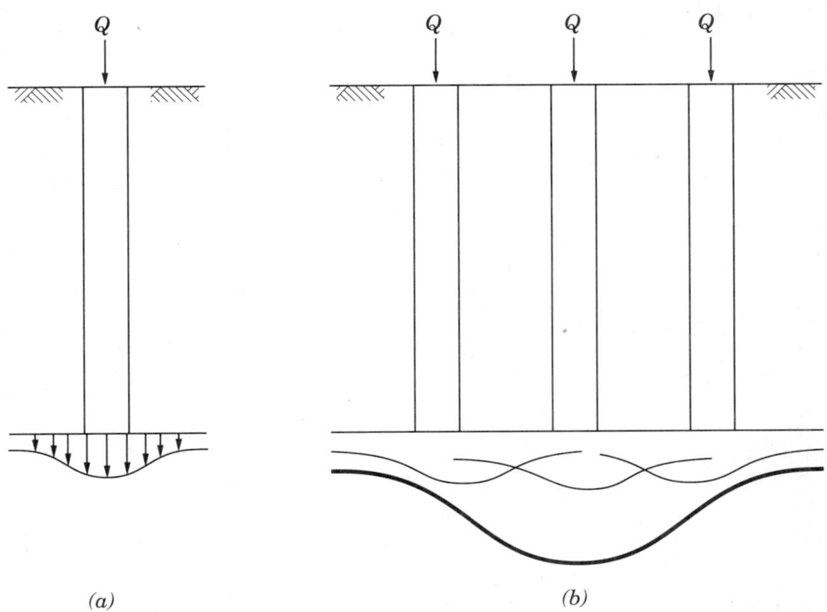


**Figure 1.8** Stress condition below tips of piles: (a) Single pile, (b) group of piles.

due to overlap of the zones of increased stress below the tip of the bearing piles and the pile group is likely to act as a unit (Figure 1.8b). The total stress shown by the heavy line may be several times greater than that under a single pile. The effective width of the group is several times that of a single pile. However, if the bearing stratum is essentially incompressible and there are no softer strata below the pile tips, the settlement of a group of bearing piles may be essentially equal to the settlements observed in loading tests on isolated piles. In this case, the piles may, if desired, be spaced about as closely as it is practicable to drive them (Taylor, 1948).

In a large group of closely spaced friction piles, the actions of the piles overlap and the distribution of load to the various piles is not uniform. In Figure 1.9, let

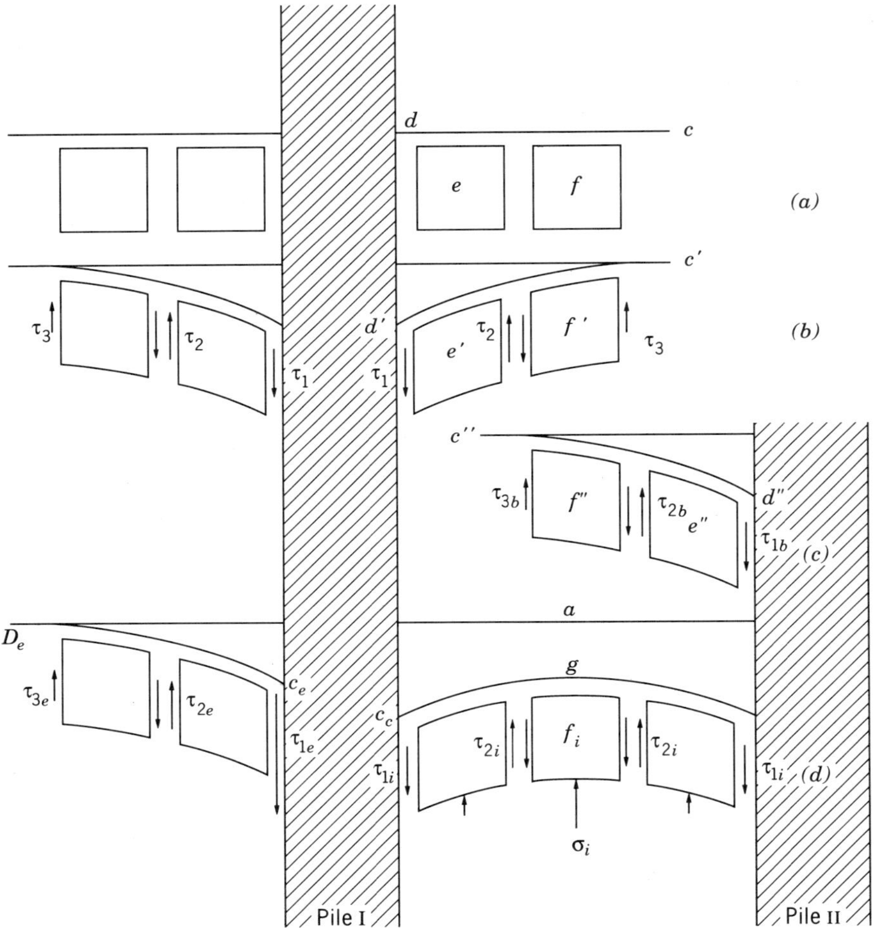


**Figure 1.9** Shearing stresses and shearing strains in the soil adjacent to loaded, single friction piles and pile groups.

piles I and II be two adjacent piles of a friction pile group and that pile I is loaded first and pile II later. Before either pile is loaded, the conditions are as shown in (a); *cd* is a horizontal reference line within the soil, and squares *e* and *f* represent reference elements within the clay. After pile I is loaded, the conditions are as shown in (b). The original reference line *cd* moves to *c′d′*. The reference elements have been distorted to the shapes *e′* and *f′*. The pile exerts a shearing stress $\tau_1$ on element *e′*. The soil on the outer side of element *f′* offers vertical support to the element by the shearing stress $\tau_3$. The distortions shown in the figure indicate that, even at fairly large radial distances from the pile, the major portion of the skin friction is transferred to the soil by shearing stresses on vertical cylindrical surfaces. It may be argued that for piles of large length, $\tau_3$ multiplied by the circumference over which it acts is nearly as large as $\tau_1$ multiplied by circumference of the pile.

Now let it be assumed that pile II is loaded. If this pile were loaded separately (c), the displacements and distortions that would be caused would be similar to those for pile I. When the two piles are loaded simultaneously, an overlapping of stresses occurs between them and gives a much more complex situation shown in (d). Element $f_i$ is symmetrically loaded by the two piles; therefore, the distortions

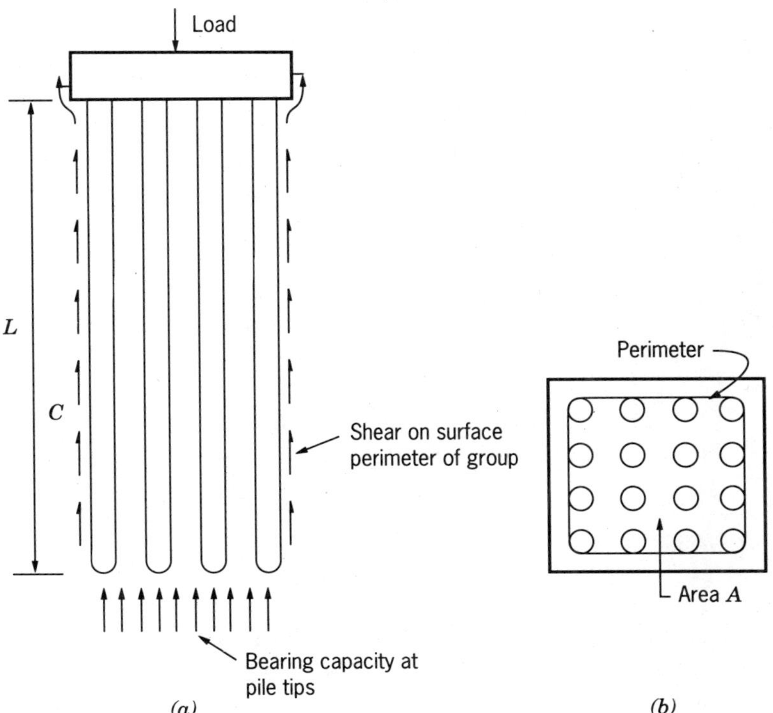


**Figure 1.10** Load-carrying capacity of a pile group in clays: (a) Section, (b) plan.

shown in $f'$ and $f''$ of (b) and (c), respectively, are not possible. Furthermore, it is not possible for shears on vertical planes to be transferred outward indefinitely, as for the single pile. Since square $f_i$ must be symmetrical after distortion, the shearing stresses it takes on its sides are much smaller than those on $f'$ and $f''$. Therefore, $\tau_{1i}$ must be much smaller than $\tau_1$. To carry the pile load, the pile must settle further. This causes larger distortions on the outer side of the piles and increases the skin friction there to $\tau_{1e}$. The frictional force represented by $\tau_{1i}$ cannot be transmitted by shear beyond point $g$. To the left of pile I, much of the skin friction is transferred by shearing stresses on vertical planes to a large distance from the pile.

The concept that two piles greatly interfere in development of skin friction around each other applies in much greater degree to large groups of closely spaced friction piles than it does to the two piles as just discussed. Thus, it may be concluded that, in foundations of friction piles, the distribution of load to the various piles is far from uniform. If the centrally located piles could settle more on loading than the exterior piles, it is possible that they may develop a slightly greater skin friction than if all piles settle equally. Since all piles settle the same amount in a pile group, each exterior pile carries a much greater load than an interior pile.

A rough estimate of the load carrying capacity $(Q_v)_{ult}$ of a friction pile may be obtained by considering the resistance to penetration along the periphery of the single pile since the contact is between soil and pile.

Usually, friction piles are driven in groups, the spacing of piles being from 3 to $4B$. A group of piles may fail under a load per pile less than the failure load of a single pile. The load-carrying capacity of group of piles (Figure 1.10) may be determined by considering failure along the perimeter of the pile groups.

The load-carrying capacity of the friction pile groups in clay is smaller of the two:

1. Sum of the failure load of the individual piles or
2. Load carried as in group action and failure as a pier along the perimeter, as in Figure 1.10

Details of the estimation of failure and working loads on pile groups in clays, are discussed in Chapter 5. Methods of load tests are described in Chapter 9.

## 1.4 NEGATIVE SKIN FRICTION

If a pile is driven in a soft clay or recently placed fill and has its tip resting in a dense stratum (see Figure 1.11), the settlement of both the pile and the soft clay or fill is taking place after the pile has been driven and loaded. During and immediately after driving, a portion of the load is resisted by adhesion of soft soil with pile (Figure 1.11a). But, as consolidation of the soft clay proceeds, it transmits all the load onto the tip of the pile.

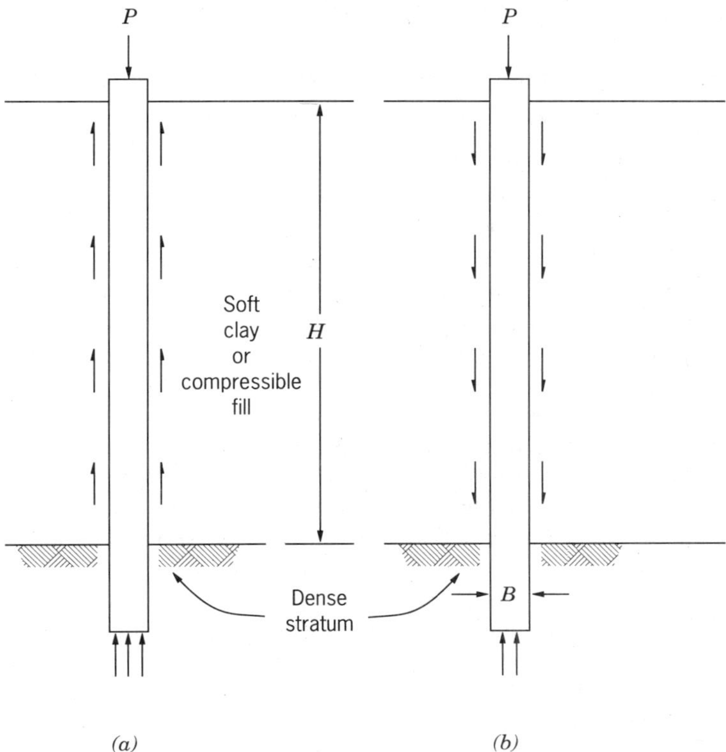


**Figure 1.11** Piles in soft soil overlying dense strata: (a) Skin friction immediately and during pile driving, (b) negative skin friction.

In case of a fill, the settlement of the fill may be greater than that of the pile. More specifically, this condition occurs in any case in which the soil subsides relative to the piles (Taylor, 1948). In the initial stages of consolidation of the fill, it transmits all the load resisted by adhesion onto the tip of the pile. A further settlement results in a downward drag on the pile. It is known as *negative skin friction* (Figure 1.11b). Both these cases should be recognized in the field in the design of bearing piles.

When this condition occurs, the pile must be capable of supporting the soil weight as well as all other loads that the pile is designed to carry. Also, if fill is to be placed around an existing pile foundation, the ability of the piles to carry the added load should be thoroughly investigated. *Load due to negative skin friction may often be large, since values of unit negative skin friction can be as large as positive values, and pile failures that are caused by such loads are not uncommon* (Taylor, 1948).

A detailed discussion on methods of computing negative skin friction loads and field techniques to reduce negative skin friction are discussed in Chapter 5.

## 1.5 SETTLEMENT OF PILE GROUPS

The settlement of a group of friction piles are considered to result from three causes (Taylor, 1948):

1. Settlement due to compression of the pile and from the movement of the piles relative to the immediately adjacent soil (Figure 1.10). When full skin friction is developed, this settlement corresponds to that observed in a loading test on a single pile.
2. Settlement due to compression occurring in the soil between the piles.
3. Settlement due to compression that occurs in compressible strata below the tips of the piles.

The settlements due to compression of the soil between piles ((2) above) and that due to compression of the strata below the tips of the piles ((3) above) are generally of much larger magnitude than that due to compression of the pile and movement of pile relative to the soil ((1) above). However, these settlements may occur very slowly in saturated soil because of consolidation and slow dissipation of pore pressure.

Since there is partial disturbance to the structure of the soil around the piles, accurate estimates of the amount of settlement occurring under item (2) are not possible. The disturbance of soil structure during pile driving may result in increased settlements after the final loading of a pile foundation. It is well known that a remolded clay, when subjected to a given load, consolidates to a considerably smaller void ratio than that reached under the same load by the same clay in undisturbed state (Taylor, 1948). Therefore, structural disturbance results in increased settlements. The magnitude of this settlement increase depends largely on such factors as (1) the distance the disturbance extends from the pile, (2) the type of soil, (3) the degree to which the soil is disturbed, and (4) the details of the action in the complicated consolidation process subsequent to driving. Definite increases in settlements may not be quantitatively defined, but it is possible that in some soils they are much larger than many engineers may suspect (Taylor, 1948). Estimates of item (3) may be made by the methods based on Terzaghi's theory of consolidation (see Chapter 5).

In loading tests, the settlements of a single friction pile are not representative of the settlements of the pile group. Therefore, such a load test will give information on failure load rather than the settlements under actual loading conditions of a friction pile. The installation of piles usually alters the deformation and compressibility characteristics of the soil mass in a different way and to a different extent as compared to that around and below the tip of the single pile although this influence extends only to a few pile diameters. Accordingly, the total settlement of a group of driven or bored piles under the safe design load not exceeding one-third to one half of the ultimate group capacity can generally be estimated roughly as for an equivalent pier foundation Terzaghi and Peck (1967) (see Chapters 5 and 9 for further details). Several simplifying assumptions are made for this computation.

## 1.6 LOAD TEST ON PILES

The amount of resistance to penetration which developed between a pile and the soil it penetrates, because of group action can be determined only by loading tests.

There are several methods of performing a load test (see Chapter 9). In the simplest case, a load is applied on the pile head and its settlement is monitored. Load settlement curves are usually plotted as in Figure 1.12. In a pile-loading test on sand, Figure 1.12a load is continuously increasing with deflection but at a decreasing rate. In a test on clay (Figure 1.12b), the plot may be practically a straight line nearly to failure. Therefore, the test in clay must be carried to failure, otherwise the magnitude of the failure load cannot be determined. In clays or fine silts, which are loaded by dead weights, the failure occurs suddenly and the pile may sink many feet into the soil without warning. When the pile is loaded by some type of jack, the actual loading curve passes a maximum load and then decreases, as shown in Figure 1.12b.

In a pile that has been driven into a clay deposit and loaded after complete consolidation of the clay around it, let the solid light horizontal lines of Figure 1.13 represent the position of surfaces within the soil before loading. These lines probably do not conform to the original strata because of disturbance during driving. Actual strengths within the clay are probably as shown by curve *D* of Figure 1.3. On application of load near failure, the horizontal surfaces are bent downward from the horizontal as shown with dotted lines close to the pile. The main portion of the load on the pile is transferred by skin friction in the form of downward vertical shearing stresses on the soil against the pile. The resulting shearing strains are represented by the deviations of the dotted lines from the horizontal in Figure 1.13. At a distance of one diameter from the pile center, the circumference is twice the pile circumference. The shearing stress at this point is

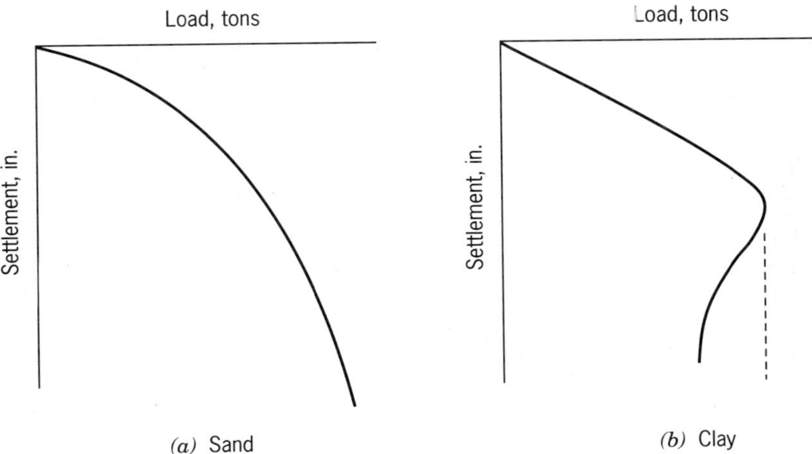


**Figure 1.12** Plots of loading tests on piles: (a) Sand, (b) clay.

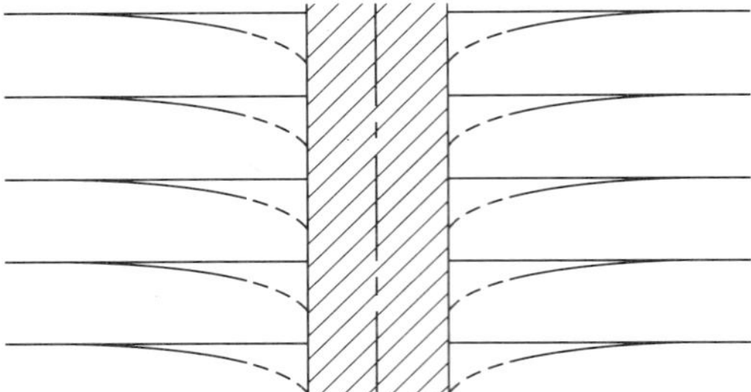

**Figure 1.13** Distortions occurring in the soil adjacent to a loaded friction pile.

only half as large as the skin friction. The shearing strains are slightly less than half of the values at the pile surface if nonlinear behavior of clay is accounted for. Thus, we see that the stresses and strains caused by the loading of one pile die out quite quickly with distance from the pile center. This explains, at least in part, the fact that settlements in loading tests on single piles are small and may be only a small fraction of the settlement the structure will undergo as a whole. Thus, the loading test furnishes the limiting value of the resisting force a soil can exert on a pile. It also gives indications relative to the strains required adjacent to the pile to develop this resistance.

## 1.7 BEHAVIOR OF PILES IN PULLOUT

For piles under tension both in sands and clays, the bearing capacity at the tip is lost. For piles of uniform diameter in sands, the ultimate uplift capacity is made up of the shaft resistance and the weight of the pile. The shaft friction in upward loading may not be of the same nature and therefore may be unequal to that in vertical downward loading.

In clays, the ultimate skin friction in pullout (adhesion $c_a$) may be quite similar to that under vertical downward loading. However, in pullout in soft clays, the failure may not necessarily occur along the perimeter of the single pile (Taylor, 1948). Also, negative pore pressures may occur in clays during pullout. The uplift capacity under sustained loading may therefore be smaller than the short-term or undrained capacity. The clays tend to soften with time, and their strength is reduced due to dissipation of negative pore pressures.

If the pile has a pedestal at the base or an enlarged tip, or plug (e.g., a Franki pile or an underreamed pile (see Chapter 5)), the failure will not take place along or near the periphery of the shaft but along failure surfaces starting from the perimeter of the base and extending up to the ground level. Several theories have been developed to compute this resistance.

On the basis of actual pullout tests of uniform diameter piles, Hegedus and Khosla (1984) found the following:

1. In overconsolidated clays, the undrained shear strength approach for ultimate pullout capacity predictions resulted in good agreement with the observed value when the *effective* pile surface was used in predictions.
2. In sands and nonplastic silts, the uplift capacity predicted on the basis of actual pile perimeter as the failure surface and the soil to pile friction, tallied well with the measured pullout load.

## 1.8 ACTION OF PILES UNDER LATERAL LOADS

Piles are generally used in groups. However, first we describe the action of a single pile under lateral load followed by discussion of pile groups.

### 1.8.1 Single Pile Under Lateral Load

In this section the behavior of fully embedded flexible vertical piles, subjected to moment ($M_g$), shear ($Q_g$), and axial loading ($Q_v$), are described (see Figure 1.14a).

Figure 1.14a illustrates the deflected shape of the pile with the soil reactions caused by the external forces. At any point $x$ along the pile, the soil reaction $p$ is taken as proportional to the deflection $y$ at that point, and $p = ky$ where $k$ *is the coefficient of subgrade reaction* for a pile of width $B$. The distribution of the soil reactions must be such that the equations of statics, when applied to the pile, are satisfied. Therefore, the distribution of soil reactions is a function of the applied loading as well as the load deformation characteristics of the soil. Obviously, all horizontal loads applied to the pile must be resisted by horizontal soil reactions.

In Figure 1.14b, the deflections are shown due to curvature of the pile. By taking moments about any point $x$ along the embedded portion of the pile, it can be shown that lateral displacements of the pile produce an eccentricity $e$ of the vertical load (Davisson, 1960). This leads to a moment ($Q_v \cdot e$) in addition to the moments caused by $Q_g$ and $M_g$. Therefore, it may be concluded that an axial load leads to a magnification of deflections, rotations, moments, and shears that would occur in the pile if $Q_g$ and $M_g$ acted alone.

The laterally loaded pile problem may be considered as an opposite extreme to the problem of an embedded flexible vertical wall subjected to horizontal forces. Figure 1.15a shows a wall whose length ($B$) may be considered infinite with respect to its embedded depth ($L_s$) and is therefore a two-dimensional problem in plane strain. A horizontal load $Q_g$ forces the wall against the soil on one side of the wall and produces an increase in pressure over the at-rest pressure, while on the other side of the wall the at-rest pressure is reduced. The maximum and minimum values that the pressures may attain are the passive and active earth pressures respectively (Davisson, 1960).

In Figure 1.15b, a pile with width $B$ is embedded length $L_s$ in the ground where the ratio of $B$ to $L_s$ is quite small. A load $Q_g$ acts at the level of the ground surface. In this case, the increase of pressure on one side of the pile (Figure 1.14a) occurs while the pressure on the other side decreases, but the pressures are now

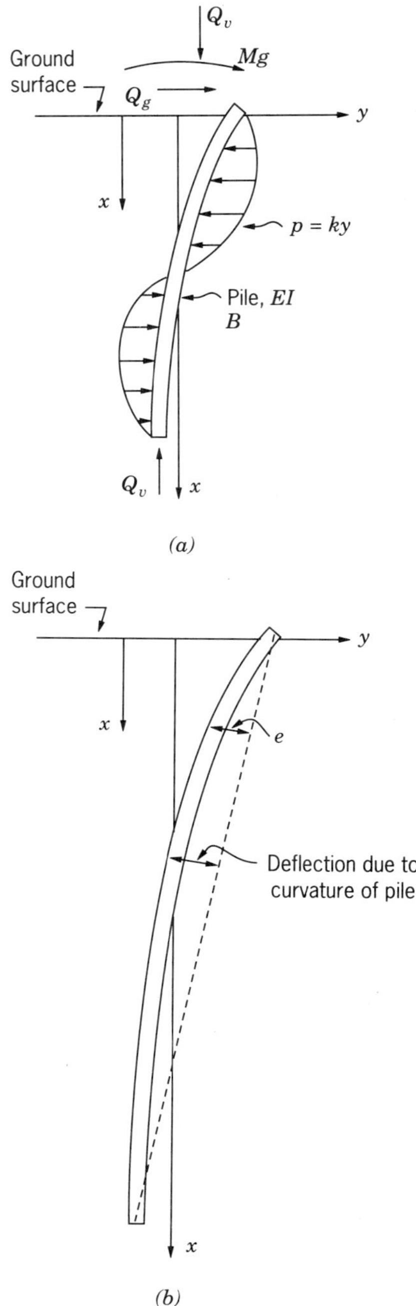


**Figure 1.14** System of forces and deflected shape of pile: (a) Forces and deflected shape, (b) deflection.

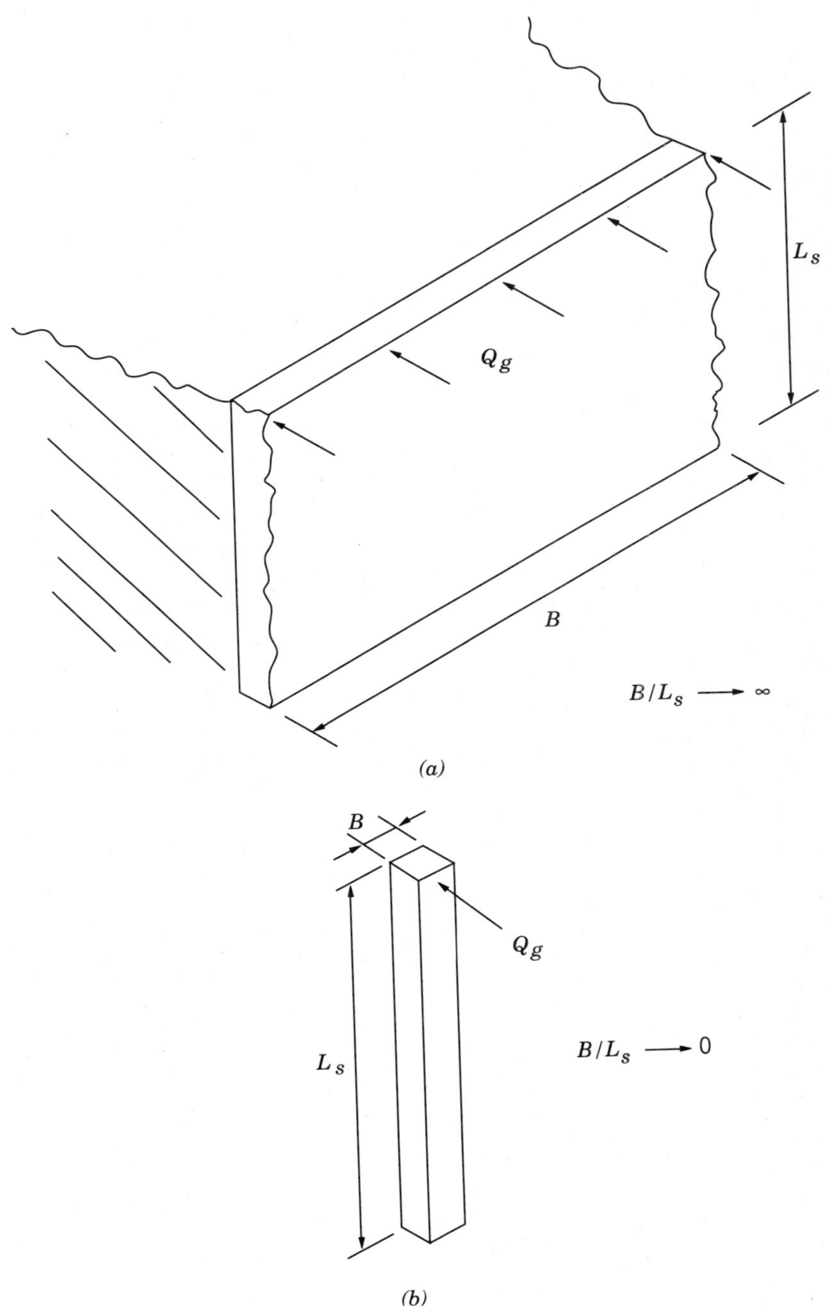


**Figure 1.15** Passive pressure and bearing capacity problem: (a) Wall, (b) pile.

influenced considerably by the shape of the pile cross-section. The shears along the sides of the pile may be negligible for walls but tend to influence the pressure changes that occur on the front and back of the pile under lateral loads. This effect on pressure changes should vary with the ratio of pile width to pile thickness. The net load $p$ is the result of several forces acting on the pile and is quite complex. Nevertheless, the expression $p = ky$ is found to be useful (Davisson, 1960).

Near the ground surface, a laterally loaded pile distorts the soil in a mode different from the mode occurring for two-dimensional active and passive pressure. The net effect is that the maximum unit soil resistance that is available to resist the deflection of a laterally loaded pile is somewhat higher than the maximum unit passive resistance computed for the two-dimensional case. At a depth of about $3B$ or greater, below the ground surface, the laterally loaded pile deforms the soil in the mode for bearing capacity of a deep footing (completely embedded in soil). The result is that the unit soil resistance, available to resist the deformation of a laterally loaded pile, is considerably greater than the unit, two-dimensional passive pressure at depths exceeding about $3B$ below the ground surface. From the ground surface to a depth of approximately $3B$, the mode of deformation makes a transition from a mode that resembles passive pressure to the mode for bearing capacity of an embedded footing (Davisson, 1960).

Reese et al. (1974) considered two cases near the ground surface (Figure 1.16a) and at large depth (Figure 1.16b). These surfaces have been used to compute soil resistance as explained above (see Chapter 6).

Tests have shown that a zone of plastic soil resistance occurs adjacent to the ground surface when piles are loaded laterally. There is a transition from plastic to elastic soil behavior at some depth below the ground surface. This behavior is in accordance with the assumed failure surfaces as in Figures 1.15 and 1.16. Only

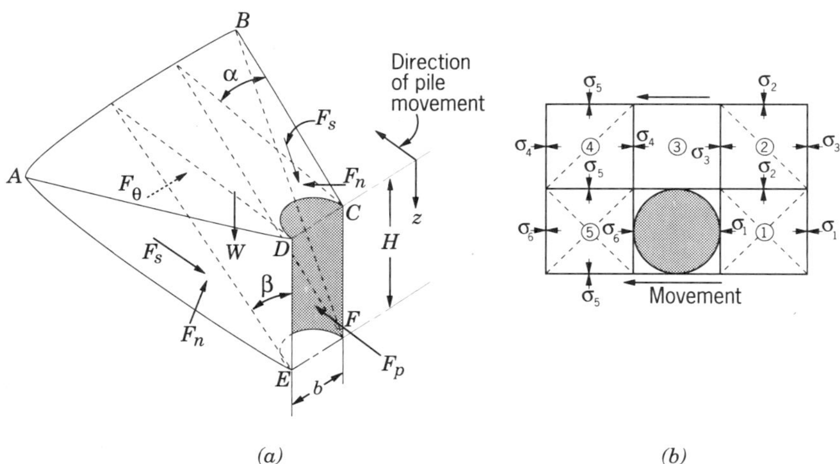


**Figure 1.16** Assumed failure surfaces around a pile under lateral load: (a) Assumed passive wedge type at shallow depth, (b) Assumed mode of soil failure by lateral flow around the pile at larger depth (After Reese et al. 1974).

rigid poles or very short piles are strong enough to produce a fully plastic state in the soil when loaded laterally. Therefore, for normal piles, a combination of elastic and plastic soil resistance must be considered.

It was shown, as early as 1880, that laterally loaded piles will fail by flexure below the ground surface. Furthermore, it was shown that increasing the embedded length of a pile beyond a certain depth had a negligible effect on the load at which a flexural failure occurred (Davisson, 1960). For piles of normal size, the point of maximum moment may occur within 2 to 3 feet of the ground surface where as, in soft soils such as bay muds, the point of maximum moment may be up to 15 feet below the mudline. Furthermore, the deflection of a pile with a fixed head will only be one-third to one-half, as for a given lateral load, as compared to that of a free head pile.

Theoretical studies, combined with the results of load tests, have shown that most piles may be analyzed for lateral loads as though they were infinitely embedded. Tests have shown that if several types of piles are embedded deep enough in the same soil deposit to be analyzed as though infinitely deep, the deflection for a given lateral load will be less for the stiffer piles than for the more flexible piles (Davisson, 1960).

If the soil is stiff close to the ground level, and soft soil occurs below this stiff layer, the beneficial effects of the stiff crust of soil at the ground surface reduces deflections and advantage can be taken of a layer of stiff dessicated clay close to the ground surface. Repeated loading of piles embedded in clay may cause a progressive deterioration of the soil resistance adjacent to the ground surface. Shearing distortion may cause a reduction in the shear strength and stiffness of clay. If a soil disturbed by repeated loading is given a rest period, an increase in strength and stiffness may occur; but such an occurrance will depend on the consolidation and thixotropic properties of the clay as for vertical loading. These effects are similar to those observed in pile driving in clay. Also, relatively stiff piles may lose contact with the pile close to the ground surface.

The lateral resistance of a pile, embedded in clay, has been shown to increase with the time after driving. In sands, repeated loading may almost double the deflection of a laterally loaded pile as compared to the deflection observed for the first application of the load. If the sand is loose, repeated loading will densify the sand deposit and subsequent increments in deflections may be reduced. It appears, that for a given repeated load, the sand will reach a state of equilibrium with regard to density. A change in the magnitude of the load will bring about a corresponding change in the density of the sand. The foregoing comments apply principally to the zone of sand near the ground surface. Upon release of a lateral load from a pile, a permanent pile deflection is usually observed. Therefore, a curvature is locked into the pile and soil pressures are necessary to sustain the curvature (Davisson, 1960).

### 1.8.2 Pile Groups Under Lateral Loads

The response of a laterally loaded pile group differs from that of a single pile because of interference with the zone of influence of the pile by adjacent piles and

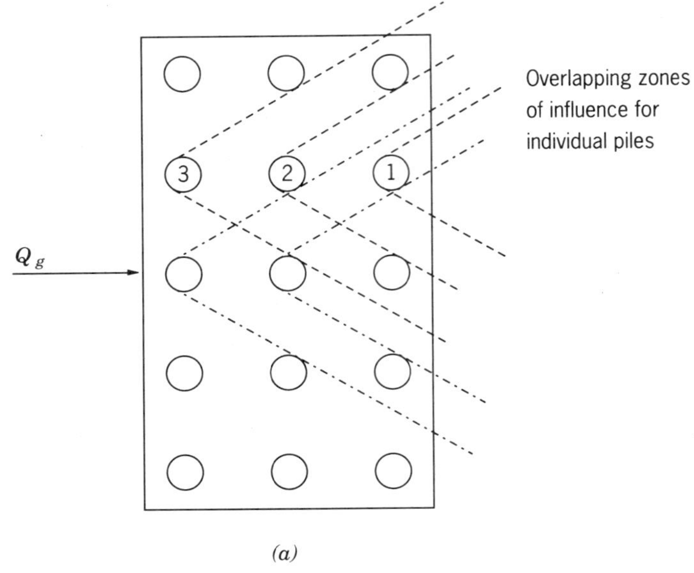


*(a)*

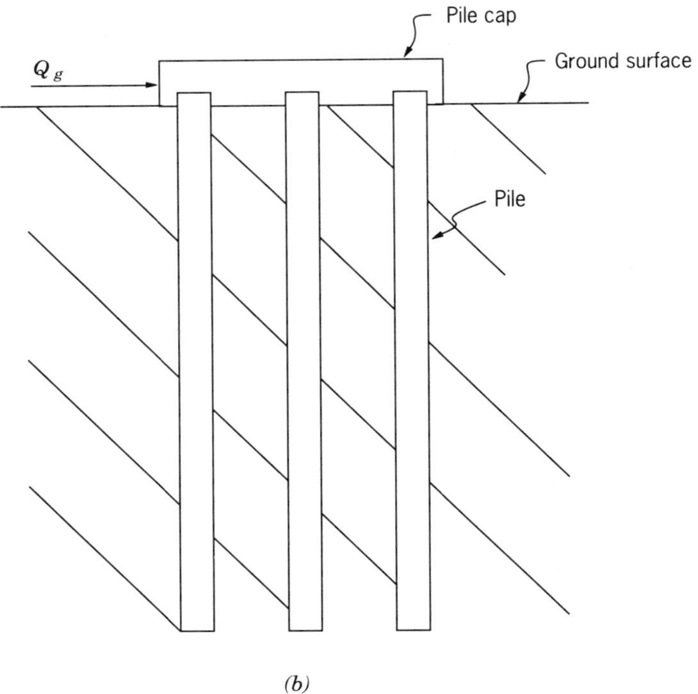


*(b)*

**Figure 1.17** Pile-group behavior (overlapping zones of stress): (a) Plan, (b) section.

their zones of influence (Figure 1.17). A difference may also exist between the degree of fixity of a single pile and a corresponding pile group; however, this is primarily a theoretical problem and not one of important behavioral differences in the soil.

***Interference of Adjoining Piles*** Figure 1.17 shows a plan and profile of a pile group loaded horizontally at the ground line by a load, $Q_g$ (Prakash 1962). The dotted lines in Figure 1.17a indicate schematically how one pile in a group may affect its neighbors. Pile 1 (Figure 1.17a) stresses the soil outside of the pile group, whereas piles 2 and 3 generally stress the soil immediately in front of their locations. This, coupled with the deflection of piles 1 and 2, causes a lower soil resistance for piles 2 and 3. Therefore, piles 2 and 3 would exhibit less stiffness than pile 1. Application of the above concepts leads to a qualitative comparison

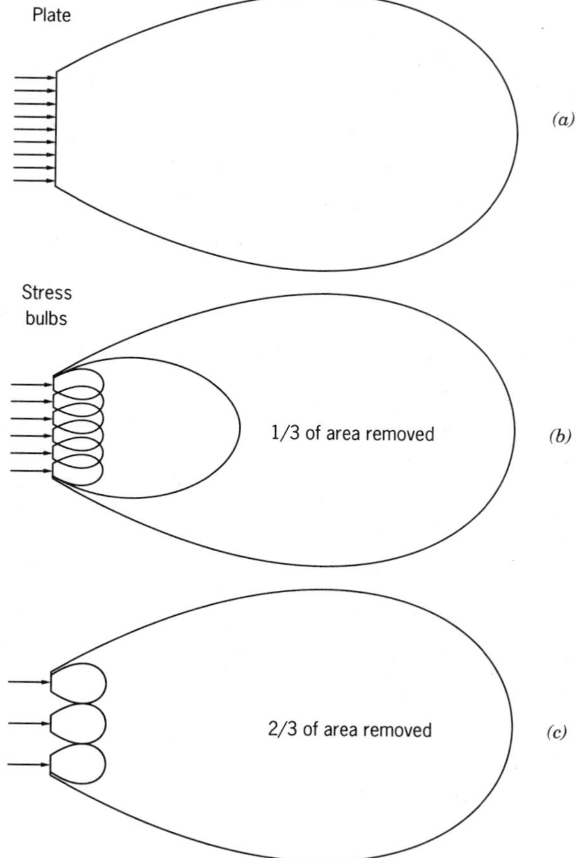


**Figure 1.18** Size effects: (a) Plate, (b) one-third of area removed, and (c) two-thirds area removed.

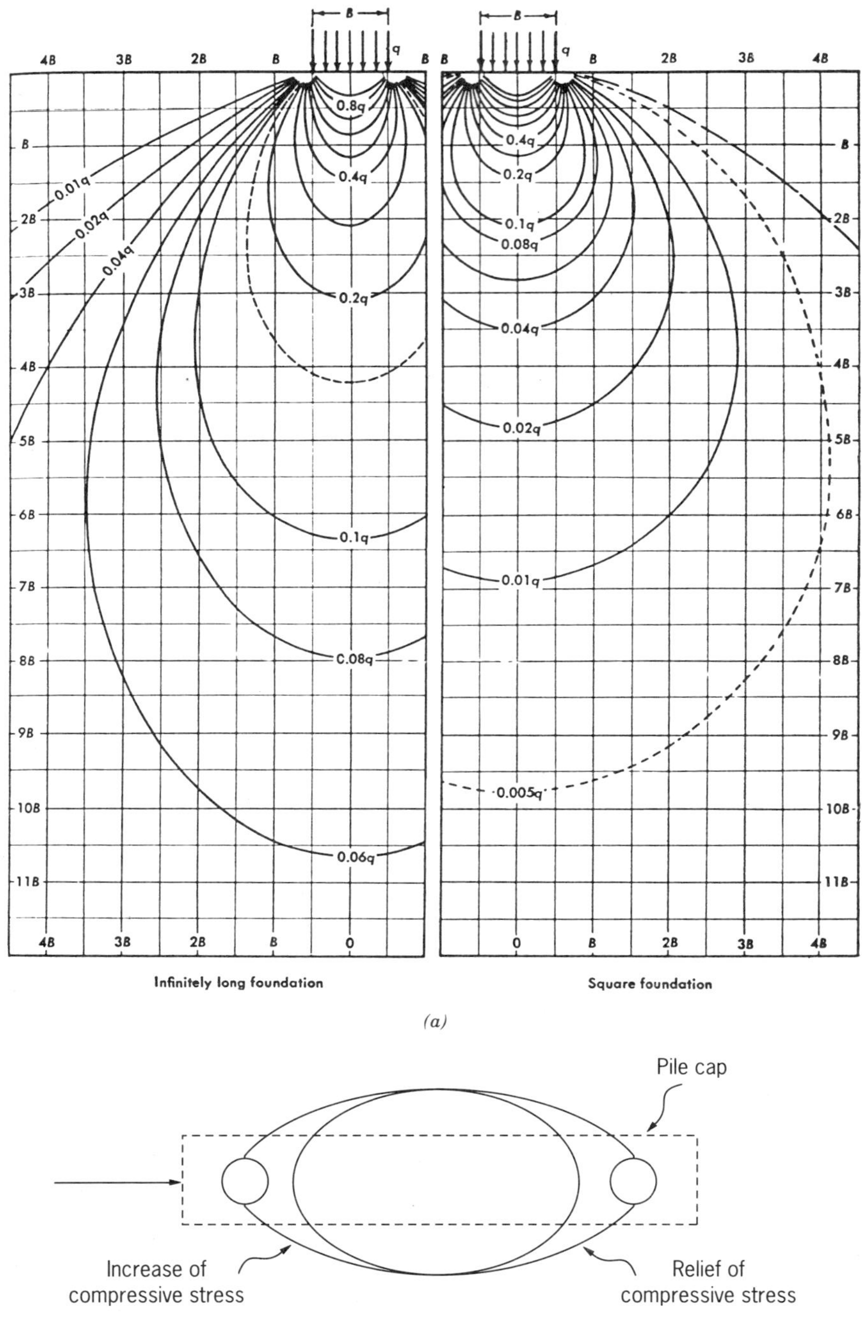


**Figure 1.19** (a) Contours of equal compressive stress intensity below infinitely long and square footing, (After Sowers 1979, Reprinted by Permission, McMillan Publishing Co. New York NY), (b) Stress zones infront and rear of 2-piles under lateral load.

of the stiffness of piles in a pile group. For example, the front corner pile should be the stiffest and an interior pile the most flexible.

The effect that a laterally loaded pile group has on the soil in front of the group may be examined further qualitatively by referring to Figure 1.18. A vertical plate loaded laterally with a uniform pressure produces an arbitrarily limited stress bulb as shown schematically in Figure 1.18a. If the plate is divided into 9 parts and 3 of the pieces are removed (Figure 1.18b), the individual pieces each have a stress bulb that overlaps with the adjacent stress bulbs. This is not a true picture because the effective stress bulb is outside the limits of the individual bulbs and is within the limits of the stress bulb outlined in Figure 1.18a. If half of the remaining pieces are removed, the individual stress bulbs do not overlap, and the effect of the pieces is essentially as defined by the individual bulbs.

The effect of pile spacing on the interference or group action of piles may also be evaluated theoretically. Figure 1.19a shows the Boussinesq compressive stress bulbs on an elastic half-space for a uniformly loaded surface area of width $B$. Both the square footing and the infinitely long footing results are shown. The following discussion assumes that the Boussinesq solution can be used for horizontal pressures exerted in the soil. Referring to the pressure bulb for an infinitely long footing, and arbitrarily regarding as negligible all compressive stresses less than 10% of the applied surface pressure, the approximate limits on pile spacing can be established so that mutual interference does not occur. In the direction of the load, a spacing in excess of $6B$ is indicated, whereas normal to the load a spacing of $4B$ appears appropriate. Note that for two piles aligned in the direction of the load, an increase of pressure on the face of the rear pile adds compressive stresses behind the front pile, whereas deflection of the front pile had relieved compressive stresses in the same zone (Figure 1.19b). Obviously, an upper limit on the spacing would be about $12B$. The effects of group action would probably disappear at pile spacings of about $8B$ in the direction of the load and $3B$ to $4B$ normal to the load. The experimental results indicate that the foregoing theoretical reasoning, although approximate, is quite satisfactory (Prakash, 1962).

***Fixity of Pile Heads*** An analogy with a familiar structure will help illustrate the prime importance of evaluating the fixity of a pile head. In a free-top, fixed-bottom column subjected to a load $Q_h$ at its top, the deflection, $y$ is $Q_hL^3/3EI$ at the top. However, in an identical column in which the top is fixed but may translate, the load $Q_h$ at the top causes a deflection at the top of $y/4$, a reduction of 75 percent when compared to the free-top column. However, for piles embedded in soils, the presence of soils will change this ratio. It has been found that the reduction in deflection of a fixed-translating head pile is reduced by a factor between 1/2 and 1/2.5 (Prakash, 1962).

## 1.9 BUCKLING OF PILES

The buckling of fully embedded piles, under the influence of vertical loads only, appears to be rare. Long, unsupported lengths of timber and H-pile sections have been shown to be vulnerable to buckling failures (Davisson, 1960). However, modern pile practices use very long unsupported lengths of piles for offshore structures; therefore, buckling of piles may become important.

Two cases must be recognized. First, the pile is perfectly vertical and there is no eccentricity in the vertical load. These are ideal situations and may not be fully realized in practice. There is eccentricity both due to pile driving as well as due to vertical load being not at the center of the section.

However, when a lateral and a vertical load are applied simultaneously the deflections due to lateral loads result in automatic eccentricity of the vertical loads. Piles with large eccentricities tend to deflect laterally quite rapidly at low loads. The lateral deflection of the pile produces soil reactions which may exceed the bearing capacity of the soil. Slender pile sections have a low ultimate bearing capacity resistance because the bearing capacity is proportional to the pile width. Buckling is not likely to occur in stiff soils unless the soil reactions become fully plastic (Davisson, 1960).

Three piers that were apparently stable, when loaded vertically, failed when a lateral load was added. It has generally been appreciated that vertical stresses and flexural stresses caused by lateral loads are cumulative. It has not been appreciated, however, that a lateral load translates a pier laterally, thereby producing an eccentricity of the vertical load (Davisson, 1960).

Two test series on single piles have been reported where axial loads were applied to a pile before a lateral load was applied. When compared to piles where only a lateral load is applied, the observed deflections were magnified because of the additional moments in the pile caused by the eccentricity of the vertical load that was produced by lateral loading (Davisson, 1960).

## 1.10 BEHAVIOR OF PILES UNDER DYNAMIC LOADS

Vibrations have been shown to have a pronounced effect on laterally loaded piles embedded in sand. A No. 1 Vulcan hammer was shown to have a negligible effect on the deflection of a laterally loaded pile when it was operated a distance of 50 feet or more from the pile. However, pile deflections increased in inverse proportions to the distance of the pile from the pile driver for distances of less than 50 feet. In the presence of vibration, piles rebounded to their initial position when unloaded. Apparently, vibration may allow the relief of any residual curvature in the pile after a cycle of lateral loading (Davisson, 1960).

However, the more important sources of dynamic loads may be earthquakes and machine foundations or a similar source. In machine foundations, loads are applied on the top of the piles, and the dynamic loads may be estimated from the unbalances in the machines (Prakash and Puri, 1988). The strains in the soils are usually of small magnitude.

The pile foundations may be excited in (1) vertical vibrations, (2) combined horizontal sliding and rocking, and (3) torsional vibrations depending on the nature of the dynamic loads. The soil pile system may be considered elastic and the relevent soil properties are the shear modulus ($G$) and Poisson's ratio ($\nu$). If $G_s$ is the shear modulus around the pile shaft and $G_b$ below the tip of the pile and the values of $G_s$ and $G_b$ are comparable, the pile behaves like a friction pile. However, if $G_b$ is much larger than $G_s$, the pile becomes a bearing pile.

Model tests on an acrylic resin pipe 2 cm in diameter and 40 cm in length were performed under two directional cyclic loading (Kishida et al. 1985). The thin lead sheet is attached inside the pile shaft and the deflection of the pile was taken by the X-ray photographs.

Two kinds of tests were made, one for the dry dense sand of relative density of 95% and the other for Kawasaki clay. The clay was remolded and reconsolidated. The pile and the lead shots were placed in the soils and the horizontal cyclic load was applied at the top of the pile and X-ray photographs were taken during the test.

The relationships between load and displacement at the top of the pile are shown in Figure 1.20a and b. The test result in the sand (Figure 1.20a) indicates that the hysteresis curves under cyclic loadings show about the same shape and that the area enclosed by the curve increases with the increment of load. The test result in the clay (Figure 1.20b), however, indicates the different shape of the hysteresis curves compared with those of sand. The areas enclosed by the curves are much smaller than the ones in the sand.

Movements of the sand and the pile in Figure 1.21a show that the sand in front of the pile is compacted due to movements of the pile and that the sand in back of the pile moves down to the pile shaft decreasing its density. No gap between the sand and the pile was observed. The sand near the pile shaft is compacted during

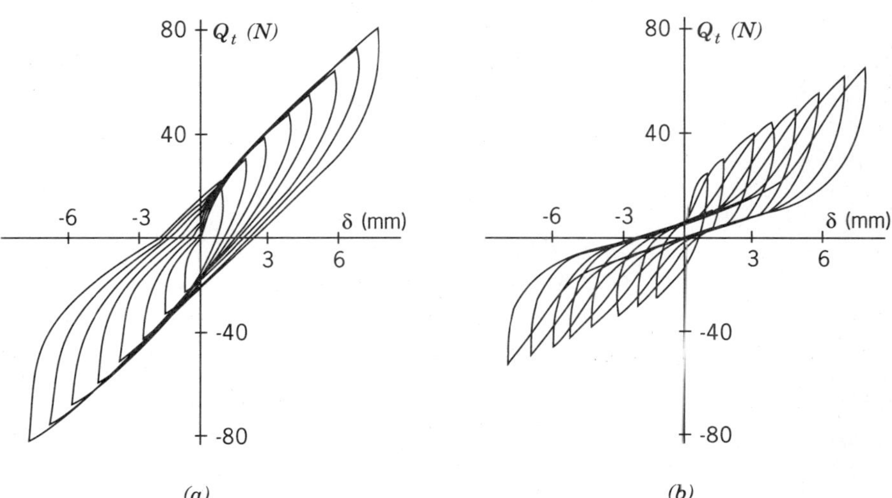


**Figure 1.20** Load displacement relationships: (a) Sand, (b) clay (After Kishida et al. 1985).

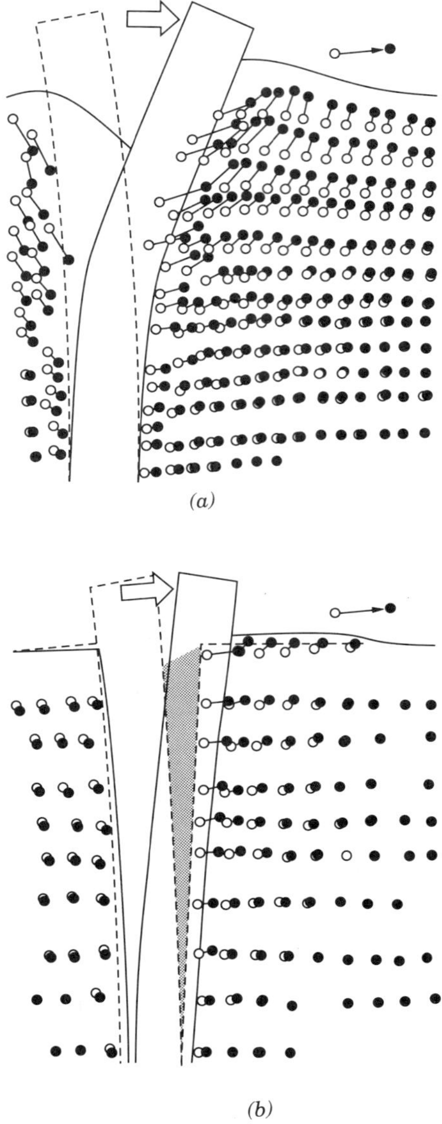


**Figure 1.21** Movements of soils: (a) Sand, (b) clay (After Kishida et al. 1985).

horizontal cyclic loading. The sand had settled to a distance of 10 in. around 10.75 in. pipe pile in lateral cyclic load tests in sand (Brown et al., 1988).

Figure 1.21b shows the gap between the clay and the pile at the back of the pile. The clay in front of the pile is remolded and may decrease its strength significantly.

The value of shear modulus $G_s$ determined from tests on undisturbed samples of soil may, therefore, not be representative of the actual values, since the soil around the pile is considerably disturbed due to remolding in all soils and the time effects in clays. This is a serious problem in practice at this time (1990).

For piles in stiff clays, there may be a loss of contact between the soil and the pile close to the ground surface as seen in Figure 1.21b.

In lateral vibrations also, the disturbance of soil around the pile due to pile driving and time effects in clay will alter the properties of the undisturbed soil. The loss of contact of the pile with the soil near the ground in clays is more serious than for vertical vibrations.

If the pile cap is resting on the ground and is embedded in backfilled soil, the group response will be affected by the contribution to stiffness and damping on both these counts. However, the contact of the pile cap with ground is not certain. A small amount of settlement of sand and shrinkage in clays may result in complete loss of this contact. However, the side soil will contribute to both the stiffness and damping in all modes of vibrations. The properties of the backfilled soil can be controlled to a degree. However, here again, the question of partial loss of contact of the pile cap with the backfilled soil would be ascertained and then only its complete contribution accepted.

In torsional vibrations of a single pile, it is shear along the shaft of the pile that is important, while in a group of piles the lateral stiffness of the pile and its distance from the mass center of gravity of the pile group controls the group stiffness and damping. In fact, depending on the stiffness of the pile in lateral direction and its distance of the center of gravity of the group, the contribution of stiffness of the individual pile to the total stiffness of the group may become negligible. If the soil pile system is considered elastic, there is no material damping in the system. However, geometrical damping will be present.

In earthquake loading, the deformations may be much larger than those in machine foundations. Therefore, the soil behavior is definitely nonlinear, which should be accounted for in an analysis. Also, the loading condition is in the form of ground motion. The response of the pile foundation to a given ground motion is a problem similar to any structural problem and can be solved from the response technique analysis of the soil–pile system. The methods of analysis based upon these concepts are described in Chapter 7.

## 1.11 ACTION OF SOIL AROUND A BORED PILE

In stiff clays and dense sands, the piles may be installed by making a hole in the soil. Such a pile is called a *bored pile*. Bored piles are used to minimize vibrations due to pile driving and reduce heave in adjacent piles and buildings. In clays, the hole may stand without support for a limited time, but in sands the hole must be supported either by a drilling fluid or by a steel casing. The action of soil around a bored pile in clays and sands is quite different than that around driven piles and is described in the following sections.

### 1.11.1 Bored Piles in Clay

Three questions are important in bored piles in clay:

1. Effect of moisture on the adhesion between soil and pile
2. Effect of boring on the bearing capacity at the tip of the pile and
3. Construction problems

***Adhesion Between Soil and Pile*** The adhesion between the pile and the soil has been found to be less than the undrained strength of soil before installation. Softening of the clay immediately adjacent to the soil surface occurs due to (1) absorption of moisture from the wet concrete, (2) migration of the water from the clay away from the pile toward the borehole on excavation of the hole; the stresses at the periphery of the whole are reduced and gradients are set up towards the hole, and (3) water poured into the boring to facilitate operation of the cutting tool.

Meyerhof and Murdock (1953) found that water contents of the clay immediately adjacent to the shaft of a bored pile in London clay increased nearly 4% at the contact surface. However, at a distance of 3 in. from the shaft, the water contents had not altered.

The larger the time taken in excavation and/or the larger the time elapsed between making a hole and its concreting, the larger the changes in moisture content.

***Bearing Capacity at the Tip*** The installation of a bored pile may cause disturbance in the clay just beneath the pile base. Softening of the clay may occur by the action of the boring tools. These effects may result in increased settlements, especially for belled piers, in which the base carries a major proportion of the load. However, base disturbance and softening should have a negligible effect on the ultimate bearing capacity of the base because of the comparatively large mass of clay involved when the base penetrates the clay (Skempton, 1959).

***Construction Problems*** Construction problems that arise with bored piles are discussed by Pandey (1967):

1. Caving of the borehole, resulting in necking or misalignment of the pile
2. Aggregate separation within the pile
3. Buckling of the pile reinforcement

### 1.11.2 Bored Piles in Sand

Bored piles in sand usually require casing or drilling fluid to support the walls of the hole. If a casing pipe is used, its withdrawal while concreting the shaft is likely to disturb and loosen the soil to some extent. Also, some loosening may occur at the bottom of the pile due to baling or "shelling-out" the hole. In underwater operation the upward surge on withdrawal of the baler or shell may loosen the soil for several feet below and around the pile. If the concrete at the base of the piles is compacted with high energy, the disturbed and loosened soil may be recompacted. However, presence of the reinforcing cage may obstruct such compaction. If drilling fluid (mud) is used to keep the hole open, then this mud forms a coating on the soil surface resulting in the reduction of skin friction between the pile and the surrounding soil and must be considered in the design (Chapter 5).

## REFERENCES


Adams, J. I. and Hanna, T. H., "Ground Movements Due to Pile Driving", Proceedings, Conf. on Behavior of Piles, I.C.E. (London) 1970.

Avery, S. B., and Wilson, S. D. "Discussion on Paper by Cummings, Kerkhoff and Peck, "Proc. ASCE, Vol. 75, pp. 1190–1199, 1950.

Brown, D. A., Morrison, C. and Reese, L. C. "Lateral Load Behavior of Pile Groups in Sand," *J. Geot., Engg. Dn.* ASCE, Vol. 114, No. 11, pp. 1261–1276, 1988.

Coyle, H. M. and Reese, L. C., "Load Transfer for Axially Loaded Piles in Clay," *J. Soil Mech and Found Dn*, ASCE, Vol. 92, No. SM-2 March 1966, pp. 1–26.

D'Appolonia D. J. and Lambe T. W., "Performance of Four Foundations on End Bearing Piles," *J. Soil Mech. & Found Dn.*, ASCE, Vol. 97, No. SM1, 1971 pp. 77–93.

Davisson, M. T., "Behavior of Flexible Vertical Piles Subjected to Moment, Shear and Axial Load," Ph.D. Thesis, University of Illinois, Urbana 1960.

Davisson, M. T., "Estimating Buckling Loads for Piles," *Proceeding of the Second Pan-American Conference of Soil Mechanics and Foundation Engineering*, São Paulo, 1963, Vol. 2 pp. 351–369.

De Mello V. F. B.,"Foundations of Buildings on Clay," State of the Art Report, *Proceedings 7th International Conference of Soil Mechanics and Foundation Engineering*, Mexico City, Vol. 2, 1969, pp. 49–136.

Ellison, R. D., D'Appolonia, E., and Thiers, G. R., "Load Deformation Mechanism of Bored Piles," *J. Soil Mech and Found Dn.*, ASCE, Vol. 97, No. SM-4, 1971, pp. 661–678.

Hagerty, D. J. and Peck, R. B., "Heave and Lateral Movements Due to Pile Driving," *J. Soil Mech. and Found. Dn.*, ASCE, Vol. 97, No. SM11, 1971, pp. 1513–1532.

Hegedus, E. and Khosla, V. K., "Pullout Resistance of H-Piles," *J. Geotech. Eng.*, ASCE, Vol. 110, No. 9 September, 1984, pp. 1274–1290.

Hoadley, P. J., Francis, A. J., and Stevens, L. J., "Load Testing of Slender Steel Piles in Soft Clay," *Proceedings of the 7th International Conference of Soil Mechanics and Foundation Engineering*, Mexico City, Vol. 2, 1969, pp. 123–130.

Horn, H. M., "Influence of Pile Driving and Pile Characteristics on Pile Foundations Performances," Notes for Lectures to New York Metropolitan Section, ASCE, *Soil Mechanics and Foundation Group*, 1966.

Kishida, H., Suzuki, Y., and Nakai, S., "Behavior of a Pile Under Horizontal Cyclic Loading," *Proceedings of the XIIth International Conference of Soil Mechanics and Foundation Engineering*, San Francisco, Vol. II, 1985, pp. 1413–1416.

Lambe, T. W. and Horn, H. M., "The Influence on an Adjacent Building of Pile Driving for the MIT Materials Center," *Proceedings of the 6th International Conference of Soil Mechanics and Foundation Engineering*, Montreal, Vol. 2, 1965, pp. 280–285.

Lambe, T. W. and Whitman, R. V., *Soil Mechanics*, Wiley, New York, 1969.

Meyerhof, G. G. and Murdock, L. J., "An Investigation of the Bearing Capacity of Some Bored and Driven Piles in London Clay," *Geotechnique*, Vol. 3, 1953, p. 267.

Nataraja, M. S. and Cook, B. E., "Increase in SPT–N Values Due to Displacement Piles," *J. Geotech, Engg. Dn.*, ASCE, Vol. 109, No. 1, January 1983, pp. 108–113.

Orrje, O. and Broms, B. B., "Effects of Pile Driving on Soil Properties," *J. Soil Mech. and Found. Dn.*, ASCE, Vol. 93, No. SM5, pp. 59–73.

Pandey, V. J., "Some Experiences with Bored Piling," *J. Soil Mech. and Found. Dn.*, ASCE, Vol. 93, No. SM5, 1967, pp. 75–87.

Peck, R. B., "A Study of the Comparative Behavior of Friction Piles," Highway Research Board, Special Report 36, 1958.

Poulos, H. G. and Davis E. H., *Pile Foundations Analysis and Design* Wiley, New York, 1979.

Prakash, S., "Behavior of Pile Groups Subjected to Lateral Loads," Ph.D. Thesis, University of Illinois, Urbana, 1962.

Prakash, S., *Soil Dynamics* McGraw-Hill Book Co., New York, 1981.

Prakash, S. and Puri V. K., *Foundations for Machines, Analysis and Design*, Wiley, New York, 1988.

Prakash, S. and Saran S., "Behavior of Laterally Loaded Piles in Cohesive Soils," *Proceeding of the 3rd Asian Regional Conference on Soil Mechanics and Foundation Engineering*, Haifa (Israel), Vol. 1, 1967, pp. 235–238.

Reddy, A. S. and Valsangker, A. J., "An Analytical Solution for Laterally Loaded Piles in Layered Soils," *Sols-Soils*, No. 21, 1968, pp. 23–28.

Reddy, A. S. and Valsangker, A. J., "Buckling of Fully and Partially Embedded Piles," *J. Soil Mech. and Found. Dn.*, ASCE, Vol. 96, No. SM6, 1970, pp. 1951–1965.

Reese, L. C., Cox, W. R., and Koop, F. D., "Analysis of Laterally Loaded Piles in Sand," *Proceeding of the 6th Off-Shore Technology*, Copy Paper No. TC 208A, 1974.

Seed, H. B. and Reese L. C., "The Action of Soft Clay Along Friction Piles," *Transaction of the American Society of Civil Engineers*, Vol. 122, 1957, pp. 731–754.

Skempton, A. W., "Cast-in-Situ Bored Piles in London Clay," *Geotechnique*, Vol. 9, 1959, pp. 158.

Soderberg, L. O., "Consolidation Theory Applied to Foundation Pile Time Effects," *Geotechnique*, Vol. XII, No. 3, 1962, pp. 217–225.

Sowers, G. F., *Introductory Soil Mechanics and Foundation Engineering*, 4th ed. Macmillan Publishing Co., New York, 1979.

Taylor, D. W., *Fundamentals of Soil Mechanics.* Wiley, New York, 1948.

Terzaghi, K. and Peck, R. B., *Soil Mechanics in Engineering Practice*, 2nd ed. Wiley, New York, 1967.

Vesic, A., "Ultimate Loads and Settlement of Deep Foundations in Sand," *Proceeding Symposium on Bearing Capacity and Settlement of Foundations*, Duke University, Durham NC, 1967, pp. 53.

# 2

# TYPES OF PILES AND PILE MATERIALS

A practicing engineer comes across various types of piles and their trade names in the technical and the contractor's literature. A unified method of classifying piles is not available. In this chapter, several pile classification criteria are identified. Then the criterion used here in a particular section to classify piles is mentioned. Based on a classification system, the use, advantages, and disadvantages of each pile type, material specifications, and the protective measures required for these piles are discussed. A comparison between various widely used pile types is also made. The discussion and data on pile types in this chapter will aid the engineer in selecting appropriate piles suitable for a particular project.

## 2.1 CLASSIFICATION CRITERIA

Information available on piles in literature reveals that they can be classified in different ways (Chellis, 1961, 1962, NAVFAC DM. 7.2, 1982, Fuller, 1983, Tomlinson, 1977, and Vesic, 1977). All these methods of classifications are grouped into one of the following five categories:

1. Pile material
2. Method of pile fabrication
3. Amount of ground disturbance during pile installation
4. Method of pile installation into ground
5. Method of load transfer

Classification of piles based on pile material identifies piles on the basis of their principal material, such as timber, concrete, steel, and composite piles. Common composite piles are either made of timber and concrete or steel and concrete. Whole trees with branches and bark removed are generally used as round timber piles. Timber piles are treated with preservatives when they are either installed above the water table or are installed in marine environment. Timber piles are always installed by driving them into the ground (see Section 2.2). Concrete piles can either be cast-in-place by pouring concrete into a predrilled hole or are precast piles installed by driving them into the ground. Precast concrete piles are either reinforced or prestressed concrete piles (see Section 2.3). Most common types of steel piles are pipe piles and H-section piles (see Section 2.4).

Piles types based on the method of pile fabrication identifies piles if they are prefabricated (i.e., precast or are cast-in-place). Timber and steel piles are always prefabricated. Concrete piles, on the other hand, can either be precast or cast-in-place. Pile types based on the amount of ground disturbance during pile installation can be placed into the following four categories:

1. *Large-displacement* (commonly known as displacement piles) piles displace soil during their installation, such as driving, jacking, or vibration, into the ground. Examples of these types of piles are timber, precast concrete, prestressed concrete, close-ended steel pipe, and fluted and tapered steel tube piles.
2. *Small-displacement* piles displace a relatively small amount of soil during installation. These piles include steel H-sections, open-ended pipe piles, steel box sections, and screw piles. These categories are based on the amount of soil disturbed during pile installation. The terms "large" or "small displacement" used are for qualitative description only, since no quantitative values of displacement have been assigned.
3. *Nondisplacement* piles do not displace soil during their installation. These piles are formed by first removing the soil by boring and then placing prefabricated or cast-in-place pile into the hole from which an equal volume of soil was removed. Their placement causes little or no change in lateral ground stress, and, consequently, such piles develop less shaft friction than displacement piles of the same size and shape. Piling operation is done by such methods, as augering (drilling, rotary boring) or by grabbing (percussion boring). Most common types of nondisplacement piles are bored and cast-in-place concrete piles.
4. *Composite* piles can be formed by combining units in above categories. An example of a displacement type composite pile is having an H-section jointed to the lower end of a precast concrete pile. An example for a displacement and nondisplacement type composite pile is by first driving an open-ended tube, then drilling out the soil and extending the drill hole to form a bored and cast-in-place pile. Numerous other combinations may be formed by combining units in each of the above categories.

Piles types based on the method of pile installation into ground can be divided into driven piles, bored (or drilled) piles and a combination of driven and bored piles. Timber, steel (both H-pile and pipe piles), and concrete (both the precast and compacted expanded base piles) are examples of driven piles. Bored piles are necessarily cast-in-place concrete piles.

Classification of piles based on the method of load transfer from the pile to the surrounding soil consists of end-bearing piles, friction piles, combining end-bearing and friction piles, and laterally loaded piles. End-bearing piles are driven through soft and loose material and their tips rest on the underlying stiff stratum, such as dense sand and gravel, clay shale, or hard rock. Friction piles primarily transfer the load to various soil layers along its shaft. Combined end-bearing and friction piles support the load partly through skin friction to the soil around them and the remaining load is transferred to the underlying denser or stiffer stratum. An example of combined end-bearing and friction piles is cited by Sharma and Joshi (1986). In this case, 24-in. shaft diameter and 36-in. bell diameter cast-in-place 40-ft-long drilled piles were installed through sand till to soft rock called *oilsand*. Full-scale load tests carried out on these piles indicated that approximately 50 percent of the pile load is transferred through skin friction to surrounding "*sand till*" and remaining 50 percent is taken by the base soft rock.

It is apparent from these classification methods that no single method is capable of providing a complete description of the types of piles. In the following paragraphs, piles are first identified based on pile material and on other characteristics, such as method of pile installation, load transfer, which will be used to further describe these piles. Piles are, therefore, classified into following five major categories:

1. Timber piles
2. Concrete piles
3. Steel piles
4. Composite piles
5. Special types of piles

Similar or a slight variation of the above classification method is also used in the literature (ASCE 1984, NAVFAC DM 7.2 Foundations and Earth Structures 1982 and Vesic 1977).

## 2.2 TIMBER PILES

Timber piles are the oldest type of pile foundations that have been used to support the structural loads even before the dawn of the recorded history. These are easy to handle, readily cut to desired lengths, and under favorable environmental conditions can last a very long period of time. Several species of timber piles are used depending on their application and availability. For example, Southern Yellow Pine can provide piles up to 75 ft (23 m) in length and West

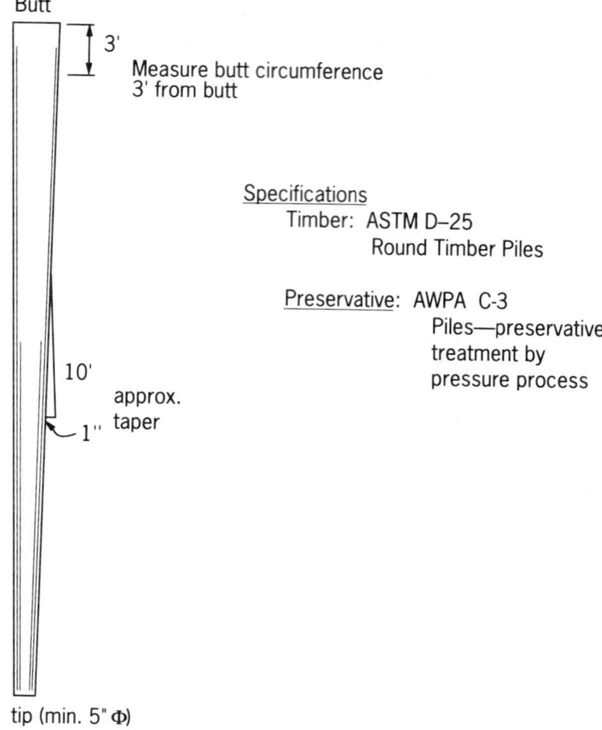


**Figure 2.1** Typical timber pile (ASCE, Committee on Deep Foundations, 1984).

Coast Douglas Fir can be used to provide piles in lengths of about 120 ft (37 m). Timber piles fully embedded below the permanent fresh groundwater level may last for many years without treatment (i.e., untreated timber piles). However, where timber piles either extend above the groundwater level or are installed under marine environment, the piles should have creosote pressure treatment to prevent decay. Figure 2.1 exhibits a typical timber pile and the applicable specifications.

In the following paragraphs, the use of timber piles, material specifications, and the material deterioration and protection methods are presented.

### 2.2.1 Use of Timber Piles

Timber piles can either be round untrimmed logs or sawed square sections. The practice of sawing can be detrimental to its durability since it removes the outer sapwood that absorbs preservatives. The most economical form of timber piles consists of round untrimmed logs.

Timber piles are best suited as friction piles in granular soils. They normally are used as friction piles in sands, silts, and clays. The piles cannot be driven against high resistances without damage; therefore, they are generally not recommended for use in dense gravel or till or as end-bearing piles to rock.

Common lengths used for these piles may range from 20 to 60 ft (6 to 20 m) for diameters of 6 to 16 in. (150 to 400 mm). These dimensions correspond to the natural sizes of available tree trunks. The design loads vary from 10 to 50 ton (89 to 441 kN). However, as a protection against damage due to high driving, timber piles are rarely used for loads in excess of 30 tons (267 kN) (Peck, Hanson, and Thornburn 1974). Capozzoli (1969) cites case histories from three projects where timber piles were successfully load tested to between 75 and 100 tons (668 kN and 890 kN). One of these projects was a seven-story men's dormitory constructed on pressure-treated timber foundation piles having 40 tons (356 kN) design load per pile for Southwestern University, Lafayette, LA. For this project, a 32-ft (9.6 m) pile had 7/16 in. (11 mm) gross movement at the top at a 100-ton (890 kN) test load. The soil at this site consisted of 30 ft (9 m) stiff clay over dense sand. The pile tip was driven into dense sand at four blows per inch with a Vulcan No. 1 hammer. This confirms the successful use of timber piles for the load and length ranges mentioned above.

### 2.2.2 Material Specifications

Timber piles should be free from large or loose knots, splits, decay, and sharp bends. It should have uniform taper from butt to tip, and the center of butt and tip within pile body should lie on a central line. Bark should be removed from timber piles where they are to be used primarily as friction piles. This should be specially removed where they are to carry uplift forces by skin friction, because if this is not done, a slip may occur between the bark and the trunk. Furthermore, when timber piles are treated with preservatives, removal of the bark will increase the depth of impregnation of the preservative.

ASTM D25 material specifications are applicable for round timber piles. Also, preaugering or preboring through hard stratum should be recommended so that the material damage due to hard driving is reduced.

Many codes of practice specify a maximum working stress in the pile material. Other codes limit the maximum load that can be carried by a pile of any diameter. This limit is applied to avoid the risk of damage to a pile during driving. Piles should, therefore, conform to the requirements of local codes. (See Stresses (Allowable) in Piles, FHWA Report, 1983.)

### 2.2.3 Material Deterioration and Protection

Material deterioration is caused by one or a combination of factors, such as decay due to fungi, insect attack, marine borer attack, and mechanical wear. Growth of fungi needs moisture, air, and favorable temperature. Decay of timber pile caused by the growth of fungi can, therefore, be prevented if the timber can be kept either dry or permanently submerged. Thus, timber piles, when situated wholly below permanent groundwater level, are resistant to fungal decay. However, the portion of piles exposed to soil or air above the permanent water table are vulnerable to decay particularly when these are subjected to lowering and raising of the water table.

Insects like termites are destructive to timber piles. Beetles may damage them above high water level. Also, no marine location is safe from causing serious damage to timber piles from marine borers. Insects have also been found above the Arctic Circle.

The life of timber piles above the permanent water table can be considerably increased by treating it with creosote, oil-borne preservatives, or salts. Creosote application by pressure treatment is the most effective method of protection for long preservation. The effective duration of this treatment has not yet been fully established, but it is known that the life of the pile is increased by about 40 years by this treatment. Effectiveness of chemical treatment to timber piles in brackish or salt water should be fully investigated before it is used as preservative. This is because various marine organisms such as teredo and limnorio may attack chemically treated piles (Peck, Hanson, and Thornburn 1974). Furthermore, treated or untreated timber piles may also lose strength under long-term effects of high temperatures when used as foundation units under structures such as blast furnaces and chemical reaction units. Therefore, timber piles are not recommended under such structures.

Timber piles may also be subjected to mechanical wear such as abrasion. Various recommended methods for protecting timber piles against such mechanical wear are to place fill around damaged piles, armor placement to provide resistance to abrasion, and concrete encasement of piles. These methods should be used in conjunction with creosote treatment.

## 2.3 CONCRETE PILES

There are numerous ways of classifying concrete piles depending on installation techniques, equipment and material used for installation, and propriety names. Information on different types of concrete piles, their uses, material specifications, and protection against material deterioration is included in this section. Details on concrete piles can also be found in ACI 543 (1980).

### 2.3.1 Types and Use of Concrete Piles

Concrete piles can be classified into following three major categories:

1. Precast concrete piles
2. Cast-in-place concrete piles
3. Composite concrete piles

Precast concrete piles can be further divided into reinforced piles and the prestressed piles. Prestressed piles can either be pretensioned or posttensioned.

Cast-in-place concrete piles can either be installed as cased or as uncased piles. The three general types of cased concrete piles are cased-driven shell piles, drilled-in-caisson piles, and the dropped-in-shell piles. Common types of uncased concrete piles are (1) uncased driven casing piles, where casing in withdrawn after the hole is filled with concrete, (2) cast-in-drilled hole piles (these piles are also called drilled piers), (3) Franki piles, which are also called compacted or expanded base compacted piles (in some engineering literature, they are also called pressure injected footings), and (4) auger grout or concrete injected piles. These pile types are further discussed in the following paragraphs.

***Precast Concrete Piles*** As the name suggests, these piles are cast, cured, and stored in a yard before they are installed in the field, mostly by driving. These piles are available in various cross-sectional shapes such as circular, octagonal, or square with chamfered corners and may have central core holes to save weight. Precast concrete piles must be designed to withstand handling and driving stresses in addition to service loads. They can be designed to carry a wide range of loads (typically up to 300 tons or 2670 kN) and can be reinforced for bending and uplift. These piles are useful in carrying fairly heavy loads through soft material to firmer strata as end-bearing piles. They are also suitable for use as friction piles when driven in sand, gravel or clay.

Precast concrete piles can be subdivided into two categories:

1. Reinforced precast concrete piles
2. Prestressed concrete piles

*Reinforced Precast Concrete Piles* Typically, these piles are of 40 to 50 ft (12 to 15 m) length. The maximum allowable stress for precast piles is 33 percent of 28-day concrete strength. These piles consist of internal cage reinforcement having four or more longitudinal bars. The lateral or tie reinforcement is provided in the form of individual hoops or a spiral. In order to resist driving forces, tie reinforcement is closely spaced at the ends. Minor cracking with crack widths up to 0.01 in. (0.25 mm) is normally considered acceptable in these piles because cracking is virtually impossible to prevent. These cracks may cause deterioration of pile under environmental conditions such as marine or freeze-thaw action. These piles have, therefore, generally been replaced by prestressed concrete piles in North America.

*Prestressed Concrete Piles* These piles are constructed by using steel rods or wires under tension to replace the longitudinal steel used in reinforced concrete piles. This steel is enclosed in a conventional spiral. These piles can be further subdivided into (1) pretensioned and (2) posttensioned piles. Pretensioned prestressed concrete piles are usually cast full length. Their lengths can be as much as 130 ft (40 m). Posttensioned prestressed piles are usually manufactured in sections and can either be assembled and prestressed to required lengths in the plant or at the site. Figure 2.2 shows a typical pretensioned prestressed pile, and

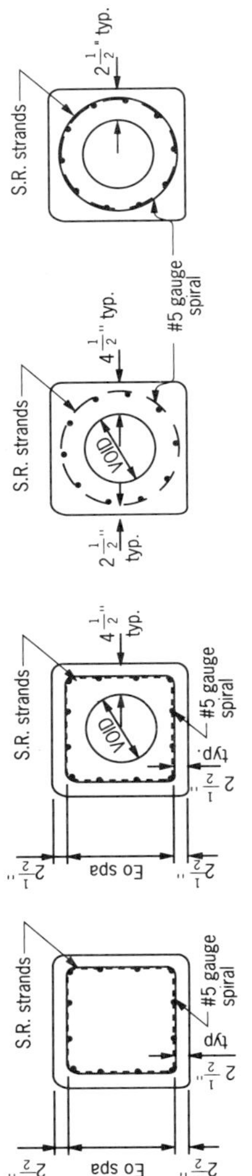


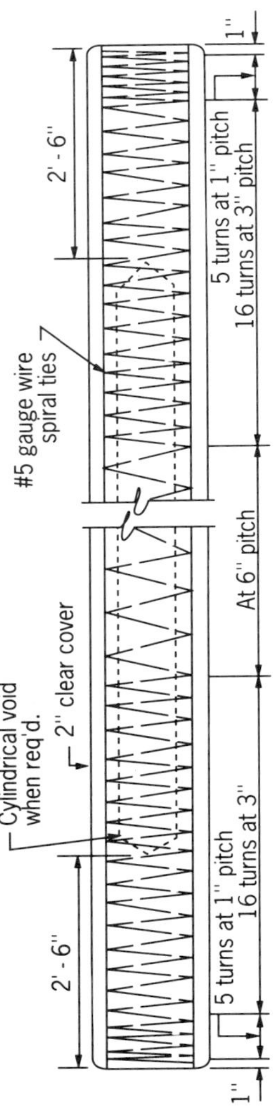


**Figure 2.2** Typical design of a prestressed (pretensioned) concrete pile (ASCE, Committee on Deep Foundations, 1984).

Figure 2.3 shows section properties and allowable loads for prestressed concrete piles.

Raymond cylinder prestressed concrete pile is an example of these piles. These piles are made up of a series of hollow-spun concrete sections reinforced with longitudinal and spiral steel. After curing, sections are assembled and high-strength steel wires are threaded through the holes, tensioned, and locked in place. The wire holes are grouted and locking devices are removed after the grout has set. These piles are then picked up as a unit and are ready for installation. Engeling et al. (1984) present a case history of the design and construction of about 1500 prestressed Raymond concrete cylinder piles that were installed for the Ju'Aymah Trestle in the Arabian Gulf. The piles ranged from 85 to 160 ft (26 to 49 m) in length, 54 to 66 in. (1350 to 1650 mm) diameter, and had an ultimate compression loads of 1400 kips (6230 kN) and ultimate tension loads as high as 560 kips (2492 kN).

Prestressed piles are well suited to soil or water conditions that require high-capacity long piles. These piles can usually be made lighter and longer than conventionally reinforced solid section concrete piles. Prestressed concrete piles are also more durable than reinforced concrete piles because the concrete is under continuous compression. This prevents spalling during driving; also, compression keeps hairline cracks closed and deleterious chemicals do not easily penetrate the concrete mass.

A case history reported by Dugan and Freed (1984) cites cases in which 14 in. (350 mm) and 16 in. (400 mm) square precast prestressed concrete piles were installed in the Boston area for buildings ranging from 5 to 40 stories high. The pile lengths varied from 90 to 160 ft (28 to 49 m) and their axial compression load capacities ranged from 140 kips (623 kN) to 350 kips (1558 kN). These piles were driven through clay into end bearing glacial till or on bedrock.

***Cast-in-Place Concrete Piles*** These piles are installed by placing concrete in a hole formed in the ground either by driving, boring, jetting, coring, or a combination of these and other methods. These piles have the following major advantages over precast piles:

1. These piles do not need casting and storage yards, do not require splicing or cutting off, and are only designed for service loads since they are not subject to driving and lifting stresses.
2. Pile lengths can be adjusted to suit field requirements; therefore, predetermination of pile length is not critical.

Various types of cast-in-place concrete piles are shown schematically in Figure 2.4. The following information on these piles is useful to the reader.

*Cased-Driven Cast-in-Place Concrete Piles* Installation procedure for "cased-driven shell piles" (Figure 2.4a) consists of (1) driving the steel casing, (2)

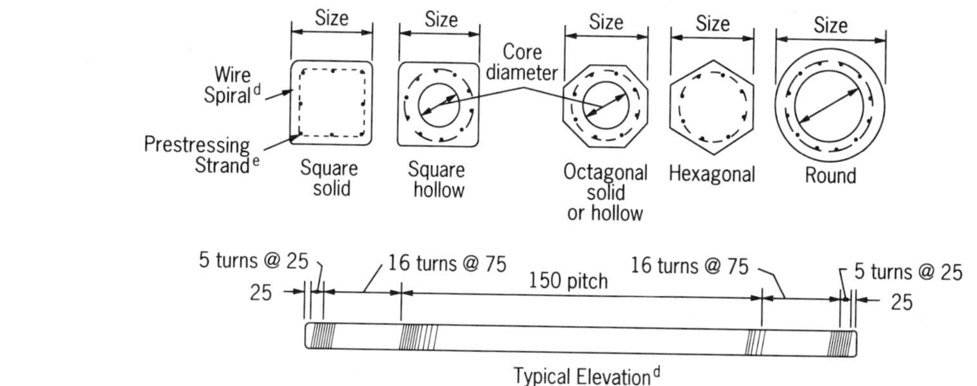

Typical Elevation[d]

| | | Section Properties[a] | | | | | | Allowable Concentric Service Load[b,c] (kN) | | | | |
|---|---|---|---|---|---|---|---|---|---|---|---|---|
| | | | | | | | | $f'_c$ (MPa) | | | | |
| Size (mm) | Core Diameter (mm) | Area (mm²) | Mass (kg/m) | Moment of Inertia ($10^6$ mm⁴) | Section Modulus ($10^3$ mm³) | Radius of Gyration (mm) | Perimeter (m) | 35 | 40 | 45 | 50 | 55 |
| Square Piles | | | | | | | | | | | | |
| 250 | Solid | 63,000 | 151 | 326 | 2,610 | 72 | 1.00 | 646 | 750 | 853 | 958 | 1,060 |
| 300 | Solid | 90,000 | 216 | 675 | 4,500 | 87 | 1.20 | 922 | 1,070 | 1,220 | 1,370 | 1,520 |
| 350 | Solid | 123,000 | 295 | 1,250 | 7,140 | 101 | 1.40 | 1,260 | 1,460 | 1,670 | 1,870 | 2,070 |
| 400 | Solid | 160,000 | 384 | 2,130 | 10,700 | 116 | 1.60 | 1,640 | 1,900 | 2,170 | 2,430 | 2,700 |
| 450 | Solid | 203,000 | 487 | 3,420 | 15,200 | 130 | 1.80 | 2,080 | 2,420 | 2,750 | 3,090 | 3,420 |
| 500 | Solid | 250,000 | 500 | 5,210 | 20,800 | 144 | 2.00 | 2,560 | 2,970 | 3,390 | 3,800 | 4,210 |
| 500 | 275 | 191,000 | 458 | 4,930 | 19,700 | 161 | 2.00 | 1,960 | 2,270 | 2,590 | 2,900 | 3,220 |
| 600 | Solid | 360,000 | 864 | 10,800 | 36,000 | 173 | 2.40 | 3,690 | 4,280 | 4,880 | 5,470 | 6,070 |
| 600 | 300 | 289,000 | 694 | 10,400 | 34,700 | 190 | 2.40 | 2,960 | 3,440 | 3,920 | 4,390 | 4,870 |
| 600 | 350 | 264,000 | 634 | 10,100 | 33,700 | 196 | 2.40 | 2,710 | 3,140 | 3,580 | 4,010 | 4,450 |
| 600 | 375 | 250,000 | 600 | 9,830 | 32,800 | 198 | 2.40 | 2,560 | 2,970 | 3,390 | 3,800 | 4,210 |

| | | | | | | | | | | | | |
|---|---|---|---|---|---|---|---|---|---|---|---|---|
| **Octagonal Piles** | | | | | | | | | | | | |
| 250 | Solid | 52,000 | 125 | 215 | 1,720 | 64 | 0.77 | 533 | 620 | 704 | 790 | 876 |
| 300 | Solid | 75,000 | 180 | 446 | 2,970 | 77 | 0.92 | 769 | 892 | 1,020 | 1,140 | 1,260 |
| 350 | Solid | 101,000 | 242 | 825 | 4,710 | 90 | 1.07 | 1,030 | 1,200 | 1,370 | 1,540 | 1,700 |
| 400 | Solid | 133,000 | 319 | 1,410 | 7,050 | 103 | 1.22 | 1,360 | 1,580 | 1,800 | 2,020 | 2,240 |
| 450 | Solid | 168,000 | 403 | 2,260 | 10,000 | 116 | 1.38 | 1,720 | 2,000 | 2,280 | 2,550 | 2,830 |
| 500 | Solid | 207,000 | 497 | 3,440 | 13,800 | 129 | 1.53 | 2,120 | 2,460 | 2,800 | 3,150 | 3,490 |
| 500 | 275 | 148,000 | 355 | 3,160 | 12,600 | 146 | 1.53 | 1,520 | 1,760 | 2,000 | 2,250 | 2,490 |
| 550 | Solid | 251,000 | 602 | 5,030 | 18,300 | 142 | 1.68 | 2,570 | 2,990 | 3,400 | 3,810 | 4,230 |
| 550 | 325 | 168,000 | 403 | 4,480 | 16,300 | 163 | 1.68 | 1,720 | 2,000 | 2,280 | 2,550 | 2,830 |
| 600 | Solid | 298,000 | 715 | 7,130 | 23,800 | 154 | 1.84 | 3,050 | 3,550 | 4,040 | 4,530 | 5,020 |
| 600 | 375 | 188,000 | 451 | 6,160 | 20,500 | 181 | 1.84 | 1,930 | 2,240 | 2,550 | 2,860 | 3,170 |
| **Round Piles** | | | | | | | | | | | | |
| 900 | 650 | 304,000 | 730 | 23,400 | 52,000 | 277 | 2.83 | 3,120 | 3,620 | 4,120 | 4,620 | 5,120 |
| 1,200 | 950 | 422,000 | 1,010 | 61,800 | 103,000 | 383 | 3.77 | 4,320 | 5,020 | 5,720 | 6,410 | 7,110 |
| 1,350 | 1,100 | 481,000 | 1,150 | 91,200 | 135,000 | 435 | 4.24 | 4,930 | 5,720 | 6,520 | 7,310 | 8,100 |
| **Hexagonal Piles** | | | | | | | | | | | | |
| 300 | Solid | 78,000 | 187 | 486 | 3,240 | 79 | 0.90 | 800 | 928 | 1,060 | 1,190 | 1,320 |
| 350 | Solid | 106,000 | 254 | 900 | 5,140 | 92 | 1.05 | 1,090 | 1,260 | 1,440 | 1,610 | 1,790 |
| 400 | Solid | 139,000 | 334 | 1,540 | 7,700 | 106 | 1.20 | 1,420 | 1,650 | 1,880 | 2,110 | 2,340 |

[a]Form dimensions may vary with producers, with corresponding variations in section properties.

[b]Allowable loads based on $N = (A/10^3)(0.33 f_c^1 - 0.27 f_{pc})$: $f_{pc} = 4.8$ MPa: Area in min$^2$.

[c]Allowable loads based on short column structural capacity only.

[d]Wire spiral varies with pile size.

[e]Strand pattern may be circular or square.

**Figure 2.3** Section properties and allowable loads for prestressed concrete piles (CPCI, 1982).

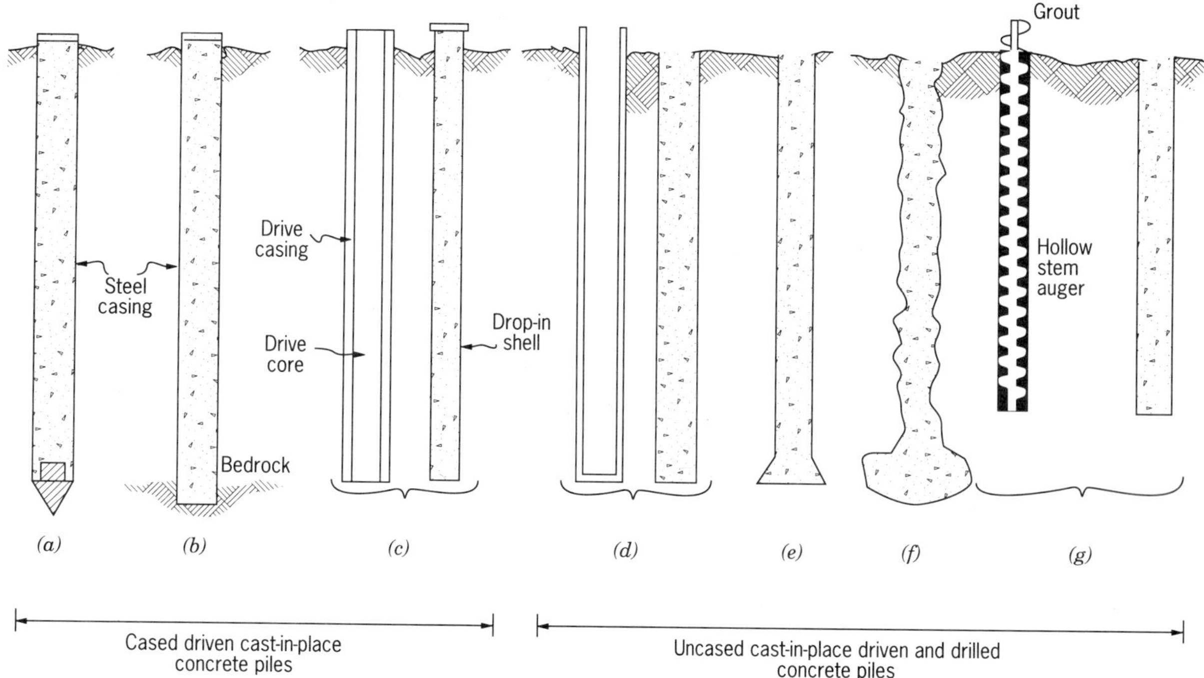


**Figure 2.4** Diagrammatic sketches of cast-in-place concrete piles. (a) Cased driven shell pile, (b) drilled-in caisson, (c) dropped-in shell pile, (d) uncased driven casing pile, (e) drilled pier (f) Franki or expanded base compacted pile, (g) auger grout injected pile.

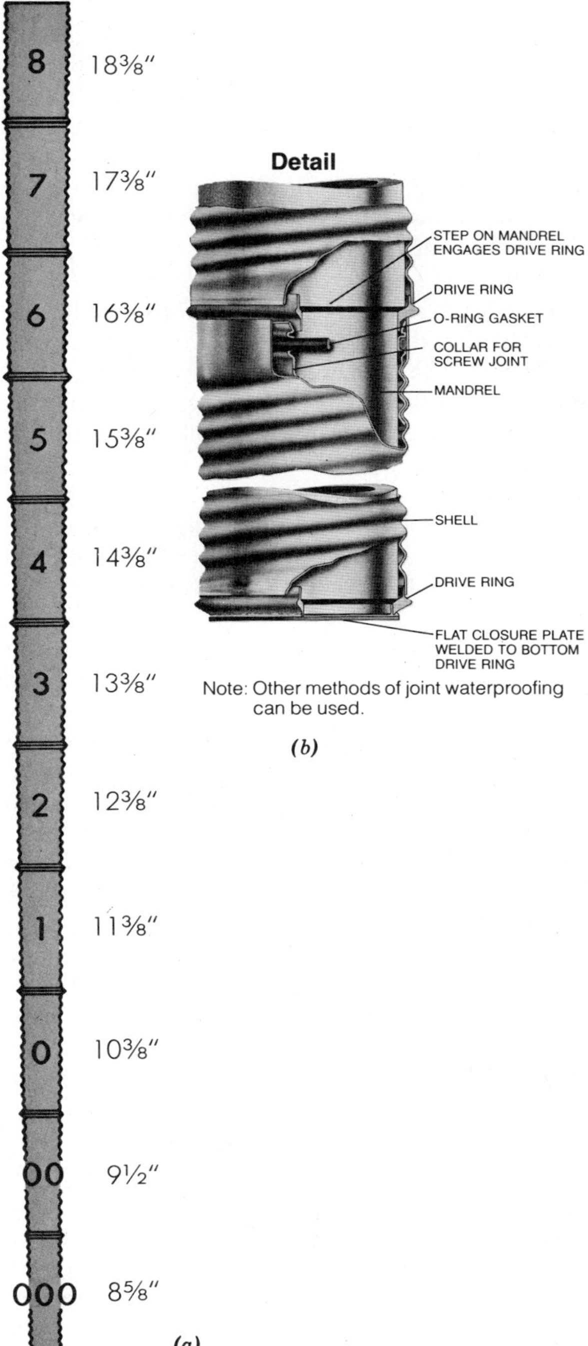


**Figure 2.5** (a) Nominal dimensions of Raymond step-taper piles (b) Detail (Raymond International, Inc., 1985).

inspecting the casing for damages, and (3) filling the driven casing with concrete. The driven steel casing can either be thin corrugated shells, or pipe (either open or close ended), or longitudinally fluted tubular shells. These piles are suitable when freshly placed concrete needs protection against ground pressures and intrusions.

"Drilled-in-caisson piles" (Figure 2.4b) are installed by (1) driving a heavy-wall open-end pipe to bedrock, (2) cleaning out the inside of the pipe by coring or jetting, (3) drilling a socket into the bedrock, and (4) filling the entire socket and pipe with concrete. This is suitable as a high-capacity pile to bedrock.

Method of installation for dropped-in-shell concrete piles (Figure 2.4c) consists of (1) driving a closed ended steel casing, (2) dropping a steel shell inside the drive casing, (3) filling the inner shell with concrete, and (4) extracting the outer steel drive casing. This pile is suitable when the concrete shaft is to be formed through unstable soil and water pressures may be high. Another type of steel driven concrete filled pile is Raymond step-taper pile. This pile is installed by driving a closed-end steel shell (Figure 2.5a) with a heavy steel mandrel (Figure 2.5b) to the required resistance to penetration. The mandrel is then withdrawn, and the shell is filled with concrete. The shell is helically corrugated to resist subsoil pressures. Typically, these piles are about 120 ft (36 m) long and maximum allowable stress is 33 percent of 28-day concrete strength.

*Uncased Cast-in-Place Driven and Drilled Concrete Piles* Uncased driven casing piles (Figure 2.4d) are installed by (1) driving a steel casing closed at the end with either an internal mandrel or unattached closure point or driven core, (2) removing the core and filling with concrete, and (3) extracting the casing. Sometimes, an enlarged base can be formed by driving out some of the concrete through the bottom. These piles need fairly close inspection and control because they can be damaged from soil pressures resulting from adjacent pile driving.

Drilled piers (Figure 2.4e) are installed (1) by mechanically drilling a hole to required depth and (2) filling the hole with reinforced or plain concrete, as required. Sometimes an enlarged base is formed by a belling tool. When the sides of the hole are unstable, either a temporary steel liner can be inserted or stabilizing bentonite slurry can be used during drilling. Sharma et al. (1984) cite a case where about 1500 cast-in-place drilled and belled concrete piles were installed at a petrochemical plant site. The pile shaft diameter varied from 20 to 40 in. (500 to 1000 mm) and their lengths varied from 20 to 40 ft (6 to 12 m). These piles were drilled through clay till into clay shale rock that was under artesian pressure. The piles were reinforced with 8–25 M vertical bars complete with 10 M ties at 12 in. (300 mm) spacing. These piles were designed to carry axial loads ranging from 50 kips (220 kN) to 110 kips (490 kN) and lateral loads of about 5 kips (20 kN). Full-scale pile load tests were carried out to confirm these pile capacities. Longer drilled piers (also called caissons) can also be installed to suit site conditions. For example, the U.S. Corps of Engineers designed three bridges for a highway to cross a proposed canal near St. Stephens, South Carolina, where 42-in. (1050 mm) diameter and 54 ft (17 m) long drilled piers were installed through sand–clay mixture into dense sand. Slurry was used to keep the hole from caving in and prevent groundwater from entering the excavation (Lane, 1984). The design axial load of piles was 470 kips (2100 kN).

Another type of concrete piles commonly known as Franki piles or expanded base compacted piles (also called pressure-injected footings) are installed by driving a steel casing into the ground. This is done by using a drop weight inside the casing and driving on a zero slump concrete at the bottom of the casing. When the required depth is reached, the casing is held in place and the plug is driven out. The base is then enlarged by ramming more dry concrete into the pile base. The pile shaft is then formed by pouring the concrete as the steel casing is withdrawn (Figure 2.4f). Another type of cast-in-place pile is the "auger grout injected pile"

**Figure 2.6** Comparison of normal drilled and belled pile and bored compaction pile (Rai and Jai Singh, 1986).

(Figure 2.4g). This pile is installed by pumping grout through hollow stem of the auger as it is withdrawn. These two pile types (expanded base compacted, and auger grout injected piles) are further discussed in Section 2.6.

A pile that combines the advantages of both bored and driven piles is called bored compaction piles. In these piles, after the pile has been bored and concreted, the reinforcement cage is driven into the freshly laid concrete. This results in compacting both the surrounding soil and the concrete. Therefore, these piles are particularly suited in loose to medium dense sandy and silty soil conditions. Figure 2.6 shows the size differences that can be achieved in bored compaction piles when compared with normal bored and belled piles installed under similar soil conditions. The extra compaction of the surrounding soil and the enlarged pile size due to driving operation may result in an increase in load carrying capacity by 1.5 to 2.0 times over the normal and belled piles (Rai and Jai Singh, 1986).

***Composite Concrete Piles*** Composite concrete piles are made either by encasing the steel or timber piles by concrete in the zone susceptible to deterioration or by making steel sections at lower part and concrete in upper areas where hard driving may be encountered. Further information on these pile types has been included in Section 2.5.

### 2.3.2 Material Specifications

Materials that are used for various concrete piles and/or their components are concrete, reinforcement, steel casing, structural steel cores, grout, anchorage, and splices. Concrete piles must conform to the requirements of national building codes (e.g., subsection 4.2.3 of the National Building Code of Canada, 1980 or ACI Code 318).

Material specifications for concrete mix should be designed as per "Recommendations for Design, Manufacture, and Installation of Concrete Piles," reported by ACI Committee 543 R-74, reaffirmed in 1980. Some of the material requirements described in these recommendations are as follows:

1. Cement content: For durability, concrete piles should have the minimum requirements as specified in Table 2.1.

**TABLE 2.1 Cement Content for Various Site Conditions**

| Site | Cement Weight/Volume of Concrete | |
|---|---|---|
| | lb/yd$^3$ | kg/m$^3$ |
| Normal environment | 564 | 335 |
| Marine environment | 658 | 390 |
| Tremie placement | 564–752 | 335–446 |

Source: ACI 543 (1980).

TABLE 2.2 **Slump for Various Pile Types**

| Conditions | Usually Specified Slump | |
|---|---|---|
| | in. | mm |
| Cast-in-place piles | 3–6 | 75–150 |
| Precast piles | 0–3 | 0–75 |
| Tremie placed concrete | 6–8 | 150–200 |

Source: ACI 543 (1980).

2. Concrete slump: Slumps indicate the workability of concrete and is related to water content of the mix, Table 2.2 lists general recommendations for usual slump values for various conditions.

### 2.3.3 Material Deterioration and Protection

Concrete piles may be subject to following deteriorating conditions (Chellis, 1962).

1. Destructive chemicals in groundwater
2. Destruction due to seawater
3. Damage due to freezing and thawing
4. Damage due to handling and driving stresses, and
5. Damage due to concrete material defects

Destructive chemicals in groundwater may cause serious damage to concrete piles. These chemicals may come from manufacturing plant wastes, leaky sewers, and other sources. The severity of these damages may also depend on the availability of air that accelerates the deterioration process. For example, in sandy soils, which permit leaching and provide more air, chemical damage is more severe than in clayey soils. Groundwater must be chemically analyzed and concrete specialist be consulted to determine the long-term impact of these chemicals on the durability of concrete.

Seawater may cause deterioration in concrete in many ways such as abrasive action (from ice, debris, wind, and waves), mechanical action, and chemical action. Mechanical action may cause deterioration if freezing water in the pores causes progressive disintegration and exposes reinforcing bars. Concrete piles often have surface cracks caused by shrinkage, temperature differences, and tension. Chemical destruction of concrete piles in seawater is promoted by these cracks that causes reinforcing bars to rust. Concrete piles that are exposed to freezing and thawing conditions should therefore contain air-entraining admixtures. These admixtures also reduce the water–cement ratio resulting in low absorption factor (low permeability). This makes concrete less susceptible to sulphate attack in environments, such as seawater because of reduced water penetration into concrete.

Many methods to protect concrete piles against destructive environments are available, which include painting, asphalt impregnation, steel points, concrete armor, shotcrete encasement, wrought iron armor, creosoted wood jackets, and Fabriform pile jacket. Hunt (1979) cites an example in which steel sections HP 14 × 102 with cast steel points were attached to prestressed 24-in. octagonal piles that were required for the Trident submarine home base. The points were installed so that the piles can penetrate glacial till.

The prestressed pile combined with steel H pile can also be a solution to the corrosion problem. For example, for piling in saline water dock facilities, strong H-steel section with cast steel point, if required, can be used below the depth of corrosion. This will also facilitate driving through waterfront debris into underlying rock. One example for such extension in saline water environment is using of Pruyn Point 75600 on HP 14 × 89 extensions of 18-in. octagonal prestressed piles at the Port of Vancouver, WA (Hunt, 1979). The Fabriform Pile Jacket System first introduced by Intrusion-Prepakt can also be used to protect piles against marine environment. A case is cited in Intrusion-Prepakt (1981) in which 16-in. (400 mm) square concrete piles were badly damaged in the 3-ft (0.9 m) tidal range and needed repair. In this case, the deteriorated concrete was removed, and the piles were encased with preassembled synthetic Fabrifoam sleeves or pile jackets. The repair was then done by pumping concrete into the voids. It is reported that, after repair, these piles have been in service without damage for at least 7 years. Selection of a protective method should depend on local experience, specific soil, water and environmental conditions, and the economic life of the structure.

## 2.4 STEEL PILES

Steel piles are strong, lightweight to handle, and capable of carrying heavy loads to deeper bearing stratum. They can be extended to any length since splicing is relatively easy, and these can also be readily cut to any required length. This makes steel piles suitable for areas where the depth of bearing strata are variable.

### 2.4.1 Types and Use of Steel Piles

Various types of steel piles in common use include pipe piles, H-section piles, box section piles, and tapered and fluted tubes (Monotubes). Pipe piles and H-section piles are the most commonly used steel piles in engineering practice.

Steel pipe piles can either be driven open ended or closed ended. Open-ended piles will experience less driving resistance and can be drilled through obstructions such as boulders and bedrock. Circular shape of the pipe piles have two main advantages: (1) the soil within the pipe can be easily taken out since there are no obstructions for cleaning out tools (e.g., no corners), and (2) the circular shape minimizes drag from waves and current forces in deep waters. Pipe piles can also be inspected for any damage and/or deviation from plumb by lowering a light source within the hollow section. As shown in Figure 2.7, pipe piles can also be fitted with end caps in areas of hard driving. Where the hard-bearing strata are inclined or sloping, the flat plate at the end of pipe may cause uneven stresses on the pipe pile resulting into stress concentration and crippling of the pile. In such situations, conical points, as shown on the pile on the right in Figure 2.7, are used to distribute the stress around the pipe.

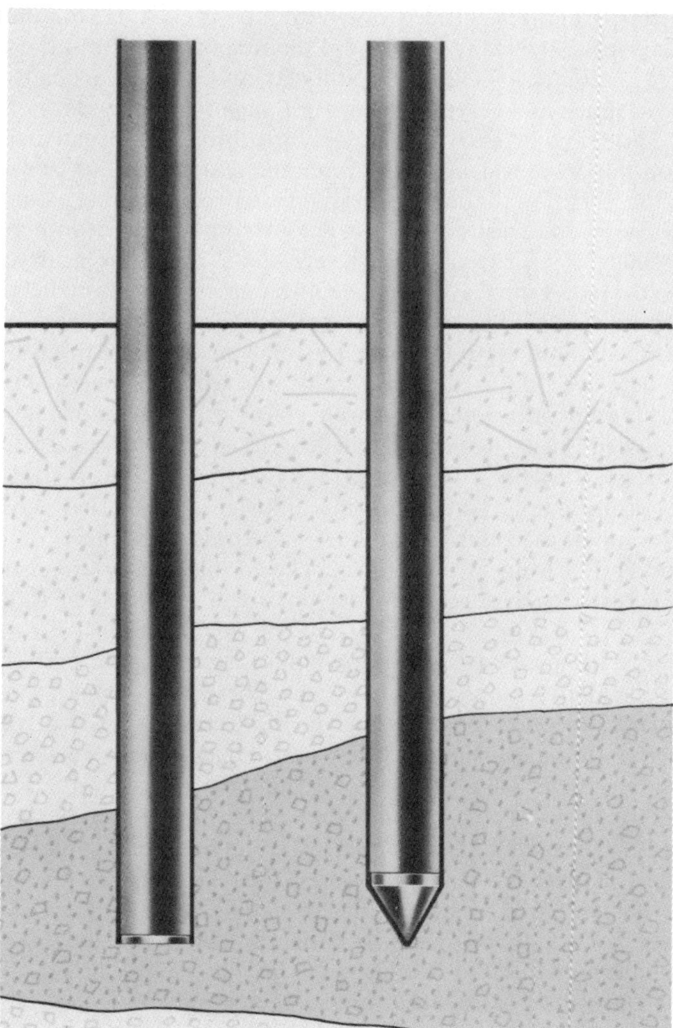

**Figure 2.7** Typical pipe pile with tip fittings (Courtesy: Associated Pile and Fitting Corp., Bulletin PP777, 1985).

Pipe piles are always filled with concrete after driving in the U.S.A. This gives the piles a higher section modulus and rigidity. The piles are generally economical in the range of 40 to 80 ft (12 to 24 m) and can carry loads as high as about 250 kips (1115 kN). Pipe piles are most suited where overburden is soft clays, silts, and loose-to-medium dense sand and is underlain by dense-bearing granular material. They also have successfully been installed in layered soils. For example, Lee et al. (1984) describe a foundation system for the Shangri-La Hotel on the bank of Chao Phraya River in Bangkok, Thailand, which, among other facilities, also consists of a 27-story tower block. The foundation soils consist of the soft Bangkok clay from the surface down to about 43 ft (13 m) underlain by alternating layers of stiff clay and sand. Pumping of water from the sand layers has reduced piezometric head in the stiff clay and sand layers causing ground subsidence as much as 4 in./yr (10 cm/yr) in Bangkok. Pile foundations designed to rest on dense sand layers will, therefore, be subjected to negative (downward) skin friction due to subsidence of surrounding clay layers that are undergoing consolidation. The foundation system consisted of installing 24-in. (600 mm) diameter open-ended steel pipe piles. The installation procedure consisted of auger-pressing the pile through clay layers and through the near surface sand layer. Then the piles were driven with a K45 hammer with a drop height of about 8 ft (2.5 m) until a set of about 0.04 in. (1 mm) per blow was achieved. At this time, the piles were at about 180 ft (55 m) to 190 ft (58 m) depth below ground surface. To reduce negative skin friction, some pipe pile sections that were to be in the settling clay layer depths were coated with a bitumen slip layer that was protected by a polyethylene layer. Remaining pile lengths (sections) were left uncoated to mobilize the skin friction. These piles thus supported the imposed loads by mobilizing skin friction and end bearing in lower stiff clay and dense sand.

Pipe piles can be used as friction piles, end-bearing piles, and a combination of friction and end-bearing or even rock-socketed piles. They are also useful for marine structures where large diameter pipes can resist lateral forces in deep waters.

Steel H piles (designated as HP) are suitable for penetrating rock as well as for driving through hard and resistant materials. These piles displace a minimum of soil mass when driven through it and, therefore, can be easily driven through dense material without causing soil heave. These piles can carry loads in the range of 80 kips (356 kN) to 240 kips (1068 kN) and have lengths in the range of 40 ft (12 m) to 100 ft (30 m). The maximum stresses in the pile section should not be more than 12,000 psi (82.7 MPa) or as per the allowable code or specificition for the job. Steel H piles are generally driven through soft soils to hard-bearing strata. The classic case of danger for these piles driven through loose materials to hard uneven rock is that these piles generally get demolished at their ends, resulting in questionable end-bearing capacity. These piles should, therefore, be protected by attaching hard steel points at their ends. Associated Pile Fitting Corp. (1985) cites a Federal Highway Administration Ohio test case where HP 10 × 42 piles were driven to hard limestone. None of the piles that had APF cast steel points experienced damage despite hard driving with up to 50,000 ft-lb

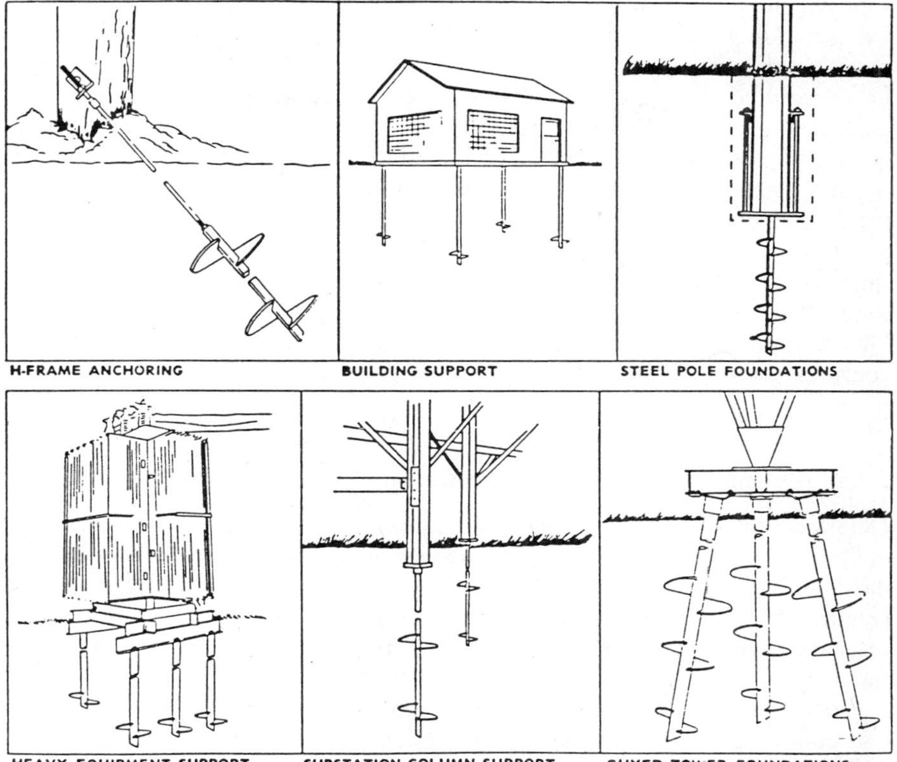


**Figure 2.8** Typical application of a screw-type pile (Courtesy: Chance Anchors, 1983).

hammer energy. In contrast, all piles driven without point protection got damaged even by driving energy of a 8700 ft-lb hammer.

Another type of steel piles that have been used to support light loads are called *screw* piles. These piles consist of installing by screwing the helix steel sections down into the ground by applying the torque without digging into the ground. Main advantage of this type of piles is that the structure or the equipment can be placed on the foundation immediately after the piles have been installed without having to return to the job site after concrete has cured. These piles can be installed in all soil types and have been used in several countries for mast and tower foundations. Figure 2.8 gives examples of some typical application of these pile types that have been used in the past. These piles are mostly used to support lightly loaded foundations.

### 2.4.2 Material Specifications

Steel piles must conform to the requirements of national codes (such as the Uniform Building Code, 1976 and National Building Code of Canada, 1985).

Pipe piles may be specified by grade with reference to ASTM-A-252. Steel H piles will generally be specified as per ASTM-A-36 or ASTM-A-572. Mill certificates or laboratory test reports should be furnished to show that the material conforms to the required specifications, including type of steel and yield strength. Steel with high yield strength should be used for piles that are to be subjected to hard-driving stresses or to be socketed into bedrock.

Steel pipe and H sections are available in various standard sizes. H piles are produced in standard mill lengths of 40 to 60 ft (12 to 18 m). Longer lengths can also be ordered. In general, the flange and web should have a minimum nominal thickness of not less than 3/8 in. (10 mm) and the flange width should not be less than 80 percent of the depth of the section. Fuller (1983) provides further information on material specifications, lengths, dimensions, fittings, special coatings, welding, handling, unloading, storage, and maintenance of material records.

### 2.4.3 Material Deterioration and Protection

Deterioration of steel piles may either occur when they may get damaged (deflected) by obstruction during driving or when they get corroded. Pipes may be damaged during driving when they encounter sloping or level hard stratum (Figure 2.9). This problem can be resolved by carefully monitoring the driving resistance and by providing driving shoe at the end of the pile. Further details on the driving shoes are included in Chapter 3.

Corrosion, on the other hand, is a complex phenomena. Only the basic concepts of corrosion mechanism are addressed here and are summarized as follows:

1. Most metals before being processed occur (in natural stable state) in their oxide form.
2. If suitable environmental conditions are permitted, metals will return to their natural state (i.e., oxide form) by reacting with oxygen and water. This may be represented as follows (Hanna, 1982).

$$\text{Metal} + O_2 \xrightarrow{H_2O} \text{Metal } (OH)_x$$

3. In the foregoing chemical reaction, the metal moves from, one region, called the anode, to another region, called the cathode, where oxygen and water are converted to hydroxyl ions.
4. This chemical reaction is considered to result from a potential difference between the anode and the cathode and depends on the chemistry of the environment.

This whole process is called corrosion. In general, all metals will return to their natural stable form and will therefore corrode. The severity of corrosion will

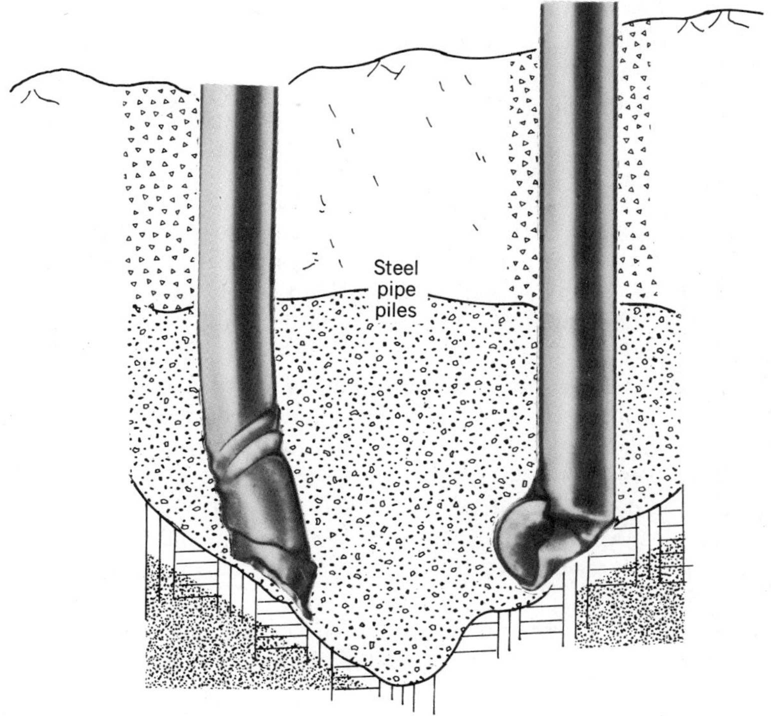


**Figure 2.9** Damage to steel pipe pile due to hard-driving conditions (Courtesy: Associated Pile and Fitting Corp., Bulletin PPP777, 1985).

depend on the nature of the environment in which the metal is placed. The rate of corrosion of a metal varies greatly with soil composition and texture, depth of embedment, and moisture content. Generally, swamps, peat bogs, and industrial and mine waste areas are corrosive environments. There are various tests such as soil resistivity and pH that will indicate if a soil has potential for corrosion. A testing laboratory should be referred in this matter. Oxygen availability is another factor that should be considered in corrosion evaluation. For example, in coarse-grained soils corrosion may approach to that of atmospheric conditions. In clays, on the other hand, the deficiency of oxygen would result in conditions approaching those in submerged corrosion and very little corrosion may occur.

From the foregoing discussion, it may be concluded that when a steel pile is embedded in ground it might corrode. The degree of corrosion will depend on the availability of moisture and oxygen in the environment and the composition of the surrounding soil. Corrosion protection alternatives would therefore require one of the following measures:

1. Provide additional metal by increasing pile section
2. Isolate pile from its surroundings by either surface coating or by encasement, and
3. Cathodic protection method

These methods are briefly described as follows:

Pile sections may either be increased by procuring a thicker pile than required or by adding plates at locations that are considered to be most susceptible to deterioration. This may be achieved by allowing a higher factor of safety in the design resulting in a thicker section. Thicker sections can either be provided locally in danger zones or along full pile lengths depending on the economics of the solution.

Surface coatings are normally applied in areas where usual maintenance can be done. There are many types of coatings available in the market, such as paints, coal tar, and other bituminous paints. In selecting a proper coating, various factors, such as weather and abrasion conditions, chemical composition of soil, and water, should be considered. Manufacturers warranty and contractors insurance against workmanship must be obtained to meet the site-specific environmental and service conditions before a surface coating on a job is specified.

Another protective measure that can be used for steel piles is providing partial or full-length encasement. These may either consist of concrete jackets or the gunite encasements. Concrete jackets may either be precast or cast-in-place. For cast-in-place jackets, steel forms having a tight closure around the pile may be driven or jetted in place. These forms may either be removed or left in place after

**TABLE 2.3 Corrosion Protection Guidelines**

| Pile Embedment Environment | Corrosion Potential | Recommended Protection |
|---|---|---|
| In impervious soils[a] | Very little | No protection required |
| In pervious soils[a] | To about 0.5 m below ground surface | Surface coating |
| Projecting into air | Atmospheric corrosion | Painting above ground |
| | Soil corrosion near ground | Concrete encasement or coal tar to 0.5 m above and below ground |
| Projecting into clean fresh water | No corrosion | No protection required |
| Projecting in sea water | Atmospheric corrosion above high tide | Painting |
| | Between high tide and mudline will corrode | Concrete encasement or coal tar |

[a]Final recommendations will depend on the results of site-specific soil tests. If soils are corrosive one of the corrosion protection methods outlined in the text should be considered.

concrete has been poured. Gunite encasement is provided before the pile is driven in place. A gunite thickness of about 2 in. (50 mm) is normally used and reinforcing bars are welded to the pile.

The basic principle behind cathodic protection is to provide sufficiently large countercurrents to the corroding metal so that the corroding currents are neutralized. This can either be provided by the use of sacrificial anodes or by impressed currents. Normally, piles in seawater or piles in the vicinity of high-voltage lines may need cathodic protection. The overall topic of cathodic protection is complex, and the recommendation regarding the need, level, and kind of protection required should be provided by a corrosion engineer.

Table 2.3 provides preliminary guidelines for corrosion potential of steel piles installed in different environments. Site specific corrosion potential and protection requirements should however be recommended by a corrosion specialist. Corrosion normally is not a practical problem for steel piles when installed into natural soil. Romanoff (1962) has documented surveys on corrosion of piles. Similar results have been reported in an investigation by Manning and Morley (1981).

## 2.5 COMPOSITE PILES

Composite piles can be made by joining sections of dissimilar materials together so that the advantages of both can be utilized.

### 2.5.1 Types and Use of Composite Piles

As shown in Figure 2.10, composite piles can be made of concrete and timber sections, concrete and steel sections, and concrete filled steel pipes. Other combinations have also been used. It is difficult to form good joints between two materials, especially concrete and timber. This type of construction (timber–concrete) has therefore been abandoned in North America. High-capacity pipe and HP–concrete composite piles do not have this problem and are used when proved economical.

### 2.5.2 Material Specifications

Material specifications for timber, concrete, and steel piles as discussed in Sections 2.2.2, 2.3.2, and 2.4.2 also apply here.

## 2.6 SPECIAL TYPES OF PILES

Pile types that have not been discussed in the previous sections are described here. These piles are special in the sense that they have special construction method and/or specialized use such as when used in permafrost areas.

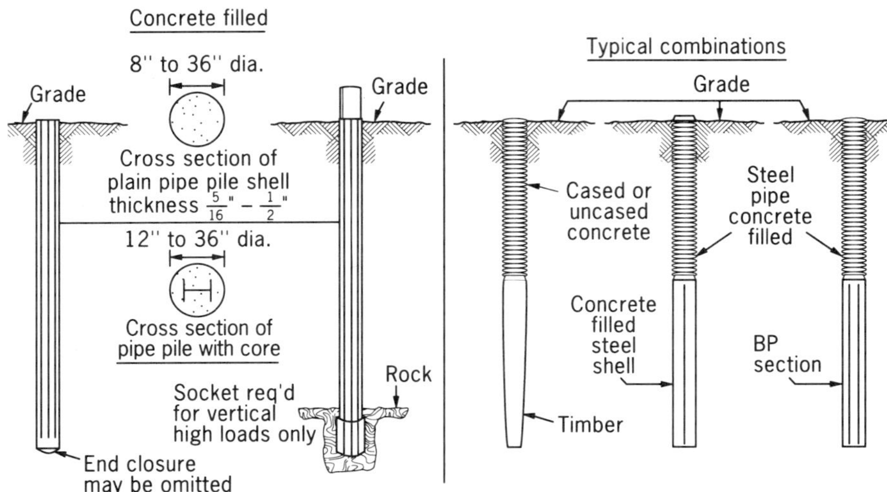


**Figure 2.10** Typical sections for some composite piles (Design Manual, NAVFAC DM 7.2, 1982).

### 2.6.1 Expanded Base Compacted Piles (Franki Piles)

These piles, also called pressure-injected footings, were originally developed and patented by the Franki Pile Company by utilizing special equipment for their installation. In these piles, a steel tube is first driven to the desired depth and then an enlarged base is formed by feeding in small charges of zero-slump concrete. Each charge is driven out into the soil with hammer blows until the required base is formed. A pile shaft is then formed by depositing zero-slump concrete charges into the drive tube. Each charge of concrete is compacted and rammed against the soil as the tube is withdrawn in short lifts. Figure 2.11 exhibits typical examples of uncased shaft and the cased shaft expanded base compacted piles. Details of equipment for pile installation are included in Chapter 3.

These piles are best suited for granular soils where bearing is achieved primarily from the densification of soil around the expanded base. These piles are not recommended in cohesive soils where compaction of the base is not possible. Commonly used pile lengths are of 20 to 60 ft (6 to 18 m) and pile shaft diameters range from 12- to 24-in. (300 to 600 mm). These piles have normal design loads of 60 to 120 tons (534 to 1068 kN). These piles provide high-capacity foundations without the necessity for excavation or dewatering.

Material used for expanded base compacted piles should also meet the specifications detailed in Section 2.3.2. Concrete for forming the base and the uncased shaft of these piles should, however, be of zero slump concrete. This concrete should have enough water to ensure hydration of the cement. Normally 3.5 gallons of water per cement bag is considered adequate for it but must be checked with concrete testing laboratory. For cased shaft expanded base

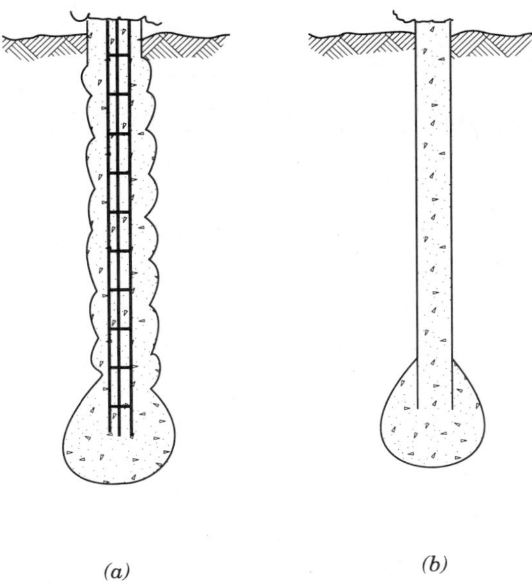


**Figure 2.11** Expanded base-compacted piles (Franki piles). (a) Uncased shaft, (b) cased shaft.

compacted piles, normal slump 6 to 8 in. (150 to 200 mm) concrete should be used.

Kozicki (1985) cites various case histories where these piles have successfully been installed through different soils. Load tests carried out on these piles confirmed that these piles could carry the design load with small settlements. For example, for Calgary Air Terminal Complex, 17 ft (5 m) long and 20 in. (500 mm) shaft diameter piles had their base on silt till. These piles were designed for a working load of 350 kips (1560 kN). When load tested to 944 kips (4200 kN), these piles exhibited a total settlement of 0.59 in. (14.7 mm). For the Outlook Manor Project in Toronto, Canada, 45 ft (14 m) long, 16 in. (400 mm) shaft diameter expanded base compacted piles bearing on dense sand were designed for a working load of 300 kips (1335 kN). When load tested to 600 kips (2670 kN), the pile showed a gross settlement of 0.585 in. (14.6 mm). For the Brickwell Bay Club Project, Miami, Florida, 27.5 ft (8.4 m) long $17\frac{5}{8}$ in. (440 mm) shaft diameter piles having their base on loose sand, shells, and limerock had a working load of 300 kips (1335 kN). When tested to 600 kips (2670 kN), these piles exhibited a gross settlement of 0.64 in. (16 mm). These examples indicate that expanded base compacted piles can provide high-capacity foundation system.

### 2.6.2 Thermal Piles

Piles in permafrost soil conditions, normally transfer their loads to ground in the following two ways:

1. The side support is provided by the development of the adfreeze* bond between soil or backfill (slurry) and the pile surface.
2. The point or end-bearing support is provided in the conventional way by firm strata (such as bedrock or dense thaw-stable sands and gravel) if encountered at suitable depths.

Adfreeze bond between the pile surface and the surrounding soil decreases as the permafrost temperature increases. Thermal piles are therefore used (1) *to ensure that long-term degradation of permafrost is prevented by removing* heat from the ground and (2) to decrease the existing ground temperature around piles that are installed in warm-temperature permafrost. Thus, thermal piles ensure the development of adequate adfreeze bond by keeping ground temperatures low and ensuring long-term thermal stability of foundations.

The two basic types of thermal piles that have been in use are natural convection system type and the forced circulation refrigeration system type. These piles are briefly discussed in the following paragraphs. Johnston (1981) provides further details on these piles.

***Natural Convection System Type Thermal Piles*** These piles remove heat from ground by natural convection system. They require no external power source and function only under conditions when air temperatures are lower than the ground temperature. These piles can either be *single-phase* (Figure 2.12a) or *two-phase* (Figure 2.12b) system. In *single-phase* system, heat from the soil surrounding the embedded portion of the pipe is absorbed by it during the winter months. This warms up the working fluid, which then rises to the above-ground radiator section of the pipe. Since the radiator section is exposed to the cooler air, it loses its heat by conduction and natural convection. This process keeps the ground cool and maintains a good adfreeze bond between pile and the surrounding soil.

In a *two-phase* system, the working fluid is part vapor, part liquid. As shown in Figure 2.12b when air temperature falls below the ground temperature, the vapor condenses. This reduces the pressure and the liquid in the lower section of the pipe starts to boil causing the vapor to flow up where it will condense again and return down. This process transfers heat from the ground up to the air and thus keeps the ground frozen. Long (1963) first suggested the use of this type of pile. Piles designed on the basis of this concept were extensively used to support the above ground section of the Trans-Alaska Oil Pipeline (Waters, 1974; Heuer, 1979). These are called vertical supported member (VSM) and are shown in Figure 2.12c).

***Forced Circulation Refrigeration System Thermal Piles*** This system of thermal piles keeps the ground frozen by forced circulation of either a liquid or cold air refrigerant system. The refrigerant is circulated by mechanical equipment operated by an external power source. Figure 2.13 illustrates schematic

*See Chapter 8 for definitions.

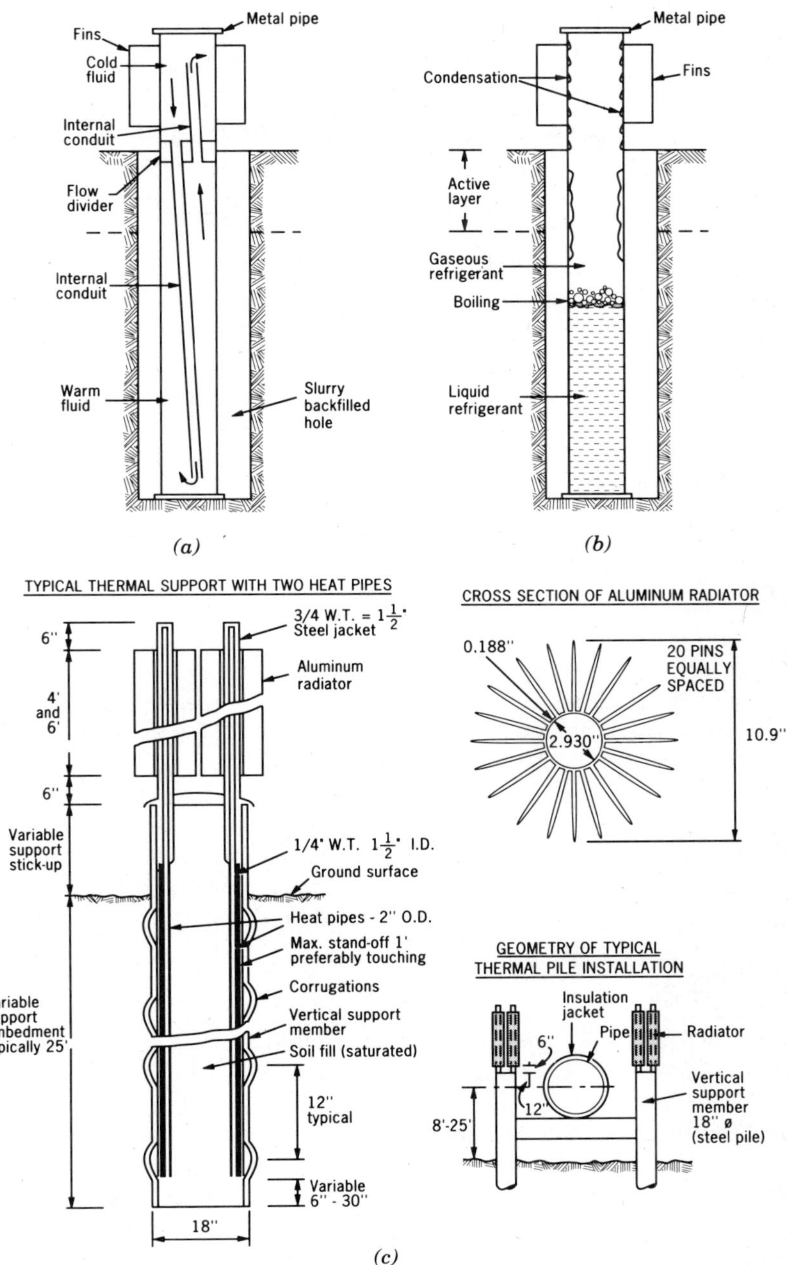


**Figure 2.12** Thermal pile types based on natural convection system (Johnston 1981). (a) Schematic representation of Single-phase and (b) two-phase system of thermal piles, (c) typical vertical support member (VSM) for Alyeska Oil Pipeline. (After Alyeska Pipe-line Service Co., 1976.)

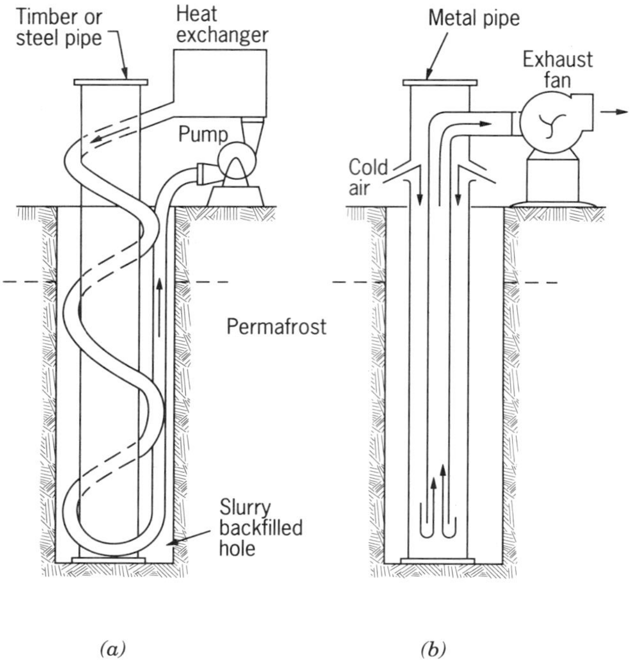


**Figure 2.13** Schematic representation of forced circulation refrigeration system thermal piles (Johnston 1981). (a) Liquid refrigerant system (Rice, 1973), (b) cold-air refrigerant system (Reed, 1966).

representation of forced circulation refrigeration system thermal piles. This system is more complex, needs external power source, and requires regular maintenance since moving parts are involved.

### 2.6.3 Other Pile Types

Some of the other pile types that have been used are auger grout or concrete injected piles, drilled-in tubular piles, and preplaced aggregate piles.

***Auger Grout Injected Piles*** As shown in Figure 2.4g, these piles are installed by first drilling a hole to the required depth by a continuous-flight, hollow-stem auger. The second step is then to raise the auger tip by about 12 in. (300 mm) and pump the grout under pressure through the hollow stem. The grout pressure is adjusted to offset the hydrostatic and lateral earth pressures as the auger is retrieved upward. These piles can also be reinforced by pushing the reinforcing cage through unset concrete/grout and can thus be designed to resist uplift and lateral loads. A temporary steel sleeve can be placed at the top of the pile before top portion of the pile is grouted and the auger is removed. This steel sleeve may not be required where the ground surface is at least 12 in. (300 mm) higher than the pile cutoff grade. These piles are suitable where ground and water conditions do not allow uncased holes without sloughing. These piles also develop excellent skin friction because the concrete/grout are injected under pressure. Further pile lengths can be adjusted during drilling if drilling operations indicate changed soil conditions.

***Drilled-in Tubular Piles*** These piles are installed by rotating heavy-gauge steel casing (tubular pile) having a cutting edge into the soil. Soil cuttings are removed with circulating drilling fluid. The hole is then filled by pumping a sand–cement grout through tremie. Reinforcing bars may be placed to resist lateral and uplift forces. The steel casing may be withdrawn during placing the grout. These piles can be used where boulders and other obstructions are encountered.

***Preplaced Aggregate Piles*** These piles are installed by first drilling a hole to the required depth. Grout pipes are installed into the hole, which is then filled with coarse aggregate. Grout is then pumped through the pipes, which are withdrawn as the grouting operation proceeds from the bottom up.

These three types of piles are nondisplacement types and can be used in and around existing foundations. They are also suitable for underpinning work.

## 2.7 SELECTION CRITERIA AND COMPARISON OF PILE TYPES

As has been discussed in the preceding sections, there are various types of piles that are widely used in engineering practice. Advantages and disadvantages of these piles will control the choice of any particular pile type for a specific project. The final selection will depend on the soil and water conditions, availability of material, local experience, construction schedule, type of the structures to be supported, and the overall economy. Cost comparison should include the estimation of the cost of the entire foundation system (e.g., pile caps, grade beams etc.) rather than comparing only the cost per pile. Important characteristics and advantages and disadvantages of several types of piles are now presented, which may help the reader to make a comparison between various pile types and aid in their selection process.

### 2.7.1 Timber Piles

Typically, these piles are used in lengths from 30 to 60 ft (9 to 18 m) and can carry loads ranging from 20 to 100 kips (89 to 445 kN). The maximum recommended stresses for cedar, Norway pine, and spruce is 870 psi (6000 kPa) and for southern pine, Douglas fir, and oak cypress in 1200 psi (8277 kPa). These piles are mostly installed by driving and are best suited as friction piles in granular material. The main advantages of timber piles are that they have low initial cost, are easy to handle, and resist decay when they are permanently submerged. The main disadvantages are that they are difficult to splice, are vulnerable to damage in hard driving, and are susceptible to decay unless treated. Treatment becomes necessary when these piles are intermittently submerged.

### 2.7.2 Concrete Piles

***Precast Concrete Piles*** Typically, precast concrete piles are 40 to 50 ft (12 to 15 m) long while the precast prestressed piles are typically 60 to 100 ft (18 to 30 m) long. They can be designed for a wide range of loads. However, a typical load range is 80 to 800 kips (356 to 3560 kN). The maximum stresses for precast sections should not exceed 33% of 28-day concrete strength ($f_c$). For prestressed sections the maximum stresses should not exceed ($0.33 f_c - 0.27 p_e$); where $p_e$ = effective prestress stress on the section. The main disadvantages of these piles are that they are difficult to handle without damage unless prestressed. They have a high initial cost, and prestressed piles are difficult to splice. The advantages of these pile types include high load capacities, corrosion resistance, and resistance to hard driving. (See also "Stresses in Piles" 1983)

***Cast-in-Place Concrete Piles*** Cast-in-place concrete piles with their shell driven with mandrel are typically 50 to 80 ft (15 to 24 m) long and can specifically be designed for a wide range of loads. Typical loads that these piles can carry are 50 to 120 kips (222 to 534 kN) provided the maximum stress in concrete, is not more than 33% of 28-day strength. The main disadvantages are that these piles are difficult to splice after concreting, their thin shells can be damaged during driving, and redriving is not recommended. Generally, stress in steel should not exceed 0.35 × yield strength of steel. The advantages are that they have low initial cost, and tapered sections can provide higher-bearing resistance in granular stratum. These piles are best suited as medium-load friction piles in granular soils.

### 2.7.3 Steel Piles

Concrete-filled steel pipe piles can be installed to any length. However, typically 40 to 120 ft (12 to 36m) lengths are commonly used. The maximum stresses in concrete should be less than 0.33 × 28-day compressive strength of concrete and the stresses in steel should not exceed 0.40 × yield strength of steel. Design load ranges for these piles are 160 to 240 kips (712 to 1068 kN) without cores and 1000 to 3000 kips (4450 to 13,350 kN) with cores. The main disadvantages of these piles are a high initial cost and soil displacement for closed-end pipe. The advantages of steel piles are that they offer best inspection control during installation, can be cleaned out and driven further, have high load capacities, and can be easily spliced. These piles also provide high bending resistance where freestanding sections are required to support lateral loads. (NAVFAC, 1982)

Steel H piles are typically installed in lengths ranging from 40 to 160 ft (12 to 49 m). However, longer lengths can also be installed to suit ground conditions.

Design loads range from 80 to 240 kips (356 to 1068 kN). The maximum stresses should not exceed the values specified in section 2.4.1 for H-piles. The disadvantages of these piles are that they may be susceptible to corrosion, and HP sections may be damaged during driving through obstructions. Advantages of these piles include that they can be easily spliced, are available in various lengths and sizes, are of high capacity, displace small amount of soil during installation, and are best suited for end bearing on rock.

### 2.7.4 Composite Piles

Composite piles are generally considered for lengths ranging from 60 to 200 ft (18 to 60 m) and for design loads of 60 to 200 kips (267 to 900 kN). The maximum stresses in timber, steel and concrete should not exceed the values specified above for various materials. The main disadvantage of these piles is that it is difficult to attain good joint between two materials. The main advantage is that considerable length can be provided at comparatively low cost.

### 2.7.5 Special Types of Piles

Expanded base compacted piles (Franki piles) are generally 20 to 60 ft (6 to 18 m) long and can carry 120 to 240 kips (534 to 1068 kN) loads. The main disadvantage is that when clay layers must be penetrated to reach suitable material, special precautions such as preboring may be required. Their installation also requires more than average dependence on quality of workmanship. Its main advantages include installation of a high-capacity pile without any excavation or dewatering and great uplift resistance if suitably reinforced. These piles are best suited for granular soils where bearing is achieved through compaction around the pile base.

Another special type of piles called Tapered Pile Tip (TPT) consists of a mandrel driven corrugated shell with an enlarged precast concrete base. A pipe mandrel inside the shell is used to drive the base and the shaft shell unit to the required bearing depth. The shaft is then filled with concrete while the annular space left around the shaft is filled with sand. The main advantage of this type of pile is spreading the load at the base thus preventing punching through the bearing layer specially when it is relatively thin.

Thermal piles are specialized used piles and are still in the development stage. These piles are used to support structures in permafrost areas. Section 2.6.2 lists their main features.

## REFERENCES

American Concrete Institute 543 (1980). "Recommendations for Design, Manufacture and Installation of Concrete Piles," *Journal American Concrete Institute*, Vol. 70, No. 8, August (1973), pp. 509–544, and revisions Vol. 71, No. 10, October 1974, pp. 477–492, reaffirmed in 1980.

American Wood Preservers Association C.3: "Piles—Preservative Treatment by Pressure Processes," Washington, DC, 1981.

American Society of Civil Engineers: Committee on Deep Foundations, "Practical Guidelines for the Selection, Design and Installation of Piles," American Society of Civil Engineers, 1984.

American Standards for Testing and Materials A572, Specification for High-Strength Low Alloy Columbian Vanadium Steel of Structural Quality, 1979.

American Standards for Testing and Materials D25, Specifications for Round Timber Piles, Philadelphia, PA, 1979.

American Standards for Testing and Materials A252, Specification for Welded and Seamless Steel Pipe Piles, 1980.

Associated Pile and Fitting, Corp *Pile Tips: Piling and Foundation News*, November–December 1985.

*Canadian Foundation Engineering Manual, Part 3, Deep Foundations*, Canadian Geotechnical Society, March 1978 and 1985.

Canadian Prestressed Concrete Institute (CPCI), *Metric Design Manual: Precast and Prestressed Concrete*, 1982.

Capozzoli, L. J., "Current Status of Timber Foundation Piles," Pile Foundations Know-how, American Wood Preservers Institute, 1969, pp. 6–9.

Chellis, R. D., *Pile Foundations*, McGraw-Hill Book Co., 2nd ed., New York, 1961.

Chellis, R. D., Pile Foundations in *Foundation Engineering*, Chapter 7, G. A. Leonards, ed. McGraw-Hill Book Co., New York, 1962.

Dugan, J. P. and Freed, D. L., "Ground Heave Due to Pile Driving," *Proc. International Conference on Case Histories in Geotechnical Engineering*, St. Louis, MO, Shamsher Prakash, ed., 1984, Vol. 1, pp. 117–122.

Engeling, P. D., Hyden, R. F., and Hawkins, R. A., "Raymond Concrete Cylinder Piles in the Arabian Gulf," *Proc. International Conference on Case Histories in Geotechnical Engineering*, St. Louis, MO, Shamsher Prakash, ed. 1984, Vol. 1, pp. 249–257.

Fuller, F. M., *Engineering of Pile Installation*, McGraw-Hill Book Co., Chapters 2, 3, and 6, 1983.

Hanna, T. H., *Foundations in Tension: Ground Anchors, Transactions Technical Publications*, McGraw-Hill Book Co., New York, 1982, p. 83.

Heuer, C. E., "The Application of Heat Pipes on the Trans-Alaska Pipeline," U.S. Army, Cold Regions Research and Engineering Laboratory, Special Report, 79–26, 1979.

Hunt, H. W., "Design and Installation of Driven Pile Foundations," Associated Pile & Fitting Corp, NJ, 1979.

Intrusion-Prepakt Inc., "Fabriform Marine Pile Jackets," Cleveland, OH, 1981.

Johnston, G. M. (ed.), *Permafrost Engineering Design and Construction*, Chapter 7, Part III, Wiley, New York, 1981.

Kozicki, P., "Expanded Base Piles," *Symposium on Deep Foundations*, Toronto, Ontario, 1985, p. 14.

Lane, D. J., "Caisson Design by Instrumented Load Test," *Proc. International Conference on Case Histories in Geotechnical Engineering*, Shamsher Prakash, ed., Vol. I, St. Louis, MO, 1984, pp. 41–50.

Lee, S. L., Karunaratne, G. P., MO, and Sithichaikasem, S., "Non-Negative Skin Friction

Piles in Layered Soil," *Proc. International Conference on Case Histories in Geotechnical Engineering*, Shamsher Prakash, ed., Vol. I, St. Louis, MO, 1984, pp. 285–288.

Long, E. L., "The Long Thermopile," *Proceedings of the 1st International Conference on Permafrost*, Lafayette, IN, NAS-NRC Publication 1287, 1963, pp. 487–491.

Manning, J. T. and Morley, J., "Corrosion of Steel Piles," Piles and Foundations, F. F. Young, ed., Tharm Telford Ud., The Institution of Civil Engineers, London, 1981, pp. 223–229.

*National Building Code of Canada*, National Research Council of Canada, Ottawa, NRCC No. 23174, 1985.

NAVFAC DM-7.2, *Foundation and Earth Structures, Design Manual 7.2*, Department of Navy, Alexandra, VA, Chapter 5, May 1982.

Peck, R. B., Hanson, W. E., and Thornburn, T. H., *Foundation Engineering*, 2nd ed., Wiley, New York, 1974.

Rai, M. and Jai Singh, M. P., *Advances in Building Materials and Construction*, Central Building Research Institute, Roorkee, India, 1986, p. 221.

Raymond Step-Taper Piles, Raymond International Inc., 1985.

Reed, R. E., "Refrigeration of a Pipe Pile by Air Circulation," U.S. Army Cold Regions Research and Engineering Laboratory, Technical Report 156, 1966.

Rice, E., "Northern Construction: Siting and Foundations," *The Northern Engineer*, Vol. 5, No. 1, 1973, pp. 11–18.

Romanoff, M., "Corrosion and Steel Piling in Soils," *Journal Soil Mechanics and Foundation Division*, ASCE, Vol. 66, No. 3, February 1962, pp. 1–22.

Sharma, H. D. and Joshi, R. C., "Comparison of In Situ and Laboratory Soil Parameters for Pile Design in Granular Deposits" *Proc. 39th Canadian Geotechnical Conference*, Ottawa, August 1986, pp. 131–138.

Sharma, H. D., Sengupta, S., and Harron G., "Cast-in-Place Bored Piles on Soft Rock Under Arterisan Pressures," *Canadian Geotechnical Journal*, Vol. 21, No. 4, 1984, pp. 684–698.

"*Stresses (Allowable) in Piles*," Federal Highway Administration Report No. FHWA/RD-83-059, McLean, VA, December 1983.

Tomlinson, M. J. "*Pile Design and Construction Practice*," A Viewpoint Publication, Cement and Concrete Association, 1977.

*Uniform Building Code*, International Conference of Building Officials, Whittier, CA, 1976.

Vesic, A. S., "Design of Pile Foundations," Transportation Research Board, NRC, Washington, DC, 1977, pp. 3–7.

Waters, E. D., "Heat Pipes to Stabilize Pilings on Elevated Alaska Pipeline Sections," *Pipeline and Gas Journal*, August 1974, pp. 46–58.

# 3

# PILING EQUIPMENT AND INSTALLATION

This chapter provides the background information on piling equipment and general pile installation requirements. Brief description of the pile driving rigs and hammers includes various rig components and their functions and the basic background on drop hammers, single-acting hammers, double-acting hammers, diesel hammers, and vibratory pile drivers (hammers). A summary of available type hammer-data sheet that provides information on rated energies for several commercially available hammers is also included. Equipment for bored piles consisting of truck, crane, and crawler-mounted drilling rigs and augers, belling, and coring tools are also presented in this chapter. Following this, procedures for installing driven and bored piles are described. Procedures for installing special type piles such as compacted expanded base (Franki) piles and thermal piles for permafrost areas are also presented. Finally, the requirements for installation records both for the driven and drilled piles are briefly outlined.

## 3.1 GENERAL INSTALLATION CRITERIA

Installation and inspection of pile foundation unit is less controllable and has more uncertainty than other foundation types due to changes in subsoil and groundwater conditions. Therefore, it is important that details of piling equipment and installation methods be fully understood by the design engineer.

The two main pile installation methods are (1) installation by driving, and (2) installation by drilling (or boring).

As shown in Figure 3.1, principal components of a pile-driving system are the pile, the hammer, and other components that transfer the hammer load to the pile and protect the pile from possible damage due to hammer impact (see Section 3.2.4 for a detailed description).

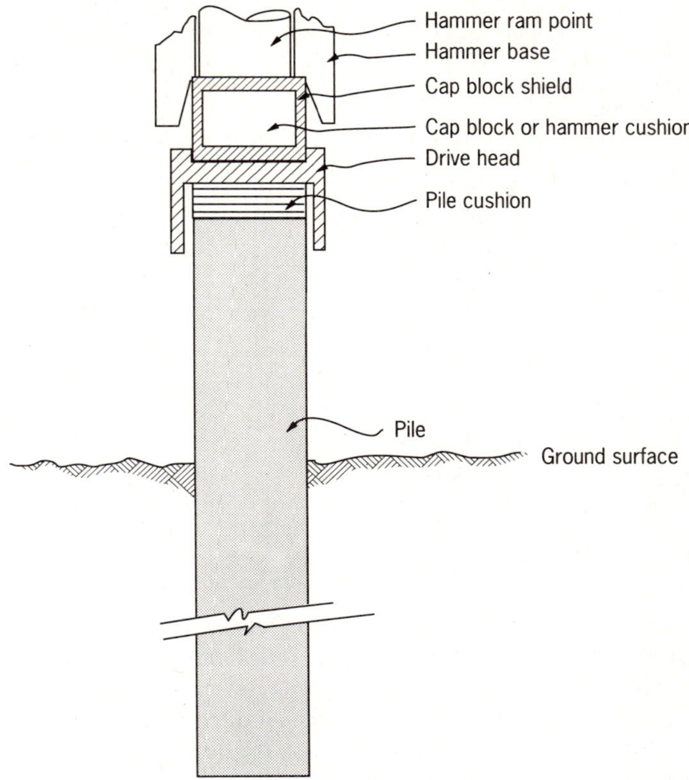


**Figure 3.1** Principal components of the pile-driving system (ASCE Deep Foundations Committee, 1984).

Figure 3.2 exhibits basic concepts of drilled (bored) pile installation into the ground. Figure 3.2a shows a bored bell pile. These piles are installed first by drilling a hole to the top of the bell with an auger. Once the bearing stratum is reached, the auger is withdrawn and the belling tool is lowered to form a bell at the base. After the bell is completed, the base is inspected, reinforcement placed, and the bell and the shaft are filled with the concrete. Figure 3.2b shows a straight-shafted bored pile with a steel casing to protect the hole from slumping and water ingress. This casing is withdrawn as the hole is filled with concrete to form the pile. Straight-shafted drilled piles are normally friction piles, but a combination of friction and end-bearing capacities can also be mobilized if pile base is properly cleaned to ensure that it is free of any slumped material. Bored and belled piles also are a combination of friction and end-bearing piles. However, in most cases, these are primarily end-bearing piles.

The general pile foundation installation criteria used in practice requires that the minimum center-to-center spacing for piles installed into the rock should be

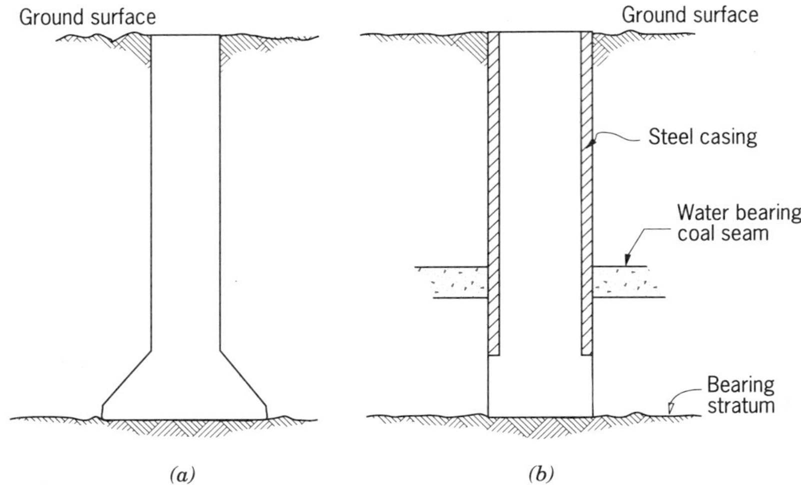


**Figure 3.2** Basic concepts of bored pile installation (Sharma et al., 1983). (a) Bored belled pile, (b) bored pile with steel casing during drilling.

at least twice the average pile diameter (or width), but not less than 24 in. (600 mm). For piles that are installed into overburden soils, the minimum center-to-center spacing should be at least 2.5 times the average diameter (or width) of the pile. This minimum spacing should also be limited by the requirements that the combined pile group load distributed into the bearing stratum shall be less than the bearing capacity of the stratum. Normally, the minimum number of piles in a group should consist of at least three piles. In cases such as floor slabs and grade beams, individual piles can be used. However when single piles are used, their shaft diameters are generally greater than 12 in. (300 mm). For proper load transfer and to maintain structural integrity of the pile and its cap, it should be ensured that the top of piles should extend at least 4 in. (100 mm) into the pile cap.

Driving sequence of piles in a group should be such that the piles are driven from the interior of the group towards the periphery. This would preclude densification and hard-driving conditions in the interior and would thus facilitate pile driving. Tolerances of pile location in horizontal direction should not exceed 4 in. (100 mm), and the vertical alignment should not vary more than 2 percent from the plumb position.

In the following sections, the pile installation equipment are presented first, followed by the details of installation procedures and the needed installation document (records).

## 3.2 EQUIPMENT FOR DRIVEN PILES

It is important to understand the basic concepts of pile installation equipment by an engineer. Also, the piling inspector must be familiar with the details of piling equipment. This is because of the fact that the equipment is not only key to obtaining an efficient construction rate but is also important to measure the adequacy of installation.

The two key players in pile driving operation are the pile and the hammer. The hammer is operated and guided on a rig. Different types of rigs are available in the

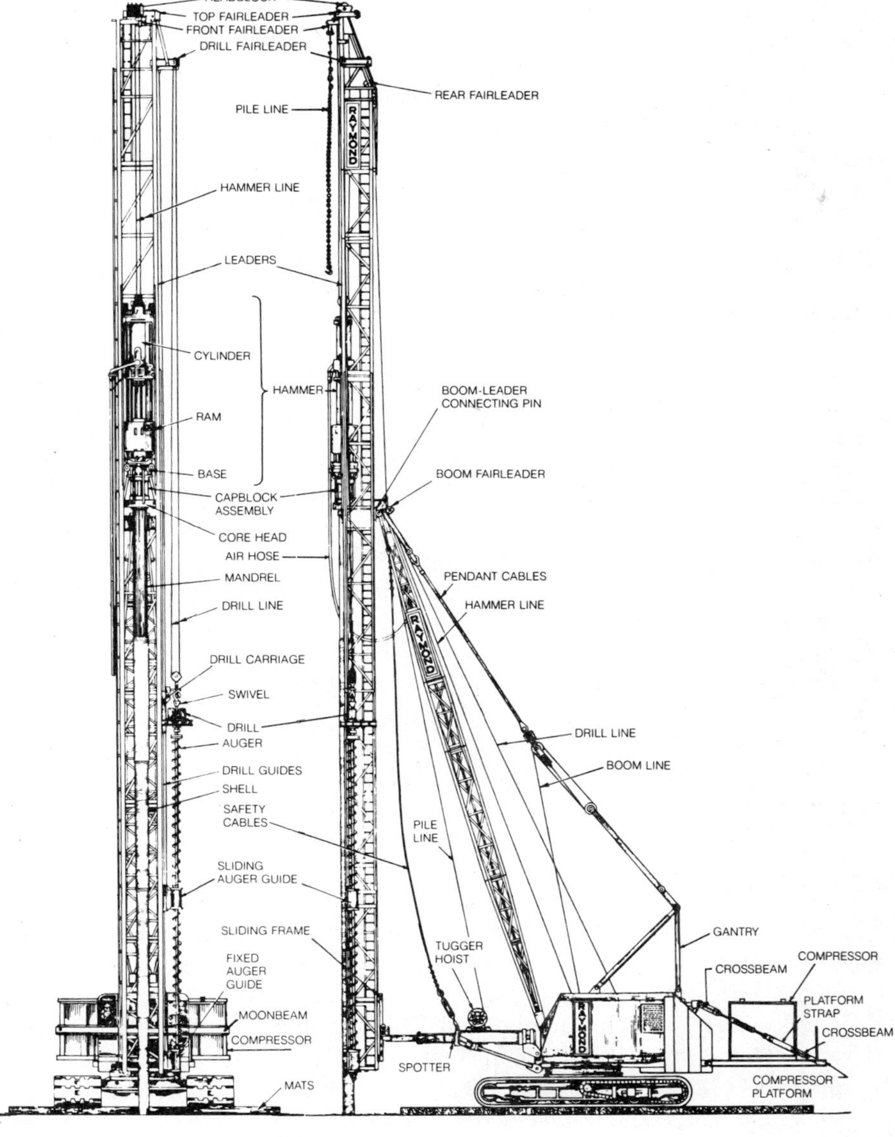


**Figure 3.3** Typical pile-driving rig: Various components labeled (Courtesy: Raymond International, Inc.).

pile-driving industry. However, the basic components of these rigs are similar. Various components of a typical rig are first identified and then their functions are outlined, followed by details of hammers used for pile driving.

### 3.2.1 Rigs

Figure 3.3 presents various components of a typical pile-driving rig. The leaders serve dual purposes of holding the pile in position and maintaining the axial alignments of the pile and the hammer. This requires that the leader should be sufficiently rigid to ensure that the pile is firmly held in its position and is in axial alignment with the hammer. Normally, leaders are fixed to the boom tip at the top and to the spotter at the bottom. The spotter, as shown in Figure 3.3, is a horizontal frame connecting the bottom of the leaders to the main body of the rig. The spotter can either be fixed in length or can be of telescopic type thus enabling an adjustable operating radius to the pile-driving rig. This permits the piles to be driven over a wide range of in-and-out batters.

Installation of piles in side-batter is done by using the moonbeam, which is a straight or curved member and is mounted at the end of the spotter. The moonbeam is located perpendicular to the longitudnal axis of both the leaders and the spotter. As shown in Figure 3.3, the bottom of the leaders is fastened to the moonbeam. This allows the bottom of the leaders to move in a lateral direction, permitting piles to be driven on a batter in any direction. In situations where prejetting or predrilling is specified for pile installation, a jet or drill may be mounted on the leaders at the same driving radius as for the hammer and pile.

Power sources such as a boiler for steam or compressor for compressed air are used to operate pile-driving hammers. Boilers are normally sized by horsepower. According to Fuller (1983), boilers should be sized according to the pounds of steam delivered per hour at the required operating pressure and compressors should be rated by the volume of compressed air delivered per minute corrected to standard conditions when operating at required pressure. Boilers or compressors should be of adequate capacity for the hammer. The hammer manufacturer's data sheet can be used to determine the appropriate boiler or compressor capacity. A summary of hammer data sheet from manufacturers' literature is included in Section 3.2.2.

### 3.2.2 Hammers

There are various types of pile-driving hammers that can be used to install piles. Drop hammers, single-acting hammers (steam or air), double-acting hammers (steam or air), differential hammers (steam, air, or hydraulic power), diesel hammers (single or double acting), and vibratory pile drivers are the principal types of hammers that have been in common use as pile drivers in the industry.

Figure 3.4 exhibits the principles of operation of a drop hammer, single-acting hammer, differential and double-acting hammers, diesel hammer, and vibratory pile driver. The drop or gravity hammers are generally raised manually and then

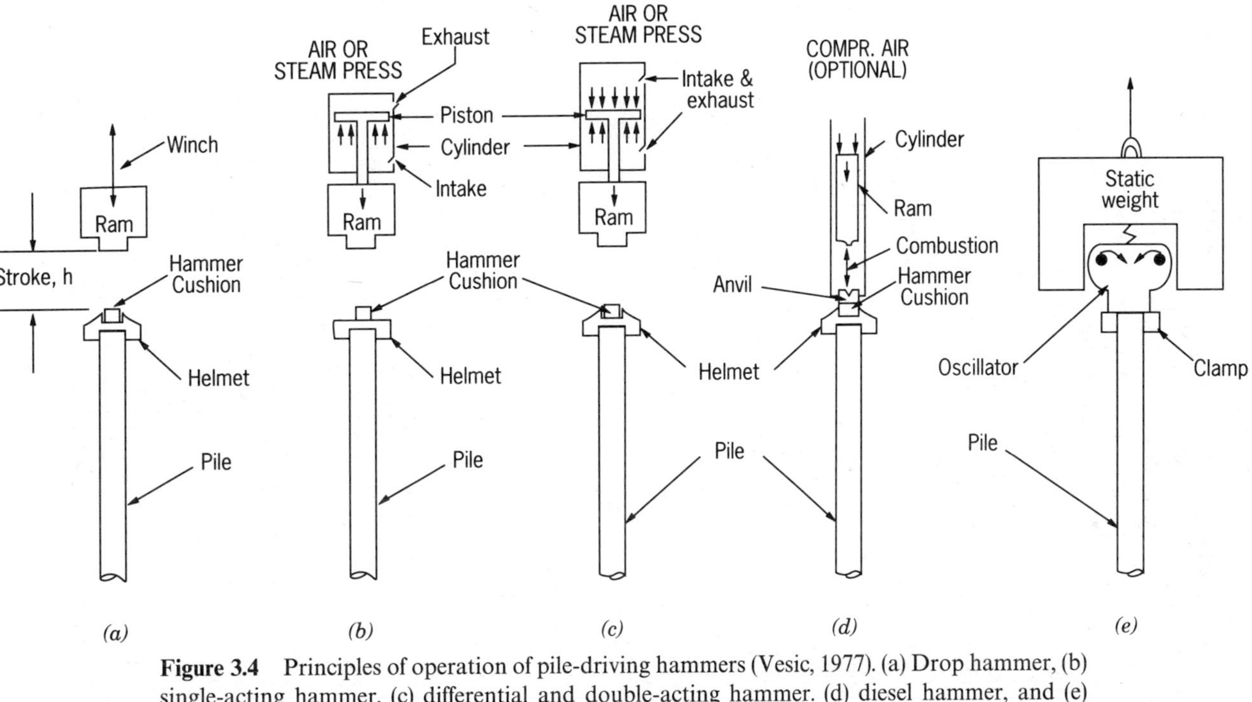


**Figure 3.4** Principles of operation of pile-driving hammers (Vesic, 1977). (a) Drop hammer, (b) single-acting hammer, (c) differential and double-acting hammer. (d) diesel hammer, and (e) vibratory driver.

impact to the pile is delivered by its free fall. The energy is calculated by multiplying the weight of the hammer by its fall. Single-acting hammers are essentially drop hammers except that the hammer is raised either by steam or air pressure. In double-acting hammers, steam or compressed air raises the ram in the upstroke, and the same pressurized fluid accelerates the ram through its downward fall. Thus, a shorter double-acting hammer with a lighter ram delivers comparable impact energy per blow at two to three times the blow rate of a longer single-acting hammer. In differential acting hammer (Figure 3.4c), the heavier ram in lifted and driven down with a lower volume of air or steam than is used in a double-acting hammer of similar energy rating. This is achieved by using a higher pressure fluid (Compton, 1981). Vibratory pile drivers (Figures 3.4e and 3.5b) employ paired rotating weights that are set eccentric from their centers of rotation. This results in a mechanical sine wave oscillator that drives the pile through the soil. Figure 3.5 illustrates the principle of a mechanical oscillator and

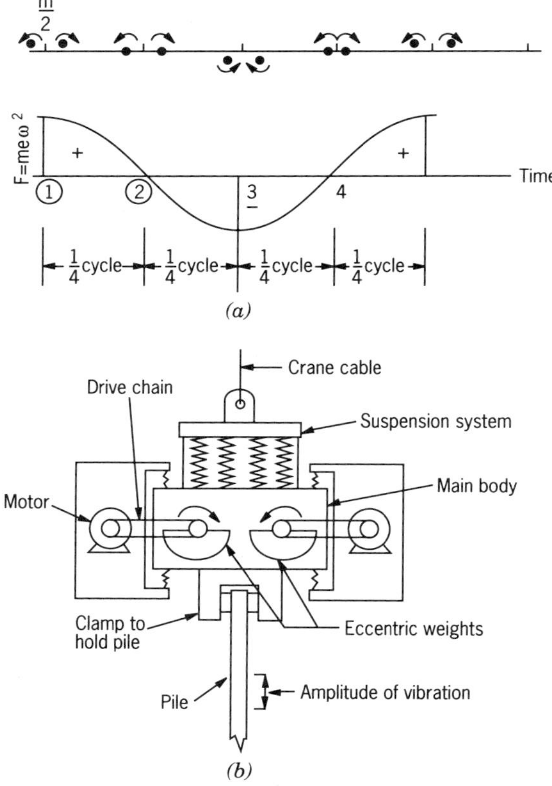


**Figure 3.5** (a) Principle of a mechanical oscillator, (b) vibratory pile Driver (Prakash 1981).

vibratory pile driver. These hammers are most effective in cohesionless or granular soils. Table 3.1 lists various makes of hammers with their rated energy in decreasing order.

In pile installation specifications, the type of pile-driving hammer and the rated energy is generally specified. This also is used as one of the parameters to determine pile capacity during driving. In order to understand clearly the actual energy delivered to the pile, one must note how consistent hammer performance is during the pile-driving operation and, most importantly, to understand the energy losses in the overall driving system. The hammer transmits the energy to the pile through various components (see Figure 3.1). There are certain losses in each component due to various reasons such as mechanical friction, valving timing, and actual stroke length in air or steam hammers. These losses reduce the actual energy delivered to the pile as compared to the theoretical rated energy of the hammer. Hammer manufacturers normally would provide maintenance and proper operation schedule to minimize energy losses. The ratio of actual energy delivered to the theoretical rated energy is called the "*hammer efficiency*." The efficiency value can range from below 50 percent for poorly maintained hammers to about 90 percent for well-maintained diesel hammers. The ram mass and the terminal velocity of ram at impact determine the actual energy delivered by the hammer. Thus, the hammer efficiency can be determined if the terminal velocity of a ram of known mass can be measured. Various measuring devices such as high-speed photography, radar, and instrumentations such as an accelerometer have been tried to measure terminal velocity with little success. Recently, pile-driving analyzers have successfully been used to monitor hammer performance (Authier and Fellenius, 1983). This consists of attaching instrumentation to the pile near its top and measuring force or energy delivered to the top of the pile. The hammer performance or efficiency can then be computed as the ratio of actual energy to the theoretical rated energy. Further description of a pile-driving analyzer is discussed in Chapter 5.

### 3.2.3 Vibratory Pile Drivers

As shown in Figure 3.4e, the basic principle of operation of vibratory pile driver consists of imparting a dynamic vertical force from a set of two rotating, eccentrically set weights. These weights are positioned so that their horizontal force components are balanced by each other while their vertical force components are added. Figure 3.5a illustrates the principles of these mechanically oscillating weights in a vibratory pile driver. The efficiency of this systems can be further improved if a static weight, as shown in Figure 3.4e, is placed on the driver.

Vibratory hammers can be categorized based on their manufacturer's brand names. In general, however, there are two types: low-frequency vibrators, and high-frequency vibrators. The low-frequency vibrators have up to 30 cycles per second operating frequency and are powered either by electric or hydraulic motors. The high-frequency vibratory pile drivers have operating frequency

**TABLE 3.1 Impact Pile-Driving Hammer Data[a]**

| Rated Energy (kip-ft) | Make of Hammer | Model Number | Type[b] | Blows per Minute (max/min) | Stroke at Rated Energy (in.) | Weight of Striking Part (kips) | Total Weight (kips) |
|---|---|---|---|---|---|---|---|
| 1800.00 | Vulcan | 6300 | S-A | 38 | 72 | 300.0 | 838.00 |
| 300.00 | Delmag | D100-13 | Dies. | 45/34 | n/a | 44.894 | 70.435 |
| 225.00 | Delmag | D80-23 | Dies. | 45/36 | n/a | 37.275 | 58.704 |
| 200.00 | Raymond | RU-200 | — | 40/30 | 40 | 60.0 | — |
| 180.00 | Vulcan | 060 | S-A | 62 | 36 | 60.0 | 121.00 |
| 165.00 | Delmag | D62-22 | Dies. | 50/36 | — | 27.077 | 42.834 |
| 150.00 | Vulcan | 530 | S-A | 42 | 60 | 30 | 141.82 |
| 149.60 | Mitsubishi | MH80B | Dies. | 60/42 | — | 17.6 | 43.9 |
| 130.00 | MKT | S-40 | S-A | 55 | 39 | 40.0 | 96.0 |
| 127.00 | MKT | DE-150 | Dies. | 50/40 | 129 | 15.0 | 29.5 |
| 120.00 | Vulcan | 040 | S-A | 60 | 36 | 40.0 | 87.5 |
| 113.5 | Vulcan | 400c | Diff. | 100 | 16.5 | 40.0 | 83.0 |
| 107.177 | Delmag | D46-32 | Dies. | 53/37 | n/a | 19.58 | 30.825 |
| 97.5 | MKT | S-30 | S-A | 60 | 39 | 30.0 | 86.0 |
| 83.88 | Delmag | D36-32 | Dies. | 53/36 | n/a | 17.375 | 26.415 |
| 79.6 | Kobe | K42 | Dies. | 52 | 98 | 9.2 | 22.0 |
| 70 | ICE | 1072 | Dies. | 68/64 | 72 | 10.0 | 25.5 |
| 68.898 | Delmag | D30-32 | Dies. | 52/36 | n/a | 13.472 | 20.704 |
| 60.0 | Vulcan | 020 | S-A | 60 | 36 | 20.0 | 39.0 |
| 60.0 | MKT | S20 | S-A | 60 | 36 | 20.0 | 38.6 |
| 58.248 | Delmag | D25-32 | Dies. | 52/37 | n/a | 12.370 | 18.50 |
| 50.2 | Vulcan | 200C | Diff. | 98 | 15.5 | 20.0 | 67.815 |
| 48.75 | Raymond | 150C | Diff. | 115/105 | 18 | 15.0 | 32.5 |
| 48.7 | Vulcan | 016 | S-A | 60 | 36 | 16.2 | 30.2 |

|  |  |  |  |  |  |  |  |
|---|---|---|---|---|---|---|---|
| 48.7 | Raymond | 0000 | S-A | 46 | 39 | 15.2 | 23.0 |
| 44.5 | Kobe | K22 | Dies. | 52 | 98 | 4.8 | 10.6 |
| 44.0 | MKT | MS-500 | S-A | 50/40 | 48 | 15.5 | — |
| 42.0 | Vulcan | 014 | S-A | 60 | 36 | 14.0 | 27.5 |
| 40.6 | Raymond | 000 | S-A | 50 | 39 | 12.5 | 21.0 |
| 39.8 | Delmag | D-22 | Dies. | 52 | n/a | 4.8 | 10.0 |
| 39.366 | Delmag | D16-32 | Dies. | 52/36 | n/a | 7.166 | 11.079 |
| 37.5 | MKT | S14 | S-A | 60 | 32 | 14.0 | 31.6 |
| 36.0 | Vulcan | 140C | Diff. | 103 | 15.5 | 14.0 | 27.9 |
| 33.0 | Vulcan | 33D | Dies. | 50/40 | 120 | 7.94 | — |
| 32.5 | MKT | S10 | S-A | 55 | 39 | 10.0 | 22.2 |
| 32.5 | Vulcan | 010 | S-A | 50 | 39 | 10.0 | 18.7 |
| 32.5 | Raymond | 00 | S-A | 50 | 39 | 10.0 | 18.5 |
| 32.0 | MKT | DE-40 | Dies. | 48 | 96 | 4.0 | 11.2 |
| 30.2 | Vulcan | OR | S-A | 50 | 39 | 9.3 | 16.7 |
| 30 | ICE | 520 | Dies. | 84/80 | 71 | 5.07 | 17.04 |
| 28.1 | Mitsubishi | MH15 | Dies. | 60/42 | — | 3.31 | 8.4 |
| 28.0 | MKT | DE-33B | Dies. | 50/40 | 126 | 3.3 | 7.75 |
| 26.3 | Link-Belt | 520 | Dies. | 82 | 43.2 | 5.0 | 12.5 |
| 26.0 | MKT | C-8 | D-A | 81 | 20 | 8.0 | 18.7 |
| 26.0 | Vulcan | 08 | S-A | 50 | 39 | 8.0 | 16.7 |
| 26.0 | MKT | S8 | S-A | 55 | 39 | 8.0 | 18.1 |
| 25 | Vulcan | 505 | S-A | 46 | 60 | 5.0 | 29.5 |
| 24.4 | Vulcan | 80C | Diff. | 111 | 16.2 | 8.0 | 17.8 |
| 24.4 | Vulcan | 8M | Diff. | 111 | n/a | 8.0 | 18.4 |
| 24.3 | Vulcan | 0 | S-A | 50 | 39 | 7.5 | 16.2 |
| 24.0 | MKT | C-826 | D-A | 90 | 18 | 8.0 | 17.7 |
| 22.6 | Delmag | D-12 | Dies. | 51 | n/a | 2.7 | 5.4 |
| 22.4 | MKT | DE-30 | Dies. | 48 | 96 | 2.8 | 9.0 |

**Table 3.1** (*Continued*)

| Rated Energy (kip-ft) | Make of Hammer | Model Number | Type[b] | Blows per Minute (max/min) | Stroke at Rated Energy (in.) | Weight of Striking Part (kips) | Total Weight (kips) |
|---|---|---|---|---|---|---|---|
| 19.8 | Union | K 13 | D-A | 110 | 24 | 3.0 | 14.5 |
| 19.8 | MKT | 11 B 3 | D-A | 95 | 19 | 5.0 | 14.5 |
| 19.5 | Vulcan | 06 | S-A | 60 | 36 | 6.5 | 11.2 |
| 19.2 | Vulcan | 65C | Diff. | 117 | 15.5 | 6.5 | 14.8 |
| 18.2 | Link-Belt | 440 | Dies. | 88 | 36.9 | 4.0 | 10.3 |
| 18.0 | Delmag | D8-22 | Dies. | 52/38 | n/a | 4.0 | 6.147 |
| 17.0 | MKT | DE-20B | Dies. | 50/40 | 126 | 2.0 | 6.4 |
| 16.2 | MKT | S5 | S-A | 60 | 39 | 5.0 | 12.3 |
| 16.0 | MKT | DE-20 | Dies | 48 | 96 | 2.0 | 6.3 |
| 16.0 | MKT | C5 | Comp. | 110 | 18 | 5.0 | 11.8 |
| 15.1 | Vulcan | 50C | Diff. | 120 | 15.5 | 5.0 | 11.7 |
| 15.1 | Vulcan | 5M | Diff. | 120 | 15.5 | 5.0 | 12.9 |
| 15.0 | Vulcan | 1 | S-A | 60 | 36 | 5.0 | 10.1 |
| 15.0 | Link-Belt | 312 | Dies. | 100 | 30.9 | 3.8 | 10.3 |
| 13.1 | MKT | 10B3 | D-A | 105 | 19 | 3.0 | 10.6 |
| 12.7 | Union | 1 | D-A | 125 | 21 | 1.6 | 10.0 |
| 9.0 | Delmag | D5 | Dies. | 51 | n/a | 1.1 | 2.4 |
| 9.0 | MKT | C-3 | D-A | 130 | 16 | 3.0 | 8.5 |
| 9.0 | MKT | S3 | S-A | 65 | 36 | 3.0 | 8.8 |
| 8.75 | MKT | 9B3 | D-A | 145 | 17 | 1.6 | 7.0 |
| 8.8 | MKT | DE-10 | Dies. | 48 | 96 | 11.0 | 3.5 |
| 8.7 | MKT | 9B3 | D-A | 145 | 17 | 1.6 | 7.0 |
| 8.2 | Union | 1.5A | D-A | 135 | 18 | 1.5 | 9.2 |
| 8.1 | Link-Belt | 180 | Dies. | 92 | 37.6 | 1.7 | 4.5 |

| | | | | | | | |
|---|---|---|---|---|---|---|---|
| 8.1 | ICE | 180 | Dies. | 95/90 | 57.0 | 1.725 | 5.208 |
| 7.2 | Vulcan | 2 | S-A | 70 | 29.7 | 3.0 | 7.1 |
| 7.2 | Vulcan | 30C | Diff. | 133 | 12.5 | 3.0 | 7.0 |
| 7.2 | Vulcan | 3M | Diff. | 133 | n/a | 3.0 | 8.4 |
| 6.5 | Link-Belt | 105 | Dies. | 94 | 35.2 | 1.4 | 3.8 |
| 4.9 | Vulcan | DGH900 | Diff. | 238 | 10 | 0.9 | 5.0 |
| 3.6 | Union | 3 | D-A | 160 | 14 | 0.7 | 4.7 |
| 3.6 | MKT | 7 | D-A | 225 | 9.5 | 0.8 | 5.0 |
| 0.4 | Union | 6 | D-A | 340 | 7 | 0.1 | 0.9 |
| 0.4 | Vulcan | DGH100A | Diff. | 303 | 6 | 0.1 | 0.8 |
| 0.4 | MKT | 3 | D-A | 400 | 5.7 | 0.06 | 0.7 |
| 0.3 | Union | 7A | D-A | 400 | 6 | 0.08 | 0.5 |

[a] Table revised and updated from the original table by Vesic (1977) based on Manufacturer's catalogue data from Pileco, Inc. of Houston, TX, Vulcan Iron Works Inc. Chattanooga, TN, International Construction Equipment (ICE), Matthews, N. C., MKT Geotechnical Systems, Dover, N. J., and Raymond International Builder Inc.

[b] S–A: Single Acting   Dies.: Diesel
D-A: Double Acting   Comp.: Compound
Diff.: Differential

**TABLE 3.2 Vibratory Pile Driver Data[a]**

| Make | Model | Total Weight (kips) | Maximum HP | Frequency (cps) | Force (kips) |
|---|---|---|---|---|---|
| Bodine (USA) | B | 22 | 1000 | 0–150 | 63/100 to 175/100 |
| Foster (France) | 2–17 | 6.2 | 34 | 18–21 | — |
| | 2–35 | 9.1 | 70 | 14–19 | 62/19 |
| | 2–50 | 11.2 | 100 | 11–17 | 101/17 |
| ICE (USA) | 1412 | 31.7 | 550 | 6.67–20 | 500 |
| | 416 | 13.1 | 200 | 6.67–25 | 200 |
| | 116 | 4.2 | 94 | 6.67–26.67 | 100 |
| Menck (Germany) | MVB22–30 | 4.8 | 50 | — | 48 |
| | MVB65–30 | 2.0 | 7.5 | — | 14 |
| | MVB44–30 | 8.6 | 100 | — | 97 |
| MKT (USA) | V-36 | 18.8 | 550 | 26.67 | 386 |
| | V-30 | 15.0 | 510 | 26.67 | 320 |
| | V-20 | 12.5 | 315 | 28.34 | 214 |
| | V-17 | 12.0 | 260 | 26.67 | 160 |
| | V-5B | 6.8 | 99 | 26.67 | 80 |
| Muller (Germany) | MS-26 | 9.6 | 72 | — | — |
| | MS-26D | 16.1 | 145 | — | — |
| (Russia) | BT-5 | 2.9 | 37 | 42 | 48/42 |
| | VPP-2 | 4.9 | 54 | 25 | 49/25 |
| | 100 | 4.0 | 37 | 13 | 44/13 |
| | VP | 11.0 | 80 | 6.7 | 35/7 |
| | VP-4 | 25.9 | 208 | | 198 |
| Tunkers (Germany/ (USA) | HVB 260.02 | 22.0 | 1072.8 | 23.34 | 573.04 |
| | HVB 130.02 | 12.6 | 547.1 | 23.34 | 286.52 |
| | HVB 60.02 | 7.05 | 288.3 | 29.17 | 132.24 |
| | HVB 30 | 2.1 | 111.7 | 30 | 65.20 |
| | MVB 10 | 2.0 | 42.9 | 35 | 23.60 |
| Uraga (Japan) | VHD-1 | 8.4 | 40 | 16–20 | 43/20 |
| | VHD-2 | 11.9 | 80 | 16–20 | 86/20 |
| | VHD-3 | 15.4 | 120 | 16–20 | 129/20 |
| Vulcan (USA) | Vulcan 1150 | 6.5 | 125 | 1600 | 85.6 |

[a]Revised and updated the original table by Vesic (1977) based on Manufacturer's Catalogue data from Pileco, Inc. of Houston, TX; Vulcan Iron Works Inc. of Chattanooga, TN; International Construction Equipment, Matthews, NC; and MKT Geotechnical systems, Dover, NJ.

range of 80 to 120 cps and are powered by internal combustion engines. Table 3.2 provides a comparison of various vibratory pile drivers.

The selection of a particular hammer type on a job depends on the past experience of the engineer and the piling contractor. It also depends on the availability of the type of hammer. However, it is generally recognized that drop hammers have a slow rate of operation and deliver inconsistent energy to the pile. These hammers are, therefore, used only on small projects or in remote areas. The single-acting hammers use either the air or the steam pressure to raise the hammer ram. The impact energy delivered to the pile by a single-acting hammer is developed by the gravity fall of the ram. Thus, heavy ram can be used to deliver impact energy to the pile by using single-acting hammer.

Double-acting hammers have light rams, and they operate at a relatively high speed. These hammers are generally used to drive small-sized and lighter piles. Differential hammers have heavy rams and higher operating speed. These hammers combine the advantages of single-acting and double-acting hammers. Vibratory hammers are most effective in installing piles in cohesionless soils. It is generally believed that vibrations generated by these hammers liquify the surrounding soil and hence reduce the skin friction along the pile surface during pile driving. A major disadvantage of these hammers is that the load-carrying capacity of a pile cannot be estimated during the pile-driving operation.

### 3.2.4 Other Driving Accessories

As shown in Figure 3.1, the principal components of a pile-driving system consist of a pile, pile cushion, drive head, hammer cushion capblock shield, and pile driver (hammer). The *hammer cushion or capblock* serves a dual purpose. First, it protects the hammer and the pile from damage by reducing peak forces. Second, its elastic properties are such that it effectively transmits the hammer energy to the pile. Various types of hammer cushions or capblocks consist of the hardwood cushion, laminated cushion, and mechanical cushions. A typical hardwood cushion is 6 in. (150 mm) thick. Its main drawbacks are that it gets crushed and burned during driving, resulting in varying elastic properties. Due to these problems, it needs frequent replacements during driving operation. Hardwood blocks were mostly used in the past. They have now generally been replaced by laminated cushions. Laminated cushions or blocks are made of alternating layers of aluminium and micarata disks. Because of their nearly constant properties during their life, they transmit approximately consistent hammer energy to the pile. Therefore, these cushions are more efficient and have a longer life than hardwood blocks. Mechanical cushions are another type of hammer cushion which are made of a cylinder, a piston and springs. They can be designed to have consistent elastic properties during driving. These cushions reduce peak forces and extend the duration of hammer blow.

A *drive head* is a steel cap on the top of a pile that is used to distribute the hammer blows uniformly to the pile. It is important that the drive-head be of correct size to provide full contact with the pile. Drive-head shape and size should preferably be similar to that of the pile (e.g., H-pile shape for steel section or the section snugly fitting on pipe pile top).

A *pile cushion* is placed between the top of a precast concrete pile and the drive head and usually consists of layers of wood such as hardwood or softwood boards or plywood. A typical cushion made of compressible material such as plywood has a minimum thickness of 6 in. (150 mm). Their purpose is to protect the pile from damage, to transmit sufficient hammer energy to the pile, and to distribute the hammer blows uniformly over the pile head. In certain situations where the pile head is to be driven below ground surface or water level, a rigid steel member (called follower) is used as an extension of the pile. The follower must have enough stiffness so that the hammer energy is transmitted to the pile without buckling during driving or without significant elastic losses.

For the installation of light-gauge steel shells such as Raymond step taper piles, a special tool such as a *madrel* is used to provide the necessary stiffness for driving (see Figure 2.5). A mandrel can either be rigid or expandable by using pneumatic, mechanical, or hydraulic methods. During driving, the mandrel pulls the shell into ground. Once the shell is in place, the mandrel is withdrawn.

## 3.3 EQUIPMENT FOR BORED PILES

A wide variety of drilling equipment for installation of bored piles is commercially available in the market. Drilling contractors and equipment manufacturers can provide detailed specifications and capabilities and suitabilities of locally available equipment for a specific project. Woodward et al. (1972) provide detailed information on various aspects of construction equipment for installing bored piles. Basic information on equipment for installing these piles is summarized in the following sections.

### 3.3.1 Drilling Rigs

Two types of drilling rigs that are used for drilling pile holes are the auger type and the rotary type. In an auger type drilling machine, boring is done by an auger of suitable capacity into soil or soft rock. In rotary rigs, fluid pressure is used to drill the hole, and the drilling fluid also carries the cuttings out of the hole.

Various types of auger drilling rigs are available and manufacturer's cataloges can be obtained from major equipment suppliers. Information, such as maximum hole size and depth, driving arrangement, rig-mounting details, maximum continuous torque, and maximum continuous downward force are the main characteristics of auger drilling rigs. For example, the Hughes LLDH model 120T is capable of drilling up to 10 ft (3 m) diameter shaft to approximately 100 ft (30 m) depths. This rig is truck mounted and is driven by a mechanically geared rotary table. This has a maximum continuous torque of 59 kips-ft at 12 rpm and can apply a maximum downward force of 50 kips (223 kN). Figure 3.6 shows a typical Hughes LLDH truck mounted drilling rig. Table 3.3 gives weights and

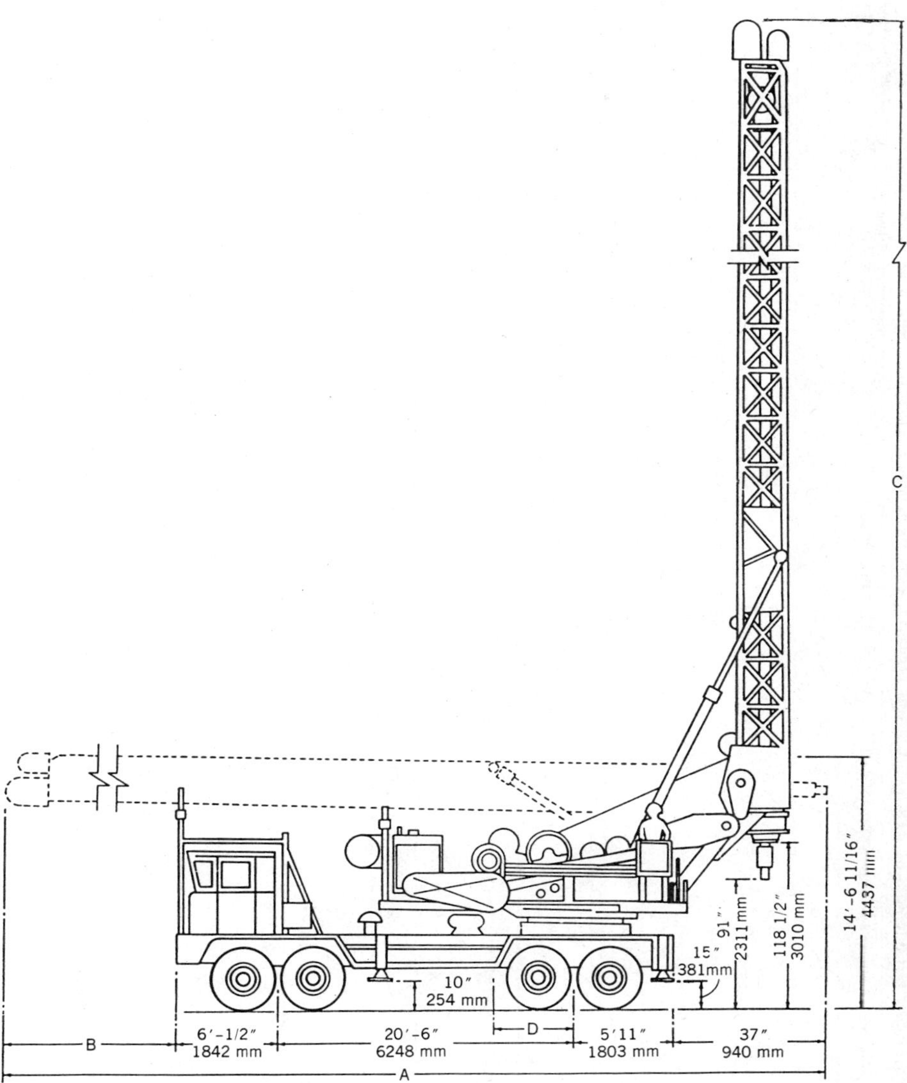


**Figure 3.6** Typical sketch of a truck-mounted LLDH-drill rig (Courtesy of Hughes-Micon Inc.).

TABLE 3.3[a] **Weights and Dimensions of Truck-mounted Drilling Rigs. (Hughes LLDH)**

| | LLDH 80 | | LLDH 100 | | LLDH 110 | | LLDH 120 | |
|---|---|---|---|---|---|---|---|---|
| | lb | kg | lb | kg | lb | kg | lb | kg |
| Total | 84 300 | 38 238 | 85 830 | 38 932 | 86 550 | 39 259 | 87 350 | 39 622 |
| Front tandem | 30 700 | 13 925 | 35 450 | 16 080 | 37 500 | 17 010 | 40 180 | 18 226 |
| Rear tandem | 53 600 | 24 313 | 50 380 | 22 852 | 49 050 | 22 249 | 47 170 | 21 396 |
| Derrick | 18 600 | 8 437 | 20 100 | 9 117 | 20 800 | 9 435 | 21 600 | 9 798 |
| Drawworks W/T.T. and jacks | 36 200 lb<br>16 420 kg | | | | | | | |
| Carrier (Purchased) | 24 729 lb<br>11 217 kg | | | | | | | |

Weights are approximate and will vary with equipment provided.

| | LLDH 80 | | LLDH 100 | | LLDH 110 | | LLDH 120 | |
|---|---|---|---|---|---|---|---|---|
| Dimensions[b] | ft | mm | ft | mm | ft | mm | ft | mm |
| A | 53′-10$\frac{3}{8}$″ | 16 418 | 63′-10$\frac{3}{8}$″ | 19 466 | 68′-10$\frac{3}{8}$″ | 20 990 | 73′-10$\frac{3}{8}$″ | 22 514 |
| B | 17′-10″ | 5 436 | 27′-10″ | 8 484 | 32′-10″ | 10 008 | 37′-10″ | 11 532 |
| C | 61′-5$\frac{1}{2}$″ | 18 729 | 71′-5$\frac{1}{2}$″ | 21 777 | 76′-5$\frac{1}{2}$″ | 23 301 | 81′-5$\frac{1}{2}$″ | 24 825 |
| Width jacks retracted | 8′-2$\frac{1}{4}$″<br>2 496 mm | | | | | | | |
| Width-front jacks extended | 10′-2$\frac{1}{4}$″ or 13′-2$\frac{1}{4}$″<br>3 105 mm or 4 020 mm | | | | | | | |
| Width-rear jacks extended | 13′-2$\frac{1}{4}$″<br>4 020 mm | | | | | | | |

Dimensions are approximate and will vary with equipment provided.

| Centre of gravity Location D | LLDH 80 | | LLDH 100 | | LLDH 110 | | LLDH 120 | |
|---|---|---|---|---|---|---|---|---|
| | in. | mm | in. | mm | in. | mm | in. | mm |
| Complete unit | 90 | 2 286 | 102 | 2 591 | 107 | 2 718 | $113\frac{9}{16}$ | 2 884 |
| Carrier and Drawworks W/T.T. and jacks | | | | $83\frac{3}{16}$ in. 2 113 mm | | | | |
| Derrick only | $114\frac{7}{16}$ | 2 907 | $163\frac{11}{16}$ | 4 158 | $182\frac{1}{4}$ | 4 629 | 206 | 5 232 |
| Complete unit minus carrier | $78\frac{3}{4}$ | 2 000 | 102 | 2 591 | $111\frac{1}{2}$ | 2 832 | $123\frac{3}{4}$ | 3 143 |

[a]Courtesy of Hughes Micon, Corsicana, Texas.

[b]See Figure 3.6.

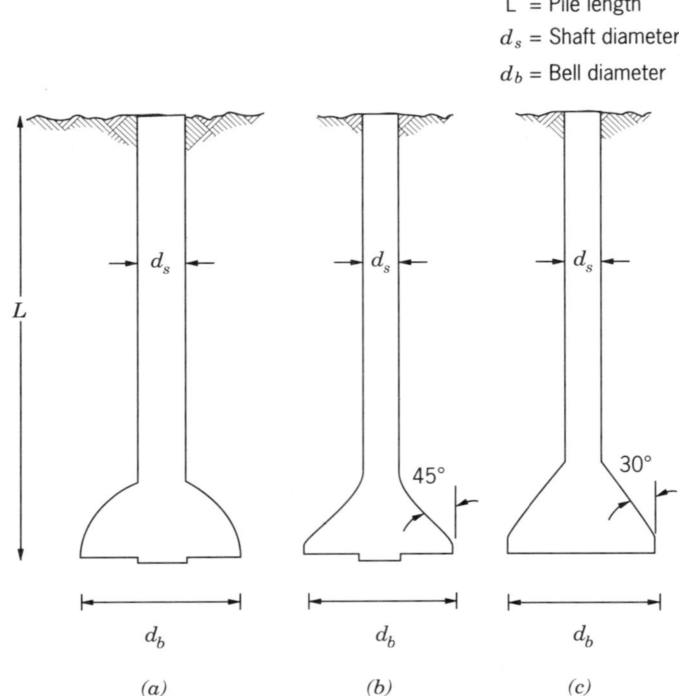


**Figure 3.7** Various drilled pile bell shapes. (a) Dome-shaped bell, (b) 45° bell, and (c) 30° bell.

dimensions of their various models. A Watson Model 3000 crawler-mounted drill rig can drill up to 110 ft deep. The Texoma model 600 drill rig has a mechanically geared mechanism that is driven by a hydraulic motor and is capable of drilling a 6-ft (1.8 mm) diameter hole to about 35 ft (11 m) deep. It can apply up to 23.5 kips-ft torque with a maximum continuous downward force of 26 kips (116 kN).

Driving arrangements can either be a kelly, usually a solid steel square driving shaft, or a hydraulic drive. A hydraulic drive arrangement will either be mounted at the turntable or on top of the drill stem and will ride up and down with it. Kelly can either be driven by a mechanically geared rotary table or by a yoke turned by a ring gear. For drilling shallower holes to about 60 ft (18 m), single-piece kellys are used. For deeper holes, two alternative methods are available. One is to add pin-connected sections of a drill shaft as required for the depth, and another is to have an inner square shaft sliding (by telescoping) in a larger hollow square section.

Drilling rigs can have one of the following mountings:

1. Truck

2. Crane
3. Crawler
4. Skid
5. Wheeled trailer

A majority of pile-boring drill rigs in North America are mounted on trucks. For larger and heavier jobs, crane-mounted rigs have proved to be more suitable. Truck-mounted rigs have the advantages in that they are very mobile, they have easy maneuverability, and many of them are also equipped with adjustments such as rotating mounts and masts that can be tilted to drill batter piles. These rigs can drill holes ranging from 4 ft (1.2 m) to as large as 10 ft (3 m) in diameter and to about 200 ft (60 m) deep. For example, the Texoma model 254 can drill a 4-ft (1.2 m) diameter hole to 20 ft (6 m) depths while a Texoma DM B100 can drill 8 ft (2.4 m) diameter holes to about a 100-ft (30 m) depth. Crane-mounted drill rigs need to be transported to the site. These rigs are less mobile and less adaptable to smaller jobs. Because of higher capacities of the cranes, these rigs are more suitable for larger jobs that require larger-diameter deeper holes. For example, crane-mounted rigs are capable of drilling holes in the range of 10 ft (3 m) to 20 ft (6 m) diameter and to depths of approximately 200 ft (60 m). These rigs can also drill holes to a maximum batter of 45°. Like crane-mounted rigs, crawler-mounted rigs also have to be transported to the site by heavy equipment trailers and therefore are less mobile. Crawler-mounted rigs, however, have excellent maneuverability and can drill large holes 20 ft (6 m) to typical depth ranges as deep as 200 ft (60 m) by applying torque as high as 350 kips and a downward force of 22 kips (Woodwards et al, 1972). A feature that may prove to be important in achieving an economical and efficient penetration rate in harder formations is the addition of downward thrust in a drilling rig. This can be done by providing one or more hydraulic cylinders that react against the weight of the machine carrying the drilling equipment. The engineer should be satisfied that sufficient thrust is available during the selection of equipment and support system. Equipment manufacturer and the drilling contractor should ensure this before a job is started to avoid undue delays during execution of the work.

### 3.3.2 Other Drilling (Boring) Accessories

The various drilling accessories that are required as tools for drilling rig to bore the pile hole may include augers, belling tools, core barrels, hole bottom cleaning tools, and casings. Other auxiliary tools, although not commonly used, may be utilized to suit specific site conditions.

***Augers*** Most of the pile shafts in soils are drilled with augers. These augers are open-helix shaped and can drill through most uniform soft to hard soils and rocks. Auger bits with hard metal cutting teeth are suitable for drilling through hard soils and soft to moderate hard rocks.

***Belling Tools*** Among the various belling tools (also called underreaming tools) to drill expanded base piles, the three main shapes of belling tools are the dome-shaped bell, the 45-degree bell, and the 30-degree bell (see Figure 3.7). Contractors' opinions vary regarding advantages and disadvantages of various types of belling tools. In general, the selection of a type will depend on their availability, contractors' preference, and past local experience.

***Coring Tools*** Coring tools such as core barrels and short barrels are used to cut through hard rocks that cannot be cut by augers-type tools. A core barrel is a cylinder with hard metal teeth at its lower part for cutting purposes. Various arrangements, spacings, and orientations of these teeth are available and their details are available with manufacturers. Another type of coring tool, which is suitable for cutting through hard rocks, is called a shot barrel. This tool does not have cutting teeth. The cutting is done by the chilled steel shots that grind the rock under the rotating edge of the barrel. The fine ground dust and cuttings of steel and rock are then washed out in suspension by water.

***Hole Bottom Cleaning Tools and Casings*** In many situations, the soils within the depths of pile installation may contain water-bearing strata. This water will seep into the hole drilled for pile installation. In most circumstances, this water is removed from the hole on completion of drilling before concrete is poured. Dewatering equipment that can be used for this purpose include airlift pumps, down-hole pumps that operate continuously until concrete pour is started, or specially designed drilling buckets that are also used as bailers to remove the water from the hole. Selection and use of these equipments should be the responsibility of the drilling contractor. However, the engineer must be satisfied that the equipment available on site is capable of handling the expected amount of water at the specific site.

In some situations, casing may be required to maintain a clean hole during or after drilling is completed. These casings could either be a temporary measure and removed after concrete has been poured or may be left in the hole as a permanent component of the pile. These casings are made of steel and should be of sufficient thickness to withstand soil and water pressures from the surrounding environment before concrete is placed into the hole.

## 3.4 PROCEDURE FOR PILE INSTALLATION

### 3.4.1 Planning Prior to Installation

Prior to proceeding with actual pile installation operation, the following steps should be followed. The first step is preparing pile specifications. These specifications are written after proper field drilling and testing, laboratory testing, geotechnical evaluation, pile load tests, and pile design and construction criteria have been established. These specifications should include soil conditions at the site, drilling methods if bored piles are used, driving method and sequence for driven piles, material specifications, tolerances, inspection and testing, and the data and record requirements.

Office planning would then consist of identifying piles by numbers and specifying pile installation sequence on construction drawings. Actual numbering of piles is important because this way any communication between design engineer, construction engineer, and pile installation contractor regarding reporting and recording is made easy when they can mention each pile by an identification tag. Methods of numbering depends on engineer's preference. They could be pure numbers or a combination of alphabets and numbers. Numbering criteria may be based on structure identification, location, coordinates, or other method to suit the specific job. Pile installation sequence is equally important because it provides guidelines to the field inspector and also furnishes criteria to the contractor to plan the activities including scheduling. Pile installation sequence should be agreed upon between the engineer and the contractor.

The next step is field planning prior to installation. This consists of surveying, pile preparation, and preexcavation, if required. A field survey is required to identify obstructions and stake out pile locations. Obstructions could either be on surface such as existing structures or they could be underground obstructions such as utility lines, electric cables, and so forth. The contract documents should clearly spell out the name of the party responsible for staking out the pile locations. Normally, a survey to identify surface obstructions and stake out pile locations is carried out by one specialist contractor, and the survey to identify underground structures is done by another specialist contractor. Proper communication and coordination between these different groups of specialist contractors should be carried out and information be transmitted on time to the pile installation contractor.

The next step in field planning consists of pile preparation. Some examples of pile preparation are providing pile protection, pile splicing, and pile coatings. Driven piles such as timber, steel, and precast concrete piles may require the attachment to protect pile tips. Information on various types of shoes that are commonly used to protect pile tips is included in Section 3.4.2. For timber piles, in addition to drive shoes, steel bands are also provided at specified intervals along the pile and at the pile butt to protect the pile from splitting during driving. Another pile protective method provided at pile preparation phase for steel piles is the attachments for cathodic protection cables (see Section 2.4.3).

Steel, precast concrete, and timber piles may require splice or joint fittings to make up the necessary pile lengths. Splicings may be welded to steel piles on ground at the job site whereas splicing for sectional precast concrete piles are attached to the pile at the shop when the pile is being cast (see Section 3.4.2).

Normally, coatings are provided on the piles for three purposes: (1) for pile material protection, (2) for pile friction reduction in the zone(s) where surrounding soft soils will drag the pile downwards causing negative skin friction, and (3) for pile friction reduction in cold regions where adfreeze forces are important. Methods of coatings for material protection were discussed in Sections 2.3.3 and 2.4.3. The most common type of coating for pile friction reduction is the application of bituman (asphaltic) coatings. The manufacturer's recommendations must be followed during coating applications. When these coatings are applied at the site, the coated piles should be protected from damage during drying.

Preexcavation or predrilling of near-surface hard soil strata may be required to facilitate installation of driven piles. The main methods of preexcavation are either by dry (augering) process or by the wet (drilling, jetting, or a combination) process. The method and the equipment used for preexcavation will depend on the site soil conditions. These methods should be approved by the engineer who should be familiar with the detailed mechanism of predrilling process and its impact on the performance of piles.

The last step is the pile installation itself. This consists of pile handling, pile alignment, and the actual installation. During pile handling, precautions should be taken so that the pile is not damaged during lifting. For example, treated timber piles should not be handled with pointed tools to avoid damage, and precast concrete piles should only be picked up at the predesignated points. For pile alignment, the plumb for vertical piles and the inclination or off-verticality for batter piles should be specified on the pile drawings. Prior to the start of pile driving, the pile and pile-driving equipment should be properly checked for alignment. The final step is actual pile installation. This is discussed in Sections 3.4.2, 3.4.3, and 3.4.4.

### 3.4.2 Installation of Driven Piles

Driven piles are either installed by the impact of a hammer or by a vibratory driver. For piles driven by impact hammers, the installation criteria is generally based on a specified penetration resistance for a driving energy that may be established either on the basis of a wave equation analysis and/or conventional driving formula. These criteria are discussed in Chapter 5. An estimated pile length is also established based on a static analysis for the soil profile and with the knowledge of the properties of the bearing strata. In Chapter 5, details of several static analyses are also discussed. Both these criteria—the specified penetration resistance and the estimated pile lengths for a desired pile capacity—provide a check that design assumptions have been realized during actual installation. If, at the estimated pile length, the specified penetration resistance is significantly different than the one determined above, the situation must be properly evaluated by the design engineer.

For piles that are to be installed by vibratory hammers, the specifications normally should establish a driving criteria and the required pile lengths. The required pile length is determined on the basis of subsoil information and by using the conventional bearing capacity formulas. The driving criteria normally should be a combination of the two factors. The first is a minimum rate of pile penetration for a specified dynamic force and operating frequency. The second criterion would require that the final penetration resistance of the pile be checked with an impact hammer. In such cases, a minimum specified penetration resistance for a driving (impact) energy applied by a hammer should be met. This is further discussed in Chapter 5.

***Driving of Timber Piles*** Timber piles do not require any special consideration for handling stresses. However, precautions are required to protect the timber pile tip and head from damage due to driving stresses. These damages may occur in the form of splitting the butt or the body or breaking the pile during driving. The pile butt and body may be protected with a steel ring, and the pile tip could be protected with steel shoes where hard driving is expected. Also, low-velocity hammer blows should be specified for driving purposes. As a guide, the hammer with rated energy per blow in foot-pounds (joules) not to exceed 3000 (1600) times the diameter in inches (cm) may be specified (*Canadian Foundation Engineering Manual*, 1978). Also, driving should be stopped when driving resistance is four to five blows per inch.

***Driving of Precast and Prestressed Concrete Piles*** Handling stresses are important for these piles, which restrict the use of very long precast concrete piles. Therefore, special splices are required to join smaller lengths. Various pile splices are produced by specialized manufacturers. The strength of the splices used must be at least equal to that of the pile. In addition, the splice should be designed so that the slack between two joined sections of a pile is less than 0.02 in (0.5 mm). This will minimize the loss of driving energy. According to the recommendation by the *Canadian Foundation Engineering Manual* (1985), splices must be cast square with the pile segment ends with maximum allowable deviation out of squareness being 1 in 100.

***Splicing of Precast Prestressed Concrete Piles*** Splicing allows extensions of piles enabling the use of shorter pile sections. This reduces the handling of weight and length of pile and thereby reducing the probability of cracking due to excessive handling stresses and other associated problems with the installation of long piles. Proper splicing methods eliminate the need to predetermine exact pile lengths and allow extensions of piles when necessary. Bruce and Hebert (1974) presented a review and evaluation on splicing methods developed and used in several parts of the world. The splices investigated were categorized as follows:

1. Welded
2. Bolted
3. Mechanical locking
4. Connector ring
5. Wedge
6. Sleeve
7. Dowel
8. Posttensioned

Table 3.4 provides a summary of these splices, and Figure 3.8 is a schematic

**TABLE 3.4 Summary of Splices (Bruce and Hebert, 1974)**

| Name of Splice | Type | Origin | Approximate Size Range (in./cm) | Approximate Field Time (min.) | Strength: Percent Compressive | Strength: Percent Tensile | Strength: Percent Flexural Cracking |
|---|---|---|---|---|---|---|---|
| Marier | Mechanical | Canada | 10–13 (25–33) | 30 | 100[a] | 100[a] | 100[a] |
| Herkules | Mechanical | Sweden | 10–20 (25–51) | 20 | 100[b] | 100[b] | 100[b] |
| ABB | Mechanical | Sweden | 10–12 (25–30) | 20 | 100[b] | 100[b] | 100[b] |
| NCS | Welded | Japan | 12–47 (30–119) | 60 | 100[b] | 100[b] | 100[b] |
| Tokyu | Welded | Japan | 12–47 (30–119) | 60 | 100[b] | 100[b] | 100[b] |
| Raymond cylinder | Welded | USA | 36–54 (91–137) | 90 | 100[b] | 100[b] | 100[b] |
| Bolognesi-Moretto | Welded | Argentina | Varied | 60 | 100[a] | 55[a] | 100[a] |
| Japanese bolted | Bolted | Japan | Varied | 30 | 100[b] | 90[b] | 90[b] |
| Brunsplice | Connector ring | USA | 12–14 (30–36) | 20 | 100[b] | 20[a] | 50[a] |
| Anderson | Sleeve | USA | Varied | 20 | 100[a] | 0[a] | 100[a] |
| Fuentes | Welded sleeve | Puerto Rico | 10–12 (25–30) | 30 | 100[b] | 100[a] | 100[a] |
| Hamilton form | Sleeve | USA | Varied | 90 | 100[a] | 75[b] | 100[b] |
| Cement dowel | Dowel | USA | Varied | 45 | 100[b] | 40[b] | 65[b] |
| Macalloy | Posttensioned | England | Varied | 120 | 100[a] | 100[a] | 100[a] |
| Mouton | Combination | USA | 10–14 (25–36) | 20 | 100[a] | 40[a] | 100[a] |
| Raymond wedge | Welded wedge | USA | Varied | 40 | 100[a] | 100[a] | 100[a] |
| Pile coupler | Connector ring | USA | 12–54 (30–137) | 20 | 100[b] | 100[b] | 100[b] |
| Nilsson | Mechanical | Sweden | Varied | 20 | 100[a] | 100[a] | 100[a] |
| Wennstrom | Wedge | Sweden | Varied | 20 | 100[a] | 100[a] | 100[a] |
| Pogonowski | Mechanical | USA | Varied | 20 | 100[a] | 100[a] | 100[a] |
| Thorburn | — | Scotland | | No information available on this splice | | | |

[a,b]Based on data furnished by Proponent.
[a]Calculated.
[b]Observed.

presentation of various splice types. Their use will depend on their availability and required characteristics as presented in Table 3.4. The pile head should be provided with a minimum of a 0.5-in. (12.5 mm)-thick steel plate to protect it against hard driving especially when pile loads are expected to exceed 100 tons (900 kN). The head should also be encased with a steel collar connected to the head plate and extending to a depth equal to half the pile diameter. Also, special steel points or shoes should be provided to protect pile tips where hard-driving conditions such as rock are encountered. For situations where driving conditions are easier, the pile head and tip need only be chamfered at the edges and corners.

Before a job is started, the pile manufacturer, the driving contractor, and the engineer must evaluate and agree with (1) the head and the tip protection methods, (2) the driving equipment to be used, and (3) all step-by-step pile installation techniques to be used at a specific job site.

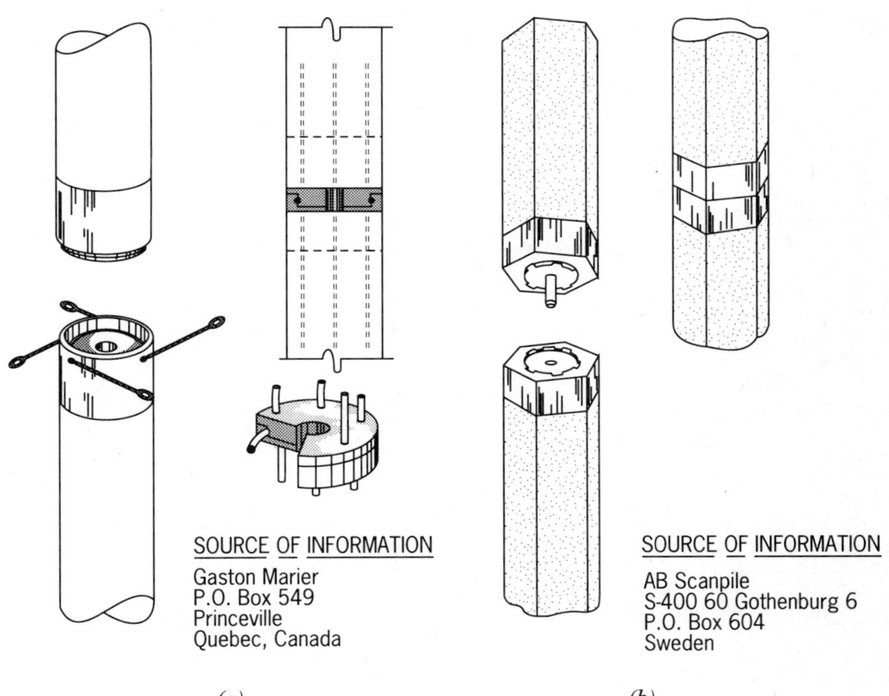


**Figure 3.8** Schematic presentation of various splice types (Bruce and Hebert, 1974). (a) Marier splice, (b) Herkules splice, (c) ABB splice, (d) NCS splice, (e) Tokyu splice, (f) Raymond cylinder pile splice, (g) Bolognesi-Moretto splice, (h) Japanese bolted splice, (i) Brunspile connector ring, (j) Anderson splice, (k) Fuentes splice, (l) Hamilton form splice (m) cement-dowel splice, (n) Macalloy splice, (o) Mouton splice, (p) Raymond wedge splice, (*q*) pile coupler splice, (r) Nilsson splice, (s) Wennstrom splice, and (t) Pogonowski splice.

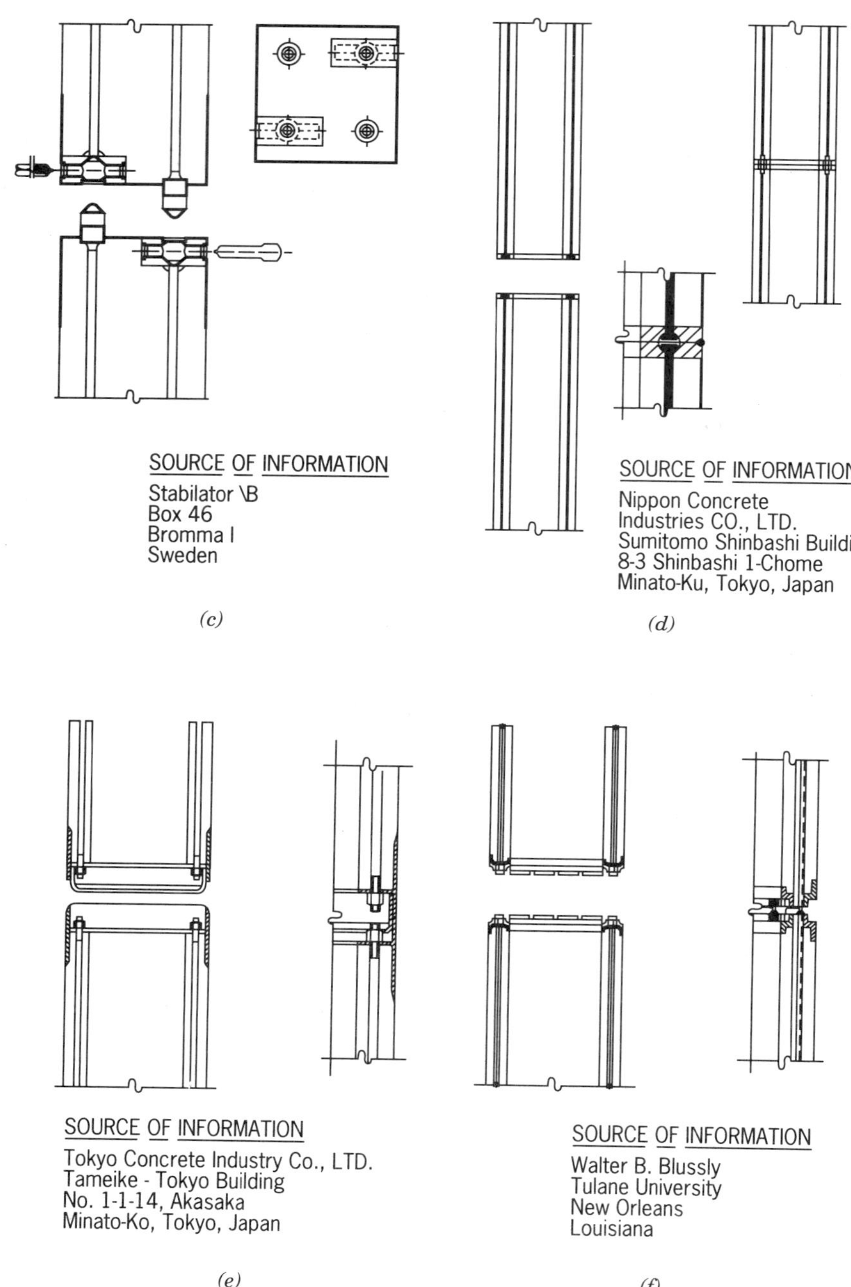


**Figure 3.8** (*Continued*)

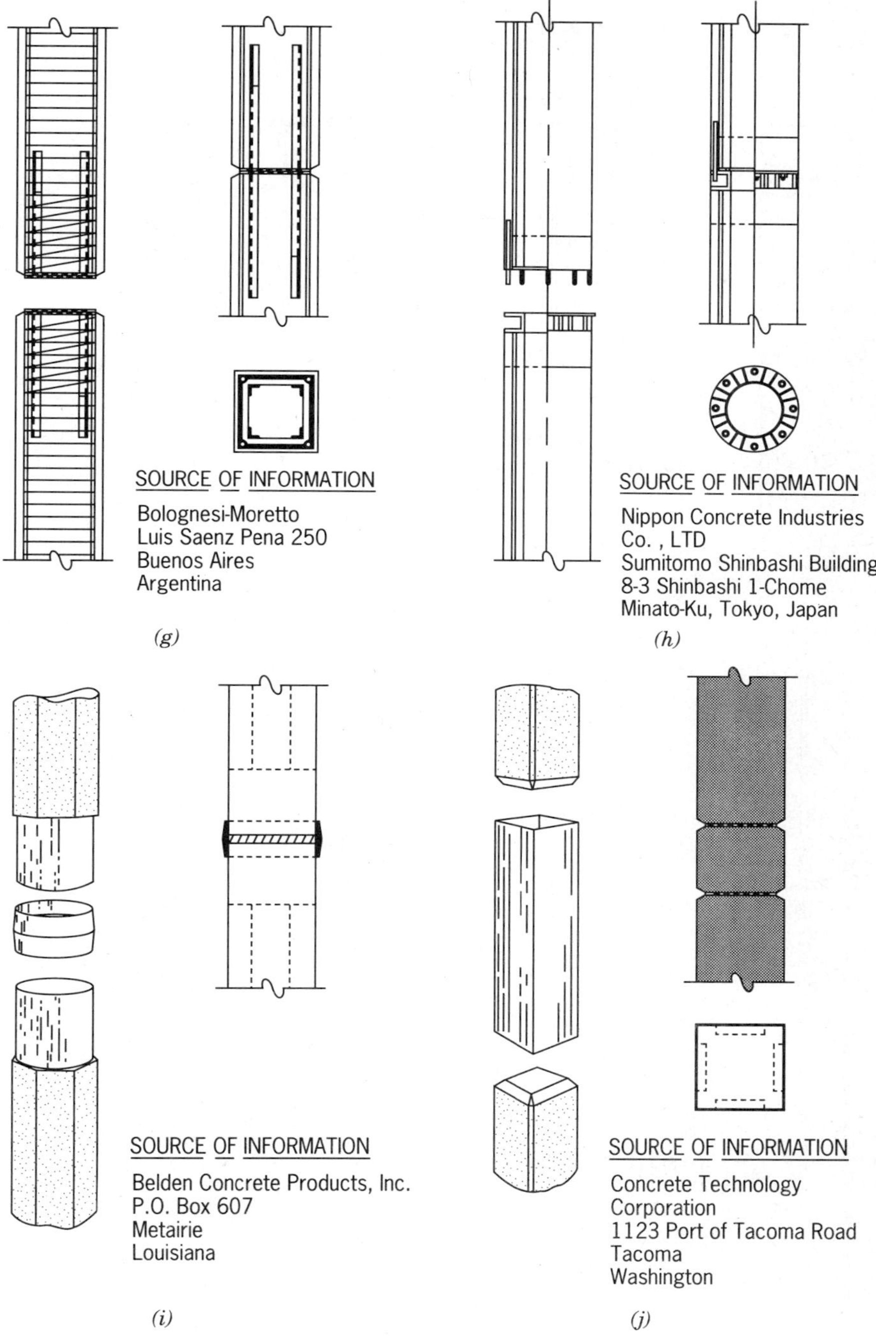


**Figure 3.8** (*Continued*)

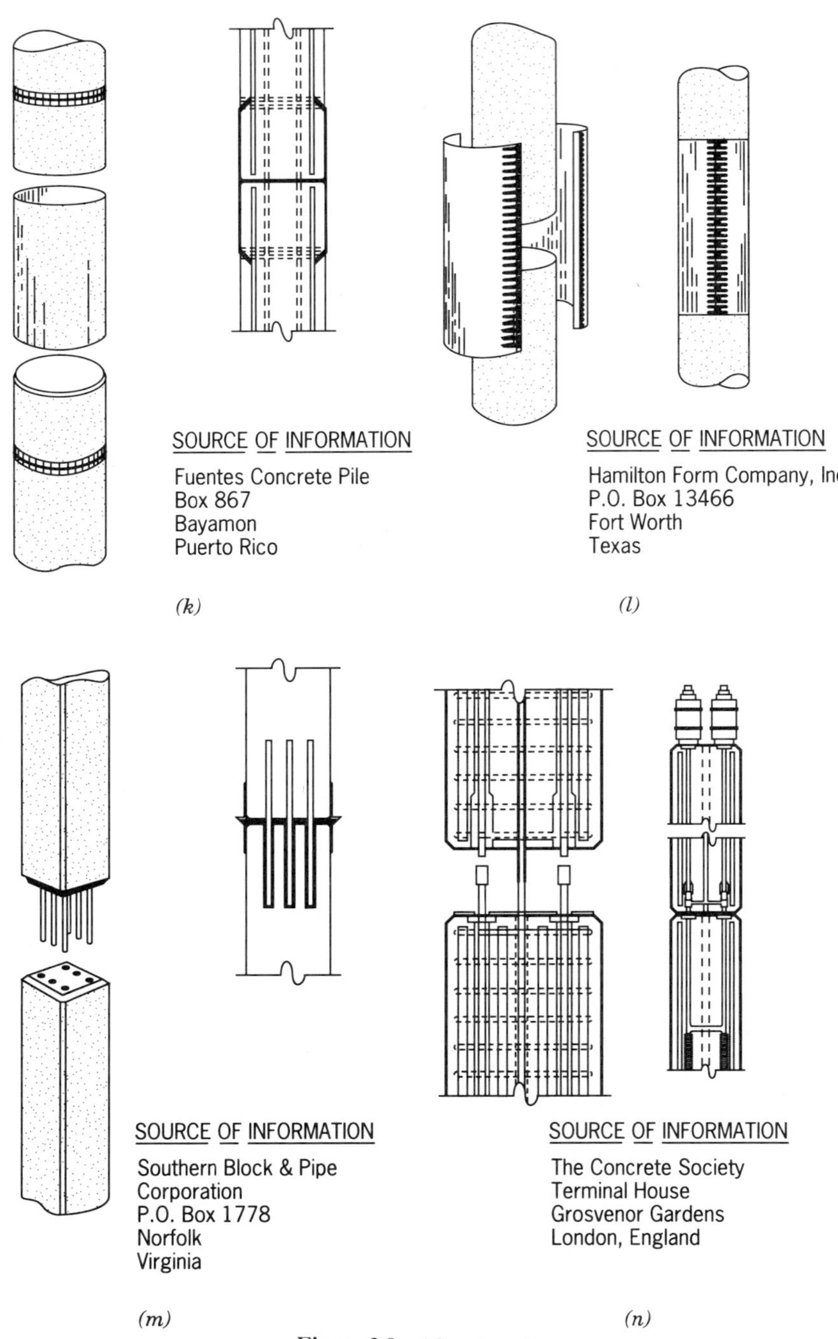


**Figure 3.8** (*Continued*)

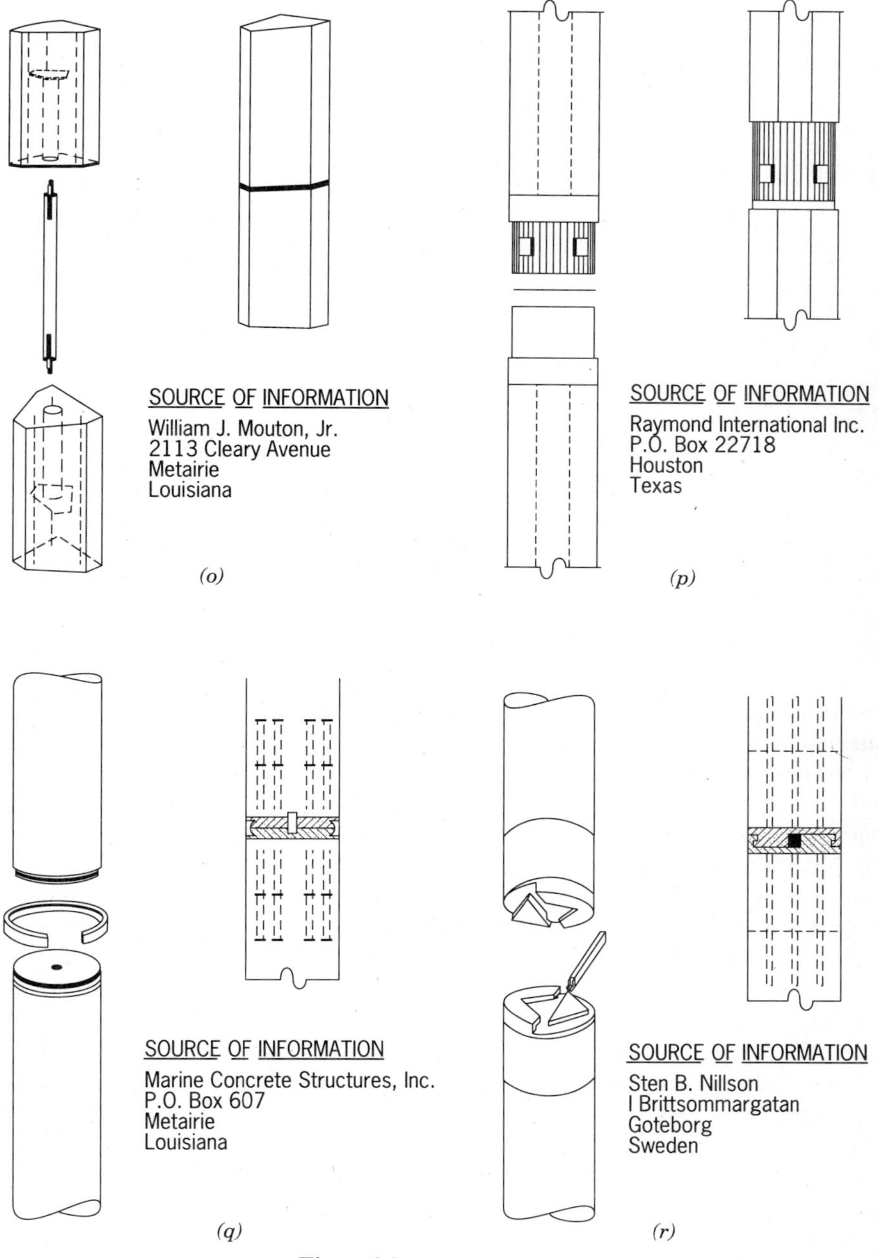


**Figure 3.8** (*Continued*)

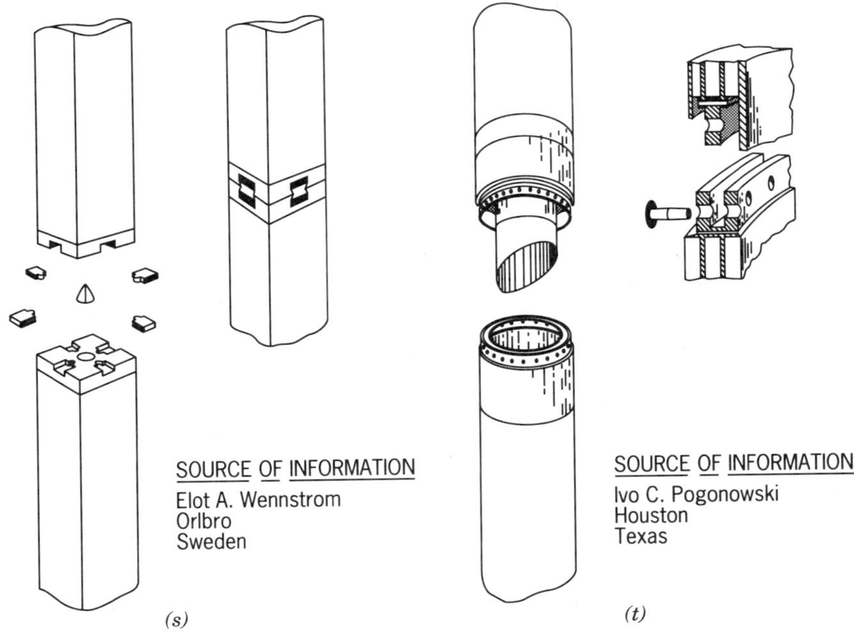


**Figure 3.8** (*Continued*)

Although single-acting and differential-acting (steam/air) hammers may be used for driving precast or prestressed concrete piles, drop hammers and diesel hammers are the most common hammer types that are used for driving these piles. As a preliminary guide, it is generally recommended that the pile-mass-to-ram-mass ratio should not exceed 2 to 1 for steam- or air-driven hammers and should not exceed 4 to 1 for diesel-powered hammers. In order to minimize higher impact velocities and unacceptable driving stresses, it is recommended that the height of freefall of drop hammers should not exceed 30 in. (75 mm). Also, driving should be stopped when driving resistance reaches 6 to 8 blows per inch.

***Driving of Steel Piles*** Normally installation of steel H piles by driving is easy except for cases where very dense gravel or tills with boulders are encountered. These conditions may damage the pile. To avoid this damage, the piles should be protected by using cast steel drive shoes and by welding steel plates to the toe of the pile to reinforce it. Driving stresses in piles may be estimated by wave equation analysis unless an acceptable local experience is available on a specific project.

Various joint fitting methods such as riveting, bolting, or welding can be used for splicing H pipes. A common practice is to make full-strength butt welds on these piles. Jointing can also be made by using available splicers in the market. Figure 3.9 shows some typical H-pile joint and point fittings that can be used as splicers and drive shoes.

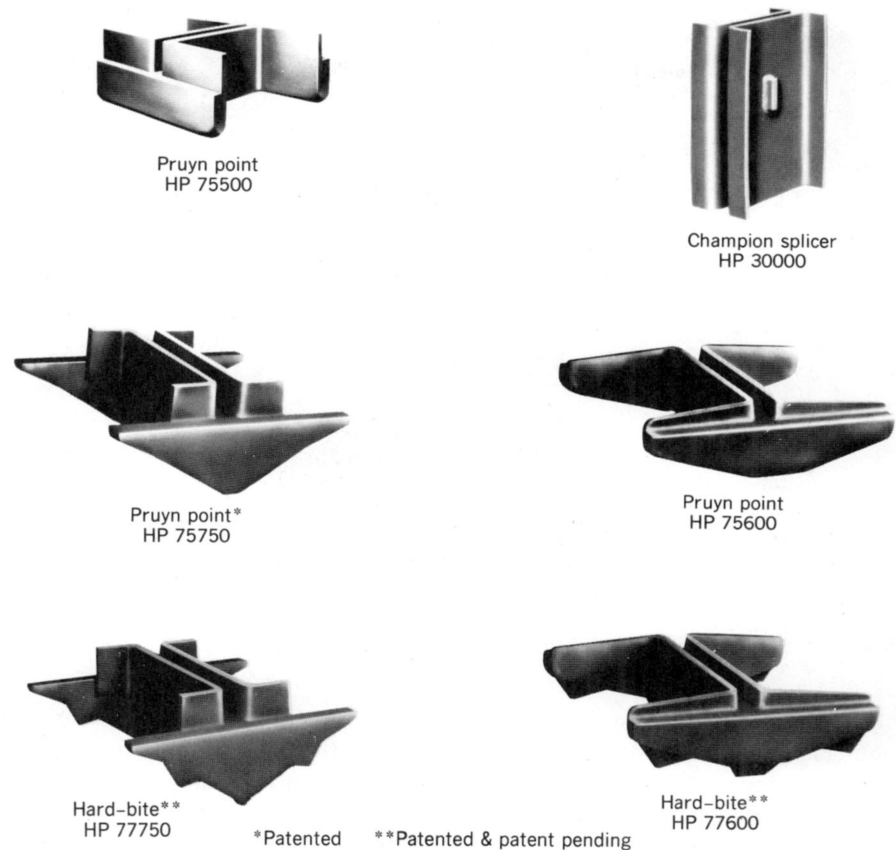


**Figure 3.9** Typical H-pile fittings (Courtesy of Associated Pile and Fitting Corp.).

Driving of open-ended pipes is easier than closed-ended pipes. If pipe piles are driven through very dense material or hard obstructions, these may be damaged at their tips. Figure 2.9 exhibited that pipe piles may deflect and/or their tips may get damaged when a hard stratum is encountered. Steel driving shoes are generally recommended to protect open-ended pipe piles against damage when hard-driving conditions such as dense gravel are encountered. Proper control and monitoring of driving energy should also be recorded on a continual basis to identify obstructions so that they can be removed when encountered.

When open-ended pipe piles are installed as nondisplacement piles, care must be taken to ensure that these piles do not pick up an immovable soil plug during driving. These piles would require periodic cleaning out by drilling or by washing, with or without jet to achieve the necessary penetration. These piles can then be filled with concrete. However, one must ensure that soil beneath the pile tip is not removed during this cleaning process. Closed-ended pipe piles normally have

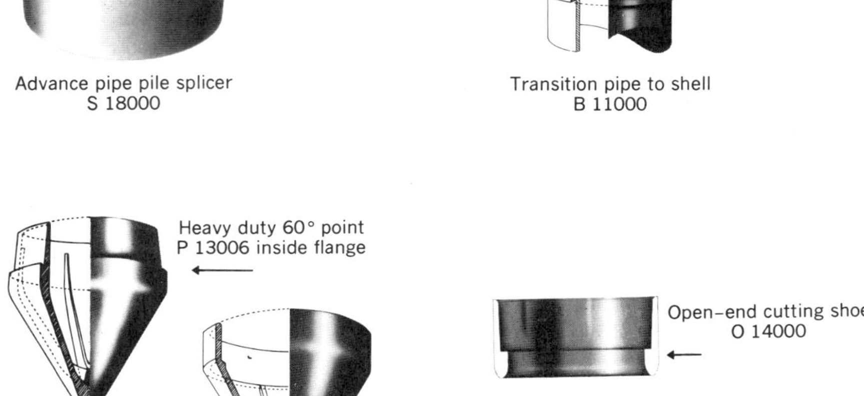


**Figure 3.10** Typical pipe pile fittings (Courtesy of Associated Pile and Fitting Crop.).

about a 0.5-in. to 0.75-in. (12.5 to 18.8 mm) thick steel plate at its tip. Special points such as conical shape points are provided at the tip when they are driven to rock or through boulders. O'Neill et al. (1982) report that 10.75-in. (272 mm)-diameter steel pipe piles with a 1.0-in. (25 mm)-thick base plate can be successfully installed into overconsolidated stiff clays. Various types of splicers and drive shoes are available for pipe piles in the market. See Figure 3.10 for typical pipe pile joint and point fittings. In addition to the foregoing protective measures, it is also recommended that the rated energy of the hammer be limited to 3000 ft-lb/in.$^2$ (630 joules/cm$^2$) both for steel H and pipe piles. Driving should be stopped when driving resistance has reached 12 to 15 blows per inch to avoid damage.

For the typical offshore platform, large high-strength steel plates up to 2.5 in. (62.5 mm) thick are rolled into tubular shapes and welded longitudinally. These tubular shapes are then welded end to end. Enough sections are fabricated so that when welded end to end, the sum of the sections will constitute as long as 400 to 800 ft (122 to 124 m) pile. As each section is added to the pile being driven, it is field welded. All welds are full penetration groove welds (Graff, 1981).

***Some Special Pile-driving Problems*** A few special problems that can be encountered during pile installation by driving are noticeable unexpected pile length variations in an area during driving, soil freeze, pile heave, and relaxation. These problems are briefly discussed as follows.

Noticeable unexpected pile length variation in an area may be due to various reasons, such as obstructions, change in soil conditions, existence of cavernous limestone formation, and a result from the gradual densification of the subsoil during pile driving. To minimize the densification of subsoil during driving, it should be specified in pile installation specifications that *driving sequence shall be such that the driving commences at the center of a group and continues outward.* Any variation in pile length in a group more than 5 ft (1.5 m) should be investigated and the causes identified. Proper corrective measure(s), if required, can then be instituted.

Soil freeze or setup occurs in cohesive soils that show a decrease in strength when remolded due to pile driving, and regain their strength with time (see Chapter 1). The amount and time required to regain their strength and hence the soil freeze will depend on the sensitivity of the cohesive soil. If driving a pile is interrupted due to uncontrollable circumstances in soils that exhibit soil freeze, these piles may experience increased resistance when redriven or retapped. Therefore, these piles should be retapped to break the soil freeze and then driving continued to reach the required bearing strata. In extreme cases, there piles may have to be abandoned and replaced because it may not be possible to reach the required bearing strata by retapping without damaging the piles (Fuller, 1983).

Already-driven piles may experience heave or upward movements when displacement-type piles are driven in cohesive soils. Piles heave should be monitored by taking elevations of tops of the already-driven piles. Contract documents should clearly specify that if heave occurs, then these piles must be redriven (Klohn, 1961). This redriving may be to the original penetration resistance or tip elevations. Pile redriving should only be commenced when pile-driving operation has progressed beyond the range that it would not cause any more heave in the already-driven piles. Koutsoftas (1982) monitored pile heave in H-piles driven into very dense sand stratum. Based on this study, it was concluded that heave up to 1.0 in. (25 mm) had no effect on load capacity because H-piles had penetrated into the bearing stratum. Also, as expected, it is recorded that heave decreases exponentially with distance from the pile-driving operations.

Certain soils exhibit high resistance to pile driving due to the apparent high soil strength resulting from the development of negative pore water pressures during shear. This normally occurs in submerged, dense, fine sands and inorganic silts. When pile driving ceases, the negative pore water pressure would dissipate resulting in decreased shear strength with time. This would cause lower pile resistance with time and is termed *relaxation.* Tapping of already-driven piles should be carried out for such soil conditions. If, after tapping, it is found that the original driving resistance has decreased, these piles should be further retapped until specified driving resistance has been achieved.

### 3.4.3 Installation of Bored Piles

There are two main steps in bored pile installation: boring the hole and then placing the concrete into the drilled hole. Selecting the method of drilling or boring the hole will depend on soil and groundwater conditions. When soil is weak and/or groundwater is encountered above the level of bearing stratum, temporary casing may be required during boring (drilling) the hole. Whatever method of boring is chosen, it must be ensured that the base is clean and is free of any slumped or loose material. As shown in Figure 3.11, bored piles can be straight shafted, underreamed, or multiunderreamed. Underreaming is done to increase the pile base size and hence provides additional load-carrying capacity. Multiunderreaming is provided to mobilize additional skin friction along the shaft. Conventional belling tools can be used to drill more than one bell along the shaft provided the ground and water conditions are suitable for making bell(s) without caving in. Martin and De Stephen (1983) confirm, by load testing an instrumented double underreamed pile and by monitoring performance of such production piles, that such piles are viable cost effective foundation in very stiff overconsolidated clays.

For soft to very firm uniform soils, an auger with a cutting blade will drill the pile hole with relative ease. For very hard, stony, and cemented soils, a toothed auger would be required to drill the hole. However, augers will be stopped by hard boulders and may require special tools designed to grab the boulders. In some cases when large stones are encountered, they may first be broken and then picked up by the tool.

Sharma et al. (1983) describe a situation where the pile-bearing stratum consisted of weathered shale bedrock under artesian pressures. Specifications required that this bearing stratum should not be punctured during the pile bell formation to avoid water flow into the hole. This was achieved by attaching a small 1-in. (25 mm) diameter, 10-in. (250 mm)-long pilot auger in front of the main auger to locate the bedrock. Once the bedrock was located, the auger was lifted

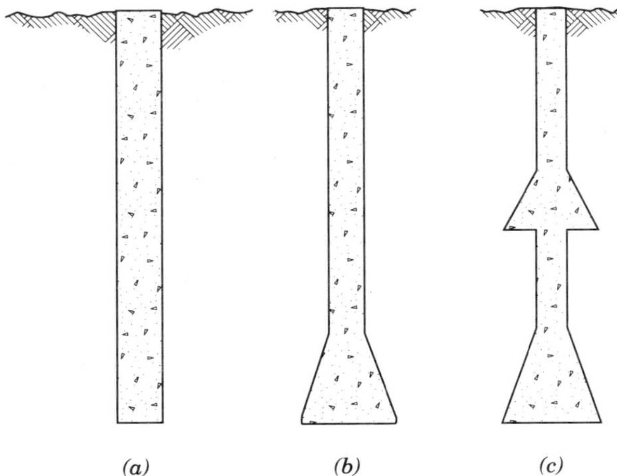


**Figure 3.11** Types of bored piles: (a) Straight-shafted bored pile (b) belled or underreamed pile, (c) multiunderreamed bored pile.

and belling commenced in such a way that the bottom of bell could be placed at the top of bedrock. Dry soil or concrete was placed in the hole to place the belling tool properly so that the bedrock was not punctured during the belling operation. This technique ensured that the bell could be formed at a desired elevation without disturbing the underlying water-bearing stratum. This technique is schematically presented in Figure 3.12.

Where the hole for drilled pile is dry the concrete may be placed by buckets, chutes or elephant trunks to avoid concrete segregation. Concrete can be placed by free fall without compaction provided the free fall height is 5 ft (1.5 m) or less.

In situations where temporary casing is used to keep the drilled hole open without the surrounding soil slumping into it, withdrawal of the casing during concreting should be carefully controlled. This should be done by maintaining a minimum of 5 ft (1.5 m) head of concrete within the casing. When the temporary casing is provided to seal out the groundwater, then the casing must not be disturbed until enough concrete has been placed so that its pressure is higher than the outside water pressure. When the flow of groundwater into the hole cannot be controlled, it may be necessary to clean out the hole and place the concrete by tremie without removing the water. Under such circumstances, it must be ensured that during concreting there is no flow of groundwater into the hole. If this happens, the hole should be filled with water to avoid dilution or segregation of concrete due to water flow into the hole. The tremie with some kind of closure at its bottom should then be inserted to the bottom of the hole and water should not be allowed to go into it otherwise it may dilute the concrete. This closure should not be opened until the concrete inside the tremie has reached the appropriate level so that inside concrete pressure is higher than the water

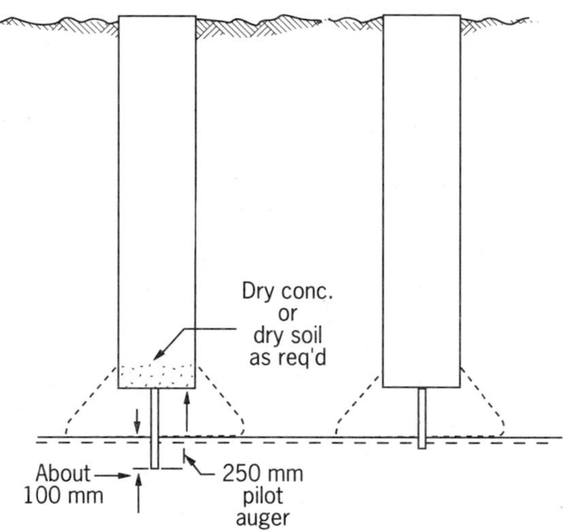


**Figure 3.12** Scheme of pile installation when bell is resting on top of rock under artesian pressures (Sharma et al., 1983). (a) Locating bedrock, (b) pile hole ready for concreting.

pressure from outside. Concreting can then proceed in normal way. Integrity of the poured shaft (concrete) shortly after construction can be checked by wave propagation method (Hearne et al., 1981).

### 3.4.4 Installation of Special Types of Piles

Various types of special piles were described in Section 2.6. Installation details of two of these piles—expanded base compacted (Franki) and thermal piles—are detailed further as follows.

***Installation of Expanded Base Compacted (Franki) Piles*** The following three major activities are required in the installation of these piles:

1. Driving the casing
2. Base or bulb construction
3. Shaft construction

#### Driving the Casing

The casing may be driven with a hammer striking its top or using a plug of dry concrete placed inside a heavy steel tube at the ground surface (Figure 3.13a). In the latter, a heavy ram is dropped on the plug. Repeated impacts of the ram on the plug will drive the concrete plug and the steel tube with it into the ground. Driving operation is stopped when the desired depth of penetration is achieved (Figure 3.13b and c).

#### Base or Bulb Construction

When the required depth has been achieved, the steel tube is clamped to the driving rig to maintain its elevation. The concrete plug is then forced out into the ground by repeated impact of the ram (Figure 3.13d). Additional dry concrete is added and forced into the ground until desired (or specified) number of blows for last 5 $ft^3$ of concrete are achieved. A reinforcing steel cage, if required, is placed inside the tube before the last batch of dry concrete is compacted in the base.

#### Shaft Construction

After the base is formed, additional small batches of dry concrete are placed at the bottom of the tube. With the ram resting on the top of each batch, the tube is withdrawn slightly and concrete is compacted by the impacts of the ram. This is repeated until the shaft is completed to the desired elevation (Figure 3.13e and f).

These piles can be subdivided into uncased shaft and cased shaft types. Up to

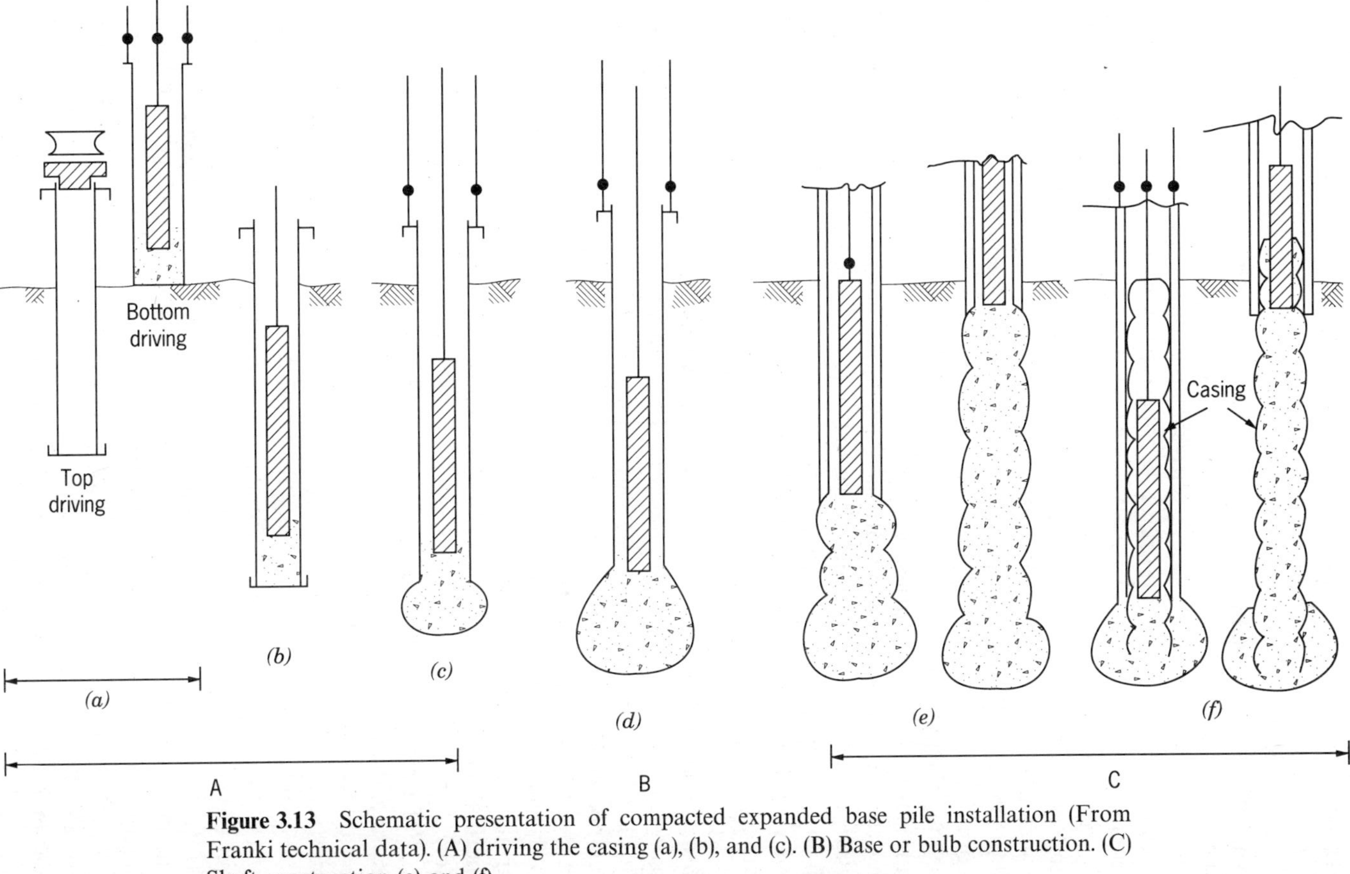


**Figure 3.13** Schematic presentation of compacted expanded base pile installation (From Franki technical data). (A) driving the casing (a), (b), and (c). (B) Base or bulb construction. (C) Shaft construction (e) and (f).

the base construction (Figure 3.13d), both these types have similar installation steps. The difference is only in the shaft construction.

In an *uncased shaft*, successive charges of zero-slump (dry) concrete are deposited in the drive tube, each charge being compacted and rammed against the soil as the tube is withdrawn in short lifts. The cycle is repeated until the design cutoff elevation is reached. The finished shaft is a rough concrete column, 1 to 3 in. greater in diameter than the drive tube. This type of pile develops maximum skin friction and maximum horizontal shear resistance (Figure 3.13e).

In a *cased shaft*, a steel pipe of suitable diameter is inserted inside the drive tube and a charge of zero-slump concrete is deposited in the pipe (Figure 3.13f). The concrete is then compacted by the ram so that the pipe and the base are in intimate contact. The drive tube is removed and the pipe is then filled with normal-slump concrete. Natural recovery of the soil fills the annular space around the pipe up to the groundwater table. Above the groundwater table, the annular space is filled with sand. These piles are useful in soft and highly plastic overburden soils. In some situations where the surficial soils consist of very stiff clays, a hole is drilled through these soils and the steel tube is dropped into it. Below this depth of drilling the driving operation is similar to as described. These piles are called *prebored compacted* shaft piles. This predrilling facilitates pile driving operations through very stiff surficial soils.

***Installation of Thermal Piles*** In order to install piles in permafrost areas, special heavy equipment is normally required. This equipment may cause disturbance in the permafrost. As good construction·practice, it is therefore, required that temporary working pads, such as gravel, metal, or timber pads in summer and compacted snow work pads in winter, be constructed to install piles. These pads will prevent undesirable disturbance to natural ground and hence minimize ground thermal disturbance (Johnston, 1981). The common methods by which thermal piles are normally installed are (1) in steam-thawed holes, (2) in augered holes, and (3) by driving piles.

For *stream-thawed holes*, piles are installed by steam jetting into the frozen ground. Because of little control over the construction details and a long waiting period required before the piles can be used due to delayed refreeze, this method is now seldom used.

For piles that are placed in *augered holes*, the usual method is to drill a hole first about 4 to 8 in. (100 to 200 mm) larger than the required pile diameter. The pile is then dropped into the hole, and the annular space between the pile and soil is filled by soil slurry. This slurry will freeze back quickly providing strength to the pile shaft. A recommended gradation range for the slurry is provided in Table 3.5, Slurry should be properly mixed with water so as to have a slump of 6 in. (150 mm). The temperature of slurry when placed into the hole should not exceed 40°F (4°C). This will ensure an adequate adfreeze bond within a reasonable period. This bond is developed between the pile and the surrounding frozen soil and provides support to the pile. Many piles are installed by this method.

**TABLE 3.5 Recommended Gradation Range for Typical Slurry Backfill**[a]

| Grain Size, in (mm) | Percentage Finer by Weight |
|---|---|
| 1 (25) | 100 |
| Sieve #4 | 90–100 |
| Sieve #10 | 70–100 |
| Sieve #20 | 30–75 |
| Sieve #40 | 15–60 |
| Sieve #100 | 5–30 |
| Sieve #200 | 0–15 |
| (0.02) | 0 |

[a]These values have been obtained from the gradation curves provided by Johnston (1981).

*Driven* open-ended steel and H piles can be installed into fine-grained frozen soils where ground temperatures may be as low as 27°F (−3°C). The main advantage of using driven piles is that freezeback, and hence adfreeze bond, can be achieved in a very short time. Once driving is begun, the work cannot be stopped because it may allow the soil to freeze or set up and may forbid further driving. When driving is to be done at close spacing into dense, hard frozen soils, piles are driven into predrilled small diameter pilot holes. For further details on pile installation techniques in permafrost, readers may refer to Johnston (1981).

## 3.5 INSTALLATION RECORDS

It is important that all piling projects are inspected by an experienced person preferably a registered professional engineer. All pile installation records for each pile should be recorded, and the design engineer should be informed of any variation in soil conditions or abrupt changes in soil resistance during driving. The requirements for such records is detailed as follows:

### 3.5.1 Driving Records

The inspector should keep a record of pile-driving logs. Basic information that should always be recorded are project name, pile type, location, size of pile, and driving system information. Driving system information could include type, size, and model of hammer; information on cap block; and pile cushion. The inspector must note all observations during driving, such as unexpected pile resistance change, any evidence of pile damage, interruption(s) in driving operation and so on. It is a good practice to record the number of blows required for each foot of pile penetration in addition to final driving resistance in blows per inch.

Each project normally has driving record format depending on the preference

DATA SHEET 1
PILE INSTALLATION RECORD
DRIVEN PILES

| | |
|---|---|
| Date: | Inspected By: |
| Project: | Contractor: |
| Pile Type: | Rig Number: |
| Pile Size: | Pile Location (No.): |
| Hammer Type: | Energy: |
| Ground Elev. | Cut-off Elev: |

DRIVING RECORD

| Depth, ft (m) | Blows | Remarks |
|---|---|---|
| 0–1 (0.3) | | |
| 1–2 (0.6) | | |
| 3 (0.9) | | |
| 4 (1.2) | | |
| 5 (1.5) | | |
| 6 (1.8) | | |
| 7 (2.1) | | |
| 8 (2.4) | | |
| 9 (2.7) | | |
| 10 (3.0) | | |
| 11 (3.3) | | |
| 12 (3.6) | | |
| 13 (3.9) | | |
| 14 (4.2) | | |
| 15 (4.5) | | |
| 16 (4.8) | | |
| 17 (5.1) | | |
| 18 (5.4) | | |
| 19 (5.7) | | |
| 20 (6.0) | | |
| 21 (6.3) | | |
| 22 (6.6) | | |
| 23 (6.9) | | |
| 24 (7.2) | | |
| 25 (7.5) | | |
| 26 (7.8) | | |
| 27 (8.1) | | |
| 28 (8.4) | | |
| 29 (8.7) | | |
| 30 (9.0) | | |

**Figure 3.14** Pile installation record: Driven piles.

DATA SHEET 2
PILE INSTALLATION RECORD
COMPACTED EXPANDED BASE CONCRETE
(FRANKI TYPE) PILES

Date:
Project:
Pile Size:
Ground Elev:
Hammer and Drop Data:
Weight of Hammer:
Hammer Drop for Driving Casing:
Hammer Drop for Forming Base:
Hammer Drop for Forming Shaft:
Elev. to Top of Base: Cutoff Elev:
Remarks:

Inspected By:
Contractor:
Rig Number:
Pile Location (No).
Specified Concrete Strength:

| Driving Record for Casing | | Base Construction | |
|---|---|---|---|
| Depth, ft (m) | Number of Blows | Concrete Volume | Number of Blows |
| 0–1 (0.3) | | | |
| 2 (0.6) | | | |
| 3 (0.9) | | | |
| 4 (1.2) | | | |
| 5 (1.5) | | | |
| 6 (1.8) | | | |
| 7 (2.1) | | | |
| 8 (2.4) | | | |
| 9 (2.7) | | | |
| 10 (3.0) | | | |
| 11 (3.3) | | | |
| 12 (3.6) | | | |
| 13 (3.9) | | | |
| 14 (4.2) | | | |
| 15 (4.5) | | | |
| 16 (4.8) | | | |
| 17 (5.1) | | | |
| 18 (5.4) | | | |
| 19 (5.7) | | | |
| 20 (6.0) | | | |
| 21 (6.3) | | | |
| 22 (6.6) | | | |
| 23 (6.9) | | | |
| 24 (7.2) | | | |
| 25 (7.5) | | | |
| 26 (7.8) | | | |
| 27 (8.1) | | | |
| 28 (8.4) | | | |
| 29 (8.7) | | | |
| 30 (9.0) | | | |

**Figure 3.15** Pile installation record: Compacted expanded base concrete (Franki type) piles.

of the engineer, client, and the contractor. Figures 3.14 and 3.15 are examples of such pile inspection record forms for driven and Franki Piles, respectively.

DATA SHEET 3
PILE INSTALLATION RECORD
DRILLED PILES

| | |
|---|---|
| Date: | Inspected By: |
| Project: | Contractor: |
| Pile Shaft Dia: | Rig Number: |
| Pile Bell Dia: | Pile Location (No.) |
| Pile Base Elev: | Ground Elev: |
| Pile Cutoff Elev: | Weather Conditions: |

Inspection Mode:
Vertical Reinforcement:
Horizontal Reinforcement:
Projection:
Time: Start of Drilling:
Completion of Drilling:
Start of Concreting:
Completion of Concreting:
Specified Concrete Strength:
Remarks:

Drilling Log

| Depth | Soil/Rock Description |
|---|---|

**Figure 3.16** Pile installation record: Drilled piles.

### 3.5.2 Drilling Records

Drilling records for each pile are kept by pile inspector. These records should note the name of the project, drill rig type, pile type and location, soil logs as observed during drilling, and observations such as sloughing, water flow, and so on.

Figure 3.16 is an example of pile installation report form for drilled pile. The examples given here are provided as a guide only. They will need to be revised to suite the requirements of a specific job.

### 3.5.3 Other Records

Internal inspection of driven pipe or drilled piles can be done from the surface by a powerful spotlight, by reflecting sunlight down the hole with a mirror, or by lowering droplight into the hole. This would help detect any damage in the pipe pile or any sloughing or seepage into the drilled hole before the concreting is done. These visual observation records should be noted in the installation report form.

For cast-in-place concrete piles, the concrete cylinders should be cast to carry out compressive strength tests in the laboratory. The concrete should also be tested for slump and air entraintment.

In some cases contract specifications require that some piles be load tested during actual installation. Load test records should be maintained (see Chapter 9 for details).

## REFERENCES


ASCE Deep Foundations Committee, "Practical Guidelines for the Selection, Design and Installation of Piles," American Society of Civil Engineers, 1984.

Authier, J. and Fellenius, B. H., "Wave Equation Analysis and Dynamic Monitoring of Pile Driving," *Civil Engineering for Practicing and Design Engineers*, Pergamon Press Ltd., Vol. 2, No. 4, 1983, pp. 387–407.

Associated Pile and Fitting Corp., Clifton, New Jersey.

Bruce, Jr., R. N. and Hebert, D. C., "Splicing of Precast Prestressed Concrete Piles: Part I—Review and Performance of Splices," *Journal of the Prestressed Concrete Institute*, Vol. 19, No. 5, 1974, pp. 70–97

*Canadian Foundation Engineering Manual, Part 3, Deep Foundations*, Canadian Geotechnical Society, March 1978 and 1985.

Compton, Jr., G. R., "Selecting Pile Installation Equipment," MKT Geotechnical Systems, 1981, 22 pp.

*Foundations and Earth Structures*, Design Manual 7.2, NAVFAC DM-7.2, Department of the Navy, Alexandria, Va., May 1982.

Franki: Technical Data Supplement, Franki Foundation Company.

Fuller, F. M., *Engineering of Pile Installation*, McGraw-Hill Book Co. New York, 1983.

Graff, W. J., *Introduction to Offshore Structures*, Gulf Publishing Company, Houston, Texas, 1981.

Hearne, T. M., Stokoe, K. H., and Reese, L. C., "Drilled-Shaft Integrity by Wave Propagation Method," *Journal of The Geotechnical Engineering Division*, ASCE, Vol. 107, No. GT 10, Oct. 1981, pp. 1327–1344.

Johnston, G. H. (Editor), *Permafrost: Engineering Design and Construction*, Wiley, New York, 1981.

Klohn, E., "Pile Heave and Redriving," *Journal of the Soil Mechanics and Foundations Divisions*, ASCE, Vol. 87, No. SM4, August 1961, pp. 125–145.

Koutsoftas, D. C., "H-Pile Heave: A Field Test," *Journal of the Geotechnical Engineering Division*, ASCE, Vol. 108, No. GT 8, Aug. 1982, pp. 999–1016.

Martin, R. E. and De Stephen, R. A., "Large Diameter Double Underreamed Drilled Shafts," *Journal of Geotechnical Engineering*, ASCE, Vol. 109, No. 8, August 1983, pp. 1082–1098.

O'Neill, M. W., Hawkins, R. A., and Audibert, J. M. E., "Installation of Pile Groups In Overconsolidated Clay." *Journal of The Geotechnical Engineering Division*, ASCE, Vol. 108, No. GT 11, November 1982, pp. 1369–1386.

Prakash, S., *Soil Dynamics*, McGraw-Hill Book Co., New York, 1981.

Sharma, H. D., Sengupta, S., and Harron, G., "Design and Construction of Pile Foundations Bearing on Top of Soft Weathered Rock Surface," 36th Canadian Geotechnical Conference, Vancouver, June 1983, pp. 1.3.1–1.3.10.

Vesic, A. S., "Design of Pile Foundations," Transportation Research Board, NRC, Washington, D.C., 1977, pp. 44–47.

Woodward, Jr., R. J., Gardner, W. S., and Greer, D. M., *Drilled Pier Foundations*, McGraw-Hill Book Co., New York, 1972.

# 4

# SOIL PARAMETERS FOR PILE ANALYSIS AND DESIGN

This chapter discusses information on methods of investigation of soils and the resulting soils parameters that are required for pile analysis and design. The chapter is divided into four sections. Section 4.1 on soil parameters for static design covers various investigation methods and the laboratory and field testing procedures. Results obtained from these tests are then used to obtain soil parameters for static design. Relationships between soil types and their properties such as friction angle, adhesion, cohesion, and elastic properties are also presented to facilitate their application for pile design. Section 4.2., which covers soil parameters for dynamic pile design, discusses laboratory and field methods for determining dynamic soil properties, followed by the selection criteria of design values. Section 4.3 provides information on soil behavior and design parameters for permafrost environment. Finally, a brief description of the modulus of horizontal subgrade is presented in Section 4.4. Soil parameters discussed in this chapter are later covered in Chapters 5 through 8 where specific pile design concepts and procedures are presented.

## 4.1 SOIL PARAMETERS FOR STATIC DESIGN

In this section, the scope of the foundation investigation, investigation methods and the resulting design parameters for static pile design are presented. The modulus of subgrade reaction ($k_h$) also a static pile design parameter, is discussed separately in Section 4.4. This parameter requires detailed coverage because of its importance in lateral load design of piles and the lack of clear understanding of this parameter by many practicing engineers.

The material presented in this section may be elementary to readers who are either experts in soil mechanics or have extensive background in geotechnical engineering. However, civil engineers who have only basic exposure to soil mechanics principles will find the material useful when designing pile foundations. This section provides them sufficient background so that design parameters for static pile design can be understood and selected without referring to other soil mechanics textbooks.

The main purpose of soils investigation is to determine the nature and sequence of soil strata, the types and properties of the soils and rock, such as gravels, sands, silts, and clays and their strength and compressibility behavior, and the groundwater conditions at the site. Soils investigations, in general, are carried out by boring and test pits, disturbed and undisturbed soil sampling, rock coring, and the measurement of groundwater levels.

ASCE (1976) provides a valuable guide for soils investigation work for design and construction of foundations. According to this document, the investigations generally proceed through following four phases:

1. **Preliminary Soils Investigations** These initial studies are conducted to establish project feasibility and preliminary design and to outline detailed investigation criteria. This work starts with reviewing available data, maps, and reports and is followed by a limited field boring and sampling work.

2. **Detailed Soils Investigations** Based on the general soils information collected from preliminary data, this stage of investigation is performed to obtain site specific soils information that is used to prepare detailed design and contract documents.

3. **Construction Verification** This is performed to identify any variation from previous investigations and assess its impact on the design and construction procedures.

4. **Postconstruction Monitoring** This stage of investigation is only conducted either on structures of major importance that are very sensitive to settlements or on structures that are built on very weak soils. Its main objective is to verify design assumptions and to monitor predicted responses.

The primary objective of this chapter is to familiarize the reader with soils investigation methods that provide soil parameters for pile design. Therefore, only the first two stages, preliminary and detailed soils investigations, are addressed in the following sections.

### 4.1.1 Scope of the Foundation Investigation

***Classification of Soils*** Soils are divided into following three major groups:

1. **Coarse-grained Soils** These soils contain more than 50 percent particles by weight retained on a No. 200 sieve (0.075 mm). They include gravels and sands and are referred to as cohesionless or noncohesive soils. Gravels contain more than 50 percent by weight of coarse fraction retained on a No. 4 sieve (4.75 mm) while sands have 50 percent or more of coarse fraction passing a No. 4 sieve.

2. **Fine-grained Soils** These soils have particles 50 percent or more passing a No. 200 sieve, which are not distinguishable to the naked eye. These soils include silts and clays. They are further classified based on both the plasticity index and the liquid limit values. This is shown in Figure 4.1. In this figure, C stands for clay, M stands for silt, O stands for organic soil, L for low-plasticity soils and H for high-plasticity soils.

3. **Organic soils** These soils have high natural organic content and are readily identified by color, odor, and spongy feel. They frequently have a fibrous texture.

For classification of soils, certain laboratory classification tests (e.g., grain size analysis (ASTM D422) and liquid-limit and plastic-limit tests (ASTM D4318)) are carried out. These tests are not discussed here; for further information refer to the ASTM Annualbook(1989).

***Soil Investigations*** The objectives of foundation soil investigations are to determine the extent, thickness, and properties of the soils and rocks and the

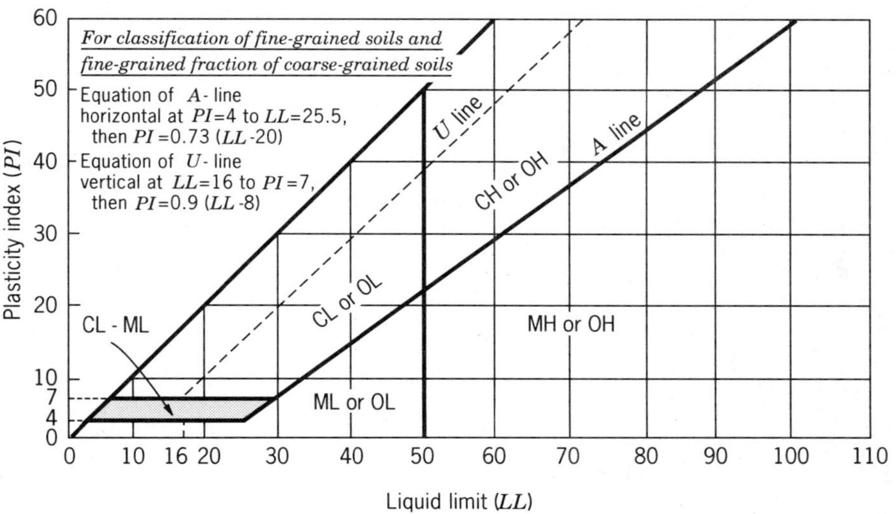


**Figure 4.1** Plasticity chart (ASTM, 1989).

groundwater levels at a site. Foundation investigations should therefore be carried out in such a manner that there are sufficient test pits and borings so that the soil stratigraphy can be described in detail. An interesting example of the importance of a site-specific detailed investigation is reported by Sharma et al. (1984). This example consisted of a pile foundation design for a major petrochemical project in Canada. Preliminary soils investigations indicated that clay shale bedrock depths varied between approximately 8.0 m (26 ft) to about 10.0 m (33 ft) across the site. The water table recorded in a few holes was reported at about 8.5 m (28.0 ft) below ground. The overburden soil was clay till. Based on

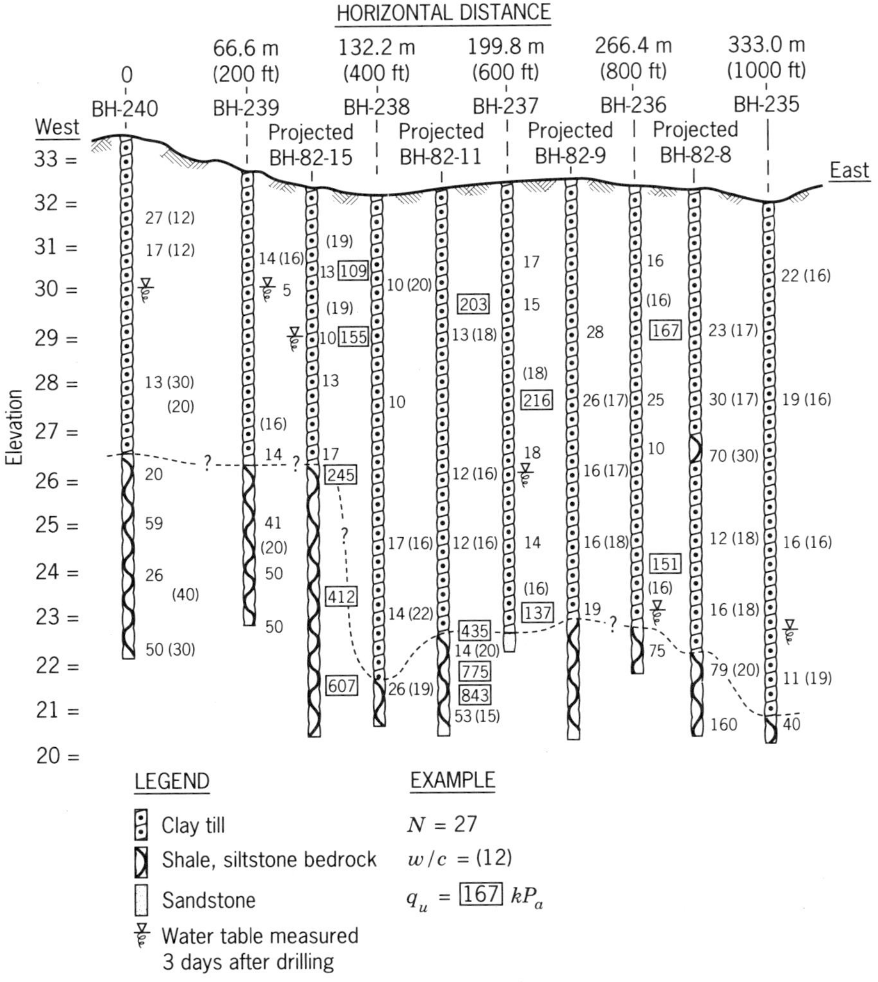


**Figure 4.2** Typical soil profile across a petrochemical project site in Alberta, Canada (Sharma et al., 1984).

this information, bored and belled concrete piles were recommended for the site. Typically, a 500-mm (20 in.) shaft diameter and 1200 mm (48 in.) bell diameter drilled pile had an ultimate capacity of about 1800 kN (405 kips). Site-specific detailed investigations later indicated that the bedrock depths at the site varied significantly as shown in Figure 4.2. Also, the existence of high artesian pressures in shallow bedrock had caused swelling of the shale bedrock. This reduced the bedrock strength. Based on the field pile load tests and laboratory strength determinations of the shale, the revised drilled pile capacities in the area of high water pressure were found to be less than half the values originally recommended at the preliminary phase. Important lessons learned from this case were that the site-specific detailed investigations must be performed to a minimum of 3 m (10 ft) into the bedrock, the equilibrium water table must be recorded, and appropriate field and laboratory tests must be conducted to determine soil and the bearing rock strengths.

In general, the soil investigations should be performed to such depths that all the soil or rock affected by the changes caused by the structure or the construction are adequately explored. Some general guidelines that should be followed by practicing engineers for soils investigation are as follows:

1. The depth of exploration should be such that the vertical stress induced by the new construction is smaller than 10 percent of the imposed stress at the level.
2. At least one borehole should be carried to bedrock unless past experience at the site has confirmed the bedrock depth.
3. The bedrock should be explored by coring into it to a minimum depth of 3 m (10 ft).
4. Groundwater levels should be recorded over a period of time to obtain equilibrium water levels. This period could vary depending on soil type. For example, it could be one day in coarse-grained soils to several weeks in cohesive fine-grained soils.

### 4.1.2 Soils Investigation and Testing Methods

***Soils Investigations*** Soil investigations consist of boring and excavating test pits to obtain soil stratigraphy and to recover samples for laboratory testing. The quality of the samples depends mainly on the boring and test pit excavation methods, the sampling equipment, and the procedure used to retrieve soil samples.

*Boring Methods* The following boring methods are commonly used for soils investigation:

1. **Auger Boring** These consist of hand- or power-operated augering with periodic removal of soil from the ground. In situations where continuous

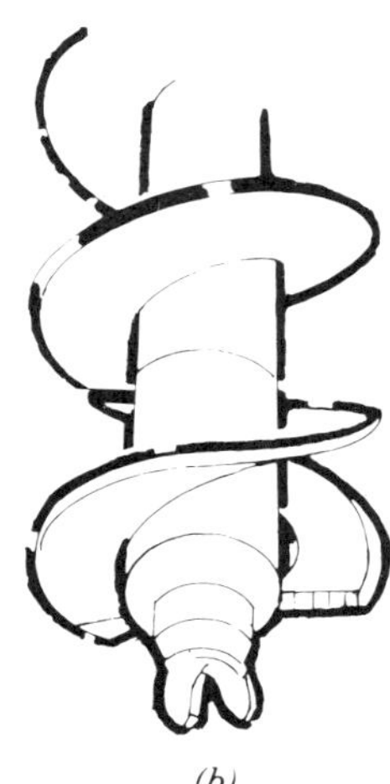

(b)

**Figure 4.3** (a)Auger boring with Mobil Model B-80, (b) auger Drill. (Courtesy: Mobil.)

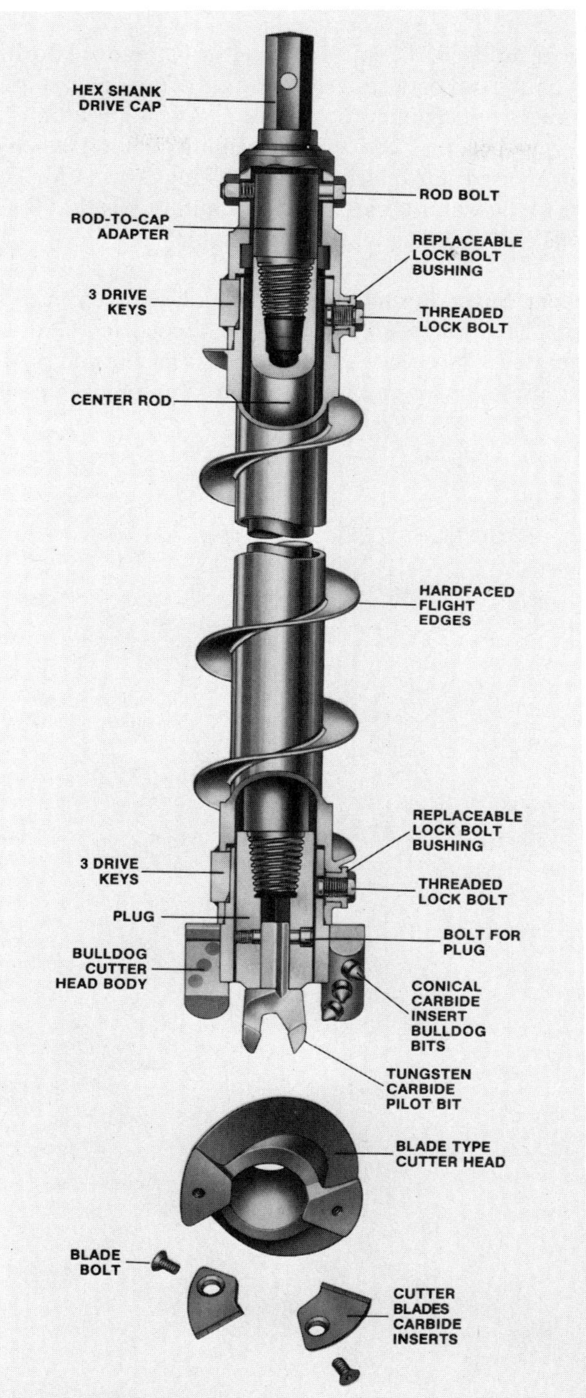


**Figure 4.4(a)** Hollow-stem auger (Courtesy: Mobil.)

flight auger is utilized, the soil is continuously removed during drilling. The material coming out of the auger is examined and noted. Figure 4.3 a shows a typical auger boring operation while Figure 4.3b exhibits an auger drill. Auger boring becomes a very fast-drilling method when power-driven equipment is used. Ordinarily, auger boring is used for shallow explorations above the water table. Its major limitation is that the hole collapses in soft soils and soils below the water table.

2. **Hollow-Stem Auger Boring** Figure 4.4a shows a typical Mobil hollow-stem auger. The hollow-stem auger is attached to the drill rig, which is power operated. The hollow stem serves as a casing and provides access for both the representative and undisturbed sampling. Figure 4.4b shows the

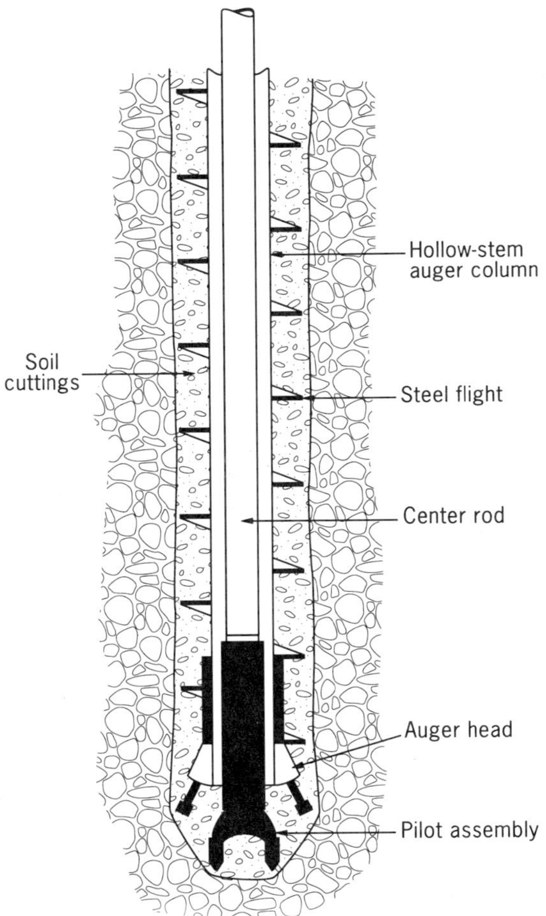


**Figure 4.4(b)** Auger drilling with hollow-stem auger (Dunnicliff, 1988).

assembly into the borehole. The figures show a pilot assembly that prevents the soil entering the hollow stem. The assembly is removed before inserting a testing or sampling device through the hollow stem. The methods of obtaining samples are presented in the following paragraphs.

3. **Wash Boring** This boring method involves chopping, twisting, and jetting action of a light drill bit as circulating fluid removes cuttings from holes. As shown in Figure 4.5, during driving the soil enters the casing at the bottom and is then removed by pumping water through a small diameter wash pipe. Casing may be used to prevent caving. Changes indicated by progress of rate of drilling and examination of cuttings in drilling fluid are used to identify soil type. It is most common method of subsoil exploration and is used in sands, sand and gravel without boulders, and soft to hard cohesive soils.

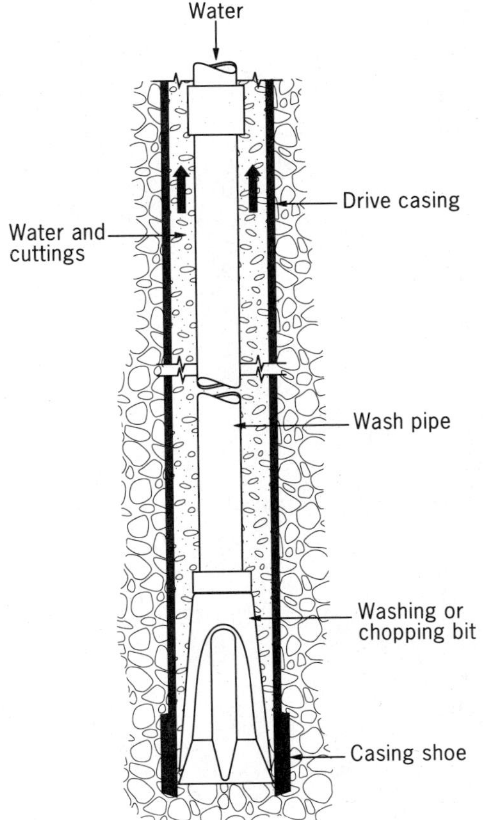


**Figure 4.5** Wash boring method (Dunnicliff, 1988).

4. **Rotary Drilling** This method utilizes power rotation of a drilling bit as circulating fluid removes cuttings from the hole. Changes indicated by rate of progress, action of drilling tools, and examination of cuttings in drilling fluid are used to identify soils. During drilling, soil samples can also be recovered at the bottom of the hole. This method is applicable to all soils except those containing large cobbles and boulders. Figure 4.6a shows a typical rotary boring drill rig in operation and Figure 4.6b shows a rotary drill in a borehole.

5. **Percussion Drilling** This is also called churn drilling and utilizes power chopping with limited amount of water at bottom of the hole. When water becomes a slurry, it is removed with bailer or sand pump. Casing is required

**Figure 4.6(a)** Typical rotary boring drill rig. (Courtesy: Mobil.)

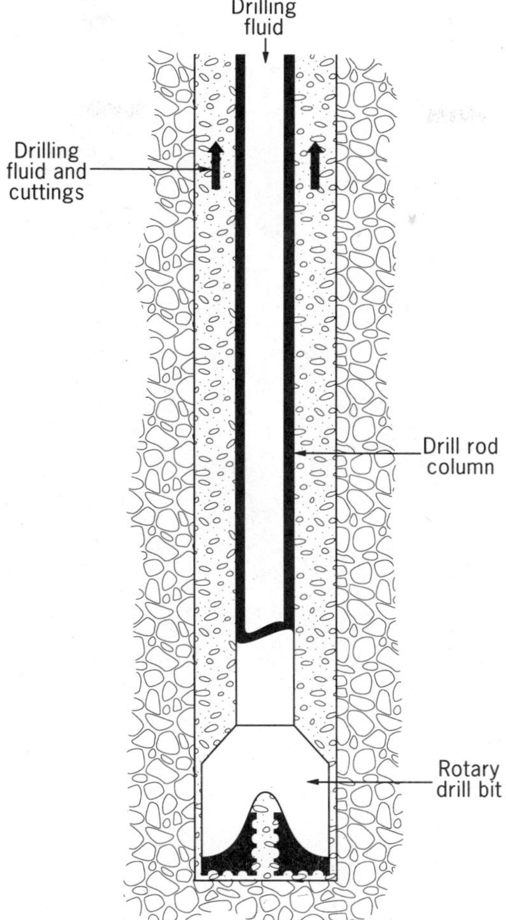


**Figure 4.6(b)** Rotary drilling operation (Dunnicliff, 1988).

in this method except in stable rock. This method is usually expensive and is used for rock drilling but not ordinarily for soil. It can be useful to probe cavities and weakness in rock by recording changes in drill rate. Percussion drilling is also used to penetrate boulders and rock formations.

6. **Rock Core Drilling** The rock core drilling operation consists of power rotation of a core barrel using a diamond-set cutting head bit as circulating water removes ground-up materials from the hole. During drilling, water also acts as a coolant for the core barrel bit. It is used to drill weathered rocks, bedrock, and boulder formation and can be either used alone or in combination with other boring methods.

(b)

**Figure 4.7** (a) Wire-line core sampling operation, (b) core/rotary drill. (Courtesy: Mobil.)

7. **Wire-line Drilling** This is a rotary-type drilling method where the coring device is an integral part of the drill rod string, which also serves as a casing. Core samples are obtained by removing the inner barrel assembly from the core barrel portion of the drill rod. It is efficient for deep-hole coring over 30 m (100 ft) depth. Figure 4.7a shows a wire-line diamond core sampling operation, while Figure 4.7(b) shows core/rotary drill bit. Figure 4.8 shows a wire-line core-drilling assembly consisting of an inner barrel that is withdrawn from the borehole on a wire-line. The bit, the outer barrel, and drill rods stay in the borehole (Dunnicliff, 1988).

*Test Pits* Test pits are either hand dug or machine excavated. The samples from the test pits are used to examine the strata and to prepare samples for soil tests. These samples are obtained at shallow depths only. Test pits are limited to depths above groundwater level. Following are some commonly used test pit methods of soils investigation:

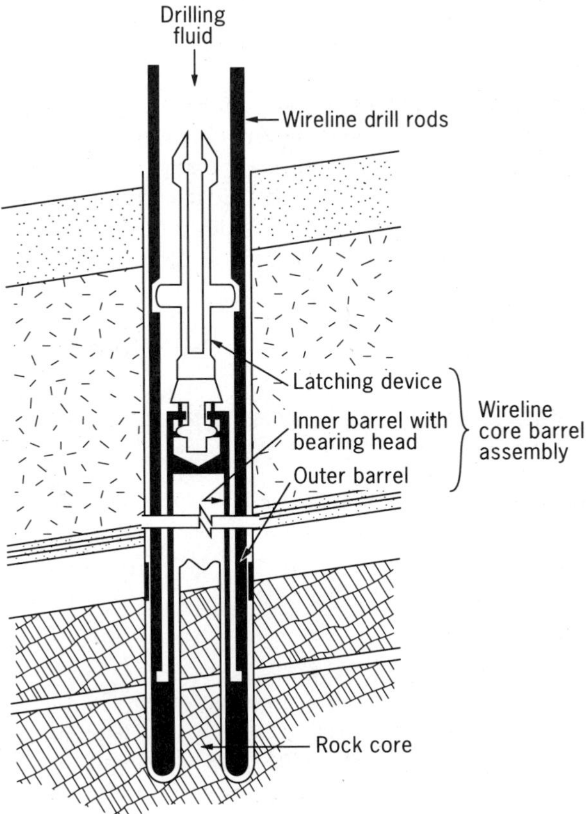


**Figure 4.8** Wire-line core drilling (Dunnicliff, 1988).

1. **Hand Excavated** These test pits and shafts are used for bulk soil sampling, in-situ testing, and visual inspection. Due to their excavation by hand, they cause less mechanical disturbance of surrounding ground. However, they are expensive and time consuming.

2. **Backhoe Excavated** These are used for bulk sampling, in-situ testing, visual inspection, and are fast and economical. They can be excavated up to 10 m (33 ft) depth.

3. **Dozer Cuts** These are used to determine the bedrock characteristics and the depth of bedrock and groundwater level. They are a relatively low-cost methods of exploration.

*Soil Sampling* Two types of soil sampling are generally carried out during soils investigation: disturbed and undisturbed soil sampling. Disturbed samples are primarily used for soil classification tests and must contain all of the constituents of the soil even though its structure is disturbed. Undisturbed samples are taken primarily for laboratory strength and compressibility tests. The soil parameters obtained serve as a basis for foundations design.

1. **Disturbed Soil Samples** Disturbed samples are generally taken at vertical intervals of no less than 5 ft (1.5 m) and at every change in strata. These soil samples are primarily used for carrying out identification and index property tests and are obtained with thick-walled samplers. The most commonly used thick-walled sampler is the *split-spoon sampler*, which is described in ASTM D1586 and is shown in Figure 4.9. It consists of a 2-in. (50 mm) outside diameter and 1.5-in.

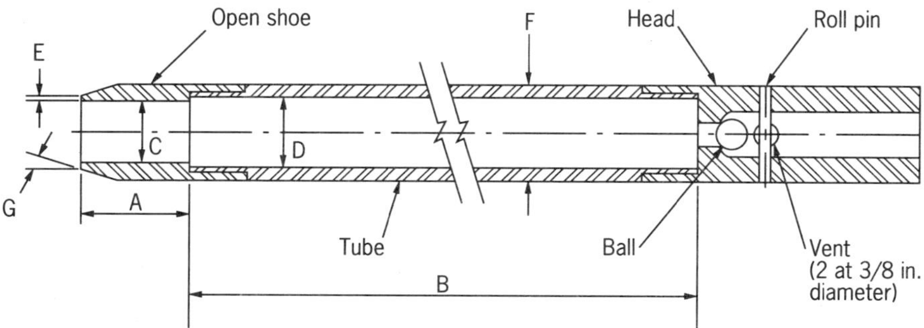


**Figure 4.9** Thick-walled split spoon (barrel) sampler ASTM D 1586 (ASTM, 1989). A = 1.0 to 2.0 in. (25 to 50 mm), B = 18 to 30 in. (457 to 762 mm), C = 1.375 ± 0.005 in. (34.93 ± 0.13 mm), D = 1.50 ± 0.05 − 0.00 in. (38.1 ± 1.3 − 0.0 mm), E = 0.10 ± 0.02 in. (2.54 ± 0.50 mm), F = 2.00 ± 0.05 − 0.00 in. (50.8 ± 1.3 − 0.00 mm), G = 16.0° to 23.0°. The $1\frac{1}{2}$ in. (38 mm) inside diameter split barrel may be used with a 16-gauge wall thickness split liner. The penetrating end of the drive shoe may be slightly rounded. Metal or plastic retainers may be used to retain soil samples.

(38 mm) inside diameter split barrel driven by a 140-lb (64 kg) weight falling 30 in. (760 mm) when the sampler is at the bottom of a borehole. The barrel must be 18 in. long or larger. The number of blows required to drive this sampler into the ground for 12 in. (300 mm) is called the *standard penetration value* and is commonly represented by $N$. The blow counts are measured for an 18-in. (450 mm) penetration of the sampler. Blows required for the first 6 in. (150 mm) are neglected because this record may be in highly disturbed and slumped material. It is, however, recommended that the blows for each 6-in. (150 mm) penetration be recorded because it furnishes additional data for interpreting the results. The blow counts for last two 6-in. penetrations are then added together to obtain the $N$ value in blows per foot (0.3 m).

This type of soil sampler is used for following two purposes:

(a) To obtain disturbed samples for laboratory identification and index property tests, and
(b) By blow count to obtain relative density and indirect strength parameters of cohesionless soils (see Section 4.1.3)

Limitations and modifications for the standard penetration test are discussed in the section, "Field Testing".

**2. Undisturbed Soil Samples** Undisturbed soil samples should not contain any visible distortion of strata nor should they have any softening of materials. The soil samples and the sampler should meet the following criteria:

$$\text{Specific recovery ratio} = \frac{\text{Length of undisturbed sample recovered}}{\text{Length of sampling push}} \times 100 \geqslant 95\%$$

$$\text{Sampler area ratio} = \frac{\text{Annular cross-sectional area of tube}}{\text{Full area of outside diameter of sampler}} \times 100 \leqslant 15\%$$

The use of a thick-walled sampler and/or taking samples by driving the sampler by falling weight usually cause disturbance in the soil. Samples obtained by this method are not suitable for density, permeability, strength, and deformation tests in the laboratory.

In order to obtain undisturbed samples of cohesive soils for laboratory strength and deformation tests, thin-walled samplers are used. Figure 4.10 shows a thin-walled sampler that can be forced into the soil smoothly and continuously. These sample tubes should be clean and free of all surface irregularities. Also, the weld seams should not project above the surface. Their outside diameter may

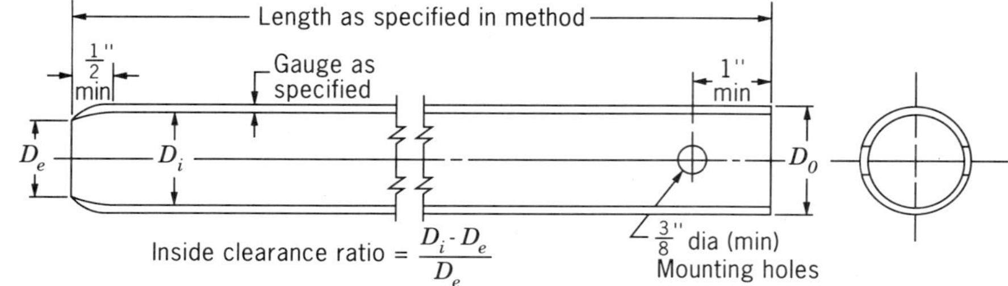


**Figure 4.10** Thin-walled soil sampler: ASTM D 1587 (ASTM 1989). Note 1: Minimum of two mounting holes on opposite sides for 2- to $3\frac{1}{2}$-in. sampler. Note 2: Minimum of four mounting holes spaced at 90° for samplers 4 in. and larger. Note 3: Tube held with hardened screws. Note 4: Two-inch outside-diameter tubes are specified with a 18-gauge wall thickness to comply with area ratio criteria accepted for "undisturbed samples." Users are advised that such tubing is difficult to locate and can be extremely expensive in small quantities. Sixteen-gauge tubes are generally readily available (ASTM, 1989).

range from 2 in. (50 mm) to 5 in. (125 mm). ASTM designation D1587-89 provides further requirements for these samplers.

In hard or dense soils, where thin-walled samplers may become damaged, either Denison or Pitcher samplers may be used. For sampling soft, sensitive clays, a Swedish foil sampler is generally recommended. Terzaghi and Peck (1967) provide further details on these samplers.

In most cohesive soils of soft to stiff consistency, good-quality samples can be obtained by pushing thin-walled tube samplers, usually referred to as Shelby tubes, about 3 in. (76 mm) or larger diameter into soil provided there is a proper cutting edge and low area ratio (10 percent) (*ASCE Manual No. 56*, 1976). Most soils investigations use this type of sampler for obtaining undisturbed soil samples.

*Rock Coring* Rocks, as opposed to soils, cannot be readily broken by hand and will not disintegrate on first drying and wetting cycles. Intact rocks are very strong, but usually blocks of rock are separated by discontinuities such as joints, faults, bedding, and shear planes. These discontinuities reduce the strength of large masses of intact rock.

The International Society of Rock Mechanics recommends that the rock be classified based on the strength of rock material, spacing, orientation, roughness, and filling in the rock mass (Deere, 1963). Therefore, in situations where foundations are to be carried to or into bedrock, investigation of the bedrock is made to determine the elevation of rock surface, rock type, depth and pattern of weathering, presence of solution channels in rocks such as limestone, and discontinuities such as bedding planes and joints.

The 3-in. (75 mm) outside diameter, double-tube, N-size core barrel drills with nonrotating inner barrels are generally used for good-quality rock coring. After the cores have been taken out of the core barrel, they should be properly placed in wooden or metal core boxes in the order in which they are taken. These cores are then examined for identification and classification in the core laboratory.

For engineering purposes, rock identification and classification are limited only to broad basic classes. Based on the range of unconfined compressive strength and field identification description, rocks can be classified with regards to their strength. According to the classification shown in Table 4.1, rocks are graded from classification R0 to R6. R0 is the extremely weak rock that can be indented by thumb nail and has an unconfined compressive strength less than 20 kips/ft$^2$ (1 MPa) while R6 is extremely strong rock having unconfined compressive strength greater than 5000 kips/ft$^2$ (250 MPa) that can only be chipped by a geological hammer.

**TABLE 4.1 Classification of Rock with Regard to Strength (Canadian Foundation Engineering Manual 1985)**

| Strength | | Field Identification Method | Range of Unconfined Compressive Strength | |
|---|---|---|---|---|
| Grade | Classification | | MPa | kip/ft$^2$ |
| R0 | Extremely weak | Indented by thumbnail | $<1$ | $<20$ |
| R1 | Very weak | Crumbles under firm blows of gelogical hammer; can be peeled with a pocket knife | 1–5 | 20–100 |
| R2 | Weak rock | Can be peeled by a pocket knife with difficulty; shallow indentations made by a firm blow with point of geological hammer | 5–25 | 100–500 |
| R3 | Medium strong | Cannot be scraped or peeled with a pocket knife; specimen can be fractured with a single firm blow of geological hammer | 25–50 | 500–1000 |
| R4 | Strong | Specimen requires more than one blow of geological hammer to fracture | 50–100 | 1000–2000 |
| R5 | Very strong | Specimen requires many blows of geological hammer to fracture | 100–250 | 2000–5000 |
| R6 | Extremely strong | Specimen can only be chipped by the geological hammer | $>250$ | $>5000$ |

The quality and the strength of rock mass is highly dependent on the spacing of discontinuities that can be measured from the outcrops, trenches, drill cores, or by viewing the boreholes with borehole cameras and periscopes. As shown in Table 4.2, the spacing of discontinuities can vary from extremely close (less than 0.06 ft (0.02 m)), to extremely wide (greater than 18 ft (6 m)). These discontinuities should be determined by measuring the distances between adjacent discontinuities over a minimum sampling length of 10 ft. Rock quality designation (RQD) is defined as:

$$\text{RQD} = \frac{\text{Length of core in pieces 4 in. and longer}}{\text{Length of core run}} \times 100 \tag{4.1}$$

RQD is an index of general quality of rock for engineering purposes. It indirectly measures the number of fractures and amount of softening or alteration in a rock mass. It is determined from the rock cores that have been obtained by using double-tube core barrels of at least NX size (54 mm in diameter) by summing up the length of core recovered and counting only those pieces of sound core that are 4 in. (100 mm) long or more. In determining RQD, if the core is broken by handling or during drilling, the fresh broken pieces should be fitted together and

**TABLE 4.2 Classification of Rock with Regard to Spacing and Discontinuities (Canadian Foundation Engineering Manual 1985)**

| | Spacing Width | |
|---|---|---|
| Spacing Classification | (m) | (ft) |
| Extremely close | <0.02 | <0.06 |
| Very close | 0.02–0.06 | 0.06–0.18 |
| Close | 0.06–0.20 | 0.18–0.6 |
| Moderately close | 0.2–0.6 | 0.6–1.8 |
| Wide | 0.6–2.0 | 1.8–6 |
| Very wide | 2–6 | 6–18 |
| Extremely wide | >6 | >18 |

**TABLE 4.3 Rock Quality Designation, RQD (Deere et al., 1967)**

| RQD Classification | RQD Value (%) |
|---|---|
| Excellent | >90 |
| Good | 75–90 |
| Fair | 50–75 |
| Poor | 25–50 |
| Very poor | <25 |

counted as an intact piece. Table 4.3 provides RQD classification and corresponding RQD values that are used as an index of rock quality for foundation engineering purposes. For further details on North American geotechnical exploration practice, readers may refer to Riggs (1986).

*Measurement of Groundwater Levels* Groundwater is a critical factor in foundation design and should be given careful attention during all stages of soil investigations. Groundwater measurements should provide information on the existence of normal, perched, hydrostatic, or artesian levels and the variations of these levels over the site and with time. Groundwater levels should be measured at the depth at which water is first encountered as well as at the level at which it stabilizes after drilling.

Groundwater level measurements generally are made by installing piezometers. The most common types of piezometers used in practice are briefly described as follows:

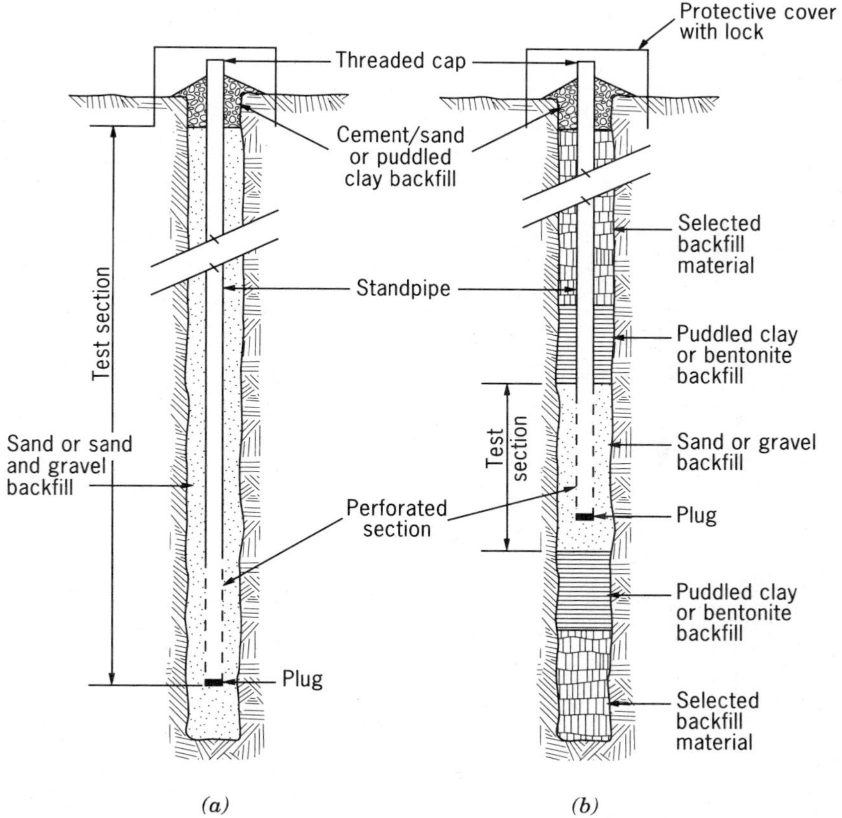


**Figure 4.11** Open standpipe piezometers (a) Long test section, (b) Isolated test section. (NAVFAC Design Manual DM 7.1, 1982).

1. **Open Standpipe Piezometers** The most common type of open standpipe piezometer consists of a perforated pipe installed in the borehole. The annular space is then backfilled with sand or gravel as shown in Figure 4.11a. The height of water in the standpipe will indicate the groundwater level at the site. The disadvantage of this system is that if

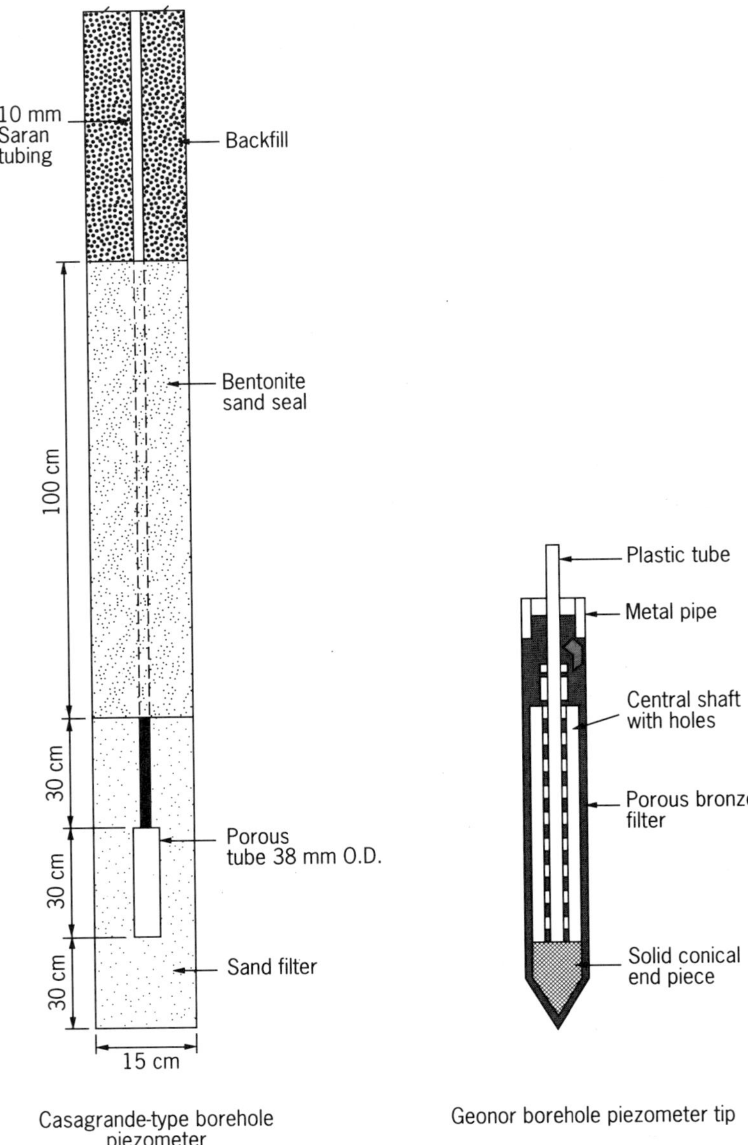


**Figure 4.12** Porous element piezometers (NAVFAC Design Manual DM 7.1, 1982).

there are different layers of soil that are under different hydrostatic pressures then groundwater levels recorded by this method will be misleading. This is because the groundwater level in the standpipe will show a combined effect of all these layers since they have not been isolated. This problem can be resolved if different strata can be isolated, as has been done in Figure 4.11b. An open standpipe piezometer system is a simple and reliable groundwater measuring installation. However, they have a slow response time and are susceptible to freezing during winter.

2. **Porous Element Piezometers** As shown in Figure 4.12, these piezometers consist of a porous element connected to the riser pipe. This pipe has a small diameter to reduce the equalization time. Porous elements are about 50 $\mu$. These tips can be used in direct contact with fine-grained soils also.

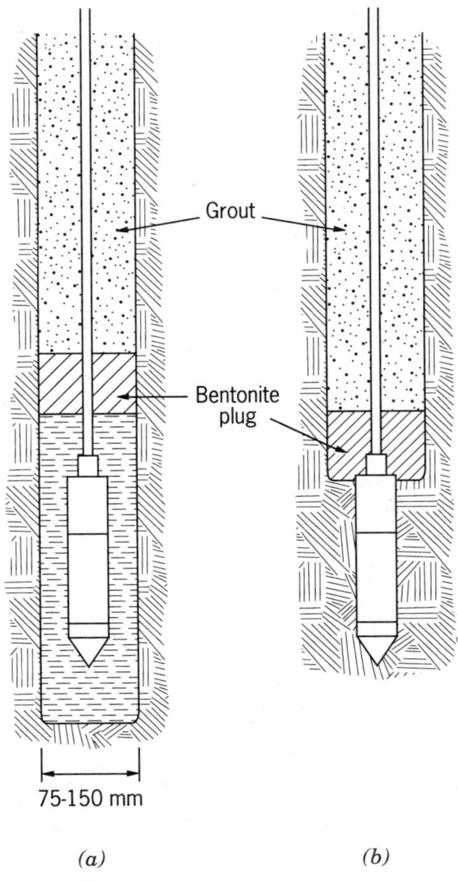


**Figure 4.13** Pneumatic-type piezometer (Solinst Canada Ltd.). (a) Sand cell installation, (b) Push-in installation.

**TABLE 4.4 In-situ Test Methods and Their Perceived Applicability (Robertson, 1986)**

| | Geotechnical Information | | | | | | | | | | | | | | Ground Conditions | | | | | | |
|---|---|---|---|---|---|---|---|---|---|---|---|---|---|---|---|---|---|---|---|---|---|
| Test Method | Soil Type | Profile | Piezometric Pressure ($u$) | Angle of Friction ($\phi$) | Undrained shear strength ($S_u$) | Density ($\gamma_t$) | Compressibility ($m_v$, $C_c$) | Rate of consolidation ($C_v$, $C_h$) | Permeability ($k$) | Modulus: shear and Young's ($G$, $E$) | In-Situ stress ($K_0$) | Stress history (OCR) | Stress–strain curve | Hard Rock | Soft Rock—till, etc. | Gravel | Sand | Silt | Clay | Peat—Organics |
| Dynamic cone (DCPT) | C | B | — | C | C | B | — | — | — | C | — | — | C | — | C | B | A | B | B | B |
| Static cone | | | | | | | | | | | | | | | | | | | | |
| Mechanical | B | A | — | B | C | B | C | — | — | C | C | C | — | — | C | — | A | A | A | A |
| Electronic friction (CPT) | B | A | — | B | C | B | C | — | — | B | C | C | — | — | C | — | A | A | A | A |
| Electronic piezo | B | A | A | B | B | B | C | A | B | B | C | B | B | — | C | — | A | A | A | A |
| Electronic piezo/friction (CPTU) | A | A | A | B | B | B | C | A | B | B | C | B | B | — | C | — | A | A | A | A |
| Electronic seismic/piezo/friction (SCPTU) | A | A | A | B | B | B | C | A | B | A | B | B | B | — | C | — | A | A | A | A |
| Acoustic probe | B | B | — | C | C | C | C | — | — | C | — | C | — | — | C | — | A | A | A | A |
| Flat plate dilatometer (DMT) | B | A | C | B | B | C | B | — | — | B | B | B | B | — | C | — | A | A | A | A |
| Field vane shear (VST) | C | C | — | — | A | — | — | — | — | — | C | B | — | — | — | — | — | B | A | B |
| Standard penetration test (SPT) | A | B | — | B | C | B | — | — | — | B | — | C | — | — | C | B | A | B | C | C |
| Resistivity probe | B | B | — | B | C | A | C | — | — | C | — | — | — | — | C | — | A | A | A | A |

| | | | | | | | | | | | | | | | | | | | | |
|---|---|---|---|---|---|---|---|---|---|---|---|---|---|---|---|---|---|---|---|---|
| Electronic conductivity probe | A | B | — | C | C | A | B | — | — | B | C | C | C | — | — | — | A | A | A | B |
| Total stress cell | — | — | — | — | — | — | — | — | — | — | B | B | — | — | — | — | — | C | A | A |
| $K_0$ stepped blade | — | — | — | — | — | — | — | — | — | — | B | B | — | — | — | — | B | A | A | B |
| Screw plate | C | C | — | C | B | B | B | C | C | A | C | B | B | — | — | — | A | A | A | A |
| Borehole permeability | C | — | A | — | — | — | — | B | A | — | — | — | — | A | A | A | A | A | A | B |
| Hydraulic fracture | — | — | A | — | — | — | — | C | C | — | B | B | — | B | B | C | C | B | A | C |
| Borehole shear | C | C | — | B | C | — | — | — | — | C | — | C | — | B | B | C | B | B | C | C |
| Prebored pressuremeter (PMT) | B | B | — | C | B | C | C | C | — | A | C | C | C | A | A | B | B | B | A | B |
| Push-in pressuremeter (PPMT) | A | B | B | C | B | C | C | A | B | A | C | C | C | — | — | — | B | A | A | B |
| Full-displacement pressuremeter (FDPMT) | C | B | B | C | B | C | C | A | B | A | C | C | C | — | — | — | A | A | A | A |
| Self-boring pressuremeter (SBPMT) | B | B | A | A | B | B | B | A | B | A | A | A | A | — | C | — | B | A | A | A |
| Self-boring devices | | | | | | | | | | | | | | | | | | | | |
| $K_0$ meter | B | B | — | — | — | — | — | — | — | — | A | A | — | — | — | — | B | A | A | A |
| Lateral penetrometer | B | B | — | B | B | B | — | — | — | B | C | C | C | — | — | — | B | A | A | A |
| Shear vane | B | B | — | — | A | — | — | — | — | — | C | B | — | — | — | — | B | A | A | A |
| Plate test | B | B | — | C | B | B | B | C | C | A | B | A | C | — | — | — | B | A | A | B |
| Seismic cross/downhole/surface | C | C | — | — | — | — | — | — | — | A | — | — | — | A | A | A | A | A | A | A |
| Nuclear probes | — | — | — | B | — | A | — | — | — | — | C | — | C | — | — | — | A | A | B | A |
| Plate load tests | C | C | — | C | B | B | B | C | C | A | C | B | B | B | A | B | B | A | A | A |

Note: A = high applicability, B = moderate applicability, C = limited applicability, — = not applicable.

3. **Electric Piezometers** In these piezometers, a waterproof chamber is separated from the porous tip by a diaphragm. The deflection of this diaphragm can be measured by a strain gage that is read by means of an electric circuit. These instruments have rapid response and high sensitivity and are suitable for automatic readout. Their disadvantages are that they are expensive and may require temperature correction. Field experience shows that the long-term performance of most of these types of devices has not been satisfactory.
4. **Pneumatic Piezometers** The diaphragm deflection of these piezometers is balanced by applying an air pressure on the backside of the diaphragm. The measure of this applied air pressure is the pore pressure (see Figure 4.13). They are the most common type of instruments used where rapid pore pressure response is required. Figure 4.13a shows where the piezometer is installed in a borehole, which is then backfilled with sand while Figure 4.13b shows where the piezometer is pushed into the natural soil.

Experience indicates that with increasing sophistication of the instruments, there is a greater probability of malfunction. Therefore, it is recommended that if water-level measuring devices such as pneumatic piezometers are installed at a site, they must be supplemented by simpler devices such as open standpipes and/or porous element piezometers.

Terzaghi and Peck (1967) provide further information on piezometers for porewater pressure measurements. Various manufacturer's catalogues, such as SINCO of Seattle, Washington; Solinst Canada Ltd., Burlington, Ontario; and others provide specific piezometer data.

***Field Testing*** The measurement of soil parameters by field testing methods has developed rapidly during the last decade primarily because of their ability to determine properties of soil that cannot be easily sampled in the undisturbed state. Field testing increases cost effectiveness of an exploration and testing programs by testing a larger volume of soil than can be tested in the laboratory. Robertson (1986) provides a comprehensive list and the application of various field (in-situ) tests. Table 4.4 summarizes these tests. Mitchell et al. (1978), Campanella and Robertson (1981), Goel (1982), Melzer and Smoltczyk (1982), Nixon (1982), Robertson (1985) and In Situ (1985) also provide information on these testing techniques and their applicability. Because of their direct applicability to pile foundation design the penetrometer tests, vane shear tests, and the pressuremeter tests will only be discussed here.

The generally known penetrometer tests are the Standard Penetration Test (SPT), Dynamic Cone Penetration Test (DCPT), Static Cone Penetration Test (CPT), and Flat Plate Dilatometer Test (DMT).

*Standard Penetration Test (SPT)* As discussed under disturbed soil sampling, SPT values can be obtained by counting the blows required to drive a standard split spoon into the soil at the bottom of a borehole. Details of the test equipment and techniques are well known as provided by Nixon (1982) and ASTM D 1586 and consists of the following steps:

1. Place the split barrel (spoon) sampler (shown in Figure 4.9) at the bottom of the borehole.
2. Drive this sampler into the soil by using a 140 1b (64 kg) weight falling 30 in. (760 mm).
3. Count the number of blows to drive the sampler every 6 in. (150 mm) for a total distance of 18 in. (450 mm). Some practitioners count blows for the first 6 in. (150 mm) and then 12 in. (300 mm) penetration only. This, however, is not the standard practice.
4. Add the blow counts for last two 6 in. (150 mm) drives of the sampler into the soil.
5. The SPT value, usually called the $N$ value, is then the number of blows required to drive the sampler the final 12 in. (300 mm) into the soil.

The blow count for the first 6 in. (150 mm) is assumed to seat the split barrel sampler into the disturbed soil in the borehole. These blows are therefore not considered in the SPT ($N$) values. In situations where the soil is very dense, it may not be possible to drive the sampler the full 12 in. (0.3 m) into the soil. In such cases, blow counts are recorded with the amount of penetration (e.g., 50/4 in.). This means that the sampler required 50 blows for 4 in of penetration into the soil or the rock as the case may be.

The SPT has several significant advantages (Robertson, 1986): (1) The equipment is relatively simple and rugged. (2) A sample of the soil is usually obtained as a part of the investigation. (3) A test can be carried out in most soil types. (4) Based on past experience (over 50 years), many useful correlations with soil parameters have been developed. It is a widely used field test method. In spite of its wide use and simple procedure, the results of SPT are greatly affected by sampling, drilling, equipment, and operator characteristics. Some of the improper drilling and sampling procedures that can affect the SPT values are as follows:

1. *Not using the standard hammer drop*: This results in nonuniform energy delivered per blow. European countries have adopted an automatic trip hammer to solve this probelm.
2. *Free fall of the drive weight is not attained*: Using more than one and one half turns of rope around the drum and/or using wire cable will restrict the fall of the drive weight.
3. *Not using the correct weight*: Driller frequently supplies drive hammers with weight varying from the standard by as much as 10 lb (5 kg).
4. *Weight does not strike the drive cap concentrically*: This reduces the impact energy, which results in increasing SPT values.

5. *Use of drill rods heavier than standard*: With heavier rods, more force is developed in the rods, which results in incorrect SPT values.
6. *Failure to maintain sufficient hydrostatic head in boring*: The water in the borehole must be at least equal to the piezometric level in the sand, otherwise the sand at the bottom of the borehole may become quick and be transformed into a loose state.

Most significant factors affecting the measured SPT (N) values are identified by Schmertmann (1977) and Kovacs and Salomone (1982). Kovacs et al. (1981) showed that the energy delivered to the rods can vary from about 30 to 80 percent of the theoretical maximum, 4200 in.-lb (475 J), with an average of about 55 percent. More recently (1987), standardization of energy of the free fall of the SPT hammer has been discussed in detail by Seed et al. (1985) and Skempton (1986).

For these reasons discussed, it is apparent that the accuracy of the Standard Penetration Test $N$ values in questionable. Therefore its correlations with soil parameters should be used with caution. However, the Standard Penetration Test, with all its problems, is still the most commonly used field test today. For example, up to 80 to 90 percent of the routine foundation designs in the United States are accomplished using the SPT, $N$ value (Robertson 1986).

A correction is required for depths in Standard Penetration values because of the greater confinement caused by increasing overburden pressure. Increasing $N$ values due to confinement may indicate larger density than the actual. The need for normalizing or correcting the results of the Standard Penetration Test $N$ values in sands was first demonstrated by Gibbs and Holtz (1957). Since then, several formulas and charts for making the correction have been published (Teng, 1962; Bazarra, 1967; Peck et al., 1974; Seed, 1979; Tokimatsu and Yoshimi, 1983). Liao and Whitman (1986) reviewed all these methods and concluded that correction factors provided by Bazarra (1967), Peck et al. (1974), and Seed (1979) will lead to fairly consistent results and any one of these can be used. Liao and Whitman (1986) also propose a simple correction factor that leads to similar results. The commonly used correction factor ($C_N$) recommended by Peck et al. (1974) is as follows:

$$C_N = 0.77 \log_{10}\frac{20}{\sigma'_V}; \quad \sigma'_v \geqslant 0.25 \text{ tsf} \tag{4.2}$$

where $\sigma'_v$ is the effective overburden vertical pressure in tsf.

*Dynamic Cone Penetration Test (DCPT)* The dynamic cone penetration is a continuous test in which an impacting weight drives a rod that is attached to a cone tip of 10 cm$^2$ cross-sectional area. The DCPT values are the number of blows for 12-in. (300 mm) penetration of the rod into the ground. In North America, the rods and impact weight are usually the same as those used for the SPT. The enlarged cone tips are used to reduce the rod friction. The DCPT is subject to all the same problems associated with energy levels as those for the SPT

(Robertson, 1986). Lack of standardization is the main reason that this test method has not been advanced more in recent years. The main advantage of this test is that it is fast and inexpensive and provides a continuous profile of qualitative soil density variation with depth. In some areas, local experience has made the DCPT a useful field test technique (*Canadian Foundation Engineering Manual*, 1985).

*Static Cone Penetration Test* (*CPT*) This test, originally developed in Europe, is now gaining acceptance in North America. It mainly consists of pushing a cone by devices such as hydraulic arrangement into the soil and measuring the corresponding resistance. The proceedings of the European symposia on penetration testing, ESOPT I (1974) and ESOPT II (1982), provide detailed information on the CPT.

The CPT system can be divided into two main groups: Mechanical and electrical (electronic). The mechanical cones require a double-rod system for their telescopic action while the electronic cones have the friction sleeve and tip advanced continuously with a single-rod system. Both systems consist of a cone with 10 cm$^2$ base area and the cone tip with apex angle is 60°. The friction sleeve, located above the tip, has a standard area of 150 cm$^2$. Although mechanical cones offer the advantage of an initial low cost for equipment and simplicity of operation, their main disadvantage is of slow incremental procedure, labor-intensive data handling, and generally poor accuracy. The electrical cones, as shown in Figure 4.14, have built-in load cells that measure continuously the end

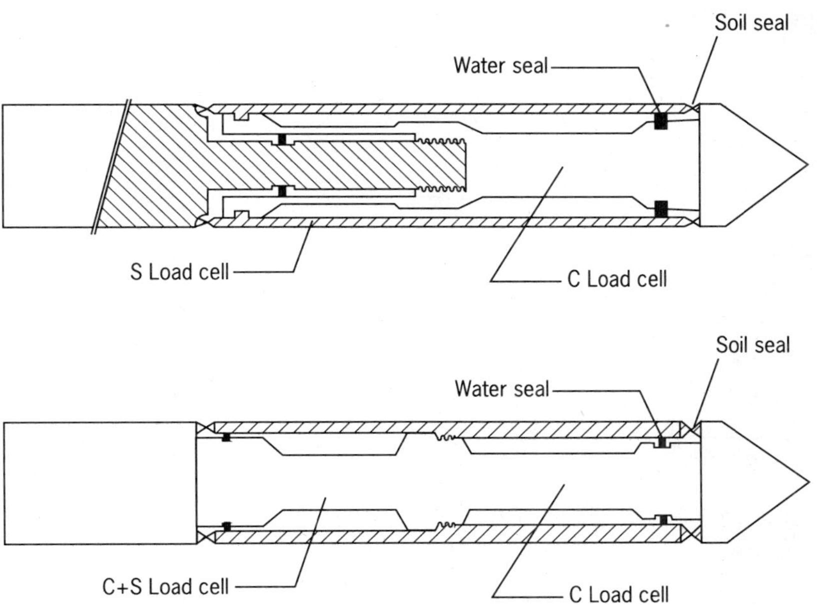


**Figure 4.14** Typical electronic friction cones (Robertson, 1986).

resistance ($q_c$) and side friction ($f_s$) (Robertson, 1986). Further information on various electronic cone designs and their uses have been described by Robertson and Campanella (1984).

The most significant advantage of electrical cones are their repeatability, accuracy, and continuous data collection. ASTM D 3441–79 provides further details of test procedure for CPT (*ASTM ANNUAL Book*, 1989). Although CPT provides a continuous soil resistance record with depth, its major drawback is that soil samples are not recovered from this test. Use of this method combined with borings is therefore recommended.

In North America, many engineers have developed considerable experience and confidence with design based on SPT correlations. Data presented in Figure 4.15 can therefore be used for converting CPT data to equivalent SPT $N$ values.

*Flat Plate Dilatometer Test (DMT)* DMT was developed by S. Marchetti in Italy and is shown in Figure 4.16. It consists of a flat plate 14 mm thick, 95 mm wide by 220 mm long. It has a flexible stainless steel membrane, 60 mm in diameter, located on one face of the blade. A measuring device is located beneath the membrane. This turns a buzzer off in the control box at the surface when the membrane starts to lift off the sensing disc and turns a buzzer on again after a

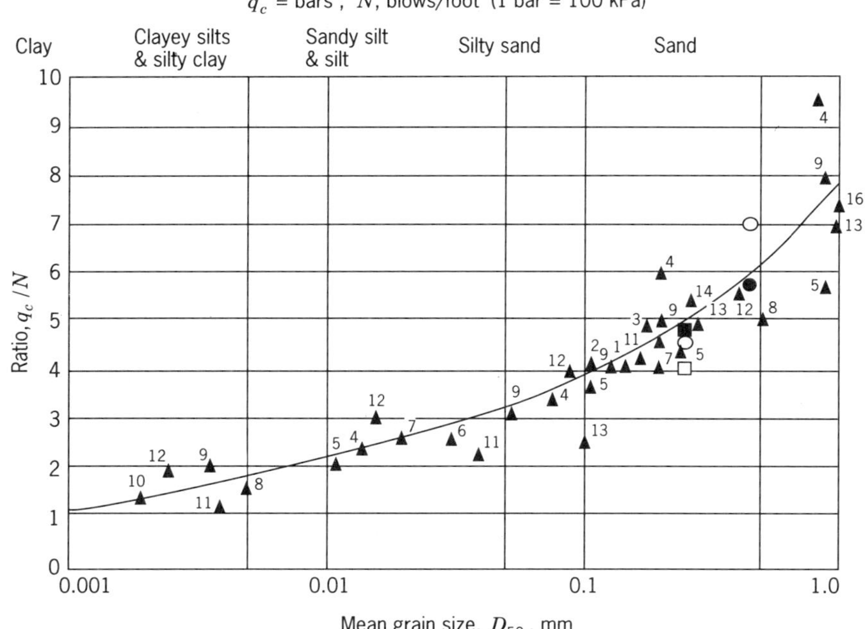


**Figure 4.15** Variation of $q_c/N$ with mean grain size (Robertson et al., 1983).

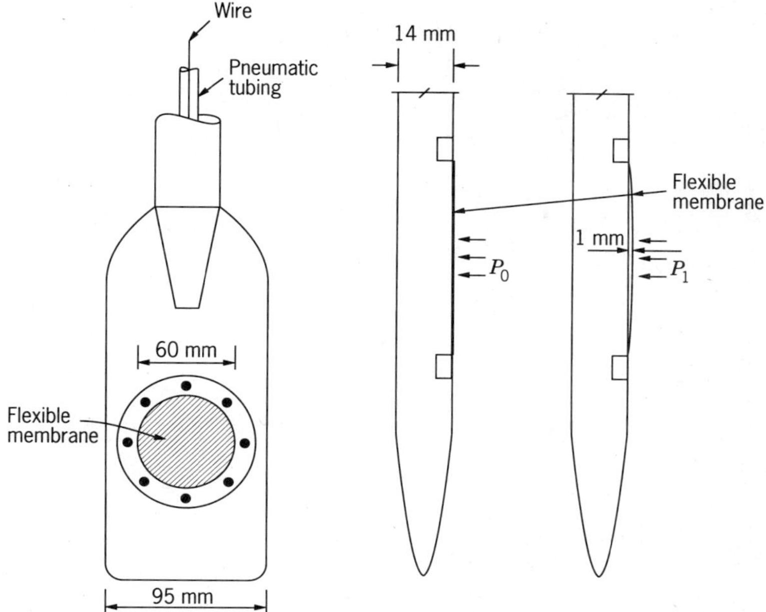


**Figure 4.16** Marchetti flat-plate dilatometer (Robertson, 1986).

deflection of 1 mm at the center of the membrane. Readings are taken every 20 cm in depth. The test measurements consist of the following:

1. Inflate the membrane by using high-pressure nitrogen gas supplied by a tube prethreaded through the rods.
2. As the membrane is inflated, the pressures required to just lift the membrane off the sensing disc is reading A. This is $P_0$.
3. The pressure required to cause 1 mm deflection at the center of the membrane is reading B. This is $P_1$.

These readings are made from a pressure gauge in the control box. Marchetti (1980) provides further details on the in-situ test details on the DMT test procedure.

The dilatometer readings A and B are corrected to pressures $P_0$ and $P_1$ to allow for offset in the measuring gauge and membrane stiffness. Using $P_0$ and $P_1$, the following three index parameters were proposed by Marchetti (Robertson 1986).

$$\text{Material index} = I_d = \frac{P_1 - P_0}{P_0 - U} \tag{4.3}$$

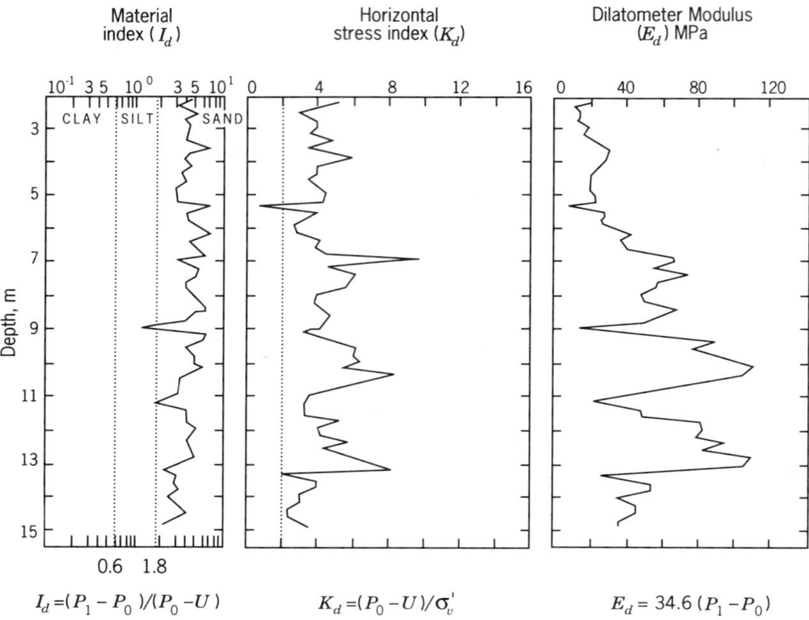


**Figure 4.17** Example of dilatometer index parameter presentation (Robertson, 1986).

$$\text{Horizontal stress index} = K_d = \frac{P_0 - U}{\sigma_V'} \tag{4.4}$$

$$\text{Dilatometer modulus} = E_d = 34.6\,(P_1 - P_0) \tag{4.5}$$

where $U$ = assumed in-situ hydrostatic water pressure and $\sigma_v'$ is the in-situ vertical effective stress. An example of DMT indices is shown in Figure 4.17.

The DMT is used primarily for stratigraphic profile determination. No general correlations exist yet for direct foundation design based on DMT data. However, Davidson and Boghrat (1983), Campanella and Robertson (1983), and Jamiolkowski et al. (1985) suggest that DMT can provide nearly continuous data for soil type identification.

*Field Vane Shear Tests* The field vane shear test, as shown in Figure 4.18, is used to obtain the shear strength of cohesive soils. The vane is best suited for soft-to-firm cohesive soils and should not be used in cohesionless soils. The main equipment parts consist of the torque assembly that is capable of producing constant angular rotation of 1 to 6° per minute, a calibrated proving ring with a dial gauge for torque measurement within 5 percent, a vane blade, and a set of rods. The vane blade should have a height-to-diameter ratio of 2. Typical vane dimensions are 2 in. (50 mm) to 4 in. (100 mm). Detailed test procedure and

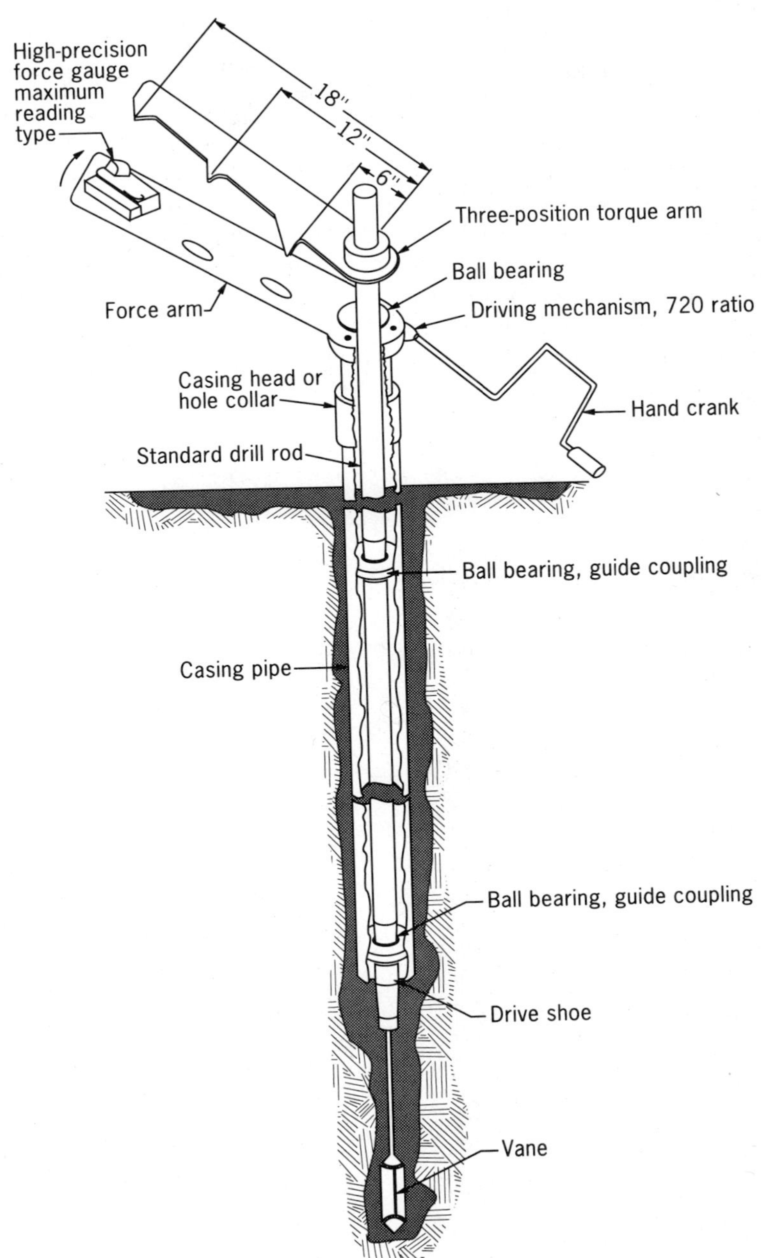


**Figure 4.18** Vane shear test arrangement (Acker Sampling Catalogue and *Design Manual* NAVFAC DM 7.1, 1982).

equipment description are also provided in ASTM D 2573 (*ASTM ANNUAL Book* 1989).

The main features of this test procedure are to push a four-bladed vane attached to the end of a rod into the undisturbed clay below the bottom of a boring. The vane is then turned by applying a torque at the top by turning the crank at a uniform rate. According to the ASTM D 2573 procedure, the torque applied to the vane should not exceed 0.1°/s. The failure mode around a vane is complex. However, test interpretation based on simplified assumptions of a cylindrical failure surface corresponding to the periphery of the blade and of a uniform strength mobilization on that surface can be made (Aas, 1965). Based on these assumptions, the undrained shear strength, $c_u = S_u$, of a clay for a measured torque $T$ can be obtained from the following relationship:

$$c_u = \frac{T}{k} \tag{4.6}$$

where

$c_u$ = undrained shear strength of clay, 1b/ft$^2$ (kN/m$^2$)
$T$ = torque 1b-ft (N-m)
$k$ = constant, depending on dimensions and shape of vane, ft$^3$ (m$^3$)

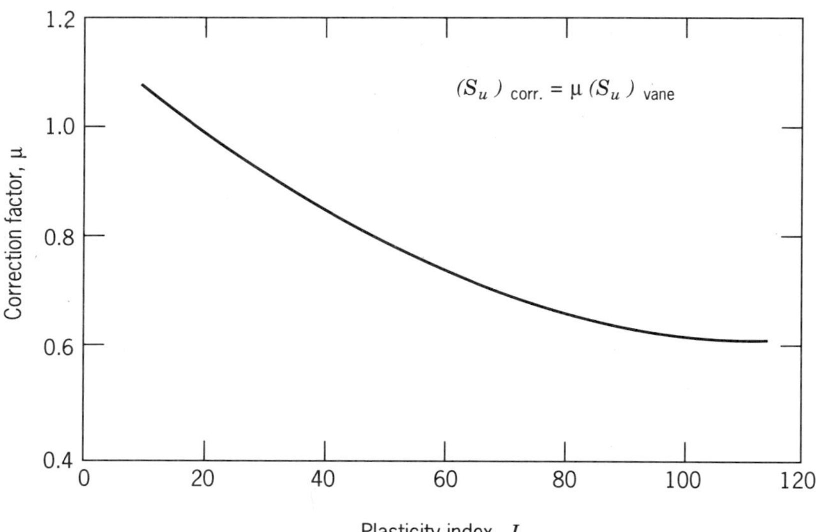


**Figure 4.19** Vane correction factor as a function of plasticity index (Bjerrum, 1973).

For a length-to-width ratio of vane of 2:1,

$k = 0.0021D^3$ in in.-1b units and

$k = 0.00000366D^3$ in metric units (ASTM D 2573, 1988).

$D$ = measured diameter of vane in inches (or centimeters).

Since the undrained shear strength of clays is known to be time dependent, the vane test results must be corrected for time effect factor, $\mu$, as shown on Figure 4.19 (Bjerrum, 1973).

As for cone penetration tests, vane shear tests should also be combined with borings so that soil samples can be recovered for laboratory testing and

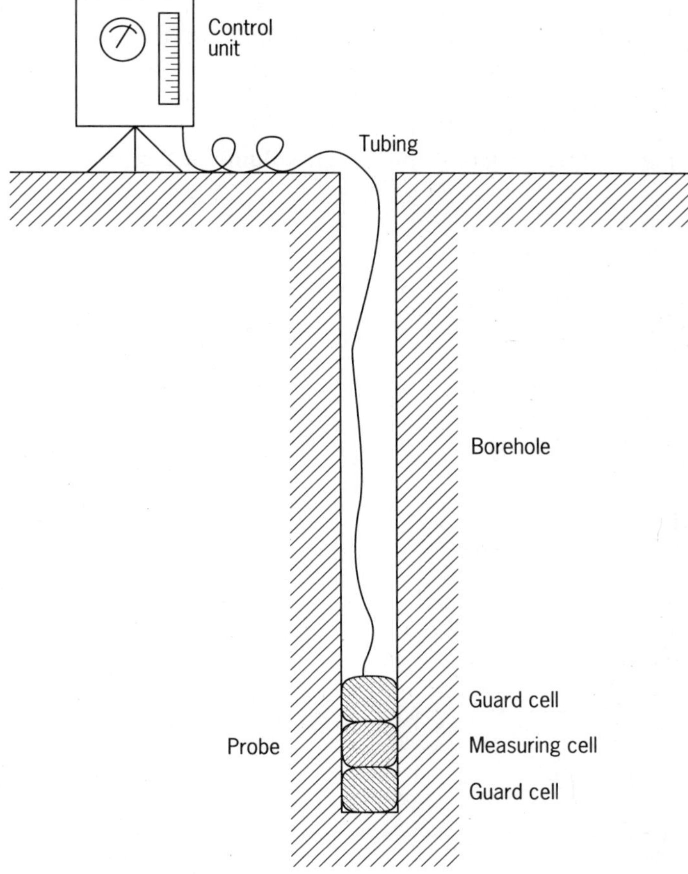


**Figure 4.20** Schematic representation of the Menard-type prebored pressuremeter (Robertson, 1986).

correlations. This test is very useful for soft sensitive soils that cannot be sampled for laboratory testing.

*Pressuremeter Tests* As indicated in Table 4.4, the pressuremeter test method is a highly rated test. This device essentially consists of an expandable cylindrical tube placed at the bottom of a borehole. This cylinder is then expanded under controlled conditions against the surrounding soil. Existing pressuremeters can be divided into three main groups: prebored, self-bored, and full displacement. The prebored pressuremeter test is performed in a predrilled hole, the self-bored pressuremeter is self-bored into the soil to minimize soil disturbance, and the full displacement pressuremeter is pushed into the soil with a solid tip (Robertson, 1985).

The most widely used pressuremeter was developed by Menard (1956). This is a prebored type pressuremeter as shown in Figure 4.20. This instrument is expanded by applying air pressure to a liquid filling the lines and the instrument. The volume expansion is measured by measuring the amount of liquid forced into the expanding central section, which is protected by two guard cells, one above and one below it. A typical pressure-volume increase curve is shown in Figure 4.21 in which *A* refers to the initial volume of the pressuremeter $V_0$. *B*

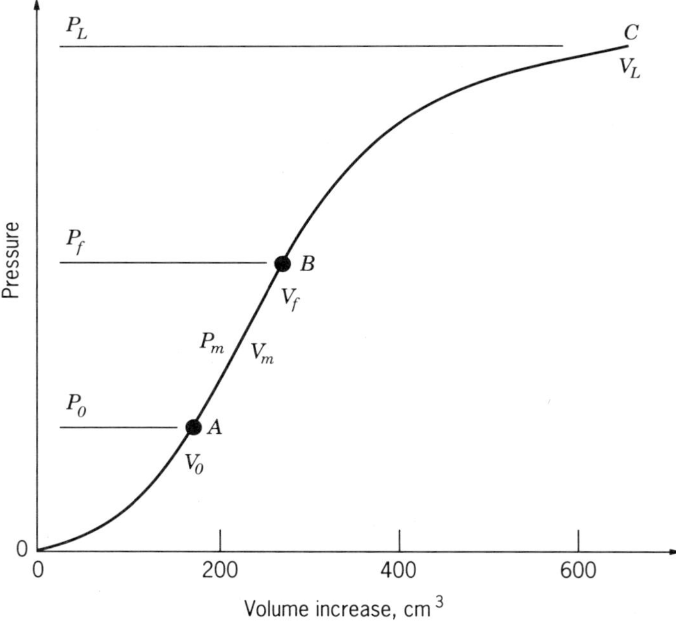


**Figure 4.21** Idealized pressure-expansion curve from Menard-type prebored pressuremeter test (Robertson, 1986).

defines the upper limit of the linear diagram. $V_m$ is the mean of volumes $V_0$ and $V_f$. The corresponding pressures are $P_0$, $P_f$, and $P_m$ respectively.

The undrained shear strength, $S_u$, of clays can be estimated from the following semiempirical relationship (Robertson 1986).

$$S_u = \frac{P_L - P_0}{5.5} \tag{4.7}$$

$P_L$ defines the maximum pressure and the corresponding volume is $V_L$ (Figure 4.21). The pressuremeter modulus, $E_m$, is obtained from the slope of the linear portion of the pressure-volume increase curve (Figure 4.21) as follows:

$$E_m = 2.66(V_0 + V_m)(P/v) \tag{4.8}$$

where

$V_0$ = initial volume of the measuring cell,
$P_0$ = Pressure corresponding to initial volume $V_0$.
$V_m$ = volume change read on the volumeter at a pressure corresponding to the mean pressure in the pseudoelastic range
$P/v$ = slope of the pressure volume curve ($AB$).

In the absence of experimental data, the values of $E_m$ for preliminary design may be estimated with the help of Table 4.5 for different soils.

Based on French experience on the Menard type pressuremeter, empirical design procedures have been developed for both the shallow and deep foundation (Baguelin et al., 1978; Mair and Wood, 1987). The pressuremeter is a useful tool for investigation and design of foundations when dealing with soils that are hard

**TABLE 4.5 Typical Menard Pressuremeter Values (*Canadian Foundation Engineering Manual*, 1985)**

| Type of Soil | $P_L$ Limit Pressure (kPa) | $E_m/P_L$ |
|---|---|---|
| Soft clay | 50–300 | 10 |
| Firm clay | 300–800 | 10 |
| Stiff clay | 600–2500 | 15 |
| Loose silty sand | 100–500 | 5 |
| Silt | 200–1500 | 8 |
| Sand and gravel | 1200–5000 | 7 |
| Till | 1000–5000 | 8 |
| Old fill | 400–1000 | 12 |
| Recent fill | 50–300 | 12 |

to investigate by conventional methods (e.g., granular soils, till, soft rock, and frozen soil).

***Laboratory Testing*** Laboratory testing is carried out to classify the soils and to provide soil parameters for design. The type and number of soil tests will depend on a number of factors such as:

1. Degree of variation of soils at the site
2. Soils information available from previous explorations in the area on similar soils
3. Character of soils
4. Requirements of structure such as importance of differential settlements

Following is a brief description of these tests. For details, consult testing manuals and other relevent publications such as Lambe (1951), Terzaghi and Peck (1967), Prakash et al. (1979), and *Annual Book of Standards*, ASTM (1989).

*Atterberg Limits* Determination of Atterberg limits for engineering purposes according to ASTM Designation D 4318–83 requires obtaining the liquid limit, plastic limit and plasticity index of soils.

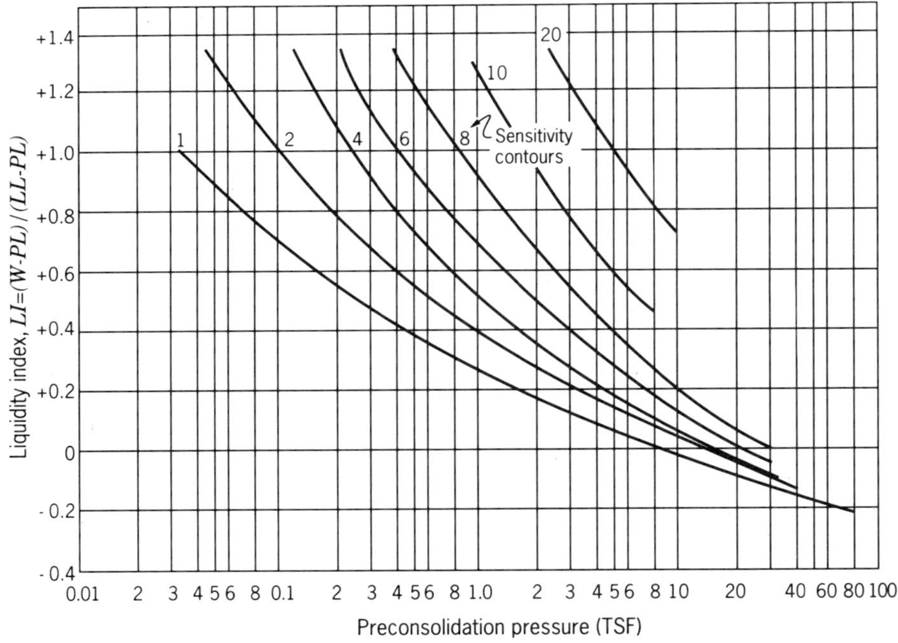


**Figure 4.22** Preconsolidation pressure vs. liquidity index (*Design Manual* NAVFAC DM 7.1, 1982).

The liquid limit (LL) of a soil is the limiting water content of a saturated soil beyond which the soil will attain a liquid state. The soil has infinitesimal strength at liquid limit.

The plastic limit (PL) is the percent water content of a wet soil below which it does not exhibit any plasticity. Thus, plastic limit defines a boundary between the plastic and nonplastic states.

The plasticity index (PI) is the difference between the liquid limit and plastic limit (PI = LL − PL) and signifies the range of water content over which the soil remains plastic.

As we present in the following paragraphs, these soil characteristics (e.g., LL, PL, and PI) can be empirically related with certain engineering soil properties.

*Unconfined Compressive Strength* The unconfined compression test is carried out on clay samples (undisturbed or remolded) to determine shear strength, $S_u$, under undrained conditions. ASTM D 2166–66 (1989) describes its detailed test procedure. The undrained strength, $S_u$, is then obtained by dividing the unconfined compressive value, $q_u$, by 2.

Approximate values of the unconfined compressive strength, $q_u$, can also be obtained from the following relationship (Design Manual NAVFAC DM 7.1, (1982).

$$q_u = 2S_u = 2\bar{p}_c(0.11 + 0.0037\ \text{PI}) \tag{4.9}$$

where

$\bar{p}_c$ = preconsolidation pressure (i.e., the maximum past effective normal stress at which the soil deposit has been consolidated). This can be obtained from consolidation test or can be approximated from Figure 4.22.

PI = plasticity index as discussed above

*Consolidation Parameters* One-dimensional consolidation tests as per ASTM D 2435–80 are conducted to determine compression (or settlement) characteristics of fine-grained cohesive soils under applied loads. The soil parameters determined by this test are compression index, $C_c$, coefficient of consolidation, $C_v$, and the preconsolidation pressure, $\bar{p}_c$.

The typical void ratio ($e$) versus log $\sigma'_v$ plot, obtained from consolidation test ASTM D 2435, is shown schematically in Figure 4.23. In this figure $e_0$ is the initial void ratio, $\bar{p}_c$ is the preconsolidation pressure, $C_r$ is the recompression index, and $C_c$ is the virgin compression index. For further details, standard textbooks on soil mechanics, such as Terzaghi and Peck (1967) should be referred to.

The preconsolidation pressure $\bar{p}_c$ is the maximum normal effective stress to which the material in situ has been consolidated by a previous loading. If the existing effective overburden pressure, $\sigma'_{v0}$, is larger than $\bar{p}_c$ then the soil is called *under consolidated*, if $\sigma'_{v0} = \bar{p}_c$ then the soil is called *normally consolidated*, and if

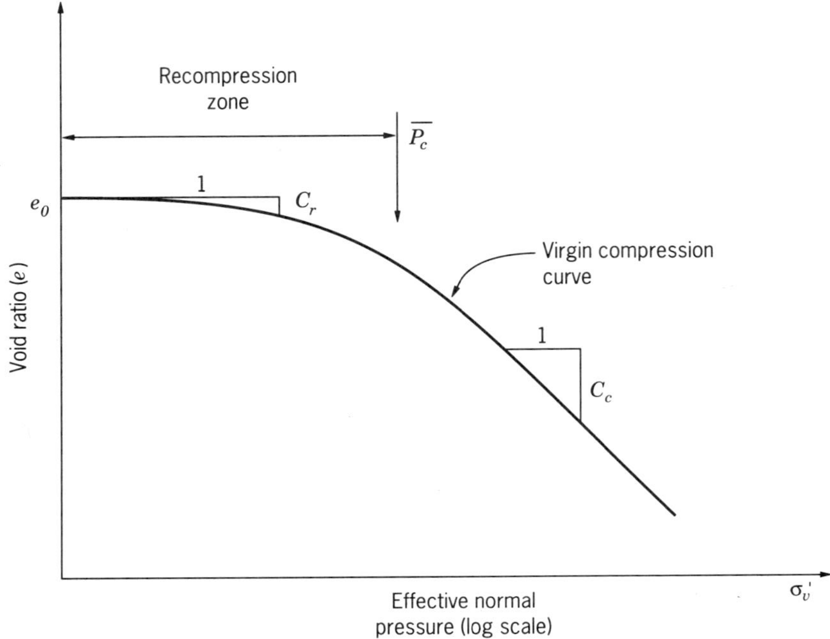


**Figure 4.23** Typical void ratio vs. log $\sigma_v'$ curve from consolidation test.

$\sigma_{v0}'$ is less than $\bar{p}_c$ then the soil is called *overconsolidated.* The ratio $(\bar{p}_c/\sigma_{v0}')$ is called the *over consolidation ratio* (OCR). If OCR is between 1 to 4, then the soils are called lightly overconsolidated while if this ratio is greater than 4, they are called heavily overconsolidated. These concepts and terms are later used in settlement calculations in Chapter 5 (Section 5.1.10).

Approximate values of compression index, $C_c$, can also be obtained from following relationships (Design Manual NAVFAC DM 7.1 1982). Similar other relationships have been proposed by Nishida (1956), Hough (1969) and Sowers (1979).

1. $C_c = 0.009$ (LL − 10 percent)
   for inorganic soils with sensitivity less than 4
2. $C_c = 0.0115\, w_n$ for organic soil
3. $C_c = 1.15\,(e_0 - 0.35)$ for all clays
4. $C_c = (1 + e_0)(0.1 + (w_n - 25)0.006)$ for varved clays where LL is the liquid limit, $w_n$ is the natural moisture content and, $e_0$ is the initial void ratio.

*Shear Parameters* The direct shear tests are carried out as per ASTM test procedure D 3080–72 (1979) on cohesionless soils to determine the angle of internal friction, $\phi'$. The triaxial test is generally not used to determine shear parameters for design of piles. The shear parameter $c = c_u = S_u$ for $\phi = 0$ for cohesive soils determined from unconfined compressive strength test has previously been discussed.

### 4.1.3 Design Parameters

This section presents the information on strength parameters, soil–pile adhesion, and elastic soil parameters both for the cohesionless and cohesive soils that are required for static pile design.

***Strength Parameters*** The two commonly used strength parameters in pile design are the angle of internal friction ($\phi'$) for cohesionless soils and the undrained shear strength ($S_u$) for cohesive soils.

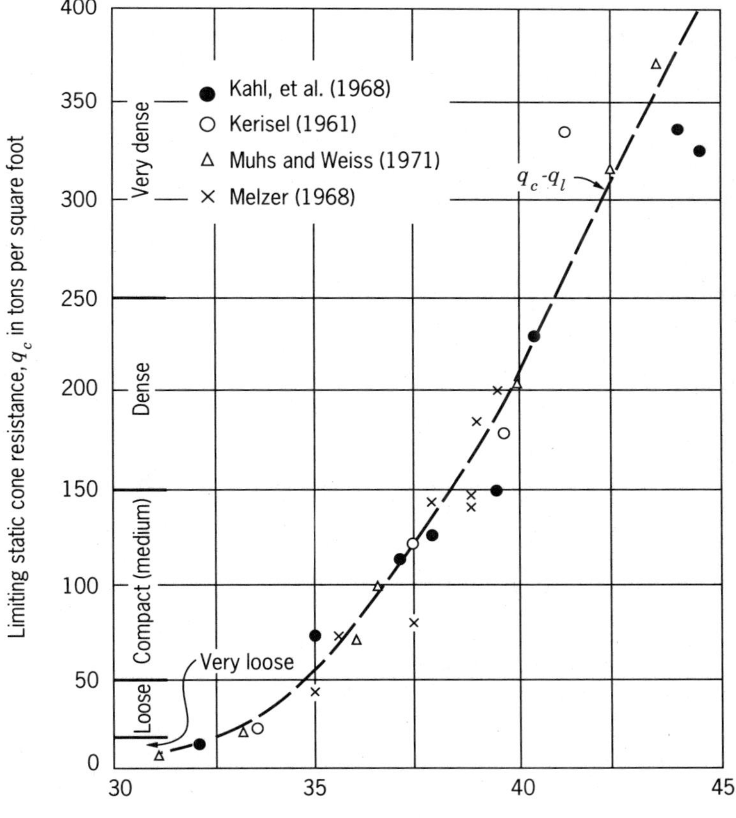


**Figure 4.24** Approximate relation between limiting static cone resistance and friction angle of sand (1 tsf = 95.8 kN/m$^2$) (Meyerhof, 1976).

The angle of internal friction can either be obtained from laboratory testing (Section 4.1.2) or from the correlations established with field penetrometer test values (e.g., $N$ or $q_c$). Figure 4.24 presents a relationship between the static cone penetration test (CPT) values, $q_c$ and the angle of internal friction, $\phi'$, values. Meyerhof (1976) recommends the use of this relationship for pile design. If only standard penetration test values, $N$ are available at a site, then Figure 4.15 should be used to first obtain the equivalent $q_c$ values. Figure 4.24 can then be used to obtain $\phi'$ values.

Another method of obtaining the angle of internal friction, as recommended in Design Manual NAVFAC, DM 7.1 (1982), consists of the following:

1. Obtain the relative density, $D_r$, for the field measured, $N$, values from Figure 4.25.
2. Then from Figure 4.26, for the known soil or dry density (or void ratio or porosity) and $D_r$ from (1) above, obtain the angle of internal friction, $\phi'$,

Example 4.1 explains the use of both the foregoing methods to estimate the $\phi'$ value from field test data for cohesionless material. The first method using the $q_c/N$ relationship and then the use of $q_c$ versus $\phi'$ relationship yields $\phi' = 36°$ while the use of the $N$, $D_r$, and $\phi'$ relationship yields $\phi' = 35°$.

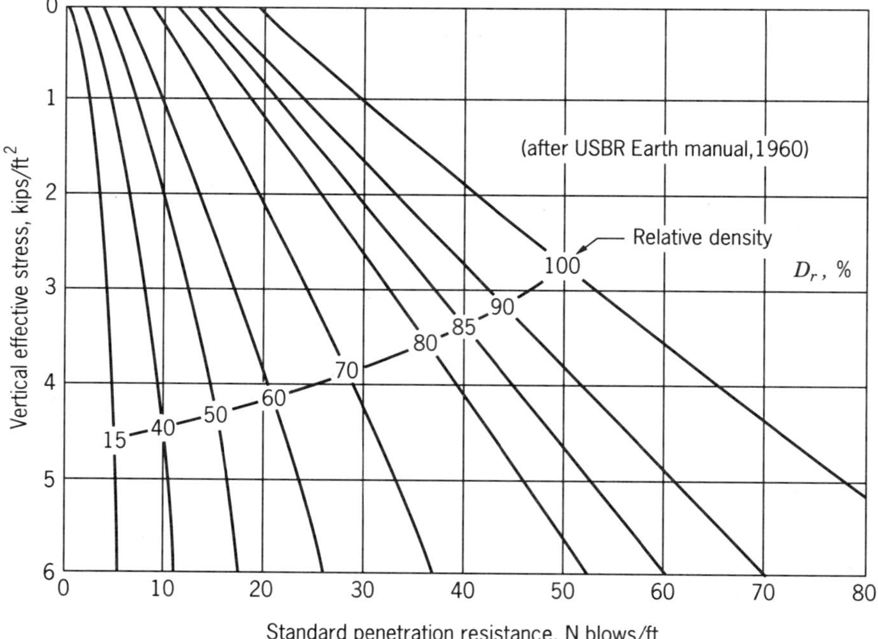


**Figure 4.25** Correlations between relative density and standard penetration resistance in accordance with Gibbs and Holtz (1957) (NAVFAC *Design Manual* DM 7.1, 1982).

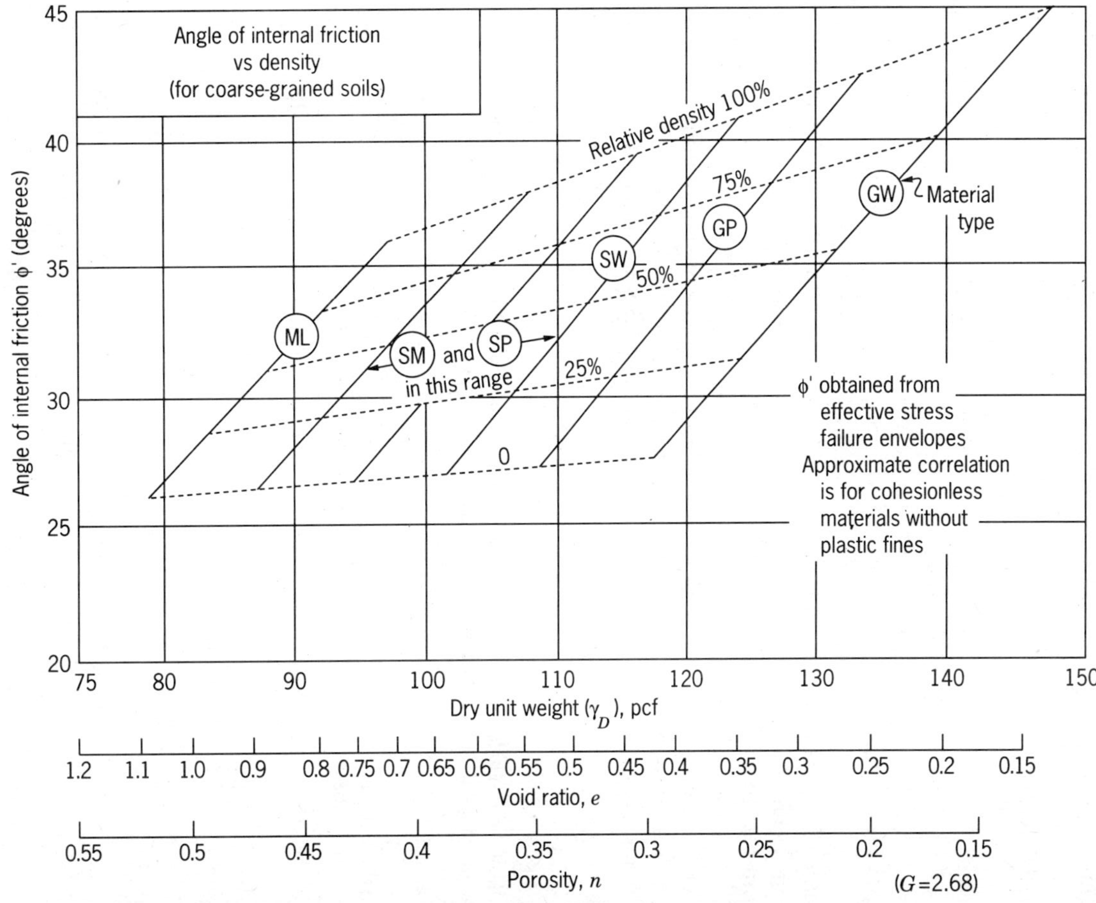


**Figure 4.26** Correlations of strength characteristics for granular soils (NAVFAC *Design Manual* DM 7.1, 1982).

***Example 4.1*** During a site investigation work, borehole logs indicated the SPT value of 20 at a depth of 25 ft in sand. Laboratory grain size analysis indicated that the sand had mean grain size, $D_{50} = 0.004$ in. (0.1 mm). The density of the overburden soil was estimated to be 125 lb/ft$^3$ and dry density of this sand was estimated at 110 lb/ft$^3$. No groundwater table was observed in the borehole. Estimate the angle of internal friction for the sand.

SOLUTION

*Method 1* From Figure 4.15, $q_c/N = 3.8$ for $D_{50} = 0.1$ mm
$q_c = 3.8 \times 20 = 76$ bar for $N = 20$
$= 7600$ kPa $= 79$ tons per square foot
From Figure 4.24, for $q_c = 79$ tons/ft$^2$, $\phi' = 36°$

*Method 2* Vertical effective stress $= \sigma'_v = 125 \times 25 = 3125$ lb/ft$^2 = 3.125$ kips/ft$^2$
From Figure 4.25 for $N = 20$, $\sigma'_v = 3.125$ kips/ft$^2$, $D_r = 64$ percent
From Figure 4.26 for $\gamma_d = 110$ psf, $D_r = 64\%$, $\phi' = 35°$

The undrained shear strength, $S_u$, of a cohesive soil can either be obtained from laboratory testing of undisturbed soil sample or by field vane shear tests, equation (4.6), on soft cohesive soils and pressuremeter tests, equation (4.7), on stiff soils.

**TABLE 4.6 Guide for Consistency of Fine-grained Soils (Terzaghi and Peck, 1967, Design Manual NAVFAC, DM 7.1, 1982, *Canadian Foundation Engineering Manual*, 1985)**

| SPT Penetration $N$ Values[a] | Estimated Consistency | Estimated Range of $S_u = c_u$ kPa | Estimated Range of $S_u = c_u$ kips/ft$^2$ |
|---|---|---|---|
| < 2 | Very soft (extruded between fingers when squeezed) | < 12 | < 0.25 |
| 2–4 | Soft (molded by light finger pressure) | 12–25 | 0.25–0.50 |
| 4–8 | Firm or medium (molded by strong finger pressure) | 25–50 | 0.50–1.00 |
| 8–15 | Stiff (readily indented by thumb but penetrated only with great effort) | 50–100 | 1.00–2.00 |
| 15–30 | Very stiff (readily indented by thumbnail) | 100–200 | 2.00–4.00 |
| > 30 | Hard (indented with difficulty by thumbnail) | > 200 | > 4.00 |

[a] *The Canadian Foundation Engineering Manual* does not recommend the relationship with $N$.

For normally consolidated natural deposits, $S_u$ can also be estimated by the following relationship (Skempton, 1948; Bjerrum and Simons 1960).

$$S_u = c_u = \sigma'_v(0.1 + 0.004\text{PI}) \tag{4.10}$$

where $\sigma'_v$ is the effective vertical overburden pressure and PI is the plasticity index. This equation is similar to equation (4.9) except that $\bar{p}_c$ has been replaced with $\sigma'_v$ for normally consolidated soils i.e., $\sigma'_v = \bar{p}_c$. Equation (4.9) is applicable for both the normally and overconsolidated soils and therefore is generalized form of equation (4.10). However, both equations would yield similar results for normally consolidated soils.

Consistency of cohesive soils and the approximate relationships with $N$ and $S_u$ can be obtained from Table 4.6. Since these relationships are approximate, they

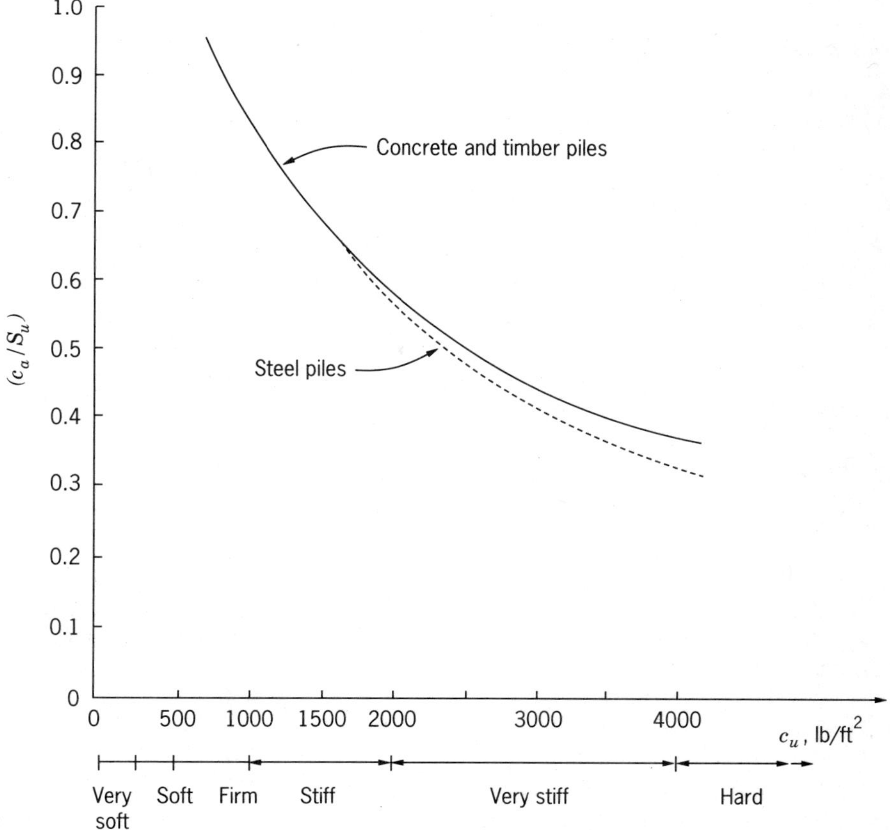


**Figure 4.27** Variation of $c_a/S_u$ with $c_u$ for different pile materials for driven piles (developed from data in Tomlinson, 1963).

should only be used in the preliminary design. For final design, field and/or laboratory determined $S_u$ values should be used.

***Soil–Pile Adhesion ($c_a$)*** Estimation of soil–pile adhesion ($c_a$) is complex. It depends on factors such as (1) soil consistency, (2) method of pile installation, (3) pile material, and (4) time. Reliable values of $c_a$ can only be obtained by performing full-scale pile load tests in the field. Figure 4.27 can be used as a guide for estimating $c_a$ values for *driven* piles in clay with different consistency (Tomlinson, 1963). These values have also been recommended by Terzaghi and Peck (1967). The soil–pile adhesion value $c_a$ is also termed as side friction. For drilled piles or piers, $c_a$ can be estimated from Table 4.7.

**TABLE 4.7 Design Parameters for Side Friction for Drilled Piers in Cohesive Soils (NAVFAC Design Manual, DM 7.2, 1982)**

| Side Resistance Design Category | $c_a/c_u$ | Limit on Side Shear—tsf | Remarks |
|---|---|---|---|
| A. Straight-sided shafts in either homogeneous or layered soil with no soil of exceptional stiffness below the base | | | |
| 1. Shafts installed dry or by the slurry displacement method | 0.6 | 2.0 | |
| 2. Shafts installed with drilling mud along some portion of the hole with possible mud entrapment | 0.3(a) | 0.5(a) | (a) $c_a/c_u$ may be increased to 0.6 and shear increased to 2.0 tons per sq. ft. for segments drilled dry |
| B. Belled shafts in either homogeneous or layered clays with no soil of exceptional stiffness below the base | | | |
| 1. Shafts installed dry or by the slurry displacement methods | 0.3 | 0.5 | |
| 2. Shafts installed with drilling mud along some portion of the hole with possible mud entrapment | 0.15(b) | 0.3(b) | (b) $c_a/c_u$ may be increased to 0.3 and side shear increased to 0.5 tons per sq. ft. for segments drilled dry |
| C. Straight-sided shafts with base resting on soil significantly stiffer than soil around stem | 0 | 0 | |
| D. Belled shafts with base resting on soil significantly stiffer than soil around stem | 0 | 0 | |

***Elastic Soil Parameter*** The most common elastic soil parameter required in pile design is the modulus of elasticity, $E_s$. In cohesionless soils, the static elastic modulus, $E_s$, may be estimated from empirical methods using relations of $E_s$ with SPT $N$ values or with static cone penetration $q_c$ values. Many studies relating $N$ values with $E_s$ indicate that such relationships are of little use because the relationships vary significantly and the ratio of predicted to observed settlements based on these $E_s$ values may range between 0.12 to 20 (Talbot, 1981; Robertson, 1986). This is due to the fact that $E_s$ depends on a large number of variables as explained in Section 4.2. Therefore, these relationships should not be used unless local experience supports them. A value of $E_s$ can, however, be estimated from results of the static cone penetration test, $q_c$, as follows (Schmertmann, 1970).

$$E_s = C_1 q_c \tag{4.11}$$

where $C_1$ is a constant and depends on the soil compactness as follows (*Canadian Foundation Engineering Manual*, 1985):

| | |
|---|---|
| Silt and sand | $C_1 = 1.5$ |
| Compact sand | $C_1 = 2.0$ |
| Dense sand | $C_1 = 3.0$ |
| Sand and gravel | $C_1 = 4.0$ |

For cohesive soils, the values of $E_s$, as recommended by the *Canadian Foundation Engineering Manual* (1985) can be estimated from the following relationship.

$$E_s = C_2 \bar{p}_c \tag{4.12}$$

where $\bar{p}_c$ is the preconsolidation pressure and $C_2$ is a constant such that $C_2 = 80$ for stiff clays, $C_2 = 60$ for firm clays, and $C_2 = 40$ for soft clays. These relationships are approximate at best and may be used only in preliminary design.

## 4.2 SOIL PARAMETERS FOR DYNAMIC DESIGN

Several problems in engineering practice require a knowledge of dynamic soil properties. In general, problems involving the dynamic loading of soils are divided into *small-* and *large-strain* amplitude responses. In a pile foundation, the amplitudes of dynamic motion and, consequently, the strains in the soil are usually small for machine foundations whereas during an earthquake or blast loading, large strains may occur. A large number of field and laboratory methods have been developed for determination of the dynamic soil properties. The principal properties that are used in dynamic soil–pile analysis include dynamic moduli, such as Young's modulus $E$ and shear modulus $G$, with corresponding spring constants; damping; and Poisson's ratio. The first two are dependent on

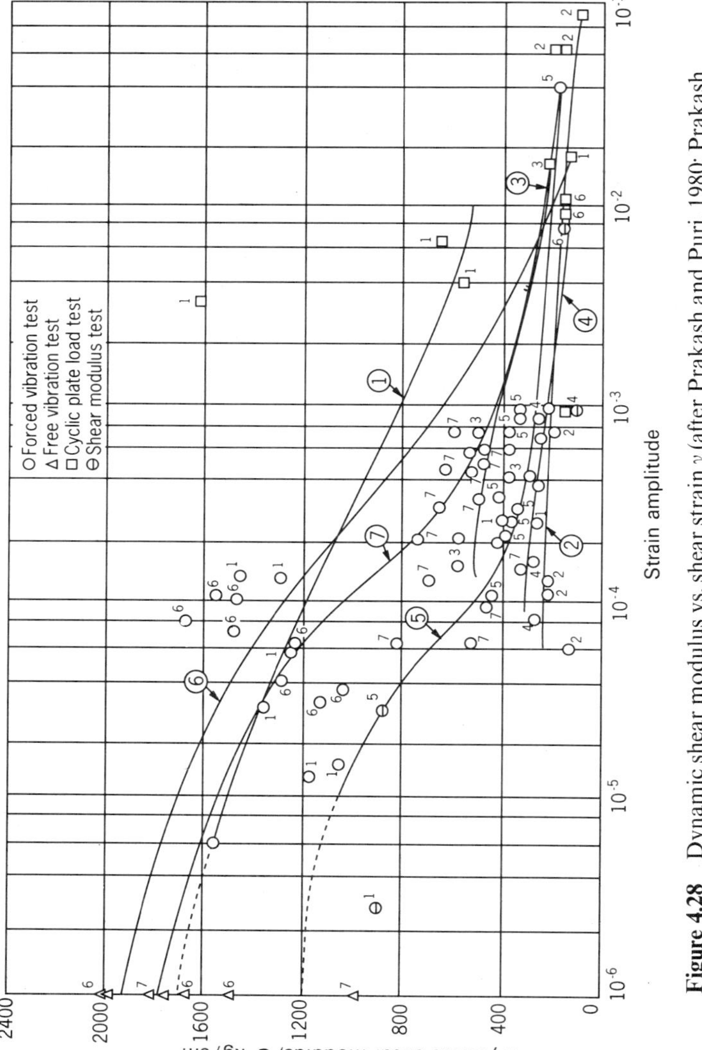


**Figure 4.28** Dynamic shear modulus vs. shear strain $\gamma$ (after Prakash and Puri, 1980; Prakash, 1981).

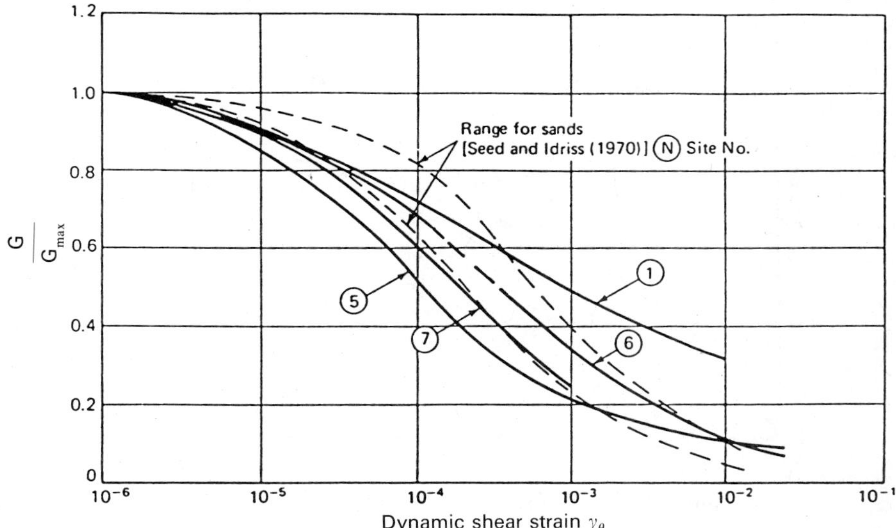


**Figure 4.29** Normalized shear modulus ($G/G_{max}$) vs. shear strain, $\gamma_\theta$.

strain amplitude ($\gamma_\theta$) since behavior of the soil is nonlinear (Figure 4.28). In Figure 4.29, the plot of $G$ vs. $\gamma_\theta$ (in Figure 4.28) has been normalized by dividing the ordinate with $G_{max}$, the value of $G$ at small strain ($10^{-6}$ or smaller).

In this section, a brief discussion of the laboratory and field methods used to determine dynamic soil moduli is presented along with typical values of dynamic soil moduli and damping.

### 4.2.1 Elastic Constants of Soils

The behavior of a soil is nonlinear from the beginning of stress application. For practical purposes, the actual nonlinear stress–strain curves of soils are *linearized.* Therefore, a modulus and a Poisson's ratio are not constants for a soil but depend on several parameters as will be explained further. Two moduli used in dynamic loading are Young's modulus and shear modulus.

If a uniaxial stress $\sigma_z$ is applied to an elastic cylinder that causes axial strain $\varepsilon_z$, then Young's modulus $E$ is defined as

$$E = \frac{\sigma_z}{\varepsilon_z} \tag{4.13}$$

The lateral strains $\varepsilon_x$ and $\varepsilon_y$ are

$$\varepsilon_x = \varepsilon_y = -\nu\varepsilon_z \tag{4.14}$$

where $\nu$ is Poisson's ratio.

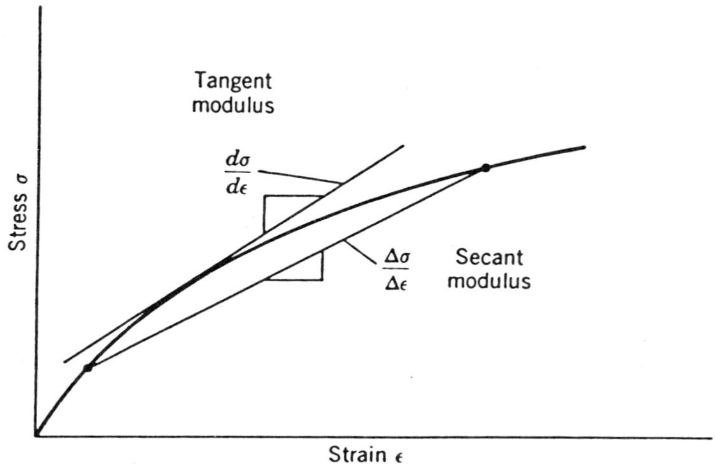


**Figure 4.30** Definitions of secant and tangent modulus.

If shear stress, $\tau$, is applied to an elastic cube, there will be a shear distortion, $\gamma_\theta$, and shear modulus $G$ is defined as

$$G = \frac{\tau}{\gamma_\theta} \quad \text{or} \quad \gamma_\theta = \frac{\tau}{G} \tag{4.15a}$$

Of the three constants ($E$, $G$, and $v$), only two are needed, because they are related as follows:

$$E = 2G(1 + v) \tag{4.15b}$$

The Young's modulus $E$ and shear modulus $G$ may be measured in terms of either tangent modulus or secant modulus. Tangent modulus is the slope of the tangent to a stress–strain curve at a particular point on the curve and is strain dependent (Figure 4.30). Secant modulus is the slope of a straight line connecting two separate points of a stress–strain curve. Based on a linear stress–strain relationship, the above *elastic constants* have been defined.

### 4.2.2 Factors Affecting Dynamic Modulus

Based on the study of dynamic elastic constants, the factors on which these depend are (Hardin and Black 1968):

1. Type of soil and its properties (e.g., water content and $\gamma_d$) and state of disturbance.
2. Initial (sustained) static stress level or confining stress

3. Strain level
4. Time effects
5. Degree of saturation
6. Frequency and number of cycles of dynamic load
7. Magnitude of dynamic stress
8. Dynamic prestrain

***Type of Soil, its Properties and Initial Static Stress Level*** Since the soil modulus is strain dependent (Figure 4.30), more than one method is needed to determine the variation of modulus with strain.

The large amount of data on the values of soil constants that had been collected was analyzed by Hardin (1978), who developed a mathematical formulation of soil elasticity and soil plasticity in terms of effective stresses. On this basis, the maximum value of the shear modulus, $G_{max}$, (at low shear strain of $10^{-6}$) is expressed by equation (4.16a) (Hardin and Black 1969):

$$G_{max} = 1230\,\mathrm{OCR}^k \frac{(2.973 - e)^2}{(1 + e)} (\bar{\sigma}_0)^{0.5} \tag{4.16a}$$

in which OCR is the overconsolidation ratio, $e$ the void ratio, and $k$ a factor that depends on the plasticity index of clays, Table 4.8, and $\bar{\sigma}_0$ the mean effective confining stress in psi, equals

$$\bar{\sigma}_0 = (\bar{\sigma}_1 + \bar{\sigma}_2 + \bar{\sigma}_3)/3 \tag{4.16b}$$

or

$$\bar{\sigma}_0 = (\bar{\sigma}_x + \bar{\sigma}_y + \bar{\sigma}_z)/3. \tag{4.16c}$$

If the shear modulus is determined at a mean effective confining pressure of $(\bar{\sigma}_0)_1$, its value at any other mean effective confining pressure $(\bar{\sigma}_0)_2$ can be determined from equation (4.17)

$$\frac{(G_{max})_1}{(G_{max})_2} = \frac{(\bar{\sigma}_0)_1^{1/2}}{(\bar{\sigma}_0)_2^{1/2}} \tag{4.17}$$

Effective overburden pressure $\bar{\sigma}_v$ may be used in place of $\bar{\sigma}_0$ in equation (4.17).

**TABLE 4.8 Values of *k* after Hardin and Drnevich, 1972**

| Plasticity Index PI | $k$ |
|---|---|
| 0 | 0 |
| 20 | 0.18 |
| 40 | 0.30 |
| 60 | 0.41 |
| 80 | 0.48 |
| 100 | 0.50 |

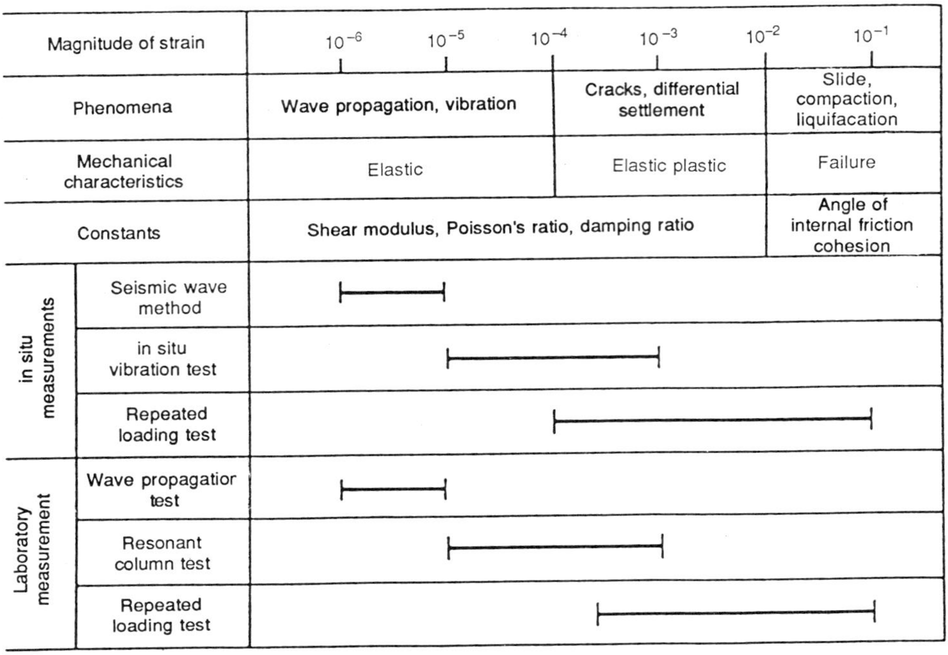

| Magnitude of strain | | $10^{-6}$ | $10^{-5}$ | $10^{-4}$ | $10^{-3}$ | $10^{-2}$ | $10^{-1}$ |
|---|---|---|---|---|---|---|---|
| Phenomena | | Wave propagation, vibration | | | Cracks, differential settlement | | Slide, compaction, liquifacation |
| Mechanical characteristics | | Elastic | | | Elastic plastic | | Failure |
| Constants | | Shear modulus, Poisson's ratio, damping ratio | | | | | Angle of internal friction cohesion |
| in situ measurements | Seismic wave method | | | | | | |
| | in situ vibration test | | | | | | |
| | Repeated loading test | | | | | | |
| Laboratory measurement | Wave propagatior test | | | | | | |
| | Resonant column test | | | | | | |
| | Repeated loading test | | | | | | |

**Figure 4.31** Strain level associated with different in-situ and laboratory tests (after Ishihara, 1971).

***Strain Level*** Figure 4.31 shows strain levels associated with different phenomenon in the field and in corresponding field and laboratory tests. Typical variations of $G$ versus shear strain amplitude for different types of in-situ tests are shown in Figure 4.28. The soil modulus values may vary by a factor of 10, depending on the strain level.

It is customary to plot a graph between normalized modulus (defined as $G$ value at a particular strain, divided by $G_{\max}$ at a strain of $10^{-6}$) and shear strain (Figure 4.29).

The shear strains induced in soil may not be precisely known (Prakash and Puri, 1981). In the case of wave propagation tests, the shear strain amplitudes are low and are assumed to be of the order of $10^{-6}$. The shear strain induced in soil essentially depends on the amplitude of vibration or settlement, which in turn depends on superimposed loads, the foundation contact area, and soil characteristics. The measured values of amplitude or settlement take care of the factors affecting them. In vertical vibrations, the shear strain amplitudes, $\gamma_\theta$, is equal to the ratio of the amplitude or settlement to width of the oscillating footing for all practical purposes, both at low and high strains (Prakash, 1975; Prakash and Puri, 1977; Prakash and Puri, 1988). For values of $\phi$ and $\nu$, in the range of interest, it is reasonable to assume, therefore, that $\gamma_\theta \cong \epsilon_z$.

***Time Effects*** The effect of duration of confinement at a constant pressure on the magnitude of shear moduli is well established both in natural and prepared soils (Anderson and Stokoe, 1977; Prakash and Puri, 1987; Richart 1961).

In Figure 4.32, the time-dependent behavior at *low strain levels* can be characterized by an initial phase when modulus changes rapidly with time, followed by a second phase when the modulus increases almost linearly with the logarithm of the time. For the most part, the initial phase results from the void ratio changes and increase in effective confinement during primary consolidation. The second phase—in which the modulus increases almost linearly with the logarithm of time—is probably due largely to the decrease in void ratio and changes in the soil structure due to a strengthening of the physicochemical bonds in the case of cohesive soils and to an increase in particle contact for cohesionless soils. This increase in modulus proceeds at a constant confining stress and is referred to as the *long-term time effects* and represents the increase in the modulus with time that occurs, after primary consolidation is completed.

The long-term time effects may be described as:

1. *Coefficient of shear modulus increase with time,* $I_G$.

$$I_G = \Delta G / \log_{10}(t_2/t_1) \tag{4.18a}$$

in which $t_1$ and $t_2$ are the times after primary consolidation, and $\Delta G$ is the change in low-amplitude shear modulus from $t_1$ to $t_2$ (Figure 4.32).

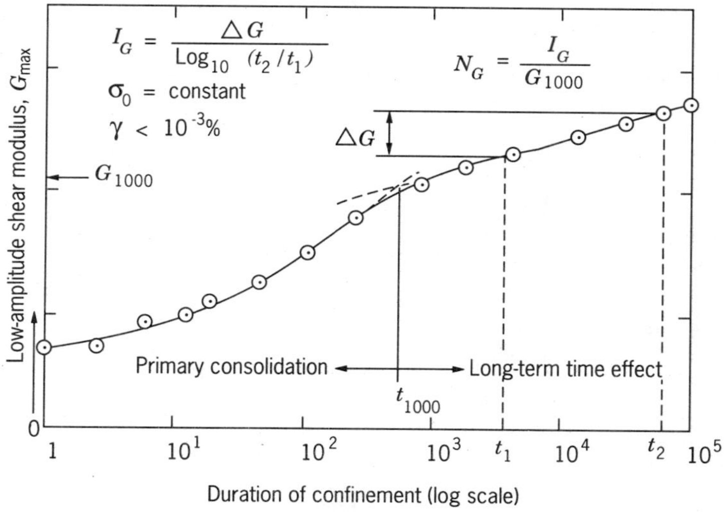


**Figure 4.32** Phases of modulus–time response in soils (after Anderson and Stokoe, 1977).

**TABLE 4.9 Typical values of $I_G$ and $N_G$**

| Soil Type | Specimen Type | Confining Pressure $(kN/m^2)^c$ | Low-Amplitude Shear Modulus $G_{1000}(kN/m^2)^c$ | Typical $I_G^a$ $(kN/m^2)^c$ | Typical $N_G^b$ (%) | Reference |
|---|---|---|---|---|---|---|
| EPK kaolinite<br>Ottawa sand<br>Quartz sand<br>Quartz silt<br>Dry clay | Vacuum extruded<br>Compacted by raining and tamping | 200–300<br>70–280 | 140,000–190,000<br>50,000–180,000 | 24,000–35,000<br>1,400–5,500 | 17–18<br>1–11 | Hardin and Black (1968)<br>Afifi and Woods (1971) |
| Kaolinite<br>Bentonite | Vacuum extruded | 70–550 | 4,000–170,000 | 1,000–8,500 | 5–25 | Marcuson and Wahls (1972) |
| Agsco sand<br>Ottawa sand<br>Air-dried EPK Kaolinite | Compacted by raining and tamping | 70–280 | 50,000–110,000 | 2,000–10,000 | 1–17 | Afifi and Richart (1973) |
| Saturated EPK Kaolinite | Vacuum extruded | | | | | |
| Silty sand<br>Sandy silt<br>Clayey silt<br>Shale | Undisturbed[d] | 70–220 | 80,000–2,600,000 | 2,000–22,900 | 1–14 | Stokoe and Richart (1973a, b) |

| | | | | | | |
|---|---|---|---|---|---|---|
| Boston blue clay | Undisturbed[d] | 70–700 | 32,500–54,000 | ≃ 7,000 | 15–18 | Trudeau et al. (1974) |
| 9 Clays 1 Silt | Undisturbed[d] | 35–415 | 13,000–235,000 | 26,000–23,500 | 2–40 | Anderson and Woods (1975, 1976) |
| Clay fills | Undisturbed[d] | 35–70 | 50,000–200,000 | 4,200–15,000 | 7–14 | Stokoe and Abdel-razzak (1972) |
| Decomposed marine limestone | Undisturbed[d] | 325–830 | 365,000–1,300,000 | 28,000–102,000 | 3–4 | Yang and Hatheway (1976) |
| San Francisco Bay mud | Undisturbed[d] | 17–550 | 7,600–150,000 | 725–32,000 | 8–22 | Lodde (1977) |
| Dense silty sand | Undisturbed[d] | 220–620 | 45,000–180,000 | 5,000–17,000 | 4–10 | Fugro, Inc. (1977) |
| Stiff OC[e] clay | Undisturbed[d] | 1,280–1,300 | 300,000–320,000 | 14,000–26,000 | 4–8 | Fugro, Inc. (1977) |

*Source:* Anderson and Stokoe, 1977, copyright ASTM. Reprinted with permission.

[a] $I_G$ defined by equation. 4.18a.
[b] $N_G$ defined by equation. 4.18b.
[c] 1 $kN/m^2 = 0.145$ psi.
[d] Nominally undisturbed.
[e] Overconsolidated.

Numerically, $I_G$ equals the value of $G$ for one logarithmic cycle of time.

2. *Normalized shear modulus increase with time, $N_G$.*

$$N_G = \left[\frac{\Delta G}{\log_{10}(t_2/t_1)}\right]\left(\frac{1}{G_{1000}}\right)100\% = \frac{I_G}{G_{1000}}100\% \qquad (4.18b)$$

in which $G_{1000}$ is the shear modulus measured after 1000 minutes of constant confining pressure (after completion of the primary consolidation).

The duration of primary consolidation and the magnitude of the long-term time effect vary with such factors as soil type, initial void ratio, undrained shearing strength, confining pressure, and strèss history. Typical values of $I_G$ and $N_G$ are given in Table 4.9

The results of a number of tests show that long-term modulus increases occur at low to intermediate strain levels (0.001 to 0.1 percent) for stiffer clays (Lodde, 1977). Preliminary results from a long-term modulus increases occur in clean, dry sands at strain amplitudes up to 0.1 percent as well.

Because of the general similarity between the increase in moduli with time at low- and high-shearing strain amplitudes, it seems reasonable to conclude that many of the factors that affect the low-amplitude modulus-time response also affect the high-amplitude modulus-time response (at the start of high-amplitude cycling) (Anderson and Stokoe, 1977). Anderson and Stokoe also proposed a method that can be used to predict the in-situ shear moduli from laboratory tests after allowing for time effects.

***Degree of Saturation*** Biot (1956) showed that the presence of fluid exerts an important influence on the longitudinal wave velocity. However, shear wave velocity change was very small. The fluid affects the shear wave velocity only by adding to the mass of the particles in motion. Therefore, for an evaluation of $V_s$ or $G$ in cohesionless soils, the in-situ unit weight and the effective pressure are considered.

***Frequency and Number of Cycles of Dynamic Load*** Hardin and Black (1969), found that for number of cycles between 1 and 100, the dynamic shear modulus of dry sands increased slightly with number of cycles whereas for cohesive soils the modulus decreased. Low strain shear modulus was found to be practically unaffected by the frequency of loading.

***Magnitude of Dynamic Stress*** The magnitude of dynamic stress controls the shear strain levels induced in the soil, and hence the dynamic shear modulus should be expected to decrease with increase in the dynamic stress.

***Dynamic Prestrain*** The test data of Drnevich, Hall, and Richart (1967) from torsional vibration type resonant column equipment show that the value of the

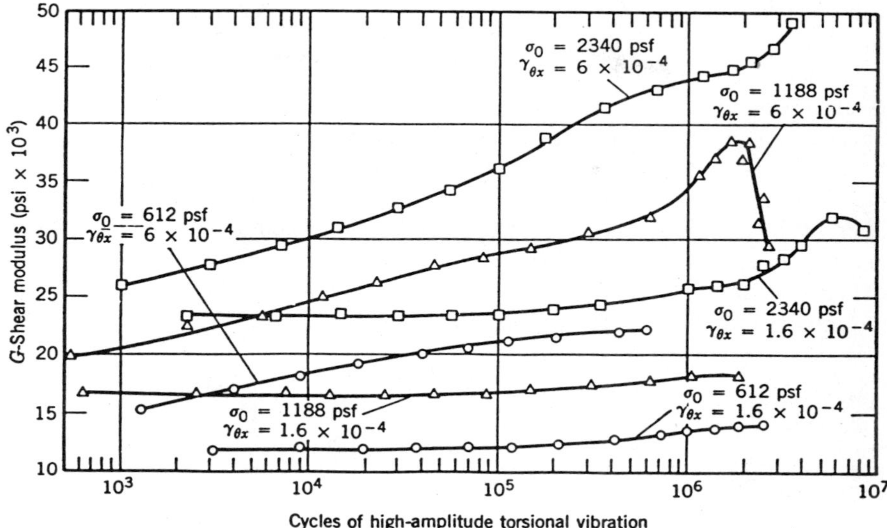


**Figure 4.33** Effect of number of cycles of high-amplitude vibration on the shear modulus at low amplitude (C-190 Ottawa sand, $e_0 = 0.46$, hollow cylindrical specimens) (after Drnevich, Hall, and Richart, 1967).

dynamic shear modulus generally increased with the number of prestrain cycles, as shown in Figure 4.33. The soil samples were first subjected to high-amplitude vibrations (dynamic prestrain) for a predetermined number of cycles and then the low-amplitude vibration modulus was determined. No data are available on the effect of dynamic prestrain on the dynamic shear modulus of clays and silts.

There are several laboratory and field methods for determination of dynamic soil properties that are described briefly as follows.

### 4.2.3 Laboratory Methods

The following laboratory methods are used to determine the dynamic elastic constants and damping values of soils:

1. Resonant column
2. Cyclic simple shear
3. Cyclic torsional simple shear
4. Cyclic triaxial compression

The resonant column test for determining the modulus and damping characteristics of soils is based on the theory of wave propagation in prismatic rods (Richart et al., 1970). Either compression waves or shear waves can be propagated through the soil specimen so that either the Young's modulus or shear modulus is determined.

In such a test, more often a soil sample is subjected to vibrations at the first-mode resonance at which the material in a cross section at every elevation vibrates in phase with the top of the specimen. The shear wave velocity and shear modulus are then determined on the basis of system constants and the size, shape, and weight of the soil specimen (Drnevich et al., 1977). In a resonant column test, different end conditions can be used to constrain the specimen (Figure 4.34). Each configuration requires a slightly different type of driving equipment and methods of data interpretation. In the fixed-free apparatus (Figure 4.34a) the distribution of angular rotation, $\theta$, along the specimen is $\frac{1}{4}$ sine wave, but by adding a mass

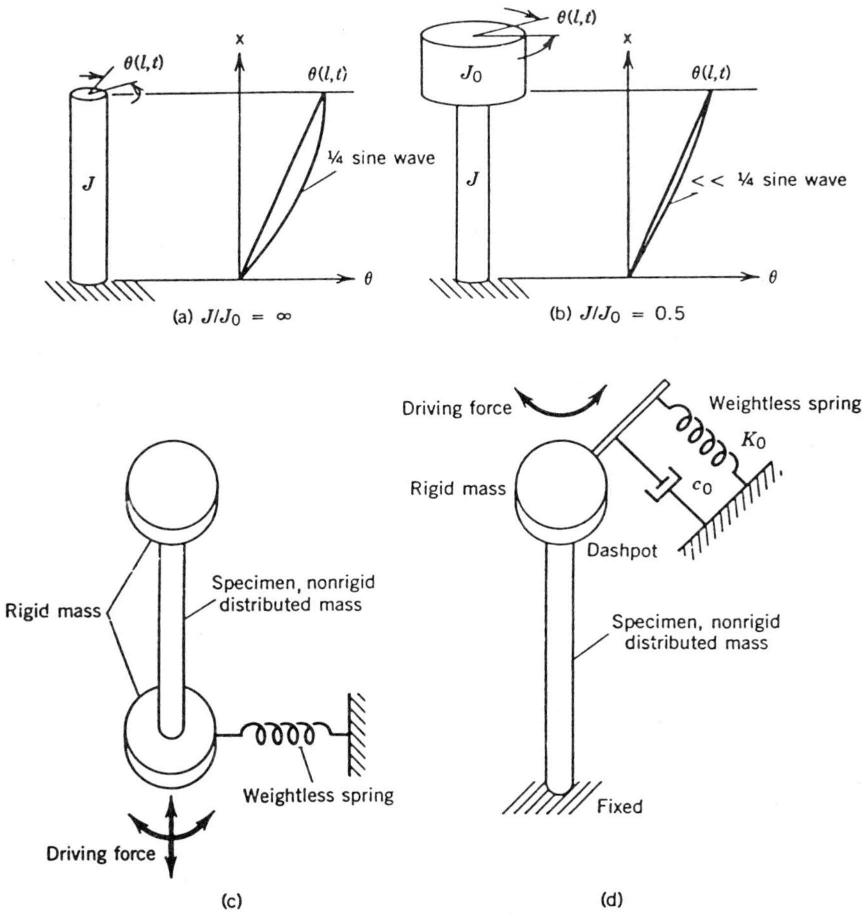


**Figure 4.34** Schematic of resonant column end conditions (after Hardin, 1965, 1970; Drnevich, 1967). (a) $J/J_0 = \infty$, (b) $J/J_0 = 0.5$, (c) free-free (d) fixed base-spring top.

with polar mass moments $J_0$. at the top of the specimen (Figure 4.34b) the variation of $\gamma_\theta$ along the sample becomes nearly linear. Later models of the fixed-free device (Drnevich, 1967) take advantage of end-mass effects to obtain uniform strain distribution throughout the length of the specimen. In Figure 4.34d, the sample has a fixed base and a top cap partially restrained by a spring, which in turn reacts against an inertial mass. If the spring in Figure 4.34c is weak compared to the specimen, this configuration could be called *free-free*. In such a case, a node will occur at midheight of the specimen, and the rotation distribution would be a $\frac{1}{2}$ sine wave. By adding end masses, the rotation distribution can also be made nearly linear. For $K_0 = 1.0$ tests, the inertial mass is balanced by a counterweight, but if one changes the counterweight, an axial load can be applied to the specimen.

In Figure 4.35, a hollow cylinder is used for test so that the shearing strain is

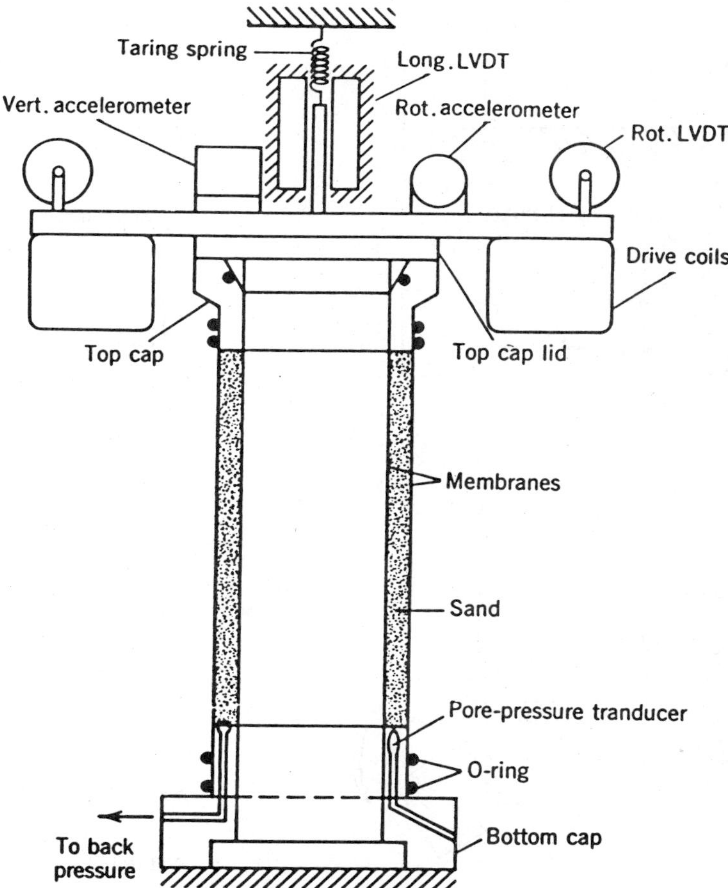


**Figure 4.35** Hollow specimen resonant column and torsional shear apparatus (after Drnevich, 1972).

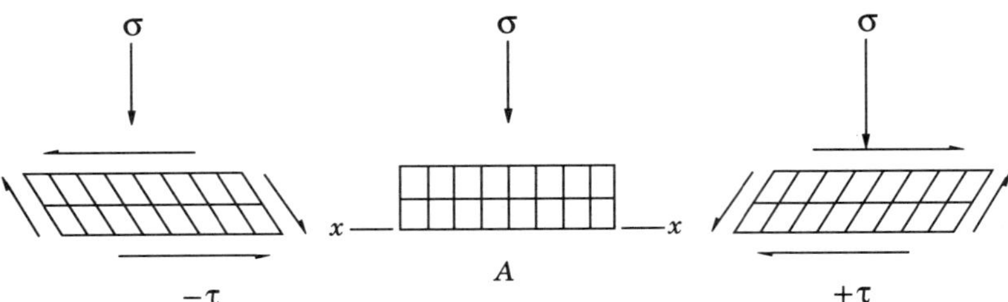


**Figure 4.36** Idealized stress conditions for element of soil below ground surface during an earthquake.

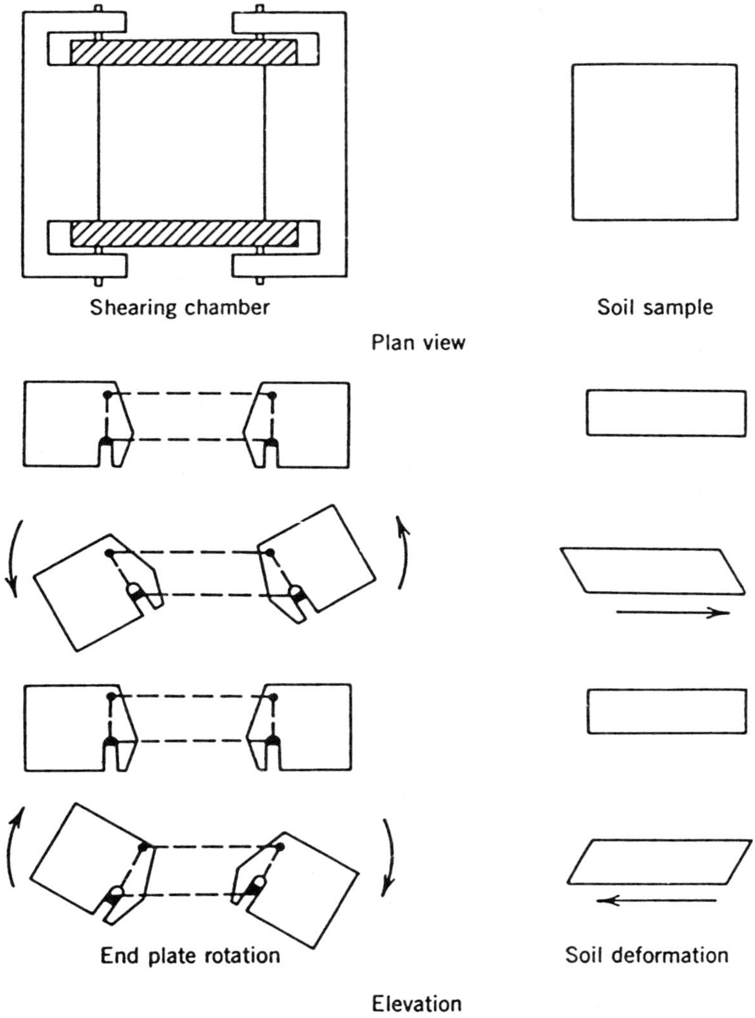

Shearing chamber

Soil sample

Plan view

End plate rotation

Soil deformation

Elevation

**Figure 4.37** Schematic diagram illustrating rotation of hinged end plates and soil deformation in oscillatory simple shear (after Peacock and Seed, 1968).

more or less uniform along the height of the specimen. Unlike the strain distribution in a solid sample with zero strain in the center and maximum at the periphery (Drnevich, 1967, 1972), the torque capacity of this device was increased to produce large shearing strain amplitudes. Anderson (1974) used a modified Drnevich apparatus to test clays at shearing strain amplitudes up to 1 percent. Woods (1978) tested dense sands on the same device at shearing strain amplitudes up to 0.5 percent at 40 psi (276 kN/m$^2$) confining pressure. Drnevich et al. (1977) described a calibration procedure and aids for reducing data of compression or shear wave propagation along a cylindrical sample.

A soil element at $xx$, as indicated in Figure 4.36, may be considered to be subjected to a series of cyclic shear stresses, which may reverse many times during dynamic loading. In the case of a horizontal ground surface, there is no initial shear stress on the horizontal plane.

In practice, initial static shear stresses are present in the soil ($k_0$-initial condition). Oscillatory shears may be introduced due to ground motion or a machine load at the surface of the ground. A simple shear device simulates all these loadings and consists of a sample box, an arrangement for applying a cyclic load to the soil, and an electronic recording system (Figure 4.37), Peacock and Seed (1968). Kjellman (1951), Hvorslev and Kaufman (1952), Bjerrum and Landra (1966), and Prakash et al. (1973) have described this type of apparatus.

Typical shear-stress, shear-strain relationships obtained during cyclic simple shear tests are shown in Figure 4.38a. A soil exhibits nonlinear stress–strain behavior. For purposes of high-stress, high-strain loading as in an earthquake, this behavior can be represented by a bilinear model (Figure 4.38b) defined by three parameters: (1) modulus $G_1$ until a limiting strain, $\gamma_y$, is reached, (2) modulus $G_2$ beyond strain $\gamma_y$, and (3) strain $\gamma_y$ (Thiers and Seed, 1968).

Typical simple shear stress–strain plots of San Francisco Bay mud for different cycles of loading are shown in Figure 4.39 for cycles 1, 50, and 200, with about 4 percent shearing strain. The decrease in peak load as the number of cycles increase is reflected by the progressive flattening of the stress–strain curves. However, corrections for confining pressure and other factors need to be applied, as described in section 4.2.2.

A major drawback of most of the cyclic simple shear apparatus is that they do not permit measurement or control of lateral confining pressures during cyclic loading. Therefore, the value of $k_0$ is not known and hence the effect of the $K_0$ condition on the behavior of soils cannot be studied.

Cyclic torsional simple shear is used to provide the capability of measuring confining pressure and controlling $K_0$ conditions. Ishihara and Li (1972) modified a triaxial apparatus to provide torsional straining capabilities. As in resonant column sample, the shear strain distribution in a hollow sample is more uniform.

The apparatus configuration (Figure 4.35) has an advantage in that both resonant column and cyclic torsional shear tests can be performed in the same device. For details refer to Woods (1978), Iwasaki et al., (1977) and Prakash and Puri (1988).

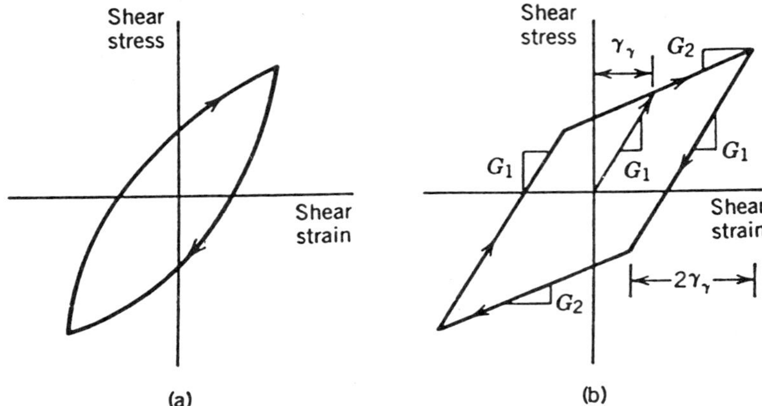


**Figure 4.38** (a) Stress–strain curve of a soil, (b) bilinear model (after Thiers and Seed, 1968).

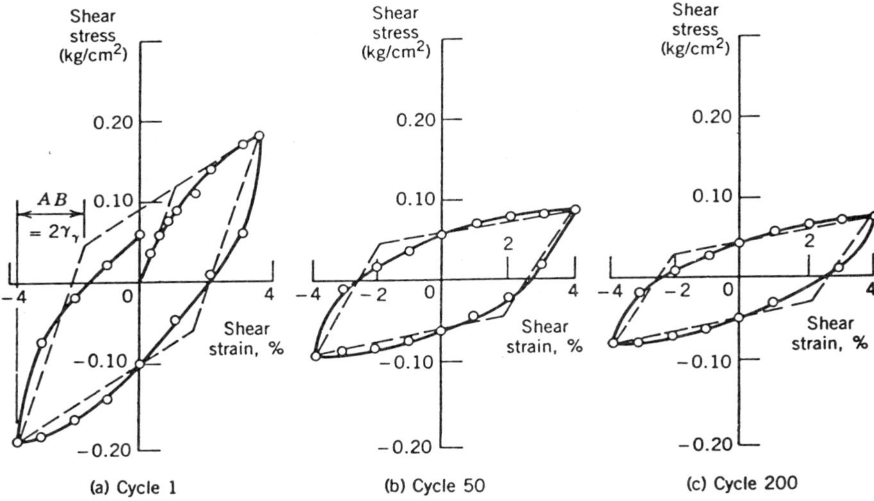


**Figure 4.39** (a) Stress–strain curves and bilinear models in San Francisco Bay mud (a) Cycle No. 1, (b) cycle No. 50, (c) cycle No. 200 (after Thiers and Seed, 1968).

Cyclic triaxial tests have been extensively used to study the stress-deformation behavior of saturated sands and silts (Puri, 1984), and Seed (1979). Also, Young's modulus, $E$, and the damping ratio, $\xi$, have often been measured in cyclic triaxial tests (Figure 4.40) when *strain-controlled* tests have been conducted. These tests are performed in essentially the same manner as the stress-controlled tests for liquefaction studies.

As in all laboratory attempts to duplicate dynamic field conditions, cyclic triaxial tests have the following limitations:

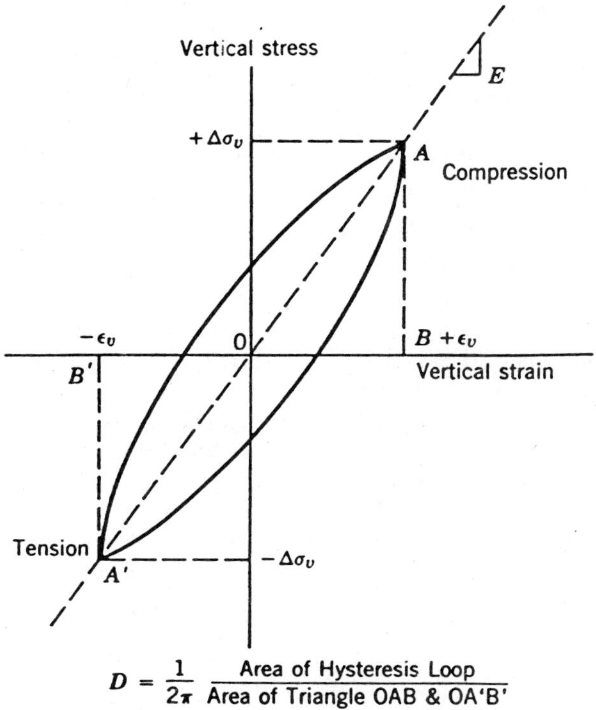


**Figure 4.40** Equivalent hysteretic stress–strain properties from cyclic triaxial test.

1. Shearing strain measurements below 1 percent are generally difficult.
2. The extension and compression phases of each cycle produce different results (Annaki and Lee, 1977); therefore, the hysteresis loops are not symmetric in strain-controlled tests. In stress-controlled tests, the samples tend to neck.
3. Void ratio redistribution occurs within the specimen during cyclic testing (Castro and Poulos, 1977).
4. Stress concentrations occur at the cap and base of the specimen being tested.
5. The principal stress changes direction by 90° during each cycle.

Void ratio redistribution is common to all cyclic shear tests, whereas the other limitations are related mostly to the cyclic triaxial test.

For details on laboratory methods, the reader is referred to Woods (1978), Silver (1981), Puri (1984), and Prakash and Puri (1988).

There are several available field methods with which the dynamic soil properties and damping of soils can be determined. Salient features of these methods will now be described.

### 4.2.4 Field Methods

The following methods for determining dynamic properties of soil are in use:

1. Cross-borehole wave propagation test
2. Up-hole or down-hole wave propagation test
3. Surface wave propagation test
4. Standard penetration test
5. Footing resonance test
6. Cyclic plate load test

Brief descriptions of these tests are presented here. For details, the reader is referred to Prakash and Puri (1988).

In the cross-borehole method, the velocity of shear wave propagation ($V_s$) is measured from one borehole to another (Stokoe and Woods, 1972). A minimum of two boreholes are required, one for generating an impulse and the other for the sensors. In Figure 4.41, the impulse rod is struck on top, causing an impulse to travel down the rod to the soil at the bottom of the hole. The shearing between the rod and the soil creates shear waves that travel through the soil to the vertical motion sensor in the second hole; and the time required for a shear wave to

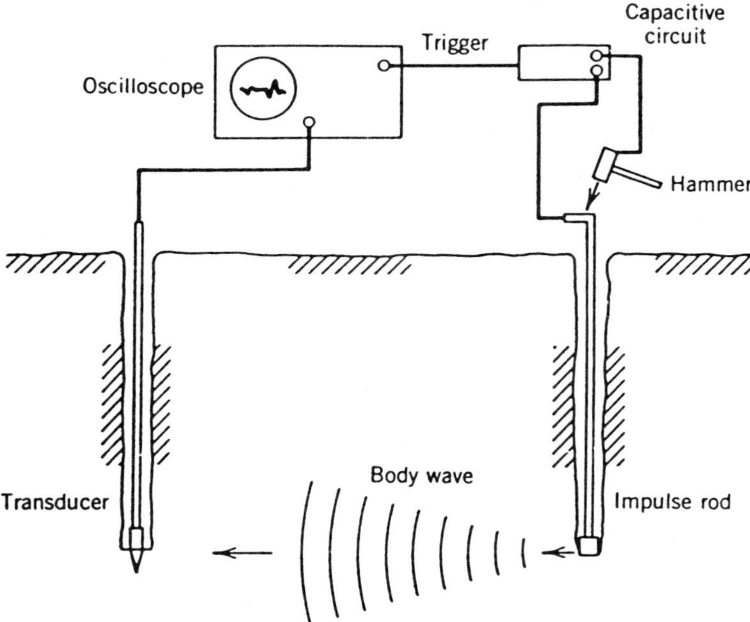


**Figure 4.41** Sketch showing cross-bore hole technique for measurement of velocity of wave propagation.

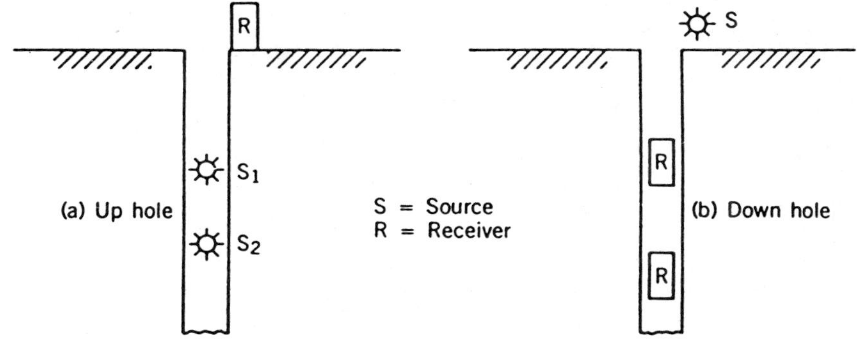


**Figure 4.42** (a) Up-hole and (b) down-hole techniques for measurement of velocity of wave propagation.

traverse the known distance is monitored. Alternatively, shear wave may be generated at any depth in a borehole with a special tool. The arrival of the shear wave is monitored at the same elevation in the second borehole (ASTM D 4428, 1989).

Up-hole and down-hole tests are performed by using only one borehole. In the up-hole method, the sensor is placed at the surface, and shear waves are generated at different depths within the borehole, while in the down-hole method, the excitation is applied at the surface, and one or more sensors are placed at different depths within the hole (Figure 4.42). Both the up-hole and the down-hole methods give average values of wave velocities for the soil between the excitation and the sensor (Prakash and Puri, 1988).

The shear modulus $G$ is then determined as

$$V_s = \sqrt{G/\rho} \tag{4.19a}$$

or

$$G = \rho V_s^2 \tag{4.19b}$$

where $\rho$ is mass density of the soil.

The Rayleigh wave (R wave) travels in a zone one-half to one-third its wavelength below the ground surface (Ballard, 1964). An impact or other harmonic vibration at the surface is used to sample soil for dynamic moduli.

The velocity of the Rayleigh waves, $V_R$, is then given by

$$V_R = \lambda_R \cdot f \tag{4.20}$$

in which $f$ is the frequency of vibration at which the wavelength ($\lambda_R$) has been measured.

It is important to note that the Rayleigh wavelength ($\lambda_R$) will vary with the frequency of excitation ($f$). For smaller $f$, the $\lambda_R$ is larger and the soil will be sampled to a larger depth (Prakash and Puri, 1988; Stokoe and Nazarian, 1985).

**TABLE 4.10 Representative Values of Poisson's Ratio**

| Type of Soil | $v$ |
|---|---|
| Clay | 0.5 |
| Sand | 0.3–0.35 |
| Rock | 0.15–0.25 |

The Rayleigh wave velocity $V_R$ and shear wave velocity $V_s$ are generally approximately equal, therefore:

$$G \simeq V_R^2 \rho \tag{4.21}$$

and

$$E \simeq 2\rho V_R^2 (1 + v) \tag{4.22}$$

in which $\rho$ is the mass density and $v$ the Poisson's ratio of the soil. Values of $v$ from Table 4.10 may be used.

More recently, the interpratication of surface wave by a method called the *spectral analysis of surface waves* (*SASW*) has been developed (Stokoe and Nazarian, 1985). In the field, two vertical velocity transducers are used as receivers. The receivers are placed securely on the ground surface symmetrically about an imaginary centerline. A transient impulse is transmitted to the soil by means of an appropriate hammer. The range of frequencies over which the receivers should function depends on the site being tested. To sample deep materials, 50 to 100 ft, the receiver should have a low natural frequency, in range of 1 to 2 Hz. In contrast, for sampling shallow layers, the receivers should be able to respond to high frequencies in the range of 1000 Hz or more.

Several tests with different receiver spacing are performed. The distance between the receivers after every test is generally doubled. The geophones are always placed symmetrically about the selected, imaginary centerline. The raw data obtained from the impact test is reduced with the help of a Dynamic Signal Analyzer (DSA) and the inversion curve is obtained.

A typical shear wave profile for a site in which the velocity profiles have been determined both by the crosshole method and SASW method show a good tally between the values measured by the two methods. The SASW method is very economical and less time consuming than the cross-borehole method and has the advantage of complete automation. The detailed description of this technique is given by Nazarian and Stokoe (1984). However, the inversion techniques applicable to soils are still not perfected (1990).

In the Standard Penetration Test (SPT), a standard split spoon sampler is driven with a 140-lb hammer that falls freely through a distance of 30 in. The number of blows for 12 in. of penetration of the split spoon sampler is designated as the $N$ value. This is $N_{measured}$. In a design problem using $N$ values, a correction for effective overburden pressure is applied (Peck et al., 1974). Although the test is designated as a standard test, there are several personal errors as well as errors that are equipment based. Therefore, the use of SPT to measure any soil property has been questioned by many engineers (Woods, 1978). Recent careful studies by Kovacs (1975), Kovacs et al. (1977a, 1977b), Palacios (1977), and Schmertmann (1975, 1977) have described the potential of SPT for obtaining consistent and useful soil properties. Seed (1979) and Seed and Idriss (1982) presented correlations between SPT and observed liquefaction.

Imai (1977) developed a correlation between (uncorrected) $N$ and shear wave velocity, $V_s$(m/sec), in 943 recordings at four urban locations in Japan and established the following relationship:

$$V_s = 91 N^{0.337} \tag{4.23}$$

Then,

$$G = \rho V_s^2 \tag{4.19b}$$

In the above relationship, he converted the $N$ values over 50 or under 1 for the penetrating length at the time of 50 or 1 blows into the number of blows for 30-cm penetration. Prakash and Puri (1981, 1984) successfully applied the above relationship in predicting dynamic soil properties at different depths.

In footing resonance tests and free vibration test, a test footing $1.5 \times 0.75 \times 0.70$ m high is cast either at the surface or in a pit $4.5 \times 2.75$ m at a suitable depth and is excited in vertical or horizontal vibrations.

From the natural frequency determined either in the forced or free footing vibration tests, the soil modulus is determined (Prakash, 1981a; Prakash and Puri, 1988).

The cyclic plate load test is a static test. There is ample evidence to show that in non-cohesive soils, the values of soil modulus from this test match with those from dynamic tests at appropriate strains and confining pressures (Prakash, 1981a; Prakash and Puri, 1988).

### 4.2.5 Selection of Design Parameters

The modulus of a given soil varies with strain and the confining pressure. It is therefore necessary to make a plot of $G$ vs. shear strain. $G$ values are determined at a mean effective confining pressure corresponding to the depth of soil and at a shear strain that may be induced in the soil when the pile is subjected to dynamic load. Prakash (1981a) and Prakash and Puri (1981) used a mean confining pressure $\bar{\sigma}_{01}$, of 1 kg/cm$^2$ or (1000 KN/m$^2$) to reduce the data from different tests to a common confining pressure for comparison purpose only using Equations (4.17) and (4.16c)

$$\frac{G_1}{G_2} = \left(\frac{\bar{\sigma}_{01}}{\bar{\sigma}_{02}}\right)^{0.5} \tag{4.17}$$

$$\bar{\sigma}_0 = \frac{\bar{\sigma}_x + \bar{\sigma}_y + \bar{\sigma}_z}{3} \tag{4.16c}$$

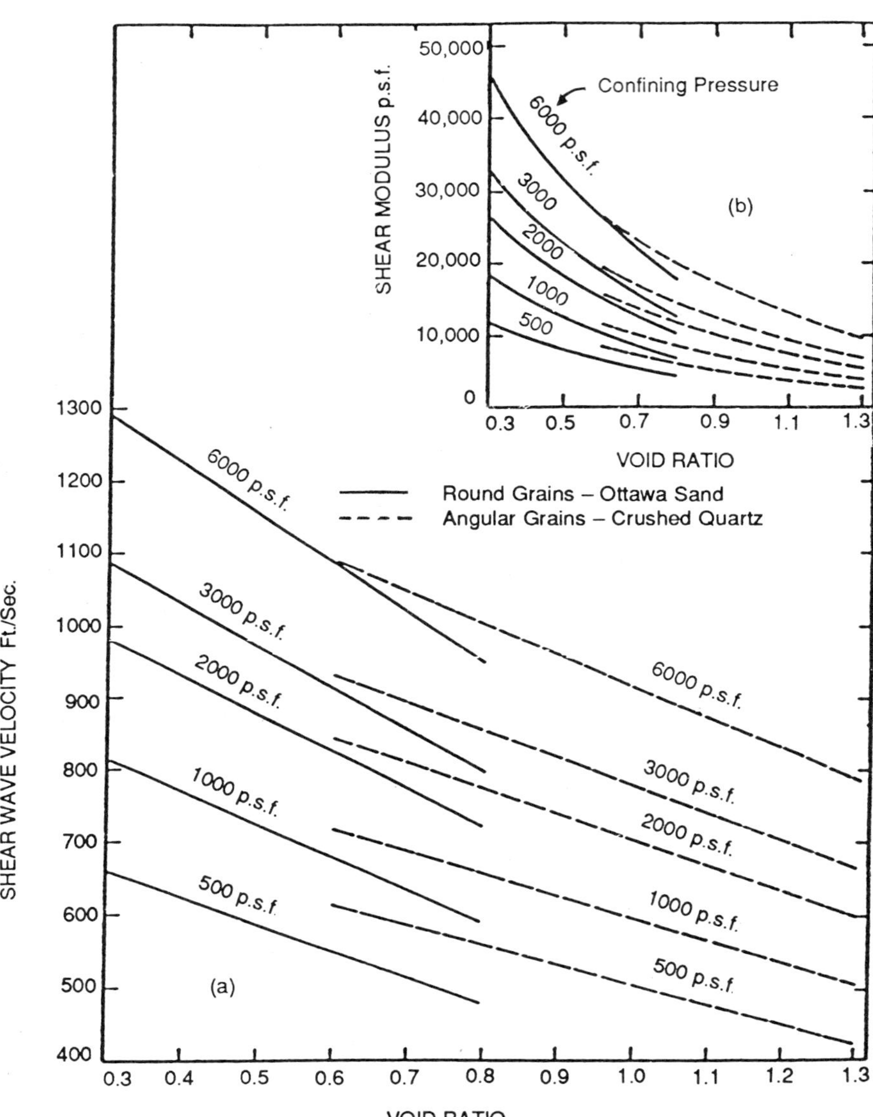


**Figure 4.43** Variation of shear wave velocity and shear modulus with void ratio and confining pressure for dry round and angular sands (After Hardin and Richart, 1963).

where

$$\bar{\sigma}_x = \bar{\sigma}_y = \frac{\nu}{1-\nu} \cdot \bar{\sigma}_z$$

The variation of modulus with strain is determined from different tests and a plot similar to that in Figure 4.28 is obtained. This plot is then used to select the design value at a predetermined strain and confining pressure. In the absence of experimental data, values of shear modulus at low strain for preliminary design may be selected from Figure 4.43 and Equation (4.16a). The following numerical examples explain the selection method of dynamic design parameters.

***Example 4.2*** In a deposit of dry sand with $G = 2.70$ and dry density of 112 lb/ft$^3$, estimate the shear wave velocity at 10, 20, and 30 ft below ground level. Also determine $G_{max}$.

SOLUTION

$$\gamma_d = \frac{G\gamma_w}{1+e} \qquad e = \frac{G\gamma_w}{\gamma_d} - 1 = \frac{2.7 \times 62.4}{112} - 1 = 0.504$$

$$V_s = \sqrt{G/\rho} \qquad \rho = \frac{112}{32.4} = 3.46 \text{ slugs}$$

SHEAR MODULUS Equation (4.16a) will be used to compute $G_{max}$

$$G_{max} = 1230 \quad \text{OCR}^k \frac{(2.973 - e)^2}{1+e} (\bar{\sigma}_0)^{1/2}$$

where $\bar{\sigma}_0$ is effective all-around stress in psi.

$$\bar{\sigma}_0 = (\bar{\sigma}_x + \bar{\sigma}_y + \bar{\sigma}_z)^{1/3} \tag{4.16c}$$

$$= \frac{\bar{\sigma}_v + 2\bar{\sigma}_h}{3}$$

Let $\qquad k_0 = 0.5$

then $\qquad \bar{\sigma}_h = 0.5\,\bar{\sigma}_v$

Also, $\qquad \bar{\sigma}_v = \gamma . z$

*At 10′ Depth*

$$\bar{\sigma}_v = 10 \times 112 = 1120 \text{ lb/ft}^2$$

$$= 0.55 \text{ kg/cm}^2$$

$$\bar{\sigma}_h = \bar{\sigma}_v \times 0.5 = 560\ \text{lb/ft}^2$$
$$= 0.273\ \text{kg/cm}^2$$

$$\bar{\sigma}_0 = 746\ \text{lb/ft}^2 = 5.1851\ \text{lb/in.}^2$$
$$= 0.364\ \text{kg/cm}^2$$

$$G_{\max} = \frac{1230(2.973 - 0.504)^2}{1 + 0.504}(5.185)^{1/2}$$
$$= 4985(5.185)^{1/2}$$
$$= 11{,}352\ \text{lb/in.}^2 = 5.543\ \text{kg/cm}^2$$
$$= 78{,}271\ \text{kN/m}^2$$

$$V_s = \sqrt{\frac{11{,}352 \times 144}{3.46}} = 687.7\ \text{ft/sec} = 209.6\ \text{m/sec}$$

*At 20′ Depth*

$$\bar{\sigma}_v = 20 \times 112 = 2240\ \text{lb/ft}^2$$
$$= 1.094\ \text{kg/cm}^2$$

$$\bar{\sigma}_h = 1120\ \text{lb/ft}^2$$
$$= 0.545\ \text{kg/cm}^2$$

$$\bar{\sigma}_0 = 1493\ \text{lb/ft}^2 = 10.37\ \text{lb/in.}^2$$
$$= 0.73\ \text{kg/cm}^2$$

$$G_{\max} = 4985(10.37)^{1/2} = 16054\ \text{lb/in.}^2 = 7.84\ \text{kg/cm}^2 = 110{,}690\ \text{kN/m}^2$$

$$V_s = \sqrt{\frac{16{,}054 \times 144}{3.46}} = 817.78\ \text{ft/sec} = 249.3\ \text{m/sec.}$$

*At 30′ Depth*

$$\bar{\sigma}_v = 30 \times 112 = 3360\ \text{lb/ft}^2$$
$$= 1.64\ \text{kg/cm}^2$$

$$\bar{\sigma}_h = 1680\ \text{lb/ft}^2$$
$$= 0.82\ \text{kg/cm}^2$$

$$\bar{\sigma}_0 = 2240\ \text{lb/ft}^2 = 15.55\ \text{lb/in.}^2$$
$$= 1.094\ \text{kg/cm}^2$$

$$G_{\max} = 4985(15.55)^{1/2} = 19{,}662\ \text{lb/in.}^2$$
$$= 9.6\ \text{kg/cm}^2 = 135{,}569\ \text{kN/m}^2$$

$$V_s = \sqrt{\frac{19{,}662 \times 144}{3.46}} = 905\ \text{ft/sec} = 275.8\ \text{m/sec}$$

***Example 4.3*** A sand layer in the field is 20 m thick. The groundwater table is located at a depth 5 m below the ground surface. Estimate the shear modulus $G_{max}$ up to a depth of 20 m below the ground surface. The sand has a void ratio of 0.6, a specific gravity of soil solids of 2.7, and Poisson's ratio of 0.3.

SOLUTION

$$\bar{\sigma}_h = \bar{\sigma}_x = \bar{\sigma}_y = \left(\frac{\nu}{1-\nu}\right)\bar{\sigma}_v = \frac{\nu}{1-\nu}\bar{\sigma}_z = 0.428\,\bar{\sigma}_z$$

$$\gamma_d = \frac{G\gamma_w}{1+e} = \frac{2.7}{1+0.6} \times 1$$

$$\gamma_d = 1.6875\ \text{g/cm}^3 = 0.0016875\ \text{kg/cm}^3 = 105.3\ \text{lb/ft}^3$$

$$e = 0.6 \qquad G = 2.7 \qquad \nu = 0.3$$

$$\gamma_t = \left(\frac{G+Se}{1+e}\right)\gamma_w = \frac{2.7+0.6}{1+0.6} \times 1$$

$$= 2.0625\ \text{g/cm}^3 = 0.0020625\ \text{kg/cm}^3 = 128.7\ \text{lb/ft}^3$$

$$\bar{\sigma}_v = \gamma \cdot z \qquad \bar{\sigma}_h = 0.428\bar{\sigma}_v \qquad \bar{\sigma}_0 = \frac{\sigma_v + 2\bar{\sigma}_h}{3}$$

*Depth $z = 5\ m$*

$$\bar{\sigma}_v = \frac{1.6875}{1000}(5)\,100 = 0.84375\ \text{kg/cm}^2$$

$$= 1728\ \text{lb/ft}^2$$

$$\bar{\sigma}_h = 0.3611\ \text{kg/cm}^2 = 739\ \text{lb/ft}^2$$

$$\bar{\sigma}_0 = 0.522\ \text{kg/cm}^2 = 1069\ \text{lb/ft}^2$$

*Depth $z = 10\ m$*

$$\bar{\sigma}_v = \frac{(2.0625-1)}{1000}(5)\,100 + 0.84375 = 1.37475\ \text{kg/cm}^2$$

$$= 2815\ \text{lb/ft}^2$$

$$\bar{\sigma}_h = 0.5884\ \text{kg/cm}^2 = 1205\ \text{lb/ft}^2$$

$$\bar{\sigma}_0 = 0.8505\ \text{kg/cm}^2 = 1742\ \text{lb/ft}^2$$

*Depth $z = 15\ m$*

$$\bar{\sigma}_v = \frac{(2.0625-1)}{1000}(10)\,100 + 0.84375 = 1.90625\ \text{kg/cm}^2$$

$$= 3904\ \text{lb/ft}^2$$

$$\bar{\sigma}_h = 0.81587\ \text{kg/cm}^2 = 1671\ \text{lb/ft}^2$$

$$\bar{\sigma}_0 = 1.1793\ \text{kg/cm}^2 = 2415\ \text{lb/ft}^2$$

*Depth z = 20 m*

$$\bar{\sigma}_v = \frac{(2.0625 - 1)}{1000}(15)\,100 + 0.84375 = 2.4375\ \text{kg/cm}^2$$

$$= 4992\ \text{lb/ft}^2$$

$$\bar{\sigma}_h = 1.04325\ \text{kg/cm}^2 = 2136\ \text{lb/ft}^2$$

$$\bar{\sigma}_0 = 1.508\ \text{kg/cm}^2 = 3088\ \text{lb/ft}^2$$

FOR CLEAN SANDS

$$G_{max} = 700\frac{(2.17 - e)^2}{1 + e}(\bar{\sigma}_0)^{1/2}$$

where $\bar{\sigma}_0$ is expressed in $\text{kg/cm}^2$

$$G_{max} \text{ at (5)m} = 700\frac{(2.17 - 0.6)^2}{1 + 0.6}(0.522)^{1/2} = 799\ \text{kg/cm}^2$$

$$= 1{,}636{,}480\ \text{lb/ft}^2$$

$$= 78{,}355\ \text{kN/m}^2$$

$$G_{max} \text{ at (10)m} = 1078.39(0.8505)^{1/2} = 994.6\ \text{kg/cm}^2$$

$$= 2{,}037{,}100\ \text{lb/ft}^2$$

$$= 97{,}537\ \text{kN/m}^2$$

$$G_{max} \text{ at (15)m} = 1078.39(1.1793)^{1/2} = 1171.05\ \text{kg/cm}^2$$

$$= 2{,}398{,}498\ \text{lb/ft}^2$$

$$= 114{,}840\ \text{kN/m}^2$$

$$G_{max} \text{ at (20)m} = 1078.39(1.508)^{1/2} = 1324.2\ \text{kg/cm}^2$$

$$= 2{,}712{,}174\ \text{lb/ft}^2$$

$$= 129{,}859\ \text{kN/m}^2$$

***Example 4.4*** A uniformly graded dry-sand specimen was tested in a resonant column device with confining pressure of 30 psi. The shear wave velocity $V_s$ determined by torsional vibration of the specimen was 776 ft/sec. The longitudinal wave velocity determined on a similar specimen in longitudinal vibrations was 1275 ft/sec. Determine:

(a) Low-amplitude Young's modulus ($E$) and shear modulus ($G$). The specific gravity of soil solids is 2.7
(b) Poisson's ratio
(c) Estimation of $G_{max}$ at a confining pressure of 15 psi.

SOLUTION

$$V_s = 750\ \text{ft/sec}$$

$$V_c = 1275\ \text{ft/sec}$$

Assuming $\gamma_d = 112\ \text{lb/ft}^3$

Mass density, $\rho = \dfrac{\gamma_d}{g} = \dfrac{112}{32.2} = 3.478 \dfrac{\text{lb} \times \text{sec}^2}{\text{ft}^4}$

(a) $E = \rho V_c^2 = (1275)^2 \times 3.478 = 5{,}654{,}348\ \text{lb/ft}^2 = 2760\ \text{kg/cm}^2$

$G = \rho V_s^2 = (776)^2 \times 3.478 = 2{,}094{,}525\ \text{lb/ft}^2 = 1022\ \text{kg/cm}^2$

(b) $E = 2G(1 + \nu)$

$$\therefore \nu = \frac{E}{2G} - 1$$

$$= \frac{5{,}654{,}348}{2(2{,}094{,}525)} - 1 = 0.35$$

(c) $\dfrac{G_1}{G_2} = \left(\dfrac{\bar{\sigma}_{01}}{\bar{\sigma}_{02}}\right)^{0.5}$

$$\frac{2{,}094{,}525}{G_2} = \left(\frac{30}{15}\right)^{0.5}$$

$$G_2 = 1{,}481{,}053\ \text{lb/ft}^2 = 723\ \text{kg/cm}^2 = 70{,}213\ \text{kN/m}^2$$

## 4.3 SOIL PARAMETERS FOR PERMAFROST

With the development of resources in cold regions of the world, the need for geotechnical information on seasonal and permanently frozen ground has been growing. A great deal of research, design, and construction activity in the past two decades has provided a lot of geotechnical information in this area. Andersland and Anderson (1978), Johnston (1981) and Morgenstern (1983) provide updated and excellent documentation on geotechnical related design and construction data for permafrost areas. This section briefly outlines the geotechnical information from these sources that are relevant for pile design in permafrost area.

### 4.3.1 Northern Engineering Basic Consideration

Permafrost is the thermal condition in soil and rock when the ground stays colder than the freezing temperature of water over at least two consecutive years. *Continuous permafrost* areas are those areas where permafrost occurs everywhere beneath the exposed land surface with the exception of widely scattered sites such as newly deposited unconsolidated sediments. These areas will eventually become permafrost. In *discontinuous permafrost* areas, some areas have permafrost while others are free of permafrost. In the *seasonally frost* areas, the top layer of the ground has temperatures below freezing during the winter and above freezing during rest of the year.

In permafrost areas, foundation loads are often transferred to frozen ground. If these frozen grounds consist of materials such as sound rock, material free of ice-filled fissures, clean well-drained sand, and gravel deposits free of ice, then no special care is required to keep them frozen. On the other hand clays, silty soils, or soils with ice may be subjected to downward movement due to dissipation of water on melting of excess ice in the soil. These soils will require special precautions to maintain them in a frozen state so that thaw degradation does not cause uneven and excessive settlements.

In seasonally frozen soils, structure loads are either taken to depths below the frost depths or the soils in frost zones are replaced with non-frost-susceptible soils (e.g., clean sands and gravel) to avoid problems due to frost action.

***Frost Action in Soils*** Frost action in soils is commonly associated with frost heave and thaw weakening, described as follows:

*Frost Heave* Frost heave occurs in frost-susceptible soils as the freezing front penetrates the soil resulting in the freezing of the pore water and the formation of ice lenses. Freezing of pore water in soils will only cause a volume expansion or heave of 9 percent, which is small. The larger part of the total heave occurs mainly due to the growth of ice lenses at the freezing front due to migration of water towards this front. The supply of moisture for ice lens formation is normally from water within the soil in either of the following ways:

1. From the groundwater table or
2. By the reduction in the water content of the soil near the zone of freezing (i.e., moisture migration from surrounding soil mass).

*Thaw Weakening* Thaw weakening or reduction in bearing capacity of these soils occurs when these ice lenses melt in the spring, and this melt water softens the soils. The situation worsens as the rate of moisture release from thawing of the ice lenses exceeds the rate at which the released moisture can escape. The phenomena of frost heave and thaw weakening is complex and depends on many factors such as soil type, permeability, and the rate of freezing. However, the following three basic conditions must exist for frost action to occur.

1. Existence of a frost-susceptible soil
2. Sufficiently low soil temperatures to cause soil water to freeze
3. Existence of a source of water supply (e.g., a water table)

Items (2) and (3) are self-explanatory. Therefore, item (1), the frost susceptibility of the soils, is discussed as follows.

**TABLE 4.11 U.S. Corps of Engineers Frost Susceptibility Criteria (Johnston, 1981) (a) Frost Design Soil Classification**

| Frost Group | Soil Type | Percentage Finer than 0.02 mm, by Weight | Typical Soil Types under Unified Soil Classification System | Remarks |
|---|---|---|---|---|
| F1 | Gravelly soils | 3 to 10 | GW, GP, GW-GM, GP-GM | |
| F2 | (a) Gravelly soils | 10 to 20 | GM, GW-GM, GP-GM | |
| | (b) Sands | 3 to 15 | SW, SP, SM, SW-SM, SP-SM | Soil types are listed approximately in order of increasing susceptibility to frost heaving and/or thaw weakening (i.e., F1 is stronger and better than F2) |
| F3 | (a) Gravelly soils | $>20$ | GM, GC | |
| | (b) Sands, except very fine silty sands | $>15$ | SM, SC | |
| | (c) Clays, $PI > 12$ | — | CL, CH | |
| F4 | (a) All silts | — | ML, MH | |
| | (b) Very fine silty sands | $>15$ | SM | |
| | (c) Clays, $PI < 12$ | — | CL, CL-ML | |
| | (d) Varved clays and other fine-grained, banded sediments | — | CL and ML<br>CL, ML, and SM<br>CL, CH, and ML<br>CL, CH, ML, and SM | |

**(b) Classification of Frost Susceptibility Based on Laboratory Test**

| Average Rate of Heave (mm/day) | Frost Susceptibility Classification |
|---|---|
| 0.0–0.5 | Negligible |
| 0.5–1.0 | Very low |
| 1.0–2.0 | Low |
| 2.0–4.0 | Medium |
| 4.0–8.0 | High |
| $>8.0$ | Very high |

FROST-SUSCEPTIBLE SOILS The most commonly used criteria for frost susceptibility of soils are based on grain size. The amount of fines and gradation are usually specified as governing criteria for frost susceptibility. For structures where frost heaving is an exceptionally critical factor, laboratory heaving tests carried out at the rate that simulate field conditions should be considered to classify frost susceptibility. Table 4.11 presents both these criteria. The foregoing frost design soil classification system is based on Casagrande's (1932) grain size criterion that suggests that frost action would occur if (1) well-graded soils contain more than 3 percent of grains smaller than 0.02 mm, and (2) uniform soils contain more than 10 percent of particles smaller than 0.02 mm.

Although there is no sharp dividing line between frost-susceptible and non-frost-susceptible soils, Table 4.11 can be used for frost design classification of soils. The mechanism of frost heaving in the active layer in permafrost areas and in seasonal frost areas are generally similar with the exceptions noted in Table 4.12.

### 4.3.2 Properties of Frozen Soils

The behavior of frozen soil under load is usually different from that of unfrozen soil because of the presence of ice and unfrozen water films surrounding the soil particles restricting the interparticle contact. Bonding of particles by ice is the dominant strength factor in frozen soils. Frozen soils therefore exhibit much more time-dependent deformation behavior under constant stress, called *creep*, and the behavior is also temperature dependent. The basic concepts related to the behavior of frozen soils consist of the following factors:

1. **Particle Bonding by Ice** This is the dominant strength factor for frozen soils. For soils that contain ice in excess to that required to fill pore spaces,

**TABLE 4.12 Differences in Frost Action in Permafrost and Seasonal Frost Areas**

| Feature | Permafrost Areas | Seasonal Frost Areas |
|---|---|---|
| Supply of water for ice segregation | Water moves in a closed system within the active layer | Water is supplied from all unfrozen soil below the freezing plane (open system) |
| Freezing mode | Although downward freezing is dominant it can also take place upwards from the permafrost tables | Freezing of soil always occurs downwards from ground surface |
| Thaw period | Lasts through the entire summer | Lasts for a month or so |

called ice rich frozen soils, frozen soil behavior under stress is similar to ice. At lower ice content, when interparticle forces begin to contribute, the unfrozen water films surrounding the soil particles play important role.

2. **Stress Transmission** On application of pressure, stress concentration on ice between soil particles develops. This causes the ice to melt and increases the amount of unfrozen water as pressure increases. This results in flow of water to lower stress areas where it refreezes.

3. **Creep Behavior** When water migrates under a stress gradient the structural and ice cementation bonds break and particles reorient themselves resulting in strength reduction with time. The movement of water causes consolidation resulting in some new ice cementation. This causes strength increase with time. If strength reduction overcomes strength increase, creep rate will accelerate eventually causing failure. If strength increase dominates then steady-state creep may exist for some period of time. This will be further discussed in the following paragraphs.

***Creep Behavior of Frozen Soils*** As shown in Figure 4.44, when stress is applied to frozen soil, the soil responds with either damped or undamped creep behavior. The stress level above which undamped creep occurs is not well defined (Savigny and Morgenstern, 1986). According to Morgenstern et al. (1980) polycrystalline

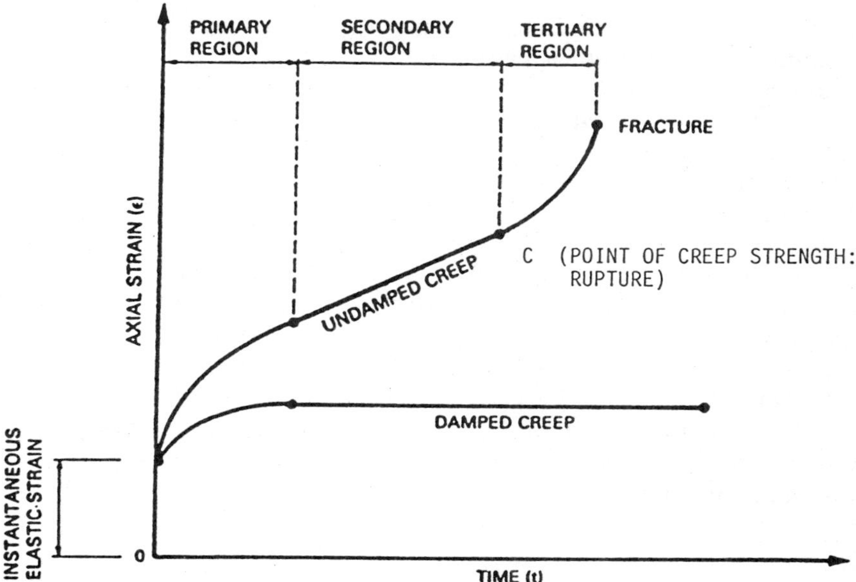


**Figure 4.44** Typical constant axial stress creep curves (after Vialov 1965; Savigny and Morgenstern, 1986).

ice creeps at small stresses; therefore, it generally forms an upper bound for undamped creep. The undamped creep curve, as shown by the upper part of Figure 4.44, has following three distinct regions.

### Primary Creep Region

This is the dominant behavior for frozen soils at low stress levels in low ice content. It means that for a stress level less than some critical value, called long-term strength, the creep rate will tend to be zero.

### Secondary or Steady-State Creep Region

The behavior is dominant in ice-rich soils under moderate stress conditions. In this case, primary creep may be neglected and entire curve may be considered linear. This situation is found in most practical problems.

### Tertiary Creep Region

At high stress levels, the soil may appear to go straight into accelerated creep and may fail after a short period of time. This stress is referred to as the *short-term strength.*

The strain–time curves are obtained from creep tests on frozen cylindrical soil samples subjected to a constant stress. One typical creep test apparatus is illustrated schematically in Figure 4.45. This apparatus consists of a cell placed inside an insulated cabinet. The cabinet can be mounted on the testing frame and the entire apparatus is then placed in a controlled-temperature laboratory. Further details of this apparatus are described by Savigny (1980). The samples are then tested under constant stress at a particular temperature and deformation with time are recorded.

The creep behavior of ice-rich frozen soil can be written by the following relationship (Nixon and McRoberts, 1976):

$$\dot{\varepsilon} = B_1\sigma^{n_1} + B_2\sigma^{n_2} \tag{4.24}$$

where $\sigma$ is uniaxial stress, and $B$ and $n$ are creep parameters dependent on temperature. Generally, the geotechnical engineer works with soil subjected to relatively low stresses (0 to 100 kPa) at temperatures within a few degrees of the melting point of permafrost. In this low-stress range the undamped behavior of ice-rich, fine-grained soils has been interpreted in a simple power law of the following form (Glen, 1952, 1975)

$$\dot{\varepsilon} = B\sigma^n \tag{4.25}$$

where $\dot{\varepsilon}$ is the uniaxial creep rate, $\sigma$ is uniaxial stress, $B$ is a coefficient with dimensions of $(\text{time})^{-1}$ and $(\text{stress})^{-n}$ and $n$ is a dimensionless exponent.

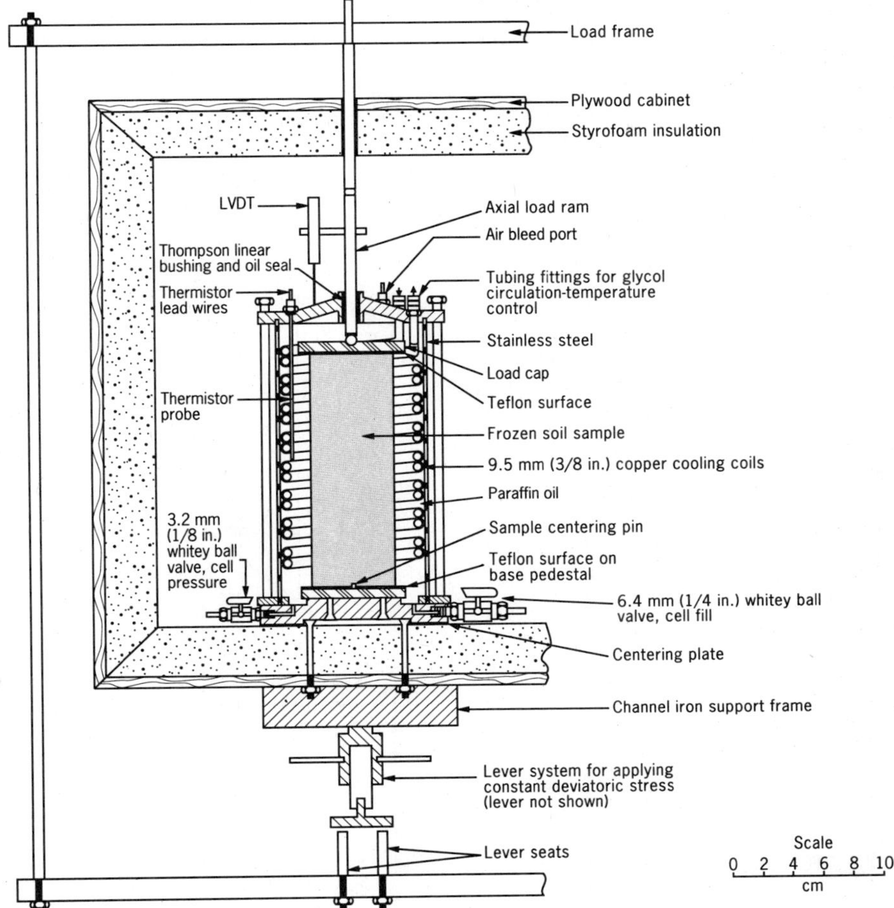


**Figure 4.45** Schematic layout of specially designed triaxial cell used for creep tests (Savigny and Morgenstern, 1986).

Morgenstern et al. (1980) reviewed the data on parameters $B$ and $n$, summarized in Table 4.13. Nixon and Lem (1984) provide the variation of $B$ with salinity and temperature, given in Figure 4.46.

***Creep Strength in Uniaxial Compression*** As shown by point $C$ in Figure 4.44, in constant stress creep testing the uniaxial compression creep strength is the stress at the time at which steady-state creep changes to accelerating creep. In a constant strain rate test, this sign of instability coincides with the first drop of strength after the peak of the stress–strain curve. Thus, creep strength is defined as the stress at which either rupture or instability leading to rupture occurs in the material after a finite time interval. There is little published data on the long-term

**TABLE 4.13 Creep Parameters $B$ and $n$ (Morgenstern et al., 1980)**

| Temperature (°C) | $B$ $(kPa^{-n})(yr^{-1})$ | $n$ |
|---|---|---|
| −1 | $4.5 \times 10^{-8}$ | 3.0 |
| −2 | $2.0 \times 10^{-8}$ | 3.0 |
| −5 | $1.0 \times 10^{-8}$ | 3.0 |
| −10 | $5.6 \times 10^{-9}$ | 3.0 |

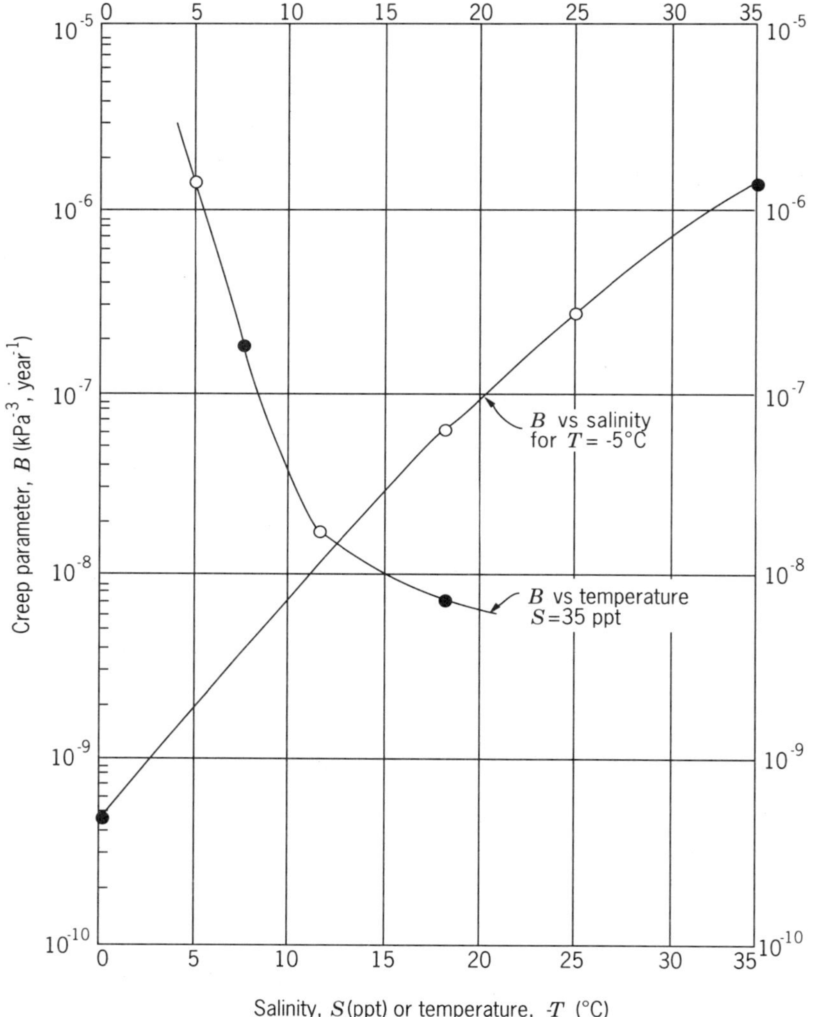


**Figure 4.46** Relationship between creep parameter $B$, salinity, and temperature (Nixon and Lem, 1984).

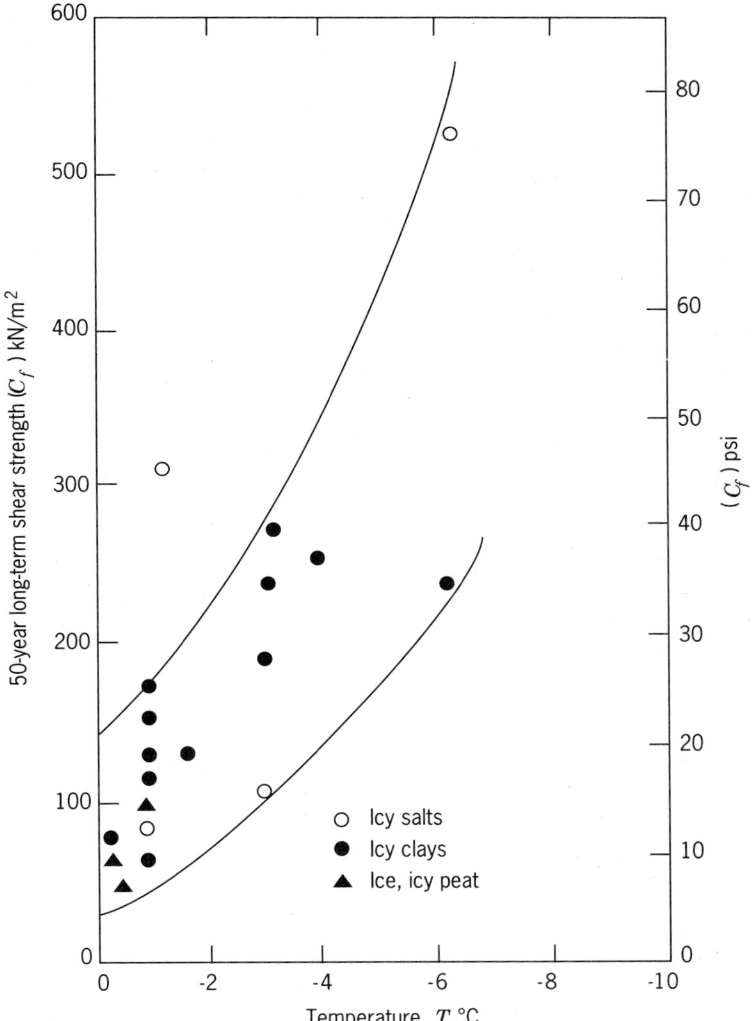


**Figure 4.47** Long-term shear strength for ice-rich soil (McRoberts, 1982).

shear strength for ice-rich soils. Various available data are presented in Figure 4.47 and can be used in absence of specific testing (McRoberts, 1982). For ice-poor soils where the friction angle is mobilized due to grain contact, it would be appropriate to use a friction angle equal to the value of similar unfrozen soil.

Creep and strength properties of frozen soils are strongly influenced by their temperature and ice content. Table 4.14 gives values of short and long-term uniaxial compressive and tensile strengths for some typical frozen soils.

**TABLE 4.14 Short-Term and Long-Term Strengths of Some Typical Frozen Soils and Ice (Voitkovskiy, 1968)[a]**

| | | | Uniaxial Strength, $kg/cm^2$ | | | |
|---|---|---|---|---|---|---|
| | | | Short Term | | Long Term | |
| Soil type | Total Water Content (%) | Temp. (°C) | Compressive | Tensile | Compressive | Tensile |
| Medium and fine Sand | 17–23 | −3 | 60–70 | 17 | 6.5 | 1.8 |
| Silty sand | 20–25 | −0.3 | 10–12 | 5–8 | 2–3 | 1.0–1.5 |
| | | −5.0 | 30–40 | 20–25 | 6–10 | 3–5 |
| | | −10.0 | 60–70 | 40–50 | 35 | 11 |
| | | −20.0 | 120–140 | 50–60 | 60 | 21 |
| Clayey silt | 20–25 | −5.0 | 23 | 20 | 20 | 9–12 |
| | | −10.0 | 39 | 30 | 25 | 12–15 |
| | | −20.0 | 66 | 40 | 40 | 16–20 |
| | 30–35 | −3.0 | 30–35 | 12–16 | 3.6 | 2.5 |
| | 35–40 | −0.5 | 8–10 | 4–6 | 2 | 1–2 |
| Clay | 25–35 | −1.0 | 15 | 5 | — | 1.6 |
| | | −5.0 | 35 | 13 | — | 5.0 |
| Polycrystalline ice | 100 | −3.0 | 16–20 | 10–12 | 0 | 0 |
| | | −10.0 | 32–40 | 17–20 | 0 | 0 |

[a] As cited by Johnston (1981).

**TABLE 4.15 Coefficient of Compressibility ($m_v$) of Frozen Soils (Tsytovich, 1975)[a]**

| Soil type | Total Water Content in Frozen Soil (%) | Unfrozen Water Content at a Given Temperature (%) | $T$ (°C) | $m_v$ ($cm^2/kg \times 10^{-4}$) Load ($kg/cm^2$) 0–1 | 1–2 | 2–4 | 4–6 | 6–8 |
|---|---|---|---|---|---|---|---|---|
| Medium sand | 21 | 0.2 | −0.6 | 12 | 9 | 6 | 4 | 3 |
| | 27 | 0.0 | −4.2 | 17 | 13 | 10 | 7 | 5 |
| | 27 | 0.2 | −0.4 | 32 | 26 | 14 | 8 | 5 |
| Silty sand, massive structure | 25 | 5.2 | −3.5 | 6 | 14 | 18 | 22 | 23 |
| | 27 | 8.0 | −0.4 | 24 | 29 | 26 | 18 | 14 |
| Medium silty clay, massive structure | 35 | 12.3 | −4.0 | 8 | 15 | 26 | 28 | 24 |
| | 32 | 17.7 | −0.4 | 36 | 42 | 37 | 21 | 14 |
| Medium silty clay, reticulate structure | 42 | 11.6 | −3.8 | 5 | 10 | 18 | 42 | 32 |
| | 38 | 16.1 | −0.4 | 56 | 59 | 39 | 24 | 16 |
| Medium silty clay, layered structure | 104 | 11.6 | −3.6 | 54 | 54 | 59 | 44 | 34 |
| | 92 | 16.1 | −0.4 | 191 | 137 | 74 | 36 | 18 |
| Varved clay | 36 | 12.9 | −3.6 | 15 | 22 | 26 | 23 | 19 |
| | 34 | 27.0 | −0.4 | 32 | 30 | 25 | 20 | 16 |

[a] As cited by Johnston (1981).

***Compressibility of Frozen Soils*** Compressibility of frozen soils can generally be neglected since they are practically incompressible when compared with creep deformation. However, in cases where large areas are loaded, the compressibility of frozen soils can be significant. Table 4.15 gives some data for the volume compressibility ($m_v$) for various soils at different temperatures. These values can be used where site specific test data are not available.

## 4.4 MODULUS OF HORIZONTAL SUBGRADE REACTION

In analysis of piles under lateral loads, two stiffness parameters are needed: (1) the flexural stiffness of the pile (EI) and (2) the horizontal stiffness of the soil, $E$, $G$, or $K_h$.

If the theory of elasticity is used, the soil stiffness is expressed by Young's modulus $E$ or shear modulus $G$ (Section 4.2.1). However, soil stiffness may also be defined by the *modulus of horizontal subgrade reaction* (lb/in.$^2$) as:

$$k_h = p/y \tag{4.26}$$

where

$p$ = soil reaction at a point on the pile per unit of length along the pile and is the *resultant for width B of the loaded face* (lb/in.)

and

$y$ = deflection at that point (in.)

The actual soil reaction thus becomes independent of the soil continuity and the soil may be assumed to be replaced by *closely spaced independent elastic springs* (Winkler, 1867, see Figure 6.6b).

Figure 4.48 shows a typical soil reaction versus deflection curve ($p$–$y$ curve) for soil surrounding a laterally loaded pile. For soil reactions less than one-third to one-half of the ultimate soil reaction, the $p$–$y$ relationship can be expressed adequately by a tangent modulus. The slope of the line is the coefficient of horizontal subgrade reaction for the pile, $k_h$. For soil reactions exceeding approximately one-third to one-half of the ultimate soil reaction, the secant modulus shown by the dashed line on Figure 4.48 should be considered; in this case, the modulus becomes a function of the deflection. Matlock and Reese (1961) presented an analysis for laterally loaded piles wherein the subgrade modulus can vary with deflection (see Chapter 6). The actual variation of the subgrade modulus with depth is shown in Figure 4.49. The dashed line in Figure 4.49a illustrates the variation of $k_h$ with depth ($k$ = constant) that has been recommended by Terzaghi (1955) for uniform preloaded cohesive soils. Because of the presence of the soil boundary, the soil at the ground surface is deformed in a mode similar to that for two-dimensional passive pressure. However, at depths

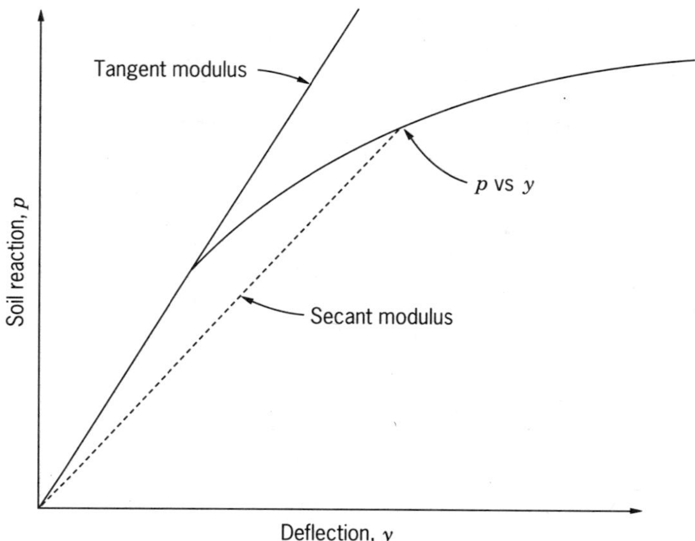


**Figure 4.48** Soil reaction vs. deflection for soil surrounding a pile.

greater than four to six pile diameters below the ground surface, the lateral deflection of the pile deforms the soil in a mode similar to that for a deep two-dimensional footing (see Chapter 1). Both the stiffness and the ultimate soil resistance are, therefore, lower near the soil boundary (Davisson, 1963).

A more realistic variation of the subgrade modulus with depth for preloaded cohesive soils is shown by the solid line in Figure 4.49a.

For granular soils, Terzaghi (1955) recommends that $k_h$ be considered directly proportional to the depth $x$ as shown on Figure 4.49b. The expression for $k_h$ in this case is

$$k_h = n_h x \tag{4.27}$$

where $n_h$ is the *constant of horizontal subgrade reaction* expressed in lb/in.$^3$ The validity of Terzaghi's recommendation for sands has been demonstrated on a model scale by Prakash (1962). The actual variation of $k_h$ with depth is indicated schematically by the solid line in Figure 4.49b, Prakash's tests also indicated that this variation is realistic.

Davisson (1960) has shown that $k_h$ is proportional to depth for normally loaded clays, whereas Peck and Davisson (1962) have shown that the assumption is also valid for normally loaded silts. It is convenient to note that $k_h$ varies with depth in a manner similar to the variation of the ultimate soil resistance with depth.

Therefore, it would appear probable that variation of $k_h$ with depth can be estimated. For example, in a desiccated normally loaded clay the variation of $k_h$

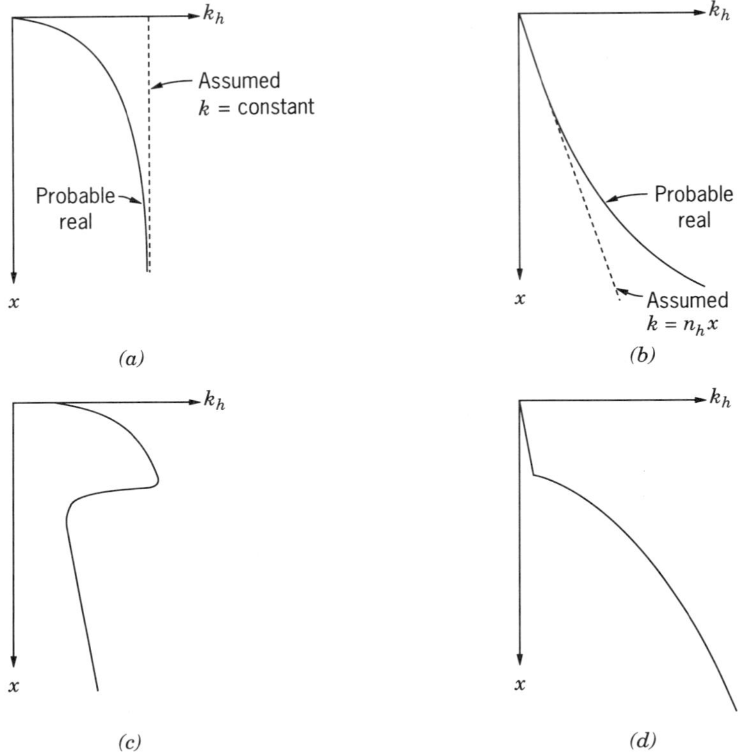


**Figure 4.49** Variation of subgrade modulus with depth. (a) preloaded cohesive soils; (b) granular soils, normally loaded silts, and clays; (c) desiccated normally loaded clay; (d) soft surface layer (Davisson, 1963).

with depth would be approximately as shown on Figure 4.49c. For a preloaded cohesive soil with a layer of soft weak soils at the surface, the variation of $k_h$ with depth shown on Figure 4.49d is appropriate. (Davisson, 1963).

### 4.4.1 Validity of Subgrade Modulus Assumption and Size Effects

In the theory of a subgrade modulus, the soil stiffness is represented by a series of independent elastic springs (Figure 6.6), while in reality they are interrelated in a complex fashion. Vesic (1961) extended Biot's (1937) work concerning a flexible beam supported on an elastic half-space and showed that, for a long relatively flexible member such as a pile, the error in the computed bending moments based on the subgrade modulus assumption is no more than a few percent when compared to the theory of elasticity solution. Therefore, the subgrade modulus concept has a reasonable theoretical foundation and has been extensively used for computing response of piles under lateral loads. Terzaghi (1955) presented an

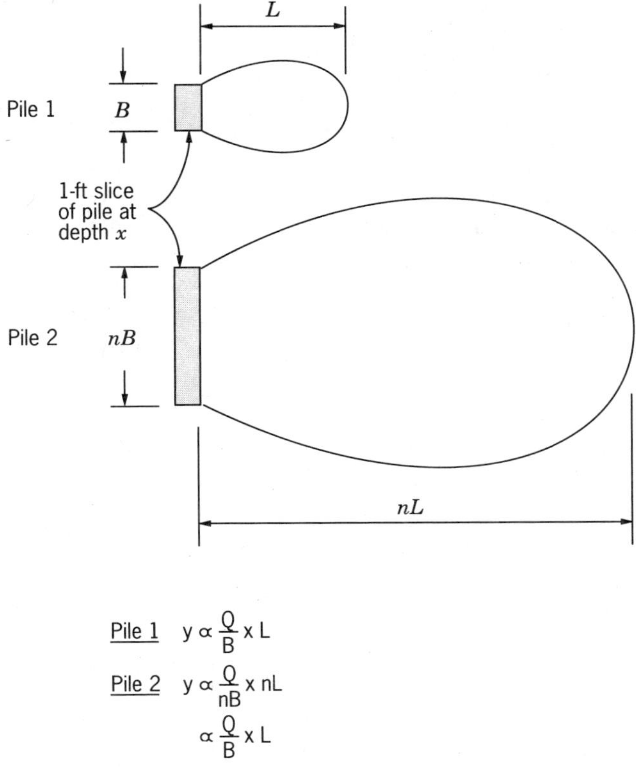


**Figure 4.50** Influence of pile width on dimensions of bulb of pressure.

extensive discussion regarding the effect of the size of the loaded area on the subgrade modulus.

Consider the bulb of pressure behind a unit length of a pile at depth $x$ below the ground surface as shown in Figure 4.50. If the pile width $B$ is increased to $nB$, then the length of the pressure bulb is also increased from $L$ to $nL$. The deflection is approximately proportional to the length of influence of the pressure bulb and the average pressure. For a given load $Q$ *per unit length of pile*, the pressure for width $B$ is $Q/B$, and for width $nB$, the pressure is $Q/nB$. Then the deflection $y$ is proportional to $Q/B \times L$ for the first case and $Q/nB \times nL = Q/B \times L$ for the second case, giving approximately the same net effect. Therefore, after $k_h$ has been determined for a given pile, its value is unchanged if the pile width $B$ is changed (Davisson, 1963). This assumption is considered reasonable for the range of pile widths used in most engineering problems that is, 20 to 90 cm, (8 in. to 3 ft) (Robinson, 1979).

This discussion concerning the size of the loaded area is based on the assumption of linear elastic behavior for the soil. Actually, plastic soil behavior will most likely be initiated at the ground surface, especially for pile heads that can translate, because the soil is weakest at that point. If a pile of width $B$ were to be widened, for example, to a width $2B$ for some depth below the ground surface, strictly elastic considerations would indicate no change in the load deflection behavior. However the ultimate soil reaction for the enlarged portion of the pile is approximately doubled because it depends on the pile width; this has the effect of increasing the secant modulus (Figure 4.48) because the soil reaction is now a smaller percentage of the ultimate soil reaction. Therefore, the observed deflections would be somewhat reduced. It may be concluded that there is some effect of the pile width on the subgrade modulus (Davisson, 1963).

It may be important to mention the following points in support and against this approach (Reese and Matlock, 1956):

1. Accuracy of $k_h$ or $n_h$ is not critical. A 32:1 variation in $n_h$ is necessary to produce a 2 to 1 variation in moment in piles.
2. The assumption of some function that results in zero pressure at the mud line is a logical one. Considering the soil to fail as a wedge that moves up and out near the mud line (Figure 1.16a), it may be concluded that the limiting pressure at the mud line must be close or equal to zero.

This is particularly valid for sand or soft muck at the mud line. However, a rational consideration of the phenomena leads to the conclusion that $k_h$ is essentially empirical in nature and may vary with a number of parameters (e.g. (1) deflection, (2) depth, (3) diameter of pile, (4) type of loading, (5) rate of loading, and (6) and number of load applications.

### 4.4.2 Recommended Design Values of Soil Modulus

Terzaghi (1955) recommended the values of $k_h$ for stiff clays ($k_h$ constant with depth Figure 4.49a) for a one-foot-square plate as 75 to 300 ton/ft$^3$. Converting these values to horizontal subgrade reactions and in turn to the terms and units used in this section, values of $k_h$ are found to range from 58 to 232 lb/in.$^2$ (Reese and Matlock, 1956). For sands, the values of the constant of horizontal subgrade reaction $n_h$ are given from 4 to 56 ton/ft$^3$. Converting these values to the units of this section, values of $n_h$ are computed to be 4.6 to 69 lb/in.$^3$.

Typical values for $k$ are available for a wide variety of soils. For a given soil, $k_h$ increases, as density increases, as would be expected. The values for $k_h$ given in Table 4.16a (Davisson, 1970) are on the basis of simple soil tests, such as the standard penetration test or the unconfined compressive strength, from which reasonable values of $K_h$ may be selected. Reese et al. (1974) have recommended values of soil modulus which are also listed in Table 4.16(b).

In most of the pile analyses according to horizontal subgrade reaction used in the United States, the preceding concepts and recommended values are used. However, the *Canadian Foundation Engineering Manual* (1985) uses the following terms:

**TABLE 4.16a Estimated Values for $k_h$[a]**

| Soil Type | Values |
|---|---|
| Granular | $n_h$ ranges from 1.5 to 200 lb/in.$^3$, is generally in the range from 10 to 100 lb/in.$^3$, and is approximately proportional to relative density |
| Normally loaded | |
| organic silt | $n_h$ ranges from 0.4 to 3.0 lb/in.$^3$ |
| Peat | $n_h$ is approximately 0.2 lb/in.$^3$ |
| Cohesive soils | $k_h$ is approximately 67 $S_u$, where $S_u$ is the undrained shear strength of the soil |

[a]After Davisson, 1970.

*Note*: The effects of group action and repeated loading are not included in these estimates.

**TABLE 4.16b Recommended Values of $n_h$ for Submerged Sand**

| Relative Density | Loose | Medium | Dense |
|---|---|---|---|
| 1. Terzaghi (1955) | | | |
| Range of values of $n_h$ (lb/in.$^3$) | 2.6–7.7 | 7.7–26 | 26–51 |
| 2. Reese et al. (1974) | | | |
| (Static and Cyclic Loading) | | | |
| Relative Density | Loose | Medium | Dense |
| Recommended $n_h$ (lb/in.$^3$) | 20 | 60 | 125 |

### *Cohesive Soils*

$$k_s = 67\frac{S_u}{B} \tag{4.28a}$$

and

$$R = 4\sqrt{\frac{EI}{K_S B}} \tag{4.28b}$$

where

$k_s$ = coefficient of horizontal subgrade reaction (force per unit volume)
$S_u$ = undrained shear strength of soil
$B$ = pile diameter
$EI$ = flexural rigidity of the pile

**TABLE 4.17 Comparison between Suggested and Observed Values of $k_h$ and $n_h$ at Load of 3 Tons Applied at the Ground Surface Against Displacement Piles**

| Soil Conditions | | | | Horizontal Subgrade Reaction | | | |
|---|---|---|---|---|---|---|---|
| | | | | | | Computed[f] | |
| Test No. | Soil Description | $N$[a] | $S_u$[b] | Horizontal[c] Movement (in.) | Type[d,e] | From Deflection | From Earlier Estimate |
| 1 | 2 | 3 | 4 | 5 | 6 | 7 | 8 |
| 1 | Amorphous peat | < 1 | — | 1.04 | $k_h$ | 100 | 70 |
| | | | | | $n_h$ | 4.2 | < 1 |
| 2 | 3 ft sand | — | — | 0.38 | $k_{h1}$ | 500 | 400 |
| | over amorphous peat | | | | $k_{h2}$ | 100 | 70 |
| 3 | 4 ft gravelly clay | — | 800[g] | 0.31 | $k_{h1}$ | 740 | 370 |
| | Over clayey silt | 1.5 | 400[g] | | $k_{h2}$ | 370 | 185 |
| 4 | 5 ft stiff clay | 3 | 1200[g] | 0.37 | $k_{h1}$ | 500 | 270 |
| | Over silt and peat | 1 | 400[g] | | $k_{h2}$ | 180 | 90 |
| 5 | Organic clayey silt | < 1 | 300 | 0.60 | $k_h$ | 300 | 140 |
| | | | | | $n_h$ | 16.5 | 1 |
| 6 | Layered silty sand and sandy silt | 3 | — | 0.90[h] | $n_h$ | 30 | 5.6 |
| 7 | Layered sand and sandy silt | 5 | — | 0.25 | $n_h$ | 62 | 8 |

| | | | | | | | |
|---|---|---|---|---|---|---|---|
| 8 | 3.5 ft sand | 10 | — | 0.11 | $n_h$ | 256 | 13 |
| | Over clayey silt | 4 | | | | | |
| 9 | Silty sand | 5 | — | 0.24 | $n_h$ | 100 | 8 |
| 10 | Slightly organic silt | 2 | — | 0.64 | $n_h$ | 15 | 3.8 |
| 11 | 3 ft organic silt | 1 | — | 0.68[h] | $n_h$ | 34 | 5.6 |
| | Over sandy silt | 3 | — | | | | |

After Robinson 1979.

[a]Average standard penetration blow count upper 10 ft.

[b]Average undrained shear strength upper 10 ft in lb/ft$^2$.

[c]Movement at load of 3 tons, except test 1 at 2 tons.

[d]$k_h$ coefficient of horizontal subgrade reaction in lb/in.$^2$, $k_{h1}$ coefficient of horizontal subgrade reaction in lb/in.$^2$ for surface layer; $k_{h2}$ coefficient of horizontal subgrade reaction in lb/in.$^2$ for lower layer.

[e]$n_h$ constant of horizontal subgrade reaction in lb/in.$^3$

[f]Computed from movements at 3 tons, except test 1 at 2 tons.

[g]Undrained shear strength estimated from $N$ value and tests on similar soil types at nearby locations.

[h]Movements measured about 2 ft above ground surface.

It must be noted that $(k_s \times B)$ in the above equations is the same quantity as $k_h$ in equation (4.26). The units of $k_h$ and $(k_sB)$ are the same as in Table 4.16.

### *(b) Cohesionless soils*

$$k_s = \frac{n_h}{B}x$$

or

$$k_sB = n_h \cdot x \tag{4.29a}$$

$$T = 5\sqrt{EI/n_h} \tag{4.29b}$$

The same remarks apply to $(k_sB)$ in equation (4.29a) as for cohesive soils. However, $n_h$ in equation (4.29b) is similar to that in equation (4.27).

Robinson (1979) described test results on timber piles at 11 sites in the lower Fraser Valley of British Columbia, Canada. The piles were load tested vertically one day prior to the lateral loading. Vertical loading tests were continued until either the test pile or the reaction piles failed, whichever occurred first. Therefore, the horizontal load tests were performed on one of the piles that did not fail during the vertical load tests.

The lateral load tests consisted of jacking apart two adjacent piles. The lateral deflections of the piles were recorded as the loads were increased. With the exceptions of tests 1 and 11, loads were applied in approximately 1-ton (907 kg or 8.9 kN) increments up to 3-tons (2720 kg or 26.7 kN) and then cycled from one to five times. Test 1 in peat was cycled at 1.5 tons (1360 kg or 13.35 kN) after excessive deflections were recorded at a load of 2 tons (1815 kg or 17.8 kN).

Based on the measured pile deflections, $k_h$ and $n_h$ were computed for each test site and are summarized in Table 4.17. These values are generally higher than those listed here in column 8 from earlier investigations. A detailed comparison has been presented by Robinson (1979). Figure 4.51 is a plot of Standard Penetration value ($N$) against $n_h$ based on Terzaghi's (1955) and Robinson's (1979) tests up to 6 ton (53.4 kN) loading. Observe that Terzaghi's recommended values are the smallest.

The recommended values of Reese et al. (1974) are about two and a half times those of Terzaghi. Robinson's values of $n_h$ are several order higher than those recommended by earlier investigations (Terzaghi, 1955; Davisson, 1970; Broms, 1965). Based on the present test results and results provided by Broms (1965), Alizadeh (1969), and Prakash (1962), repetitive loading on very loose soil could double the deflection, while for $N$ of 10 or more, a 25 percent increase over the first cycle deflection would be a conservative assumption.

Smith (1987) suggested that significant side shear and front pressures are mobilized to resist the translating pile. The equivalent uniform pressure $p$ cannot be measured directly in a field load test, but it can be calculated by double differentiation of measured bending strains. In addition, the actual distributions of pressure are far from linear. Briaud, Smith, and Meyer (1983) illustrated from

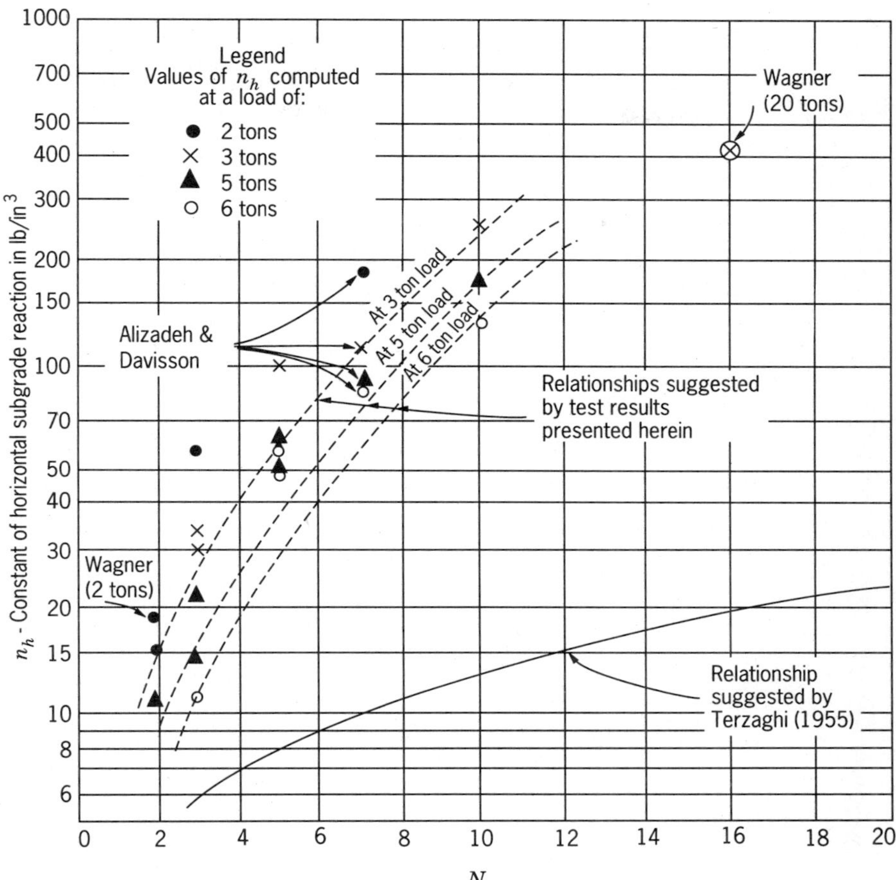


**Figure 4.51** *N* Versus $n_h$ for cohesionless soil where loads are applied at the ground surface against displacement piles (Robinson, 1979). ($N$ = average standard penetration resistance blows per foot (upper 10 ft). 1 ft = 30.48 cm, 1 lb/in.$^3$ = 0.02768 kg/cm$^3$, 1 ton = 907.2 kg .)

an earth pressure cell instrumented field load test result that the measured distributions of mobilized front pressure $Q$ around the circumference of a pile approximately followed the theoretical predictions of Baguelin et al. (1977). This distribution, and that of side shear $F$, is illustrated in Figure 4.52 at a safety factor of 8, to ultimate load during this test. Smith (1987) reports that side shear contributed 88 percent of the soil reaction from horizontal equilibrium. It will be interesting to see the contribution of shear as above at a factor of safety of 1.5 or 2.

On the basis of above discussion, it is recommended that:

1. For preliminary analysis, values of $k_h$ or $n_n$ be taken from Table 4.16.

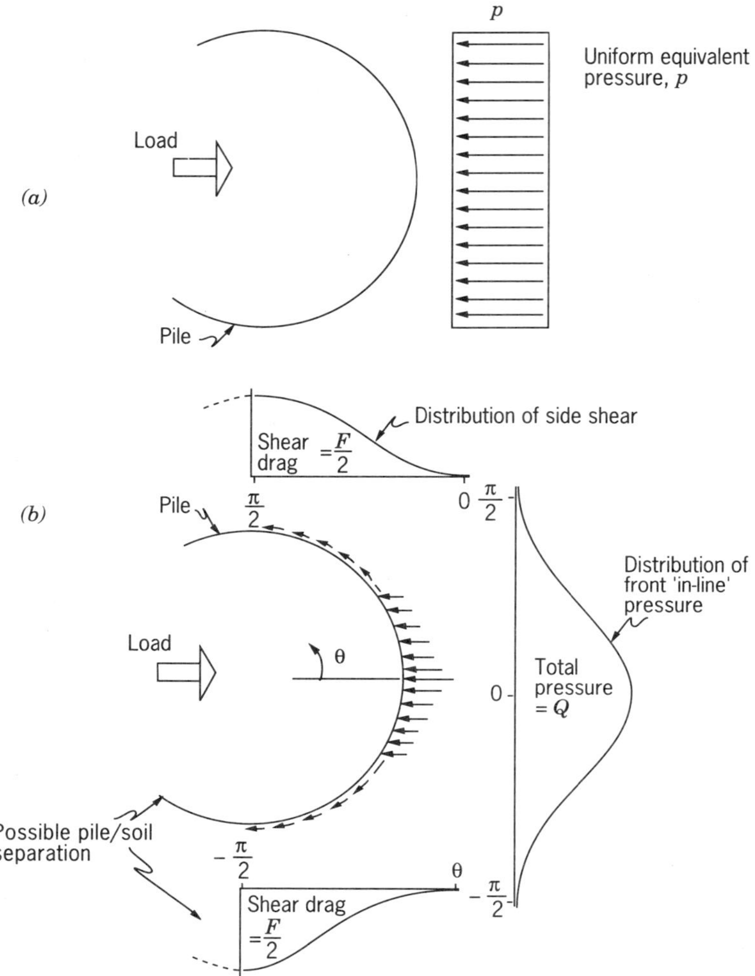


**Figure 4.52** Comparison of simplified modulus and theoretical stress distributions around pile. (a) Modulus approach; (b) actual theoretical (Smith, 1987).

2. For more realistic analysis, typical lateral pile load tests be performed to estimate the values of $k_h$ or $n_n$ for a given site and the piles to be used in that project.

## 4.5 OVERVIEW

In this chapter, various methods of soil investigation, field and laboratory testing techniques and the resulting design parameters for piles subjected to static and dynamics loads and piles installed in permafrost environment have been presented. Results from field tests—such as $N$, $q_c$, field-vane shear, pressuremeter values, and shear wave velocity—are also related to laboratory tested parameters—such as, $c_u$, $\phi'$, and dynamic shear modulus $G$. A brief description of soil behavior and design parameters for permafrost environment was also presented. Design parameters developed and discussed in this chapter will be later used in Chpater 5 (for vertical static loads), Chapter 6 (for lateral loads), Chapter 7 (for dynamic loads), and Chapter 8 (for permafrost environment). Data provided herein will therefore supplement chapters on design of piles.

Soils investigations consist of boring and test pits to obtain soil stratigraphy and to recover samples for laboratory testing. The generally used boring methods consist of auger boring, hollow stem boring, wash boring, rotary drilling, percussion drilling, rock core drilling, and wire-line drilling. Auger boring is normally used for investigations to shallow depths and above groundwater levels. This is not suitable for loose and soft materials unless a casing is used to support the borehole. Hollow-stem auger boring serves as a casing during boring and also provides access both for disturbed and undisturbed soil sampling. This method is therefore most common method of investigation where soil consistency allows augering. In soils that consist of harder layers of gravels and very stiff cohesive soils, wash boring is used. The major disadvantage of this method is the difficulty in obtaining undisturbed soil samples. Rotary and percussion drilling methods are used to advance holes into harder soil stratum. Rock core drilling and wire-line drilling are used to advance and obtain rock core when boring through rocks. Wire-line drilling is an efficient method for deep hole rock coring over 30 m (100 ft) depth.

Measurement of groundwater (normal, perched, and artesian) is an important aspect of the soil investigation program. Groundwater levels are generally monitored by piezometers, the most common types of which are the open standpipe piezometer, the porous element piezometer, the electric piezometer, and the pneumatic piezometer. Open standpipe piezometers are simple and reliable means of groundwater monitoring system but have a slow response time. Porous element piezometers are an improvement over open standpipes but still require equalization time in fine-grained soils. Experience indicates that electric piezometers are not satisfactory on a long-term basis. Pneumatic piezometers have proved to be successful in fine-grained soils and are commonly used where rapid pore pressure response is required.

In recent years, the measurement of soil parameters by field testing methods has gained wide acceptance. Field testing provides properties for soils that cannot be sampled in undisturbed state (e.g., saturated sands below a water table). In addition, field testing increases the cost effectiveness of an exploration and testing program because larger volume of soil can be tested in the field when compared with the laboratory testing. Penetrometer tests, vane shear tests, and the pressurementer tests have direct applicability to the pile foundation design (Robertson, 1986). Among the various penetrometer tests, the Standard Penetration Test (SPT) and the Static Cone Penetration Test (CPT) are the widely used field testing techniques for pile design. The major advantages of SPT are that it provides soil samples along with the SPT values and has been used for pile design for over 50 years. Its major disadvantage is that it is affected by many factors such as operator, drilling, equipment, and sampler driving characteristics (Schmertmann, 1977, and Kovacs and Salomone, 1982). CPT, originally developed in Europe, is now gaining acceptance in North America. The major advantages of CPT are that the results are repeatable, accurate, and provide a continuous record of soil stratigraphy. Its major drawback is that the soil samples are not recovered from this test. This method should be used in conjuction with borings.

Laboratory tests are carried out to classify the soils and to provide soil parameters for pile design. The common type of laboratory tests are the Atterberg limits, the unconfined compressive strength, the consolidation tests, and the direct shear test.

The soil parameters for static pile design are the friction angle ($\phi'$) in cohesionless soils, the undrained strength ($c_u = S_u$), and the soil-pile adhesion factor ($c_a$) in cohesive soils. The $\phi'$ value is generally obtained from field tests relationships with $N$ or $q_c$ values. The relationship between $\phi'$ and $q_c$ as cited by Meyerhof (1976) appears to be the most reliable means of obtaining an in-situ $\phi'$ value. The best method of obtaining an $S_u$ value for cohesive soil is the unconfined compressive tests on undisturbed samples in the laboratory. However, for soft sensitive clays, undisturbed sampling is difficult. Therefore, for such cases, field vane shear tests should be carried out to obtain $S_u$ values. Determination of soil parameters for permafrost is still in the development stage. The major factor that controls the behavior of ice-rich frozen soils is creep. The uniaxial creep rate ($\dot{\varepsilon}$) is given by equation (4.25). The creep parameters $n$ and $B$, in this equation, can be taken from Table 4.13. Tables 4.14 and 4.15 can be used to estimate the strength (short and long term) and compressibility parameters of frozen soils for preliminary design. Further laboratory and field testing of frozen soils and the back-calculated design parameters from field pile load testing are required to provide a better understanding of design parameters for piles in permafrost areas.

Soil moduli under dynamic loads depend on soil characteristics, such as void ratio, relative density, stress history, preconsolidation pressure, confining pressure, and strain level. Simple equations have been developed for use with available data to make preliminary estimates of soil moduli at low strain amplitudes for sands and clays equation (4.16a). For estimations of $\bar{\sigma}_0$ the value of $K_0$, the coefficient of earth pressure at rest, which is a function of the plasticity index and overconsolidation ratio of clays is needed, which may be determined from Figure 4.53. If the soil modulus is determined at one confining pressure, the corresponding value at any other confining pressure can be determined with the help of equation (4.17). For preliminary design, values may be determined as above.

As already explained, different tests in the laboratory and field result in different strains (Figure 4.31). In triaxial tests, generally intermediate strains can be developed. Efforts have been made to extend the strain ranges in resonant column apparatus from small to intermediate values and in triaxial tests from intermediate to small values. The determination of shear wave profile with depth

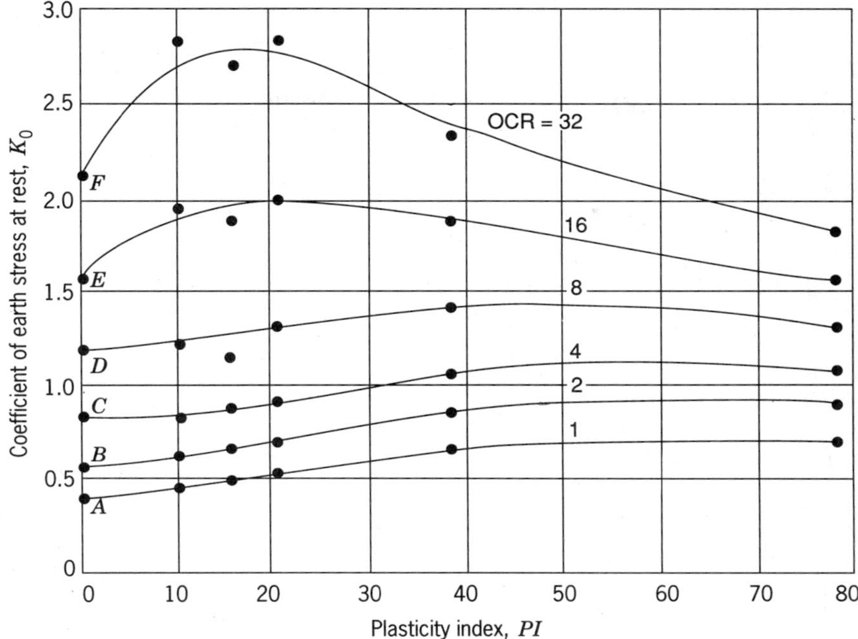


**Figure 4.53** $K_0$ as function of overconsolidation ratio and plasticity index for clays, (after Brooker and Ireland, 1965; Lambe and Whitman, 1969). (Reprinted by permission of Wiley, New York.)

from the spectral analysis of surface waves (SASW) appears to be a promising field method for the future (Prakash and Puri, 1988).

In lateral loading of piles, the modulus of subgrade reaction is used in most analyses. For preliminary estimate, $k_h$ or $n_h$ may be taken from Tables 4.16a and b. For a more realistic determination of the modulus of subgrade reaction, a lateral pile load test is recommended.

As for soil modulus, the $k_h$ or $n_h$ values depend on strain in the soil or pile displacement. For piles supporting buildings, 1/4 in. to 1/2 in. deflection of pile head may be admissible. In pile-supported machine foundations, the lateral displacements are too small. However, for offshore piles, large displacements may be permissible. Appropriate corrections to $k_h$ or $n_h$ values need be applied. There is no universal relationship and agreement on the variations of $k_h$ or $n_h$ with displacement. There is, however, a need to develop one.

## REFERENCES

Aas, G.,"A Study of the Effect of Wave Shape and Rate of Strain on the Measured Values of In-Situ Shear Strength of Clays," *Proceedings, 6th International Conference on Soil Mechanics and Foundation Engineering*, Montreal, Canada, Vol. 1, 1965, pp. 141–145.

Afifi, S. E. A. and Richart, F. E. Jr., "Stress History Effects on Shear Modulus of Soils," *Soils and Foundations (Japan)*, Vol. 13, No. 1, 1973, pp. 77–95.

Afifi, S. E. A. and Woods, R. D., "Long-Term Pressure Effects on Shear Modulus of Soils," *J. Soil Mech. and Found. Dn.*, ASCE, Vol. 97, No. SM10, 1971, pp. 1445–1460.

Alizadeh, M. Lateral Load Tests on Instrumented Timber Piles, *Performance of Deep Foundations*, ASTM STP444, 1969, p. 379.

Alizadeh, M. and Davisson, M. T., "Lateral Load Tests on Piles—Arkansas River Project," *J. Soil Mech. and Found Dn.*, ASCE, Vol. 96, No. SM5, 1970, pp. 1583–1604.

American Society for Testing and Materials, Annual Book of Standards, Soil and Rock, Building Stones; Geotextiles, Vol. 4.08, 1989.

American Society for Testing and Materials, Designation D 1589–67, Standard Practice for Penetration Test and Split-Barrel Sampling of Soils (Reapproved 1974), Annual Book of ASTM Standards, 1989.

American Society for Testing and Materials, Designation D 1581–83, Standard Practice for Thin-Walled Tube Sampling of Soils, Annual Book of ASTM Standards, 1989.

American Society for Testing and Materials, ASTM D 2487–85, Standard Test Method for Classification of Soils, for Engineering Purposes, Annual Book of ASTM Standards, 1989.

American Society of Civil Engineering, ASTM Designation D 3080–72(1979) Direct Shear Test of Soils Under Consolidated Drained Conditions, Annual Book of ASTM Standards 1989.

American Society of Civil Engineers, Manual No. 56, Subsurface Investigation for Design and Construction of Foundations of Buildings, ASCE—Manuals and Reports on Engineering Practice No. 56, 1976, p. 61.

Anderson, D. G., "Dynamic Modulus of Cohesive Soils," Ph.D. Dissertation Univ. of Mich. Ann Arbor.

Anderson, D. G. and Stokoe K. H. II "Shear Modulus: A Time-Dependent Soil Property," *Dynamic Geotechnical Testing*, ASTM Spec. Tech. Pub. 654, Denver, CO, June 1977, pp. 66–89.

Anderson, D. G. and Woods, R. D., "Comparison of Field and Laboratory Shear Moduli," *Proceedings, In Situ Measurement of Soil Properties*, ASCE, Raleigh, N.C. Vol. 1, 1975, pp. 69–92.

Anderson, D. G. and Woods, R. D., "Time Dependent Increase in Shear Modulus of Clay," *J. Geotech. Eng. Div.*, ASCE, Vol. 102 (GT5), May 1976, pp. 525–537.

Annaki, M. and Lee, K. L., "Equivalent Uniform Cycle Concept of Soil Dynamics," *J. Geotech. Eng. Div.*, ASCE, Vol. 103 (GT6), June 1977, pp. 549–564.

Andersland, O. B. and Anderson, D. M. (ed.), *Geotechnical Engineering for Cold Regions*, McGraw-Hill Book Co., New York, 1978.

*Annual Book* of *American Society for Testing and Materials Standards*, Philadelphia, Vol. 04–08, 1989.

Baguelin, F., Frank, R., and Said, Y., "Theoretical Study of Lateral Reaction Mechanism of Piles," *Geotechnique*, Vol. 27, No.3, September 1977, pp. 405–434.

Baguelin, F., Jezequel, J. F., and Sheilds, D. H., "The Pressuremeter and Foundation Engineering," Transtech Publication, Rockport, MA, 1978.

Ballard, R. F. Jr., "Determination of Soil Shear Moduli at Depth by In-Situ Vibration Techniques," Waterways Experiment Station Misc. Paper No. 4–691, December 1964.

Bazarra, A. R., "Use of the Standard Penetration Test for Estimating Settlements of Shallow Foundations on Sand," Ph.D. Thesis, University of Illinois, Urbana, 1967, p. 379.

Biot, A. M. "Theory of Propagation of Elastic Waves in Fluid Saturated Porous Solid" *J. Accoust. Soc. Am.* Vol. 28, 1956, pp. 168–191.

Bjerrum, L., "Problems of Soil Mechanics and Construction on Soft Clays," *Proceedings 8th International Conference on Soil Mechanics and Foundation Engineering*, Moscow, State-of-the-Art Report, Vol. 3, 1973, pp. 111–159.

Bjerrum, L. and Landra, A., "Direct Simple Shear Tests on a Norwegian Quick Clay," *Geotechnique*, Vol. 26 No. 1, 1966, pp. 1–20.

Bjerrum, L. and Simons, N. E., "Comparison of Shear Strength Characteristics of Normally Consolidated Clays," *Proceedings of the American Society of Engineers*, Research Conference on Shear Strength of Cohesive Soils, Boulder, 1960, pp. 711–726.

Bowles, J. E., *Foundation Analysis and Design*, 2nd ed. McGraw-Hill Book Co., New York, 1977.

Briaud, J. L., Smith, T. D., and Meyer, B. J., "Pressuremeter Gives Elementary Model for Laterally Loaded Piles," *Proceedings on International Symposium on In-Situ Testing of Soil and Rock*, Paris, France, Vol. II May 1983, pp. 2–221

Broms, B. B., Discussion to Paper by R. Roshimi, *J. Soil Mech. and Found. Dn*, ASCE, Vol. 91, No. SM 4, 1965 pp. 199–205.

Brooker, E. W. and Ireland, H. O., "Earth Pressure at Rest Related to Stress History," *Can. Geot. J.* Vol. 11, Nos (Feb)

Campanella, R. G., and Robertson, P. K., "Applied Cone Research," Symposium on Cone Penetration Testing and Experience, Geotechnical Engineering Division, ASCE, 1981, pp. 343–362.

*Canadian Foundation Engineering Manual*, National Research Council of Canada, Ottawa, 1978 and 1985.

Castro, G. and Poulos, S. J., "Factors Affecting Liquefaction and Cyclic Mobility," *J. Geot. Eng. ASCE*, Vol. 103, No. GT6, 1977, pp. 501–506.

Casagrande, A., "A New Theory of Frost Heaving: Discussion," *Proceedings U.S. Highway Research Board*, Vol. 11, Part I, 1932, pp. 168–172.

Davidson, J. L. and Boghrat, A. G., "Flat Dilatometer Testing in Florida," *Proceedings of the International Symposium on Soil and Rock Investigation by In-Situ Testing*, Paris, Vol. II, 1983, pp. 251–255.

Davisson, M. T. "Behavior of Flexible Vertical Piles Subjected to Moment, Shear and Axial Load" Ph.D. Thesis, University of Illinois, Urbana, 1960.

Davisson, M. T. "Estimating Buckling Loads for Piles" *Proceedings of the Second Pan American Conference on Soil Mechanics and Foundation Engineering*, Vol. I, 1963, pp. 351–369.

Davisson, M. T. "Lateral Load Capacity of Piles" *Highway Research Record No. 333*, Washington, DC, 1970, pp. 104–112.

Deere, D. U., "Technical Description of Rock Cores for Engineering Purposes," *Felsmechanik und Ingeniergeologie*, Vol. 1, 1963, pp. 17–22.

Deere, D. U., Hendron, A. J., Jr., Datton, F. D., and Cording, E. J., "Design of Surface and

Near Surface Construction in Rock," *Proceedings of the 8th Symposium of Rock Mechanics*, American Institute of Mining, Metallurgical and Petroleum Engineers, New York, 1967, pp. 237–302.

Drnevich, V. P. "Effects of Strain History on the Dynamic Properties of Sand," Ph.D. Thesis, Department of Civil Engineering University of Michigan, ann Arbor, 1967.

Drnevich, V. P. "Undrained Cyclic Shear of Saturated Sand," *J. Soil Mech. Found. Div.*, ASCE, Vol. 98 No. SM8, August 1972, pp. 807–825.

Drnevich, V. P., "Resonant Column Testing—Problems and Solutions," *ASTM Symposium on Dynamic Soil and Rock Testing in the Field and Laboratory for Seismic Studies*, Denver, CO, June 1977, pp. 384–398.

Drnevich, V. P., Hall, J. R., and Richart, F. E., "Effects of Amplitude of Vibration on the Shear Modulus of Sand," *Proceedings of the International Symposium on Wave Propagation and Dynamic Properties of Earth Materials*, Albuquerque, NM 1967, pp. 189–199.

Drnevich, V. P., Hardin, B. O., and Shippy, D. J. "Modulus and Damping of Soils by the Resonant-Column Method," *Dynamic Geotech. Testing*, ASTM Spec. Tech. Pub. No. 654, Denver, June 1977, pp. 91–125.

Dunnicliff, J. L., *Geotechnical Instrumentation for Monitoring Field Performances* Wiley, New York, 1988.

ESOPT I, *Proceedings of the First European Symposium on Penetration Testing*, Stockholm, Sweden, 1974.

ESOPT II, *Proceedings of the Second European Symposium on Penetration Testing*, Amsterdam, Netherlands, A. A. Balkema 1982.

*Foundations and Earth Structures, Design Manuals 7.1 and 7.2*, NAVFAC DM-7.1 and 7.2, Department of the Navy, Alexandria, VA, May 1982.

Furgo, Inc., "Sustained-Pressure Studies," Furgo Technical Development Program, 1977, unpublished.

Gibbs, H. J. and Holtz, W. G., "Research on Determining the Density of Sands by Spoon Penetration Testing," *Proc. 4th International Conference on Soil Mechanics and Foundation Engineering*, Vol. 1, London, 1957, pp. 35–39.

Glen, J. W., "Experiments on the Deformation of Ice" *J. Glaciology*, Vol. 2, 1952, pp. 111–114.

Glen, J. W., "The Mechanics of Ice", U.S. Army Cold Regions Research and Engineering Laboratory, Monograph II-C2, 1975.

Goel, M. C., "Correlation of Static Core Resistance with Bearing Capacity," *Proceedings of the Second European Symposium on Penetration Testing*, Amsterdam, Netherlands, Vol. 2, 1982.

Hardin, B. O., "The Nature of Stress–Strain Behavior of Soils," State-of-the-Art Report, *Proceedings of the ASCE Speciality Conference on Earthquake Engineering and Soil Dynamics*, Pasadena, CA, June 1978, pp. 3–90.

Hardin, B. O. and Black, W. L., "Vibration on Modulus of Normally Consolidated Clays," *J. Soil Mech. and Found. Dn.*, ASCE, Vol. 92, No. SM2, March 1968, pp. 353–369.

Hardin, B. O. and Black, W. L., "Closure to Vibration Modulus of Normally Consolidated Clays," *J. Soil Mech. and Found. Dn.*, ASCE, Vol. 95, No. SM6, November 1969, pp. 1531–1537.

Hardin, B. O. and Drnevich, V. P., "Shear Modulus and Damping in Soils Design Equations and Curves," *J. Soil Mech. Found. Dn.*, ASCE, Vol. 98, No. SM7, July 1972, pp. 667–692.

Hardin, B. O. and F. E. Richart Jr, "Elastic Wave Velocities in Granular Soils" *J. Soil Mech. and Found. Dn.*, ASCE, Vol. 89, No. SM1, 1963, pp. 33–65.

Hough, B. K., *Basic Soils Engineering*, 2nd ed. The Ronald Press Company, New York, 1969.

Hvorslev, M. J. and Kaufman, R. I., "Torsion Shear Apparatus and Testing Procedure," *USAE Waterways Exper. Station Bull. 38*, May 1952, p. 76.

Imai, T., "Velocities of P- and S-Waves in Subsurface Layers of Ground in Japan," *Proceedings 9th International Conference on Soil Mechanics and Foundation Engineering*, Vol. 2, Tokyo, Japan, 1977, pp. 257–260.

In Situ '86: A Speciality Conference, *Geotechnical Engineering Division*, ASCE, Blacksburg, VA, 1986.

Ishihara, K., "Factors Affecting Dynamic Properties of Soils," *Proceedings 4th Asian Regional Conference on Soil Mechanics and Foundation Engineering*, Bangkok, Thailand, Vol. 2, August 1971.

Ishihara, K. and Li, S., "Liquefaction of Saturated Sands in Triaxial Torsion Shear Test," *Soils Found.* (Jpn) Vol. 12, No. 2, pp. 19–39.

Iwasaki, T., Tatsuoka, F., and Takagi, Y., "Shear Moduli of Sands Under Cyclic Torsional Shear Loading," Technical Memo Public Works Research Institute, Ministry of Construction, Chiba-Shi, Japan, No. 1264, July 1977.

Jamiolkowski, M., Ladd, C. C., Germaire, J. T., and Lancellotta, R., "New Developments in Field and Laboratory Testing of Soils," State-of-Art Paper, *Proc. 11th International Conference on Soil Mechanics and Foundation Engineering*, San Francisco, CA, 1985, Vol. 1, pp. 57–153.

Johnston, G. H. (ed.), *Permafrost: Engineering Design and Construction.* Wiley, New York, 1981.

Kjellman, W., "Testing of Shear Strength in Sweden," *Geotechnique*, Vol. 2, 1951, pp. 225–232.

Kovacs, W. D., Discussion of "On Dynamic Shear Moduli and Poisson's Ratios of Soil Deposits," *Soils Found* (Jpn) Vol. 15, No. 1, March 1975.

Kovacs, W. D., Evans, J. C., and Griffith, A. M., "Towards a More Standardized SPT," *Proceedings Ninth International Conference on Soil Mechanics and Foundation Engineering*, Tokyo, Japan, 1977a, Vol. 2, pp. 269–276.

Kovacs, W. D., Griffith, A. H., and Evans, J. C., "Alternate to the Cathead and Rope for the SPT," *ASTM Symposium on Dynamic Field and Laboratory Testing of Soil and Rock*, Denver, CO., June 29, 1977b.

Kovacs, W. D. and Salomone, L. A., "SPT Hammer Energy Measurements," *J. Geotech. Eng. Dn.*, ASCE, Vol. 108, No. GT4, 1982, pp. 599–620.

Kovacs, W. D., Salomone, L. A., and Yokel, F. Y., "Energy Measurements in the Standard Penetration Test" Building Science Series, National Bureau of Standards (U.S.), No. 135, 1981.

Lambea, T. W. and Whitman, R. V., *Soil Mechanics*, Wiley and Sons, NY 1969.

Liao, S. and Whitman, R. V. "Overburden Corrections Factors for SPT in Sand," *J. Geo. Eng. Dn. ASCE.* Vol. 112, No. 3, pp. 373–377, 1986.

Lodde, P. J., "Shear Moduli and Material Damping Ratios of San Francisco Bay Mud," Master's Thesis, University of Texas, Austin, 1977.

Mair, R. J. and Wood, D. M. *Pressuremeter Testing*, Butterworths, London, 1987.

Marchetti, S., "In Situ Tests by Flat Dilatometer," *J. Geot. Eng. Dn.*, ASCE, Vol. 106, No. GT3, 1980, pp. 299–321.

Marcuson, W. F., III and Wahls, H. E., "Time Effects on Dynamic Shear Modulus of Clays," *J. Soil Mech. and Found. Dn.*, ASCE, Vol. 98, No. SM12, December 1972, pp. 1359–1373.

Matlock, H. and Reese, L. C. "Foundation Analysis of Offshore Pile-Supported Structures," *Proceedings 5th International Conference of Soil Mechanics and Foundation Engineering*, Paris, 1961, Vol. 2, pp. 91–97.

McRoberts, E. C., "Shallow Foundation in Cold Regions: Design," *J. Geot. Eng.*, ASCE, Vol. 108, No. GT10, 1982, pp. 1338–1349.

Melzer, K. J. and Smoltczyk, U. "Dynamic Penetration Testing—State-of-the-Art Report," *Proceedings, Second European Symposium on Penetration Testing*, Amsterdam Vol. 1, 1982, pp. 191–202.

Menard, L., "An Apparatus for Measuring the Strength of Soils in Place," M.Sc. Thesis, Department of Civil Engineering, University of Illinois, Urbana, 1956.

Meyerhof, G. G., "Bearing Capacity and Settlement of Pile Foundations," *J. Geot. Eng.*, ASCE, Vol. 102, No. GT3, March 1976, pp. 197–228.

Mitchell, J. K., Guzikowski, F., and Villet, W. C. B., "The Measurement of Soil Properties in Situ," Report Prepared for U.S. Department of Energy Contract W-7405-ENG-48. Lawrence Berkely Laboratory, University of California, Berkely, CA, 1978, pages 67.

Morgenstern, N. R., "Geotechnical Contributions to Arctic Resource Development," *Proceedings 7th Pan American Conference on Soil Mechanics and Foundation Engineering*, Vancouver, Canada, June 1983, Vol. III, pp. 889–913.

Morgenstern, N. R., Roggensack, W. D., and Weaver, J. S., "The Behavior of Friction Piles in Ice and Ice-Rich Soils," *Can. Geotech. J.*, No. 17, Vol. 3, 1980, pp. 405–415.

NAVFAC Design Manuals 7.1 and 7.2, *Foundations and Earth Structures*, DM-7.1 and 7.2, Department of the Navy, Alexandria, VA, May 1982.

Nazarian, S. and Stokoe, K. H. II, "In Situ Shear Wave Velocities Form Spectral Analysis of Surface Waves," *Proceedings Eighth World Conference on Earth Engineering*, San Francisco, CA, Vol. III, July 1984, pp. 31–38.

Nixon, I. K., "Standard Penetration Test, State-of-the-Art Report," *Proceedings Second European Symposium on Penetration Testing*, Amsterdam, Vol. 1, 1982, pp. 3–24.

Nixon, J. F. and McRoberts, E. C., "A Design Approach for Pile Foundations in Permafrost," *Can. Geotech. J.*, Vol. 13, 1976, pp. 40–57.

Nixon, J. F., and Lem, G., "Creep and Strength Testing of Frozen Saline Fine-Grained Soils," *Canadian Geotechnical Journal*, Vol. 21, No. 3, 1984, pp. 518–529.

Nishida, Y., "Brief Note on Compression Index of Soil," *J. Soil Mech. Found. Dn.*, ASCE, Vol. 82, No. SM3, February, 1956.

Palacios, A., "The Theory and Measurement of Energy Transfer during Standard Penetration Test Sampling," Ph.D. Thesis, Department of Civil Engineering, University of Florida, August 1977.

Peacock, W. H. and Seed, H. B., "Sand Liquefaction Under Cyclic Loading Simple Shear

Condtions," *J. Soil Mech. Found. Div.*, ASCE, Vol. 94, No. SM3, May 1968, pp. 689–708.

Peck, R. B., Hansen, W. E., and Thornburn, T. H., *Foundation Engineering*, 2nd ed. Wiley, New York, 1974.

Prakash, S., "Behaviour of Pile Groups Subjected to Lateral Loads". Ph.D. thesis University of Illinois, Urbana, 1962.

Prakash, S., "Analysis and Design of Vibrating Footings," *Symposium on Recent Developments in the Analysis of Soil Behavior*," Sydney, Australia, July 1975, pp. 295–326.

Prakash, S., "Geotechnical Earthquake Problems of the Future," *International Convention of ASCE*, Preprint No. 81–087, 1981b, pp. 14, New York.

Prakash, S., *Soil Dynamics.* McGraw-Hill Book Co., New York, 1981a.

Prakash, S., Nandkumaran, P., and Joshi, V. H., "Design and Performance of an Oscillatory Shear Box," *J. Indian Geotech. Soc.*, Vol. 3 No. 2, April 1973, pp. 101–112.

Prakash, S. and Puri, V. K., "Critical Evaluation of IS-5249-1969," *J. Indian Geotech. Soc.*, Vol. VI, No. 1, January 1977, pp. 43–56.

Prakash, S. and Puri, V. K., "Dynamic Properties of Soils from In Situ Tests," unpublished report, University of Missouri-Rolla, Rolla, July 1980.

Prakash, S. and Puri, V. K., "Dynamic Properties of Soils from In Situ Test," *J. Geotech. Eng. Div.*, ASCE, Vol. 107, No. GT7, July 1981, pp. 943–963.

Prakash, S. and Puri, V. K., "Design of Compressor Foundations: Predictions and Observations," *Proc. International Conference on Case Histories in Geotechnical Engineering*, St. Louis, MO, Vol. IV, May 1984, pp. 1705–1710.

Prakash, S. and Puri, V. K., *Foundation for Machines.* Wiley, New York, 1988.

Prakash, S., Ranjan, G., and Saran, S., *Analysis and Design of Foundations* and Retaining Structures. Sarita Prakashan, Meerut (UP) India, 1979.

Puri, V. K., "Liquefaction Behavior and Dynamic Properties of Loessial (Silty) Soils," Ph.D. Thesis, Civil Engineering Department, University of Missouri-Rolla, Rolla, 1984.

Reese, L. C., Cox, W. R., and Koop, F. B., "Analysis of Laterally Loaded Piles in Sand," *Proceedings Offshore Technology Conference*, Houston, TX, Paper No. OTC 2080, 1974, pp. 473–483.

Reese, L. C. and Matlock, H. "Non-Dimensional Solutions for Laterally Loaded Piles with Soil Modulus Assumed Proportional to Depth," *Proceedings Eighth Texas Conference on Soil Mechanics and Foundations Engineering*, Austin, 1956, pp.

Richart, F. E., Jr., Closure to "Foundation Vibrations," *J. Mech. Found. Dn.*, ASCE, Vol. 84, No. SM4, August 1961, pp. 169–178.

Richart, F. E., Jr., Hall, J. R., and Woods R. D., *Vibrations of Soils and Foundations.* Prentice-Hall, Inc., Englewood Cliffs, NJ, 1970.

Riggs, C. I. "North American Geotechnical Exploration Practice," *Proceedings International Conference on Deep Foundations*, Beijing, September 1986, Vol. 2, pp. 2–146, 2–161.

Robertson, P. K., "In Situ Testing and Its Application To Foundation Engineering," *Can. Geotech. J.*, Vol. 23, No. 4, November 1986, pp. 573–594.

Robertson, P. K., "In Situ Testing and its Application to Foundation Engineerging," Soil

Mechanics Series No. 91 University of British Columbia, Department of Civil Engineering, Vancouver, B.C., 1985, 121 Pages.

Robertson, P. K. and Campanella, R. G., "Guidelines for Use and Interpretation of the Electronic Core Penetration Test," Soil Mechanics Series No. 69, 2nd Ed., Department of Civil Engineering, University of British Columbia, Vancouver, BC, 1984.

Robertson, P. K., Campanella, R. G., and Wightman, A., "SPT-CPT Correlations," *J. Geotech. Eng.*, ASCE, Vol. 109, No. 11, November 1983, pp. 1449–1459.

Robinson, K. E., "Horizontal Subgrade Reaction Estimated from Lateral Loading Tests on Timber Piles," *Behavior of Deep Foundations*, ASTM STP 670, Raymond Lundgren, ed. American Society for Testing and Materials, 1979, pp. 520–536.

Savigny, K. W., "In Situ Analysis of Naturally Occurring Creep in Ice-Rich Permafrost Soil," Ph.D. Thesis, University of Alberta, Edmonton, Alta, 439p.

Savigny, K. W. and Morgenstern, N. R., "Creep Behavior of Undisturbed Clay Permafrost," *Can. Geotech. J.*, Vol. 23, No. 4, November 1986, pp. 515–527.

Schmertmann, J. H., "Static Cone to Compute Static Settlements over Sand," *J. Soil Mech. Found Dn.*, ASCE, Vol. 96, No. SM3, 1970, pp. 1011–1043.

Schmertmann, J. H., "Measurement of In Situ Shear Strength," *Proceedings ASCE Specialty Conference on In Situ Measurement of Soil Properties*, Raleigh, NC, Vol. 2, 1975, pp. 56–138.

Schmertmann, J. H., "Use the SPT to Measure Dynamic Soil Properties? Yes, But . . .," *Dynamic Geotech. Testing*, ASTM Spec. Tech. Pub. 654, Denver June 1977, pp. 341–355.

Seed, H. B., "Soil Liquefaction and Cyclic Mobility Evaluation for Level Ground During Earthquakes," *J, Geotech. Eng. Div.*, ASCE, Vol. 105, No. GT2, 1979 pp. 201–255.

Seed, H. B., Tokimatsu, K., Harder, L. F. and Chung, R. M., "Influence of SPT Procedures in Soil Liquefaction Resistance Evaluations," J. Geot. Eng. Dn., ASCE, Vol. 111, No. 12, 1985, pp. 1425–1445.

Sharma, H. D., Sengupta, S. and Harron, G., "Cast-in-Place Bored Piles on Soft Rock under Artesian Pressures," *Can. Geotech. J.*, Vol. 21, No. 4 November 1984, pp. 684–698.

Silver, M. L., "Load Deformation and Strength Behaviour of Soils Under Dynamic Loading," State-of-the-Art Paper, *Proc. International Conference on Recent Advances in Geotechnical Earthquake Engineering and Soil Dynamics*, St. Louis, MO, Shamsher Prakash, ed. Vol. 3, 1981, pp. 873–896.

Skempton, A. W., "Vane Tests in the Alluvial Plain of the River Forth Near Grangemouth," *Geotechnique*, Vol. 1, No. 2, 1948, pp. 111–124.

Skempton, A. W., "Standard Penetration Test Procedures and the Effects of Overburden Pressure, Relative Density, Particle Size, Ageing and Over Consolidations," *Geotechnique*, Vol. XXXVI, No. 3, 1986, pp. 425–498.

Smith, T. D., "Pile Horizontal Soil Mudulus Values," *J. Geotech. Eng. Div.*, ASCE, Vol. 113, No. 9, 1987, pp. 1040–1043.

Sowers, G. F., *Introductory Soil Mechanics and Foundations*, 4th ed. Macmillan, New York, 1979.

Stokoe, K. H. II, and Abdel-razzak, K. G., "Shear Moduli of Two Compacted Fills," *Proceedings In Situ Measurement of Soil Properties*, American Society of Civil Engineers, Raleigh, NC, 1975, Vol. 1, pp. 422–447.

Stokoe, K. H., Lee, S. H. H., and Knox, D. P., "Shear Moduli Measurements Under True

Triaxial Stresses," *Proceeding Advances in the Art of Testing Soils Under Cyclic Dynamic Conditions*, ASCE, Detroit Convention, October 1985, pp. 166–185.

Stokoe, K. H. II and Nazarian, S., "Use of Rayleigh Waves in Liquefaction Studies," *Measurement and Use of Shear Wave Velocity, Proceedings of Geotechnical Engineering Division Session at Denver Spring ASCE Convention*, May 1985, pp. 1–17.

Stokoe, K. H. II and Richart, F. E. Jr., "In Situ and Laboratory Shear Wave Velocities," *Proc. 8th Intl. Conf. on SM and FE*, Moscow, Vol. 1, pp. 403–409, 1973b.

Stokoe, K. H. II and Richart, F. E. Jr., "Shear Moduli of Soils: *In Situ* and From Laboratory Measurements," *Proceedings Fifth World Conference on Earthquake Engineering*, Rome, Vol. 1, 1973a, pp. 356–359.

Stokoe, K. H. II and Woods, R. D., "*In Situ* Shear Wave Velocity by Cross-Hole Method," *J. Soil Mech. Found. Div.*, ASCE, Vol. 98, No. SM5, 1972, pp. 443–460.

Talbot, J. C. S., "The Prediction of Settlements Using In Situ Penetration Test Data," M.Sc. Dissertation, University of Surrey, Surrey, UK, 1981.

Terzahgi, K., "Evaluation of Coefficient of Subgrade Reaction," *Geotechnique*, Vol. 5, No. 4, 1955, pp. 297–326.

Terzaghi, K. and Peck, R. B., *Soil Mechanics in Engineering Practice*, 2nd ed., Wiley, New York, 1967.

Thiers, G. R. and Seed, H. B., "Cyclic Stress-Strain Characteristics of Clay", *J. Soil Mech. Found. Div.*, ASCE, Vol. 94, No. SM6, March 1968, pp. 555–569.

Tokimatsu, K. and Yoshimi, Y., "Empirical Correlations of Soil Liquefaction Based on Soil Liquefaction and Fine Contents," *Soils and Foundations, Jpn*, Vol. 23, No. 4, 1983.

Tomlinson, M. J., *Foundation Design and Construction.* Wiley, New York, 1963.

Trudeau, P. J., Whitman, R. V., and Christian, J. T., "Shear Wave Velocity and Modulus of a Marine Clay", *J. Boston Society of Civil Eng.*, Vol. 61, No. 1, January 1974, pp. 12–25.

Tsytovich, N. A., *The Mechanics of Frozen Ground*, Vysshaya Shakola Press, Moscow (in Russian), Transl. by Scripta Technica (G. K. Swinzow and G. P. Tschebotarioff, eds.), Scripta/McGraw-Hill, New York, 1975, p. 426.

Vesic, A. B. (1961) "Beams on Elastic Subgrade and Winkler's Hypothesis," *Proceedings 5th Internat. Conf. Soil Mechanics and Foundation Engineering*, Paris, Vol. 1, pp. 845–850.

Vialov, S. S., "Rheological Properties and Bearing Capacity of Frozen Soils, U.S. Army Cold Region Research and Engineering Laboratory, Technical Translation, No. 74, 1965, 219 Pages.

Williams, A. A. B., "Discussion of J. E. B. Jenning's and K. Knight's paper: The Prediction of Total Heave from the Double Oedometer Test," *Transactions of the Society of Afr. Instn. Civil Engineers*, Vol. 8, No. 6, 1958.

Wilson, S. D., Brown, F. R., Jr., and Schwarz, S. D., "In Situ Determination of Dynamic Soil Properties," *Dynamic Geotechnical Testing*, ASTM, STP 654, Denver, CO., June 1977, pp. 295–317.

Winkler E., *Die Lehre von Elastizitat und Festigkeit.* Publisher, Prague, 1867.

Woods, R. D., "Measurement of Dynamic Soil Properties—State-of-the-Art," *Proceedings ASCE Speciality Conference on Earthquake Engineering and Soil Dynamics*, Pasadena, CA, Vol. 1, June 1978, pp. 91–180.

Yang, Z. and Hathaway, A. W., "Dynamic Response of Tropical Marine Limestone," *J. Geotech. Eng.*, ASCE, Vol. 102, No. GT2, February 1976, pp. 123–138.

# 5

# ANALYSIS AND DESIGN OF PILE FOUNDATIONS FOR VERTICAL STATIC LOADS

Piles are generally used in groups. However, the allowable or design load is always determined for a single pile. The design load may be determined either from considerations of shear failure or settlement and is the lower of the following two values:

1. Allowable load obtained by dividing the ultimate failure load with a factor of safety and
2. Load corresponding to an allowable settlement of the pile

In most situations, behavior of a single pile is different from that of a pile group. Therefore, procedures will be developed to determine the allowable loads of a pile group from that of the single pile.

This chapter discusses the methods used in practice to calculate ultimate loads and settlements of pile groups in (1) cohesionless soils, (2) cohesive soils, and (3) rock. These methods will be different for piles subjected to axial compression, pullout, and lateral loads. In this chapter, piles under axial compression and pullout are discussed. Piles under lateral loads are discussed in Chapter 6.

As an aid to design engineers, theoretical concepts are explained first, followed by a design procedure. Numerical examples are included to illustrate the design procedure.

## 5.1 PILES SUBJECTED TO AXIAL COMPRESSION LOADS

Figure 5.1 shows a pile under vertical load. This load is shared between the bearing at its tip and in shaft friction around its perimeter. If $(Q_v)_{ult}$ is the axial

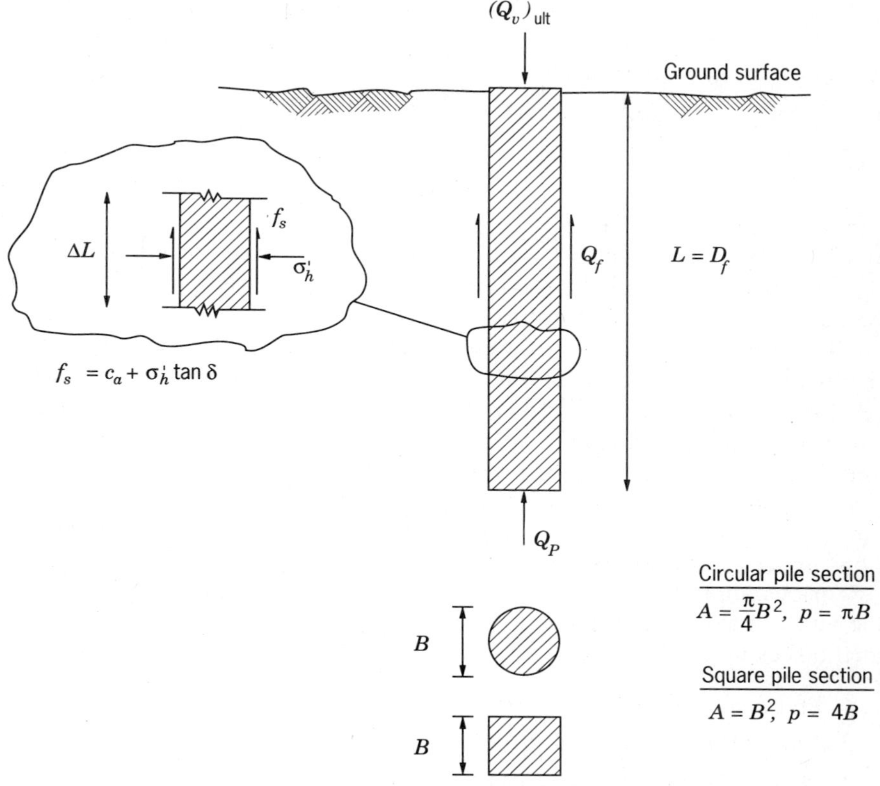


**Figure 5.1** Basic concept of load support by pile foundations.

compressive ultimate load applied on the top of a pile, it is shared by the pile tip ($Q_p$) and by the frictional resistance ($Q_f$) around the pile shaft. This can be represented by the following relationship.

$$(Q_v)_{\text{ult}} = Q_p + Q_f \tag{5.1}$$

where $(Q_v)_{\text{ult}}$ is the ultimate bearing capacity of pile, $Q_p$ is the end-bearing capacity, and $Q_f$ is the frictional capacity along the pile perimeter.

These terms can be further expanded as follows:

$$Q_p = A_p[cN_c + 1/2\gamma BN_\gamma + \gamma D_f N_q] \tag{5.2}$$

where

$A_p$ = pile end (point) area,
$c$ = cohesion of the underlying soil,
$\gamma$ = unit weight of soil,

$N_c, N_\gamma$, and $N_q$ = nondimensional bearing capacity parameters and are dependent on the angle of internal friction of the soil,
$B$ = the pile width or diameter and
$D_f$ = depth of pile tip below ground

$$Q_f = p \sum_{L=0}^{L=L} f_s \Delta L \tag{5.3}$$

where $p$ is the pile perimeter, $f_s$ is the unit shaft friction over a length $\Delta L$, and $L$ is the pile length over which shaft (skin) friction is assumed to have mobilized. These terms are further discussed in this chapter.

There are a number of state-of-the-art design manuals and review publications that provide a good documentation and background on the bearing capacity of pile foundations. Kezdi (1975), Meyerhof (1976), Tomlinson (1977), Vesic (1977), *Canadian Foundation Engineering Manual* (1978, 1985), Poulos and Davis, (1980), and *Foundations and Earth Structures Design Manual* (1982) are some of the publications. Design-related information from these and other practice-oriented publications has been used to develop rational design methods for pile foundations under axial compressive loads.

Various empirical and semiempirical methods are also available for evaluating the ultimate axial compression load capacity of piles. In general, the following methods estimate pile capacity:

1. Static analysis by utilizing soil strength
2. Empirical analysis by utilizing standard field tests
   (a) Standard penetration test values
   (b) Cone penetration values
   (c) Pressuremeter tests
3. Dynamic driving resistance
   (a) By pile driving formulas
   (b) By wave equation
4. Full-scale pile load tests

In the following sections, these methods are described listing clearly the merits and demerits of each one in different types of soils and for different situations in the field. Soil–pile interaction is complex and depends on such factors as soil types, types of loads, and pile installation methods (see Chapter 1). On account of this behavior, pile load capacity can only be estimated roughly from soil tests and semiempirical methods. Full-scale field load tests should therefore be made as a check of these estimates. This also would provide information on potential construction problems, if any, during actual pile installation. Such tests are also useful for construction control of driven piles by correlating the results of such tests with dynamic driving analysis. This question is discussed in detail in Chapter 9.

### 5.1.1 Bearing Capacity of a Single Pile in Cohesionless Soils

As mentioned previously, an allowable load on a single pile may be determined by several methods. Their use for cohesionless soils is illustrated here.

***Static Analysis by Utilizing Soil Strength (Theoretical Capacity)*** When piles are installed in homogeneous cohesionless soils, the soil near the pile gets disturbed to a distance of a few pile diameters. In driven piles, this disturbance results into compaction and increased density of the surrounding soil. In drilled or bored piles the disturbance may result into some loosening of the surrounding soil. This has already been discussed in Chapter 1. Since installation of piles results into changes in the surrounding soil density, the bearing capacity of piles should be estimated based on the changed soil properties. However, it is difficult to predict these changes in surrounding soil properties due to pile installation because of the complex interaction between the soil and pile during and after construction. The bearing capacity of piles is, therefore, estimated based on initial strength and deformation characteristics of the soil. The effect of changed soil conditions are reflected in the nondimensional empirical coefficient $N_q$ and mobilized shaft friction, $f_s$, in cohesionless soils.

In the following paragraphs, the end-bearing capacity and friction capacity based on initial soil strength and empirical coefficients are discussed separately.

*End-bearing Capacity* ($Q_p$) For cohesionless soils, $c = 0$ and the term $(1/2\gamma B N_\gamma)$ of equation (5.2) is small compared to $\gamma D_f N_q$. Equation (5.2) for cohesionless soils can then be rewritten as:

$$Q_p = A_p \gamma D_f N_q \tag{5.4a}$$

or

$$Q_p = A_p \sigma'_v N_q \tag{5.4b}$$

where $\sigma'_v$ is the effective overburden pressure at the pile tip.

Large-scale experiments and field observations show that both the point resistance and skin friction increase up to a certain critical depth, $D_c$. Beyond this depth, these values practically remain constant. This observation on critical depth was reported by Meyerhof (1976). He analyzed 33 pile load test data reported by various investigators. The tested piles were driven in sand and their depth to pile width ratio ($D_f/B$) varied from 2 to 40. Similar conclusions can be drawn from 16 load test data presented by Coyle and Castello (1981) where the $D_f/B$ ratio varied from 11 to 57 for driven piles in sand. For most design purposes, this critical depth may be taken as 20 times the pile width or diameter ($B$) although it may range between 10 to 30 times the pile diameter (Meyerhof, 1976; Coyle and Castello, 1981).

The semiempirical bearing capacity factor $N_q$ depends on (1) the $D_f/B$ ratio, (2) the angle of internal friction $\phi$ of the bearing stratum, and (3) the pile installation method. Furthermore, $N_q$ value may also change according to the theory and

**TABLE 5.1 Bearing Capacity Factors for Piles in Cohesionless Soils (Coyle and Castello, 1981)**

| | Approximate $N_q$ Values for Various Friction Angles, $\phi'$, in Degrees | | | | |
|---|---|---|---|---|---|
| Theories[a] | 25 | 30 | 35 | 40 | 45 |
| De Beer (1945) | 59 | 155 | 380 | 1150 | 4000 |
| Meyerhof (1953) | | | | | |
| Driven piles | 38 | 89 | 255 | 880 | 4000 |
| Caquot-Kerisel (1956) | 26 | 55 | 140 | 350 | 1050 |
| Brinch Hansen (1961) | 23 | 46 | 115 | 350 | 1650 |
| Skempton, Yassin, and Gibson (1953) | 46 | 66 | 110 | 220 | 570 |
| Brinch Hansen (1951) | 32 | 54 | 97 | 190 | 400 |
| Berezantsev (1961) | 16 | 33 | 75 | 186 | — |
| Vesic (1963) | 15 | 28 | 58 | 130 | 315 |
| Vesic (1972): $I_r = 60$[b] | 20 | 27 | 40 | 59 | 85 |
| $I_r = 200$[b] | 29 | 46 | 72 | 110 | 165 |
| Terzaghi (1943) | | | | | |
| General shear | 12.7 | 22.5 | 41.4 | 81.3 | 173.3 |
| Localized shear | 5.6 | 8.3 | 12.6 | 20.5 | 35.1 |

[a] Various references are cited by Vesic (1972, 1977).
[b] Rigidity factor.

**TABLE 5.2 Values for $N_q$ and $\phi$[a]**

| $\phi°$ | 20 | 25 | 28 | 30 | 32 | 34 | 36 | 38 | 40 | 42 | 45 |
|---|---|---|---|---|---|---|---|---|---|---|---|
| $N_q$ (driven) | 8 | 12 | 20 | 25 | 35 | 45 | 60 | 80 | 120 | 160 | 230 |
| $N_q$ (drilled) | 4 | 5 | 8 | 12 | 17 | 22 | 30 | 40 | 60 | 80 | 115 |

[a] These values have been obtained from the curves provided by Meyerhof (1976).

corresponding failure surfaces assumed in the basic theoretical model. Some of these assumed failure surfaces have been presented in Chapter 1 (Figure 1.7). Table 5.1 summarizes the range of $N_q$ values according to the different investigators. A review of this table indicates that the true failure mechanism of a pile is not yet well understood. However, the analysis of the pile load test data presented by Meyerhof (1976) and Coyle and Castello (1981) indicate that, for design purposes, the $N_q$ values presented in Table 5.2 can be used. These values appear to be primarily affected by the $\phi$ values and the method of pile installation.

*Friction Capacity* ($Q_f$) In equation (5.3), the unit shaft friction $f_s$ needs to be determined to calculate $Q_f$. As shown in Figure 5.1, and from basic soil mechanics principles, $f_s$ can be written as

$$f_s = c_a + \sigma'_h \tan\delta \tag{5.5a}$$

where $c_a$ is the unit adhesion, $\delta$ is the angle of friction between soil and pile, and $\sigma'_h$ is the normal effective stress along the pile.

For cohesionless soils $c_a = 0$, then

$$f_s = \sigma'_h \tan\delta \tag{5.5b}$$

Also, if $K_s = \sigma'_{hl}/\sigma'_{vl}$, where $K_s$ is an earth pressure coefficient, $\sigma'_{vl}$ is the effective vertical pressure on an element at a depth $l$ along the pile, and $\sigma'_{hl}$ is the normal effective stress along the pile at a depth $l$, then

$$f_s = K_s \sigma'_{vl} \tan\delta \tag{5.5c}$$

Equation (5.3) may then be rewritten as:

$$Q_f = pK_s \tan\delta \sum_{L=0}^{L=L} \sigma'_{vl} \Delta L \tag{5.6}$$

For most design purposes, $\delta = 2/3\phi$. Meyerhof (1976) also analyzed the load test data to estimate $K_s$ values. (See Table 1.1 also.)

The results of the analysis show that the value of $K_s$ and, hence the skin friction, increases with the volume of displaced soil. Therefore, small displacement piles such as H piles and nondisplacement piles such as bored piles will have lower $K_s$ values than large displacement piles (Meyerhof, 1976). Table 5.3 lists values for $K_s$ for design purposes. As discussed, it should be recognized that, like end bearing, the shaft (skin) friction also increases up to the critical depth ($= 20B$) beyond which it can be assumed constant.

The final expression for ultimate load capacity, $(Q_v)_{\text{ult}}$, of a pile then becomes

$$(Q_v)_{\text{ult}} = Q_p + Q_f = A_p \sigma'_v N_q + pK_s \tan\delta \sum_{L=0}^{L=L} \sigma'_{vl} \Delta L \tag{5.7}$$

**TABLE 5.3 Values for $K_s$ for Various Pile Types in Sands[a]**

| Pile Type | $K_s$ |
|---|---|
| Bored pile | 0.5 |
| Driven H pile | 0.5–1.0 |
| Driven displacement pile | 1.0–2.0 |

[a] These values are based on the data presented by Meyerhof (1976). Similar values have been recommended in *Foundations and Earth Structures Design Mannual* 7.2 (1982).

where

$A_p$ = pile tip area
$\sigma'_v$ = effective overburden pressure at the pile tip
$\sigma'_{vl}$ = effective vertical stress at a point along the pile length
$p$ = pile perimeter
$K_s$ = earth pressure coefficient, determined from Table 5.3
$N_q$ = bearing capacity factor, determined from Table 5.2
$\delta = 2/3\phi$
$L$ = pile length

***Example 5.1*** A closed-ended 12-in. (300 mm) diameter steel pipe pile is driven into sand to 30-ft (9 m), depth. The water table is at ground surface and sand has $\phi = 36^\circ$ and unit weight ($\gamma$) is 125 lb/ft$^3$ (19.8 kN/m$^3$). Estimate the pipe pile's allowable load.

SOLUTION For circular pile, $A_p = \pi/4(1)^2 = 0.785$ ft, $p = \pi(1) = 3.14$ ft.

$$N_q = 60 \text{ from Table 5.2}$$
$$K_s = 1.0 \text{ from Table 5.3}$$
$$\delta = 2/3\phi = 2/3 \times 36^\circ = 24^\circ$$

$$\sum_{L=0}^{L=L} \sigma'_{vl}\Delta L = (\gamma_{\text{sub}}20B/2)(20B) + (\gamma_{\text{sub}}20B)(L - 20B)$$

This assumes that $\sigma'_{vl}$ increases with depth up to $20B$. Below this depth, $\sigma'_{vl}$ remains constant.
Where $\gamma_{\text{sub}} = 125 - 62.5 = 62.5$ lb/ft$^3$, $B = 1$ ft, $L = 30$ ft. Then:

$$\sum_{L=0}^{L=L} \sigma'_{vl}L = (62.5 \times 10 \times 1)(20 \times 1) + (62.5 \times 20 \times 1)(30 - 20 \times 1)\,\text{lb}$$
$$= 12{,}500 + 12{,}500 = 25 \text{ kips } (111.25\text{ kN})$$

Then, from equation (5.7):

$$(Q_v)_{\text{ult}} = Q_p + Q_f$$
$$(Q_v)_{\text{ult}} = 0.785(\gamma_{\text{sub}}20B)(60) + 3.14 \times 1 \times \tan 24 \times 25 \quad \text{kips}$$
$$= 58.88 + 34.95 = 93.83 \text{ kips}$$

where

$$Q_p = 58.88 \quad \text{and} \quad Q_f = 34.95$$
$$(Q_v)_{\text{all}} = (Q_v)_{\text{ult}}/FS = 93.83/3 = 31 \text{ kips } (137.95\text{ kN}),$$

using a factor of safety, $FS$, equal to 3.

***Empirical Analysis by Utilizing Standard Field Tests*** The three empirical methods that can be used to estimate bearing capacity of piles based on field soil tests are based on (1) standard penetration tests, (2) static cone penetrometer (Dutch cone with friction sleeve), and (3) pressuremeter tests.

*Standard Penetration Tests* This method should only be considered as a guide to estimate bearing capacity of pile foundation in cohesionless soils.

**1. End-bearing Capacity ($Q_p$).** According to Meyerhof (1976), the ultimate end (point or tip) resistance $Q_p$ in tons of driven piles can be estimated by the following relationships:

For sand:

$$Q_p^* = (0.4\bar{N}/B)D_f A_p \leqslant 4\bar{N} A_p \tag{5.8}$$

For cohesionless or nonplastic silt:

$$Q_p^* = (0.4\bar{N}/B)D_f A_p \leqslant 3\bar{N} A_p \tag{5.9}$$

where $\bar{N}$ is the average corrected Standard Penetration Test value near the pile tip and can be obtained from the following relationship:

$$\bar{N} = C_N N \tag{5.10}$$

where $C_N$ is obtained from equation (4.2) and $N$ is the average of the observed Standard Penetration Test value near the pile tip. This correction is made for the overburden pressure and has been further discussed in Chapter 4 (Section 4.1). $D_f$ is the depth of pile into granular stratum, which is the pile length ($L$) in homogeneous cohesionless soils. $B$ is the pile width or diameter, and $A_p$ is the pile tip area in square feet.

**2. Friction Capacity on Perimeter Surface ($Q_f$).** The friction capacity of a pile can be estimated by using the following relationship:

$$Q_f = (f_s)(\text{perimeter})(\text{embedment length}) \tag{5.11}$$

where $f_s$ is the ultimate unit shaft friction in tons per square feet. For driven piles, this value is given by the following relationship (Meyerhof, 1976, 1983):

$$f_s^* = \bar{N}/50 \leqslant 1 \text{ tsf} \tag{5.12}$$

where $\bar{N}$ is average corrected Standard Penetration Test value.

*$Q_p$ value is in tons. This value should be multiplied by a conversion factor of 8.9 to obtain $Q_p$ in kN.
*$f_s$ value is in tons/ft$^2$. This value should be multiplied by a conversion factor of 95.8 to obtain $f_s$ in kN/m$^2$.

The ultimate bearing capacity of a driven pile in cohesionless soil will then be the summation of $Q_p$ and $Q_f$ from equations (5.8) and (5.12). The allowable bearing capacity can then be obtained by applying a factor of safety of 3.

For drilled piles, $Q_p$ is one-third of the values given by equations (5.8) and (5.9) and $Q_f$ is one-half the values given by equation (5.12). These reductions reflect soil density reductions in the surrounding soil due to drilling.

***Example 5.2*** Using data of Example 5.1 find allowable bearing capacity based on standard penetration data as given in Fig. 5.2.

SOLUTION

(a) Average $N$ value near pile tip is 12 $(=(10 + 12 + 14)/3)$ (see Figure 5.2)

(b) Point Bearing ($Q_p$)

$$\sigma'_v \text{ near pile tip} = (125 - 62.5)\ 30\,\text{lb/ft}^2 = 1875\,\text{lb/ft}^2 = 0.938\,\text{tsf}$$

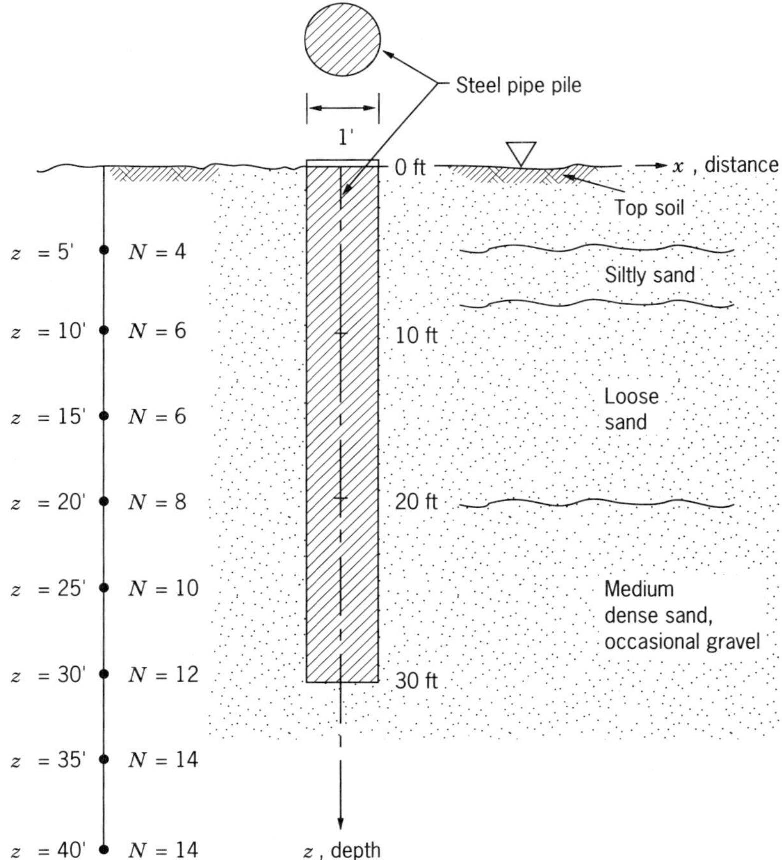


**Figure 5.2** Pile dimensions and soil properties with depth (for Example 5.2).

The correction for depth in $N$ values is applied by using equation (4.2) as follows:

$$C_N = 0.77 \log_{10} (20/0.938) = 1.02$$

Therefore, $\bar{N} = C_N N$. Then $\bar{N} = 1.02 \times 12 \simeq 12$
For driven piles from equation (5.8):

$$Q_p = (0.4\bar{N}/B)D_f A_p \leqslant 4\bar{N} A_p$$

where $0.4\bar{N}D_f A_p/B = 0.4 \times 12 \times 30 \times 0.785/1 = 113$ tons
$4\bar{N}A_p = 4 \times 12 \times 0.785 = 37.7$ tons
The lower of these two values will be $Q_p = 37.7$ tons
(c) Shaft Friction ($Q_f$)

Average $N$ value along pile shaft $= (4 + 6 + 6 + 8 + 10)/5 = 6.8$. Use an effective overburden pressure $\sigma'_v$ for average depth of $L/2 = 30/2$ ft. Then $\sigma'_v$ will be half the above value (i.e., $\sigma'_v = 0.938/2 = 0.469$ tsf). Then $C_N = 0.77 \log_{10} (20/0.469) = 1.25$. Thus, $\bar{N} = 6.8 \times 1.25 = 8.5$.

$$f_s = \bar{N}/50 = 8.5/50 = 0.17 \text{ tsf (which is less than 1 tsf (equation (5.12))}$$

$$Q_f = f_s \times p \times L = 0.17 \times \pi \times 1 \times 30 = 16 \text{ tons}$$

(d) Allowable Bearing Capacity ($Q_{all}$)

$$(Q_v)_{ult} = Q_p + Q_f = 37.7 + 16 = 53.7 \text{ tons}$$

$$(Q_v)_{all} = (Q_v)_{ult}/FS = 53.7/3 = 17.9 \text{ tons} = 35.8 \text{ say } 35 \text{ kips } (156 \text{ kN})$$

*Static Cone Penetration Values* Ultimate bearing capacity can also be estimated by using static cone penetration values as follows:

**1. End bearing ($Q_p$)** The static cone penetrometer is a model pile. The penetration resistance of a static cone, when pushed into a homogeneous cohesionless soil, can be correlated with a similarly installed full-sized pile.

According to Meyerhof (1976), the ultimate tip resistance of driven piles, $Q_p$, can also be obtained from the static cone resistance ($q_c$) value and is given by the following relationship.

$$Q_p = A_p q_c \tag{5.13}$$

where $A_p$ is the pile tip area and $q_c$ is the cone penetration resistance. Experience indicates that equation (5.13) is only applicable when pile embedment into the granular soil (i.e., pile length) is at least 10 times the pile width or diameter (Meyerhof, 1976; *Foundations and Earth Structures Design Manual* 7.2, 1982). For shallower depths, this relationship can not be used. De Ruiter and Beringen (1979) provide a modified version of equation (5.13) to estimate ultimate tip resistance of driven piles. This modified relationship needs further field verification. Readers should refer to the original paper for further details of this modified formula.

**2. Friction Capacity on Perimeter Surface ($Q_f$)** The $Q_f$ for driven piles can be estimated by the following relationship:

$$Q_f = (f_s)(\text{perimeter})(\text{embedment length}) \tag{5.14}$$

$$f_s = \text{ultimate shaft friction of driven pile}$$

The $f_s$ can be approximately given by the unit resistance of local friction sleeve, $f_c$, of static cone penetrometer (Meyerhof, 1976). Chapter 4 (Section 4.1) provides further details of the static cone penetration test method.

The ultimate bearing capacity, $(Q_v)_{ult}$, of a driven pile in cohesionless soils will then be the sum of $Q_p$ and $Q_f$ from equations (5.13) and (5.14). A factor of safety of 3 should be used to obtain the allowable bearing capacity from the above equations. In drilled or bored piles, the drilling operation may result in density reduction of the surrounding soil. Therefore, for drilled piles, use one-half of the allowable bearing capacity as obtained above (Meyerhof, 1976).

*Pressuremeter Tests* As discussed in Chapter 4, (Section 4.1.2), a pressuremeter is a very effective and useful tool to measure in-situ soil properties. Empirical relationships have been developed to relate these in-situ soil parameters by pressuremeter for pile foundation design (Baguelin et al., 1978 and *Canadian Foundation Engineering Manual* 1978,* 1985). These relationships are presented as follows for the end-bearing and the skin friction capacities.

**1. End bearing capacity ($Q_p$)** The following empirical relationship can be used to estimate the end-bearing capacity of a pile if the limit pressures $P_L$ and $P_0$ are obtained from pressuremeter tests. Methods of obtaining $P_L$ and $P_0$ have been discussed in Chapter 4 (Section 4.1).

$$Q_p = A_p[q_0 + K_q(P_L - P_0)] \tag{5.15}$$

where

$Q_p$ = ultimate end-bearing capacity

$A_p$ = pile point (end) area

$q_0$ = horizontal at rest stress in soil at the elevation of the pile tip

$k_q$ = bearing capacity factor determined from Figure 5.3. In this figure, class 1 curves are for clays and silts; class 2 curves are for hard clay, dense silt, loose sand, and soft or altered (weathered) rock; class 3 curves are for sand and gravel and rock; and class 4 curves are for very dense sand and gravel

*The 1978 edition of the *Canadian Foundation Engineering Manual* provides more data on the pressuremeter tests than 1985 edition of the manual.

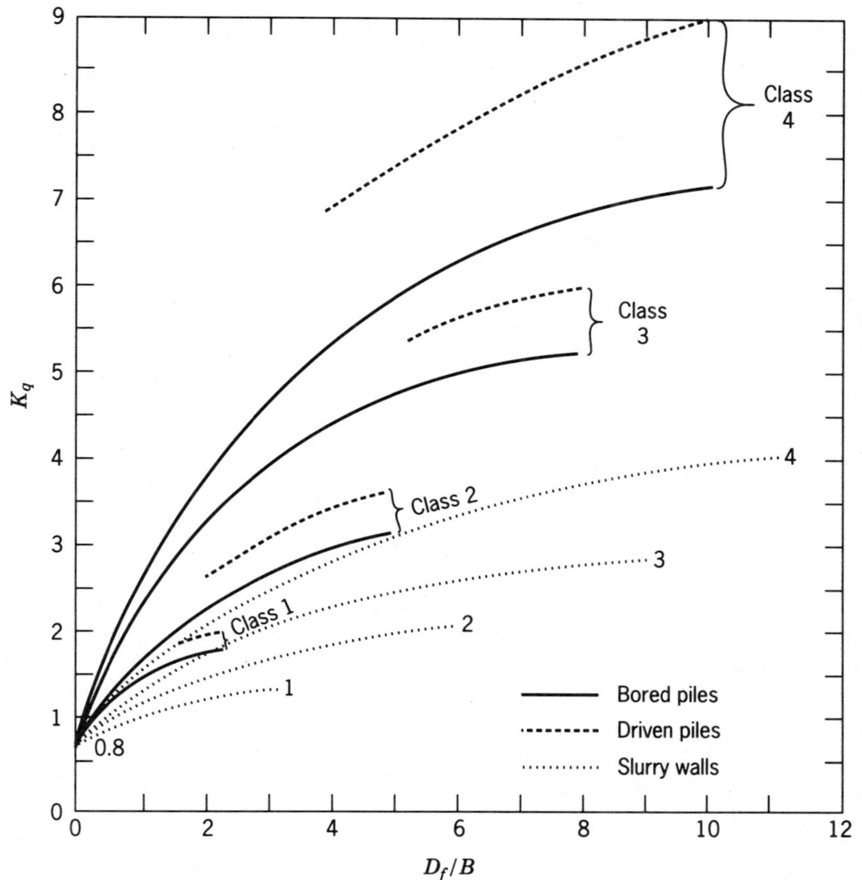


**Figure 5.3** Bearing capacity factor $K_q$ for pile foundation design by pressuremeter (*Canadian Foundation Engineering Manual*, 1978).

(Shields, 1987). A factor of safety of 3 should be used to obtain the allowable end-bearing capacity from equation (5.15).

**2. Skin friction capacity ($Q_f$)** The ultimate skin friction capacity $Q_f$ can be obtained by multiplying the ultimate skin friction, $\tau_f = f_s$, with the pile length and the pile perimeter. The ultimate skin friction can be related to the ultimate pressure $P_L$ and can be obtained from Figure 5.4. This figure is based on the empirical relationships presented by Baguelin et al. (1978). For piles embedded in cohesive soils, curve $A$ should be used directly for timber and concrete piles. These values should be multiplied by 0.75 for steel piles. For cohesionless soils, curve $A$ should be used for nondisplacement concrete piles and displacement steel piles. These values should be multiplied by a factor of 0.5 for nondisplacement steel piles. Curve $B$ should be used for displacement concrete

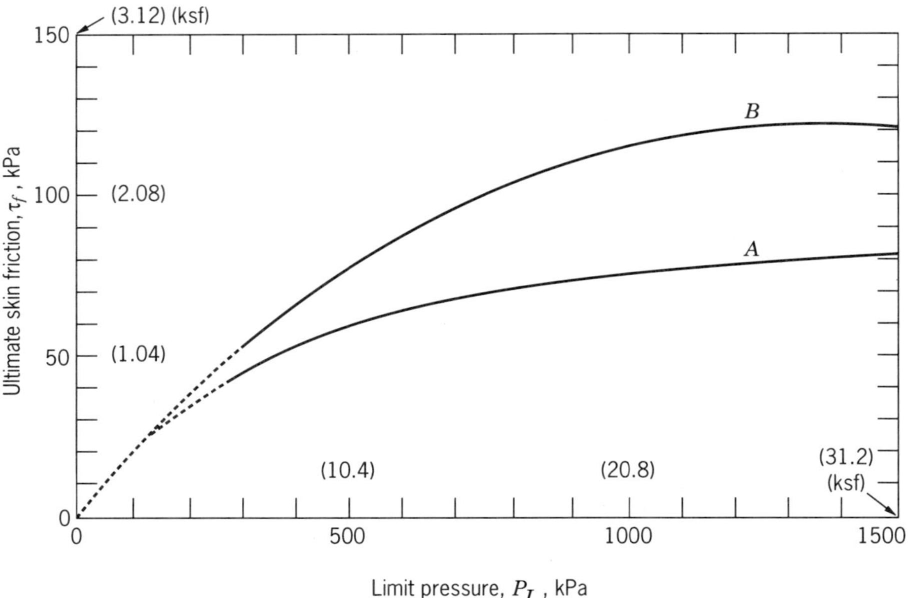


**Figure 5.4** Ultimate skin friction, $\tau_f$, on piles by pressuremeter method (*Canadian Foundation Engineering Manual*, 1978).

piles. A factor of safety of 2 is recommended to ultimate skin friction value to arrive at the allowable skin friction values (*Canadian Foundation Engineering Manual*, 1978).

***Dynamic Driving Resistance*** The two methods of estimating ultimate capacity of piles on the basis of dynamic driving resistance are pile-driving formulas and wave equation analysis. Pile capacities based on pile-driving formulas are not always reliable. They should therefore be supported by local experience or testing and should be used with caution. Pile capacities estimated on the basis of wave equation analysis have more rational approach than the estimation on the basis of pile driving formulas.

*Pile-driving Formulas* Figure 5.5 shows the basic concept behind the derivation of pile driving formula. In this figure, $xx$ shows the position of pile cap just before a ram or a hammer of weight $W$ strikes the pile cap after falling through a height $H$. The pile cap then moves a distance $(S + S_e)$. The term $S$ is the distance pile point penetrates per blow of the hammer, and $S_e$ is the elastic compression of the pile and pile cap. The work done by a falling hammer and the work required to penetrate pile point by $S$ can then be related as follows:

$$WH = Q_{\text{dyn}}S + \Delta E \tag{5.16}$$

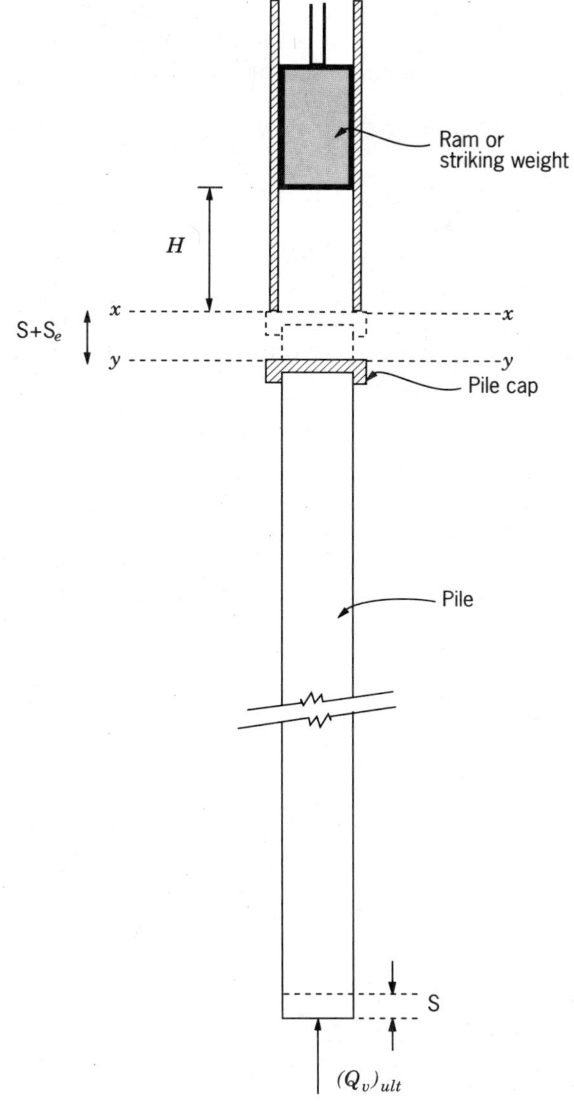


**Figure 5.5** Basic concept behind derivation of pile-driving formulas.

where

$Q_{\mathrm{dyn}}$ = dynamic resistance of soil to pile driving
$\Delta E$ = energy losses

Energy losses result when all of the energy of the falling hammer is not used in pile penetration into the soil but is converted into losses such as heat, noise, and the elastic compression of the pile ($S_e$). If we assume that $C$ represents the additional penetration of pile that would have occurred if energy losses during pile driving were zero, then equation (5.16) can be modified as follows:

$$WH = Q_{\text{dyn}} \times S + Q_{\text{dyn}} C \tag{5.17}$$

Then

$$Q_{\text{dyn}} = WH/(S + C) \tag{5.18}$$

The allowable static-bearing capacity, $(Q_v)_{\text{all}}$, can then be obtained by applying appropriate factor of safety. Numerous attempts have been made by various investigators to obtain $Q_{\text{dyn}}$ by taking into account the energy losses. These attempts have resulted in many complicated expressions. Smith (1962) reports that the editors of *Engineering News-Record* have 450 such formulas on file. All these formulas are based on equation (5.16) and ignore the dynamic aspect of pile driving, which will be discussed in the following paragraphs. Experience shows that complicated pile-driving formulas do not possess any advantages over the simpler ones (Terzaghi and Peck, 1967). In view of this, the use of elaborate formulas is not warranted. Formulas summarized in Table 5.4 can be used as a guide to estimate allowable pile capacities. These formulas have been widely used for preliminary estimates of pile capacities and for construction control when

**TABLE 5.4 Basic Pile-driving Formulas (Design Manual DM 7.2, 1982)**

| For Drop Hammer | For Single-Acting Hammer | For Double-Acting Differential Hammer |
|---|---|---|
| | $(Q_v)_{\text{all}}^{\text{a,c}} = \dfrac{2WH}{S + 0.1}$ | $(Q_v)_{\text{all}}^{\text{a}} = \dfrac{2E}{S + 0.1}$ |
| $(Q_v)_{\text{all}}^{\text{c}} = \dfrac{2WH}{S + 1}$ | | |
| | $(Q_v)_{\text{all}}^{\text{b,c}} = \dfrac{2WH}{S + 0.1\dfrac{W_D}{W}}$ | $(Q_v)_{\text{all}}^{\text{b}} = \dfrac{2E}{S + 0.1\dfrac{W_D}{W}}$ |

[a] Use when driven weights are smaller than striking weights.

[b] Use when driven weights are larger than striking weights.

[c] This is based on the most commonly used pile-driving formula, known as the Engineering New formula.

$(Q_v)_{\text{all}}$ = allowable pile load in pounds.

$W$ = weight of striking parts of hammer in pounds.

$H$ = effective height of fall in feet.

$E$ = actual energy delivered by hammer per blow in foot-pounds.

$S$ = average net penetration in inches per blow for the last 6 in. of driving set.

$W_D$ = driven weights including pile.

*Note:* 1. Ratio of driven weights to striking weights should not exceed 3.

supplemented by full-scale field load tests. The *Foundation and Earth Structures Design Manual* (1982) recommends the use of these formulas. Engineering literature provides many case histories that show problems arising from a naive dependence on such formulas. The formulas are still widely used because they provide an invaluable guide to field personnel. The engineer uses them as a guide to determine when to instruct the contractor to stop driving a pile and move on to the next one. The reliability of a dynamic driving formula can be greatly improved if the load test is first performed at the site and the dynamic formula is modified to fit the results of the load test. This adjusted dynamic driving formula can then be used as a field control. These formulas must be supplemented by an adequate site specific soil exploration program, and a minimum of three test piles should be driven if site conditions are uniform. For erratic subsurface conditions, more test piles are required. The shortcomings of dynamic pile-driving formulas can be categorized into the following three areas (Goble and Rausche, 1980):

1. **Driving System Representation** In dynamic formulas, only the rated energy and estimated losses are included. Driving system loss representations are oversimplified, and the formulas do not attempt to deal realistically with poor equipment performance. Thus, driving system representation in dynamic formulas are only approximate at best.
2. **Pile Flexibility** In dynamic formula derivation, the pile is assumed to be rigid and all effects of flexibility are neglected.
3. **Soil Resistance** The soil model approximation in the model is far from the real soil because the formulation assumes constant soil resistance. Also static pile resistance may not be equal to dynamic pile resistance (See Chapter 1)

In spite of the above shortcomings of pile-driving formulas, the use of the blow count is still widely used to assess the quality of the pile installation because it is convenient and simple to observe in the field. An alternative improved approach based on a one-dimensional wave propagation was developed for pile driving (Smith, 1962). This is called the *Wave Equation Approach* and is now described.

***Example 5.3*** Find the allowable load on a steel pipe pile that was driven by a 5000-lb drop hammer having a 6.5 feet free fall. The pile-driving record showed 12 blows for the last foot of driving into the cohesionless soil. Of these 12 blows the last 6 inches had 7 blows. Determine the allowable load on the pile.

SOLUTION

$$W = 5000\,\text{lb}$$
$$H = 6.5\,\text{ft}$$
$$S = \text{penetration in inches per blow}$$
$$= 6/7 = 0.86\,\text{in./blow}$$

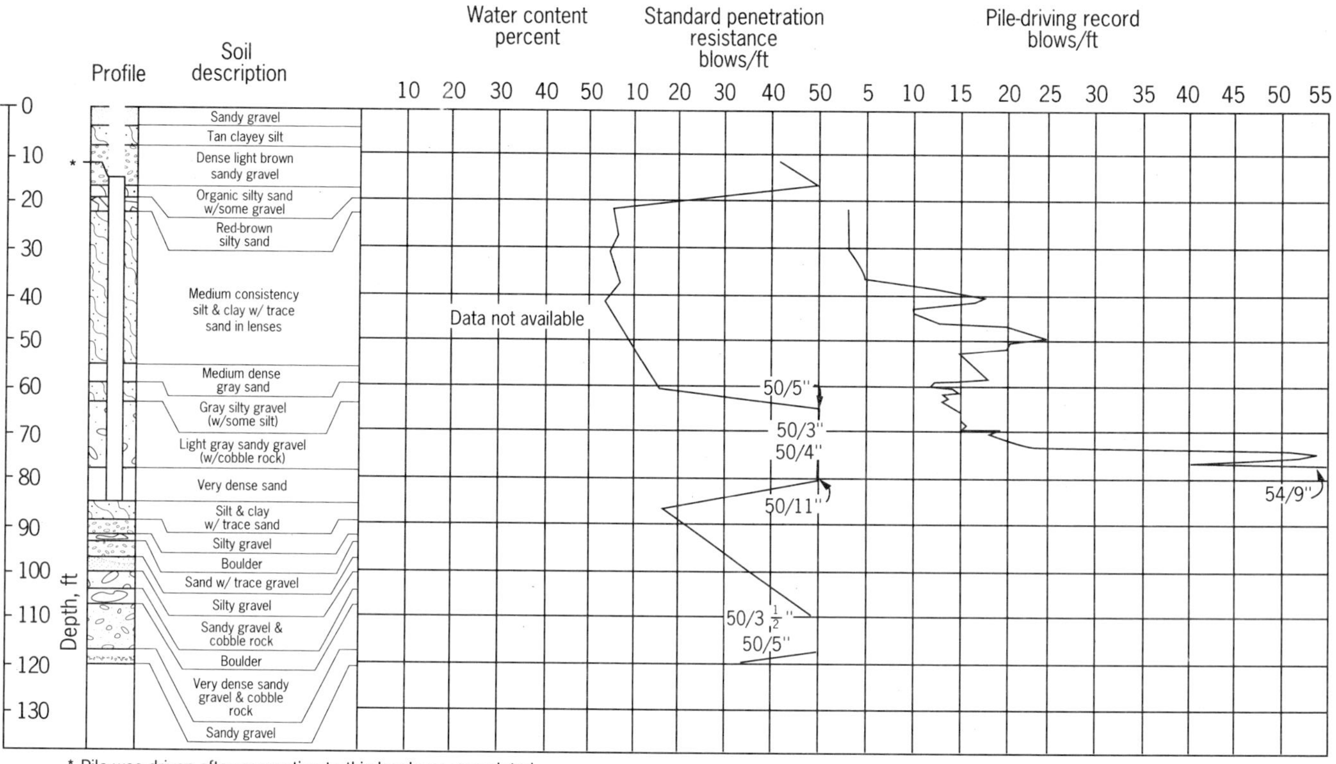


* Pile was driven after excavation to this level was completed.

**Figure 5.6** Soil stratigraphy and pile-driving records for a HP/14 × 73 pile used in Example 5.4 (American Iron and Steel Institute, 1985).

$$(Q_v)_{\text{all}} = 2WH/(S+1)(\text{from Table 5.4 for drop hammer})$$
$$= 2 \times 5000 \times 6.5/(0.86+1) = 34.8\,\text{kips}\ (155\,\text{kN})$$

***Example 5.4*** A steel HP 14 × 73 pile was installed by using a Delmag D-22 double acting hammer having a rated energy of 39,800 ft-lb. The pile was driven through various soil layers and was finally resting on very dense sand as shown in Figure 5.6. Driving records showed 54 blows for the last 9 in. of driving. Estimate the allowable load on this pile.

SOLUTION From Table 5.4, the allowable load is given by:

$(Q_v)_{\text{all}} = 2E/(S+0.1)\,\text{lb}$ (from Table 5.4 for Delmag double-acting hammer)

$E = 39{,}800\,\text{ft-lb}$

$S = 9/54 = 0.167\,\text{in./blow}$ (from Figure 5.6, pile-driving record)

$(Q_v)_{\text{all}} = 2 \times 39{,}800/(0.167+0.1) = 298\,\text{kips}\ (1326\,\text{kN})$

### 5.1.2 Wave Equation Analysis and Dynamic Pile Drivability

The wave equation analysis is based on using the theory of one-dimensional stress wave propagation. The stress wave is generated from the hammer impact on the pile head. The analysis is used to obtain the following:

1. **Pile Capacity** A plot of ultimate pile capacity, $(Q_v)_{\text{ult}}$ versus set, S, can be developed.
2. **Driving Stresses** Plots of stress versus set can be obtained to assess the potential for pile overstress.
3. **Equipment Compatibility** Appropriate hammer sizes and cushions for a particular pile are evaluated.

As will be discussed in the following paragraphs, this analysis requires certain soil and pile input parameters. These parameters are assumed and cannot be related to routinely measured soil parameters. This problem has been resolved by using wave equation analysis in conjunction with field measurements. This is called *dynamic monitoring* and will be discussed under the heading *Case Method* following the wave equation analysis.

***Wave Equation Analysis*** This method was first put to practical use for pile foundations by Smith (1962). According to this method the ram (hammer), the capblock, the pile cap, and the pile are represented as a series of weights and springs as shown in Figure 5.7. The time element is chosen sufficiently small so that the stress wave travels from one pile element of length $\Delta L$ into the next lower element during $\Delta t$. Smith (1962) recommends that for all practical applications, the following $\Delta L$ and $\Delta t$ values can be used in the analysis:

1. Steel pile: $\Delta L = 8$ to 10 ft, $\Delta t = 0.00025$ sec

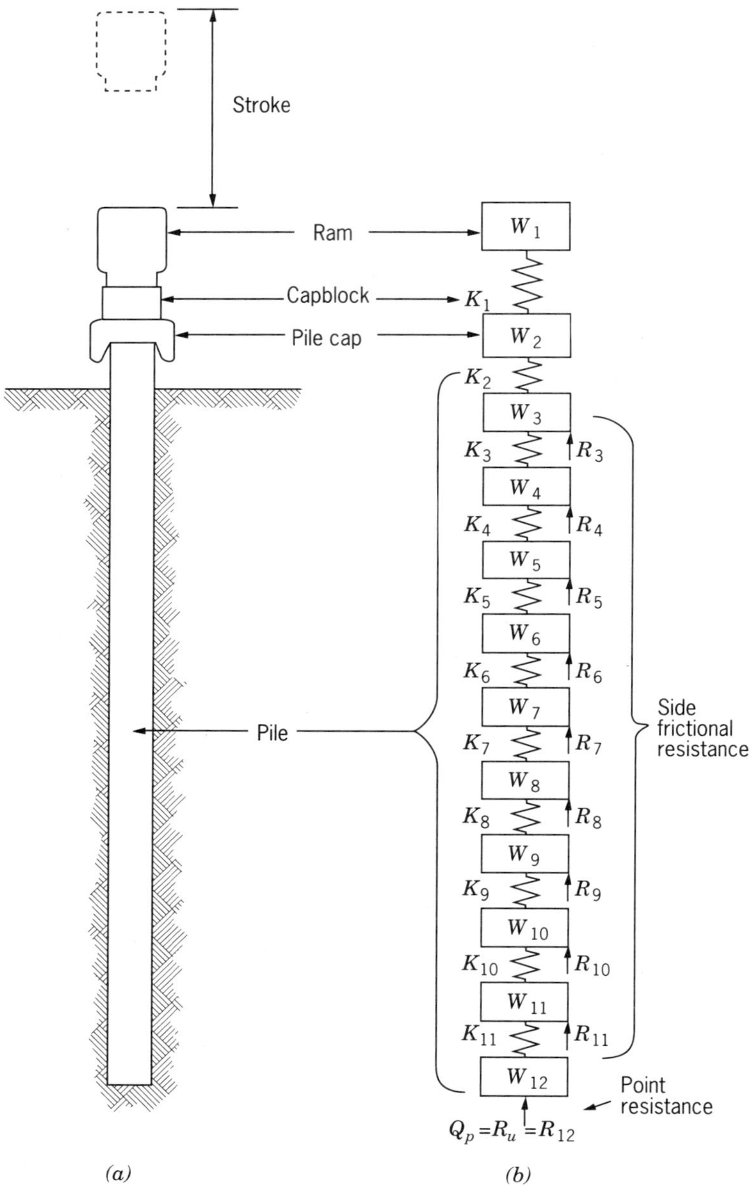


**Figure 5.7** Wave equation analysis: Method of representation of pile and other parts of model. (a) Actual, (b) as represented (after Smith, 1962).

2. Concrete pile: $\Delta L = 8$ to 10 ft, $\Delta t = 0.00033$ sec
3. Wood pile: $\Delta L = 8$ to 10 ft, $\Delta t = 0.00025$ sec

*Soil Behavior Model* Smith (1962) assumed that the soil pile response under vertical load is as shown in Figure 5.8 and is described as follows:

SOIL RESISTANCE AT PILE POINT It is assumed that when a driving force (load) is applied to a pile, the soil compresses elastically to a certain distance termed as *quake* ($Q$), and then the soil fails plastically with a constant ultimate resistance ($R_u$). On removal of the load, an elastic rebound equal to $Q$ occurs. A permanent set $S$ of the pile occurs as shown by $\overline{AB} = \overline{OC}$ (Figure 5.8). This description does not consider the element of time (i.e., soil offers more instantaneous resistance to rapid motion than to slow motion). This has been represented by introducing a factor called *viscous damping* ($J$). The damping resistance is instantaneous or temporary and does not contribute to the bearing capacity of the pile. The constant $J$ refers only to the point resistance $R_{12}$ of the pile point (see Figure 5.7).

SOIL RESISTANCE ALONG THE PILE SHAFT The resistance along the pile shaft ($R_3$ to $R_{11}$ inclusive in Figure 5.7) are calculated by using a side resistance factor called the damping constant $J'$ instead of the factor $J$, which has been used for point bearing. As the pile is driven, the soil along the shaft remains in place while the soil at the pile point is displaced rapidly. Therefore, $J'$ should be smaller than $J$. For example, Smith (1962) recommends $J' = 0.05$ and $J = 0.15$. This is further discussed in the following paragraphs.

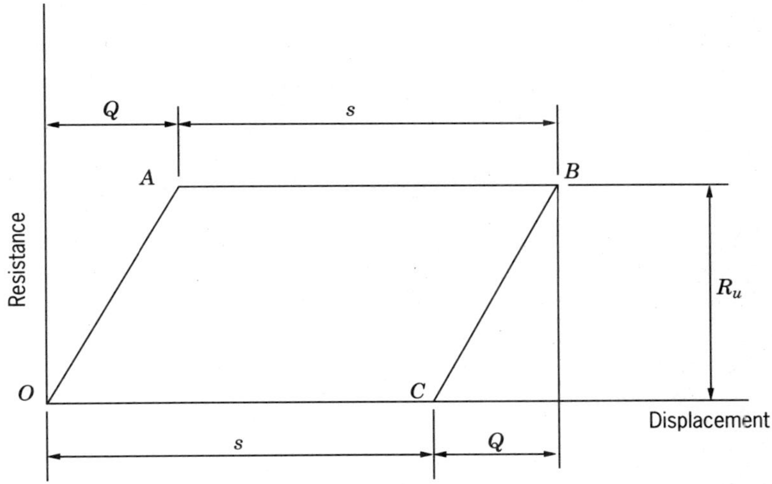


**Figure 5.8** Load displacement diagram at pile point (after smith, 1962)

*Pile and Driving Component Model* In addition to the soil, the various components such as the hammer, capblock, pile cap or follower, and pile characteristics should also be considered. The hammer ram can be represented by a single weight $W_1$ (Figure 5.7) without elasticity. This assumption is reasonable for short, heavy, and rigid hammers. In special cases where the ram is long and slender, it can be represented by a series of weights and springs. The velocity of the pile cap at the instant of impact is the same as the velocity of ram. This velocity of impact, $v_1$, can be computed as follows:

$$v_1 = \sqrt{\frac{\text{(Rated energy, in ft-lb)(efficiency)(64.4)}}{\text{Weight of ram}}} \tag{5.19}$$

The rated energy is obtained from the manufacturer's catalogue while the efficiency is sometimes given by the manufacturer or it may be assumed. The capblock is represented by spring $K_1$. The coefficient of restitution $e_1$, for capblock in accordance with the Newton's law of impact, is defined as:

$$e_1 = \sqrt{\frac{\text{Energy output}}{\text{Energy input}}} \tag{5.20}$$

Smith (1962) recommends following $e_1$ values for capblocks:

1. For a hardwood capblock with vertical grain, 6 in. in original height and with a horizontal area of $A$ in.$^2$, the following $K_1$ and $e_1$ values can be used.

$$e_1 = 50\%$$
$$K_1 = 20{,}000A \text{ lb/in. of compression}$$

2. For a 12-in. high Micarta capblock with a horizontal area of $A$ in.$^2$, the following $K_1$ and $e_1$ values can be used:

$$e_1 = 80\%$$
$$K_1 = 45{,}000A \text{ lb/in. of compression}$$

The pile cap or follower or the helmet are ordinarily short and rigid objects and can be represented by a single weight, such as $W_2$ in Figure 5.7. If the pile cap is long and slender, then it should be represented by a series of weights and springs. In general, the elastic constant, $K$, of any object of uniform cross sectional area $A$, length $\Delta L$, and modulus of elasticity $E_p$ can be determined from the following:

$$K = AE_p/\Delta L \tag{5.21}$$

The 10 springs $K_2$ to $K_{11}$ inclusive (Figure 5.7) represent the elasticity of the pile. The springs $K_3$ to $K_{11}$ can transmit tension because of the continuity of the pile material. However, springs $K_1$ and $K_2$ cannot transmit tension because the ram, the pile cap, and the pile are separate objects.

*Mathematical Formulation* Let us assign the subscript $m$ to denote the general case. For example, $W_m$ will denote any weight in Figure 5.7. Thus, if $m = 4$ then $W_4$, $K_4$, and $R_4$ will denote the element properties. The letter $t$ will be used to denote time. The instantaneous spring compression, element displacement, force, resistance, velocity, and accelerating force of this element for any time interval $t$ will be denoted by $C_m$, $D_m$, $F_m$, $R_m$, $V_m$, and $Z_m$, respectively. The letters $c_m$, $d_m$, and $v_m$ will denote spring compression, displacement, and velocity in time interval $t-1$. The letter $d^*$ will refer to a displacement value in time interval $(t-2)$.

According to Smith (1962), the following basic equations can be used for the numerical solution of the wave equation:

$$D_m = d_m + v_m(12\Delta t) \tag{5.22}$$

$$C_m = D_m - D_{m+1} \tag{5.23}$$

$$F_m = C_m K_m \tag{5.24}$$

$$Z_m = F_{m-1} - F_m - R_m \tag{5.25}$$

$$V_m = v_m + Z_m \frac{\Delta t g}{W_m} \tag{5.26}$$

where $D_m$, $d_m$, and $C_m$ are in inches; $V_m$ and $v_m$ are in feet per second; the time interval $\Delta t$ is in seconds; the spring constant $K$ is in pounds per inch; $F_m$ and $R_m$ are in pounds; and acceleration due to gravity $g$ is in feet per second per second. The subscript $m$ denotes that all these equations are applicable to the pile element $m$.

The soil spring constant along element $m$ is given by the following:

$$K'_m = R_{um}/Q \tag{5.27}$$

$R_{um}$ is portion of $R_u$ (Figure 5.8) applicable to weight $W_m$. The frictional resistance, $R_m$, alongside the pile is given by the following:

$$R_m = (D_m - D'_m)K'_m(1 + J'v_m) \tag{5.28}$$

where $D'_m$ is the soil plastic displacement around element $m$ in time interval $t$ in inches and $J'$, as described earlier, is the damping constant applicable to resistance at side of pile.

The soil resistance at the pile point, $R_p$, is then given by

$$R_p = (D_p - D'_p)K'_p(1 + Jv_p) \tag{5.29}$$

where subscript $p$ denotes values of parameters at pile point and $J$ is damping constant applicable to pile point.

Equations (5.22) through (5.26) can be combined to a obtain wave equation converted into a difference equation suitable for numerical computations and is given by the following (Smith, 1962).

$$D_m = 2d_m - d_m^* + \frac{12g(\Delta t)^2}{W_m}[(d_{m-1} - d_m)k_{m-1} - (d_m - d_{m+1})k_m - R_m] \quad (5.30)$$

*Wave Equation Analysis Computations*

INPUT DATA

1. Obtain the pile cap velocity at the instant of impact from equation [(5.19)].
2. Obtain the weight of ram $W_1$, capblock spring constant $K_1$, pile cap weight $W_2$, and the modulus of elasticity of the pile material.
3. The coefficient of restitution $e_1$ can be obtained from the data provided above for capblocks, and the pile spring constant can be computed from equation (5.21).
4. Assign soil properties $Q$, $J$, and $J'$. Smith (1962) recommended $Q = 0.1$ in., $J = 0.15$ and $J' = 0.05$.

*Computational Steps*

1. Compute the displacements of each element $D_1$ through $D_p$ by using equation (5.22). Then compute the soil plastic displacement $D'_p$, Smith (1962) gives a computer SUBROUTINE for such calculations.
2. Compute $R_m$ and $R_p$ by using equations (5.28) and (5.29).
3. Compute $C_m$ from equation (5.23).
4. Compute the forces in each element by using equation (5.24).
5. Compute the velocity of each element by using equation (5.26).
6. Place the just-computed values of $D_m$ and $V_m$ in storage (i.e., $D_1$ through $D_p$ and $V_1$ through $V_p$ are placed in computer storage).
7. Set one time interval back and repeat the calculations to compute new $D_m$ and $V_m$. The computer is programmed to stop automatically when the following two conditions are reached (Smith, 1962).

(a) All the velocities $V_1$ through $V_p$ inclusive become negative.
(b) The ground plastic displacement at pile point $(D'_p - d'_p)$ becomes zero.

The foregoing two conditions indicate that the pile will not penetrate into the soil and will begin to rebound if driving is continued.

Thus, the plots of $R_u$ versus the blows/inch are made by assuming several values of $R_u$. The blows/inch is the inverse of set, which is inch/blow. The wave equation computer program is used to obtain the set for the particular assumed $R_u$. For each $R_u$ versus blows/inch, a percent of $R_u$ is assumed to be carried by the pile point. This percent $R_u$ is constant for one set of calculations (i.e., one curve may be obtained when it is assumed that pile point carries 60 percent of the ultimate load while another curve will be obtained if it is assumed that pile point carries 50 percent of the ultimate load).

Figure 5.9 presents the results of an analysis from wave equation carried out on an 18-in. (450 mm) outside diameter, 0.375 in (9.4 mm) wall thickness, and 75-ft (22.5 m)-long steel pile that had 35 ft (10.5 m) length embedded into the soil. The pile was driven with a No. 1 Vulcan hammer having 70 percent efficiency. The hammer cushion used was a standard aluminum-micarta stack. In this analysis, it was assumed that 50 percent of the ultimate load capacity was moblized uniformly over the embedded portion of the pile, and the remaining 50 percent was mobilized at the tip.

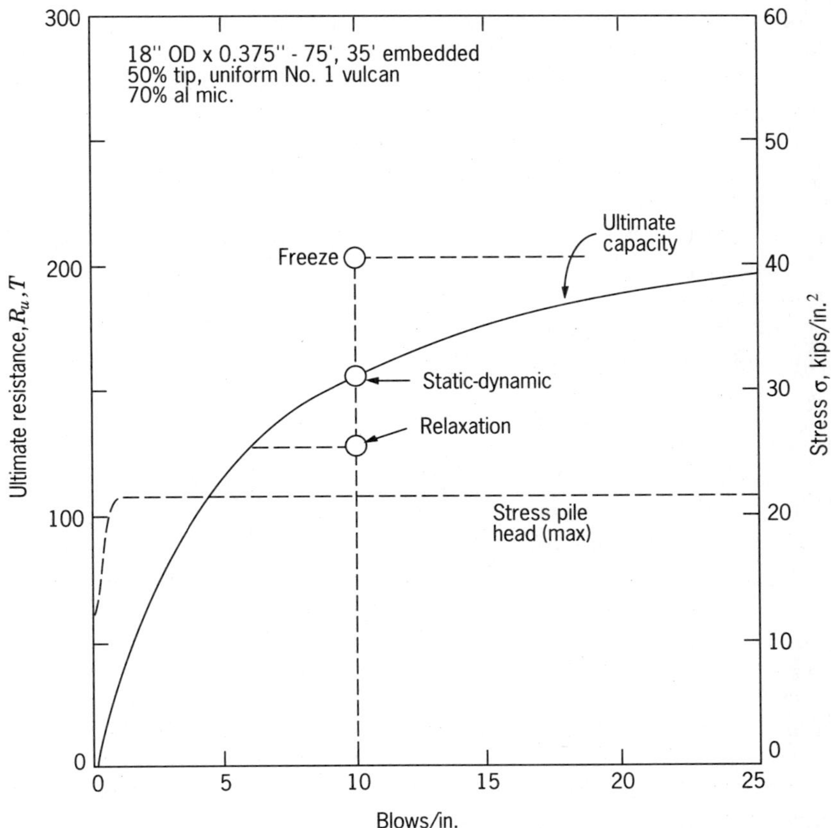


**Figure 5.9** An example of ultimate resistance versus blows per inch by wave equation analysis (Davisson, 1975, as cited in Prakash, 1981).

Based on the wave equation analysis on the previously mentioned pile, a plot of ultimate resistance $R_u$ and the driving record in blows per inch (bpi) can be obtained as shown by solid line in Figure 5.9. This resistance should then represent the ultimate static pile capacity. In case a *setup* or *freeze* occurs after driving is completed, then the static capacity will be higher than the predicted $R_u$ value. On the other hand, when *relaxation* occurs after driving is completed, then the static load capacity will fall below the predicted $R_u$ in the plot. For example, in Figure 5.9 the wave equation analysis predicts that for 10 blows/inch of driving the ultimate static pile capacity is 150 tons, while two cases have been exemplified in which this capacity would be 200 tons if *freeze* occurs and would be 125 tons if *relaxation* occurs. The terms *soil freeze* and *relaxation* were discussed in Chapter 3 (Section 3.4.2). The analysis should therefore be carried out for restriking conditions. The dashed line in this figure shows the pile head stress versus blows per inch. It shows that at 10 blows/inch, the pile is subjected to an approximate stress of 21 kips/in.$^2$, which is below the ultimate stress of 36 kips/in.$^2$ of steel pile. This plot is used to ensure that the pile is not overstressed. The wave equation analysis is also used to determine the equipment compatibility. The pile and the driving equipment are not considered compatible if the solutions to wave equations are not obtained (i.e., the equipment is either too small to provide enough driving energy or is too big so that the energy is being wasted during driving).

Wave equation analysis is easily carried out on the personal computers today (1990). The computer program mostly known in North America for the wave equation analysis of pile driving are the TT1 program (Hirsch et al., 1976) and the WEAP program (Goble and Rausche, 1980) or WEAP-86. The TT1 program was primarily developed for analysis of piles driven with air/steam hammers or drop hammers. The WEAP program, in addition, models the actual combustion sequence of the diesel hammer and also calculates the ram rebound of the hammer (Authier and Fellenius, 1983; Goble and Rausche, 1980).

The reliability of the wave equation analysis depends on the accurate estimation of various parameters such as damping factors, quake values, the hammer efficiency, capblock, and cushion properties. Thus the analysis requires reliable soil–pile parameters and an experienced operator with knowledge in both the computer analysis and the piling practice. In addition, several computer runs may be required to account for variability in the field. Rausche et al. (1985) have developed a simplified solution to wave equation to obtain axial static pile capacity by using dynamic force and acceleration measurements during pile driving and by utilizing empirical correlations to static pile load tests. This method saves computational time and effort significantly. The method is called the *Case Method* and is presented below.

***Case Method: Static Capacity from Dynamic Monitoring*** In dynamic pile formulas, the only measurement taken is the permanent set per hammer blow. In spite of many modifications made on dynamic pile formulas over a period of 100 years, there has not been better results in predicting pile capacities. However, with the development of electronics, it is now practical to measure parameters during pile driving in addition to the pile set. Based on the analysis of extensive pile testing and the force and acceleration measurements made during pile driving first reported by the Michigan State Highway Commission (1965) and later by Goble et al. (1975), it is now feasible to make force and acceleration measurements as a routine on a piling project. These data can then be used in wave equation analysis and predict pile behavior.

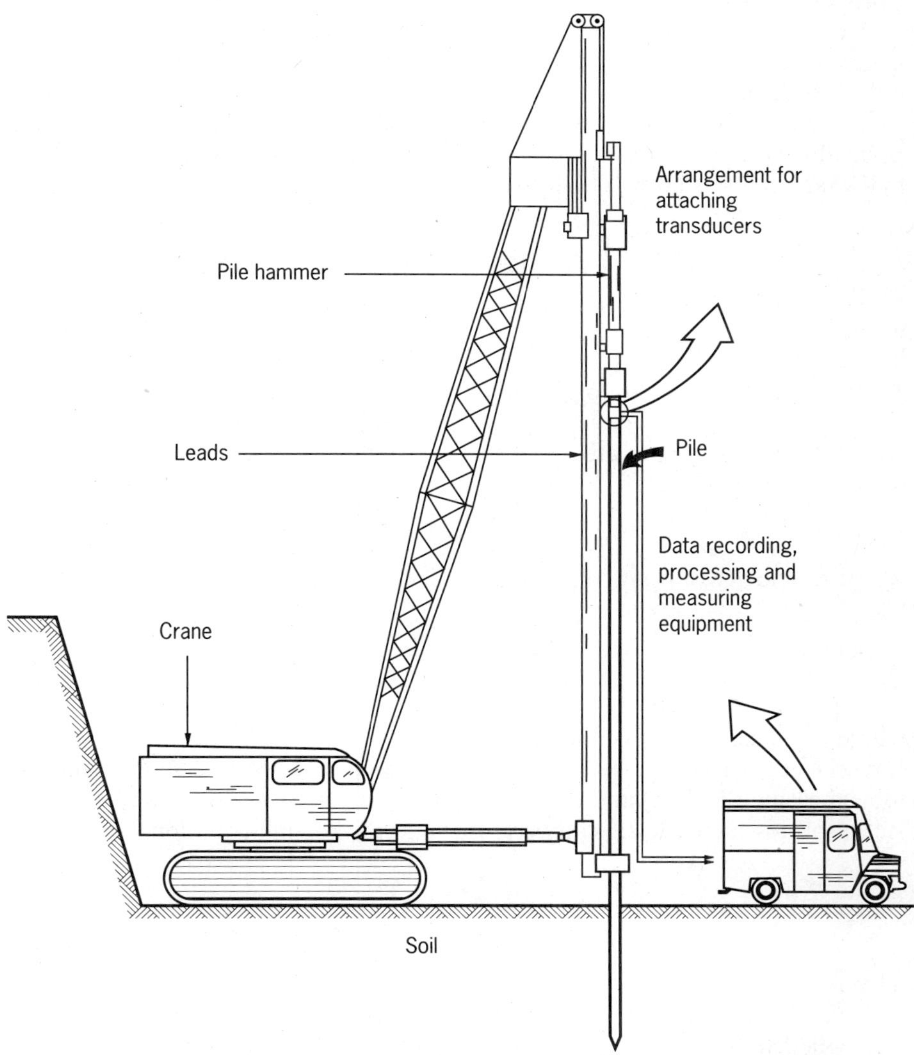


**Figure 5.10** Typical setup for dynamic monitoring (after Thompson, 1986).

The force and acceleration measurements are recorded by means of a pile-driving analyzer system. In this system, one pair of strain transducers and one pair of accelerometers with built-in amplifiers are normally bolted onto the pile below the pile head. During pile driving, the signals from the transducers are transmitted by a connector box hung below the pile head to the analyzer that is kept in a monitoring station on the ground. The overall arrangement for this monitoring is shown in Figure 5.10. The analyzer, receives the signals from the transducers and calculates and prints the values of impact force, maximum force, developed energy, and a computed estimate of the mobilized soil resistance (Authier and Fellenius, 1983).

The determination of this mobilized static soil resistance is based on the principle that when a pile head is struck with a pile-driving hammer, an axial force is suddenly applied to it resulting in a stress wave that travels down the pile away from the pile head. If we measure particle velocity $v(t)$ and force $F(t)$ at a point along this pile at time $t$, then as long as no reflections arrive at that point, the following relationship will hold (Rausche et al., (1985):

$$v(t) = \frac{V_r}{E_p A} F(t) \tag{5.31}$$

where

$E_p$ = modulus of elasticity of pile material

$A$ = Pile cross-sectional area

and

$V_r$ = the velocity of wave in pile and is given by the following equation:

$$V_r = \sqrt{\frac{E_p}{\rho}} \tag{5.32}$$

where $\rho$ = mass density of the pile material.

Rausche et al. (1985) have shown that if a pile is subjected to a sudden applied force measured as $F_m(t)$ and the measured velocity $v_m(t)$ at any time $t_m$, then the static soil resistance $R_s(t_m)$ can be given by the following equation:

$$R_s(t_m) = \tfrac{1}{2}(1 - j_c)[F(t_m) + (MV_r/L)v_t(t_m)] + \tfrac{1}{2}(1 + j_c)[F(t_m + 2L/V_r) - (MV_r/L)v_t(t_m + 2L/V_r)] \tag{5.33}$$

where

$L$ = pile length
$M$ = pile mass

**TABLE 5.5 Suggested Values for Case Method Damping Constant ($j_c$) (Rausche et al., 1985)**

| Soil Type in Bearing Strata | Suggested Range, $j_c$ |
|---|---|
| Sand | 0.05–0.20 |
| Silty sand or sandy silt | 0.15–0.30 |
| Silt | 0.20–0.45 |
| Silty clay and clayey silt | 0.40–0.70 |
| Clay | 0.60–1.10 |

$j_c$ = the Case Method damping constant *and is equal to* $(JV_r/(E_pA)$
$J$ = the damping constant.

Table 5.5 gives the suggested values for $j_c$ for various soil types. $R_s$ is the ultimate soil capacity and $F(t_m)$ and $v_t(t_m)$ are measured force and velocity at time $t_m$.

Figure 5.11a shows an example of the measured force and velocity plots for a 15-in (381 mm) diameter, 80 ft (24 m) long, 1/2. in (13 mm) wall thickness steel pile. Figure 5.11b shows predicted static resistance values from equation (5.33) for $j_c = 0.4$ and $j_c = 0$, respectively. Since the method gives capacity at the time of testing, testing should be carried out on restrike to include soil strength changes due to setup or relaxation. The ultimate static pile capacity determination by measuring force and velocity during pile driving is presented in Example 5.5. Chapter 11 presents critical evaluation of this method.

***Example 5.5*** Figure 5.11a shows the record of measured force and velocity for an 80-ft (24 m) long, 15-in. (381 mm) diameter and 0.5-in. (13 mm) wall thickness steel pipe pile driven into silty clay. Calculate the ultimate static axial pile capacity for the measured force and velocity record.

SOLUTION The modulus of elasticity $E_p$ for steel is $30 \times 10^6$ psi. The unit weight for steel is 490 lb/ft$^3$. The measured force and velocity plots are as provided in Figure 5.11a.

$$A = \pi/4\,(15^2 - 1^2) = 176\,\text{in.}^2$$

$$\rho = \gamma/g = 490/32.2 = 15.2\,\text{lb sec}^2/\text{ft}^4 = 0.733 \times 10^{-6}\,\text{kip sec}^2/\text{in.}^4$$

From equation (5.32):

$$V_r = \sqrt{E_p/\rho} = \sqrt{30 \times 10^3/0.733 \times 10^{-6}} = 202.3 \times 10^3\,\text{in./sec}$$

$$M = \rho AL$$

$$M = 0.733 \times 10^{-6} \times 176 \times (80 \times 12) = 0.1238\,\text{kips sec}^2/\text{in.}$$

$$MV_r/L = 0.1238 \times 202.3 \times 10^3/80 \times 12 = 26.08$$

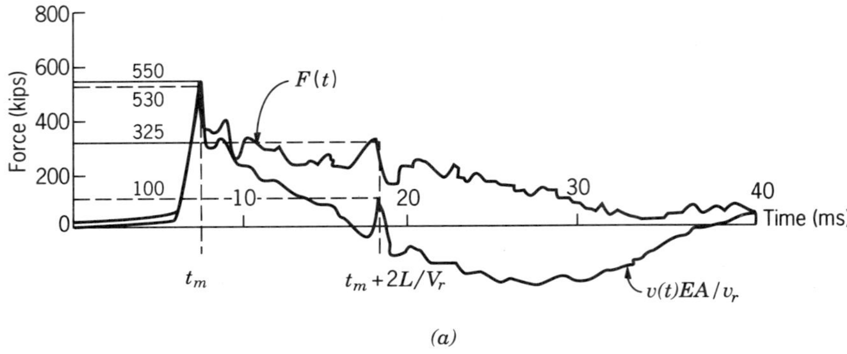


*(a)*

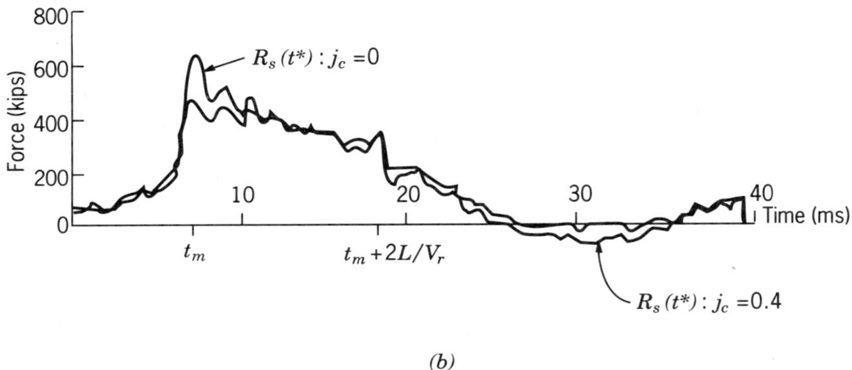


*(b)*

**Figure 5.11** Measured force and velocity and static capacity prediction plot for a steel pile. (a) Measured force and velocity, (b) resistance to penetration and static prediction (after Rausche et al., 1985).

From Figure 5.11a, the following measured values are obtained:

$$t_m = 7.5 \times 10^{-3} \text{ sec}$$

$$F(t_m) = 550 \text{ kips}$$

$$v(t_m)EA/V_r = 530 \text{ kips}$$

$$F(t_m + 2L/V_r) = 325 \text{ kips}$$

$$v(t_m + 2L/V_r)EA/V_r = 100 \text{ kips}$$

From Table 5.5 for silty clay, $j_c$ can be taken as 0.4.

From equation (5.33), the ultimate static axial capacity $R_s(t_m)$ is given by:

$$R_s(t_m) = 1/2(1 - j_c)(F(t_m) + (MV_r/L)v_t(t_m)) \\ + 1/2(1 + j_c)(F(t_m + 2L/V_r) - (MV_r/L)v_t(t_m + 2L/V_r))$$

Then, substituting various values in equation (5.33), we get:

$$R_s(t_m) = 1/2(1-0.4)\left(550 + \frac{26.08 \times 530 \times 202.3 \times 10^3}{30 \times 10^3 \times 176}\right)$$
$$+ 1/2(1+0.4)\left(325 - \frac{26.08 \times 100 \times 202.3 \times 10^3}{30 \times 10^3 \times 176}\right)$$
$$= 0.3\,(1079.424) + 0.7\,(225.12)$$
$$= 323.82 + 157.58 = 481.4\,\text{kips}\,(2142\,\text{kN})$$

### 5.1.3 Bearing Capacity of Pile Groups in Cohesionless Soils

Pile foundations in many situations are constructed as groups of closely spaced piles with a reinforced concrete pile cap or other joining systems such as cross-beams or frames. Based on economy and practicality, the optimal pile spacing normally ranges between 3 to 3.5 times the pile diameter ($B$). Normally, piles are not installed at less than 3 times the pile shaft diameter to avoid interference during installation. The following criteria may normally be used for piles to be considered as a group or acting as individual piles.

| Pile Spacing (s) | Pile Action |
|---|---|
| $3B$ to $7B$ | Group |
| Greater than $7B$ | Individual |

There is no acceptable rational theory of bearing capacity of pile groups. For cohesionless soils, the following criteria may be used for bearing capacity of pile groups, as long as the center-to-center pile spacing is more than 3 times the pile diameter (Vesic, 1977).

1. $(Q_p)_G = nQ_p$, where $(Q_p)_G$ is ultimate point load of a pile group, $(Q_p)$ is the ultimate point load of a single pile and $n$ is the number of piles.
2. The ultimate shaft friction load of a pile group may be greater than the sum of individual shaft friction load due to increased compaction and lateral compression caused by driving within a relatively small area. It is difficult to forecast this increased capacity quantitatively.

Therefore, such increases are not recommended unless demonstrated by a full-scale load test at a site. One can thus conclude that the ultimate bearing capacity of a pile group in cohesionless soils is at least equal to the sum of individual pile capacities.

However, group action of piles in cohesionless soils increases the settlement of the group. This will be discussed later in this section. Based on this, the ultimate bearing capacity of a pile group is simply the sum of individual capacities unless the pile group is founded on dense cohesionless soil of limited thickness underlain by a weak soil deposit. In such situations, the pile group capacity is lower of (1) the sum of individual pile capacities and (2) the capacity of the block failure of an equivalent base with width $\bar{b}$ punching through the dense deposit into the underlying weak deposit (Meyerhof, 1974; Terzaghi and Peck, 1967). This will be further discussed in Section 5.1.8.

There is an increase in pile group capacity due to the pile cap resting on ground (Garg, 1979). Due to uncertainties in construction, this increase in pile capacity can be neglected.

***Example 5.6*** Using the data of Example 5.1, calculate the pile group bearing capacity if the piles are placed 4 ft center to center and joined at the top by a square pile cap supported by nine piles.

SOLUTION Using the arrangement shown in Figure 5.12,

$$B = 1\ \text{ft}, s = 4\ \text{ft}, \bar{b} = 4 + 4 + 1 = 9\ \text{ft}, b = 10\ \text{ft}, n = 9$$

$$(Q_v)_{\text{ult}} = 93.83\ \text{kips for a single pile (Example 5.1)}$$

$$\begin{aligned}(Q_{vG})_{\text{ult}} &= n(Q_v)_{\text{ult}} \\ &= 9 \times 93.83 = 844.47\ \text{kips}\end{aligned}$$

$$(Q_{vG})_{\text{all}} = \frac{9 \times 93.83}{3} = 281\ \text{kips (1250 kN) with FOS of 3.}$$

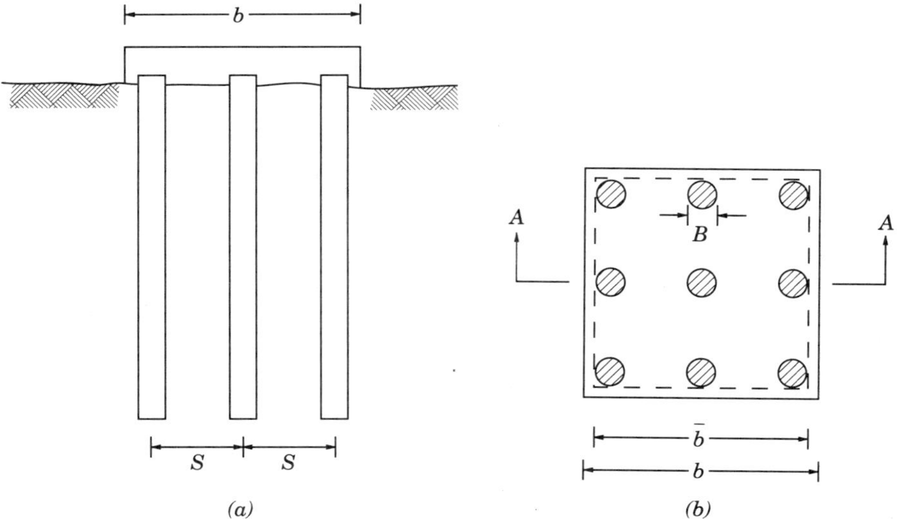


**Figure 5.12** Bearing capacity of pile group in cohesionless soils (for Example 5.6). (a) Section AA, (b) plan.

### 5.1.4 Settlement of a Single Pile in Cohesionless Soils

The settlement prediction of pile foundation is complex because of (1) disturbance and changes in the state of soil stress due to pile installation operation and (2) the uncertainty about the distribution and the exact position of load transfer from the pile to the soil. The disturbance and changes in the soil stress due to pile installation were discussed in Chapter 1. The displacement required to mobilize skin friction is small and may not exceed 0.2 in. regardless of soil and pile type and pile dimensions. However, Vesic, 1977; Sharma and Joshi, 1988 found that this value may not exceed 0.4 in. (10 mm). The displacement required to mobilize pile point resistance is, however, large and depends on the soil type and the pile type and size. Thus, the ultimate skin friction is mobilized much sooner than the point bearing. In addition, the load transfer mechanism also depends on the pile length and the load levels. This is illustrated by Figure 5.13 where at loads up to 40 kips the entire load was being taken by the shaft. Load test results presented by Sharma and Joshi (1988), however, indicated that on 24 in. (600 mm) diameter, 40 ft (12.2 m) long piles bored through sandy till, about 60 percent load was taken by the shaft and the remaining 40 percent is taken by the tip. Niyama et al. (1989)

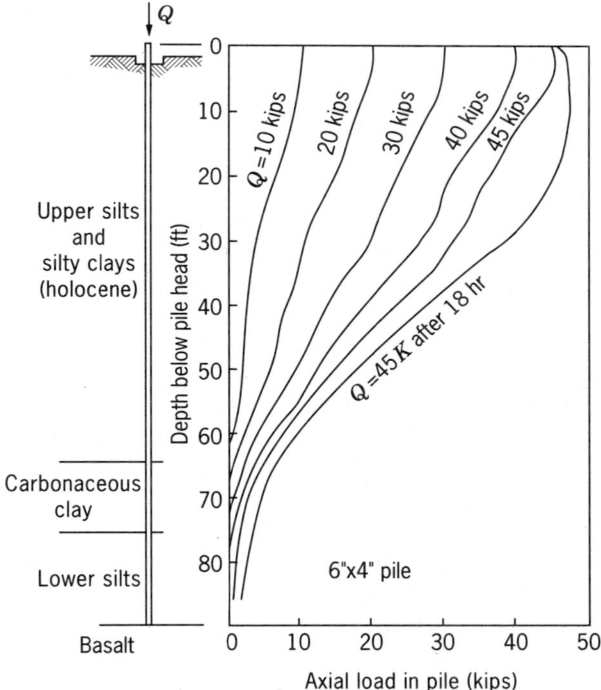


**Figure 5.13** Load transfer from a steel pile driven through compressible silt to rock (after Francis et al., 1961, reported by Vesic 1977).

estimated skin friction as 46% of the total load by CAPWAP method and 10% from static computations. These and other data presented by Vesic (1977) indicate that the load transfer mechanism in piles is not well understood. Since settlements are influenced by load transfer mechanism, only approximate solutions of this problem are available. The following three methods are recommended for estimating pile settlement in cohesionless soils. (1) semiempirical method, (2) empirical method, and (3) pile load test.

1. **Semiempirical Method** For design purposes, the settlement of a pile can be broken down into the following three components (Vesic, 1977).

$$S_t = S_s + S_p + S_{ps} \tag{5.34}$$

where

$S_t$ = total pile top settlement for a single pile
$S_s$ = settlement due to axial deformation of a pile shaft
$S_p$ = settlement of pile base or point caused by load transmitted at the base
$S_{ps}$ = settlement of pile caused by load transmitted along the pile shaft

These three components are determined separately and then are added together.

$$S_s = (Q_{pa} + \alpha_s Q_{fa})L/(A_p E_p) \tag{5.35}$$

where

$Q_{pa}$ = actual base or point load transmitted to the pile base in working stress range (force units)
$Q_{fa}$ = actual shaft friction load transmitted by the pile in the working stress range (force units)
$L$ = pile length
$A_p$ = pile cross-sectional area
$E_p$ = modulus of elasticity of the pile
$\alpha_s$ = a number that depends on distribution of skin friction along the pile shaft

Vesic (1977) recommended that $\alpha_s = 0.5$ for the uniform or the parabolic skin friction distribution along the pile shaft. For triangular (zero at pile head and maximum at pile base) skin friction distribution, the $\alpha_s = 0.67$. The shape of the skin friction distribution can only be obtained by monitoring the shaft friction during load tests. Sharma and Joshi (1988) indicated that the total settlements estimated based on uniform or triangular distribution are not sensitive to $\alpha_s$ values. Therefore, for practical purposes, either value of $\alpha_s$ will provide reasonable settlement estimates.

**TABLE 5.6 Typical Values of Coefficient $C_p$ (Vesic, 1977)**

| Soil Type | Driven Piles | Bored Piles |
|---|---|---|
| Sand (dense to loose) | 0.02–0.04 | 0.09–0.18 |
| Clay (stiff to soft) | 0.02–0.03 | 0.03–0.06 |
| Silt (dense to loose) | 0.03–0.05 | 0.09–0.12 |

The following relationships have been established based on theoretical analyses and empirical correlations between soil properties and ultimate point resistance ($q_p$) for a number of construction sites as reported by Vesic (1977).

$$S_p = C_p Q_{pa}/(B q_p) \tag{5.36}$$

$$S_{ps} = C_s Q_{fa}/(D_f q_p) \tag{5.37}$$

where

$C_p$ = empirical coefficient (typical values provided in Table 5.6).
$C_s = 0.93 + 0.16\sqrt{D_f/B} \cdot C_p$ (5.38)
$Q_{pa}$ = net point load under working conditions or allowable
$Q_{fa}$ = pile shaft load under working conditions or allowable
$q_p$ = ultimate end (point)-bearing capacity (force/area)
$B$ = pile diameter
$D_f = L$ = embedded pile length

In these estimates, it has been assumed that the bearing stratum under the pile tip extends at least 10 pile diameters below its base (tip). Also, the soil below is of comparable or higher stiffness.

2. **Empirical Method** The settlement of a displacement pile for working loads may be estimated by the following relationship (Vesic, 1970):

$$S_t = B/100 + (Q_{va} L)/(A_p E_p) \tag{5.39}$$

where

$S_t$ = settlement of pile head, in.
$B$ = pile diameter, in.
$Q_{va}$ = applied pile load, lb
$A_p$ = area of cross-section of pile in.$^2$
$L$ = pile length, in.
$E_p$ = modulus of elasticity of pile material, lb/in.$^2$

3. **Pile Load Test** If a pile load test is carried out by the standard method described in ASTM D1143-81 and discussed in Chapter 9, then the settlement observed during a load test can be considered to be representative for long-term behavior of a pile in cohesionless soils. A load test is the only accurate method of estimating pile settlements.

***Example 5.7*** For the pile described in example 5.1, estimate the pile settlement. The pile has 3/4 in. (18.75 mm) wall thickness and is closed at the bottom.

SOLUTION

$$B = 12\,\text{in. (outside diameter)}$$

$$L = 30 \times 12 = 360\,\text{in.}$$

$$(Q_v)_{\text{all}} = 31{,}000\,\text{lb (from Example 5.1)}$$

where

$$\text{Area of base} = \frac{\pi}{4}(12)^2 = 113\,\text{in.}^2$$

$$\text{Pipe inside diameter} = 12 - 2 \times 3/4 = 10.5\,\text{in.}$$
$$\text{Area of steel section} = \pi(12^2 - 10.5^2)/4 \times 144 = 0.184\,\text{ft}^2 = 26.496\,\text{in.}^2$$

1. **Semiempirical Method** From equation (5.34) $S_t = S_s + S_p + S_{ps}$: Assuming that skin friction has uniform distribution along pile shaft, then from equations (5.3) and (5.6):

$$Q_f = p\int_0^L f_s \Delta L = pk_s \tan\delta \sum_{L=0}^{L=L} \sigma'_{vl}\Delta L$$

$$Q_f = 34.95 \text{ from Example 5.1}$$

$$Q_p = 58.88 \text{ from Example 5.1}$$

$$(Q_v)_{\text{ult}} = Q_p + Q_{\text{f}} = 58.88 + 34.95 = 93.83$$

$$(Q_v)_{\text{all}} = (Q_v)_{\text{ult}}/FS = 93.83/3 = 31.2\,(\text{say } 31)\,\text{kips}$$

Assuming allowable loads are the actual loads, then

$$Q_{pa} = (Q_p)_{\text{all}} = 58.83/3 = 19.6\,\text{kips}$$

$$Q_{fa} = (Q_f)_{\text{all}} = 31 - 19.6 = 11.4\,(= 34.95/3)\,\text{kips}$$

Then

$$S_s = \frac{(Q_{pa} + \alpha_s Q_{fa})L}{A_p E_p} = \frac{(19.6 + 0.5 \times 11.4)\,1000 \times 360}{26.496 \times 30 \times 10^6} \text{ from equation (5.35)}$$

$$= \frac{25.3 \times 36 \times 10^4}{26.496 \times 3 \times 10^7} = 0.011\,\text{in.}$$

where modulus of elasticity of steel $= E_p = 30 \times 10^6$ psi $\alpha_s = 0.5$ for uniform distribution of skin friction.

$$S_p = \frac{C_p Q_{pa}}{Bq_p} \quad \text{from equation (5.36)}$$

$$= \frac{0.03 \times 19.6 \times 113}{12 \times 58.88} \quad \text{where } C_p = 0.03 \text{ from Table 5.6 and}$$

$$q_p = \frac{Q_p}{A_p} = \frac{58.88}{113} \text{ kips/in.}^2$$

$$S_p = 0.094 \text{ in.}$$

$$S_{ps} = \frac{C_s Q_{fa}}{D_f q_p} \quad \text{from equation (5.37)}$$

$$C_s = 0.93 + 0.16\sqrt{D_f/B} \cdot C_p \quad \text{from equation (5.38)}$$

$$= 0.93 + 0.16\sqrt{360/12} \cdot 0.03 = 0.054$$

$$S_{ps} = \frac{0.054 \times 11.4 \times 113}{360 \times 58.88} = 0.0033 \text{ in.}$$

$$S_t = S_s + S_p + S_{ps}$$
$$= 0.011 + 0.094 + 0.0033$$
$$= 0.108 \text{ in. } (2.7 \text{ mm})$$

**2. Empirical Method**

$$S_t = \frac{B}{100} + \frac{Q_{va} L}{A_p E_p} \quad \text{from equation (5.39)}$$

$$= \frac{12}{100} + \frac{31 \times 360 \times 1000}{26.496 \times 30 \times 10^6}$$

$$= 0.12 + 0.014 = 0.134 \text{ in. } (3.35 \text{ mm})$$

### 5.1.5 Settlement of Pile Groups in Cohesionless Soils

The settlement of a pile group ($S_G$) is normally greater than the settlement of a single pile ($S_t$) *at equal load per pile* because of the larger depth of influence ($D_e$) of a group as compared to that of a single pile ($D_{e'}$) (shown conceptually in Figure 5.14). No general theory to predict pile group settlements in cohesionless soils is available. Many empirical and semiempirical methods with gross approximations are available but cannot be recommended without reservations.

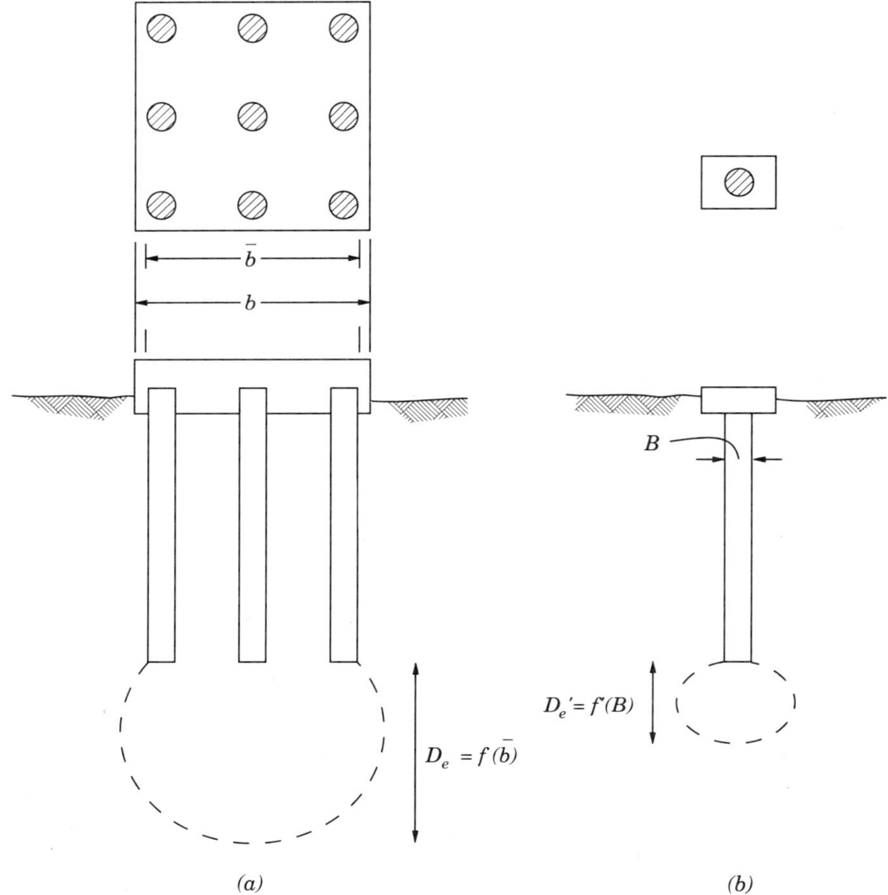


**Figure 5.14** Zone of influences for a pile group and a single pile. (a) Zone of influence for settlement of pile group. (b) Zone of influence for settlement of a single pile.

For design purposes, the simplest of these methods is recommended as follows (Vesic, 1977), according to which,

$$S_G = S_t\sqrt{(\bar{b}/B)} \tag{5.40}$$

where

$S_G$ = group settlement at load per pile equal to that of the single pile
$S_t$ = settlement of a single pile estimated or determined from pile load test
$\bar{b}$ = width of pile group (smaller dimension)
$B$ = individual pile diameter

A similar empirical relationship for estimating the settlement of pile groups has been provided by Skempton (1953). This is a very conservative approach and is not generally used in practice. In the absence of field load test data on pile groups, equation (5.40) is generally recommended in engineering practice (*Foundation and Earth Structures Design Manual* DM 7-2, 1982 and *Canadian Foundation Engineering Manual*, 1985) and can be used to obtain pile group settlements in cohesionless soils.

Meyerhof (1976) presented conservative empirical expressions for preliminary estimates of the total settlements of pile foundations in cohesionless soil using the results of standard penetration test ($N$) and static cone penetration ($q_c$) values as follows:

1. Based on standard penetration ($N$) values:

$$S_G = 2p\sqrt{(\bar{b}I/\bar{N})} \tag{5.41}$$

where

$p$ = net foundation pressure, in tons/ft$^2$
$\bar{b}$ = the width of pile group, in feet
$\bar{N}$ = the average corrected standard penetration test values, in blows/ft (blows per 0.3 m) within the seat of settlement (roughly equal to $\bar{b}$ in homogeneous soils)
$I = [1 - D_f/8\bar{b}] \geqslant 0.5$
$D_f$ = effective depth in the bearing stratum = pile length

For silty sand, the values of $S_G$ obtained from equation (5.41) should be doubled.

2. Based on static cone penetration ($q_c$) values:

$$S_G = p\bar{b}I/(2q_c) \tag{5.42}$$

All these preliminary estimates are based on the assumption that the soil is uniform within the zone of influence. None of these methods provides an accurate value for settlements of pile groups. Only load tests on a pile group can provide representative settlement estimates.

***Example 5.8*** A pile group consisting of nine 12-in. (300 mm) diameter steel piles driven 4 ft center to center to 30 ft into sand at a site is shown in Figure 5.2. The sand had $\phi = 36°$ and $\gamma = 125$ lb ft$^3$ (19.8 kN/m$^3$). Similar data were also used in Examples 5.1, 5.2, 5.6, and 5.7. Estimate the pile group settlement.

SOLUTION From Example 5.6:

$$B = 1\text{ ft}$$
$$\bar{b} = 9\text{ ft (square arrangement)}$$
$$n = 9\text{ piles}$$
$$(Q_G)_{\text{all}} = 281\text{ kips}$$

From Example 5.2: within the zone of influence, 9 ft, (equal to $\bar{b}$ depth below group base) the average $N = (12 + 14 + 14)/3 \simeq 13$. From Example 5.7: $S_t = 0.134$ in.

1. Based on Vesic's method:
   From equation (5.40):

$$S_G = S_t\sqrt{(\bar{b}/B)} = 0.134\sqrt{9/1} = 0.40 \text{ in. } (10 \text{ mm})$$

2. Based on Meyerhof's method ($N$ values)

$$p = \frac{(Q_G)_{\text{all}}}{\bar{b} \times \bar{b}} = \frac{281}{9 \times 9} = 3.47 \text{ kips/ft}^2 = 1.74 \text{ tons/ft}^2$$

$$I = (1 - D_f/8\bar{b}) \text{ where } D_f = \text{pile length} = 30 \text{ ft}$$

$$= [1 - 30/(8 \times 9)] = 0.58 > 0.5$$

Then, from equation (5.41):

$$S_G = 2p\sqrt{\bar{b}}(I/\bar{N}) = 2 \times 1.74\sqrt{9} \times 0.58/13 = 0.5 \text{ in. } (13 \text{ mm})$$

### 5.1.6 Design Procedure for Piles in Cohesionless Soils

The design procedure consists of the following six steps:

1. **Soil Profile.** From proper soils investigations, establish the soil profile and groundwater levels, and note soil properties on the soil profile based on the field and laboratory tests (see Chapter 4 for details).
2. **Pile Dimensions and Allowable Bearing Capacity.** Select a pile type, length, and diameter and calculate allowable bearing capacity based on the formulas used for the available soil parameters as follows:
   (a) *Static analysis by utilizing soil strength*

$$(Q_v)_{\text{ult}} = A_p\sigma'_v N_q + pK_s \tan\delta \sum_{L=0}^{L=L} \sigma'_{vl}\Delta L \tag{5.7}$$

The values of $N_q$ and $K_s$ are provided in Tables 5.2 and 5.3, respectively.
(*b*) *Empirical analysis utilizing the Standard Penetration Test values*
***For Sands***

$$Q_p \text{ (tons)} = \frac{0.4\bar{N}}{B} D_f A_p \leqslant 4\bar{N}A_p \tag{5.8}$$

***For Nonplastic Silt***

$$Q_p\ (\text{tons}) = \frac{0.4\bar{N}}{B} D_f A_p \leqslant 3\bar{N} A_p \tag{5.9}$$

$$Q_f = (f_s)(\text{perimeter})(\text{embedment length}) \tag{5.11}$$

where $f_s$ in tons per square foot is given by the following equation:

$$f_s = \bar{N}/50 \leqslant 1\ \text{tsf} \tag{5.12}$$

The ultimate capacity $(Q_v)_{\text{ult}}$ is then the summation of $Q_p$ and $Q_f$ from the above. These equations are for driven piles. For drilled piles use one-third of $Q_p$ and one-half of $Q_f$ from these equations.

*Empirical Analysis Utilizing the Static Cone Penetration Test Values*

$$Q_p = A_p q_c \tag{5.13}$$

$$Q_f = (f_s)(\text{perimeter})(\text{embedment length}) \tag{5.14}$$

The $(Q_v)_{\text{ult}}$ is then the summation of $Q_p$ and $Q_f$. These equations are for driven piles. For drilled piles, use one-half of the above values. Because of the uncertainties in soil parameters and the semiempirical nature of bearing capacity formulas, a factor of safety of 3 should be used to obtain the allowable bearing capacity from the foregoing equations. The allowable bearing capacity used in the design is then the lowest of these values.

3. **Number of Piles and Their Arrangement.** Determine the number of piles required by dividing the column load with the allowable bearing capacity of a pile and arrange the piles in the group so that pile spacing is three to four times the pile diameter. Establish pile cap size with reference to column spacing and other space restrictions. If the pile cap size becomes too large, increase pile length and/or pile diameter and repeat step (2) to obtain reasonable pile dimensions and capacity. Determine pile group capacity by simply adding the individual pile capacities.

4. **Settlement of a Single Pile.** Estimate the settlement of a single pile by the following methods:

   (a) *Semiempirical method*

$$S_t = S_s + S_p + S_{ps} \tag{5.34}$$

where

$$S_s = (Q_{pa} + \alpha_s Q_{fa}) \frac{L}{A_p E_p} \tag{5.35}$$

$$S_p = C_p Q_{pa}/Bq_p \tag{5.36}$$

$$S_{ps} = C_s Q_{fa}/D_f q_p \tag{5.37}$$

(b) *Empirical method*

$$S_t = B/100 + Q_{va}L/A_p E_p \tag{5.39}$$

The settlement is then higher of the values obtained from the foregoing methods.

5. **Settlement of Pile Group and Check on Design.** Estimate pile group settlement by using the following methods:
(*a*) *Vesic's method*

$$S_G = S_t \sqrt{(\bar{b})/B} \tag{5.40}$$

(b) *Meyerhof's method*

1. If Standard Pentration ($N$) values are available:

$$S_G = 2p\sqrt{\frac{\bar{b}I}{\bar{N}}} \tag{5.41}$$

where

$$I = \left(1 - \frac{D_f}{8\bar{b}}\right) \geqslant 0.5$$

2. If Static Cone Penetration ($q_c$) values are available:

$$S_G = p\bar{b}I/2q_c \tag{5.42}$$

The largest of the values obtained from Vesic and Meyerhof's methods should be equal to or less than the allowable settlement values.

6. **Pile Load Test and Pile-Driving Criteria.** Recommend a pile load test to fine tune the allowable bearing capacity. If a driven pile is selected, specify the driving criteria that should be supplemented with pile load test and dynamic monitoring. On large projects the pile load test should be carried out on a test pile that is loaded to failure. On smaller projects, one of the actual piles should be tested by loading it to two times the design load. For details of a pile load test, see Chapter 9.

***Example 5.9*** A 236-kip (1050 kN) vessel is to be supported on a pile foundation in an area where soil investigations indicated soil profile (shown in Figure 5.15).

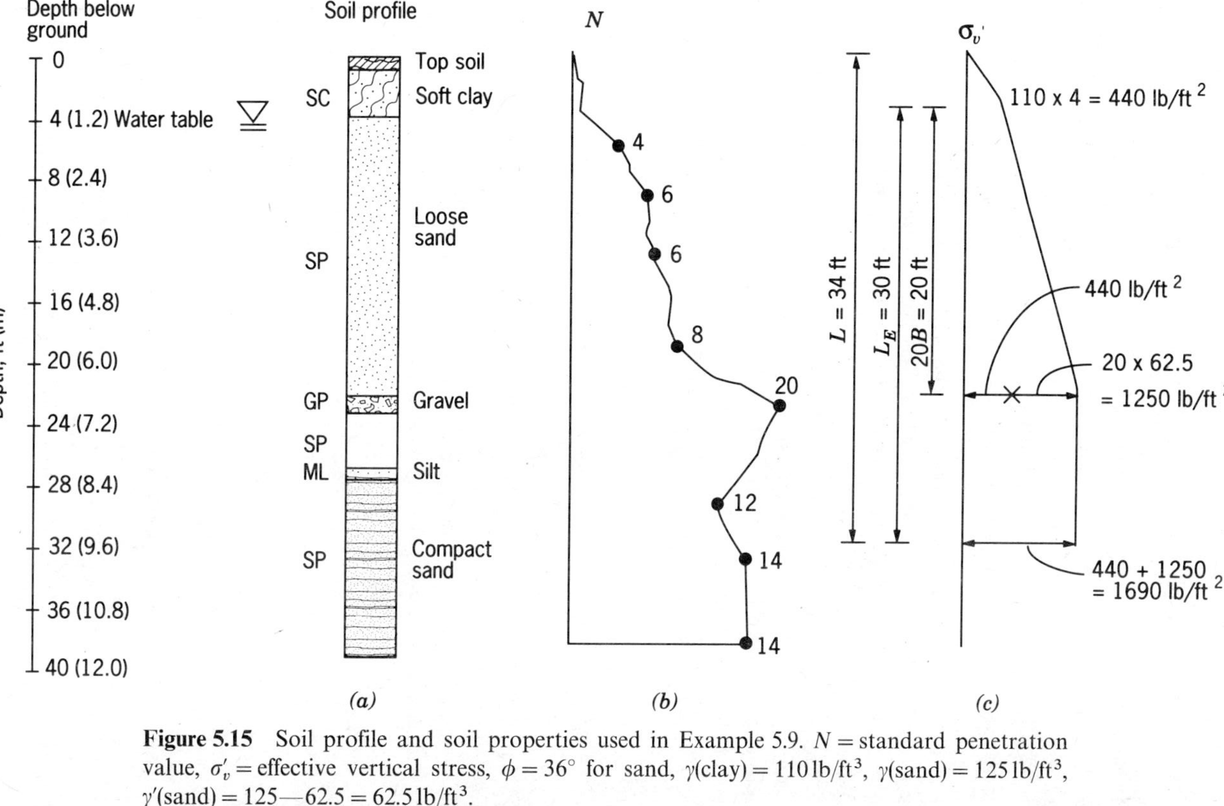


**Figure 5.15** Soil profile and soil properties used in Example 5.9. $N$ = standard penetration value, $\sigma'_v$ = effective vertical stress, $\phi = 36°$ for sand, $\gamma$(clay) = 110 lb/ft$^3$, $\gamma$(sand) = 125 lb/ft$^3$, $\gamma'$(sand) = 125—62.5 = 62.5 lb/ft$^3$.

Design a pile foundation so that the maximum allowable settlement for the group does not exceed 0.6 in. (15 mm).

SOLUTION

1. **Soil Profile** (Shown in Figure 5.15)
2. **Pile Dimensions and Allowable Bearing Capacity**. Top 4 ft of soil consists of top soil and soft clay. These are low strength materials and lie in the zone of seasonal variation. Contribution of this depth to pile side frictional resistance can therefore be neglected.

The cohesionless soil, as shown by the soil profile, is exhibiting uniformly increasing $N$ values with depth except at 24 ft depth where $N = 20$. This anomaly results because of gravel particles that cause obstruction and higher resistance to split spoon. This higher $N$ value at 24 ft can therefore be neglected.

Try a 34-ft (10.3 m) long with 30-ft (9.1 m) net penetration into sand and 12-in. (305 mm) diameter steel-driven frictional pile. This pile will have 0.75-in. (19 mm) wall thickness and is closed at the bottom. Allowable bearing capacity of this pile can be calculated as follows:

*Static Analysis by Utilizing Soil Strength*

$$(Q_v)_{\text{ult}} = A_p \sigma'_v N_q + pK_s \tan\delta \sum_{L=0}^{L=L} \sigma'_{vl} \Delta L \tag{5.7}$$

$$A_p = \pi/4(1)^2 = 0.785\ \text{ft}^2$$

$\sigma'_v$ is equal to $\gamma z$ above groundwater level and is $\gamma' z$ below groundwater level. Values of $\sigma'_v$ at various depths are shown in Figure 5.15c. In this figure, $\sigma'_v$ as discussed earlier, has been considered increasing linearly up to (20B) pile length. Below this depth, the $\sigma'_v$ value is assumed to be constant for pile design.

$$N_q = 60 \text{ for } \phi = 36^\circ \text{ from Table 5.2}$$

$$p = \pi B = \pi(1) = 3.14\ \text{ft}$$

$$K_s = 1.0 \text{ from Table 5.3}$$

$$\delta = (2/3)\phi = (2/3) \times 36^\circ = 24^\circ$$

Substituting these values in equation (5.7), we get:

$$(Q_v)_{\text{ult}} = 0.785 \times 1690 \times 60 + 3.14 \times 1.0 \tan 24 \times \left[\frac{(440 + 1690)}{2} \times 20 + 1690 \times 10\right] \text{lb}$$

$$= 79.6 + 43.7 = 123.3\ \text{kips}$$

$$(Q_v)_{\text{all}} = (Q_v)_{\text{ult}}/FS = 123.3/3 = 41.1\ \text{kips (say 41 kips)(182.5 kN)}$$

*Empirical Analysis Utilizing Standard Penetration Tests*
***Point Bearing ($Q_p$)***

$$\text{Average } N \text{ value near pile tip} = (8 + 12 + 14 + 14)/4 = 12$$

$$\sigma'_v \text{ near pile tip} = 440 + (125 - 62.5)30 = 2315\,\text{lb/ft}^2 = 1.15\text{t/ft}^2$$

From equation (4.2), $C_N = 0.77\log_{10}(20/1.15) \simeq 1$, therefore, $\bar{N} = C_N \cdot N$ $\bar{N} \simeq 12$.

$$Q_p = (0.4\bar{N}/B)D_f A_p \leqslant 4\bar{N}A_p = 0.4 \times 12/1 \times 30 \times 0.785 = 113\,\text{tons} \qquad (5.8)$$

which is greater than $4 \times 12 \times 0.785 = 37.7$ tons (say 38 tons), therefore, use $Q_p = 38$ tons.

***Shaft friction ($Q_f$)***

$$\text{Average } N \text{ value along pile shaft} = (4 + 6 + 6 + 8 + 12)/5 = 7.2 \text{ (say 7)}$$

$$f_s = (\bar{N}/50) \leqslant 1\,\text{tsf} = 7/50 = 0.14\,\text{tsf} \qquad (5.12)$$

$$Q_f = f_s pL = 0.14 \times 3.14 \times 30 = 13.2\,\text{tons}$$

Therefore,

$$(Q_v)_{\text{ult}} = Q_p + Q_f = (38 + 13.2)\,\text{tons} = 102.4\,\text{kips}$$

$$(Q_v)_{\text{all}} = (Q_v)_{\text{ult}}/3 = 34\,\text{kips (151.3 kN)}$$

where $(Q_p)_{\text{all}} = 25.3$ kips and $(Q_f)_{\text{all}} = 8.8$ kips; these values will be later used in settlement estimation.

*Empirical Analysis Utilizing Cone Penetration Values*

Cone penetration values are not available therefore allowable bearing capacity on this basis has not been calculated.

The allowable bearing capacity will be the lower of the values obtained previously. Therefore, $(Q_v)_{\text{all}} = 34$ kips (151.3 kN).

### 3. Number of Piles and Their Arrangement

The number of piles required to support 236 kips vessel load will be:

$$n = Q_{va}/(Q_v)_{\text{all}} = 236/34 = 6.9$$

Try a group of nine piles arranged in a square pattern with 3 piles on each side. Place the piles at 4-ft center to center spacing. Thus, a 10 ft × 10 ft square concrete pile cap will be required. Assume that the pile cap is 3 ft thick. This means that the pile cap width $b$ is 10 ft, and the square surrounding outer periphery of piles has $\bar{b} = b - 1 = 10 - 1 = 9$ ft ($b$ and $\bar{b}$ dimensions have been explained in Figure 5.14).

$$\text{Pile cap weight} = 3 \times 10 \times 10 \times 0.15 = 45\,\text{kips}$$

$$\text{Total weight on pile group} = 236 + 45 = 281\,\text{kips}$$

$$\text{Load per pile} = 281/9 = 31\,\text{kips} < 34\,\text{kips}$$

$$\text{Pile group capacity} = 34 \times 9 = 306\,\text{kips} > 281\,\text{kips}$$

**4. Settlement of a Single Pile**

*Settlement by Semiempirical Method*

$$S_t = S_s + S_p + S_{ps} \qquad (5.34)$$

where

$$S_s = (Q_{pa} + \alpha_s Q_{fa})L/(A_p E_p) \qquad (5.35)$$

Since the allowable load on each pile is 34 kips while the actual load is 31 kips, the point resistance and skin friction can be proportionally reduced without any significant error in calculations. Therefore,

$$(Q_p)_{\text{actal}} = 25.3(31/34) = 23\,\text{kips} = Q_{pa}$$

$$(Q_f)_{\text{actual}} = 8.8(31/34) = 8\,\text{kips} = Q_{fa}$$

The modulus of elasticity of steel, $E_p = 30 \times 10^6$ psi and $\alpha_s = 0.5$ by assuming a uniform distribution of skin friction. This is reasonable since, as discussed in Section 5.1.4, the total settlement calculated based on uniform or triangular distribution are not sensitive to $\alpha_s$ values. Substituting these values in the equation for $S_s$ we get $S_s = (23 + 0.5 \times 8)30 \times 12 \times 1000)/(\pi/4)(12^2 - 10.5^2)$ $(30 \times 10^6) = 0.012$ in.

$$S_p = C_p Q_{pa}/(B \times q_p) \qquad (5.36)$$

where

$$c_p = 0.03 \quad \text{from Table 5.6}$$

$$Q_{pa} = 23\,\text{kips}$$

$$B = 12\,\text{in.}$$

$q_p = Q_p/A_p = 76/113.09$

where

$$Q_p = 76\,\text{kips} \qquad \text{from above and}$$

$$A_p = (\pi/4)(12)^2 = 113.09\,\text{in.}^2$$

Therefore,

$$S_p = [(0.03 \times 23)/(12 \times 76/113.09)] = 0.086\,\text{in.}$$

$$S_{ps} = (C_s Q_{fa})/(D_f q_p) \qquad (5.37)$$

where

$$C_s = 0.93 + 0.16\sqrt{D_f/B} \cdot C_p = (0.93 + 0.16 \times \sqrt{30 \times 12/12})0.03$$
$$= 0.054 \text{ from equation 5.38}$$
$$Q_{fa} = 8 \text{ kips}$$
$$D_f = 30 \times 12 \text{ in.}$$
$$q_p = Q_p/A_p = 76/113.09 = 0.67 \text{ kips/in.}^2$$

Then,

$$S_{ps} = (0.054 \times 8)/(30 \times 12 \times 0.67) = 0.0018 \text{ in.}$$

Total settlement $= S_t$ is, then

$$S_t = S_s + S_p + S_{ps} = 0.012 + 0.086 + 0.0018$$
$$= 0.0998 \text{ in. } (2.53 \text{ mm}) \text{ say } 0.1 \text{ in. } (2.5 \text{ mm})$$

*Settlement by Empirical Method*

$$S_t = B/100 + Q_{va}L(A_pE_p) \tag{5.39}$$
$$= 12/100 + (31 \times 30 \times 12 \times 1000)/(\pi/4)(12^2 - 10.5^2)30 \times 10^6$$
$$= 0.12 + 0.014$$
$$= 0.134 \text{ in. } (3.4 \text{ mm})$$

From above, consider the larger of the two settlement values for a single pile that is equal to 0.134 in.

5. **Settlement of Pile Group and Check on Design** As mentioned earlier, $B = 1$ ft, $\bar{b} = 9$ ft square arrangement $n = 9$ piles within the zone of influence of 9 ft (equal to $\bar{b}$ depth below group base) the average $N$ value is $N = (12 + 14 + 14)/3 \simeq 13$ actual load on group, $Q_G = 281$ kips. Total settlement of a single pile, $S_t = 0.134$ in.

*Group Settlement Based on Vesic's Method*

$$S_G = S_t\sqrt{\bar{b}/B} \tag{5.40}$$
$$= 0.134\sqrt{9/1} = 0.402 \text{ (say 0.4 in.)(10 mm)}$$

*Group Settlement Based on Meyerhof's Method (N Values)*

$$S_G = 2p\sqrt{\bar{b}I/\bar{N}} \tag{5.41}$$

where

$p = Q_G/(\bar{b} \times \bar{b}) = 281/9 \times 9 = 3.47\ \text{kips/ft}^2$

$= 1.74\ \text{tons/ft}^2$

$I = [1 - D_f/(8\bar{b})]$, $D_f$ = effective depth in bearing stratum = 30 ft (Figure 5.15)

$= (1 - 30/8 \times 9) = 0.58$

Then,

$$S_G = 2 \times 1.74\sqrt{9 \times 0.58/13} = 0.47\ \text{in.} \quad \text{(say 0.5 in.)(13 mm)}$$

From above, take $S_G = 0.5$ in. (13 mm). This is less than the allowable settlement of 0.6 in. Therefore, the designed pile diameter, length, and group arrangement is acceptable.

**6. Pile Load Test and Pile-driving Criteria**

*Driving Criteria*

From Table 5.4, using a drop hammer,

$$Q_{\text{all}} = 2WH/(S + 1)$$

Using $W = 5000$ lb, $H = 6.5$ ft, $Q_{\text{all}} = 34$ kips

$S = 2 \times 5000 \times 6.5/34000 - 1 = 0.9$ in./blow

$\simeq 6/7$ (i.e., for last 6 in. of driving it would require 7 blows for a drop hammer with a driving energy of 32,500 ft-lb = (5000 lb × 6.5 ft))

For a 12-in. diameter closed-end steel pipe pile, driven to 34 ft below ground or driven with a 32,500 ft-lb energy requiring 7 blows for the last 6 in. of driving, carry out a compression pile load test as per ASTM D 1143-81 to confirm the design load and settlement values. The load test shall be carried to two times the design load. On small-sized projects, this load test can be carried out during actual installation of the piles to confirm that the design criteria are being met. For large projects, a full-scale pile load test (testing a pile to failure) should be conducted. This will permit the selection of optimum pile type and design load. Pile load test methods and related details are provided in Chapter 9.

### 5.1.7 Bearing Capacity of a Single Pile in Cohesive Soils

As discussed in the beginning of this chapter and shown by equation (5.1) the ultimate axial compression load capacity $(Q_v)_{\text{ult}}$ of a pile is the sum of end-bearing capacity $(Q_p)$ and the frictional capacity $(Q_f)$. These two components $Q_p$ and $Q_f$ for cohesive soils are further discussed as follows.

***End-bearing Capacity ($Q_p$)*** For cohesive soils, the bearing capacity of piles is critical on a short-term basis because clay strength will increase due to consolidation or strength regain of disturbed soils in the long term. This was discussed in Chapter 1. Therefore, for piles in clays $\phi = 0$ concept applies for bearing capacity evaluation. Thus, undrained strength, $S_u = c_u = c$ and $\phi = 0$* and bearing capacity factors $N_\gamma = 0$ and $N_q = 1$. Equation (5.2) then becomes:

$$Q_p = A_p[c_u N_c + \gamma D_f N_q] \tag{5.43}$$

When adjustment for pile weight is made then equation (5.43) can be approximated to the following:

$$Q_p = A_p[c_u N_c + \gamma D_f N_q] - \gamma D_f A_p \tag{5.44}$$

Since $N_q = 1$ for $\phi = 0$, then equation (5.44) becomes:

$$Q_p = A_p c_u N_c \tag{5.45}$$

***Friction Capacity ($Q_f$)*** For cohesive soils, applying the concept of $\phi = 0$, shaft friction $f_s$ can be written as follows (See Figure 5.1):

$$f_s = c + \sigma_h \tan \delta$$

where

$c = c_a =$ adhesion between soil and pile, $\delta = 2/3\phi = 0$
$f_s = c_a$

Then equation (5.3) becomes

$$Q_f = p \sum_{L=0}^{L=L_e} c_a \Delta L \tag{5.46}$$

Ultimate bearing capacity $(Q_v)_{\text{ult}}$ for a pile in cohesive soil can then be expressed in the following form:

$$(Q_v)_{\text{ult}} = A_p c_u N_c + p \sum_{L=0}^{L=L_e} c_a \Delta L \tag{5.47}$$

where

$A_p =$ pile point (base) area
$c_u =$ the minimum undrained shear strength of clay at pile point level (i.e., cohesion of the bearing stratum ($c = c_u = S_u = q_u/2$))
$N_c =$ the bearing capacity factor (obtained from Tables 5.7 and 5.8)
$p =$ pile paremeter

*Total stress parameters will be used for the $\phi = 0$ case.

$L_e$ = effective pile length
$c_a$ = soil–pile adhesion (obtained from Figure 4.27)

Since the unit weight of soil does not appear in this expression, the position of groundwater has no effect on pile capacity.

*Undrained Shear Strength of Bearing Stratum ($c = c_u$)* The soil near the driven pile is displaced and may get remolded to a distance of about one pile diameter. Within this disturbed zone, the pore water pressure caused by the pile-driving operation dissipates quickly and after consolidation the soil may be stronger. However, in very sensitive clays or stiff, overconsolidated clays due to the loss of soil structure, the final shear strength may be smaller than that in the undisturbed state. Near bored piles, the clay is usually softened to a distance of about 1 in. (25 mm) due to pile installation, and experience has shown that there is no significant shear strength change of the soil with time (Meyerhof, 1976).

For most practical purposes, it can be assumed that the shear strength of a bearing stratum consisting of low to medium sensitivity homogeneous clay remains unchanged during pile installation. Shear strength ($c_u$) values for bearing capacity estimation should be obtained from laboratory tests done on undisturbed clay samples. The $c_u$ value from laboratory tests is generally obtained by testing 1.5 in. (37.5 mm)-diameter intact clay samples. However, in stiff, fissured clays, the undrained shear strength ($c_u$) decreases as the size of test specimen increases. This reduction is primarily due to the greater involvement of fissured material in controlling soil strength on larger soil samples than on the smaller-sized 1.5 in. (37.5mm) diameter laboratory samples. For stiff, fissured clay, the undrained shear strength ($c_u$) should therefore be corrected for scale effects (Meyerhof, 1983). This is given by the following relationship.

$$c_u = (c_u)_{\text{lab}} R_c \tag{5.48}$$

where $(c_u)_{\text{lab}}$ is the undrained shear strength obtained from conventional triaxial compression tests. $R_c$ is the reduction factor and is obtained from following relationships.

1. For driven piles into stiff, fissured clay, $R_c$ is given by:

$$R_c = (B + 0.5)/(2B) \leqslant 1 \text{ for } B \geqslant 0.5\text{ m} \tag{5.49}$$

where $B$ is the pile base diameter in meters.

2. For bored piles into stiff, fissured clay, $R_c$ is given by:

$$R_c = (B + 1)/(2B + 1) \leqslant 1 \tag{5.50}$$

For intact clay $R_c = 1$. Meyerhof (1983) provides further information on this reduction factor, $R_c$.

In cases where bearing stratum is under high artesian pressures and drilling during pile installation has caused the base clay to swell resulting in decreased shear strength, swelled soil samples should be tested in the laboratory for shear strength determination. Undrained shear strength ($c_u$) from these results should

**TABLE 5.7 Values of $N_c$ for Various Depth to Pile Diameter ($D_f/B$) Ratios**[a]

| $D_f/B$ | $N_c$ |
|---|---|
| 0 | 6.2 |
| 1 | 7.8 |
| 2 | 8.5 |
| $\geqslant 4$ | 9 |

[a]These values have been obtained from the graph presented in the *Foundations and Earth Structures Design Manual* NAVFAC, DM 7.2, 1982.

**TABLE 5.8 Values of $N_c$ for Various Pile Diameters ($B$) (Canadian Foundation Engineering Design Manual, 1985)**

| Drilled Pile Base Diameter | $N_c$ |
|---|---|
| Less than 0.5 m ($\simeq$ 1.5 ft) | 9 |
| Between 0.5 to 1 m ($\simeq$ 1.5 to 3 ft) | 7 |
| Greater than 1 m ($\simeq$ 3 ft) | 6 |

then be used for bearing capacity estimation (Sharma et al., 1984). In highly plastic soft clays, the undrained shear strength should be obtained from field vane tests. Bengtsson and Sallfors (1983) present a method of determining the bearing capacity of axially loaded floating piles in such soils.

***Bearing Capacity Factor ($N_c$)*** As shown in Table 5.7, $N_c$ values increase as the depth-to-pile-diameter ratio increases until it reaches a value of 9 for $D_f/B \geqslant 4$ (Skempton, 1951). For most pile foundations, the depth-to-diameter ratio ($D_f/B$) is greater than 4; $N_c = 9$ may therefore be used for such cases. Table 5.8 provides recommendations for $N_c$ values for various drilled pile base diameters. $N_c$ values provided in Table 5.7 and 5.8 can therefore be used for design purposes, as applicable.

*Soil–pile Adhesion ($c_a$)* The average value of soil–pile adhesion ($c_a$) for homogeneous saturated clay is usually related to the average undrained shear strength ($c_u$) of undisturbed clay within pile embedment length. The ratio ($c_a/c_u$) depends on various factors such as (1) nature and strength of clay (2) dimensions and method of installation of pile, and (3) time effects (Meyerhof, 1976). This has also been discussed in Chapter 4 (Section 4.1.2). Kraft et al. (1981) provide correlations to relate soil–pile adhesion to (1) pile length (2) relative soil–pile stiffness, and (3) soil stress history. These correlations need further field test confirmation before they can be used in practice. Figure 4.27 provides the ($c_a/c_u$) values for various soil consistency and unconfined strength values for driven piles. Where a pile penetrates several different layers the soil–pile adhesion can be approximated by the weighted average value of $c_a$ for individual layers. For drilled piles, the values provided for ($c_a$) in Table 4.7 may be used for preliminary design calculations.

***Effective Pile Length ($L_e$)*** Effective pile length is the length that is assumed to contribute to frictional capacity of the pile. This may be different from actual pile embedment length ($L$) because, for most piles, the upper part of the pile may not be in close contact with soil due to such factors, as disturbances caused by humans and machines and softening and cracking caused by seasonal variations. This length should be evaluated for specific geographical location or job site. For most situations, this may vary from about 3 ft (1 m) to 5 ft (1.5 m).

For drilled-belled piles, the author's (Sharma) experience indicates that in addition to the above seasonal depths, soil around the shaft-bell neck gets disturbed due to a tendency for the soil to move down in that area. This disturbed length is about two times the shaft diameter. Therefore, soil–pile adhesion along this length should be neglected and effective pile length ($L_e$) should be calculated accordingly (Tomlinson, 1977; Sharma et al., 1984). In general, the criteria given in Table 5.9 may be used for estimating effective pile lengths ($L_e$) when $L$ is the total pile embedment or length. In equation (5.47), the length $L$ should therefore be replaced with $L_e$.

***Example 5.10*** A straight-shafted drilled pile was installed through clay till to bear on clay shale. The pile had a 20-in. (500 mm) shaft diameter and was 31 ft (9.5 m) long. Undrained shear strength ($c_u$) for clay till was 950 lb/ft$^2$(45.5 kN/m$^2$) and for clay shale was 6576 lb/ft$^2$ (315 kN/m$^2$). Estimate the allowable bearing capacity of this pile.

SOLUTION

$$B = 20\text{ in.} \qquad D_f = L = 31 \times 12\text{ in.}$$
$$A_p = (\pi/4)B^2 = \pi/4(20/12)^2 = 2.18\text{ ft}^2$$
$$D_f/B = 31 \times 12/48 = 7.75$$

From Table 5.7, for $(D_f/B) = 7.75$, $N_c = 9$

**TABLE 5.9 Effective Pile Length ($L_e$) of Driven and Drilled Piles[a]**

| Type of Piles | $L_e$ |
|---|---|
| Driven and<br>Straight shaft drilled | $L$ – (depth of seasonal variation) |
| Drilled and belled | $L$ – (depth of seasonal variation + 2 × pile shaft diameter) |

[a]Based on the experience documented by Tomlinson (1977) and Sharma et al. (1984).

From Table 5.8, for $B = 20/12 = 1.67$ ft, $N_c = 7$
The lower of the above two $N_c$ values is 7 and will be used for these calculations.

$c_u = 6576$ lb/ft$^2$ for the clay shale on which the pile tip will bear
$p = \pi B = \pi \times 20/12 = 5.24$ ft

From Table 4.7 for drilled concrete piles for

$c_u = 950$ lb/ft$^2$, $c_a/c_u = 0.6$, $c_a = 0.6 \times 950 = 570$ lb/ft$^2$
$L_e = 31 - 5 = 26$ ft (assuming that 5 ft is the depth of seasonal variation)

$$(Q_v)_{\text{ult}} = A_p c_u N_c + p \sum_{L=0}^{L=L} c_a \Delta L \tag{5.47}$$

$(Q_v)_{\text{ult}} = 2.18 \times 6576 \times 7 + 5.24 \times 570 \times 26$ lb
$(Q_v)_{\text{ult}} = 178$ kips
$(Q_v)_{\text{all}} = 178/3 = 59$ kips (262 kN), if a factor of safety of 3 is used

### 5.1.8 Bearing Capacity of Pile Groups in Cohesive Soils

If $(Q_v)_{\text{ult}}$ is the ultimate capacity of a single pile and $(Q_{vG})_{\text{ult}}$ is the ultimate capacity of a pile group in cohesive soils then, in general, the following applies:

$$(Q_{vG})_{\text{ult}} \neq n(Q_v)_{\text{ult}} \tag{5.51}$$

where $n$ is the number of piles in the group.

There is, at present, no acceptable rational theory of bearing capacity of pile groups (Vesic, 1977). The basic mechanism of group action of piles was discussed in Chapter 1 (Section 1.3). For most practical purposes, the ultimate load of pile group, $(Q_{vG})_{\text{ult}}$, can be estimated from the smaller of the following two values:

***Group Action*** Block failure of pile group by breaking into the ground along an imaginary perimeter and bearing at the base as shown on Figure 5.16 (Terzaghi and Peck, 1967; Meyerhof, 1976). Using equation (5.47), the ultimate capacity for the group failure of Figure 5.16 can be estimated from the following relationship:

$$(Q_{vG})_{\text{ult}} = c_u N_c(\bar{b}^2) + 4c_u(\bar{b})L_e \tag{5.52a}$$

***Individual Action*** If there is no group action, the total load the group can take is $n$ times the load of the single pile

$$(Q_{vG})_{\text{ult}} = n \times (Q_v)_{\text{ult}} \tag{5.52b}$$

If the piles are spaced closely enough, the load in group action is smaller than that in individual action. The ratio of ultimate load capacity of the group to the

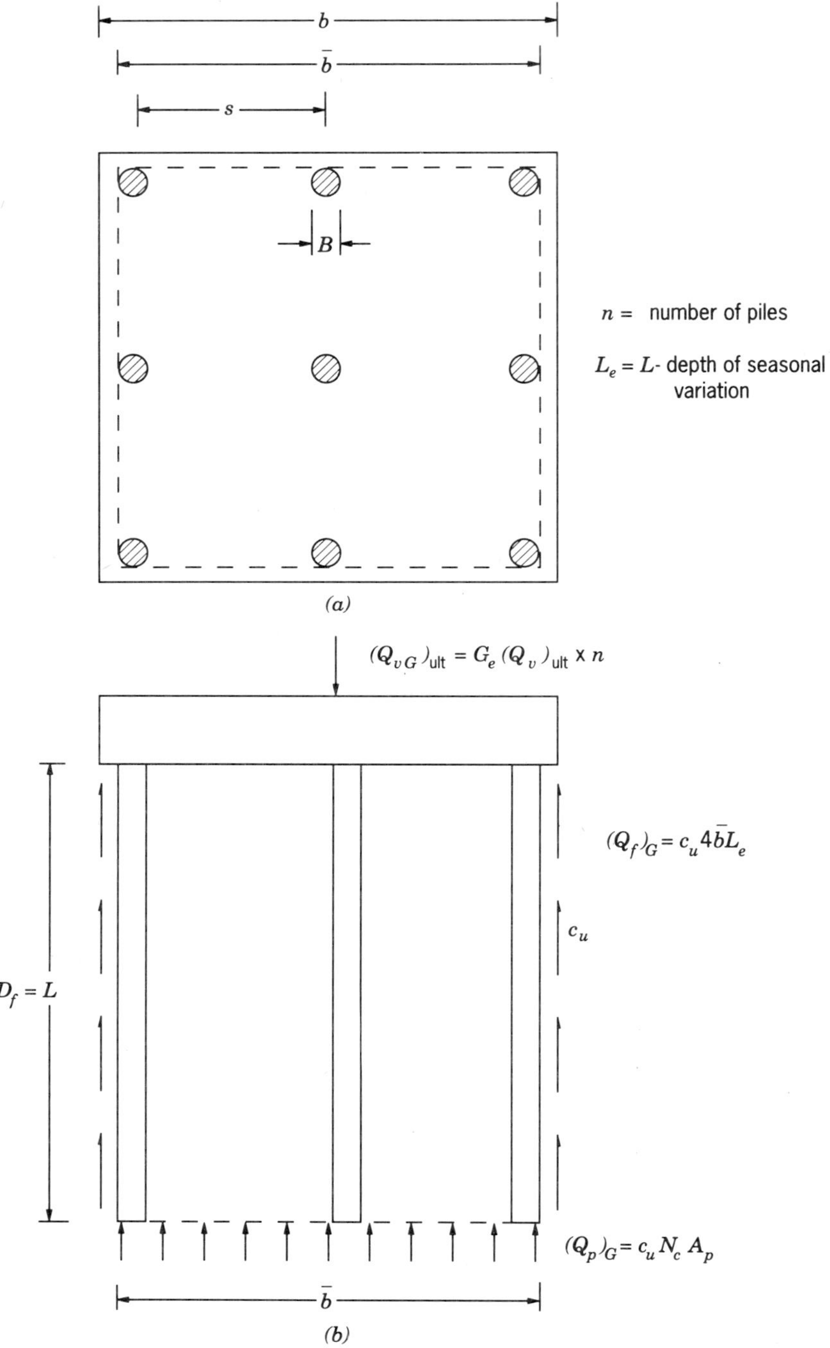


**Figure 5.16** Bearing capacity of pile group in cohesive soils. (a) Plan (b) Section

total individual capacity is defined as the pile group efficiency $G_e$. Therefore,

$$G_e = (Q_{vG})_{ult}/[n \times (Q_v)_{ult}] \tag{5.53a}$$

Thus

$$(Q_{vG})_{ult} = G_e \times n \times (Q_v)_{ult} \tag{5.53b}$$

The value of group efficiency ($G_e$) depends on (1) soil parameters, (2) size and shape of pile group, (3) pile length, and (4) pile spacing (Whitaker, 1957; Kerisel, 1967). A number of efficiency formulas are available in the literature (Chellis, 1961; Moorhouse and Sheehan, 1968). The following group efficiency or reduction factors ($G_e$) as given in Table 5.10 can be used for practical design estimates of pile group capacities in cohesive soils. Thus, $(Q_{vG})_{ult}$ will be the lower of the values estimated from equations (5.52a) and (5.53b).

***Example 5.11*** Using the data of Example 5.10, estimate the pile group bearing capacity if the piles are placed 5 ft (1500 mm) center to center and joined at the top by a square pile cap supported by nine piles.

SOLUTION Assuming the arrangements of Figure 5.16, $B = 20$ in. (500 mm), $s = 5$ ft (1500 mm), $\bar{b} = (5 + 5 + 20/12)$ ft $= 10.67$ ft, $n = 9$.

(a) Block Failure of Pile Group

$$c_u = 6576 \text{ lb/ft}^2$$

$$D_f/\bar{b} = 31/10.67 = 2.9$$

From Table 5.7: $N_c \simeq 9$ for $D_f/B = 2.9$
From Table 5.8: $N_c = 6$ for base width $\bar{b} > 3$ ft
The lower of these $N_c$ values is 6 and will be used in these calculations
$c_a = 570 \text{ lb/ft}^2$ along the shaft, from Example 5.10.
$B = 5$ ft, $\bar{b} = 10.67$ ft
$L_e = 31 - 5 = 26$ ft (assuming 5 ft is the seasonal variation depth)

Then

$$(Q_{vG})_{ult} = c_u N_c (\bar{b})^2 + 4c_a(\bar{b})L_e \tag{5.52a}$$
$$(Q_{vG})_{ult} = 6576 \times 6(10.67^2) + 570(4 \times 10.67)26 \text{ lb} = 5124 \text{ kips}$$

**TABLE 5.10 Group Efficiency Values for Various Pile Spacing**[a]

| Pile spacing(s) | $3B$ | $4B$ | $5B$ | $6B$ | $8B$ |
|---|---|---|---|---|---|
| Group efficiency ($G_e$) | 0.7 | 0.75 | 0.85 | 0.9 | 1.0 |

[a]These values are based on the experimental data obtained by Whitaker (1957) and presented in graphs in *Foundations and Earth Structures Design Manual*, DM-7.2 (1982).

(b) Sum of Ultimate Loads of Single Piles

$$\text{Pile spacing} = s = 5\,\text{ft} = 3B$$

From $s$ and $G_e$ relationship,

$$G_e = 0.7 \text{ from Table 5.10}$$

Also, from Example 5.10,

$$(Q_v)_{\text{ult}} = 178\,\text{kips}$$

Then, from equation (5.53b)

$$(Q_{vG})_{\text{ult}} = 0.7 \times 9 \times 178 = 1121\,\text{kips}$$

The smaller of $(Q_{vG})_{\text{ult}}$ from (a) and (b) is 1121 kips.

$$(Q_{vG})_{\text{all}} = (Q_{vG})_{\text{ult}}/FS = 1121/3 = 374\,\text{kips}\ (1663\,\text{kN})$$

### 5.1.9 Settlement of a Single Pile in Cohesive Soils

The settlement of piles in cohesive soils primarily consists of the sum of the following two components:

1. Short-term settlement occurring as the load is applied.
2. Long-term consolidation settlement occurring gradually as the excess pore pressures generated by loads are dissipated.

Generally, the short-term settlement results from elastic compression of cohesive soils. This component of settlement constitutes a significant portion of the total settlement for partially saturated and overconsolidated saturated cohesive soils. The overconsolidated soils are soils whose past effective vertical overburden pressures are larger than the present effective vertical overburden pressures. Methods of settlement estimation discussed in Section 5.1.4 also apply here to calculate short-term settlements when pertinent soil properties for clays are used.

The method of estimating long-term consolidation settlement of a pile group is presented in the Section 5.1.10 and Example 5.12.

### 5.1.10 Settlement of Pile Groups in Cohesive Soils

The settlement estimation of pile groups in cohesive soils is complex. Figure 5.17 shows a simple method that can be used for settlement estimation of pile groups in cohesive soils.

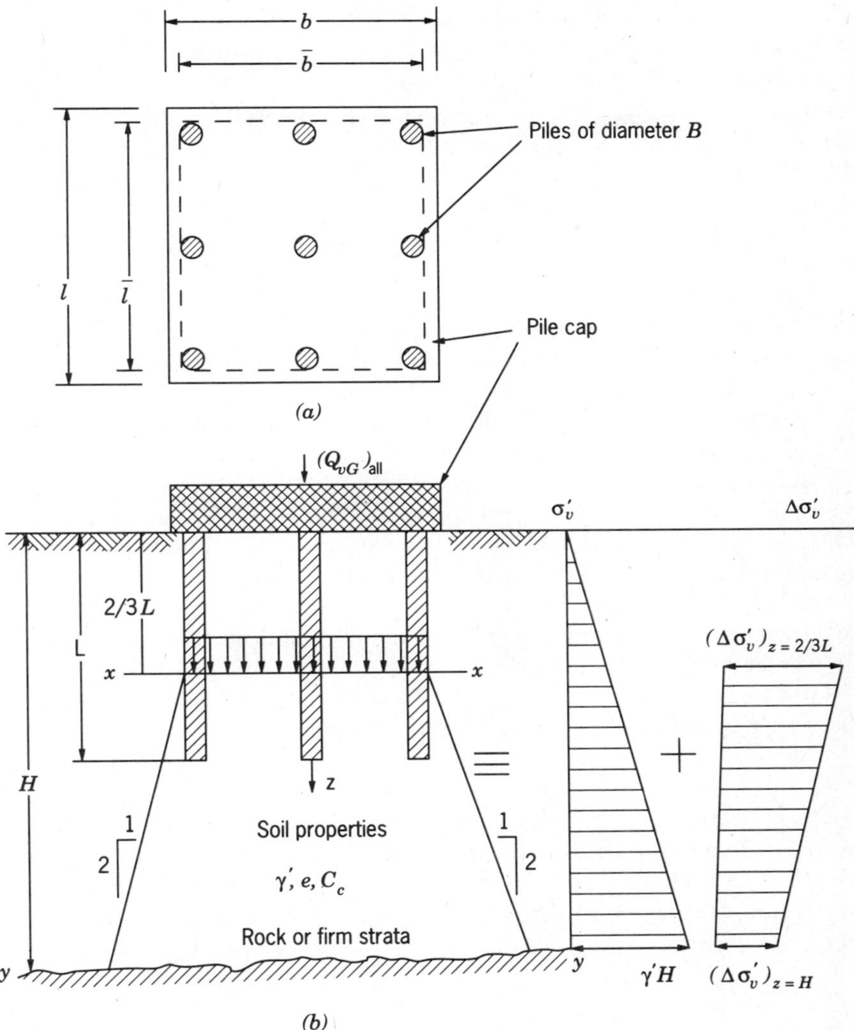


**Figure 5.17** Stress distribution for settlement estimation for friction piles in clay. (a) Plan (b) Section.

This approximate method is based on the following assumptions:

1. The allowable soil pressure $= q_{all} = (Q_{vG})_{all}/(\bar{b} \times \bar{l})$, where $\bar{b}$ and $\bar{l}$ are the base dimensions of pile group at the tip.
2. The pressure $q_{all}$ is transferred to, $(2/3) \times L$, depth below ground surface. The settlement of the soil above this depth is assumed to be small and therefore is neglected.

3. Then estimate the settlement as if a footing of dimensions $\bar{b} \times \bar{l}$ with pressure $q_{\text{all}}$ is placed at $\frac{2}{3}L$ depth below ground surface. The presence of pile below this depth is disregarded.
4. Once the pressure $q_{\text{all}}$ is applied on top of a clay layer of thickness $(H - \frac{2}{3}L)$, then consolidation settlement, $\Delta H$, can be calculated from the following relationship:

$$\Delta H = [C_c/(1 + e_0)][H - \tfrac{2}{3}L] \times \log_{10} [(\sigma'_v + \Delta\sigma'_v)/\sigma'_v] \tag{5.54}$$

where

$\Delta H$ = consolidation settlement

$\sigma'_v$ = present effective (vertical) overburden pressure at the middle of the layer $(H - 2/3L)$, determined as shown in Figure 5.17.

$\Delta\sigma'_v$ = increased pressure from pile load at the middle of the layer $(H - 2/3L)$.

$C_c$ = coefficient of consolidation

$e_0$ = initial void ratio of the soil

Figure 5.17 and Example 5.12 further explain these terms and the method of calculating consolidation settlement. Figure 5.17 shows a pile group having $b \times l$ size pile cap. There are 9 piles having $\bar{b} \times \bar{l}$ rectangular dimension at the base of the pile group. The piles are of length $L$. The soil conditions assumed are clay to a depth $H$ below ground underlain by rock.

If it is assumed that the load $(Q_{vG})_{\text{all}}$ is transferred to $\frac{2}{3}L$ depth below ground. The increased stress at his depth (level $xx$) is then:

$$(\Delta\sigma'_v)_{z=\frac{2}{3}L} = (Q_{vG})_{\text{all}}/(\bar{b} \times \bar{l}) \tag{5.55}$$

For stress distribution below this level ($xx$), it is assumed that the pressure (or stress) is distributed at $2V{:}1H$ slope as shown in Figure 5.17. Based on this assumption, the increased stress on plane $yy$ at depth $z = H$ can be obtained from the following:

$$(\Delta\sigma'_v)_{z=H} = (Q_{vG})_{\text{all}}/(\bar{b} + H - \tfrac{2}{3}L)(\bar{l} + H - \tfrac{2}{3}L) \tag{5.56}$$

The increased stress at any intermediate level between $xx$ and $yy$ can then be obtained by interpolation.

The consolidation settlement $(\Delta H)$ of this pile group due to an applied load of $(Q_{vG})_a = (Q_{vG})_{\text{all}}$ can then be estimated by using equation (5.54). In this equation $\sigma'_v$ is the present effective vertical pressure, $\Delta\sigma'_v$ is obtained by using equations (5.55) and (5.56), $C_c$ and $e_0$ are laboratory-determined soil parameters, and $H$ is the thickness of the clay stratum. Empirical relations for estimating $C_c$ are presented in Chapter 4 (Section 4.1.2).

Equation (5.54) is used when the clays are normally consolidated. For overconsolidated clays, the settlement calculation requires that the settlement be divided into two components as follows:

$$H = \Delta H_1 + \Delta H_2 \tag{5.57}$$

where

$\Delta H_1$ = settlement due to applied load in the recompression zone
$\Delta H_2$ = settlement due to applied load in the virgin curve zone
Thus $\Delta H_1$ and $\Delta H_2$ can be estimated from the following:

$$\Delta H_1 = [C_r/(1+e_0)][H - (\tfrac{2}{3})L]\log_{10}[p_c/\sigma'_v] \tag{5.58}$$

$$\Delta H_2 = [C_c/(1+e_0)][H - (\tfrac{2}{3})L]\log_{10}\left(\frac{\bar{\sigma}'_v + \Delta\sigma'_v}{p'_c}\right). \tag{5.59}$$

For highly overconsolidated clays, long-term consolidation settlements do not occur. Therefore, only short-term settlements are calculated. This is because their $p_c$ is very high and additional pressure due to $\Delta\sigma'_v$ will not result in consolidation. When the soils are underconsolidated, they settle due to their own weight and result in imposing downward loads along the pile shaft. This is discussed in Section 5.1.12. The definitions of underconsolidated, normally consolidated, overconsolidated, $C_r$, $C_c$, and $p_c$, were presented in Chapter 4 (Section 4.1.2).

***Example 5.12*** For the pile arrangement shown in Figure 5.17, let:

$(Q_{vG})_{\text{all}} = 323$ kips
$\bar{b} = 10.67\text{ ft} = \bar{l}$
$L = 30$ ft
$H = 50$ ft, unit weight of soil, $\gamma = 125\text{ lb/ft}^3$

Initial soil void ratio, $e_0 = 0.7$, compression index, $C_c = 0.17$, soil is normally consolidated, water level is at ground surface, and $e_0$ remains constant for the entire soil mass. Estimate the total settlement of the pile group.

SOLUTION

(a) Effective Overburden Pressure ($\sigma'_v$)
$\sigma'_v$ (at depth $= \frac{2}{3}L = 20$ ft) $= (125 - 62.5)20/1000 = 1.25\text{ kips/ft}^2$
$\sigma'_v$ (at depth $= H = 50$ ft) $= (125 - 62.5)\,50/1000 = 3.125\text{ kips/ft}^2$
(b) Increased Pressure Due to Loads on Pile ($\Delta\sigma'_v$)
From equation (5.55):

$$\Delta\sigma'_v \text{ (at depth } = \tfrac{2}{3}L = 20\text{ ft}) = (Q_{vG})_{\text{all}}/(10.67 \times 10.67)$$
$$= 323/(10.67)^2 = 2.83\text{ ksf}$$

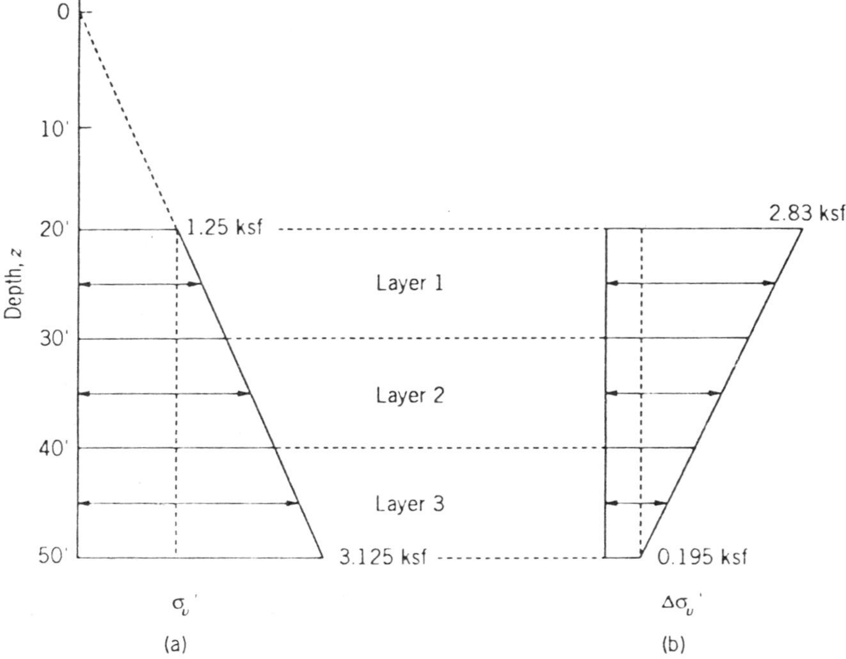


**Figure 5.18** (a) Existing vertical stress, $\sigma'_v$, and (b) the increased stress, $\Delta\sigma'_v$, for consolidation settlement calculations in Example 5.12.

From equation (5.56):

$$\begin{aligned}\Delta\sigma'_v(\text{at depth} = H = 50\,\text{ft}) &= (Q_{vG})_{\text{all}}/(\bar{b} + H - \tfrac{2}{3}L)^2 \\ &= 323/(10.67 + 50 - 20)^2 = 0.195\,\text{ksf}\end{aligned}$$

These values are plotted in Figure 5.18b. As shown in this figure, the total clay layer between 20 to 50 ft depth is then divided into three layers. The pressures at the middle of each layer are then calculated as follows:

$$\text{Layer 1: } \sigma'_v = 1.25 + \frac{1.875}{30} \times 5 = 1.56; \qquad \Delta\sigma'_v = 0.195 + \frac{2.635}{30} \times 25 = 2.39$$

$$\text{Layer 2: } \sigma'_v = 1.25 + \frac{1.875}{30} \times 15 = 2.19; \qquad \Delta\sigma'_v = 0.195 + \frac{2.635}{30} \times 15 = 1.52$$

$$\text{Layer 3: } \sigma'_v = 1.25 + \frac{1.875}{30} \times 25 = 2.81; \qquad \Delta\sigma'_v = 0.195 + \frac{2.635}{30} \times 5 = 0.63$$

The soil thickness below depth $2/3L$ (= 20 ft) is 30 ft. This soil will be consolidated under the increased pressure of 2.83 kips/ft$^2$. For settlement calculation purposes as already mentioned, this 30-ft-thick clay has been divided into three equal layers. $\sigma'_v$ and $\Delta\sigma'_v$ are determined at their mid-depths, and the settlements ($\Delta H$) are then estimated as follows. In equation (5.54) the total depth of clay that will undergo consolidation is $(H - 2/3L) = 30'$. If we divide this total thickness into three equal layers each of thickness $H_0 = 30/3 = 10$ ft, the equation (5.54) can be modified as follows:

$$\Delta H = [C_c/(1 + e_0)H_0 \log_{10} [(\sigma'_v + \Delta\sigma'_v)/\sigma'_v]$$

$$H_0 = 10 \times 12 = 120 \text{ in.}$$

$$\text{Layer 1: } \Delta H_1 = \frac{0.17}{1 + 0.7} \times 120 \log_{10} \frac{1.56 + 2.39}{1.56} = 4.84 \text{ in.}$$

$$\text{Layer 2: } \Delta H_2 = \frac{0.17}{1 + 0.7} \times 120 \log_{10} \frac{2.19 + 1.52}{2.19} = 2.74 \text{ in.}$$

$$\text{Layer 3: } \Delta H_3 = \frac{0.17}{1 + 0.7} \times 120 \log_{10} \frac{2.81 + 0.63}{2.81} = 1.05 \text{ in.}$$

Total settlement $= \Delta H = \Delta H_1 + \Delta H_2 + \Delta H_3$, $\Delta H = 4.84 + 2.74 + 1.05 = 8.63$ in. (say 9 in.).

### 5.1.11 Design Procedure for Piles in Cohesive Soils

The design procedure consists of the following five steps:

1. **Soil Profile.** From proper soils investigations, establish the soil profile and groundwater levels and note soil properties on the profile based on field and laboratory tests.
2. **Pile Dimensions and Allowable Bearing Capacity.** Select a pile type, length, and diameter and calculate allowable bearing capacity of a single pile based on the following equation:

$$(Q_v)_{\text{ult}} = A_p c_u N_c + p \sum_{L=0}^{L=L_e} c_a \Delta L \tag{5.47}$$

3. **Number of Piles.** Determine the number of piles required by dividing the column load with the allowable load or bearing capacity of the single pile. Arrange the piles in the group such that pile spacing is three to four times the pile diameter. Establish pile cap size with reference to column spacing and other space restrictions. If it becomes too large, increase pile length and/or pile diameter and repeat item (2) to obtain reasonable pile dimensions and arrangement. The pile group capacity is then the lower of the values

obtained from the following equations:

$$(Q_{vG})_{\text{ult}} = c_u N_c (\bar{b})^2 + 4c_u(\bar{b})L_e \qquad (5.52a)$$

$$(Q_{vG})_{\text{ult}} = G_e \times n \times (Q_v)_{\text{ult}} \qquad (5.53b)$$

4. **Settlement of Piles.** The settlement of piles in cohesive soils is the sum of the short-term and the long-term settlements. For short-term settlements the settlement of a single pile is first calculated. Then this value is used to estimate the short-term settlement of pile group.

*(a) Short-term settlement*

The short-term settlement of a single pile is determined as follows:

(i) Semiempirical Method

$$S_t = S_s + S_p + S_{ps} \qquad (5.34)$$

where

$$S_s = (Q_{pa} + \alpha_s Q_{fa})L/(A_p E_p) \qquad (5.35)$$

$$S_p = C_p Q_{pa}/(Bq_p) \qquad (5.36)$$

$$S_{ps} = C_s Q_{fa}/(D_f q_p) \qquad (5.37)$$

(ii) Empirical Method

$$S_t = (B/100) + (Q_{va}L)/(A_p E_p) \qquad (5.39)$$

The settlement is then higher of the values obtained from (i) and (ii) above. The settlement of a pile group is then determined from the following:

$$S_G = S_t \sqrt{(\bar{b}/B)} \qquad (5.40)$$

*(b) Long-term (consolidation) settlement*

(i) The long-term (consolidation) settlement for normally consolidated clays is determined from the following:

$$\Delta H = [C_c/(1 + e_0)][H - \tfrac{2}{3}L] \times \log_{10}[(\sigma'_v + \Delta\sigma'_v)/\sigma'_v] \qquad (5.54)$$

(ii) The long-term (consolidation) settlement for overconsolidated clays is determined from the following:

$$\Delta H = \Delta H_1 + \Delta H_2 \qquad (5.57)$$

where

$$\Delta H_1 = (C_r/(1 + e_0))(H - \tfrac{2}{3}L)\log_{10}(\bar{p}_c/\sigma'_v) \qquad (5.58)$$

$$\Delta H_2 = (C_c/1 + e_0))(H - \tfrac{2}{3}L)\log_{10}((\sigma'_v + \Delta\sigma'_v)/\bar{p}_c) \qquad (5.59)$$

The $\Delta\sigma'_v$ is calculated at depth $z = \frac{2}{3}L$ and at $z = H$ by using the following equations. The $\Delta\sigma'_v$ values at any intermediate depth can then be obtained by interpolation.

$$(\Delta\sigma_v)_{z=\frac{2}{3}L} = (Q_{vG})_{all}/(\bar{b} \times \bar{l}) \tag{5.55}$$

$$(\Delta\sigma_v)_{z=H} = (Q_{vG})_{all}/(\bar{b} + H - \tfrac{2}{3}L)(\bar{l} + H - \tfrac{2}{3}L) \tag{5.56}$$

5. **Pile Load Test and Driving Criteria.** Recommend a pile load test to fine tune the allowable bearing capacity. If driven piles are selected, specify the driving criteria that should be supplemented with the pile load test.

***Example 5.13*** In an industrial project one column of a steel frame supporting a heavy equipment carries an axial load of 500 kips (2225 kN). Soils investigation indicated the soil profile as shown in Figure 5.19a. Design a pile foundation such that the maximum settlement of the group does not exceed 0.75 inch (19 mm).

SOLUTION

1. **Soil Profile.** Soil profile and test values with depth are shown in Figure 5.19a.

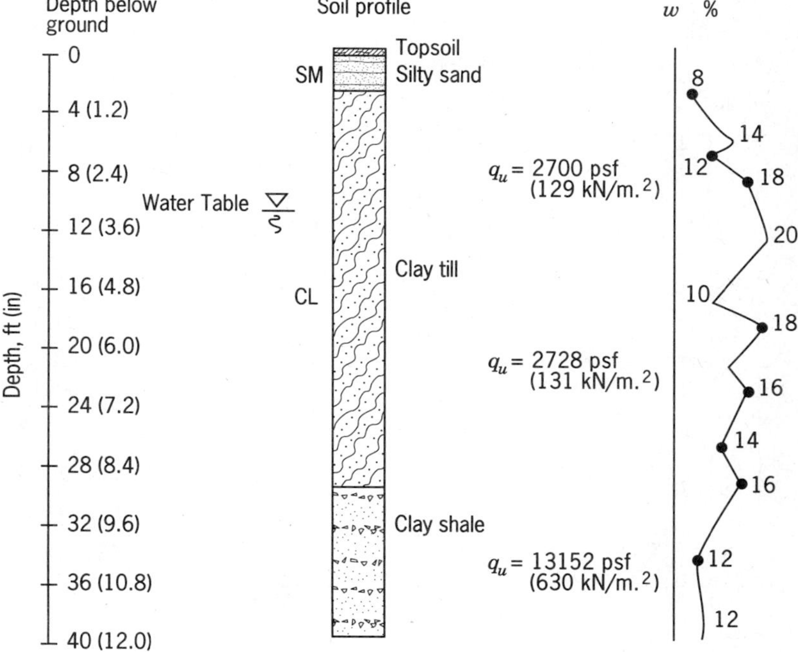


**Figure 5.19a** Soil profile and soil properties along the depth used in Example 5.13. $q_u$ = unconfined compressive strength; over consolidation ratio for the clay till = 4 to 5; over consolidation ratio for clay shale 6 to 8. $w$ = natural moisture content.

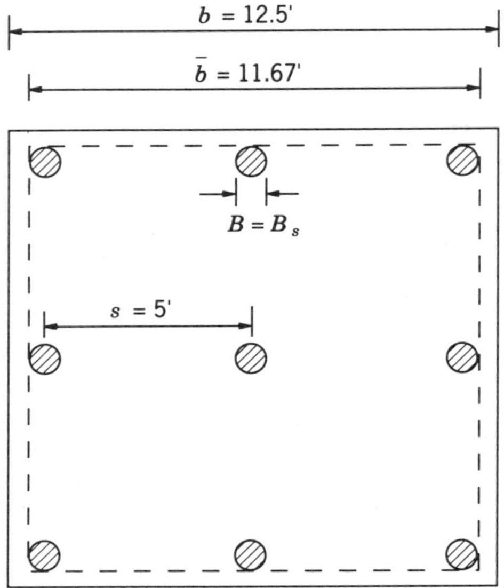


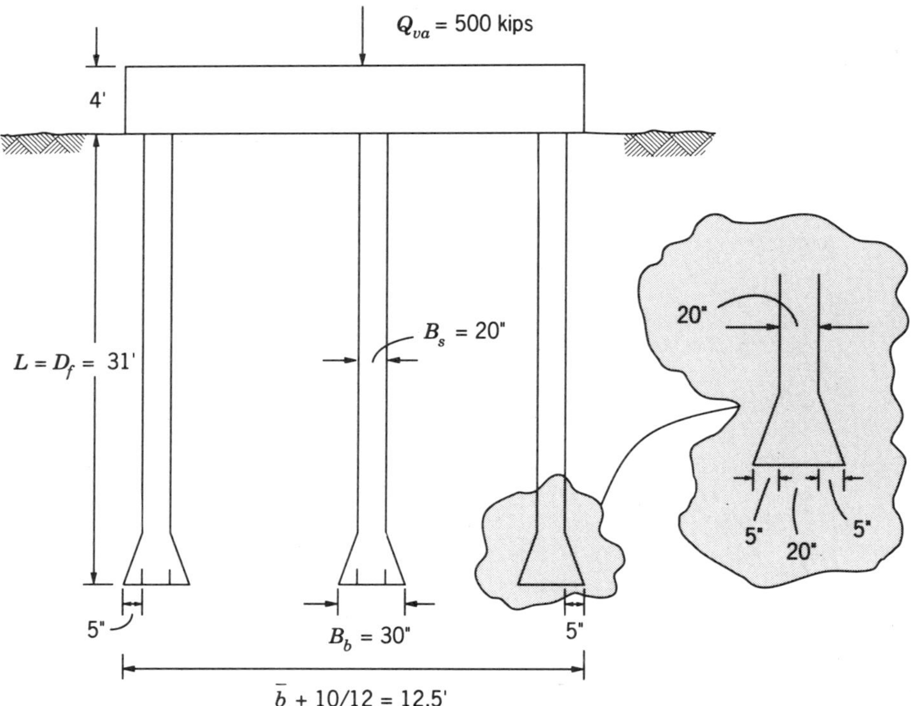


**Figure 5.19b** Pile group arrangement used in Example 5.13.

2. **Pile Dimensions and Allowable Bearing Capacity.** The top 5 ft of the soil lies in the zone of seasonal variation. Below this depth, the clay till appears to have a uniform moisture content and shear strength. The average undrained shear strength $c_u = \frac{1}{2}\{(2700 + 2728)/2\} = 1357$ psf.
   Consider a 31-ft (9.5 m) long, 20-in. (500 mm) shaft diameter cast-in-place bored concrete pile. Assume that a 30-in. diameter bell is made at the pile base. From equation (5.47) the ultimate bearing capacity is:

$$(Q_v)_{\text{ult}} = A_p c_u N_c + p \sum_{L=0}^{L=L_e} c_a \Delta L \tag{5.47}$$

where

$$A_p = \pi/4(B_b)^2 = \pi/4(30/12)^2 = 4.9\,\text{ft}^2, \text{ where } B_b \text{ is the bell diameter}$$
$$c_u = 13152/2 = 6576\,\text{psf at pile base in clay shale}$$
$$D_f/B_b = 31 \times 12/30 = 12.4$$

Then, from Table 5.7, $N_c = 9$. The pile base diameter of 30 in. = 2.5 ft. Then, from Table 5.8, $N_c = 7$. The lower of these two $N_c$ values is 7 and will be used here.

$$p = \pi B_s = \pi \times 20/12 = 5.24\,\text{ft},$$

where $B_s$ is the shaft diameter.

From Table 4.7, for drilled concrete pile, if $c_u$ for clay along pile shaft is 1357 lb/ft$^2$, then $c_a/c_u = 0.30$.

$$c_a = 0.3\,c_u = 0.3 \times 1357 = 407\,\text{lb/ft}^2$$

From Table 5.9:

$$L_e = L\text{–(depth of seasonal variation + 2 pile shaft diameter)}$$
$$= 31 - (5 + 2 \times 20/12) = 22.7\,\text{ft}$$
$$(Q_v)_{\text{ult}} = 4.9 \times 6576 \times 7 + 5.24 \times 407 \times 22.7\,\text{lb}$$
$$= 225.56 + 48.41 = 273.97\,\text{kips, say } 274\,\text{kips}$$
$$= Q_p + Q_f; \qquad Q_p = 225.56\,\text{kips}; \qquad Q_f = 48.41\,\text{kips}$$
$$(Q_v)_{\text{all}} = (Q_v)_{\text{ult}}/FS = 274/3 = 91\,\text{kips (405 kN)}$$

3. **Number of Piles and Their Arrangement.** The number of piles required to support 500 kips equipment load will be:

$$n = Q_{vG}/(Q_v)_{\text{all}} = 500/91 = 5.5\ \text{(say 6)}$$

Since in cohesive soils, group capacity is normally less than the sum of individual pile capacity, a greater number of piles should be used. Try a group of nine piles arranged in a square pattern with three piles on each side. Place the piles at 5 ft (125 mm) center-to-center spacing with 12.5 ft × 12.5 ft × 4 ft thick concrete cap. The pile group arrangement is shown in Figure 5.19b.

The allowable bearing capacity of this pile group can be calculated by the following two methods:

$$(Q_{vG})_{\text{ult}} = c_u N_c(\bar{b})^2 + 4c_u(\bar{b})L_e \tag{5.52a}$$

For straight-shafted piles, the pile base width will be $\bar{b}$. For belled piles this will be increased to $(\bar{b} + 10/12) = 12.5'$ (shown in Figure 5.19b).

$$D_f/B = D_f/(\bar{b} + 10/12) = 31/12.5 = 2.48 \qquad N_c = 8.5 \text{ from Table 5.7}$$

Also,

$$(\bar{b} + 10/12) = \text{base width} = 12.5', \qquad N_c = 6 \text{ from Table 5.8}$$

The lower of the above two $N_c$ values is 6 and will be used in these calculations. Values $c_u$, $c_a$, and $L_e$ were obtained above.

$$(Q_{vG})_{\text{ult}} = 6576 \times 6\,(12.5)^2 + 4 \times 1357 \times 11.67 \times 22.7\text{ lb}$$

$$= 6165 + 1438 = 7603\text{ kips}$$

$$(Q_{vG})_{\text{all}} = 2534\text{ kips (11278 kN) when applying a safety factor of 3}$$

$$(Q_{vG})_{\text{ult}} = G_e \times n \times (Q_v)_{\text{ult}} \tag{5.53b}$$

$$s = 5\text{ ft}, \qquad B = 20/12 = 1.67\text{ ft}, \qquad s/B = 3$$

From Table 5.10, $G_e = 0.7$. The number of piles $= n = 9$.

$$(Q_v)_{\text{ult}} = 274\text{ kips for a single pile as calculated above}$$

$$(Q_{vG})_{\text{ult}} = 0.7 \times 9 \times 274 = 1726\text{ kips}$$

$$(Q_{vG})_{\text{all}} = 1726/3 = 575\text{ kips (2559 kN) when a safety factor of 3 is used}$$

The lower of $(Q_{vG})_{\text{all}}$ calculated from equations (5.52a) and (5.53b) above is 575 kips (2559 kN).
Therefore,

$$(Q_{vG})_{\text{all}} = 575\text{ kips.}$$

$$\text{Pile cap weight} = 12.5 \times 12.5 \times 4 \times 0.15 = 93\text{ kips (417 kN)}$$

$$\text{Total load on pile group} = 500 + 93 = 593\text{ kips}$$

The group capacity is 575 kips, which is approximately equal to the load 593 kips on the group. Therefore, it is acceptable from a bearing capacity point of view.

**4. Settlement of Single Pile and Pile Group**

*(a) Short-term settlement*

(i) Semiempirical Method

$$S_t = S_s + S_p + S_{ps} \tag{5.34}$$

$$S_t = (Q_{pa} + \alpha_s Q_{fa})L/(A_p E_p) + C_p Q_{pa}/(B_p q_p) + C_s Q_{fa}/(D_f q_p)$$

by combining equations (5.34) through (5.37).

Total load on pile group = 593 kips. Therefore, the load per pile = 593/9 = 66 kips. From section (2) above, $Q_p = 225.56$, $(Q_p)_{\text{all}} = 225.56/3 = 75$ kips. Also, $Q_f = 48.41$, $(Q_f)_{\text{all}} = 48.4/3 = 16$ kips and total allowable load is 75 + 16 = 91 kips while the actual load on each pile is 66 kips. The values of actual $Q_p$ and $Q_f$ can be proportioned as shown without any significant error in calculations.

$$(Q_p)_{\text{actual}} = 75(66/91) = 54 \text{ kips} = Q_{pa}$$

$$(Q_f)_{\text{actual}} = 16(66/91) = 12 \text{ kips} = Q_{fa}$$

$L = 31$ ft, $A_p = [\pi(20/12)^2/4] = 2.18\text{ ft}^2$, $E_p = 3.6 \times 10^6$ psi (for concrete) $\alpha_s = 0.5$ by assuming uniform distribution of skin friction. This is a reasonable assumption. As discussed in Section 5.1.4, the total settlement calculated based on uniform or triangular distribution are not sensitive to $\alpha_s$ values. From Table 5.6, $C_p = 0.03$, $B_p = B_b = 30$ in. $q_p = Q_p/A_{\text{base}} = 225.56/(\pi/4)(30/12)^2 = 46\text{ kips/ft}^2$, $D_f = 31$ ft, and from equation (5.38) $C_s = (0.93 + 0.16\sqrt{D_f/B})\,C_p = (0.93 + 0.16\sqrt{31 \times 12/20}) \times 0.03 = 0.048$.

Substituting above values in the expression for $S_t$, we get:

$$S_t = \frac{(54 + 0.5 \times 12)\,31 \times 12 \times 10^3}{2.18 \times 144 \times 3.6 \times 10^6} + \frac{0.03 \times 54 \times 144}{30 \times 46} + \frac{0.048 \times 12 \times 144}{31 \times 12 \times 46}\text{ in.}$$

$$S_t = 0.019 + 0.168 + 0.005 = 0.192\text{ in. (4.8 mm)}$$

(ii) Empirical Method

$$S_t = (B/100) + Q_{va}L/(A_p E_p) \tag{5.39}$$

where $B = 20$ in., $Q_{va} = 66$ kips, $L = 31$ ft, $A_p = (\pi/4)(20)^2 = 314.16\,\text{in}^2$.

$$E_p = E_c = 3.6 \times 10^6\ \text{psi}$$

$$S_t = 20/100 + (66 \times 31 \times 12 \times 1000)/(314.16 \times 3.6 \times 10^6)$$

$$S_t = 0.2 + 0.02 = 0.22\ \text{in.}\ (5.5\ \text{mm})$$

The higher of the above two values estimated by the semiempirical and empirical methods is 0.22 in. (5.5 mm)
Settlement of pile group can be calculated by using equation (5.40)

$$S_G = S_t\sqrt{b/B} \tag{5.40}$$

$$S_G = 0.22\sqrt{(12.5 \times 12/20)} = 0.60\ \text{in.}\ (15.3\ \text{mm})$$

*(b) Long-term (consolidation) settlement.* As shown in soil profile (Figure 5.19a), both the clay till and clay shales are highly overconsolidated since their overconsolidation ratio is 4 or more. As, an example $\Delta\sigma'_v$ at pile base is equal to 11 ksf while $p'_c$ at that level is 14.6 ksf. Therefore, the consolidation settlement due to loads on pile foundations would not occur. This has been discussed in Section 5.1.10. That Section and Example 5.12 also provide the details of estimating consolidation settlement for normally consolidated soil. The calculated settlement of pile group is 0.60 in. (15.3 mm). This is less than the allowable settlement of 0.75 in. (19 mm). Therefore, the designed pile diameter, length, and group arrangement is acceptable.

5. **Pile Load Test and Driving Criteria.** These are cast-in-place bored concrete piles, therefore, no driving criteria are required. Pile load tests as per ASTM D1143–81 should, however, be recommended to confirm the design load and settlement values estimated above. Conservative design values should be used where the cost of pile load tests cannot be justified. On small-size projects, a pile load test can be performed to two times the design load on an actual foundation pile. On large projects, where economics justifies it, pile load test should be carried out to failure on a test pile that shall not be used as a part of the actual foundation and will be abandoned after the test.

### 5.1.12 Pile Design for Negative Skin Friction

In Figure 5.20a, a pile embedded in layered clay is loaded axially. The pile has a tendency to move downward with reference to the surrounding soil. This would result in the mobilization of upward (positive) resistance or friction along the pile shaft. This upward or positive resistance $Q_f$ along with point bearing $Q_p$ act in the same direction and thus help support the external load $(Q_{vG})_{\text{all}}$. In Figure 5.20b, the pile is driven through a recent fill resting over an old deposit of clay. The recent fill is underconsolidated and is consolidating under its own

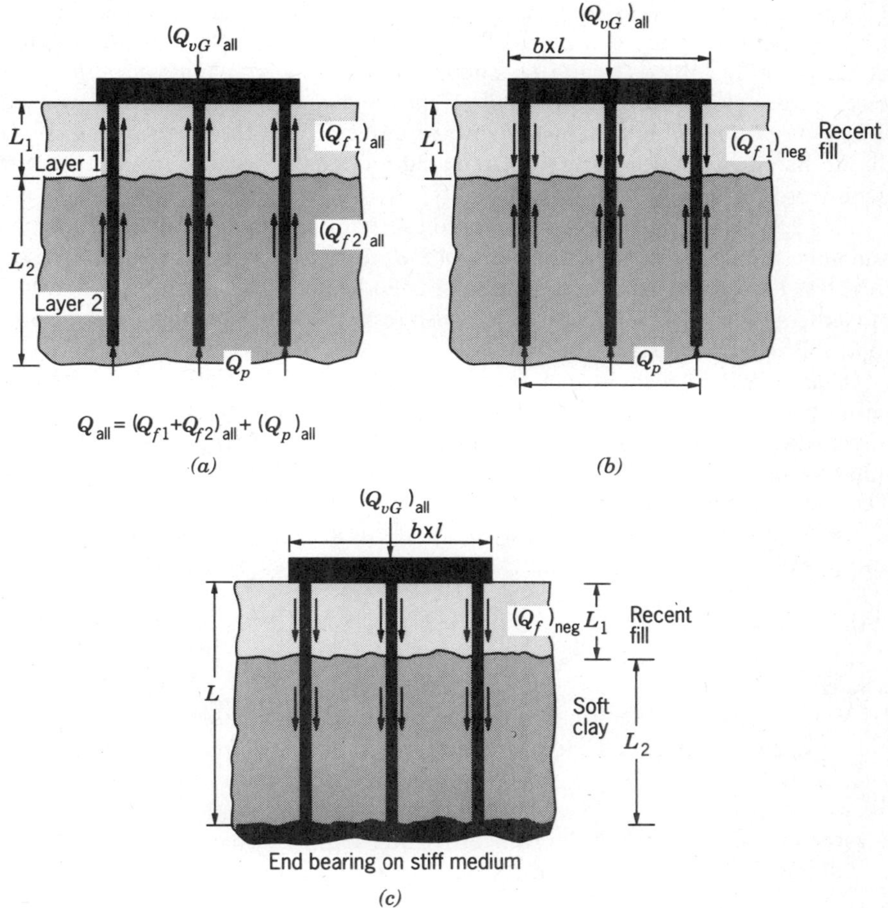


**Figure 5.20** Conceptual comparison between positive and negative skin friction development. (a) Positive friction development. (b) Negative skin friction when recent fill consolidates due to its self-weight. (c) Negative skin friction when soft clay consolidates due to dewatering and/or addition of recent fill.

weight after the pile has been installed. Therefore, the recent fill may settle more than the pile. Chapter 4 (Section 4.1.2) provides the definition for underconsolidated clays.

In Figure 5.20c, the pile has been driven through the recent fill overlying a soft clay and its tip rests on a stiff medium clay. A dewatering operation may lower the water table, and the soft clay starts consolidating under the increased effective stresses. The recent fill is also underconsolidated. The pile tip is not settling.

The foregoing two and other similar situations may cause the pile side friction or resistance to act in the same direction as the externally applied load $(Q_{vG})_{all}$.

This is called *negative skin friction* because the soil resistance along the pile surface operates in the downward direction, which is opposite to the conventional resisting forces $Q_f$ and $Q_p$ shown in Figure 5.20a. *Thus, negative skin friction develops when the settlement of surrounding soil exceeds the downward movement of the pile shaft.* This downward movement of the surrounding soil is due to its consolidation. This is also called *down drag* because it pulls the pile downward.

This downward pull on the pile would exert additional axial force on the pile and may result in excessive settlements of the pile or even failure in extreme cases. Vesic (1977) reported that observations had indicated that a relative downward movement of 0.6 in. (15 mm) of the soil with respect to the pile might be sufficient to mobilize full negative skin friction.

In areas where there is a potential for negative skin friction development, batter piles should be avoided. This is primarily due to (1) the magnitude of this down drag on the outer side of batter piles being significantly larger than the inner vertical piles and (2) the settling soil moving away from the inner piles. These phenomena can induce excessive bending on piles.

According to Vesic (1977), negative skin friction, $(Q_f)_{\text{neg}}$, for both cohesionless and cohesive soils can be estimated by the following expression:

$$(Q_f)_{\text{neg}} = \bar{N}_0 \bar{p}_0 A \tag{5.60a}$$

where

$\bar{N}_0$ = nondimensional factor that can be obtained from Table 5.11
$\bar{p}_0$ = the mean normal effective stress

and

$A$ = the area of the shaft in the zone of settling soil (e.g., $A = \pi BL$ for a pile with diameter $B$ and length $L$ in the zone of settling soil).

In Table 5.11 uncoated and coated piles have been mentioned. Uncoated piles are the regular piles that have no surface treatment. The coated piles are piles that have been coated with bitumen or bentonite. These coatings reduce the adhesion

**TABLE 5.11 Values of $\bar{N}_0$ for Various Conditions**[a]

| Soil and Pile Condition | $\bar{N}_0$ |
|---|---|
| (a) Uncoated pile | |
| (i) In soft compressible layers of silt and clay | 0.15–0.3 |
| (ii) In loose sand | 0.3–0.8 |
| (b) Pile coated with bitumen or bentonite | 0.01–0.05 |

[a]Values provided by Vesic (1977).

**TABLE 5.12 Empirical Values of Negative Skin Friction**

| Soil Type | Negative Skin Friction |
|---|---|
| Sand | 0.35 to 0.5 $\sigma'_v A$ |
| Silt | 0.25 to 0.35 $\sigma'_v A$ |
| Clay | 0.20 to 0.25 $\sigma'_v A$ |

*Note*: $\sigma'_v$ = average effective vertical stress. $A$ = area of shaft in the zone of settling soil. The units of $\sigma'_v$ and $A$ have to be consistent to yield a force unit for the negative skin friction.

or bond between the soil and the pile surface. Thus the negative skin friction is lower for coated piles than the uncoated piles.

Negative skin friction can also be estimated by an empirical relationship originally proposed by Garlanger (1973) and recommended for use in the *Foundation and Earth Structures Design Manual* (1982). According to this, negative skin friction, $(Q_f)_{\text{neg}}$, can also be estimated from Table 5.12.

A simple method to estimate negative skin friction is by using positive skin friction values in downward direction in the zones of settling soils. Negative skin friction can be estimated as follows:

1. For cohesionless soils: From equation (5.6):

$$(Q_f)_{\text{neg}} = pK_s \tan\delta \sum_{L=0}^{L=L_e} \sigma'_{vl} \Delta L \tag{5.6}$$

2. For cohesive soils: from equation (5.46):

$$(Q_f)_{\text{neg}} = p \sum_{L=0}^{L=L_e} c_a \Delta L \tag{5.46}$$

In both these relations, $L$ or $L_e$ is the pile length in the zones of settling soils. These relations normally yield higher (conservative) values.

Observations suggest that approximately 0.75 times the pile length ($L$) in compressible layer should be considered as contributing to negative skin friction (Endo et al., 1969). This is based on the observation that at about $0.75L$ a neutral point exists below which there is no relative movement between the pile and the adjacent soil. However, other investigations show that neutral point can be located higher or lower than $0.75L$ (Vesic, 1977). Until there is a definitive method of determining the depth of this neutral point, it is recommended that total pile length in the zone of settling soil be used for such calculations.

Estimated value of $(Q_f)_{\text{neg}}$ should be subtracted from the allowable pile load for the design. The mechanics of negative skin friction is complex. The estimation method for negative skin friction on pile group is still not well understood. At the present time, the negative skin friction on a pile group can be conservatively

calculated by taking the total weight of fill and/or compressible soil enclosed by the piles in the group as follows:
In Figure 5.20b:

$$(Q_f)_{\text{neg}} = (b \times l)\gamma' L_1 \tag{5.61}$$

In Figure 5.20c:

$$(Q_f)_{\text{neg}} = (b \times l)\gamma' L \tag{5.62}$$

where $\gamma'$ is the effective unit weight of settling fill and $b$, $l$, and $L_1$ and $L$ are shown in Figure 5.20.

***Example 5.14*** In Figure 5.20c, consider that each pile is spaced such that they act individually and piles are end bearing. Further assume the following: steel pile, $B = 12$ in., $L_1 = 5$ ft, $L_2 = 10$ ft, groundwater is at ground surface and soil properties for the two layers are:

| Layer | $c'$ | $\phi'$ | $\gamma$ lb/ft$^3$ | Soil Type |
|---|---|---|---|---|
| 1 | 0 | 30° | 110 (17.5 kN/m$^3$) | Sand |
| 2 | $c_u = 300$ lb/ft$^2$ (14.37 kN/m$^2$) | 0 | 120 (19 kN/m$^3$) | Clay |

Estimate the negative skin friction along pile for the above case.

SOLUTION

*Method 1: Empirical Relations*

$$(Q_f)_{\text{neg}} = 0.5\sigma'_v A + 0.25\sigma'_v A \text{ from Table 5.12}$$

$$\sigma'_v \text{ at } 5' \text{ depth} = (110 - 62.5)\ 5 = 237.5 \text{ psf}$$

$$\sigma_v \text{ at } 15' \text{ depth} = 237.5 + (120 - 62.5)\ 10 = 237.5 + 575 = 812.5 \text{ psf}$$

$$(Q_f)_{\text{neg}} = 0.5\frac{(237.5 + 0)}{2}\pi \times 1 \times 5 + 0.25\frac{(237.5 + 812.5)}{2}\pi \times 1 \times 10$$

$$= 932.66 + 4123.34 = 5 \text{ kips } (22.25 \text{ kN})$$

*Method 2: Assuming Skin Friction Values in Downward Direction*

$$(Q_f)_{\text{neg}} = pK_s \tan\delta \sum_{L=0}^{L=L_1} \Delta\sigma'_{vl}\Delta L + p \sum_{L=0}^{L=L_2} c_a \Delta L \text{(5.6) and (5.46)}$$

$K_s = 1.0$ from Table 5.3, for layer 1 $\qquad p = \pi B = \pi \times 1, \qquad \delta = 2/3\phi = 20°$

$$c_a = 300\,\text{lb/ft}^2 \text{ from Figure 4.27 for layer 2}$$

$$(Q_f)_{\text{neg}} = \pi \times 1 \tan 20 \frac{(237.5 + 0)}{2} 5 + \pi \times 300 \times 10$$

$$= 678.58 + 9424.77 = 10\,\text{kips } (44.5\,\text{kN})$$

These two methods give upper and lower bound of the negative skin friction values.

***Example 5.15*** Assume that in Figure 5.20 (b) the piles now act as a group. Assume b = 10.67 ft (3 m). Then the negative skin friction can be calculated as follows:

SOLUTION As discussed in this Section and explained by equations (5.61) and (5.62).

$$(Q_f)_{\text{neg}} = (b \times l)(\gamma'_1 L_1 + \gamma'_2 L_2)$$

$$= 10.67 \times 10.67[(110 - 62.5)5 + (120 - 62.5)10]$$

$$= 92.5\,\text{kips } (411.63\,\text{kN})$$

### 5.1.13 Piles in Swelling and Shrinking Soils

Soils that contain substantial proportions of clay minerals (e.g., montmorillonite) exhibit a high-volume increase when they are above the water table and come in contact with moisture. This volume increase is called *swelling* of clays. When this moisture is removed by drying, these soils exhibit a high-volume decrease. This phenomena of volume decrease is termed as *shrinkage*. The magnitude of this volume change will depend on many factors (e.g., mineralogy of clays, the initial moisture content, soil particle structure) and the new environmental conditions imposed on the soil (e.g., a building that imparts heat or addition of moisture due to watering the lawn). Williams (1958) provides a guide to classify the swelling and shrinking potential of clay-rich soils based on Atterberg limits and grain-size test data. Another method of determining swelling and shrinking potential of a soil is by running laboratory swelling tests. These tests consist of placing the soil in a consolidation ring and subjecting it to the pressure equivalent to its field pressures. The sample is then submerged in water and allowed to swell for 24 hours. If the increase in volume under the anticipated vertical pressure is more than 5 percent of the original volume then the soil is considered to have swelling and shrinking potential.

The foregoing methods could either become time consuming or interpretations of swelling potential may get difficult. For most practical purposes, soils with a plasticity index greater than 30 may be classified has having high swelling and shrinking potential (Seed et al., 1962). The depth of soil that contributes to swelling and shrinking at a particular site mainly depends on (1) the thickness of

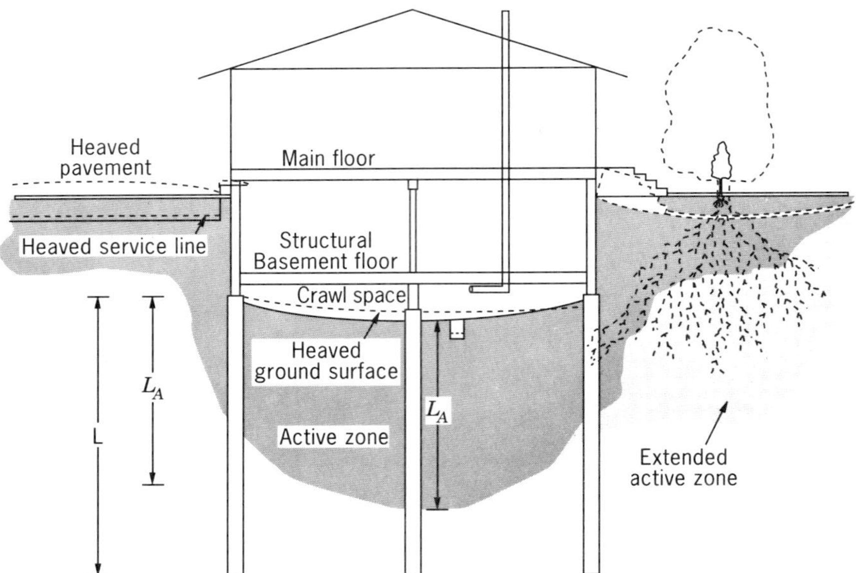


**Figure 5.21** Typical pile foundation performance on deep deposit of swelling and shrinking clays (*Canadian Foundation Engineering Manual*, 1978).

swelling and shrinking clays, (2) the depth of water table, and (3) the local environmental conditions that will influence the depth of seasonal changes. The depth of seasonal changes in soil moisture is mainly responsible for swelling and shrinking behavior of the clays. This depth is called the *active zone.* This depth can also be affected by the existence of a structure. For example, the excavation of soil below a structure and/or the heat transmitted by the structure to the underlying soil may alter the depth of active zone (Figure 5.21). The depth of active zone is generally evaluated and identified during the soils investigations work and based on the local experience.

It is a common engineering practice to utilize pile foundations in swelling and shrinking soils so that the foundations develop their bearing capacity in stable ground conditions below the active zone (Figure 5.21). Piles installed in such soils may, however, be subjected to uplift forces in the zone where swelling process due to moisture change occurs. Design considerations for such situation consists of either one or a combination of the following two methods.

***Preventive Methods*** These methods consist of eliminating uplift forces along the pile surface by isolating piles from the swelling clays in the active zone. The following methods can be used for such purposes:

1. Coating the pile surface in the active zone with bitumen
2. Separating the pile from swelling soils in the active zone by the use of floating sleeves that move up and down with the surrounding soil

***Design of Piles to Resist Uplift Swelling Forces*** The basic concept for the design of piles to resist upward swelling forces along pile surface should consist of the following:

1. The piles should have structural strength to resist these upward forces.
2. The uplift resistance to the pile in the soil should be provided from the soil below the zone that is not subjected to soil moisture changes (i.e., below the active zone).

The magnitude of uplift forces, $Q_{\text{up}}$, to be resisted by the pile can be approximated from equation (5.46) when equating $c_a = c_u$ as follows:

$$Q_{\text{up}} = p \sum_{L=0}^{L=L_A} c_u \Delta L \tag{5.63}$$

In this equation, the pile length $L$ has been equated to the pile length, $L_A$, which is the length of pile in the active zone as shown in Figure 5.21.

Thus, this $Q_{\text{up}}$ shound be resisted by the length of the pile below the active zone. This would require estimation of pullout capacities of a single pile and pile groups, as the case may be. This has been discussed in Sections 5.2.1 through 5.2.5 both for piles in cohesionless and cohesive soils, whichever are encountered below the depth of active zone.

Another alternative design to resist these uplift swelling pressures is to provide drilled and underreamed (belled) piles founded below the active zone. The estimation of pullout capacities and design formulas for such piles are discussed in Section 5.2.8. In such piles, the shaft should be designed to carry the tensile forces exerted by the uplift forces and the pile reinforcement should be carried into the bell to a point 4 in. (100 mm) above the base. Methods of estimating pullout resistance of piles have been discussed in detail in Article 5.2. Chen (1975) provides information for foundations on expansive (swelling) soils.

### 5.1.14 Piles in a Layered Soil System

A simple method of estimating bearing capacity of piles in a multilayered soil system would be to estimate frictional resistance in the strata where the shaft is located and end bearing in the strata where the tip is resting. This situation, in general form, is exhibited in Figure 5.22. In a situation where the pile shaft is mainly through clay and is resting on a sand layer, as shown in Figure 5.22a, the ultimate bearing capacity can be estimated by the following relationship:

$$(Q_v)_{\text{ult}} = p \sum_{L=0}^{L=L_1} c_a \Delta L + A_p \sigma'_v N_q \tag{5.64}$$

In estimating bearing capacities of layered soils, the relative stiffnesses and strengths of different layers penetrated by the piles should be considered. For

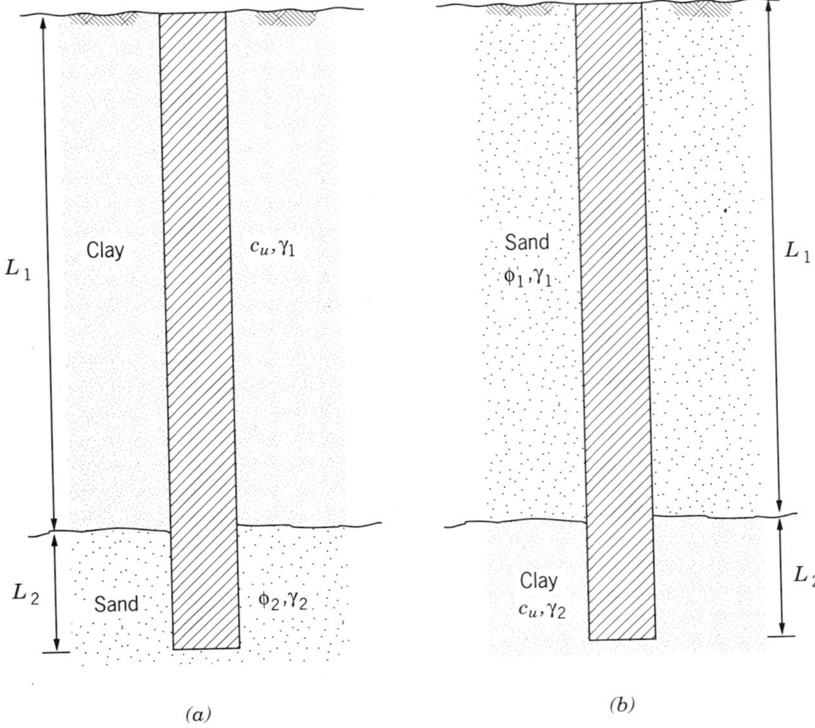


**Figure 5.22** Bearing capacity of a single pile in layered soil system. (a) Pile bearing on sand, (b) pile bearing on clay

example, if piles penetrate through a layer of soft soil into a deep deposit of competent material such as sand, the bearing capacity of this system would be derived only from frictional resistance and end-bearing capacities of the lower competent soil Figure 5.22a. $(Q_v)_{\text{ult}}$ for such cases can be obtained from the following equation:

$$(Q_v)_{\text{ult}} = pK_s \tan\delta \sum_{L=0}^{L=L_2} \sigma'_{vl}\Delta L + A_p\sigma'_v N_q \tag{5.65}$$

The critical depth, as discussed in Section 5.1.1, should be taken from the upper surface of granular stratum. The definitions of various terms in equations (5.64) and (5.65) and the concept of critical depth have already been discussed in Sections 5.1.1 and 5.1.7.

In the situation where the pile shaft is mainly through sand and is resting on the clay layer, Figure 5.22b, the ultimate bearing capacity can be estimated by the following relationship:

$$(Q_v)_{\text{ult}} = pk_s \tan\delta \sum_{L=0}^{L=L_1} \sigma'_{vl}\Delta L + A_p c_u N_c \tag{5.66}$$

Various terms in these equations have already been defined in sections 5.1.1 and 5.1.7.

In cases where a pile group is transferring load through a multilayer system to a sand stratum underlain by a weaker clay, the pile group safety at the base should be checked as follows:

1. Assume that the total applied load, $Q_{va}$, on the pile group is transferred to the soil through a theoretical footing located at the base of the pile group (shown in Figure 5.23).
2. Assume that this load $Q_{va}$ is now distributed at 2 V:1$H$ below the base of the pile group. At level $xx$, which is the sand–clay interface, the vertical stress, $\Delta\sigma'_v$, due to $Q_{va}$ will then be given by the following:

$$\Delta\sigma'_v = Q_{va}/(\bar{b} + H)(\bar{l} + H) \tag{5.67a}$$

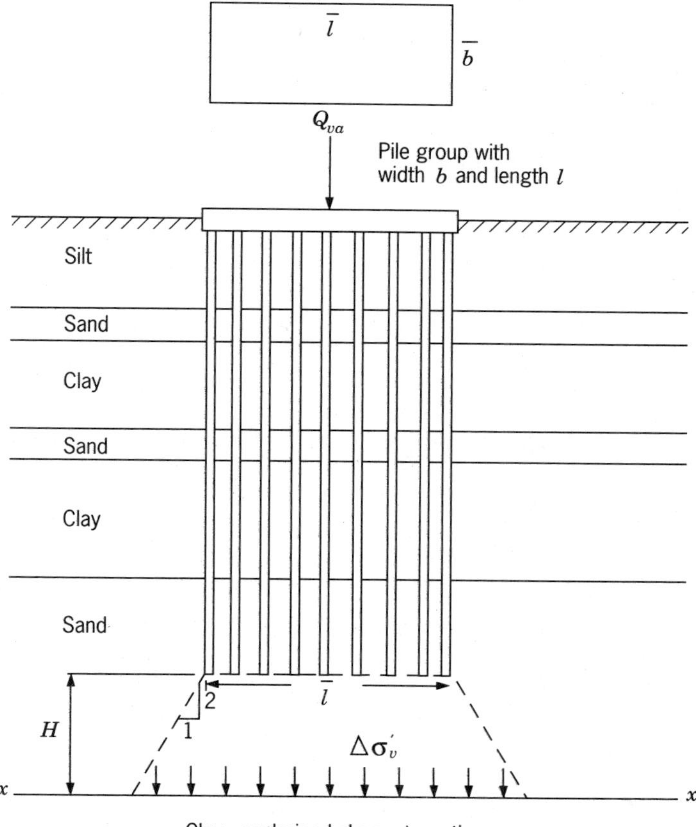


**Figure 5.23** Safety of pile groups against punching shear in layered soil (*Canadian Foundation Engineering Manual*, 1985).

where $\bar{l}$ and $\bar{b}$ are the dimensions of pile base, and $H$ is the depth of the sand–clay interface below the pile base as shown in Figure 5.23.

3. The pile group will then be safe against failure in the lower clay if following condition is met:

$$\Delta\sigma_v' \leqslant 3c_u \tag{5.67b}$$

where $c_u$ is the undrained strength of clay. This relationship ensures that the additional stress $\Delta\sigma_v'$ will not cause failure in the lower clay.

The settlement estimation of piles in layered soil system is complex and cannot be obtained accurately. Rough estimates may be made by using methods described in Sections 5.1.4, 5.1.5, 5.1.9, and 5.1.10.

### 5.1.15 Design of Franki Piles

Franki piles are also called *expanded base-compacted piles* and *pressure-injected footings*. These piles were discussed in Chapter 2 (Section 2.6.1) and Chapter 3 (Section 3.4.4).

As discussed in Chapters 2 and 3, Franki piles are special. Their installation method primarily consists of (1) driving a pipe into the ground by the impact of a drop hammer on a zero-slump concrete plug located inside and at the bottom of

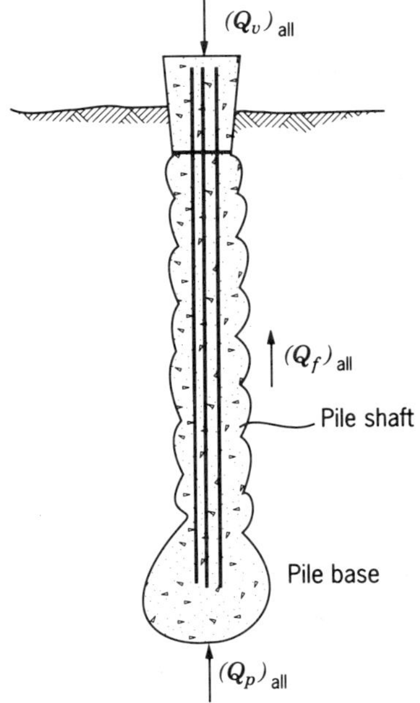


**Figure 5.24** Typical Franki pile.

the pipe, (2) after reaching the desired depth holding the drive pipe in position and expelling the concrete plug into the soil by the repeated blows of the drop hammer, (3) after expelling the plug the pile base being formed by adding and ramming zero-slump concrete out of the end of the drive pipe with drop hammer. The total number of blows of drop hammer are recorded, and the total quantity of zero-slump concrete rammed into the base is also noted when the driving is stopped. (4) The drive pipe is then withdrawn in a series of short steps while ramming the zero-slump concrete into the drive pipe to form the shaft. All these steps were detailed in Section 3.4.4.

The foregoing procedure results in a pile that has a bulb-shaped base. Since the base formation requires ramming many cubic feet (typically 10 to 30 $\mathrm{ft}^3$) of concrete into the soil, this procedure significantly improves the soil conditions by compacting the soil around the base (see Figure 5.24). The estimation of allowable capacity of these piles has not yet been completely developed. Therefore, these piles are designed on the basis of empirical relations only. Their capacities should always be confirmed by field pile load tests.

The allowable load at the pile base, $(Q_p)_{\mathrm{all}}$, can be estimated from the following empirical relationship (Nordlund, 1982):

$$(Q_p)_{\mathrm{all}} = W \times H \times \frac{N_b(V)^{2/3}}{K} \tag{5.68}$$

**TABLE 5.13 Recommended Values of *K* (Nordlund, 1982)**

| Soil Type | *K* for a Compacted Concrete Shaft | *K* for a Cased Concrete Shaft[a] |
|---|---|---|
| Gravel | 9 | 12 |
| Medium to coarse sand | 11 | 14 |
| Fine to medium sand | 14 | 18 |
| Coarse sand | 18 | 23 |
| Medium sand | 22 | 28 |
| Fine samd | 27 | 35 |
| Very fine sand | 32 | 40 |
| Silty medium to coarse sand | 14 | 18 |
| Silty fine to medium sand | 17 | 22 |
| Silty fine sand | 24 | 30 |
| Residual | $600 \div N$(but $K \nless 18$) | $1800 \div N$(but $K \nless 50$) |
| Fine sand with limerock fragments or shells, or both | 18 | 25 |
| Till with granular matrix | 20 | 27 |
| Till with clayey matrix | 30 | 40 |

*Note.* $N$ = number of blows from the Standard Penetration Test.

[a]Terminologies are described in Chapter 3 (Section 3.4.4).

where

$W$ = weight of hammer to install the pile base (lb)
$H$ = height of fall (drop) of the hammer during pile base formation (ft)
$N_b$ = number of blows of $W \times H$ energy needed to ram 1-cft of concrete into the base
$V$ = bulk volume of the base (ft$^3$)
$K$ = a dimensionless constant

Equation (5.68) has a factor of safety of 2.5.

Values of $K$ can be obtained from Table 5.13. Where standard penetration test data are available, the values of $K$ can also be estimated from Table 5.14. These values have been obtained by analyzing 10 field pile load test and pile installation data (Sharma, 1988). Example 5.15 further explains the application of equation (5.68).

Frictional capacity ($Q_f$) can be obtained by using equation (5.6) if the shaft is in cohesionless soils and equation (5.46) if the pile shaft is through cohesive soils. A factor of safety of 3 should be applied to $Q_f$ values in these equations in order to obtain $(Q_f)_{all}$. These have been discussed in Sections 5.1.1 and 5.1.7. The allowable pile load capacity $(Q_v)_{all}$ will then be the sum of $(Q_p)_{all}$ obtained from equation (5.68) and the $(Q_f)_{all}$ obtained either from equations (5.6) or from equation (5.46) as discussed above.

***Example 5.15*** A Franki-type piling system was installed at a site. The piles were installed with a 7000-lb. drop hammer and a height of fall of 20 ft. The total volume of concrete in the base was 10 ft$^3$. It required 15 blows of this drop hammer to ram out the last 5 ft$^3$ of dry concrete into the base. The general soil conditions at the site consisted of fine to medium sand. The pile was of compacted concrete shaft.

(a) Determine the allowable pile base capacities.

**TABLE 5.14 Recommended *K* versus *N* for Various Soil Types (Sharma, 1988)**

| Soil Type | $K$ |
|---|---|
| Residual soil | (i) $600/N$ but $\nless 18$ for compacted concrete shaft<br>(ii) $1800/N$ but $\nless 50$ for cased concrete shaft |
| Very fine silty sand | $2.5N$ for prebored compacted shaft |
| Silty fine sand | $3N$ for cased pile shaft |
| Coarse to medium sand | $3.5N$ for cased pile shaft |

*Note.* various terminologies such as *compacted concrete shaft*, *prebored compacted shaft*, and *cased pile shaft* are described in Chapter 3 (Section 3.4.4).

(b) Two pile load tests were carried out at the site that proved that the pile base allowable capacity is 150 kips. Provide a general formula for the site so that various capacity piles can be installed.

SOLUTION

(a) $W = 7000\,\text{lb}$
$H = 20\,\text{ft}$
$V = 10\,\text{ft}^3$
$N_b = 15/5 = 3\,\text{blows/ft}^3$

From Table 5.13, $K = 14$ for fine to medium sand and for compacted concrete shaft pile.

$$(Q_p)_{\text{all}} = W \times H \times N_b(V)^{2/3}/K \quad (5.68)$$
$$= 7000 \times 20 \times 3(10)^{2/3}/14 = 140\,\text{kips}$$

(b) Rearranging equation (5.68).
$K = W \times H \times N_b(V)^{2/3}/(Q_p)_{\text{all}} = 7000 \times 20 \times 3(10)^{2/3}/150{,}000 = 13.1$

Assume that the height and the drop of the driving hammer is the same as detailed above. Then

$$W \times H = 7000 \times 20 = 140{,}000\,\text{ft-lb} = 140\,\text{kip ft}$$
$$K = 13.1$$

Substituting these values in equation (5.68) yields the following relationship.
$$(Q_p)_{\text{all}} = 140(N_b)(V)^{2/3}/13.1 = 10.7(N_b)(V)^{2/3}\,\text{kips}$$

The required $(Q_p)_{\text{all}}$ can then be obtained by adjusting the values of $N_b$ and $V$ during the pile installation. For example, a pile with $(Q_p)_{\text{all}} = 100\,\text{kips}$ should be installed with $10\,\text{ft}^3$ concrete in the base and with 10 blows required to ram out last $5\,\text{ft}^3$ of dry concrete into the base (i.e., $N_b = 10/5 = 2$). On the other hand, a pile with $(Q_p)_{\text{all}} = 250\,\text{kips}$ should be installed with $15\,\text{ft}^3$ concrete in the base and with 19 blows required to ram out last $5\,\text{ft}^3$ of dry concrete into the base (i.e., $N_b = 19/5 = 3.8$).

### 5.1.16 Piles on Rock

This section discusses the load capacities of drilled and driven piles on rock, their settlement estimates, and a simple design procedure and two illustrative examples. Rocks may either be unweathered and intact or may be in weathered state. Pile design criteria will be different for unweathered and weathered rocks. This section is divided into following parts:

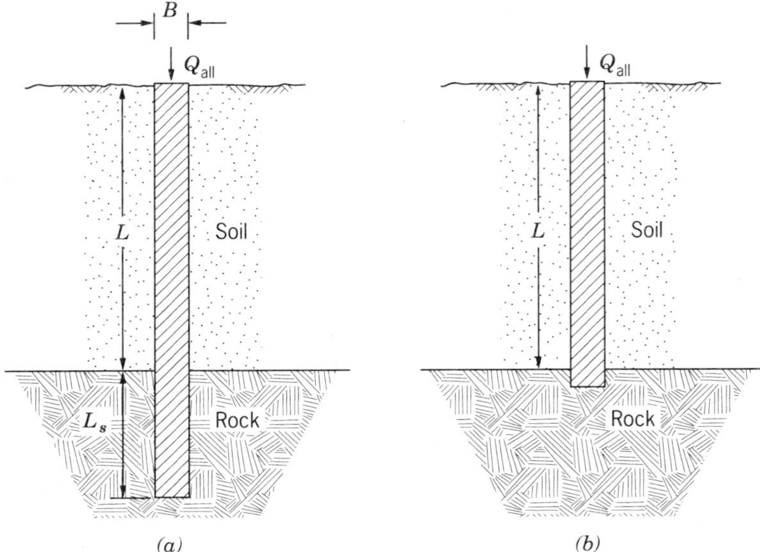


**Figure 5.25** Pile foundations on rock. (a) Bored and rock socketed pile. (b) Piles driven to rock.

1. Piles on intact (unweathered) rocks
2. Piles on weathered rocks
3. Settlement of piles on rock
4. Piles groups on rock
5. Design procedure

***Piles On Intact (Unweathered) Rocks*** As shown in Figure 5.25, two types of piles are generally installed on rock.

1. Bored cast-in-place piles: These are also called bored and rock socketed piles when they are drilled through soil and extend more than a nominal depth (typically more than 5 ft) into rock. (Figure 5.25a)
2. Piles driven to rock. (Figure 5.25b)

Methods for estimating allowable bearing capacity are different for bored (drilled) cast-in-place piles and driven piles, explained as follows:

*Bored Cast-in-Place Piles* Allowable bearing pressure on unweathered rock should normally be based on the strength of intact rock and on the influence of joints and, shear zones. Table 5.15 provides estimates of allowable bearing pressures for various types and conditions of rocks. The allowable bearing

**TABLE 5.15 Estimates of Allowable Bearing Pressure (Canadian Foundation Engineering Manual, 1985)**

| Group | Types and Conditions of Rocks | Strength of Rock Material | Presumed Allowable Bearing Pressure Kilo pascals (ton/ft$^2$) | Remarks |
|---|---|---|---|---|
| (a) | Massive igneous and metamorphic rocks (granite, diorite, basalt, gneiss) in sound condition (2) | High to very high | 10,000 (100) | These values are based on the assumption that the foundations are carried down to unweathered rock. |
| (b) | Foliated metamorphic rocks (slate, schist) in sound condition (1) (2) | Medium to high | 3,000 (30) | |
| (c) | Sedimentary rocks: shale, siltstone, sandstone, limestone without cavities, thoroughly cemented conglomerates, all in sound condition (1) (2) | Medium to high | 1,000–4,000 (10–40) | |
| (d) | Compaction shale and other argillaceous rocks in sound condition (2) (4) | Low to medium | 500 (5) | |
| (e) | Broken rocks of any kind with moderately close spacing of discontinuities (1 ft or greater), except argillaceous rocks (shale) | | 1,000 (10) | |
| (f) | Thinly bedded limestone, sandstones, shale | | See note (3) | |
| (g) | Heavily shattered or weathered rocks | | See note (3) | |

These presumed values of the allowable bearing pressure are estimates and may need alteration upwards or downwards. No addition has been made for the depth of embedment of the foundation.

*Notes*

1. The foregoing values for sedimentary or foliated rocks apply where the strata or foliation are level or nearly so, and, then only if the area has ample lateral support. Tilted strata and their relation to nearby slopes or excavations shall be assessed by a person knowledgeable in this field of work.
2. Sound rock conditions allow minor cracks at spacing not less than 1 m.
3. To be assessed by examination in situ, including loading tests if necessary, by a person knowledgeable in this field of work.
4. These rocks are apt to swell on release of stress and are apt to soften and swell appreciably on exposure to water.

**TABLE 5.16 Allowable Contact Pressure ($q_a$) on Jointed Rock (Peck, Hanson, and Thornburn, 1974)**

| Rock Quality Designation (RQD) | Rock Quality | $q_a$[a] | |
|---|---|---|---|
| | | kN/m$^2$ | tons/ft$^2$ |
| 100 | Excellent | 28,000 | 300 |
| 90 | Good | 19,000 | 200 |
| 75 | Fair | 11,000 | 120 |
| 50 | Poor | 6,000 | 65 |
| 25 | Very poor | 2,800 | 30 |
| 0 | | 900 | 10 |

[a] If values of $q_a$ exceed unconfined compressive strength ($q_u$) of intact samples of the rock, as it might in the case of some clay shales, for instance, take $q_a = q_u$.

capacity of piles on rock will be governed by (1) rock strength and (2) the settlements associated with the defects in the rock.

For tight joints or joints smaller than a fraction of an inch, the rock compressibility is reflected by the *Rock Quality Designation* (RQD) and allowable pressures on rock can be estimated as shown in Table 5.16. The RQD used to obtain $q_a$ from Table 5.16 should be averaged within a depth below foundation level equal to the width of the foundation. For these contact pressures, the settlement of foundation should not exceed 0.5 in. (12.5 mm) (Peck, Hanson, and Thornburn, 1974). The method of determination of RQD was presented in chapter 4 (Section 4.1.1).

ALLOWABLE BEARING CAPACITY FROM PROPERTIES OF ROCK CORES The allowable bearing capacity ($q_a$) for cast-in-place drilled or socketed piles in rock can be evaluated by relating it to the rock core strength as given by equation (5.69). This method is not applicable to soft stratified rock, such as shales or limestones (*Canadian Foundation Engineering Manual*, 1985; Ladanyi and Roy, 1971).

$$q_a = (q_u)_{\text{core}} K_{sp} d \tag{5.69}$$

where

$(q_u)_{\text{core}}$ = average unconfined compressive strength of rock core from ASTM D2938-79

$K_{sp}$ = an empirical factor given in Figure 5.26

$d$ = a depth factor given by equation (5.70)

$$d = [0.8 + 0.2(L_s/B)] \leqslant 2 \tag{5.70}$$

where

$L_s$ = pile length that is socketed in rock having a strength ($q_u$) and $B$ is the diameter as shown in Figure 5.25a

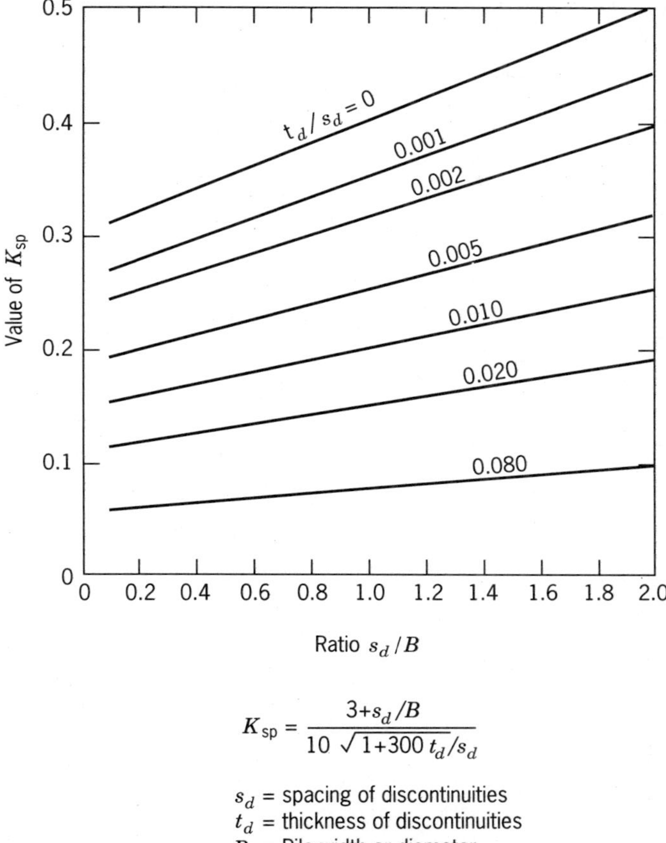


**Figure 5.26** Values of empirical coefficient, $K_{sp}$. The coefficient $K_{sp}$ takes into account the size effect and presence of discontinuities and contains a nominal factor of safety of 3 against general foundation failure. (*Canadian Foundation Engineering Manual*, 1985).

$s_d$ = spacing of discontinuities }
$t_d$ = thickness of discontinuities } Figure 5.26
$B$ = pile width or diameter

ALLOWABLE BEARING CAPACITY DERIVED FROM THE BOND BETWEEN ROCK AND CONCRETE The allowable bearing capacity, $(Q_v)_{\text{all}}$, based on the bond along the socketed surface can be expressed by the following equation:

$$(Q_v)_{\text{all}} = pL_sC_{bs} \tag{5.71a}$$

or

$$q_a = (Q_v)_{\text{all}}/(\pi B^2/4) \tag{5.71b}$$

**TABLE 5.17 Allowable Rock-Socket, Skin-Friction $C_{bs}$, and End-bearing Values for Bored Piles in Rocks (Tomlinson, 1977)**

| | Allowable Skin-Friction on Rock Socket | | Allowable End-bearing Pressure | |
|---|---|---|---|---|
| Type of Rock | (kN/m$^2$) | (tons/ft$^2$) | (kN/m$^2$) | (tons/ft$^2$) |
| Manhattan schist | 1330 | 13.9 | — | — |
| Black Utica shale (Montreal) | 1120 | 11.7 | 2,620 | 24.4 |
| Black Billings shale (Ottawa) | 1120 | 11.7 | — | — |
| Dundas shale (Toronto) | 1120 | 11.7 | 7,850 | 73.2 |
| Limestone (Chicago) | 1716 | 17.9 | 10,468 | 96 |
| Fragmented shale | 107 | 1.1 | — | — |
| Widely fissured hard sandstone | 429 | 4.5 | — | — |

where

$p$ = pile perimeter ( $= \pi B$ for circular pile)
$L_s$ = socketed pile length in the rock
$C_{bs}$ = allowable bond strength between concrete and rock

Values of bond strength $C_{bs}$ are highly dependent on the quality of contact area attained during excavation process. Table 5.17 gives values of allowable skin friction (bond strength) for some rocks. These values should always be used with caution because of the difficulty in achieving a clean hole during construction. Site-specific $C_{bs}$ values should be obtained from load tests for detailed design. In some sound rocks, maximum mobilized shear stress can exceed the allowable values given by local codes. One such instance is cited by Koutsoftas (1981). Load tests should therefore be done for detailed design. Pile load capacity can be improved by cutting grooves into the rock wall to roughen the pile rock interface (Horvath et al., 1983). This alternative should be considered where possible.

*Piles Driven to Rock* In cases where steel H piles, pipe piles, or precast concrete piles are driven to rock, their exact area of contact with rock is not known. Their bearing capacity will depend on the type and nature of rock and the depth of penetration of pile into the rock. Estimation of allowable bearing capacity of such piles by analytical method cannot be made. Load capacity of these type of piles should be estimated based on local experience and driving resistance supported by pile load tests. When driving piles to rock there is potential for damage to the pile tip due to hard driving. This will have adverse effect on pile capacity. Pile tips should therefore be fitted with proper protective features, such as, shoes or plates. This has been discussed in section 3.4.2.

***Piles on Weathered Rocks*** Weathered rocks exhibit a great variety of physical properties. From detailed field investigations, an engineer should first evaluate if weathered rock has a matrix in which the rock fragments play a minor role or a major role. In situations where weathering is extensive and rock fragments are surrounded by *decomposed soil*, the foundation should be designed as if it were supported on soil matrix. Even in relatively unweathered shale, foundation can be designed as if it were supported on a heavily overconsolidated clay.

If thin seams of compressible material are present within the mass that is predominately rock, drilled piles can be taken to depths where these seams are minimized and foundation can be designed as if it was supported on rock. Evaluation of foundation parameters for such soils is difficult to assess and requires extensive local experience supported by pile load tests.

***Settlement of Piles on Rock*** For conventional structures, that are founded on rock, settlements are small and need not be evaluated provided allowable bearing capacity is not exceeded. Full-scale pile load tests may be required for estimation of settlements for piles on rock for extremely settlement sensitive structures.

***Pile Groups on Rock*** Normally for piles on rock, pile group capacity is simply the sum of individual allowable pile capacity.

***Design Procedure***

1. Carry out field investigation and determine soil and rock profile, depth of water table, and depth and type of bedrock.
2. Carry out measurements and tests on rock to determine spacing and thickness of discontinuities and RQD.
3. If the rock is unweathered calculate allowable bearing capacity by the following methods.

   (a) A range of allowable bearing capacity from Table 5.15.

   (b) From rock core values

$$q_a = (q_u)_{\text{core}} k_{sp} d \tag{5.69}$$

   (c) From bond between rock and concrete

$$(Q_v)_{\text{all}} = pL_s C_{bs} \tag{5.71a}$$

$$q_a = (Q_v)_{\text{all}}/(\pi B^2/4) \tag{5.71b}$$

   Use Table 5.17 for $C_{bs}$ values.

   (d) From RQD values using Table 5.16

   Allowable bearing value is the lower of (b), (c) and (d) and should fall in the range given by (a).

4. In conventional structures that are founded on rock, settlements are small

provided the allowable bearing capacity is not exceeded. The settlements, therefore, need not be evaluated.

5. Carry out load tests to five time these values to fine tune the design.

***Example 5.16*** A 36-in. (900 mm) diameter drilled pile is supported on unweathered rock by socketing 6 ft into it. The rock was sandstone with ($q_u$) core of 90 tons/ft$^2$. Estimate the allowable bearing capacity for the pile.

SOLUTIONS

(a) Allowable bearing capacity from Table 5.15: For group (c) in Table 5.15, presumed allowable bearing capacity for medium to high sandstone is = 10 to 40 tons/ft$^2$ (1000 to 4000 kN/m$^2$)
(b) Allowable bearing capacity from properties of rock cores:

$$L_s = 6\,\text{ft}$$

$$B = 3\,\text{ft}$$

$$d = 0.8 + 0.2\,(6/3) = 1.2 \tag{5.70}$$

In the absence of information on sizes and spacing of discontinuities, assume $K_{sp} = 0.3$ from Figure 5.26.

Then, from equation (5.69),

$$\begin{aligned} q_a &= (q_u)_{\text{core}} K_{sp} d \\ &= 90 \times 0.3 \times 1.2\,\text{tons/ft}^2 = 32\,\text{tons/ft}^2 \end{aligned} \tag{5.69}$$

(c) Allowable bearing capacity derived from the bond between rock and concrete.

$$(Q_v)_{\text{all}} = pL_s C_{\text{bs}} \tag{5.71a}$$

The value for a allowable bond stress $C_{\text{bs}}$ is not available for unweathered sandstone and the pile material (concrete). A conservative value of 4.5 tons/ft$^2$ for sandstone can be estimated from Table 5.17.

$$C_{\text{bs}} = 4.5\,\text{tons/ft}^2$$

$$p = \pi B,\ L_s = 6\,\text{ft}$$

$$(Q_v)_{\text{all}} = \pi \times 3 \times 6 \times 4.5$$

$$q_a = (Q_v)_{\text{all}}/(\pi B^2/4) \tag{5.71b}$$

$$q_a = (\pi \times 3 \times 6 \times 4.5)/(\pi/4)(3^2) = 36\,\text{tons/ft}^2$$

From cases (b) and (c), the lower allowable bearing pressure = 32 tons/ft$^2$. This falls in the range specified in case (a). Therefore, $q_a = 32$ tons/ft$^2$.

***Example 5.17*** The pile described in Example 5.16 is supported on clay shale with $(q_u)_{\text{core}} = 60$ tons/ft$^2$. Core recovery along depth indicated the following:

Core 1, 5.0 ft recovery 2.5 ft, RQD = 2.5/5 = 50 percent from 0 to 5 ft into the rock
Core 2, 5.0 ft recovery 4.0 ft, RQD = 4.0/5 = 80 percent from 5 to 10 ft into the rock
Core 3, 5.0 ft recovery 4.4 ft, RQD = 4.4/5 = 88 percent from 10 to 15 ft into the rock

Recovery was considered by pieces that were of sizes 4 in. or larger. Estimate the allowable bearing capacity of the pile.

SOLUTION

*The RQD Method* Since pile was socketed 6 ft into the rock and pile width is 3 ft, the RQD used to obtain $q_a$ from Table 5.16 will require the average RQD within a depth below foundation level equal to the width of the foundation. Then RQD for depth 6 ft to $(6 + B) = 6 + 3 = 9$ ft will be 80 percent.

From Table 5.16 for RQD = 80 percent, $q_a = 147$ tons/ft$^2$. This value is obtained from Table 5.16, by interpolating RQD between 75 and 90 percent. Since $(q_u)_{\text{core}} = 60$ tons/ft$^2$ $< 147$ tons/ft$^2$, take $q_a = 60$ tons/ft$^2$.

*Allowable Bearing Capacity Derived from the Bond between Rock and Concrete* From equation (5.71a and b).

$q_a = pL_sC_{\text{bs}}/\text{Area of base} = 36$ ton/ft$^2$ as calculated earlier in Example 5.16. The lower of the two values gives $q_a = 36$ tons/ft$^2$.

## 5.2 PILES SUBJECTED TO PULLOUT LOADS

The ultimate pullout capacity $P_u$ of piles can be estimated in a similar manner to ultimate compression capacity. The only difference will be that the end-bearing capacity ($Q_p$) is ignored except for belled piles, which will be discussed later in Section 5.2.8. As shown in Figure 5.27, the pullout force $P_u$ is resisted by the side frictional resistance $Q_{fp}$ and the weight of the pile $W_p$. The general relationship for estimating pullout capacity will then be as follows:

$$P_u = Q_{fp} + W_p \tag{5.72}$$

where

$P_u$ = ultimate pullout capacity
$Q_{fp}$ = ultimate shaft friction in pullout
$W_p$ = pile weight

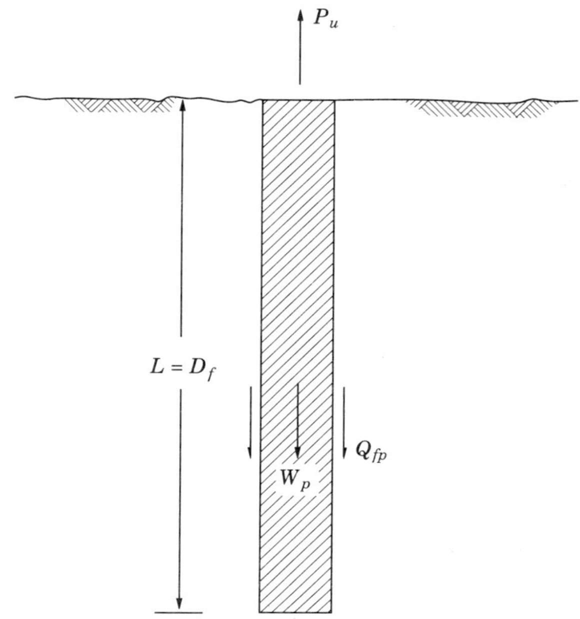


**Figure 5.27** Basic concept of pullout resistance by pile foundations.

The estimation of $Q_{fp}$ and hence $P_u$ will be discussed for cohesionless and cohesive soils separately in the following sections.

### 5.2.1 Pullout Capacity of a Single Pile in Cohesionless Soils

As discussed in Section 5.1.1, the ultimate shaft friction $Q_f$ for axial compression loads is given by equation (5.6) as follows:

$$Q_f = pK_s \tan\delta \sum_{L=0}^{L=L} \sigma'_{vl}\Delta L \tag{5.6}$$

where

$p$ = pile perimeter
$K_s$ = coefficient of earth pressure as given in Table 5.3.
$\delta = 2/3\phi$ = friction between soil and pile
$\phi$ = angle of internal friction for the soil
$L$ = pile length
$\sigma'_{vl}$ = effective vertical stress over pile length
$\Delta L$ = a small pile element

Experience indicates that the value of $K_s$ taken from Table 5.3 should be multiplied by two-thirds if equation (5.6) is to be used for uplift or tensile loads

(*Foundations and Earth Structures Design Manual* DM-7.2, 1982). The ultimate shaft friction in pullout, $Q_{fp}$, will then be given by the following:

$$Q_{fp} = 2/3\,pK_s \tan\delta \sum_{L=0}^{L=L} \sigma'_{vl}\Delta L \tag{5.73}$$

As discussed in Section 5.1.1 the $\sigma'_{vl}$ value increases with depth until the depth equals 20 times the pile diameter. Beyond this depth, $\sigma'_{vl}$ is assumed to be constant. From equations (5.72) and (5.73) the ultimate pullout capacity becomes:

$$P_u = 2/3\,pK_s \tan\delta \sum_{L=0}^{L=L} \sigma'_{vl}\Delta L + W_p \tag{5.74}$$

The allowable pullout capacity $P_{all}$ can then be written as follows:

$$P_{all} = 1/FS\left[2/3\,pK_s \tan\delta \sum_{L=0}^{L=L} \sigma'_{vl}\Delta L\right] + W_p \tag{5.75}$$

where

$FS$ = factor of safety (usually taken as 3)
$W_p$ = weight of the pile

The submerged weight of the pile should be considered in the zone where the pile length is below the water table.

### 5.2.2 Pullout Capacity of Pile Groups in Cohesionless Soils

For a pile group in soils with friction, at ultimate condition, the block of soil around the group is lifted. Exact size and shape of this block depends on the manner in which pullout load is transferred from the piles to the soil. This is a complex mechanism and depends on factors such as method of pile installation, pile properties, and soil properties including the degree of layering. A simplified method for estimating pullout resistance of pile group, in cohesionless soils, consists of using the lower of the following two values:

1. Estimate allowable pullout resistance of individual piles by the method described in Section 5.2.1 and multiply this by the number of piles. Thus, $(P_G)_{all}$ = number of piles × $P_{all}$.

2. Calculate the effective weight of the soil bound by the trapezoid from base to the top with sides inclined at 75° from the horizontal (see Example 5.19).

As shown in Figure 5.29, the effective weight of the soil bound by the trapezoid can be calculated by the following:

Effective weight of soil = effective weight of soil bound by $(xyx_1y_1)$

$$\text{Effective weight of soil} = (\tfrac{1}{3}A_1h - \tfrac{1}{3}A_2h_2)\gamma' \tag{5.76}$$

where

$A_1 = (b' + 2h_1 \tan 15°)^2$

$A_2 = b' \times b'$

$h = h_1 + h_2$

$\gamma'$ = effective unit weight of the soil

The various terms are explained in Figure 5.29.

Weights of the piles can be assumed approximately equal to the weight of displaced soil to simplify calculations. For both these cases, the weight of the pile cap should be added to the allowable pullout capacity.

### 5.2.3 Design Computations for Pullout in Cohesionless Soils

Design computations consist of the following steps:

1. From proper soil investigations, establish the soil profile and ground water levels and note soil properties on the soil profile based on field and laboratory tests. Normally, a pile type and its dimensions are already selected based on axial compression load requirements. Pullout capacity of this selected pile is then calculated.
2. Calculate allowable pullout capacity by using equation (5.75)

$$P_{\text{all}} = (1/FS)\left(2/3\, pK_s \tan\delta \sum_{L=0}^{L=L} \sigma'_{vl}\Delta L\right) + W_p \tag{5.75}$$

3. If the piles have been placed in a group then group capacity is calculated by the two methods described in an Section 5.2.2.
4. Confirm pullout capacity by pile load test.

Steps 1 and 2 are further explained in Example 5.18 and step 3 is explained in Example 5.19.

***Example 5.18*** A 12-in. (300 mm) diameter steel pipe pile was driven in a cohesionless soil. The pile was 30 ft (9 m) long. Soil properties are given in Figure 5.28. Estimate its allowable pullout capacity.

SOLUTION

1. Soil Properties: Soil properties and pressures are shown in Figure 5.28.

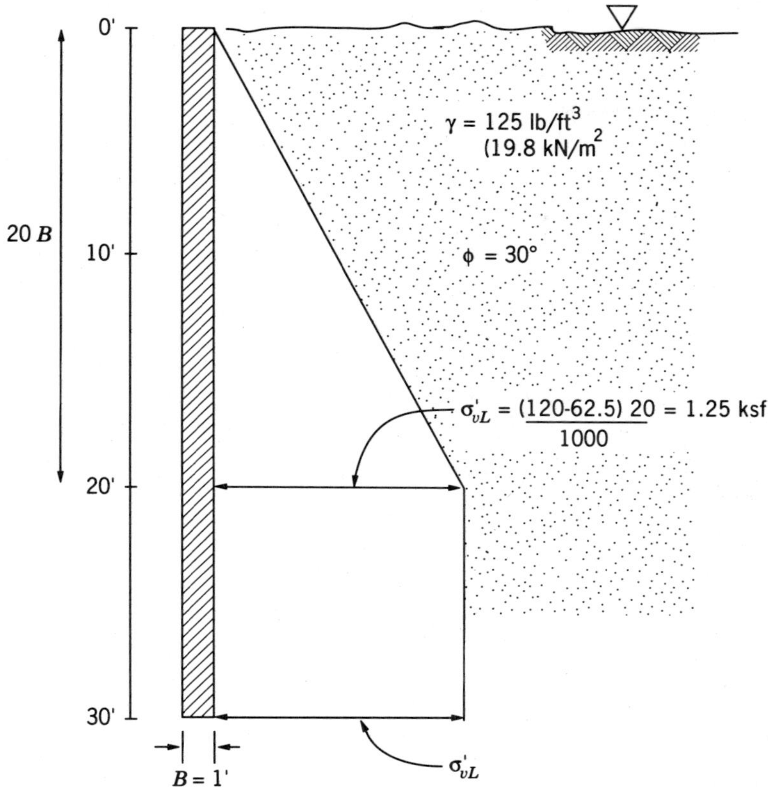


**Figure 5.28** Soil properties and pressure diagram for Example 5.18.

2. Allowable Pullout Capacity

$$K_s = 1 \text{ from Table 5.3}$$

$$\delta = \tfrac{2}{3}\phi = 20^\circ$$

For 12-in. diameter, 0.25-in. thickness of the pile, the pile weight = 31.37 lb/ft. From equation (5.75):

$$P_{\text{all}} = 1/FS\left[\, p(\tfrac{2}{3}K_s)\tan\delta \sum_{L=0}^{L=L} \sigma'_{vl}\Delta L \,\right] + W_p \tag{5.75}$$

$$= \tfrac{1}{3}(\pi \times 1)(\tfrac{2}{3} \times 1)\tan 20\left[\frac{(1.25+0)}{2} \times 20 + 1.25 \times 10\right]$$

$$+ 31.37 \times 30/1000$$

$$= 6.35 + 0.94 = 7.29 \text{ kips (say 7 kips)}$$

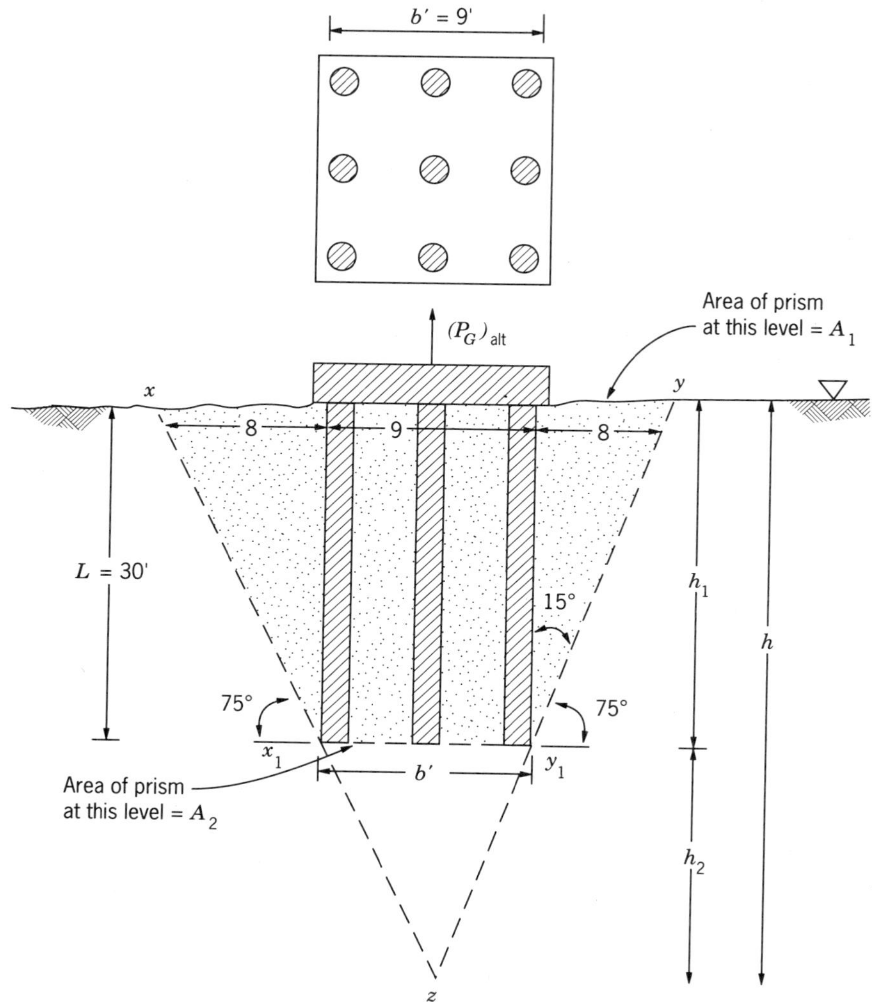


**Figure 5.29** Pile group configuration and soil weight contribution for pile group capacity for Example 5.19.

***Example 5.19*** From Example 5.18, assume that there are nine similar piles arranged in a group as shown in Figure 5.29. Estimate the pullout capacity of the group.

SOLUTION

*Method* (*a*) From example 5.18, $P_{all} = 7$ kips. Number of piles = 9. Therefore,

$$(P_G)_{all} = 9 \times 7 = 63 \text{ kips}$$

*Method* (*b*) From Figure 5.29: Effective weight of the soil inside the wedge

$$= (\tfrac{1}{3}A_1 h - \tfrac{1}{3}A_2 h_2)\gamma' \tag{5.76}$$

$$A_1 = (b' + 2 \times 30 \tan 15)^2$$

$$= (9 + 2 \times 8)^2 = 625\,\text{ft}^2$$

$$A_2 = 9 \times 9 = 81\,\text{ft}^2$$

$$h_2 = 9/2 \tan 75 = 16.8\,\text{ft}$$

$$h = h_1 + h_2 = 30 + 16.8 = 46.8\,\text{ft}$$

Then, the effective weight of the soil inside the wedge

$$= (\tfrac{1}{3} \times 625 \times 46.8 - \tfrac{1}{3} \times 81 \times 16.8)(125 - 62.5)/1000\,\text{kips}$$

$$= 581\,\text{kips} = (P_G)_{\text{ult}}$$

$$(P_G)_{\text{all}} = 581/3 = 194\,\text{kips}$$

The lower of the two methods (a) and (b) is 63 kips.

$$(P_G)_{\text{all}} = 63 + \text{weight of the pile cap}$$

### 5.2.4 Pullout Capacity of a Single Pile in Cohesive Soils

For cohesive soils, the ultimate skin friction $Q_f$ is given by equation (5.46) as follows:

$$Q_f = p \sum_{L=0}^{L=L} c_a \Delta L \tag{5.46}$$

This equation can also be used to estimate ultimate shaft friction in pullout, $Q_{fp}$. Thus, the ultimate pullout capacity in cohesive soils can be given by the following relationship:

$$P_u = p \sum_{L=0}^{L=L_e} c_a \Delta L + W_p \tag{5.77}$$

The allowable pullout capacity will then be as follows:

$$P_{\text{all}} = 1/FS\left(p \sum_{L=0}^{L=L_e} c_a \Delta L\right) + W_p \tag{5.78}$$

where $L_e = (L - \text{depth of seasonal change})$. Typically, the depth of seasonal change is 5 ft.

$p$ = pile perimeter and

$c_a$ = soil–pile adhesion obtained from Figure 4.27 or Table 4.7 as applicable

$L_e$ is pile length that is normally estimated by subtracting the zone of seasonal variation and any other soft zones that may not contribute to skin friction mobilization from $L$, the actual pile length. Zone of seasonal variation will depend on local conditions; a depth of about 5 ft (1.5 m) is normally assumed where local information is not available.

For estimating allowable pullout capacity a factor of safety ($FS$) of 3 is generally applied except for pile weight ($W_p$).

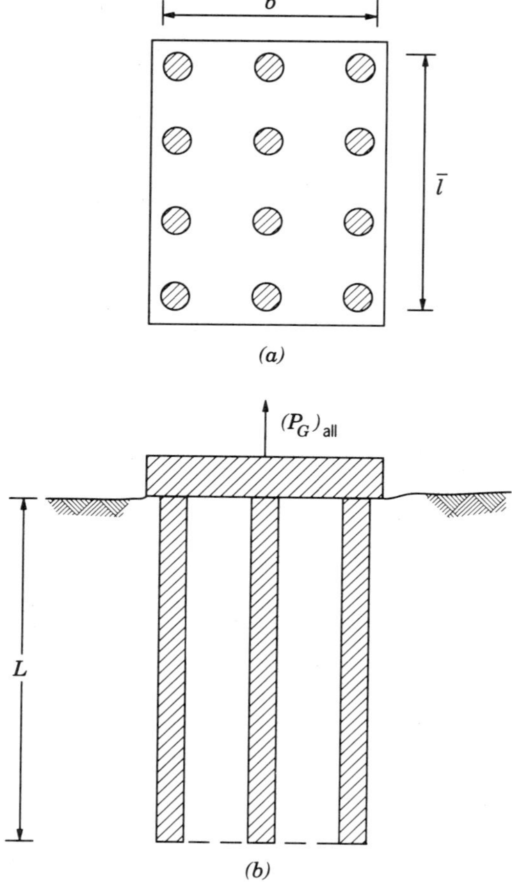


**Figure 5.30** Pullout capacity of pile group in cohesive soils. (a) Plan (b) Section.

### 5.2.5 Pullout Capacity of Pile Groups in Cohesive Soils

A simple method for estimating allowable pullout capacity of a pile group, in cohesive soils, consists of using the lower of the following two values:

1. Allowable group capacity, $(P_G)_{\text{all}} = nP_{\text{all}}$ where $n$ is number of piles and $P_{\text{all}}$ is allowable pullout capacity of a single pile.
2. Allowable group capacity is the uplift resistance of the block of soil enclosed by pile group. This is shown in Figure 5.30. In this Figure the perimeter (p) for the block of soil will be given by the following:

$$p = 2(\bar{b} + \bar{l})$$

The weight of soil, $W_s$, within the pile group is

$$W_s = (\bar{b} \times \bar{l})L_e\gamma'$$

where

$L_e$ = (pile length – the depth of seasonal changes)
$\gamma'$ = effective unit weight of soil (i.e., total weight above water table and submerged below the water table). The allowable pullout capacity of the group will then be given by the following equation:

$$(P_G)_{\text{all}} = 1/FS[2(\bar{b} + \bar{l})L_e c_u] + W_s \tag{5.79}$$

All terms have been defined earlier. In this equation, it has been assumed that the weight of piles will be approximately equal to the weight of the soil that was displaced with the piles. For all practical purposes, this assumption is reasonable.

### 5.2.6 Design Computations for Pullout in Cohesive Soils

Design Computations for pullout resistance consists of the following steps:

1. From soils investigations establish the soil profile and soil parameters from field and laboratory tests.
2. Calculate allowable pullout capacity by using equation (5.78)

$$P_{\text{all}} = 1/FS\left(p \sum_{L=0}^{L=L_e} c_a \Delta L\right) + W_p \tag{5.78}$$

3. If the piles have been placed in a group, then group capacity is calculated by the two methods described in Section 5.2.5.
4. Confirm pullout capacity by pile load test.

Steps 1 and 2 are further explained in Example 5.20 and step 3 is explained in Example 5.21.

***Example 5.20*** Estimate allowable pullout capacity for a 12 in. (300 mm) diameter, 30 ft (9 m) long, driven steel pipe pile. The $c_u$ for the soil is 1030 psf. Assume that seasonal variation is to 5 ft below ground. The weight of pile is 0.94 kips.

SOLUTION Cohesive soil with $c_u = 1400$ psf.

From Figure 4.27, $c_a/c_u = 0.68$ $\therefore c_a = 700$ psf

$$L_e = 30 - 5 = 25\text{ ft}$$

$$W_p = 0.94\text{ kips}$$

$$p = \pi \times 1 = 3.14\text{ ft}$$

From equation (5.78):

$$P_{\text{all}} = 1/FS\left(p \sum_{L=0}^{L=L_e} c_a \Delta L\right) + W_p \tag{5.78}$$

$$= \tfrac{1}{3}[3.14 \times 700 \times 25/1000] + 0.94 = 19\text{ kips}$$

***Example 5.21*** In Example 5.20 now assume that piles are in a group. Assume that the group has a square pattern with $\bar{b} = \bar{l} = 9$ ft. Assume that the total unit weight of soil = 125 lb/cu ft and water table is near ground surface.

SOLUTION

*Method (a)*

$$(P_G)_{\text{all}} = nP_{\text{all}}$$

$$= 9 \times 19 = 171\text{ kips}$$

*Method (b)*

$$(P_G)_{\text{all}} = (1/FS)[2(\bar{b} + \bar{l})L_e c_u + W_s] \tag{5.79}$$

$$= \tfrac{1}{3}(2 \times 18 \times 25 \times 1030/1000) + (9 \times 9 \times 30 \times 62.5/1000)$$

$$= 309 + 151.87 = 461\text{ kips}$$

This assumes that $W_s$ is approximately equal to the weight of soil enclosed within 9 ft × 9 ft area. The lower of the two values is 171 kips. This is then the allowable pullout pile group capacity. The weight of pile cap should be added to this capacity.

### 5.2.7 Pullout Capacity of H Piles

Methods discussed in Sections 5.2.1 and 5.2.4 can also be used to estimate pullout capacities for H piles. For such piles a soil plug is assumed to develop between the flanges. The perimeter ($p$) is then determined as $p = 2(a + b)$, where $a$ is the flange width and $b$ is the web height for the H pile.

Hegedus and Khosla (1984) experimentally determined pullout capacities of driven H piles in stiff clays, dense sands, silts, and stratified soils. Test results showed that earth pressure parameters and adhesion values were generally consistent with the values used in Sections 5.2.1 and 5.2.4 for estimating pullout capacities of circular or rectangular piles. It is therefore recommended that the H pile be treated as a rectangular pile and procedures described in Sections 5.2.1 and 5.2.4 be applied in this case also.

### 5.2.8 Pullout Capacity of Belled Piles

Enlarged (belled) bases are formed in many cases at the pile bottom for increased end-bearing capacities. Details of pile bell such as size and shape formed in

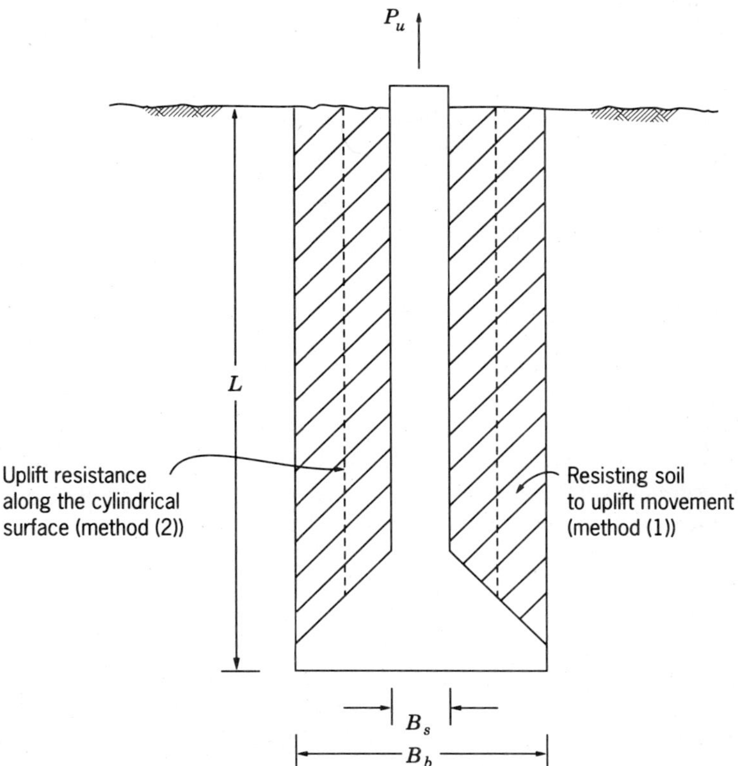


**Figure 5.31** Uplift resistance of belled piles in cohesive soil.

cohesionless soils by driving or by bentonite slurry methods cannot be controlled with reliability. Therefore, uplift capacity of such bells is difficult to estimate. Pullout tests are the only reliable methods for such estimates. Determination of uplift capacity of piles with bells formed in clay by belling tools are now described. The uplift capacity of a belled pile in cohesive soils can be estimated by using the lower of the following two values.

1. The base resistance of the pile will be the ultimate uplift bearing capacity at the annular area between the bell and the shaft (Figure 5.31). This is given by the following relationship (Tomlinson, 1977):

$$P_u = \frac{\pi}{4}(B_b^2 - B_s^2)c_u \times 9 + W_p \tag{5.80}$$

where

$B_b$ = pile bell diameter
$B_s$ = pile shaft diameter
$c_u$ = undrained strength
$N_c$ = nondimensional bearing capacity parameter; its value equals 9
$W_p$ = the weight of the pile

2. The shaft resistance along a cylindrical surface with diameter average of bell and shaft and is given by following relationship (Sharma et al., 1984).

$$P_u = [\pi(B_b + B_s)/2] \sum_{L=0}^{L=L_e} c_u L_e + W_p \tag{5.81}$$

where $c_u$ is undrained soil strength along pile length, and $L_e$ is effective pile length.

The $P_{all} = (P_u/FS) + W_p$ and will be the lower of the two values obtained from equations (5.80) and (5.81). A factor of safety ($FS$) of 3 should be used for sustained loading. Meyerhof and Adams (1968) present the uplift resistance of a circular plate embedded in $\phi = 0$ soil. The method established in this investigation can also be used for estimating uplift capacity of piles in $\phi = 0$ soils. This method needs further field verification. For final design a full-scale pile load test should be carried out to determine uplift capacity of belled piles.

## 5.3 OVERVIEW

In this chapter, bearing capacity and settlements of single pile and pile groups in cohesionless soils, cohesive soils, and on rocks under axial loads were discussed. Problem of negative skin friction and the design of piles in swelling and shrinking soils have also been discussed. Piles subjected to pullout loads both in cohesionless and cohesive soils have also been described. Following the theoretical background, design procedures and numerical examples for pile design, both for axial compression and pullout, were outlined.

Bearing capacity of piles in cohesionless soils can be estimated by utilizing soil strength, standard penetration tests, dynamic driving resistance, and the full-scale pile load tests. The end-bearing capacity of piles varies significantly depending on the theoretical model used. The bearing capacity factor $N_q$ also varies with the depth of pile penetration, soil strength, and soil compressibility. This has been discussed in detail by Meyerhof (1976) and Coyle and Castello (1981). The wide variation in $N_q$ values (Section 5.1.1) suggests that its conservative values be used in design (Table 5.2). Furthermore, the end-bearing capacity should be increased with overburden pressure only upto a depth of $20B$ (Section 5.1.1). Below $20B$ depth, the end-bearing should be considered constant. This behavior has been confirmed by field load tests.

The estimation of friction capacity of piles in cohesionless soils is based on the coefficient $K_s$ (equation 5.6)). Review of test data indicate that $K_s$ values for driven piles vary from 0.3 to 3 (Table 1.1). However, for design, maximum value of 2 is recommended (Table 5.3).

Semiempirical analysis of pile capacity in cohesionless soils by Standard Penetration Tests and the Static Cone Penetration Tests and their comparison with field load tests indicates a reasonable agreement. (Meyerhof 1976, 1983; Sharma and Joshi, 1986). These relationships can therefore be used for preliminary design.

The dynamic driving methods for estimating pile capacities are (1) pile-driving formulas and (2) wave equation analysis . Pile-driving formulas are not reliable and therefore should only be used as a field control technique when supported by full-scale pile load tests at the specific site. Wave equation analysis, originally recommended by Smith (1962), provides a better rational approach for estimating pile capacities. However, considerable judgment is needed in selecting the input parameters and interpretation of results (*Wu* et al., 1989). Davisson (1989) has demonstrated with several case histories that there may be problems in use of pile driving analyser results (See Chapter 11).

The bearing capacity of piles in cohesive soils depends on the bearing capacity factor $N_c$, which can be estimated with reasonable accuracy from Tables 5.7 and 5.8 (Skempton, 1951, 1959; Meyerhof, 1976, 1983). However, tests indicate a significant variation in soil pile adhesion $c_a$, which has been related with undrained strength of soil $c_u$. The $c_a$ value depends on soil consistency, pile material, and the method of pile installation (McClelland, 1974; Meyerhof, 1976; Vesic, 1977). Values of $c_a$ obtained from Figure 4.27 and Table 4.7, when used in equation (5.4b), provide rough estimates of friction capacity of piles.

The bearing capacity of pile groups in cohesionless and cohesive soils is not well understood. There are conflicting recommendations for group capacities specially in cohesive soils. For example the *Foundations and Earth Structures Design Manual* DM 7.2 (NAVFAC, 1982) recommends a group reduction factor while the *Canadian Foundation Engineering Manual* (1985) recommends that no group reduction factor be used for pile group capacity. Because of the limited field test data, it is suggested that group efficiency $G_e$ be taken as unity for cohesionless soils and values from Table 5.10 be used for estimating $G_e$ in cohesive soils. Also, the block failure of pile group by breaking into the ground should also be considered (Section 5.1.7) (Terzaghi and Peck, 1967; Meyerhof 1976).

The three practical methods of estimating short term or immediate settlements of pile are (1) the semiempirical method, (2) the empirical method, and (3) the pile load tests. Experience indicates that settlement prediction of piles is very complex. The only reliable method of immediate settlement prediction is the pile load test. Equation (5.34) can, however, be used for preliminary estimates of settlements (Vesic, 1977; NAVFAC, 1982). There is a need for further analytical and experimental research work in this area. Long-term settlement predictions require further work.

Pullout capacity of piles in cohesionless soils is estimated by using equation (5.73). Available test data when compared with this equation indicate wide variations (Ireland, 1957; Sowa, 1970; Hegedus and Khosla, 1984).

Equation (5.73) should be used as a guide for estimating pullout capacities in cohesionless soils. Pullout resistance for piles, in cohesive soils by using equation (5.78), on the other hand, appears to provide more reliable values when compared with test data (Sowa, 1970). This equation can therefore be used for preliminary design. Uplift capacity estimates of drilled and belled piles is not yet well understood and needs further investigation and testing.

The foregoing discussions indicate that pile capacities and settlements can be estimated conservatively by the methods provided in this chapter. These methods, however, are approximate because the bearing capacity and settlements depend on factors such as soil type, soil consistency, soil density, method of pile installation, load transfer mechanism, state of disturbance during pile installation, and soil stratigraphy. All these factors cannot be accurately modeled in an analytical formula. Therefore, the best method to predict pile capacity and short term settlement is the field pile load test. This is discussed in Chapter 9.

## REFERENCES

American Iron and Steel Institute, *Steel Pile Load Test Data*, AISI, Washington, DC, May 1985, p. 82.

Authier, J. and Fellenius, B. H., *Civil Engineering for Practicing and Design Engineers*, Vol. 2, No. 4. Pergamon Press Ltd., Oxford, England, 1983, pp. 387–407.

Baguelin, F., Jexequal, J. F., and Shields, D. H., "The Pressuremeter and Foundation Engineering," Trans. Tech. Publication, 1978.

Bengtsson, P. and Sallfors, G., "Floating Piles in Soft, Highly Plastic Clays," *Can. Geotech. J.*, Vol. 20, No. 1, February 1983, pp. 159–168.

*Canadian Foundation Engineering Manual*, Canadian Geotechnical Society, Bi Tech. Publication, 1978 and 1985.

Chellis, R. D., *Pile Foundations*, 2nd ed. McGraw-Hill Book Co., New York, 1961.

Chen, F. H. *Foundation on Expansive Soils.* Elsevier Scientific Publishing Co., New York, 1975.

Coyle, H. M. and Castello, R. R., "New Design Correlations for Piles in Sand," *J. Geotech. Eng. Div.*, Proc. ASCE, Vol. 107, No. GT7, July 1981, pp. 965–986.

Davisson, M. T., "Pile Load Capacity, Design, Construction and Performance of Deep Foundations," ASCE, University of California, Berkeley, August 1975.

Davisson, M. T. "Foundations in Difficult Soils–State of the Practice Deep Foundations", *Seminar on Foundations in Difficult Soils,* Metropolitan Section, ASCE, April 1989, New York, N.Y.

De Ruiter, J. and Beringen, F. L., "Pile Foundations for Large North Sea Structures," *Marine Geotechnology,* Vol. 3, No. 3, 1979, pp. 267–314.

Endo, M., Minou, A., Kawasaki, T., and Shibata, T., "Negative Skin Friction Acting on Steel Pipe Pile in Clay", *Proceedings 7th International Conference on Soil Mechanics and Foundation Engineering,* Vol. 2, Mexico City, 1969, pp. 85–92.

*Foundations and Earth Structures Design Manual 7.2,* NAVFAC DM-7.2, Department of the Navy, Alexandria, VA, May 1982.

Francis, A. J., Savory, N. R., Stevens, L. K., and Trollope, D. H., "The Behavior of Slender Point-Bearing Piles in Soft Soil," *Symposium on Design of Tall Buildings,* University of Hong Kong, September 1961, pp. 25–50.

Garg, K. G., "Bored Pile Groups Under Vertical Load in Sand," *J. Geotech. Eng. Div.,* ASCE, Vol. 105, No. GT 8, August 1979, pp. 939–956.

Garlanger, J. E., "Prediction of the Downdrag Load at Culter Circle Bridge," *Symposium on Downdrag of Piles,* Massachusetts Institute of Technology, Cambridge, MA, 1973.

Goble, G. G., Lickins, G. E., and Rausche, F., "Bearing Capacity of Piles From Dynamic Measurements", Final Report, Department of Civil Engineering, Case Western Reserve University, Cleveland, OH, March 1975.

Goble, G. G. and Rausche, F., "Wave Equation Analysis of Pile Driving—WEAP Program," submitted to U.S. Department of Transportation by Goble and Associates, Inc., September 1980, 4 Volumes.

Hegedus, E. and Khosla, V. K., "Pollout Resistance of 'H' Piles," *J. Geotech. Eng. Div.,* ASCE, Vol. 110, No. 9, September 1984, pp. 1274–1290.

Hirsch, T. J., Carr, L., and Lowery, L. L., "Pile Driving Analysis. Wave Equation User's Manual. TTI Program," Vol. 1—Background, Vol. 2—Computer Program, Vol. 3-Program Documentation. U.S. Dept. of Transportation, Federal Highway Administration Office of Research and Development, Washington, DC, 1976, 308p.

Horvath, R. G., Kenny, T. C., and Kozicki, P., "Methods of Improving the Performance of Drilled Piers in Weak Rock," *Can. Geotech. J.* Vol. 20, No. 4, November 1983, pp. 758–772.

Ireland, H. O., "Pullout Test on Piles in Sand," *Proceedings 4th International Conference Soil Mechanics and Foundation Engineering,* Vol. 2, London 1957, pp. 43–54.

Kerisel, J. L., "Vertical and Horizontal Bearing Capacity of Deep Foundations in Clay," *Proceeding Symposium of Bearing Capacity and Settlement of Foundations,* Duke University, Durham, NC, 1967, p. 45.

Kezdi, A., "Pile Foundation," *Foundation Engineering Handbooks,* Editors H. F. Winterkorn and H. Y. Fang. Van Nostrand Reinhold Co., New York, 1975.

Koutsoftas, D. C., "Caissons Socketed in Sand Mica Schist," *J. Geotech. Eng. Div.*, ASCE, Vol. 107, No. GT6, June 1981, pp. 743–757.

Kraft, L. M., Focht, J. A., and Amerasinghe, S. F., "Friction Capacity of Piles Driven Into Clay," *J. Geotech. Eng. Div.*, ASCE, Vol. 107, No. GT 11, November 1981, pp. 1521–1541.

Ladanyi, B. and Roy, A., "Some Aspects of Bearing Capacity of Rock Mass," *Proceedings 7th Canadian Symposium on Rock Mechanics*, Edmonton, Alberta, Canada, 1971, pp. 161–190.

McClelland, B., "Design of Deep Penetration Piles for Ocean Structures," *J. Geotech. Eng. Div.*, ASCE, Vol. 100, No. GT 7, 1974, pp. 705–747.

Meyerhof, G. G., "Ultimate Bearing Capacity of Footings on Sand Layer Overlying Clay," *Can. Geotech. J.*, Vol. 11, No. 2, 1974, pp. 223–229.

Meyerhof, G. G., "Bearing Capacity and Settlement of Pile Foundations," *J. Geotech. Div.*, ASCE, Vol. 102, No. GT 3, March 1976, pp. 197–228.

Meyerhof, G. G., "Scale Effects of Ultimate Pile Capacity," *J. Geotech. Eng. Div.*, ASCE, Vol. 109, No. 6, June 1983, pp. 797–806.

Meyerhof, G. G. and Adams, J. J., "The Ultimate Uplift Capacity of Foundations," *Can. Geotech. J.*, Vol. 5, No. 4, Novemeber 1968, pp. 225–244.

Michigan State Highway Commission, "A Performance Investigation of Pile Driving Hammers and Piles," Final Report, Lansing, MI, March 1965.

Moorhouse, D. C. and Sheehan, J. V., "Predicting Safe Capacity of Pile Groups," *Civil Engineering*, Vol. 38, No. 10, October 1968, pp. 44–48.

Niyama, S., Azevedo, N., Polla, C. M. and Dechichi M. A." Load Transfer in Dynamically and Statically Tested Pile," *Proc. 12th Intern. Conf. on Soil Mech. and Found. Eng.* Rio de Janeiro (Brazil) August 1989, Vol. II, pp. 1167–1170.

Nordlund, R. L., "Dynamic Formula for Pressure Injected Footings," *J. Geotech. Eng. Div.*, ASCE, Vol. 108, No. GT 3, March 1982, pp. 419–437.

Peck, R. B., Hansen, W. E., and Thornburn, T. H., *Foundation Engineering*, 2nd ed. Wiley, New York, 1974.

Poulos, H. G. and Davis, E. H., *Pile Foundation Analysis and Design*, Wiley, New York, 1980.

Prakash, S., *Soil Dynamics*, McGraw-Hill Book Co., New York, 1981.

Rausche, F., Goble, G. G., and Likins, G. E., "Dynamic Determination of Pile Capacity," *J. Geotech. Eng. Div.*, ASCE, Vol. 111, No. 3, March 1985, pp. 367–383.

Seed, H. B., Woodward, Jr., R. J., and Lundgren, R., "Prediction of Swelling Potential for Compacted Clays," *J. Soil Mech. and Found On.* ASCE, Vol. 88, No. SM 3, 1962, pp. 53–87.

Sharma, H. D. and Joshi, R. C., "Comparison of In Situ and Laboratory Soil Parameters for Pile Design in Granular Deposits," *39th Canadian Geotechnical Conference*, Ottawa, August 1986, pp. 131–138.

Sharma, H. D. and Joshi, R. C., "Drilled Pile Behavior in Granular Deposits," *Can. Geotech. J.*, Vol. 25, No. 2, May 1988, pp. 222–232.

Sharma, H. D., Sengupta, S., and Harron, G., "Cast-In-Place Bored Piles on Soft Rock Under Artesian Pressures," *Can. Geotech. J.*, Vol. 21, No. 4, November 1984, pp. 684–698.

Shields, D. H. Private communication on pressuremeter data, June 1987.

Skempton, A. W., "The Bearing Capacity of Clays," *Proceedings of the British Building Research Congress*, London, 1951, pp. 180–189.

Skempton, A. W., "Discussion on Piles and Pile Foundation," *Proceedings 3rd International Conference on Soil Mechanics and Foundation Engineering*, Zurich, Switzerland, Vol. 3, 1953, p. 172.

Skempton, A. W., "Cast-In-Situ Bored Piles in London Clay," *Geotechnique*, Vol. 9, 1959, p. 158.

Skempton, A. W., Yassin, A. S., and Gibson, R. E., "Theorie De La Force Portante Des Pieux," *Annales De L'Institute Technique Du Batiment Et Des Travaux Publics*, Vol. 6, Nos. 63–64, 1953, pp. 285–290.

Smith, E. A., " Pile Driving Analysis by the Wave Equation," *Transactions*, ASCE, Vol. 127, Part I, 1962, pp. 1145–1193.

Sowa, V. A., "Pulling Capacity of Concrete Cast in Situ Piles," *Can. Geotech. J.*, Vol. 17, 1970, pp. 482–493.

Standard Test Method for Unconfined Compressive Strength of Intact Rock Core Specimens, ASTM D 2938–86 pp. 345–346.

Terzaghi, K. and Peck, R. B., *Soil Mechanics in Engineering Practice*, 2nd ed. Wiley, New York, 1967.

Thompson, C. D., "New Standard Method of High-Strain Dynamic Testing of Piles," Submitted to ASTM, June 1986.

Tomlinson, M. J., "The Adhesion of Piles Driven in Clay Soils," *Proceedings 4th International Conference on Soil Mechanics and Foundation Engineering*, London, 1957, pp. 66–71.

Tomlinson, M. J., "Some Effects of Pile Driving on Skin Friction," Conference on Behavior of Piles, Institution of Civil Engineers, London, 1970, pp. 59–66.

Tomlinson, M. J.,"*Pile Design and Construction Practice*," A Viewpoint Publication, Cement and Concrete Association, London, 1977.

Vesic, A. S., "Ultimate Loads and Settlements of Deep Foundations in Sands," *Proceedings of the Symposium on Bearing Capacity and Settlement of Foundations*, Duke University, Durham, NC, April 1965, pp. 53–68.

Vesic, A. S., "Load Transfer in Pile-Soil Systems," *Proceedings Conference on Design Installation of Pile Foundations*, Lehigh University, Bethlehem, PA, 1970a, pp. 47–73.

Vesic, A. S., "Tests on Instrumented Piles, Ogeeche River Site," *J. Soil Mechanics and Foundation Div.*, ASCE, Vol. 96, No. SM2, March 1970, pp. 561–584.

Vesic, A. S., "Expansion of Cavities in Infinite Soil Mass," *J. Soil Mech. Found. Div.*, ASCE, Vol. 98, No. SM3, Proceeding Paper 8790, March 1972, pp. 265–290.

Vesic, A. S., "Design of Pile Foundations," Transportation Research Board, National Research Council, Washington, DC, 1977.

Whitaker, T., "Experiments With Model Piles In Groups," *Geotechnique*, Vol. 7, 1957, pp. 147–167.

Williams, A. A. B., "The Prediction of Total Heave from the Double Oedometer Tests," Discussion, *Transactions of the South African Institution of Civil Engineers*, Vol. 8, No. 6, 1958, pp. 123–124.

Wu, A. K. H., Kuhlemeyer, R. L., and To, C. S. W., "Validity of Smith Model in Pile Driving Analysis," *J. Geot. Engg. Div.*, ASCE, Vol. 115, No. 9, September 1989, pp. 1285–1302.

# 6

# ANALYSIS AND DESIGN OF PILE FOUNDATIONS UNDER LATERAL LOADS

Lateral loads and moments may act on piles in addition to the axial loads. The two pile head fixity conditions—free-head and fixed headed*—may occur in practice. Figure 6.1 shows three cases where such loading conditions may occur. In Figure 6.1a, piles with a free head are subjected to vertical and lateral loads. Axial downward loads are due to gravity effects. Upward loads, lateral loads, and moments are generally due to forces such as wind, waves and earthquake. In Figure 6.1b, piles with a free head are shown under vertical and lateral loads and moments, while in Figure 6.1c, fixed-headed piles ($Ft$) under similar loads are shown. The extent to which a pile head will act as free headed or fixed headed will depend on the relative stiffness of the pile and pile cap and the type of connections specified. In Figure 6.1 the deformation modes of piles have been shown under various loading conditions by dotted lines.

The allowable lateral loads on piles is determined from the following two criteria:

1. Allowable lateral load is obtained by dividing the ultimate (failure) load by an adequate factor of safety
2. Allowable lateral load is corresponding to an acceptable lateral deflection. The smaller of the two above values is the one actually adopted as the design lateral load

Methods of calculating lateral resistance of vertical piles can be broadly divided into two categories:

*Fixed against rotation but free to translate, therefore, fixed-translating headed ($Ft$).

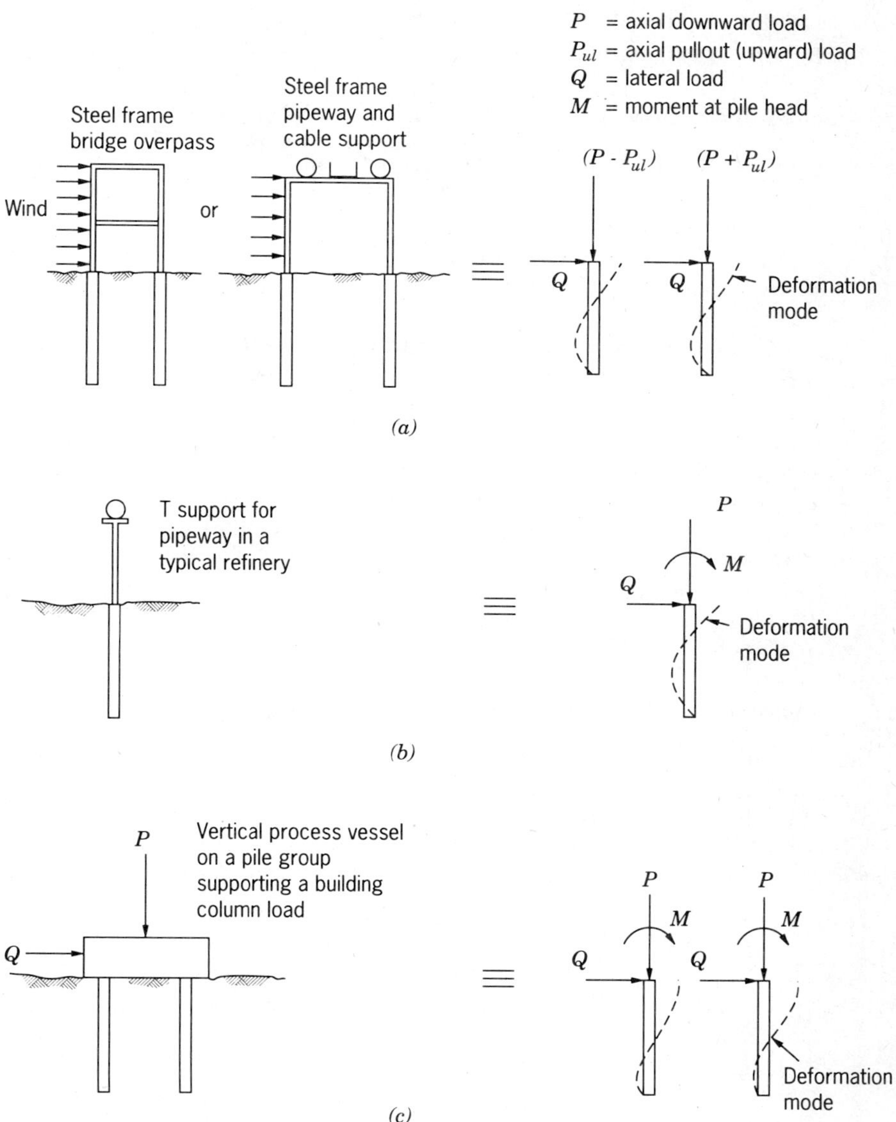


**Figure 6.1** Piles subjected to lateral loads. (a) Piles subjected to vertical and lateral loads (free head), (b) piles subjected to vertical and lateral loads and moment (free head), (c) piles subjected to vertical and lateral loads and moment (fixed head).

1. Methods of calculating ultimate lateral resistance
2. Methods of calculating acceptable deflection at working lateral load

*I. Methods of Calculating Lateral Resistance of Vertical Piles*

A. Brinch Hansen's Method (1961): This method is based on earth pressure theory and has the advantage that it is:
   1. Applicable for $c$–$\phi$ soils
   2. Applicable for layered system

   However, this method suffers from disadvantages that it is
   1. Applicable only for short piles
   2. Requires trial-and-error solution to locate point of rotation

B. Broms' Method (1964a, b): This also is based on earth pressure theory, but simplifying assumptions are made for distribution of ultimate soil resistance along the pile length. This method has the advantage that it is:
   1. Applicable for short and long piles
   2. Considers both purely cohesive and cohensionless soils
   3. Considers both free-head and fixed-head piles that can be analyzed separately

   However, this method suffers from disadvantages that:
   1. It is not applicable to layered system
   2. It does not consider $c$–$\phi$ soils

*II. Methods of Calculating Acceptable Deflection at Working Load*

A. Modulus of Subgrade Reaction Approach (Reese and Matlock, 1956): In this method it is assumed that soil acts as a series of independent linearly elastic springs. This method has the advantage that:
   1. It is relatively simple
   2. It can incorporate factors such as nonlinearity, variation of subgrade reaction with depth, and layered systems
   3. It has been used in the practice for a long time

   Therefore, a considerable amount of experience has been gained in applying the theory to practical problems. However, this method suffers from disadvantages that:
   1. It ignores continuity of the soil
   2. Modulus of subgrade reaction is not a unique soil property but depends on the foundation size and deflections.

B. Elastic Approach (Poulos, 1971a and b):
   In this method, the soil is assumed as an ideal elastic continuum. The method has the advantage that:
   1. It is based on a theoretically more realistic approach,
   2. It can give solutions for varying modulus with depth and layered system. However, this method suffers from disadvantages that:
   1. It is difficult to determine appropriate strains in a field problem and the corresponding soil moduli

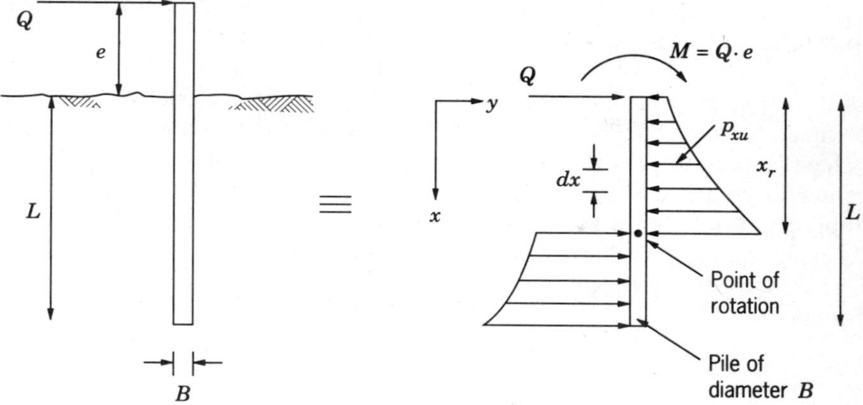


**Figure 6.2** Mobilization of lateral resistance for a free-head laterally loaded rigid pile.

2. It needs more field verification by applying theory to practical problems

***Ultimate Lateral Resistance*** Figure 6.2 shows the mechanism in which the ultimate soil resistance is mobilized to resist a combination of lateral force $Q$ and moment $M$ applied at the top of a free-head pile. The ultimate lateral resistance $Q_u$ and the corresponding moment $M_u$ can then be related with the ultimate soil resistance $p_u$ by considering the equilibrium conditions as follows:

Sum of Forces in horizontal direction $= \Sigma F_y = 0$

$$Q_u - \int_{x=0}^{x=x_r} p_{xu} B \, dx + \int_{x=x_r}^{x=L} p_{xu} B \, dx = 0 \tag{6.1}$$

$\Sigma$ Moments $= 0$

$$Q_u e + \int_{x=0}^{x=x_r} p_{xu} B x \, dx - \int_{x=x_r}^{x=L} p_{xu} B x \, dx = 0 \tag{6.2}$$

where

$B =$ width of pile
$x_r =$ depth of point of rotation

If the distribution of ultimate unit soil resistance $p_{xu}$ with depth $x$ along the pile is known, then the values of $x_r$ (the depth of the point of rotation) and $Q_u$ (the ultimate lateral resistance) can be obtained from equations (6.1) and (6.2).

This basic concept has been used by Brinch Hansen (1961) and Broms (1964a, b) to determine the ultimate lateral resistance of vertical piles.

*Brinch Hansen's Method* For short rigid piles, Brinch Hansen (1961) recommended a method for any general distribution of soil resistance. The method is based on earth pressure theory for $c$–$\phi$ soils. It consists of determining the center of rotation by taking moment of all forces about the point of load application and equating it to zero. The ultimate resistance can then be calculated by using equation similar to equation (6.1) such that the sum of horizontal forces is zero. Accordingly, the ultimate soil resistance at any depth is given by following equation.

$$p_{xu} = \bar{\sigma}_{vx} K_q + c K_c \tag{6.3}$$

where

$\bar{\sigma}_{vx}$ = vertical effective overburden pressure
$c$ = cohesion of soil
$K_c$ and $K_q$ = factors that are function of $\phi$ and $x/B$ as shown in Figure 6.3

The method is applicable to both uniform and layered soils. For short-term loading conditions such as wave forces, undrained strength $c_u$ and $\phi = 0$ can be used. For long-term sustained loading conditions, the drained effective strength values ($c'$, $\phi'$) can be used in this analysis.

*Broms' Method* The method proposed by Broms (1964a, b) for lateral resistance of vertical piles is basically similar to the mechanism outlined above. The following simplifying assumptions have been made in this method:

1. Soil is either purely cohesionless ($c = 0$) or purely cohesive ($\phi = 0$). Piles in each type of soil have been analyzed separately.
2. Short rigid and long flexible piles are considered separately. The criteria for short rigid piles is that $L/T \leqslant 2$ or $L/R \leqslant 2$

   where

$$T = \left(\frac{EI}{n_h}\right)^{1/5} \tag{6.4a}$$

$$R = \left(\frac{EI}{k_h}\right)^{1/4} \tag{6.4b}$$

$E$ = modulus of elasticity of pile material
$I$ = moment of inertia of pile section
$k_h = n_h x$ for linearly increasing soil modulus $k_h$ with depth($x$)

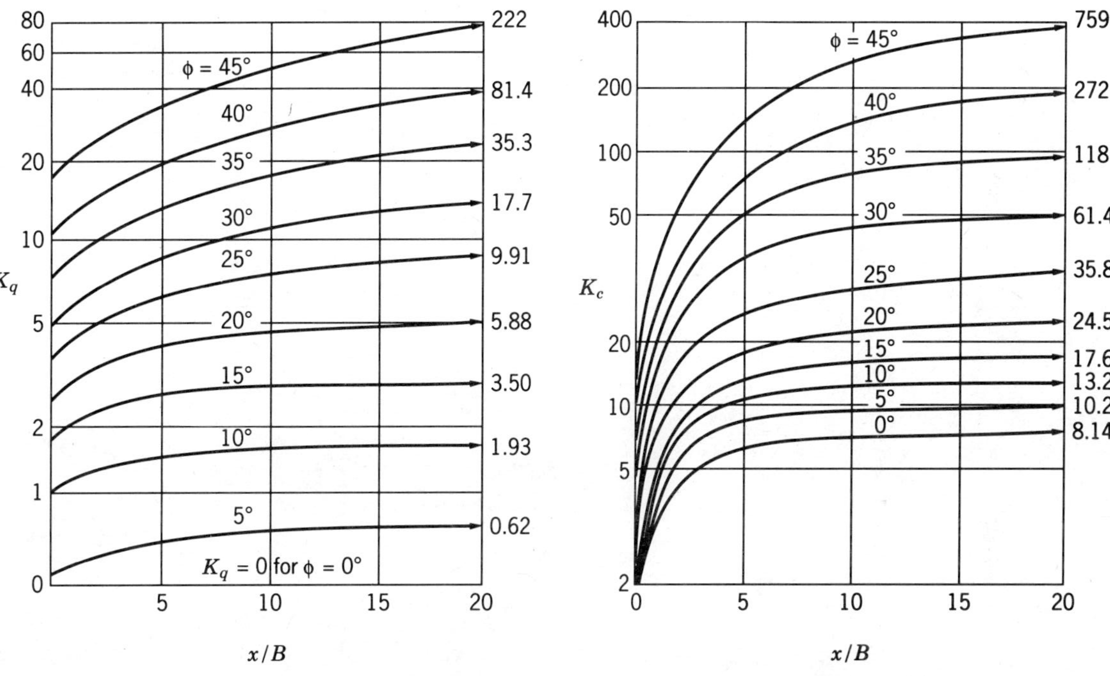


**Figure 6.3** Coefficients $K_q$ and $K_c$ (Brinch Hansen, 1961).

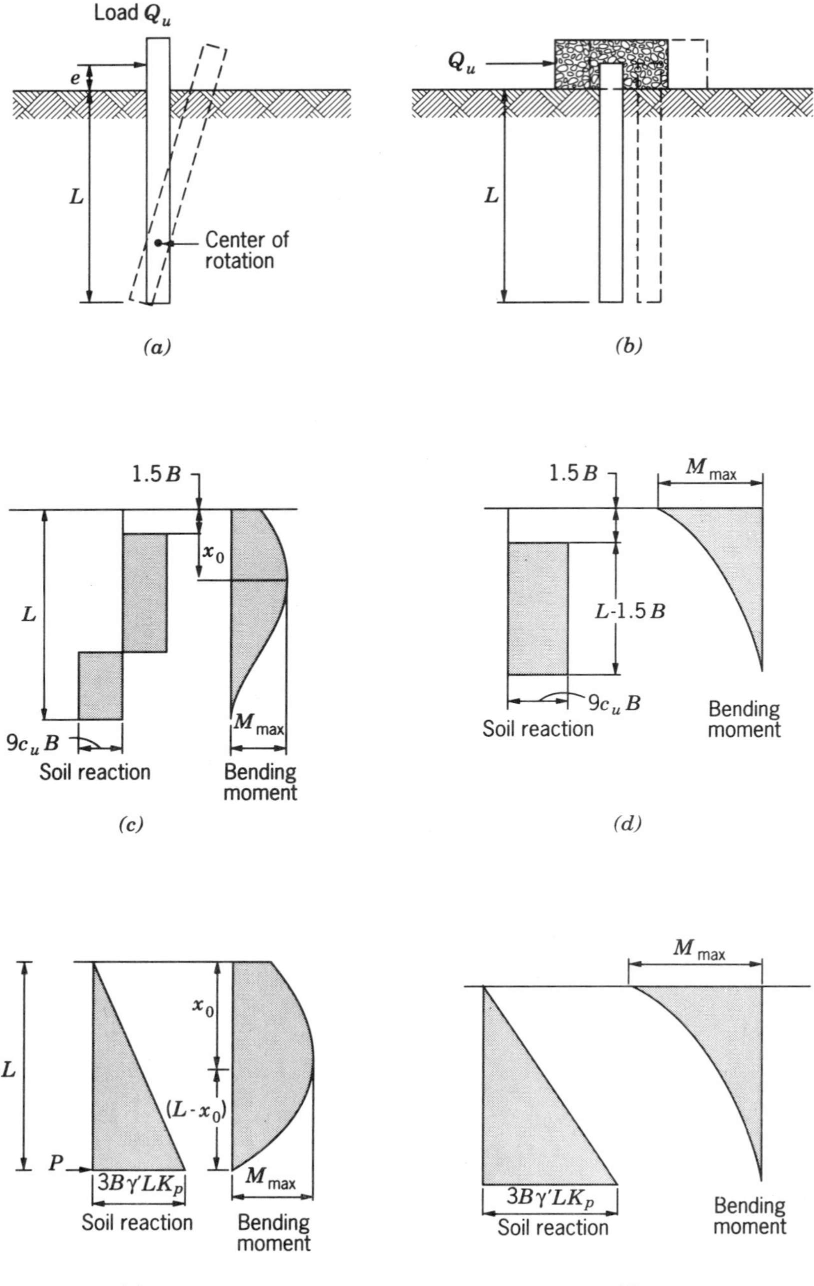


**Figure 6.4** Rotational and translational movements and corresponding ultimate soil resistances for *short piles* under lateral loads. Deformation modes: (a) Free head, (b) fixed-head. Soil reactions and bending moment in *cohesive soils*: (c) Free head, (d) fixed-head. Soil reactions and bending moments in *cohesionless soils*: (e) Free head, (f) fixed head. (After Broms, 1964a and b).

$n_h$ = constant of modulus of subgrade reaction
$k$ = modulus value in cohesive soils that is constant with depth

The criteria for long flexible pile will be $L/T \geqslant 4$ or $L/R \geqslant 3.5$, as applicable.

3. Free-head short piles are expected to rotate around a center of rotation while fixed-head piles move laterally in translation mode (Figure 6.4a, b). Deformation modes of long piles are different from short piles because the rotation and translation of long piles cannot occur due to very high passive soil resistance at the lower part of the pile (Figure 6.5a, b). Lateral load capacity of short and long piles have therefore been evaluated by different methods.
4. Distribution of ultimate soil resistance along the pile for different end conditions is shown in Figure 6.4 for short piles and in Figure 6.5 for long piles.

*Short Piles in Cohesionless Soils*

(a) The active earth pressure on the back of the pile is neglected and the distribution of passive pressure along the front of the pile at any depth is (Figure 6.4e, f)

$$p = 3B\sigma'_v K_p = 3\gamma' LBK_p \tag{6.5}$$

where

$p$ = Unit soil pressure (reaction)
$\sigma'_v$ = effective overburden pressure at any depth
$\gamma'$ = effective unit weight of soil
$L$ = embedded length of pile
$B$ = width of pile
$K_p = (1 + \sin\phi)/(1 - \sin\phi)$ = Rankine's passive earth pressure coefficient
$\phi'$ = angle of internal friction (effective)

This pressure is independent of the shape of the pile section.

(b) Full lateral resistance is mobilized at the movement considered.

*Short Piles in Cohesive Soils*

The ultimate resistance of piles in cohesive soil is assumed to be zero at ground surface to a depth of $1.5B$ and then a constant value of $9c_uB$(below this depth (Figures 6.4c, d))

In *long piles*, $L$ is replaced by $x_0$ in equation 6.5 in cohesionless soils beyond which the soil reaction decreases. In cohesive soils, the soil reaction decreases beyond $(1.5B + x_0)$. The soil reaction distribution with depth for long piles, is shown in Figure 6.5.

***Acceptable Deflection at Working Lateral Load*** In most situations, the design of piles to resist lateral loads is based on acceptable lateral deflection rather than the

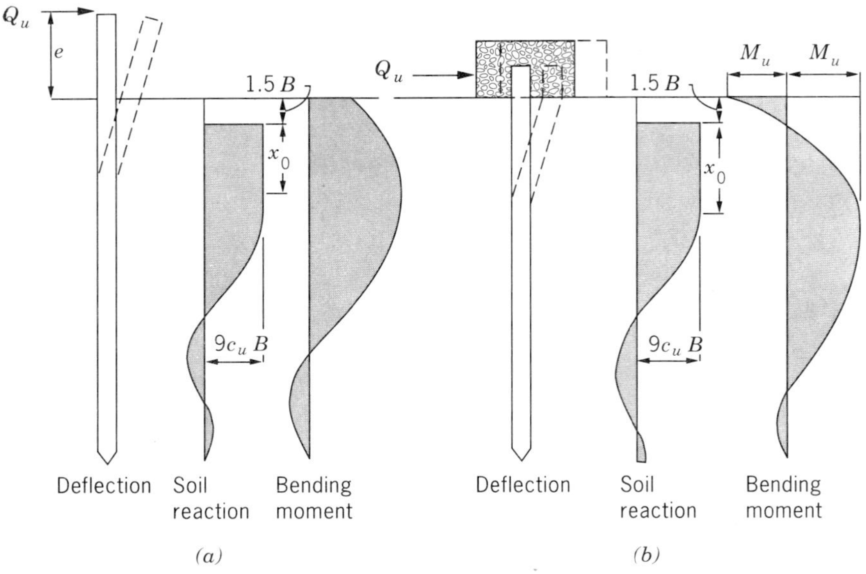


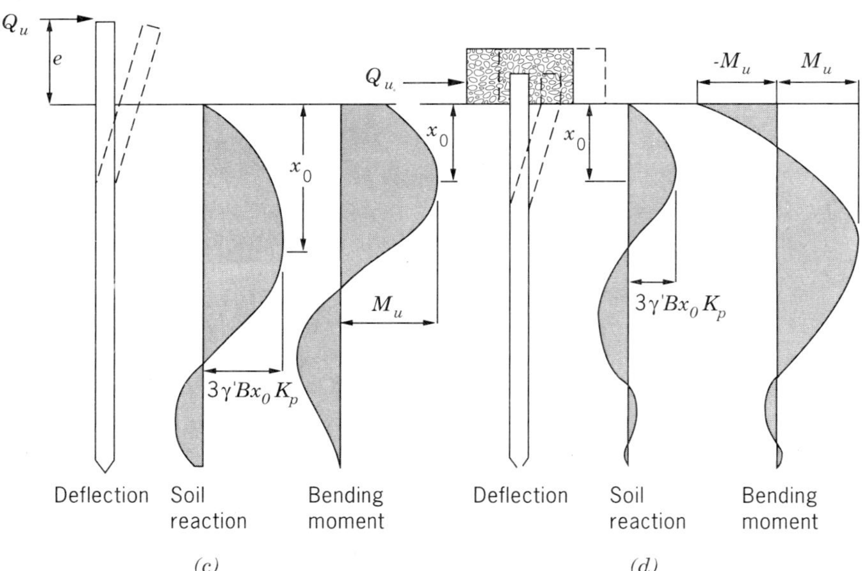


**Figure 6.5** Rotational and translational movements and corresponding ultimate soil resistances for *long piles* under lateral loads. Piles in *cohesive soil*: (a) Free-head, (b) fixed-head (*Ft*). Piles in *cohesionless soil*: (c) Free-head, (d) fixed-head (*Ft*) (After Broms 1964a and b).

ultimate lateral capacity. The two generally used approaches of calculating lateral deflections are:

1. Subgrade reaction approach (Reese and Matlock, 1956; Matlock and Reese 1960)
2. Elastic continuum approach (Poulos, 1971a and b)

*Subgrade Reaction Approach* This approach treats a laterally loaded pile as a beam on elastic foundation (Figure 6.6b, c). It is assumed that the beam is supported by a Winkler soil model according to which the elastic soil medium is replaced by a series of infinitely closely spaced independent and elastic springs. The stiffness of these springs $k_h$ (also called the modulus of horizontal subgrade reaction) can be expressed as follows (Figure 6.6d):

$$k_h = \frac{p}{y} \tag{6.6}$$

where

$p =$ the soil reaction per unit length of pile
$y =$ the pile deformation and $k_h$ has the units of force/length$^2$

Palmer and Thompson (1948) employed the following form to express the modulus of a horizontal subgrade reaction:

$$k_x = k_h\left(\frac{x}{L}\right)^n \tag{6.7a}$$

where

$k_h =$ value of $k_x$ at $x = L$ or tip of the pile
$x =$ any point along pile depth
$n =$ a coefficient equal to or greater than zero

The most commonly used value of $n$ for sands and normally consolidated clays under long-term loading is unity. For overconsolidated clays, $n$ is taken zero. According to Davisson and Prakash (1963), a more appropriate value of $n$ will be 1.5 for sands and 0.15 for clays under undrained conditions.

For the value of $n = 1$, the variation of $k_h$ with depth is expressed by the following relationship:

$$k_h = n_h x \tag{6.7b}$$

where $n_h$ is the constant of modulus of subgrade reaction (see Section 4.4). This applies to cohesionless soils and normally consolidated clays where these soils indicate increased strength with depth due to overburden pressures and the consolidation process of the deposition. Typical values are listed in Table 4.16.

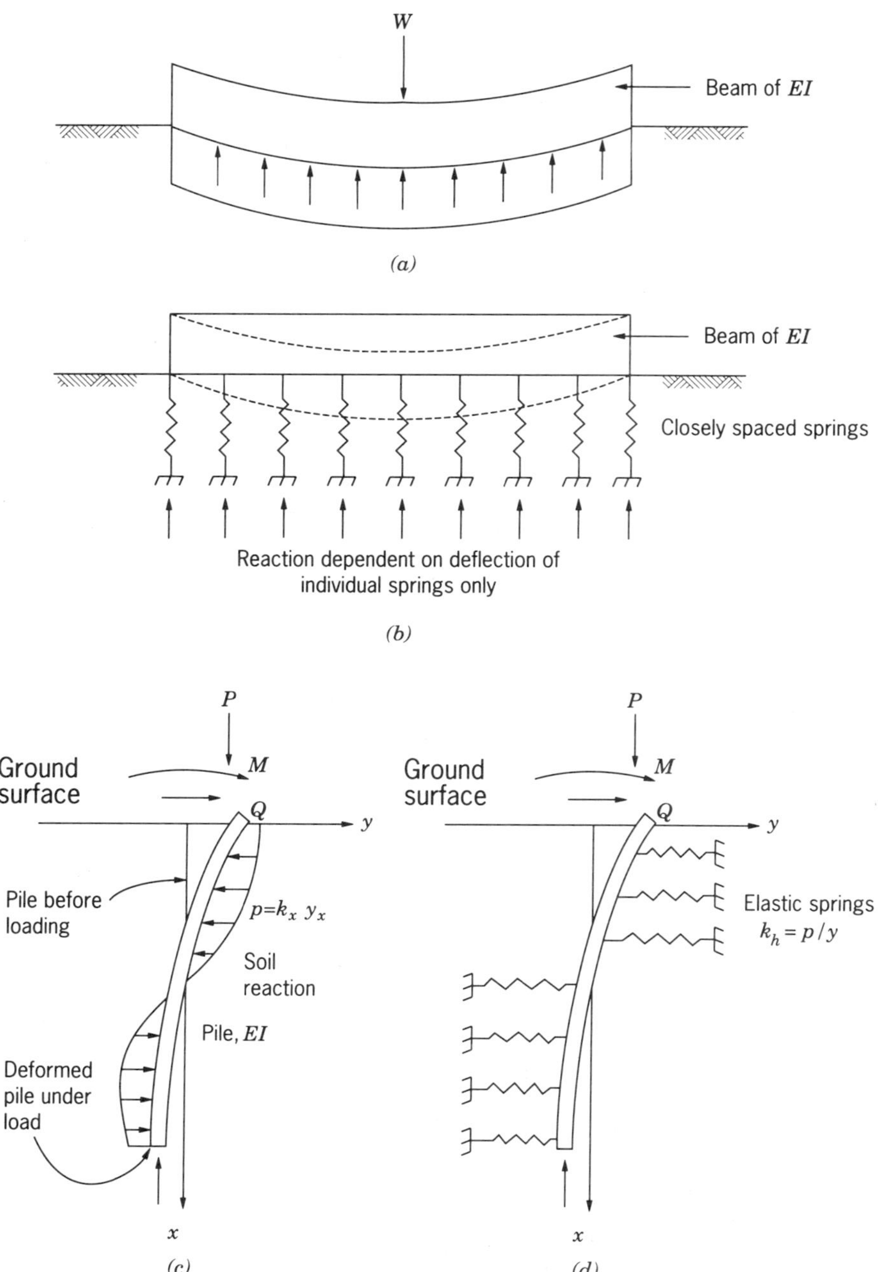


**Figure 6.6** Behavior of laterally loaded pile: subgrade reaction approach. (a) Beam on elastic foundation, (b) Winkler's idealization, (c) laterally loaded pile in soil, (d) laterally loaded pile on springs.

For the value of $n = 0$, the modulus will be constant with depth and this assumption is most appropriate for piles in overconsolidated clays.

The soil reaction–deflection relationship for real soils is nonlinear and Winkler's idealization would require modification. This can be done by using $p$–$y$ curves approach, discussed in Sections 6.1 and 6.6.

The behavior of a pile can thus be analyzed by using the equation of an elastic beam supported on an elastic foundation and is given by the following equation:

$$EI\frac{d^4y}{dx^4} + p = 0 \tag{6.8}$$

where

$E$ = modulus of elasticity of pile
$I$ = moment of inertia of pile section
$p$ = soil reaction which is equal to $(k_h y)$

Equation (6.8) can be rewritten as follows:

$$\frac{d^4y}{dx^4} + \frac{k_h y}{EI} = 0 \tag{6.9}$$

Solutions for equation (6.9) to determine deflection and maximum moments are given in Section 6.1 for cohesionless soils and Section 6.6 for cohesive soils. The extension of these solutions to incorporate nonlinear soil behavior by using $p$–$y$ curves are also described there.

*Elastic Continuum Approach* The determination of deflections and moments of piles subjected to lateral loads and moments based on the theory of subgrade reaction is unsatisfactory as the continuity of the soil mass is not taken into account. The behavior of laterally loaded piles for soil as an elastic continuum has been examined by Poulos (1971a, and b). Although this approach is theoretically more realistic, one of the major obstacles in its application to the practical problem is the realistic determination of soil modulus $E_s^*$. Also, the approach needs more field verification by applying the theoretical concept to practical problems. Therefore, only the basic theoretical concepts and some solutions, for this approach will be described here. These concepts will be helpful in comparing this approach with the subgrade reaction approach.

*$K_h$ and $E_s$ are *sometimes* used interchangeably.

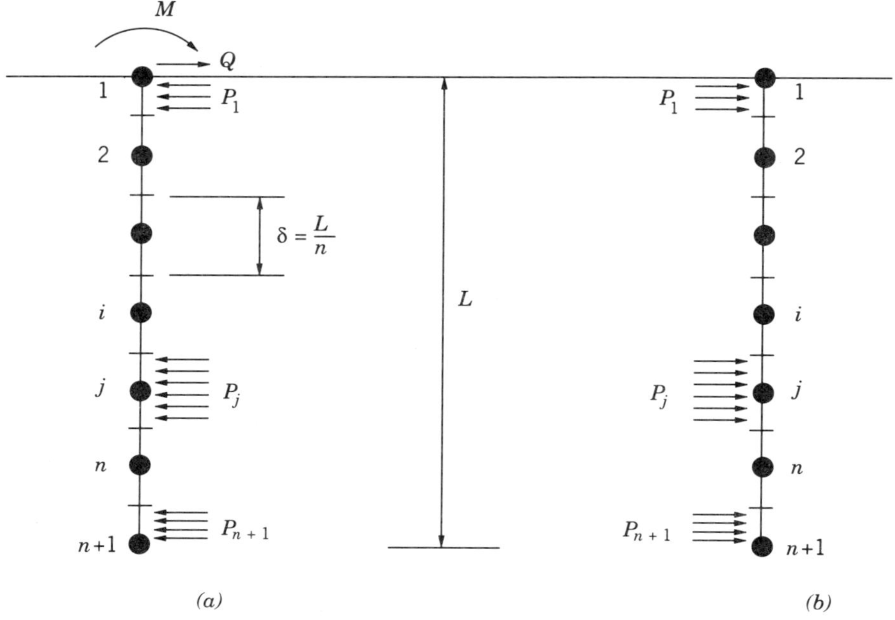


**Figure 6.7** Stresses acting on (a) Pile, (b) soil adjacent to pile (Poulos, 1971a).

*Theoretical Basis* Theoretical basis for the elastic continuum approach solution is as follows:

1. As shown in Figure 6.7, the pile is assumed to be a thin rectangular vertical strip of width $B$, length $L$, and constant flexibility $EI$. The pile is divided into $(n + 1)$ elements of equal lengths except those at the top and tip of the pile, which are of length $(\delta/2)$.
2. To simplify the analysis, possible horizontal shear stresses developed between the soil and the sides of the pile are not taken into account.
3. Each element is assumed to be acted on by a uniform horizontal force $P$, which is assumed constant across the width of the pile.
4. The soil is assumed to be an ideal, homogeneous, isotropic, semi-infinite elastic material, having a Young's modulus $E_s$ and Poisson's ratio $\nu_s$, which are unaffected by the presence of the pile.

In the purely elastic conditions within the soil, the horizontal displacements of the soil and of the pile are equal along the pile. In this analysis, Poulos (1971) equates soil and pile displacements at the element centers. For the two extreme elements (the top and the tip), the displacements are calculated. By equating soil and pile displacements at each uniformly spaced points along the pile and by using appropriate equilibrium conditions, an unknown horizontal displacement at each element can be obtained.

Solutions to obtain deflection and moments on pile for fixed- and free-head conditions are described in Section 6.1.5 for cohesionless soils and Section 6.6.3 for cohesive soil.

## 6.1 VERTICAL PILE UNDER LATERAL LOAD IN COHESIONLESS SOIL

This section presents the application of general approaches to the analysis of vertical piles subjected to lateral loads.

### 6.1.1 Ultimate Lateral Load Resistance of a Single Pile in Cohesionless Soil

The two methods that can be used to determine the ultimate lateral load resistance of a single pile are by Brinch Hansen (1961) and by Broms (1964b). Basic theory and assumptions behind these methods have already been discussed. This section stresses the application aspect of the concept discussed earlier.

***Brinch Hansen's Method*** For cohesionless soils where $c = 0$, the ultimate soil reaction at any depth is given by equation (6.3), which then becomes:

$$p_{xu} = \bar{\sigma}_{vx} K_q \tag{6.10}$$

where $\bar{\sigma}_{vx}$ is the effective vertical overburden pressure at depth $x$ and coefficient $K_q$ is determined from Figure 6.3. The procedure for calculating ultimate lateral resistance consists of the following steps:

1. Divide the soil profile into a number of layers.
2. Determine $\bar{\sigma}_{vx}$ and $k_q$ for each layer and then calculate $p_{xu}$ for each layer and plot it with depth.
3. Assume a point of rotation at a depth $x_r$ below ground and take the moment about the point of application of lateral load $Q_u$ (Figure 6.2).
4. If this moment is small or near zero, then $x_r$ is the right value. If not, repeat steps (1) through (3) until the moment is near zero.
5. Once $x_r$ (the depth of the point of rotation) is known, take moment about the point (center) of rotation and calculate $Q_u$.

This method is illustrated in Example 6.1.

***Example 6.1*** A 20-ft (6.0 m) long, 20-in. (500 mm)-diameter concrete pile is installed into sand that has $\phi' = 30°$ and $\gamma = 120\,\text{lb/ft}^3$ (1920 kg/m$^3$). The modulus of elasticity of concrete is $5 \times 10^5$ kips/ft$^2$ ($24 \times 10^6$ kN/m$^2$). The pile is 15 ft

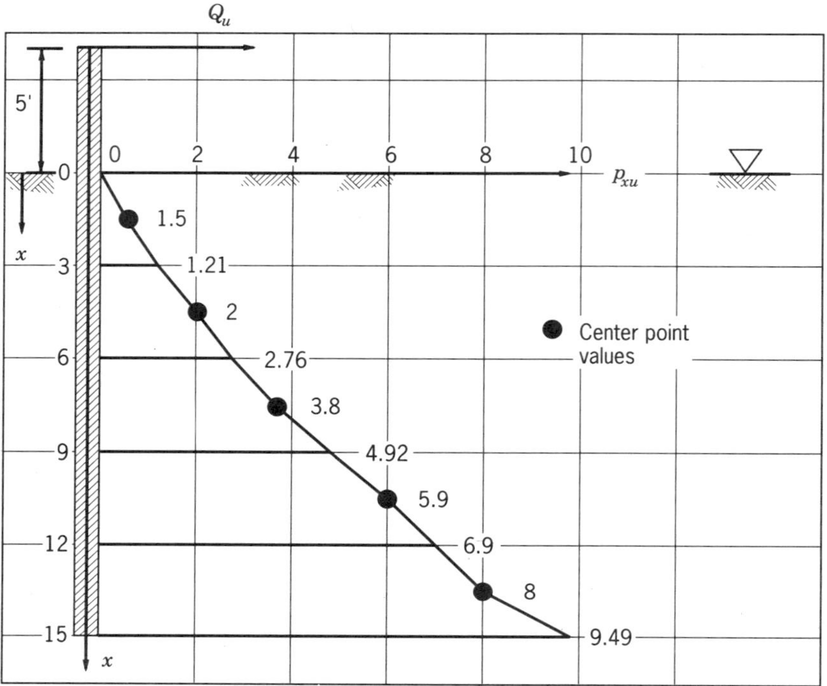


**Figure 6.8** Solution of Example 6.1.

(4.5 m) into the ground and 5 ft (1.5 m) above ground. The water table is near ground surface. Calculate the *ultimate* and the *allowable lateral* resistance by Brinch Hansen's method.

SOLUTION

(a) Divide the soil profile in five equal layers, 3 ft long each (Figure 6.8).
(b) Determine $\bar{\sigma}_{vx}$:

$$\bar{\sigma}_{vx} = \gamma' x = \frac{(120 - 62.5)}{1000} x = 0.0575\, x \text{ kips/ft}^2$$

where $x$ is measured downwards from the ground level.
For each of the five soil layers, calculations for $\bar{\sigma}_{vx}$ and $p_{xu}$ are carried out as shown in Table 6.1. $p_{xu}$ is plotted with depth in Figure 6.8. The values for $p_{xu}$ at the middle of each layer are shown by a solid dot.
(c) Assume the point of rotation at 9.0 ft below ground level and take moment about the point of application of lateral load, $Q_u$. Each layer is 3 ft thick, which

**TABLE 6.1 Calculation of $p_{xu}$ with Depth**

| $x$(ft) | $x/B$[a] | $\bar{\sigma}_{vx}$(kips/ft$^2$) | $Kq$[b] | $P_{xu} = \bar{\sigma}_{vx} K_q$ (Equation (6.10)) |
|---|---|---|---|---|
| 1 | 2 | 3 | 4 | 5 |
| 0 | 0 | 0 | 4.9 | 0 |
| 3 | 1.79 | 0.1725 | 7.0 | 1.21 |
| 6 | 3.59 | 0.3450 | 8.0 | 2.76 |
| 9 | 5.39 | 0.5175 | 9.5 | 4.92 |
| 12 | 7.19 | 0.6900 | 10.0 | 6.90 |
| 15 | 8.98 | 0.8625 | 11.0 | 9.49 |

[a] $B = 20/12 = 1.67$ ft, $\bar{\sigma}_{vx} = 0.0575x$ kips/ft$^2$.
[b] $K_q$ is obtained from Figure 6.3 for $\phi = 30°$ and for $(x/B)$ values in column 2.

gives

$$\begin{aligned}\sum M &= 1.5 \times 3 \times 6.5 + 2 \times 3 \times 9.5 + 3.8 \times 3 \times 12.5 \\ &\quad - 5.9 \times 3 \times 15.5 - 8 \times 3 \times 18.5 \\ &= 29.25 + 57 + 142.50 - 274.35 - 444 = 228.75 - 718.35 = -489.6 \\ &\text{kip-ft/ft width}\end{aligned}$$

(d) This is not near zero; therefore, carry out a second trial by assuming a point of rotation at 12 ft below ground. Then, using the above numbers,

$$\sum M = 29.25 + 57 + 142.50 + 274.35 - 444 = 59.1 \text{ kip ft/ft}$$

The remainder is now a small number and is closer to zero. Therefore, the point of rotation $x_r$ can be taken at 12 ft below ground.
(e) Take the moment about the center of rotation to determine $Q_u$:

$$\begin{aligned}Q_u(5+12) &= 1.5 \times 3 \times 10.5 + 2 \times 3 \times 7.5 + 3.8 \times 3 \times 4.5 + 5.9 \times 3 \times 1.5 - 8 \times 3 \times 1.5 \\ &= 47.25 + 45 + 51.3 + 26.55 - 36 = 134.1 \\ &= 7.89 \text{ kips/ft width} \\ &= 7.89 \times B = 7.89 \times 1.67 = 13.2 \text{ kips (where } B = 20 \text{ in.} = 1.67 \text{ ft)}\end{aligned}$$

$$Q_{all} = \frac{13.2}{2.5} = 5.3 \text{ kips using a factor of safety 2.5}$$

***Brom's Method*** As discussed earlier, Broms (1964b) made certain simplifying assumptions regarding distribution of ultimate resistance with depth, considered short rigid and long flexible piles separately, and also dealt with free-head and fixed (restrained)-head cases separately. In the following section, first the free-head piles are discussed followed by the fixed-head case.

*Free-Head (Unrestrained) Piles*

SHORT PILES For short piles ($L/T \leqslant 2$), the possible failure mode and the distribution of ultimate soil resistance and bending moments are shown in Figure 6.4 (a) and (e), respectively. Since the point of rotation is assumed to be near the tip of the pile, the high pressure acting near tip (Figure 6.4e for cohesionless soils) can be replaced with a concentrated force. Taking the moment about the toe gives the following relationship:

$$Q_u = \frac{0.5\gamma' L^3 B K_p}{(e + L)} \tag{6.11}$$

This relationship is plotted using nondimensional terms $L/B$ versus $Q_u/K_p B^3 \gamma'$ in Figure 6.9a. From this figure, $Q_u$ can be calculated if the values of $L$, $e$, $B$, $K_p = (1 + \sin\phi')/(1 - \sin\phi')$ and $\gamma'$ are known. As shown in Figure 6.4e, the maximum moment ($M_{max}$) occurs at a depth of $x_0$ below ground. At this point, the shear force equals zero, which gives:

$$Q_u = 1.5\gamma' B x_0^2 K_p \tag{6.12}$$

From this expression, we get

$$x_0 = 0.82\left(\frac{Q_u}{\gamma' B K_p}\right)^{0.5} \tag{6.13}$$

The maximum moment is:

$$M_{max} = Q_u(e + 1.5x_0) \tag{6.14}$$

LONG PILES For long piles ($L/T > 4$), the possible failure mode and the distribution of ultimate soil resistance and bending moments are shown in Figure 6.5c for cohesionless soils. Since the maximum bending moment coincides with the point of zero shear, the value of ($x_0$) is given by equation (6.13). The corresponding maximum moment ($M_{max}$) and $Q_u$ (at the point of zero moment) are given by the following equations:

$$M_{max} = Q(e + 0.67x_0) \tag{6.15}$$

$$Q_u = \frac{M_u}{e + 0.54\left(\dfrac{Q_u}{\gamma' B K_p}\right)^{0.5}} \tag{6.16}$$

where $M_u$ = the ultimate moment capacity of the pile shaft. Figure 6.9b can be used to determine the $Q_u$ value by using $Q_u/K_p B^3 \gamma'$ versus $M_u/B^4 \gamma' K_p$ plot.

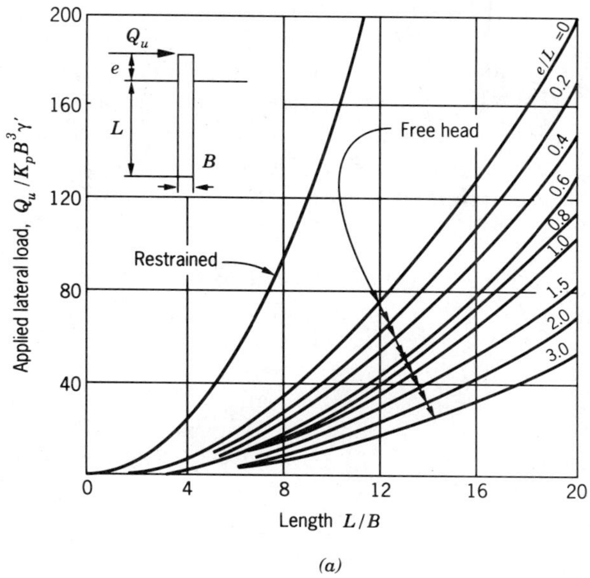


*(a)*

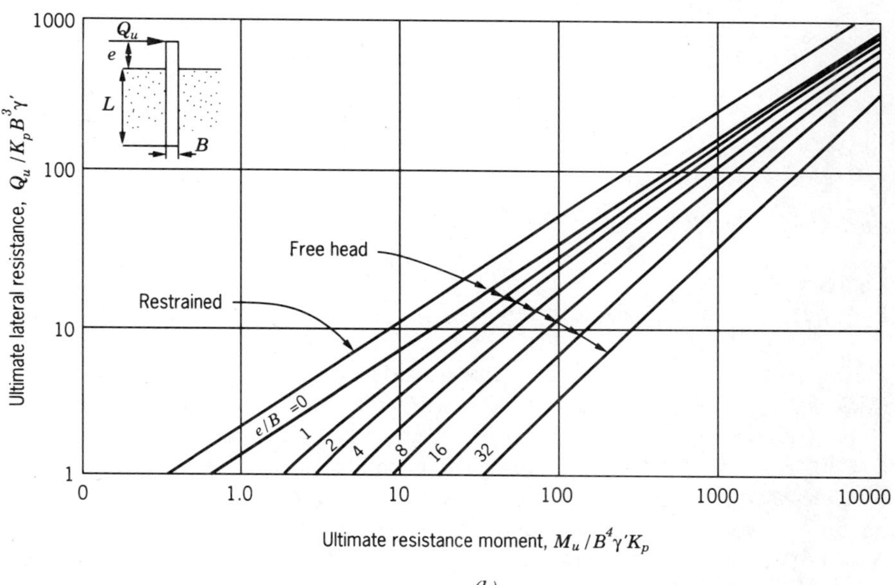


*(b)*

**Figure 6.9** Ultimate lateral load capacity of short and long piles in cohesionless soils (Broms, 1964b). (a) Ultimate lateral resistance of *short piles in cohesionless* soil related to embedded length, (b) ultimate lateral resistance of *long piles in cohesionless* soil related to ultimate resistance moment.

*Fixed-Head (Restrained) Piles*

SHORT PILES For these piles, the possible failure mode is shown on top right-hand corner of Figure 6.4b. The bottom right-hand side of Figure 6.4f shows the distribution of ultimate soil resistance and bending moments for fixed-head short piles. Since failure of these piles is assumed in simple translation, $Q_u$ and $M_{\max}$ for cohesionless soils are computed by using horizontal equilibrium conditions, which give

$$Q_u = 1.5\gamma' L^2 B K_p \tag{6.17}$$

$$M_{\max} = \gamma' L^3 B K_p \tag{6.18}$$

LONG PILES Figure 6.5 (d) shows the failure mode, the distribution of ultimate soil resistance, and bending moments for fixed head long piles in cohesionless soils. $Q_u$ and $M_{\max}$ for cohesionless soils can be determined from following relationships:

$$Q_u = \frac{2M_u}{(e + 0.67x_0)} \tag{6.19}$$

$$x_0 = 0.82\left(\frac{Q_u}{\gamma' B K_p}\right)^{0.5} \tag{6.20}$$

$$M_{\max} = Q_u(e + 0.67x_0) \tag{6.21}$$

where

$x_0$ = depth below ground level where soil reaction becomes maximum

Figure 6.9 (a) and (b) provide graphical solutions for fixed (restrained) short and long piles in cohesionless soils.

***Example 6.2*** A 10.75-inch (273 mm) outside diameter, 0.25 in. (6.4 mm) wall thickness, 30 ft (9.1 m) long steel pile (with free head) is driven into a medium dense sand with standard penetration values ranging between 20 to 28 blows/ft, $\phi = 30°$ and $\gamma = 125\,\text{lb/ft}^3$. Calculate the ultimate failure lateral load at the top of a free-head pile. Find the allowable lateral load and corresponding maximum bending moment, assuming a factor of safety against the ultimate load as 2.5. Assume Young's modulus for steel $(E) = 29000$ ksi (20 MN/m$^2$), yield strength $(f_y) = 35$ ksi (241 MPa), and $n_h = 30$ kips/ft$^3$.

SOLUTION

$$E = 29{,}000 \times 144\,\text{ksf} = 4176 \times 10^3\,\text{ksf}$$

$$I = \frac{\pi}{64}(10.75^4 - 10.25^4) = 113.7\,\text{in.}^4 = 0.0055\,\text{ft}^4$$

$Z = I/(B/2) = \dfrac{113.7 \times 2}{10.75} = 21.2\,\text{in.}^3 = 0.0122\,\text{ft}^3$, $B/2$ is the distance of farthest fiber under bending

$M_u$ = ultimate moment resistance for the section $= Zf_b$

$f_b$ = allowable bending stress $= 0.6f_y = 0.6 \times 35 = 21$ ksi $= 21 \times 144$ ksf $= 3024$ ksf

$M_u = 0.0122 \times 3024 = 37.1$ kip-ft

$$T = \left(\frac{EI}{n_h}\right)^{0.2}$$

$$= \left(\frac{4176 \times 10^3 \times 0.0055}{30}\right)^{0.2} = 3.8\,\text{ft}$$

$L/T = 30/3.8 = 7.9 > 4$. This means that it behaves as a long pile. Then using Figure 6.9,

$$M_u/B^4\gamma' K_p = \frac{37.1}{\left(\dfrac{10.75}{12}\right)^4 \times 125\left(\dfrac{1 + \sin 30}{1 - \sin 30}\right)}$$

$$= \frac{37.1 \times 1000}{0.64 \times 125 \times 3} = 154.6$$

$e/B = 0$

$Q_u/k_pB^3\gamma = 50$ from Figure 6.9b and $e/B = 0$ for free-head pile

$$Q_u = 50 \times 3 \times \left(\frac{10.75}{12}\right)^3 \times \frac{125}{1000} = 13.48\,\text{kips}$$

where $K_p = (1 + \sin\phi)/(1 - \sin\phi) = 3$

Using a safety factor of 2.5,

$$Q_{\text{all}} = \frac{13.48}{2.5} = 5.4\,\text{kips}$$

$$M_{\max} = Q_u(e + 0.67x_0) \tag{6.21}$$

$$e = 0, x_0 = 0.82\left(\frac{Q_u}{\gamma' Bk_p}\right)^{0.5} \tag{6.20}$$

$$= 0.82\left(\frac{5.4 \times 1000}{\dfrac{125 \times 10.75 \times 3}{12}}\right)^{3.5} = 3.3\,\text{ft}$$

$$M_{\max} = 5.4(0.67 \times 3.3) = 11.9\,\text{kips-ft}$$

Since we want to calculate allowable lateral load and corresponding maximum bending moment $Q_{\text{all}}$ should be substituted in equation (6.20) and (6.21).

The section is safe since the maximum moment is less than the ultimate movement resistance of 37.1 kips-ft.

### 6.1.2 Ultimate Lateral Load Resistance of Pile Groups in Cohesionless Soil

The group capacity of laterally loaded piles can be estimated by using the lower of the two values obtained from (1) the ultimate lateral capacity of a single pile multiplied by the number of piles in the group and (2) the ultimate lateral capacity of a block equivalent to the area containing the piles in the group and the soil between these piles. While the value in (1) can be obtained from methods discussed in Section 6.1.1, there is no proven method to obtain ultimate value for case (2).

A more reasonable method, one that is supported by limited tests, is based on the concept of *group efficiency* $G_e$, which is defined as follows:

$$G_e = \frac{(Q_u)_G}{nQ_u} \tag{6.22}$$

where

$(Q_u)_G$ = the ultimate lateral load capacity of a group
$n$ = the number of piles in the group
$Q_u$ = the ultimate lateral load capacity of a single pile

A series of model pile groups were tested for lateral loads by Oteo (1972) and group efficiency $G_e$ values can be obtained from the results of these tests. Interpolated values from his graph are provided in Table 6.2

**TABLE 6.2 Group Efficiency $G_e$ for Cohesionless Soils**[a]

| $S/B$[b] | $G_e$ |
|---|---|
| 3 | 0.50 |
| 4 | 0.60 |
| 5 | 0.68 |
| 6 | 0.70 |

[a] These are interpolated values from graphs provided by Oteo (1972).
[b] $S$ = center-to-center pile spacing.
$B$ = pile diameter or width.

Table 6.2 shows that group efficiency for cohesionless soils decreases as $(S/B)$ of a pile group decreases. Ultimate lateral resistance $(Q_u)_G$ of a pile group can be estimated from equation (6.22) and Table 6.2. There is a need to carry out further laboratory and confirmatory field tests in this area.

### 6.1.3 Lateral Deflection of a Single Pile in Cohesionless Soil: Subgrade Reaction Approach

As discussed earlier, the design of piles to resist lateral loads in most situations is based on acceptable lateral deflections rather than the ultimate lateral load capacity. The two methods that can be used for calculating lateral deflections are the subgrade reaction approach and the elastic approach. The basic theoretical principles behind these two approaches were discussed in the beginning of this section. The application of subgrade reaction approach is discussed here. The elastic approach is discussed later in Section 6.1.5.

***Free-Head Pile*** Figure 6.10 shows the distribution of pile deflection $y$, pile slope variation $dy/dx$, moment, shear, and soil reaction along the pile length due to a lateral load $Q_g$ and a moment $M_g$, applied at the pile head. The behavior of this pile can be expressed by equation (6.9). In general, the solution for this equation can be expressed by the following formulation:

$$y = f(x, T, L, k_h, EI, Q_g, M_g) \tag{6.23}$$

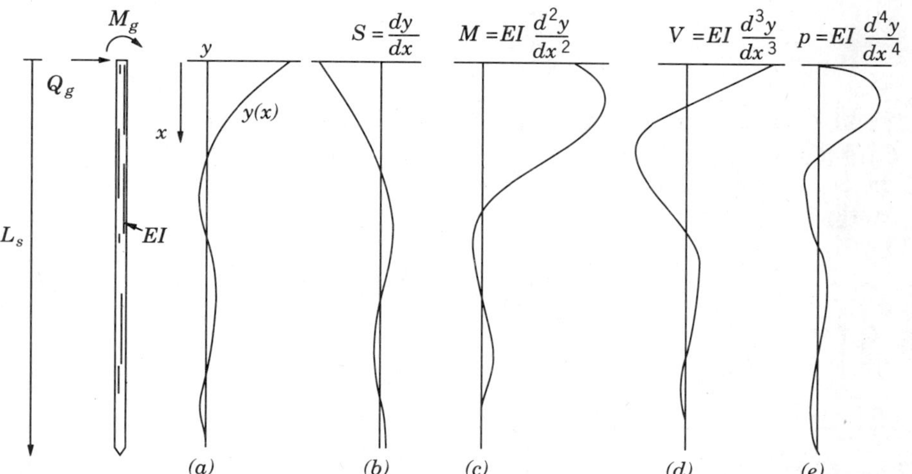


**Figure 6.10** A pile of length $L$ fully embedded in soil and acted by loads $Q_g$ and $M_g$ (a) Deflection, $y$; (b) slope, $dy/dx$; (c) moment, $EI(d^2y/dx^2)$; (d) shear, $EI\ (d^3y/dx^3)$; (e) soil reaction, $EI\ (d^4y/dx^4)$ (Reese and Matlock, 1956).

where

$x$ = depth below ground
$T$ = relative stiffness factor
$L$ = pile length
$k_h = n_h x$ is modulus of horizontal subgrade reaction
$n_h$ = constant of subgrade reaction
$B$ = pile width
$EI$ = pile stiffness
$Q_g$ = lateral load applied at the pile head
$M_g$ = the moment applied at the pile head

Elastic behavior can be assumed for small deflections relative to the pile dimensions. For such a behavior, the principle of superposition may be applied. As we discuss later, for large deformations this analysis can be used with modifications by using the concept of $p$–$y$ curves. By utilizing the principle of superposition, the effects of lateral load $Q_g$ on deformation $y_A$ and the effect of moment $M_g$ on deformation $y_B$ can be considered separately. Then the total deflection $y_x$ at depth $x$ can be given by the following:

$$y_x = y_A + y_B \tag{6.24}$$

where

$$\frac{y_A}{Q_g} = f_1(x, T, L, k_h, EI) \tag{6.25}$$

and

$$\frac{y_B}{M_g} = f_2(x, T, L, k_h, EI) \tag{6.26}$$

$f_1$ and $f_2$ are two different functions of the same terms. In equations (6.25) and (6.26) there are six terms and two dimensions; force and length are involved. Therefore, following four independent nondimensional terms can be determined (Matlock and Reese, 1962).

$$\frac{y_A EI}{Q_g T^3}, \frac{x}{T}, \frac{L}{T}, \frac{k_h T^4}{EI} \tag{6.27}$$

$$\frac{y_B EI}{M_g T^2}, \frac{x}{T}, \frac{L}{T}, \frac{k_h T^4}{EI} \tag{6.28}$$

Furthermore, the following symbols can be assigned to these nondimensional terms:

$$\frac{y_A EI}{Q_g T^3} = A_y \text{ (deflection coefficient for lateral load)} \tag{6.29}$$

$$\frac{y_B EI}{M_g T^2} = B_y \text{ (deflection coefficient for moment)} \tag{6.30}$$

$$\frac{x}{T} = Z \text{ (depth coefficient)} \tag{6.31}$$

$$\frac{L}{T} = Z_{\max} \text{ (maximum depth coefficient)} \tag{6.32}$$

$$\frac{k_h T^4}{EI} = \phi(x) \text{ (soil modulus function)} \tag{6.33}$$

From equations (6.29) and (6.30), one can obtain:

$$y_x = y_A + y_B = A_y \frac{Q_g T^3}{EI} + B_y \frac{M_g T^2}{EI} \tag{6.34}$$

Similarly, one can obtain expressions for moment $M_x$, slope $S_x$, shear $V_x$, and soil reaction $p_x$ as follows:

$$M_x = M_A + M_B = A_m Q_g T + B_m M_g \tag{6.35}$$

$$S_x = S_A + S_B = A_s \frac{Q_g T^2}{EI} + B_s \frac{M_g T}{EI} \tag{6.36}$$

$$V_x = V_A + V_B = A_v Q_g + B_v \frac{M_g}{T} \tag{6.37}$$

$$p_x = p_A + p_B = A_p \frac{Q_g}{T} + B_p \frac{M_g}{T^2} \tag{6.38}$$

Referring to the basic differential equation (6.9) of beam on elastic foundation and utilizing the principle of superposition, we get:

$$\frac{d^4 y_A}{dx^4} + \frac{k_h y_A}{EI} = 0 \tag{6.39}$$

$$\frac{d^4 y_B}{dx^4} + \frac{k_h}{EI} y_B = 0 \tag{6.40}$$

Substituting for $y_A$ and $y_B$ from equations (6.29) and (6.30), $k_h/EI$ from equation (6.33) and $x/T$ from equation (6.31), we get:

$$\frac{d^4 A_y}{dz^4} + \phi(x)A_y = 0 \tag{6.41}$$

$$\frac{d^4 B_y}{dz^4} + \phi(x)B_y = 0 \tag{6.42}$$

For cohesionless soils where soil modulus is assumed to increase with depth $k_h = n_h x$, $\phi(x)$ may be equated to $Z = x/T$. Therefore, equation (6.33) becomes

$$\frac{n_h x T^4}{EI} = \frac{x}{T} \tag{6.43}$$

This gives

$$T = \left(\frac{EI}{n_h}\right)^{1/5} \tag{6.44}$$

Solutions for equations (6.41) and (6.42), by using finite-difference methods, were obtained by Reese and Matlock (1956) for values of $A_y$, $A_s$, $A_m$, $A_v$, $A_p$, $B_y$, $B_s$, $B_m$, $B_v$, and $B_p$ for various $Z = X/T$.

It has been found that pile deformation is like a rigid body (small curvature) for $Z_{max} = 2$. Therefore, piles with $Z_{max} \leqslant 2$ will behave as rigid piles or poles. Also,

**TABLE 6.3 Coefficient $A$ for Long Piles ($Z_{max} \geqslant 5$): Free Head (Matlock and Reese, 1961, 1962)**

| $Z$ | $A_y$ | $A_s$ | $A_m$ | $A_V$ | $A_P$ |
|---|---|---|---|---|---|
| 0.0 | 2.435 | −1.623 | 0.000 | 1.000 | 0.000 |
| 0.1 | 2.273 | −1.618 | 0.100 | 0.989 | −0.227 |
| 0.2 | 2.112 | −1.603 | 0.198 | 0.956 | −0.422 |
| 0.3 | 1.952 | −1.578 | 0.291 | 0.906 | −0.586 |
| 0.4 | 1.796 | −1.545 | 0.379 | 0.840 | −0.718 |
| 0.5 | 1.644 | −1.503 | 0.459 | 0.764 | −0.822 |
| 0.6 | 1.496 | −1.454 | 0.532 | 0.677 | −0.897 |
| 0.7 | 1.353 | −1.397 | 0.595 | 0.585 | −0.947 |
| 0.8 | 1.216 | −1.335 | 0.649 | 0.489 | −0.973 |
| 0.9 | 1.086 | −1.268 | 0.693 | 0.392 | −0.977 |
| 1.0 | 0.962 | −1.197 | 0.727 | 0.295 | −0.962 |
| 1.2 | 0.738 | −1.047 | 0.767 | 0.109 | −0.885 |
| 1.4 | 0.544 | −0.893 | 0.772 | −0.056 | −0.761 |
| 1.6 | 0.381 | −0.741 | 0.746 | −0.193 | −0.609 |
| 1.8 | 0.247 | −0.596 | 0.696 | −0.298 | −0.445 |
| 2.0 | 0.142 | −0.464 | 0.628 | −0.371 | −0.283 |
| 3.0 | −0.075 | −0.040 | 0.225 | −0.349 | 0.226 |
| 4.0 | −0.050 | 0.052 | 0.000 | −0.106 | 0.201 |
| 5.0 | −0.009 | 0.025 | −0.033 | 0.013 | 0.046 |

deflection coefficients are same for $Z_{max} = 5$ and 10. Therefore, pile length beyond $Z_{max} = 5$ does not change the deflection. In practice, in most cases pile length is greater than $5T$; therefore, coefficients given in Tables 6.3 and 6.4 can be used. Figure 6.11 provides values of $A_y$, $A_m$, and $B_y$ and $B_m$ for different $Z_{max} = L/T$ values.

***Fixed-Head Pile*** For a fixed-head pile, the slope $(S)$ at the ground surface is zero. Therefore, from equation (6.36),

$$S_x = S_A + S_B = A_s \frac{Q_g T^2}{EI} + B_s \frac{M_g T}{EI} = 0 \tag{6.45}$$

Therefore,

$$\frac{M_g}{Q_g T} = -\frac{A_s}{B_s} \quad \text{at } x = 0$$

From Tables 6.3 and 6.4 for $Z = x/T = 0$;

$$A_s/B_s = -\frac{1.623}{1.75} = -0.93$$

Therefore, $M_g/Q_g T = -0.93$. The term $M_g/Q_g T$ has been defined as the *nondimensional fixity factor* by Prakash (1962). Then the equations for deflection

**TABLE 6.4 Coefficient $B$ for Long Piles ($Z_{max} \geqslant 5$): Free Head (Matlock and Reese, 1961, 1962)**

| $Z$ | $B_y$ | $B_S$ | $B_m$ | $B_V$ | $B_P$ |
|---|---|---|---|---|---|
| 0.0 | 1.623 | −1.750 | 1.000 | 0.000 | 0.000 |
| 0.1 | 1.453 | −1.650 | 1.000 | −0.007 | −0.145 |
| 0.2 | 1.293 | −1.550 | 0.999 | −0.028 | −0.259 |
| 0.3 | 1.143 | −1.450 | 0.994 | −0.058 | −0.343 |
| 0.4 | 1.003 | −1.351 | 0.987 | −0.095 | −0.401 |
| 0.5 | 0.873 | −1.253 | 0.976 | −0.137 | −0.436 |
| 0.6 | 0.752 | −1.156 | 0.960 | −0.181 | −0.451 |
| 0.7 | 0.642 | −1.061 | 0.939 | −0.226 | −0.449 |
| 0.8 | 0.540 | −0.968 | 0.914 | −0.270 | −0.432 |
| 0.9 | 0.448 | −0.878 | 0.885 | −0.312 | −0.403 |
| 1.0 | 0.364 | −0.792 | 0.852 | −0.350 | −0.364 |
| 1.2 | 0.223 | −0.629 | 0.775 | −0.414 | −0.268 |
| 1.4 | 0.112 | −0.482 | 0.688 | −0.456 | −0.157 |
| 1.6 | 0.029 | −0.354 | 0.594 | −0.477 | −0.047 |
| 1.8 | −0.030 | −0.245 | 0.498 | −0.476 | 0.054 |
| 2.0 | −0.070 | −0.155 | 0.404 | −0.456 | 0.140 |
| 3.0 | −0.089 | 0.057 | 0.059 | −0.213 | 0.268 |
| 4.0 | −0.028 | 0.049 | −0.042 | 0.017 | 0.112 |
| 5.0 | 0.000 | 0.011 | −0.026 | 0.029 | −0.002 |

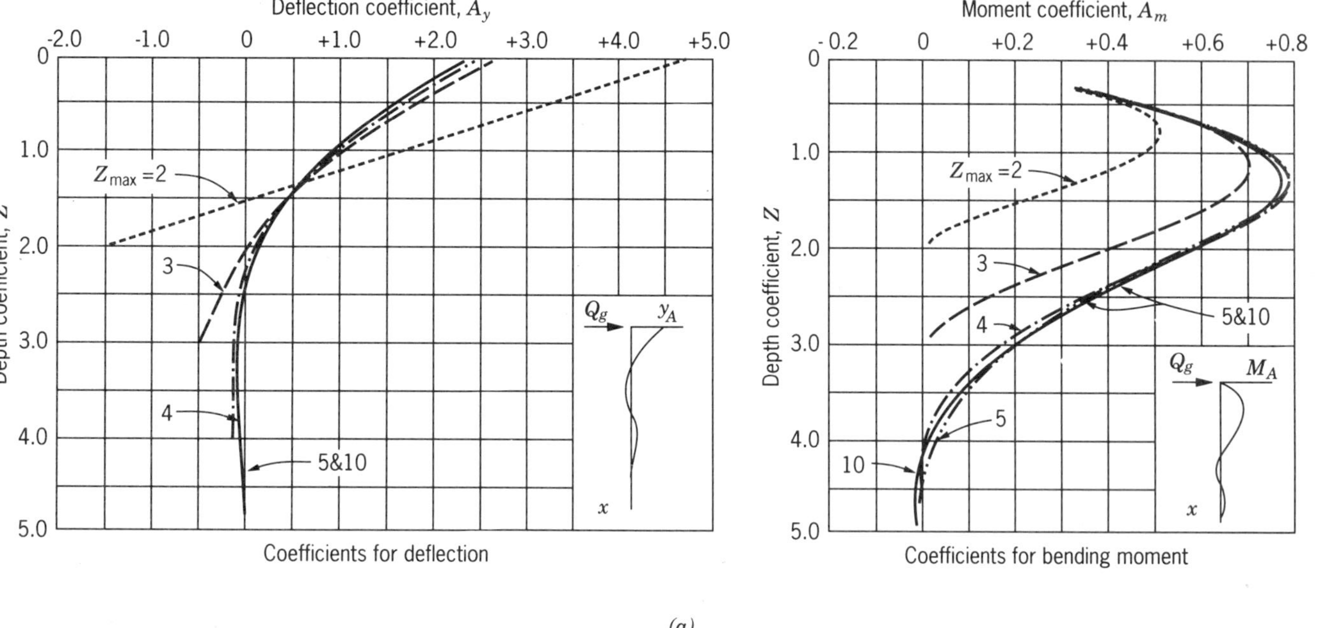
Deflection coefficient, $A_y$
-2.0
-1.0
0
+1.0
+2.0
+3.0
+4.0
+5.0
0
1.0
2.0
3.0
4.0
5.0
Depth coefficient, $Z$
$Z_{max}$ =2
3
4
5&10
$Q_g$
$y_A$
$x$
Coefficients for deflection
Moment coefficient, $A_m$
- 0.2
0
+0.2
+0.4
+0.6
+0.8
0
1.0
2.0
3.0
4.0
5.0
Depth coefficient, $Z$
$Z_{max}$ =2
3
4
5&10
5
10
$Q_g$
$M_A$
$x$
Coefficients for bending moment

*(a)*

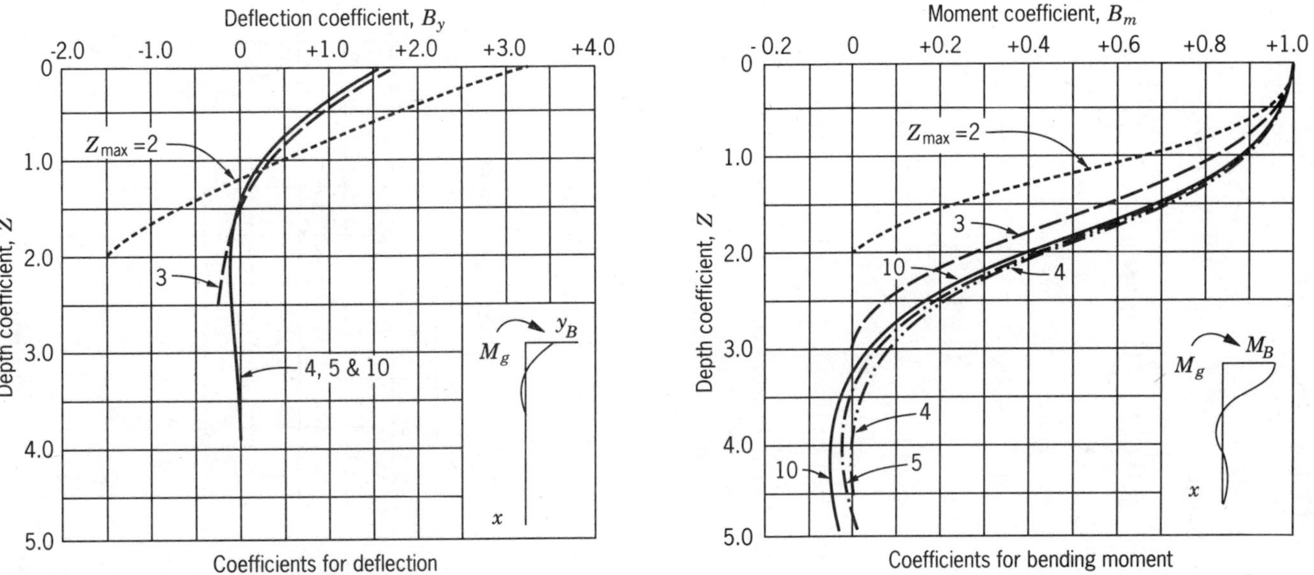


*(b)*

**Figure 6.11** Coefficients for free-headed piles in cohesionless soil (a) Free head, (b) fixed (Ft) head (Reese and Matlock, 1956).

and moment for fixed head can be modified as follows:
From equation (6.34),

$$y_x = A_y \frac{Q_g T^3}{EI} + B_y \frac{M_g T^2}{EI}$$

substituting $M_g = -0.93\, Q_g T$ for fixed head, we get

$$y_x = (A_y - 0.93 B_y) \frac{Q_g T^3}{EI}$$

or

$$y_x = C_y \frac{Q_g T^3}{EI} \tag{6.46}$$

similarly,

$$M_x = C_m Q_g T \tag{6.47}$$

values of $C_y$ and $C_m$ can be obtained from Figure 6.12.

***Partially Fixed Pile Head*** In cases where the piles undergo some rotation at the joints of their head and the cap, these are called partially fixed piles. In such a situation, the coefficient $C$ needs modification as follows:

$$C_y = (A_y - 0.93\lambda B_y) \tag{6.48}$$

$$C_m = (A_m - 0.93\lambda B_m) \tag{6.49}$$

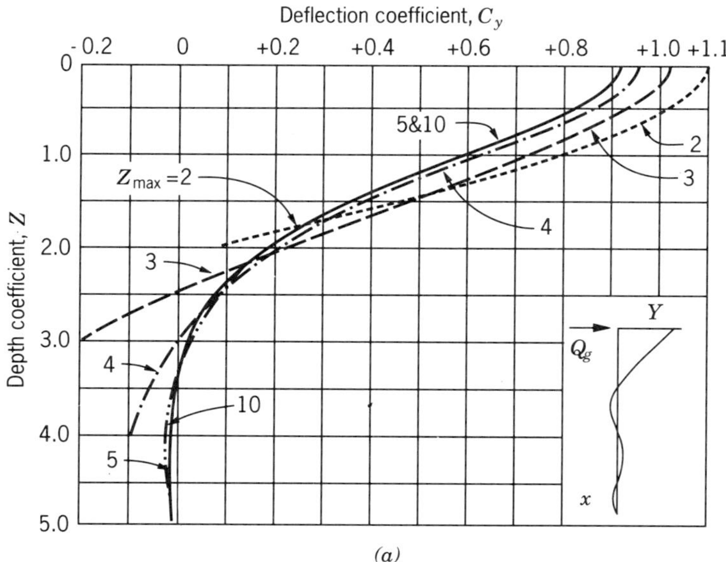


(a)

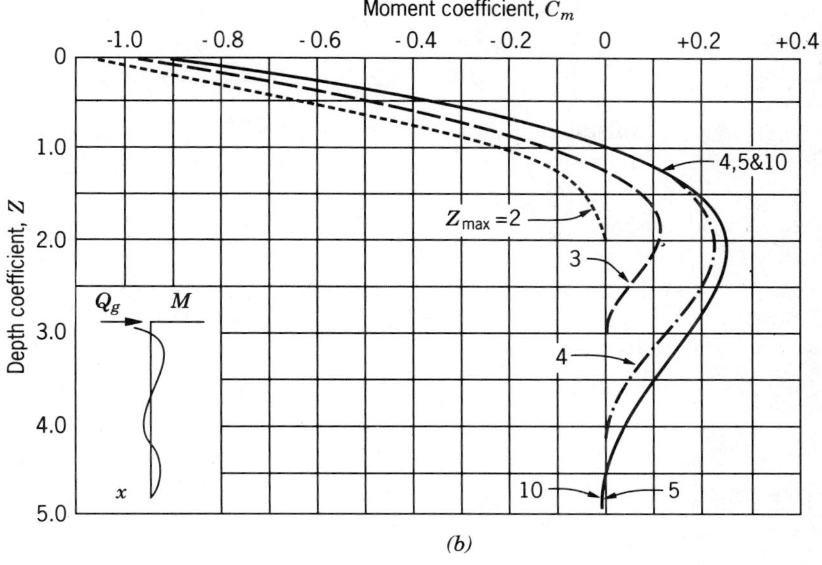


(b)

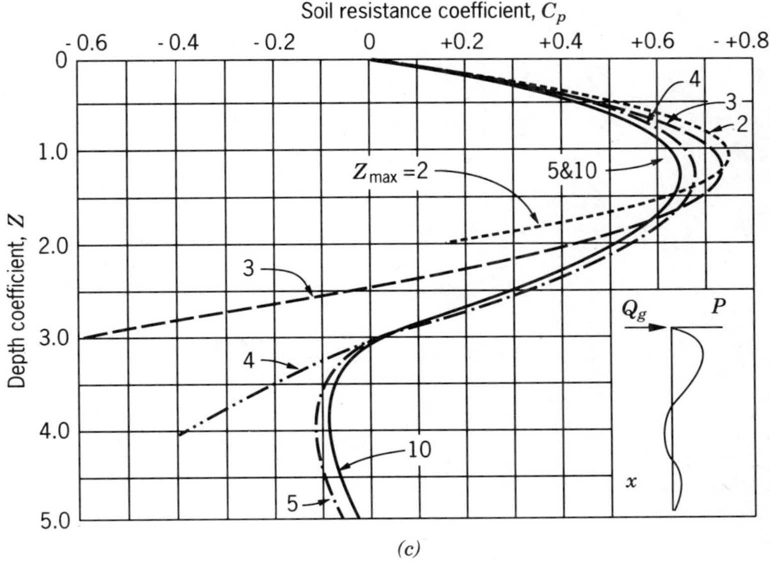


(c)

**Figure 6.12** Deflection, moment, and soil reaction coefficients for fixed-head (Ft) piles subjected to lateral load (a) Deflections, (b) bending moments, (c) soil reaction. (Reese and Matlock, 1956).

where $\lambda$ is percent fixity (i.e., $\lambda = 1$ for 100 percent fixity or fully restrained pile head and $\lambda = 0$ for fully free pile head). At intermediate fixity levels, proper $\lambda$ can be taken (e.g., $\lambda = 0.5$ for 50 percent fixity and $\lambda = 0.25$ for 25 percent fixity).

***Example 6.3*** A 3/4-in. (19.0 mm) thick, 10-in. (254 mm) inside diameter, concrete filled, 56.25-ft (17.15 m)-long pipe pile was installed as a closed-ended friction pile in loose sand. Calculate the following:

(a) Allowable lateral load for 0.25 in. (6.35 mm) deflection at the pile head, which is free to rotate

(b) Maximum bending moment for this load

(c) Allowable load if the pile head is (i) fully fixed and (ii) 50 percent fixed.

Assume that the modulus of elasticity $E$ for concrete is $3.6 \times 10^6$ psi (25,00 MPa) and for steel is $30 \times 10^6$ psi (208,334 MPa).

SOLUTION

Calculation of $T$:

Since the pile is made of two materials steel pipe and the concrete core, we will need to transform the section into the equivalent of one material. Let us transform all of the materials into concrete. Concrete thickness $t_c = n \times$ steel thickness $t_s$, where $n$ is modular ratio $(E_s/E_c)$

$$t_c = \frac{E_s}{E_c} t_s = \frac{30 \times 10^6}{3.6 \times 10^6} \times 3/4 = 6.2 \text{ in.}$$

Equivalent diameter of composite section in terms of concrete $= 10 + 6.2 + 6.2 = 22.4$ inch.

$$I = \frac{\pi B^4}{64} = \frac{\pi (22.4)^4}{64} = 12358.4 \text{ in.}^4$$

$$EI = 3.6 \times 10^6 \times 12358.4 = 44.49 \times 10^9 \text{ lb-in.}^2 (\equiv 308.96 \times 10^3 \text{ kips-ft}^2)$$

From Table 4.16a, $n_h = 20 \text{ lb/in.}^3$ for loose sand

$$T = \left(\frac{EI}{n_h}\right)^{0.2} = 73.44 \text{ in.} \ (\equiv 6.12 \text{ ft})$$

$$\frac{L}{T} = \frac{56.25}{6.12} = 9.2 > 4, \quad \text{therefore it is a long pile}$$

(a) Allowable lateral load for a 0.25-in. deflection at the top of a free-head pile: From equation (6.34)

$$y_x = A_y \frac{Q_g T^3}{EI} + B_y \frac{M_g T^2}{EI} \tag{6.34}$$

where

$M = 0$, since there is no moment on pile head
$T = 6.12\,\text{ft}$
$y = 0.25/12 = 0.02\,\text{ft}$
$EI = 308.96 \times 10^3\,\text{kips-ft}^2$

Also, since $L/T > 5$, Table 6.3 can be used. $A_y = 2.435$ for $Z = 0$ at ground level. Substituting these values in equation (6.34), we get:

$$0.02 = 2.435 \frac{Q_g(6.12)^3}{308.96 \times 10^3}$$

$$Q_g = 11\,\text{kips}$$

(b) Maximum bending moment for this lateral load:
From equation (6.35)

$$M_x = A_m Q_g T + B_m M_g \tag{6.35}$$

From Table 6.3, the maximum $A_m = 0.772$ at $Z = 1.4$, $Q_g = 11\,\text{kips}$, $T = 6.12\,\text{ft}$, $M_g = 0$.

$$M_{\text{max}} = 0.772 \times 11 \times 6.12 = 51.9\,\text{kips-ft at a depth of } x = 1.4 \times 6.12$$

or $x/T = 1.4$ equal to 8.6 ft below ground level

(c) Allowable lateral load if pile is fully fixed and 50% fixed at its head:

*Fully Fixed Head*
From Equation (6.46)

$$y_x = C_y \frac{Q_g T^3}{EI} \tag{6.46}$$

where $C_y$ can either be obtained from Figure 6.12 or $C_y = (A_y - 0.93\lambda B_y)$. $\lambda = 1$ for 100% fixity values of $A_y$ and $B_y$ at the ground surface are:

$$A_y = 2.435 \quad \text{from Table 6.3}$$

$$B_y = 1.623 \quad \text{from Table 6.4}$$

Then,

$$C_y = (2.435 - 0.93 \times 1.623) = 0.926$$

As a check from Figure 6.12a for $z = x/T = 0$, $L/T = 9.2$, $C_y = 0.93$, which is close to above. Then substituting the values of $y = 0.02\,\text{ft}$, $C_y = 0.926$, $T = 6.12\,\text{ft}$,

$EI = 308.96 \times 10^3$ in equation (6.46), we get

$$Q_g = \frac{0.02 \times 308.96 \times 10^3}{0.926(6.12)^3} = 29.1 \text{ kips}$$

*50% Fixity, $\lambda = 0.5$*

$$C_y = (2.435 - 0.93 \times 0.5 \times 1.623) = 1.68$$

Then, following the procedure for the fully fixed head,

$$Q_g = \frac{0.02 \times 308.96 \times 10^3}{1.68(6.12)^3} = 16 \text{ kips}$$

### 6.1.4 Application of $p$–$y$ Curves to Cohesionless Soils

Lateral capacity of piles calculated by the subgrade reaction approach can be extended beyond the elastic range where soil yields plastically. This can be done by employing $p$–$y$ curves (Matlock, 1970; Reese et al., 1974; Reese and Welch, 1975; Bhushan et al., 1979). In the following paragraphs, first the theoretical basis for the use of $p$–$y$ curves are explained, then the procedure of establishing $p$–$y$ curves is be described. A step-by-step iterative design procedure for a pile under lateral load is then developed.

***Theoretical Basis*** The differential equation for the laterally loaded piles, assuming that the pile is a linearly elastic beam, is as follows:

$$EI\frac{d^4y}{dx^4} + P\frac{d^2y}{dx^2} - p = 0 \tag{6.50a}$$

where $EI$ is flexural rigidity of the pile, $y$ is the lateral deflection of the pile at point $x$ along the pile length, $P$ is axial load on pile, and $p$ is soil reaction per unit length. $p$ is expressed by equation (6.50b).

$$p = ky \tag{6.50b}$$

where $k$ is the soil modulus.

The solution for equation (6.50a) can be obtained if the soil modulus $k$ can be expressed as a function of $x$ and $y$. The numerical description of the soil modulus is best accomplished by a family of curves that show the soil reaction $p$ as a function of deflection $y$ (Reese and Welch, 1975). In general, these curves are nonlinear and depend on several parameters, including depth, soil shear strength, and number of load cycles (Reese, 1977).

A concept of $p$–$y$ curves is presented in Figure 6.13. These curves are assumed to have the following characteristics:

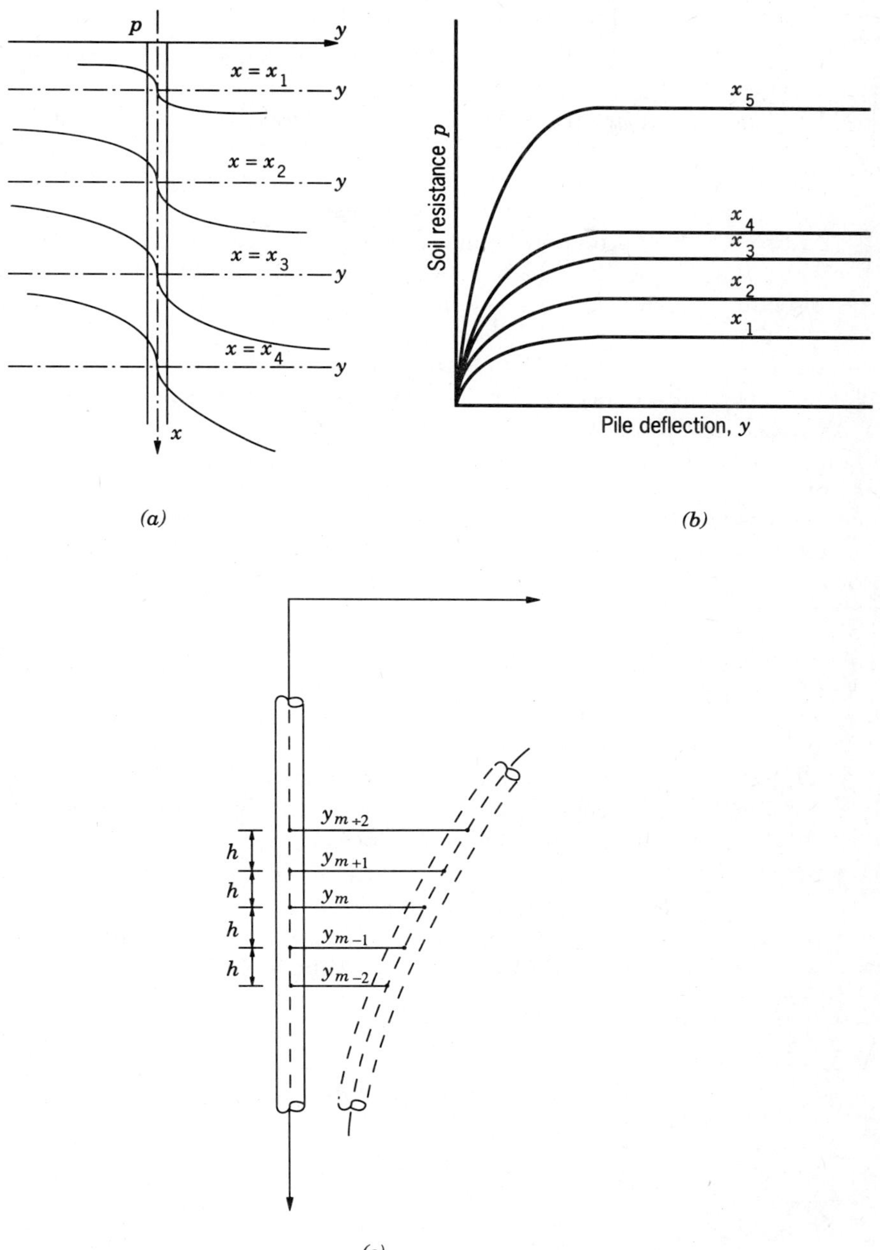


**Figure 6.13** Set of $p$–$y$ curves and representation of deflected pile. (a) Shape of curves at various depths $x$ below soil surface, (b) curves plotted on common axes, (c) representation of deflected pile.

1. A set of $p$–$y$ curves represent the lateral deformation of soil under a horizontally applied pressure on a discrete vertical section of pile at any depth.
2. The curve is independent of the shape and stiffness of the pile and is not affected by loading above and below the discrete vertical area of soil at that depth. This assumption, of course, is not strictly true. However, experience indicates that pile deflection at a depth can, for practical purposes, be assumed to be essentially dependent only on soil reaction at that depth. Thus, the soil can be replaced by a mechanism represented by a set of discrete $p$–$y$ characteristics as shown in figure 6.13b.

Thus, as shown in Figure 6.13a, a series of $p$–$y$ curves would represent the deformation of soil with depth for a range of lateral pressures varying from zero to the yield strength of soil. This figure also presents deflected pile shape (Figure 6.13c) and $p$–$y$ curves when plotted on a common axis (Figure 6.13b). At present, the application of $p$–$y$ curves is widely used to design laterally loaded piles and has been adopted in API Recommended Practice (1982).

Once a set of $p$–$y$ curves has been established for a soil–pile system, the problem of laterally loaded piles can be solved by an iterative procedure consisting of the following steps:

1. As described earlier, calculate $T$ or $R$, as the case may be, for the soil–pile system with an estimated or given value of $n_h$ or $k$. $T$ will apply for cohesionless soils and normally consolidated clays, and $R$ will apply to overconsolidated clays.
2. With the calculated $T$ or $R$ and the imposed lateral force $Q_g$ and moment $M_g$, determine deflection $y$ along the pile length by Reese and Matlock (1956) or Davisson and Gill (1963) procedures, as applicable. These procedures have been described in Section 6.1.3 and 6.6.1, respectively.
3. For these calculated deflections (step (2) above), determine the lateral pressure $p$ with depth from the earlier established $p$–$y$ curves. The soil modulus and relative stiffness ($R$ or $T$) will then be determined as:

$$k = p/y$$

(a) $n_h = \dfrac{k}{x}$, $\quad T = \left(\dfrac{EI}{n_h}\right)^{1/5}$ for modulus increasing with depth

(b) $k_1 = k$, $\quad R = \left(\dfrac{EI}{k}\right)^{1/4}$ for modulus constant with depth

Compare the ($R$ or $T$) value with those calculated in step (1). If these values do not match carry out a second trial as outlined in the following steps.

4. Assume $k$ or $n_h$ value closer to the one in step (3). Then repeat steps (2) and (3) and obtain new $R$ or $T$. Continue the process until calculated and

assumed values agree. Then, deflections and moments along the pile section can be established for the final $R$ or $T$ value.

Reese (1977) provides a computer program documentation that solves for deflection and bending moment for a pile under lateral loading. A step-by-step procedure has been provided here to establish $p$–$y$ curves for cohesionless soils. A numerical example has also been given to explain the procedure to establish $p$–$y$ curves. This step-by-step procedure and numerical example will help design engineers to solve such problems either manually or by using electronic calculators or microcomputers.

Methods to establish $p$–$y$ curves for cohesionless soils will now be presented. Methods of $p$–$y$ determination for soft and stiff overconsolidated clays are discussed in Section 6.6.2.

***Procedure for Establishing p–y Curves for Laterally Loaded Piles in Cohesionless Soils*** For the solution of the problem of a laterally loaded pile, it is necessary to predict a set of $p$–$y$ curves. If such a set of curves can be predicted, Equation 6.50 can readily be solved to yield pile deflection, pile rotation, bending moment, and shear and soil reaction for any load capable of being sustained by the pile.

The set of curves shown in Figure 6.13a would seem to imply that the behavior of the soil at a particular depth is independent of the soil behavior at all other depths. This is not strictly true. However, Matlock (1970) showed that for the patterns of pile deflections that can occur in practice, the soil reaction at a point is essentially dependent on the pile deflection at that point only. Thus, for purposes of analysis, the soil can be removed and replaced by a set of discrete closely spaced independent and elastic springs with load-deflection characteristics as in Figure 6.6b.

Cox et al. (1971) performed lateral loads tests in the field on full-sized piles, which were instrumented for the measurement of bending moment along the length of the piles. In addition to the measurement of the load at the ground line, measurements were made of pile-head deflection and pile-head rotation. Loadings were static and cyclic. For each type of loading, a series of lateral loads were applied, beginning with a load of small magnitude, and a bending moment curve was obtained for each load.

The sand at the test site varied from clean fine sand to silty fine sand, both having high relative densities. The sand particles were subangular with a large percentage of flaky grains. The angle of internal friction $\phi'$ was 39° and $\gamma'$ was 66 lb/ft$^3$ (1057 kg/m$^3$).

From the sets of experimental bending moment curves, values of $p$ and $y$ at points along the pile can be obtained by integrating and differentiating the bending moment curves twice to obtain deflections and soil reactions, respectively. Appropriate boundary conditions were used and the equations were solved numerically.

The $p$–$y$ curves so obtained were critically studied and form the basis for the following procedure for developing $p$–$y$ curves in cohesionless soils (Reese et al., 1974).

Step 1 Carry out field or laboratory tests to estimate the angle of internal friction ($\phi$) and unit weight ($\gamma$) for the soil at the site.

Step 2 Calculate the following factors:

$$\alpha = \tfrac{1}{2}\phi \tag{6.51}$$

$$\beta = 45 + \alpha \tag{6.52}$$

$$K_0 = 0.4 \tag{6.53}$$

$$K_A = \tan^2(45 - \tfrac{1}{2}\phi) \tag{6.54}$$

$$p_{cr} = \gamma x \left[ \frac{k_0 x \tan\phi \sin\beta}{\tan(\beta - \phi)\cos\alpha} + \frac{\tan\beta}{\tan(\beta - \phi)}(B + x\tan\beta\tan\alpha) + K_0 x \tan\beta(\tan\phi\sin\beta - \tan\alpha) - K_A B \right] \tag{6.55}$$

$$p_{cd} = K_A B\gamma x(\tan^8\beta - 1) + K_0 B\gamma x \tan\phi \tan^4\beta \tag{6.56}$$

$p_{cr}$ is applicable for depths from ground surface to a *critical depth* $x_r$ and $p_{cd}$ is applicable below the critical depth. The value of critical depth is obtained by plotting $p_{cr}$ and $p_{cd}$ with depth ($x$) on a common scale. The point of intersection of these two curves will give $x_r$ as shown on Figure 6.14a.

Equations 6.55 and 6.56 are derived for failure surface in front of a pile shown in Figure 1.16a for shallow depth and 1.16b for depths below the critical depth ($x_r$).

Step 3 First select a particular depth at which a $p$–$y$ curve will be drawn. Compare this depth ($x$) with the critical depth ($x_r$) obtained in step (2) above and then find if the value of $p_{cr}$ or $p_{cd}$ is applicable. Then carry out calculations for a $p$–$y$ curve discussed as follows. Refer to Figure 6.14b when following these steps.

Step 4 Select appropriate $n_h$ from Table 4.16a for the soil. Calculate the following items:

$$p_m = B_1 p_c \tag{6.57}$$

where $B_1$ is taken from Table 6.5 and $p_c$ is from equation (6.55) for depths above critical point and from equation (6.56) for depths below the critical point

$$y_m = \frac{B}{60} \tag{6.58}$$

where $B$ is the pile width

$$p_u = A_1 p_c \tag{6.59}$$

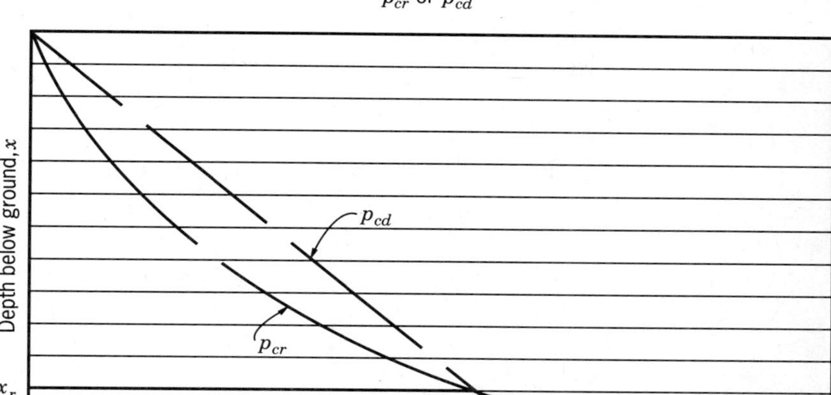


(a)

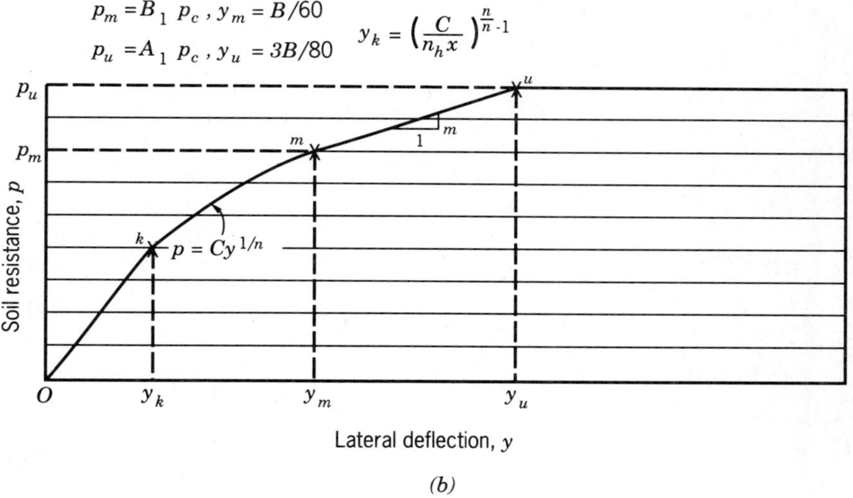


(b)

**Figure 6.14** Obtaining the value of $x_r$ and establishing $p$–$y$ curve. (a) Obtaining the value of $x_r$ at the intersection of $p_{cr}$ and $p_{cd}$, (b) establishing the $p$–$y$ curve.

and where $A_1$ is taken from Table 6.5

$$y_u = \frac{3B}{80} \tag{6.60}$$

$$m = \frac{p_u - p_m}{y_u - y_m} \tag{6.61}$$

$$n = \frac{p_m}{m y_m} \tag{6.62}$$

**TABLE 6.5 Values for Coefficients $A_1$ and $B_1$**

| $\frac{x}{B}$ | $A_1$ | | $B_1$ | |
|---|---|---|---|---|
| | Static | Cyclic | Static | Cyclic |
| 1 | 2 | 3 | 4 | 5 |
| 0 | 2.85 | 0.77 | 2.18 | 0.50 |
| 0.2 | 2.72 | 0.85 | 2.02 | 0.60 |
| 0.4 | 2.60 | 0.93 | 1.90 | 0.70 |
| 0.6 | 2.42 | 0.98 | 1.80 | 0.78 |
| 0.8 | 2.20 | 1.02 | 1.70 | 0.80 |
| 1.0 | 2.10 | 1.08 | 1.56 | 0.84 |
| 1.2 | 1.96 | 1.10 | 1.46 | 0.86 |
| 1.4 | 1.85 | 1.11 | 1.38 | 0.86 |
| 1.6 | 1.74 | 1.08 | 1.24 | 0.86 |
| 1.8 | 1.62 | 1.06 | 1.15 | 0.84 |
| 2.0 | 1.50 | 1.05 | 1.04 | 0.83 |
| 2.2 | 1.40 | 1.02 | 0.96 | 0.82 |
| 2.4 | 1.32 | 1.00 | 0.88 | 0.81 |
| 2.6 | 1.22 | 0.97 | 0.85 | 0.80 |
| 2.8 | 1.15 | 0.96 | 0.80 | 0.78 |
| 3.0 | 1.05 | 0.95 | 0.75 | 0.72 |
| 3.2 | 1.00 | 0.93 | 0.68 | 0.68 |
| 3.4 | 0.95 | 0.92 | 0.64 | 0.64 |
| 3.6 | 0.94 | 0.91 | 0.61 | 0.62 |
| 3.8 | 0.91 | 0.90 | 0.56 | 0.60 |
| 4.0 | 0.90 | 0.90 | 0.53 | 0.58 |
| 4.2 | 0.89 | 0.89 | 0.52 | 0.57 |
| 4.4 to 4.8 | 0.89 | 0.89 | 0.51 | 0.56 |
| 5 and more | 0.88 | 0.88 | 0.50 | 0.55 |

[a] All these values have been obtained from the curves provided by Reese et al. (1974).

$$C = \frac{p_m}{(y_m)^{1/n}} \tag{6.63}$$

$$y_k = \left(\frac{C}{n_h x}\right)^{n/(n-1)} \tag{6.64}$$

$$p = C y^{1/n} \tag{6.65}$$

Step 5 (i) Locate $y_k$ on the $y$ axis in Figure 6.14b. Substitute this value of $y_k$ as $y$ in equation (6.65) to determine the corresponding $p$ value. This $p$ value will define the $k$ point. Joint point $k$ with origin $O$; thus establishing line $OK$ (Figure 6.14b)

(ii) Locate the point $m$ for the values of $y_m$ and $p_m$ from equations 6.58 and 6.57 respectively.

(iii) Then plot the parabola between the points $k$ and $m$ by using equation (6.55).

(iv) Locate point $u$ from the values of $y_u$ and $p_u$ from equations (6.60) and (6.59), respectively

(v) Join points $m$ and $u$ with a straight line.

Step 6 Repeat the above procedure for various depths to obtain $p$–$y$ curves at each depth below ground.

***Example 6.4*** A 40-ft (12.2 m) long, 30-in. (762 mm) outside diameter and 1-in. (25.4 mm) wall thickness steel pipe pile is driven into compact sand with $\phi = 36°$ and unit weight $(\gamma) = 125\,\text{lb/ft}^3$ $(2000\,\text{kg/m}^3)$ and $n_h = 52\,\text{lb/in}^3$. $(14.13 \times 10^3\,\text{kN/m}^3)$. Draw the $p$–$y$ curves at 2 ft (0.6 m), 4 ft (1.2 m), and 10 ft (3.0 m) below ground surface.

SOLUTIONS

Step 1 As already given, $\phi = 36°$ and $\gamma = 125\,\text{lb/ft}^3$

Step 2 $\alpha = \dfrac{36}{2} = 18°$ (equation (6.51))

$$\beta = 45 + 18 = 63 \quad \text{(equation (6.52))}$$

$$K_0 = 0.4 \quad \text{(equation (6.53))}$$

$$K_A = \tan^2(45 - 18) = 0.259 \quad \text{(equation (6.54))}$$

$$p_{cr} = 125x\left[\frac{0.4x \tan 36 \sin 63}{\tan(63-36)\cos 18} + \frac{\tan 63}{\tan(63-36)}\left(\frac{30}{12} + x\tan 63 \tan 18\right)\right.$$

$$\left. + 0.4x \tan 63(\tan 36 \sin 63 - \tan 18) - 0.259 \times \frac{30}{12}\right] \quad \text{(equation (6.55))}$$

$$= 125x[0.534x + 9.636 + 2.457x + 0.252x - 0.647]$$
$$= 405.375x^2 + 1123.625x$$

Then, various values of $x$ and $p_{cr}$ can be calculated as given below:

$$x = 0, \qquad p_{cr} = 0$$
$$= 2', \qquad p_{cr} = 3.867 \text{ kips/ft}$$
$$= 4', \qquad p_{cr} = 10.976 \text{ kips/ft}$$
$$= 10', \qquad p_{cr} = 51.76 \text{ kips/ft}$$
$$= 20', \qquad p_{cr} = 184.46 \text{ kips/ft}$$

$$p_{cd} = 0.259 \times \frac{30}{12} \times 125x(\tan^8 63 - 1) + 0.4 \times \frac{30}{12} \times 125x \tan 36 \tan^4 63 \qquad \text{(equation (6.56))}$$
$$= 17{,}735.592x + 1346.367x = 19{,}081.959x$$

For various values of $x, p_{cd}$ can be calculated as follows:

$$x = 0, \qquad p_{cd} = 0$$
$$= 4', \qquad p_{cd} = 76.327 \text{ kips/ft}$$
$$= 10', \qquad p_{cd} = 190.819 \text{ kips/ft}$$
$$= 20', \qquad p_{cd} = 381.639 \text{ kips/ft}$$

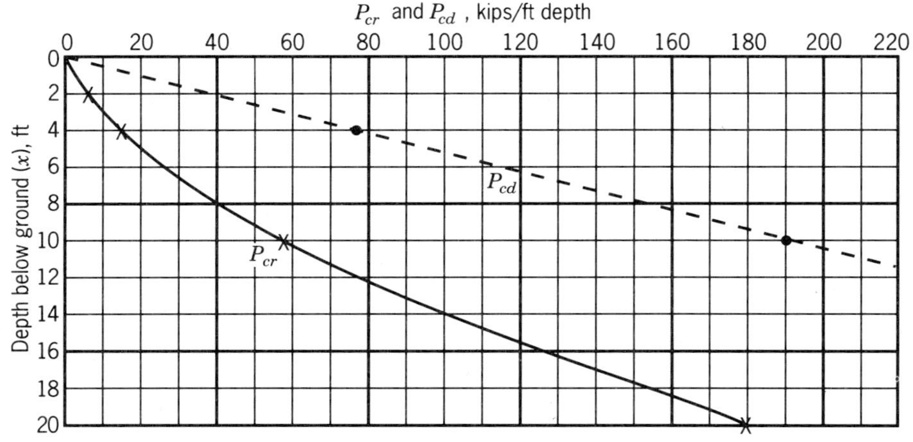


**Figure 6.15** Values of $p_{cr}$ and $p_{cd}$ with depth (Example 6.4).

Values of $p_{cr}$ and $p_{cd}$ are plotted against depth in Figure 6.15. These do not intersect up to 20 ft depth. Therefore, over the range of depth considered here (up to 20 ft), only the values of $p_{cr}$ will be applicable to the $p$–$y$ curves.

Step 3 Select the depth $x = 2\,\text{ft}$
Step 4 $n_n = 52\,\text{lb/in.}^3 = 90\,\text{kips/ft}^3$

From Table 6.5, $B_1 = 1.7$ for $\frac{x}{B} = \frac{2 \times 12}{30} = 0.8$ and for static loading condition.
From step (2), $p_c = 3.867$ kips/ft depth of pile. Substituting these values in equation (6.57), we get:

$$p_m = 1.7 \times 3.867 = 6.574\,\text{kips/ft depth of pile}$$

$$y_m = \frac{B}{60} = \frac{30}{12 \times 60} = 0.0416\,\text{ft} = 41.6 \times 10^{-3}\,\text{ft} \quad \text{(equation (6.58))}$$

Also, from Table 6.5, $A_l = 2.2$ for $x/B = 0.8$ and static conditions. Then

$$p_u = 2.2 \times 3.867 = 8.507\,\text{kips/ft} \quad \text{(equation (6.59))}$$

$$y_u = \frac{3B}{80} = \frac{3 \times 30}{12 \times 80} = 0.0937\,\text{ft} = 93.7 \times 10^{-3}\,\text{ft} \quad \text{(equation (6.60))}$$

$$m = \frac{8.507 - 6.574}{0.0937 - 0.0416} = \frac{1.933}{0.0521} = 37.1 \quad \text{(using equation (6.61))}$$

$$n = \frac{6.574}{37.1 \times .0416} = 4.26 \quad \text{(using equation (6.62))}$$

$$C = \frac{6.574}{(0.0416)^{1/4.26}} = \frac{6.574}{0.474} = 13.869 \quad \text{(From equation (6.63))}$$

$$y_k = \left(\frac{13.869}{90 \times 2}\right)^{4.26/3.26} = (0.077)^{1.306} = 35.16 \times 10^{-3}\,\text{ft} \quad \text{(equation (6.64))}$$

$$p = 13.869\,(y)^{1/4.26} = 13.869\,y^{0.2347} \quad \text{(from equation (6.65))}$$

Select two values of $y$ in between $y_k$ and $y_m$ and obtain $p$ value from above relationship of $p$ and $y$.

$$y = 37 \times 10^{-3}\,\text{ft}, \qquad p = 6.397\,\text{kips/ft}$$

$$= 40 \times 10^{-3}\,\text{ft}, \qquad p = 6.516\,\text{kips/ft}$$

$$y_m = 41.6 \times 10^{-3}\,\text{ft}, \qquad p_m = 6.574\,\text{kips/ft}$$

$$y_u = 93.7 \times 10^{-3}\,\text{ft}, \qquad p_u = 8.507\,\text{kips/ft}$$

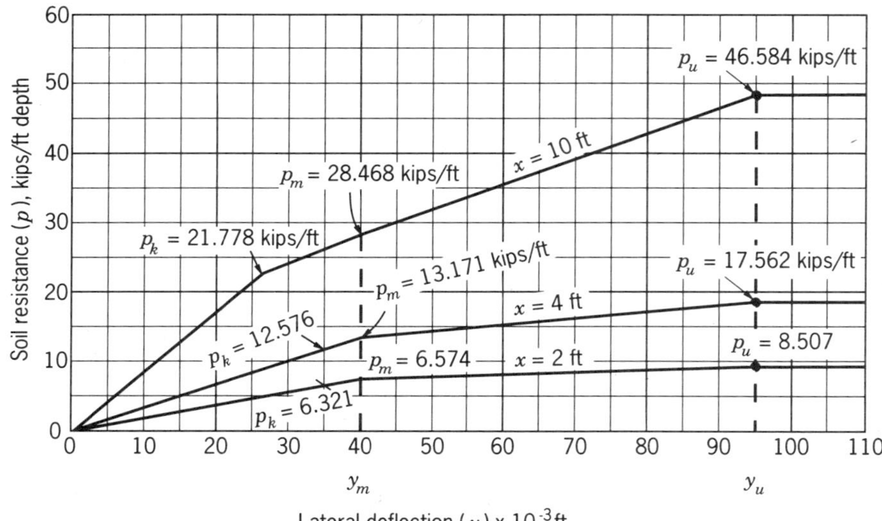


**Figure 6.16** $p$–$y$ curves at different depths (Example 6.4).

Step 5

(i) Locate $y_k = 35.16 \times 10^{-3}$ ft in Figure 6.16. Corresponding $p$ value from equation 6.65 is $p_k = 13.869(35.16 \times 10^{-3})^{0.2347} = 6.321$ kips/ft. Join this $p_k, y_k$ point to (0.0).

(ii) Locate point $m$ for $y_m = 41.6 \times 10^{-3}$ and $p_m = 6.574$ kips/ft.

(iii) Plot the parabola between points $k$ amd $m$ by using $y$ and $p$ values calculated in setp (4).

(iv) Locate point $u$ at $y_u = 93.7 \times 10^{-3}$ ft and $p_u = 8.507$ kips/ft.

(v) Join points $m$ and $u$ with a straight line. The $p$–$y$ curve for $x = 2$ ft is plotted on Figure 6.16.

Step 6 For $x = 4$ ft, $x/B = \dfrac{4 \times 12}{30} = 1.6, B_1 = 1.24$ (Table 6.5)

$$p_c = 10.976 \text{ kips/ft}, p_m = 1.24 \times 10.976 = 13.171 \text{ kips/ft}$$

$$y_m = B/60 = 41.6 \times 10^{-3} \text{ ft}, A_1 = 1.74 \text{ (Table 6.5)}$$

$$p_u = 1.74 \times 10.976 = 17.562 \text{ kips/ft}, y_u = 93.7 \times 10^{-3} \text{ ft}$$

$$m = \frac{(17.562 - 13.171)}{(93.7 - 41.6)10^{-3}} = \frac{4.391}{52.1 \times 10^{-3}} = 84.28$$

$$n = \frac{13.171}{84.28 \times 41.6 \times 10^{-3}} = 3.756 \qquad C = \frac{13.171}{(41.6 \times 10^{-3})^{1/3.756}} = 30.70$$

$$y_k = \left(\frac{30.7}{90 \times 4}\right)^{3.756/2.756} = 34.9 \times 10^{-3}$$

$$p = 30.70(y)^{1/3.756} = 30.70(y)^{0.266}$$

$y = y_k = 34.9 \times 10^{-3}$ ft $\qquad p_k = 12.576$ kips/ft

$\quad = 37 \times 10^{-3}$ ft $\qquad p = 12.773$ kips/ft

$y_m = 41.6 \times 10^{-3}$ ft $\qquad p_m = 13.171$ kips/ft

$y_u = 93.7 \times 10^{-3}$ ft $\qquad p_u = 17.562$ kips/ft

For $x = 10$ ft $\qquad x/B = \dfrac{10 \times 12}{30} = 4 \qquad B_1 = 0.53$ (Table 6.5)

$p_c = 51.76$ kips/ft $\qquad p_m = 0.53 \times 51.76 = 28.468$ kips/ft

$y_m = 41.6 \times 10^{-3}$ ft $\qquad A_1 = 0.9 \qquad p_u = 0.9 \times 51.76 = 46.584$ kips/ft

$y_u = 93.7 \times 10^{-3}$ ft $\qquad m = \dfrac{(46.584 - 28.468)}{(93.7 - 41.6)10^{-3}} = 343.757$

$$n = \frac{28.468}{343.757 \times 41.6 \times 10^{-3}} = 1.991 \qquad C = \frac{28.468}{(41.6 \times 10^{-3})^{0.502}} = 141.632$$

$$y_k = \left(\frac{141.632}{90 \times 10}\right)^{1.991/0.991} = 0.0247 \text{ ft} = 24.7 \times 10^{-3} \text{ ft}$$

$$p = 141.632(y)^{1/1.991} = 141.632(y)^{0.502}$$

$y = y_k = 24.7 \times 10^{-3}$ ft $\qquad p_k = 21.778$ kips/ft

$\quad = 30 \times 10^{-3}$ ft $\qquad p = 24.359$ kips/ft

$\quad = 35 \times 10^{-3}$ ft $\qquad p = 26.319$ kips/ft

$\quad = y_m = 41.6 \times 10^{-3}$ ft $\qquad p_m = 28.468$ kips/ft

$y_u = 93.7 \times 10^{-3}$ ft $\qquad p_u = 46.584$ kips/ft

Figure 6.16 shows the $p$–$y$ curves for these three depths $x = 2', 4'$, and $10'$, respectively.

### 6.1.5 Lateral Deflection of a Single Pile in Cohesionless Soil: Elastic Approach

As discussed earlier, the elastic approach to determine deflections and moments of piles subjected to lateral loads and moments is theoretically more realistic since it assumes the surrounding soil as an elastic continuum. However, the principles of this approach need more field verification before this approach can be used with confidence. At this time, therefore, the application aspects of this approach will be briefly presented. The information presented herein should, however, provide enough background for design engineers to use this approach in practical applications.

In this approach, the soil displacements have been evaluated from the Mindlin equation for horizontal loads within a semiinfinite mass, and the pile displacements have been obtained by using the equation (6.9), a beam on elastic foundation. Then the solutions for lateral deflections and maximum moment, described below, were obtained by assuming soil modulus $E_s$ increasing linearly with depth expressed as follows:

$$E_s = N_h x \tag{6.66}$$

where $N_h$ is the rate of increase of $E_s$ with depth and is analogous to $n_h$ in the subgrade reaction approach. If $E_s$ and $k_h$ are assumed to increase with depth at the same rate then $N_h = n_h$. The ground level deflections $y_g$ and maximum moments for a free-head and a fixed-head pile can then be given by the following relationships (Poulos and Davis, 1980).

***Free-Head Pile***

$$y_g = \frac{Q_g}{N_h L^2}\left( I'_{\rho H} + \frac{e}{L} I'_{\rho M} \right) \div F'_\rho \tag{6.67}$$

where $I'_{\rho H}$, $I'_{\rho M}$ and $F'_\rho$ are given by Figures 6.17, 6.18, and 6.19, respectively. The $Q_u$ for Figures 6.19 can be obtained from Brom's method discussed in Section 6.1.1. The maximum moment can be obtained from Figure 6.20.

***Fixed-Head Pile***

$$y_g = \frac{Q_g}{N_h L^2} I'_{pF}/F'_{pF} \tag{6.68}$$

values of $I'_{pF}$ and $F'_{pF}$ can be obtained from Figure 6.21. Again, $Q_u$ can be obtained from Broms' method (Section 6.1.1). The fixing moment ($M_f$) at the head of a fixed-head pile can be obtained from Figure 6.22.

***Example 6.5*** A 10.75-in. (273 mm) outside diameter steel pile is driven 30 ft (9.1 mm) into a medium dense sand with $\phi = 30°$, $\gamma = 125\ \text{lb/ft}^3$ and $N_h = 17.4\ \text{lb/in.}^3$. The pile has a free head, and the wall thickness is 0.25 in. (6.4 mm). The modulus of elasticity for steel is 29,000 ksi ($200 \times 10^3$ MPa) and $f_y = 35$ ksi (241 MPa). Calculate the pile head deflection and maximum moment for an applied lateral load of 5.0 kips at its head.

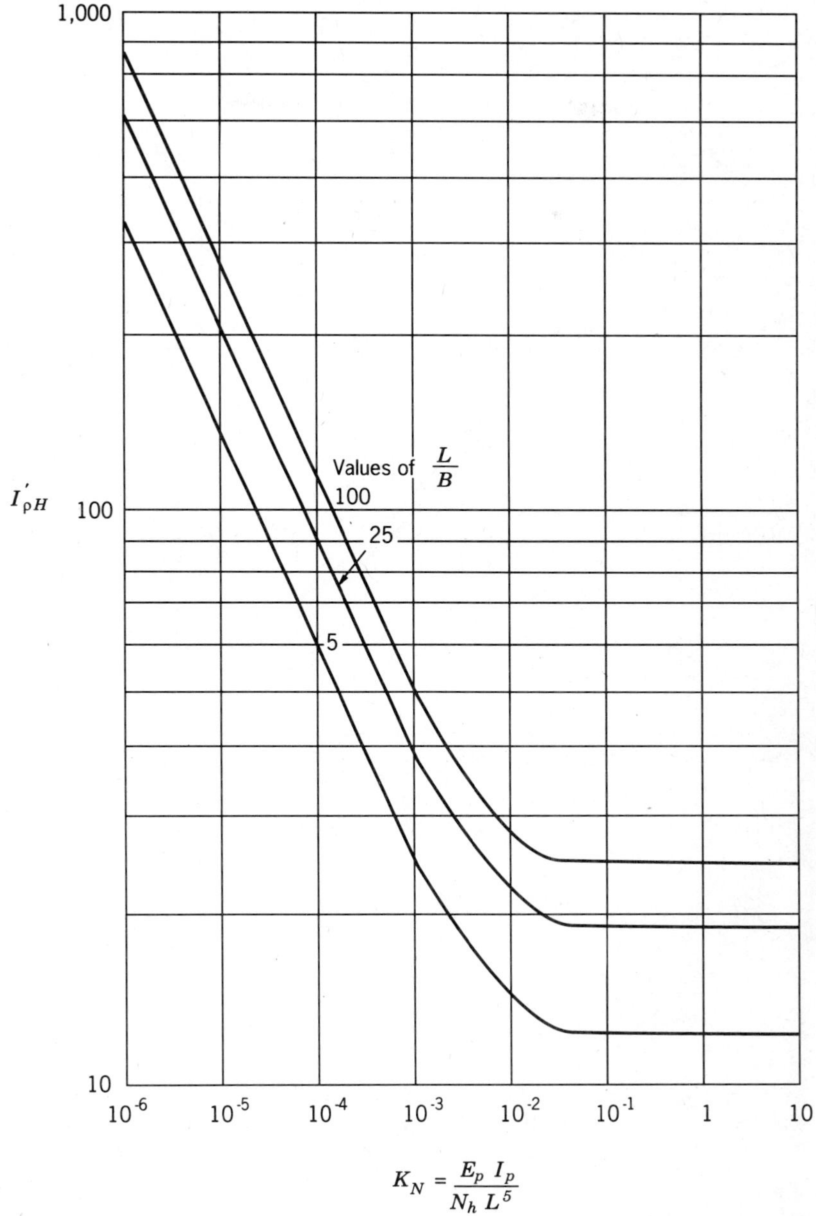


**Figure 6.17** Values of $I'_{\rho H}$: free-head pile with linearly varying soil modulus (Poulos and Davis, 1980).

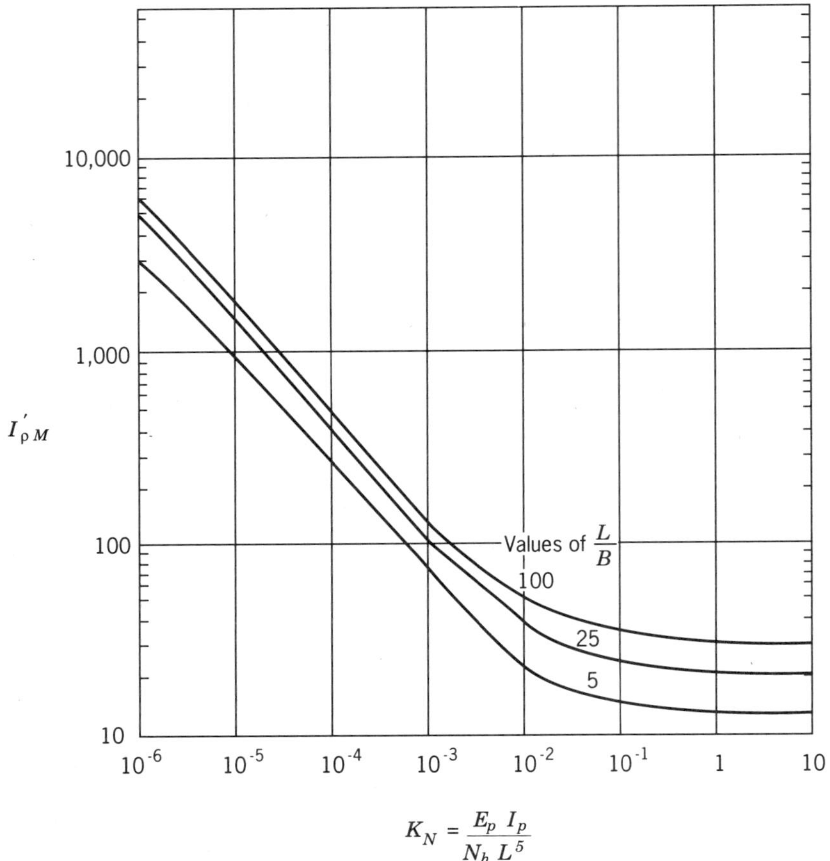


**Figure 6.18** Values of $I'_{\rho M}$: free-head pile with linearly varying soil modulus (Poulos and Davis, 1980).

SOLUTION

$K_N$ can be calculated from the following relationship.

$$K_N = \frac{E_p I_p}{N_h L^5}$$

$$N_h = n_h = 17.4\,\text{lb/in.}^3 = 30\,\text{kips/ft}^3$$

$$L = 30\,\text{ft}$$

$$E_p = 29000 \times 144\,\text{ksf} = 4176 \times 10^3\,\text{ksf}$$

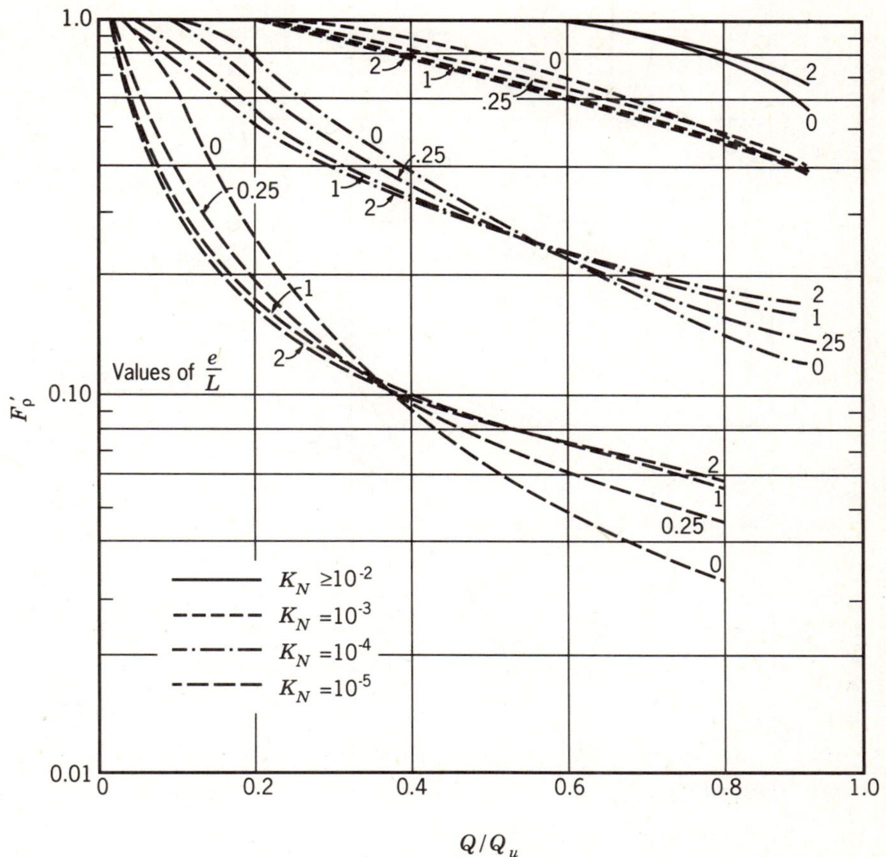


**Figure 6.19** Yield displacement factor $F'_\rho$: free-head pile, linearly varying soil modulus, and soil yield strength (Poulos and Davis, 1980).

$$I_p = \frac{\pi}{64}(10.75^4 - 10.25^4)\frac{1}{12^4} = 0.0055\,\text{ft}^4$$

$$K_N = \frac{4176 \times 10^3 \times 0.0055}{30(30)^5} = 3.15 \times 10^{-5}$$

$$\frac{e}{L} = 0 \qquad \frac{L}{B} = \frac{30 \times 12}{10.75} = 33.49$$

From Figures 6.17 and 6.18, we get:

$$I'_{\rho H} = 185 \qquad I'_{\rho M} = 700$$

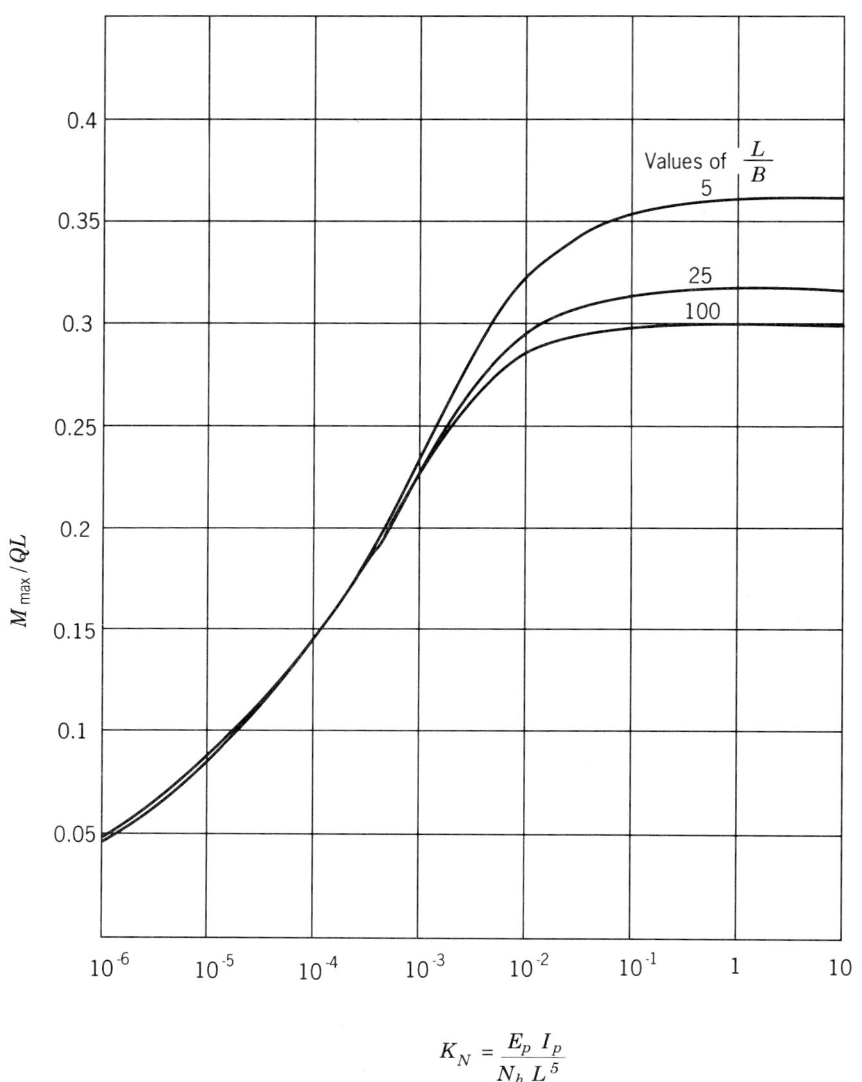


**Figure 6.20** Maximum moment in free-head pile with linearly varying soil modulus (Poulos and Davis, 1980).

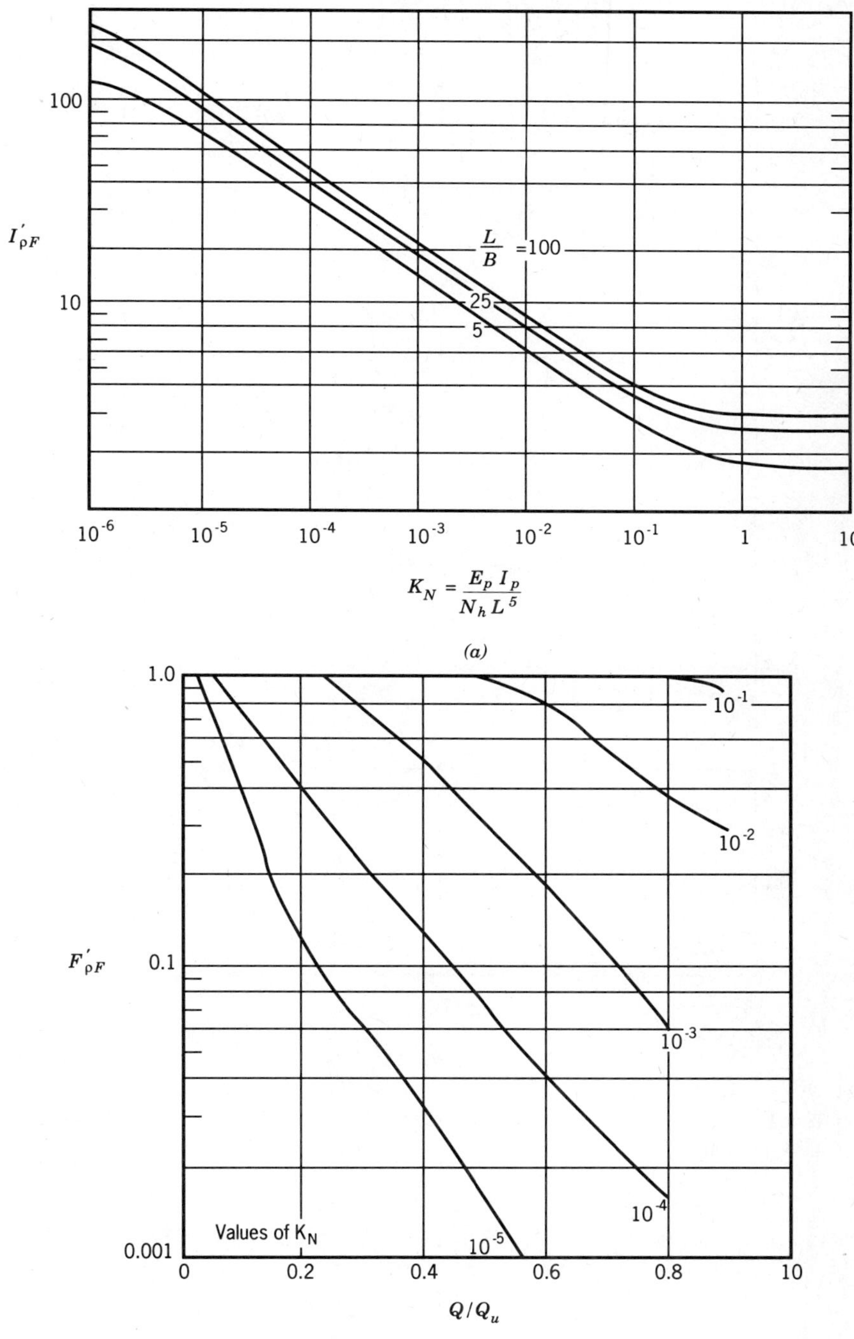


**Figure 6.21** (a) Values of $I'_{\rho F}$ (b) yield displacement factor $F'_{\rho F}$ fixed-head floating pile, linearly-varying soil modulus with depth (Poulos and Davis, 1980).

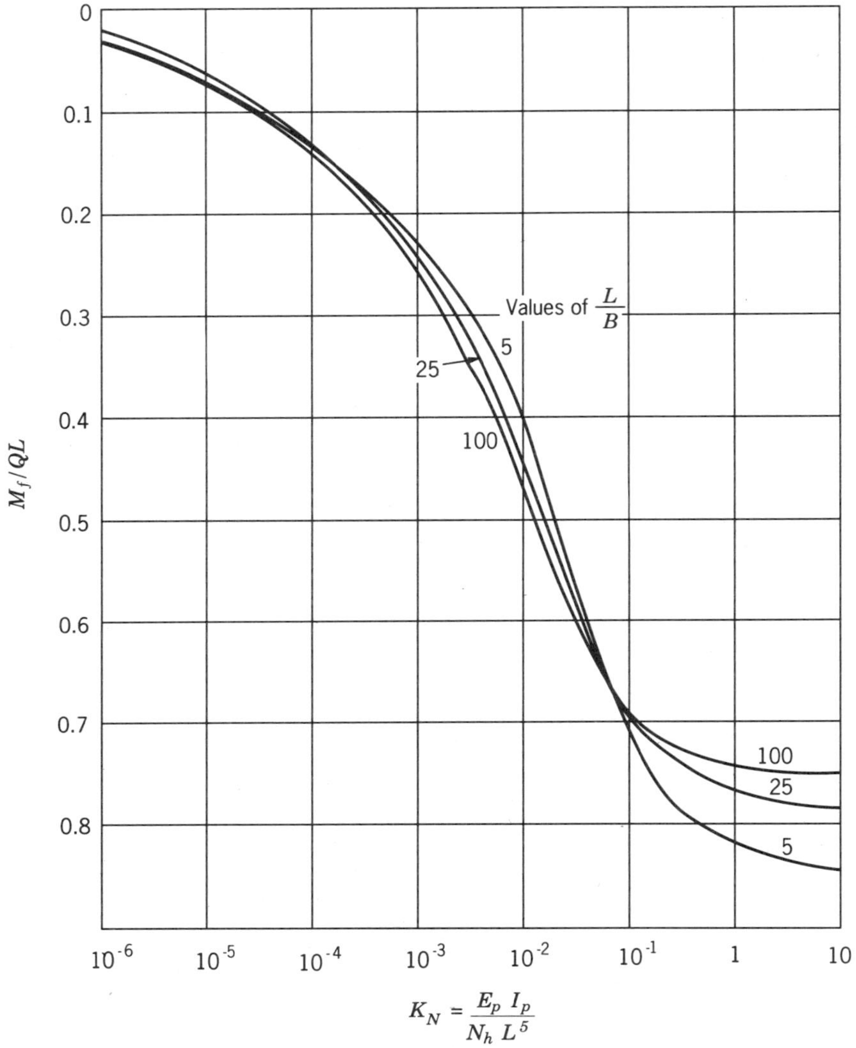


**Figure 6.22** Fixing moment in fixed-head pile: linearly varying soil modulus (Poulos and Davis, 1980).

Also,

$$T = \left(\frac{EI}{n_h}\right)^{0.2} = \left(\frac{4176 \times 10^3 \times 0.0055}{30}\right)^{0.2} = 3.8$$

$\frac{L}{T} = \frac{30}{3.8} = 7.9 > 4.$ This means that the pile is a long pile.

$$M_u = Zf_b = \frac{2I}{B}(0.6\,f_y) = 0.0122 \times 0.6 \times 35 \times 144 = 37.1\ \text{kips-ft}$$

$$\frac{M_u}{B^4 k_p \gamma} = \frac{37.1 \times 1000}{\left(\frac{10.75}{12}\right)^4 125\left(\frac{1+\sin 30}{1-\sin 30}\right)} = 154.6$$

Using Brom's method from Figure 6.9b, for

$$\frac{M_u}{B^4 \gamma k_p} = 154.6 \qquad \frac{e}{B} = 0 \qquad \frac{Q_u}{k_p B^3 \gamma} = 50, \text{ which yields}$$

$$Q_u = 50\frac{(1+\sin 30)}{(1-\sin 30)}\left(\frac{10.75}{12}\right)^3 \frac{125}{1000} = 1348\ \text{kips}$$

$$\frac{Q}{Q_u} = \frac{5}{13.48} = 0.37$$

Then, from Figure 6.19 for $Q/Q_u = 0.37, e/L = 0, K_N = 3.15 \times 10^{-5}$, we get:

$F'_\rho = 0.18$, substituting these values in equation (6.67), we get:

$$y_g = \frac{5}{30(30)^2}\frac{(185+0)}{0.18} = 0.19\ \text{ft} \equiv 2.3\ \text{in.}$$

Also, from Figure 6.20, for $k_N = 3.15 \times 10^{-5}$, $\frac{L}{B} = \frac{30 \times 12}{10.75} = 33.49$, we get:

$$\frac{M_{max}}{QL} = 0.09$$

$M_{max} = 0.09\,(5)\ (30) = 13.5$ kips-ft for an applied lateral load of 5.0 kips.

## 6.2 LATERAL DEFLECTION OF PILE GROUPS IN COHESIONLESS SOIL

Piles are mostly used in groups to support the imposed loads. As in vertical loading, there are also interaction effects in horizontal and lateral loading. Tests on groups of piles showed that piles behave as individual units if they are spaced at more than 6 to 8 diameters ($B$) parallel to the direction of lateral load application (Prakash, 1962) (see chapter 1). In order to act as individual units in a direction perpendicular to the lateral load direction, their center-to-center spacing should be at least 2.5 diameters (Prakash, 1981). In order to determine lateral load capacity of a pile group, reduction in the coefficient of subgrade

**TABLE 6.6 Group Reduction Factor for the Coefficient of Subgrade Reaction (Davisson 1970)[a]**

| Pile Spacing in the Direction of Loading | Group Reduction Factor for $n_h$ or $k^b$ |
|---|---|
| $3B$ | 0.25 |
| $4B$ | 0.40 |
| $6B$ | 0.70 |
| $8B$ | 1.00 |

[a] Also adopted in *Canadian Foundation Engineering Manual*, 1985. *Foundation and Earth Structures, Design Manual 7.2*, NAVFAC, DM 7.2 (1982) also recommends these values.
[b] $n_h$ is applicable for soil modulus linearly increasing with depth, and $k$ is applicable for soil modulus constant with depth.

reaction, $n_h$ should be made (Davisson, 1970). These reduction factors are given in Table 6.6. With an appropriately reduced $n_h$ value, the lateral load capacity of individual piles in a group can then be determined by the procedures discussed in Section 6.1.3. Pile group capacity will then be the sum of individual pile capacities calculated on the basis of reduced $n_h$ value.

Poulos (1971b) presents the behavior of laterally loaded pile groups by assuming soil as an elastic continuum having elastic parameters $E_s$ and $v_s$. At the present time, this method of analysis is not widely used in practice and needs further field verification (Poulos and Davis, 1980). The effect of the soil in contact with the cap can result in higher pile capacities (Kim et al., 1979). However, due to uncertainties in construction methods, it is safe to neglect this increased capacity.

## 6.3 DESIGN PROCEDURE FOR PILES IN COHESIONLESS SOIL

Based on the discussion of behavior and analysis of a single pile and pile group under lateral loads, a step-by-step design procedure is proposed.

### Design Procedure

The design procedure consists of the following steps:

#### 1. Soil Profile

From proper soils investigations, establish the soil profile and groundwater levels and note soil properties on the soil profile based on the field and laboratory tests. In Chapter 4, proper procedures for field investigations and relevent soil property determination were discussed.

**2. Pile Dimensions and Arrangement**

Normally, pile dimensions and arrangements are established from axial compression loading requirements. The ability of these pile dimensions and their arrangement to resist imposed lateral loads and moments is then checked by following procedure.

**3. Calculation of Ultimate Lateral Resistance and Maximum Bending Moment**

*a. Single Piles*

(i) Determine $n_h$ from Table 4.16. Calculate the relative stiffness $T = (EI/n_h)^{1/5}$. Determine the $L/T$ ratio and check if it is a short ($L/T \leqslant 2$) or long ($L/T \geqslant 4$) pile.

(ii) Calculate the ultimate lateral resistance $Q_u$, the allowable lateral resistance, $Q_{\text{all}}$, and maximum bending moment $M$ for the applied loads by Broms' method outlined in Section 6.1.1.

*b. Pile Group* From Table 6.2 determine $G_e$ for $(S/B)$ ratio of the group. The allowable lateral resistance of the group $(Q_{\text{all}})_G$ is then calculated by following equation:

$$(Q_{\text{all}})_G = nG_e Q_{\text{all}} \tag{6.32}$$

where $n$ is number of piles in the group, and $Q_{\text{all}}$ is obtained as described in step 3(a(ii)).

**4. Calculation of Lateral Resistance and Maximum Moment for Allowable Lateral Deflection**

*a. Single Piles*

(i) Determine $n_h$ from soil parameters as in step 3(a(i)). Calculate the relative stiffness, $T = (EI/n_h)^{1/5}$. Determine $L/T$ ratio.

(ii) Calculate the allowable lateral load for the specified lateral deflection and maximum bending moment for the design loading conditions by the subgrade reaction approach as outlined in Section 6.1.3.

*b. Pile Group*

(i) From Table 6.6, determine the group reduction factor for $n_h$ for the $S/B$ ratio of the group. Then determine the new $n_h$ and, as outlined in 4(a), calculate the allowable lateral load capacity of a single pile based on this new $n_h$.

(ii) The pile group capacity is the allowable lateral load capacity of single pile, obtained in 4b(i), multiplied by the number of piles $n$. The maximum bending moment for a pile is calculated by the method outlined in Section 6.1.3 except that the $Q$ value used is obtained for a single pile in the group.

**5. Allowable Lateral Load and Maximum Bending Moment**

Allowable lateral load is the lower of the values obtained in steps 3 and 4. The maximum bending moment is corresponding to the allowable lateral load.

**6. Special Design Feature: Calculation of Deflection and Moment Beyond the Elastic Range (where soil is allowed to yield plastically) for Given Lateral Load and Moment**

a. Establish the $p$–$y$ curve by the procedure outlined in Section 6.1.4.
b. Determine the $n_h$ from soil parameters. Calculate the $T = (EI/n_h)^{1/5}$. Determine the deflections along pile depth for the given lateral load and moment. The $T$ value calculated here will be first trial value and will be referred as $(T)_{\text{trial}}$ in following steps.
c. For the deflections determined in step 6(b), obtain the corresponding pressure from the $p$–$y$ curve established in step 6(a). Then obtain the soil modulus $k = (p/y)$, where $p$ is the soil reaction, and $y$ is the pile deflection. This is *first trial* value for $k$. Plot the value of $k$ with depth.
d. From $k$ obtained in step 6(c), calculate new $n_h = (k/x)$ where $x$ is the depth below ground. Then compute $T = (EI/n_h)^{1/5}$. Compare this $(T)_{\text{obtained}}$ from the $(T)_{\text{trial}}$ value calculated in step 6(b). If these values do not match, proceed with the second trial as follows.
e. Assume a $T$ value closer to the value obtained in step 6(d). Repeat steps 6(b), 6(c), and 6(d) and obtain a new $T$.
f. Plot $(T)_{\text{obtained}}$ values on the ordinate and $(T)_{\text{trial}}$ on the abscissa and join the points. Draw a line at 45° from the origin. The intersection of this line with the trial line will give actual $T$.
g. With the finally obtained $T$ value, calculate deflections $y$, soil reactions $p$, and moments $M$ along the pile length by the method outlined in Section 6.1.3.

This procedure is applicable for a single pile only.

***Example 6.6*** A group of nine piles, each with a 36-in. (914.4 mm) outside diameter and 1-in. (25.4 mm) wall thickness steel pipe piles driven 60 ft (18.3 m) into dense sand with average $N = 38$, $\phi = 36°$ and unit weight $\gamma = 120\ \text{lb/ft}^3$ ($1920\ \text{kg/m}^3$), is supporting a module. The piles are spaced at 18 ft (5.5 m) center-to-center distance and can be assumed to be free headed. Yield strength for the steel, $f_y = 44$ ksi ($303.5 \times 10^3\ \text{kN/m}^2$) and the modulus of elasticity for the steel, $E = 29{,}000$ ksi ($200 \times 10^3$ MPa). Other piles in the area around this group are 18 ft away. The constant of subgrade reaction for the soil, $n_h = 52\ \text{lb/in.}^3$.

(a) Calculate the allowable lateral load on each pile. Due to sensitive nature of the structure, the maximum allowable lateral deformation on pile head is 0.25 in. (6.35 mm).

(b) Calculate the maximum bending moment along the pile length for an applied lateral load equal to the allowable value obtained in (a).

(c) If the pile is subjected to a 50-kip (222.5 kN) cyclic lateral load and a 90-kip-ft (122 kN-m) moment at its head, calculate the maximum deflection and maximum bending moment on the pile. Assume that the soil is allowed to yield beyond the elastic range and piles are acting as single piles (i.e., no group effect).

SOLUTION

1. *Soil Profile* This is shown in Figure 6.23.
2. *Pile Dimensions and Arrangement* Piles are placed in a group of nine from axial compression loading and the space requirements. Each pile is of 36 in. or 3 ft outside diameter and spaced at 18 ft center-to-center distance. Therefore, $S/B = 18/3 = 6$; when the pile group is arranged in a square pattern, three piles are on each side of the square. Also, other piles in the area are placed 18 ft away from a pile in the group. Therefore, this $S/B = 6$ will apply for group effect in all directions.

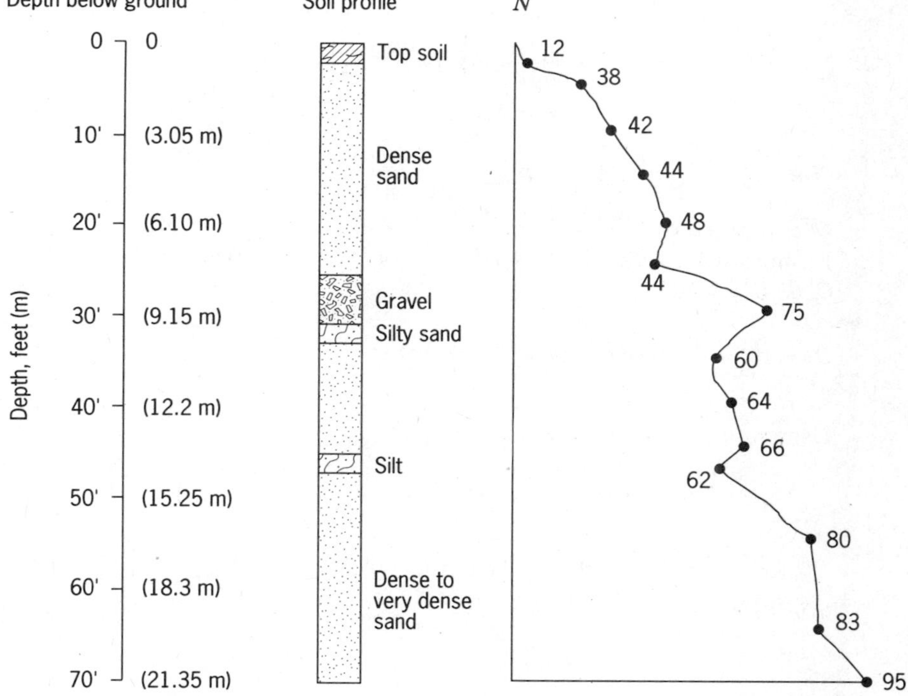


**Figure 6.23** Soil profile and soil properties along the pile depth (Example 6.6). $N$ = standard penetration value. $\phi = 36°$, $\gamma = 120\,\text{lb/ft}^3$.

3. *Calculation of Ultimate Lateral Resistance and Maximum Bending Moment*

3(a) *Single piles* The average $N$ value in the top half part of the pile (upto 30 ft depth) is $=(12+38+42+44+48+44)/6=38$.

$$n_h = 52\,\text{lb/in.}^3 = 90\,\text{kips/ft}^3$$

$$E = (29 \times 10^3 \times 144)\,\text{kips/ft}^2$$

$$I = \frac{\pi}{64}\left[\left(\frac{36}{12}\right)^4 - \left(\frac{34}{12}\right)^4\right] = 0.81\,\text{ft}^4$$

$$EI = (29 \times 144 \times 10^3)(0.81) = 3.38 \times 10^6\,\text{kips-ft}^2$$

$$T = \left[\frac{3.38 \times 10^6}{90}\right]^{1/5} = 8.2\,\text{ft}$$

$$\frac{L}{T} = \frac{60}{8.2} = 7.3 > 4, \text{ piles behave as long piles.}$$

Using Brom's method, consider the free-head long pile.

$$M_u = Zf_b = \left(\frac{I}{B/2}\right)(0.6f_y)$$

$$= \frac{0.81}{1.5} \times 0.6 \times 44 \times 144\,\text{kips-ft} = 2.05 \times 10^3\,\text{kips-ft}$$

$$\frac{M_u}{B^4\gamma k_p} = \frac{2.05 \times 10^3}{(3)^4 \times \dfrac{120}{1000}\left(\dfrac{1+\sin 36}{1-\sin 36}\right)} = \frac{2.05 \times 10^3}{81 \times 0.12 \times \dfrac{1.58}{0.42}} = 56.06$$

From Figure 6.9b, for $M_u/B^4\gamma k_p = 56.06$, $e/B = 0$, free-head pile, $Q_u/k_pB^3\gamma = 10$

$$Q_u = 10\left(\frac{1.58}{0.42}\right)(3)^3\,0.120 = 121.89\,\text{kips}$$

Using a factor of safety = 2.5

$$Q_{\text{all}} = \frac{121.89}{2.5} = 48.8\,\text{kips}$$

$$M_{\max} = Q_g(e + 0.67x_0) \text{ from equation (6.15), } e = 0,\ x_0 = 0.82\left(\frac{Q_u}{\gamma Bk_p}\right)^{0.5}$$

$$= 0.82\left(\frac{121.89}{0.12 \times 3 \times \dfrac{1.58}{0.42}}\right)^{0.5} = 7.78 \qquad \text{from equation (6.13)}$$

$$= 48.8 \times 0.67 \times 7.78 = 254.35\,\text{kips-ft}$$

3(b) *Pile group action*

$S/B = 18/3 = 6$

From Table 6.2, for $S/B = 6$, $G_e = 0.7$

$Q_{\text{all}} = 0.7 \times 48.4 = 34$ kips for each pile

$M_{\text{max}} = 254.35 \times 0.7 = 178$ kips-ft for each pile

4. *Calculation of Lateral Resistance and Maximum Moment for Allowable Lateral Deflection* Since the piles are spaced at $S/B = 6$, they will act as a group, and group reduction factor for $n_h$ is 0.7 (Table 6.6).

$n_h = 0.7 \times 90 = 63$ kips/ft$^3$

$$T = \left(\frac{3.38 \times 10^6}{63}\right)^{1/5} = 8.8 \text{ ft}$$

$\dfrac{L}{T} = 60/8.8 = 6.8 > 5$. Therefore, coefficients $A_y$ and $B_y$ from Tables 6.3 and 6.4 can be used.

From equation (6.34):

$$y = A_y \frac{Q_g T^3}{EI} + B_y \frac{M_g T^2}{EI}$$

At ground level, $Z = X/T = 0$, $A_y = 2.435$, and $B_y = 1.623$

$$y = \frac{0.25}{12} \text{ ft} \qquad T = 8.8 \text{ ft} \qquad EI = 3.38 \times 10^6 \text{ kips-ft}^2$$

$M_g = 0$

Substituting in the foregoing equation, we get

$$\frac{0.25}{12} = 2.435 \frac{Q_g (8.8)^3}{3.38 \times 10^6}$$

$$Q_g = \frac{0.25 \times 3.38 \times 10^6}{12 \times 2.435(8.8)^3} = 42.4 \text{ kips}$$

Maximum bending moment for this $Q_g$ is:

$$M = A_m Q_g T + B_m M_g \tag{6.35}$$

From Table 6.3, $(A_m)_{max} = 0.772$

$$\therefore M_{max} = 0.772 \times 42.4 \times 8.8 = 288 \text{ kips-ft}$$

5. *Allowable Lateral Load and Maximum Bending Moment* From steps 3 and 4, the allowable lateral load for a single pile of a group is the lower of the two values.

$Q_{all} = 34$ kips and corresponding deflection of pile head
$y_g = 2.435 \times 34(8.8)^3/3.38 \times 10^6 = 0.2$ in
$M_{max} = 178$ kips-ft

6. *Special Design Feature: Calculation of Deflection and Moment Beyond the Elastic Range*

6(a) *Establish the p–y curve* In order to establish the $p$–$y$ curve, refer to the steps for laterally loaded piles in Section 6.1.4.

As given above,

$$\phi = 36^\circ \qquad \gamma = 120 \text{ lb s/ft}^3$$

$$\alpha = 18^\circ \tag{6.51}$$

$$\beta = 45 + 18 = 63^\circ \tag{6.52}$$

$$k_0 = 0.4 \tag{6.53}$$

$$K_A = \tan^2(45 - 18) = 0.259 \tag{6.54}$$

$$p_{cr} = 120 \times \left[ \frac{0.4x \tan 36 \sin 63}{\tan(63-36)\cos 18} + \frac{\tan 63}{\tan(63-36)} \left( \frac{36}{12} + x \tan 63 \tan 18 \right) \right.$$

$$\left. + 0.4x \tan 63 \,(\tan 36 \sin 63 - \tan 18) - 0.259x \frac{36}{12} \right] \tag{6.55}$$

$$= 120x(0.534x + 11.563 + 2.457x + 0.252x - 0.776)$$

$$= 389.16x^2 + 1294.44x$$

Then, values of $x$ and $P_{cr}$ can be calculated as follows

$x = 0$ $\quad P_{cr} = 0$
$= 2$ ft $\quad = 4.144$ kips/ft
$= 4$ ft $\quad = 11.40$ kips/ft
$= 10$ ft $\quad = 51.84$ kips/ft
$= 15$ ft $\quad = 106.935$ kips/ft
$= 20$ ft $\quad = 181.480$ kips/ft
$= 30$ ft $\quad = 388.920$ kips/ft

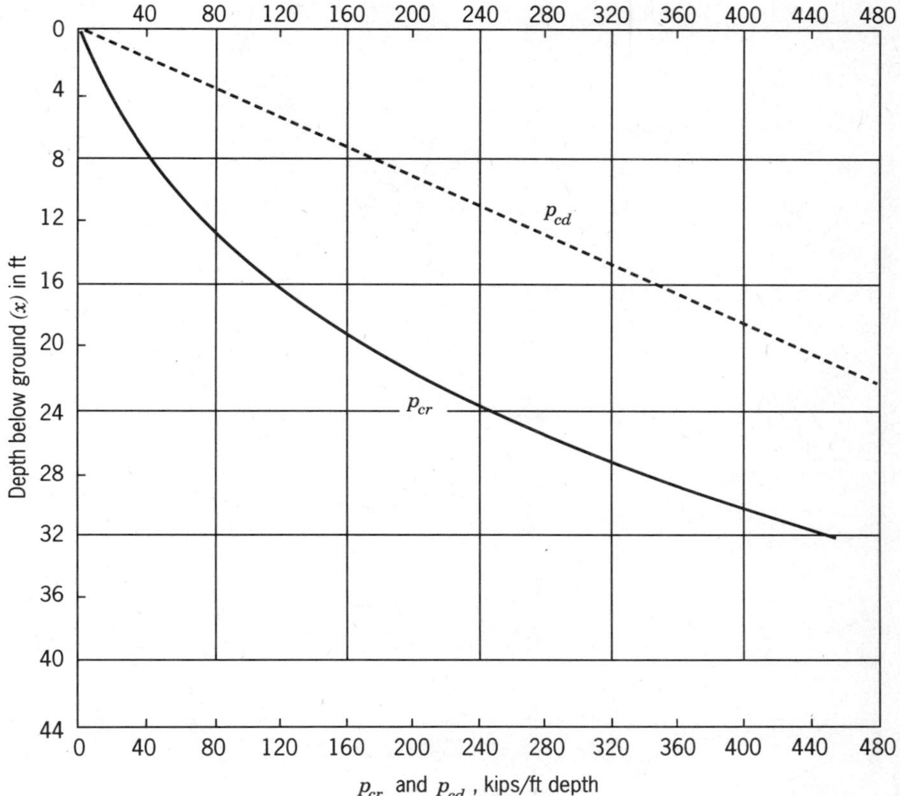


**Figure 6.24** Values of $p_{cr}$ and $p_{cd}$ with depth (Example 6.6).

$$P_{cd} = 0.259 \times \frac{36}{12} \times 120x\,(\tan^8 63 - 1) + 0.4 \times \frac{36}{12} \times 120 \times \tan 36 \tan^4 \quad 63$$

from equation (6.56)

$$= 21.982x \text{ kip/ft}$$

$$\begin{aligned} x &= 0 & P_{cd} &= 0 \\ &= 2' & &= 43.964 \text{ kips/ft} \\ &= 4' & &= 87.928 \text{ kips/ft} \\ &= 10' & &= 219.82 \text{ kips/ft} \\ &= 15' & &= 329.73 \text{ kips/ft} \\ &= 20' & &= 439.64 \text{ kips/ft} \\ &= 30' & &= 659.46 \text{ kips/ft} \end{aligned}$$

Values of $P_{cr}$ and $P_{cd}$ are plotted against depth in Figure 6.24. $P_{cr}$ and $P_{cd}$ do not intersect, therefore over the range of depth that is important for

**TABLE 6.7 Calculations for Establishing the *p–y* curve**

| $X$ (ft) | $\frac{X}{B}$ | $A_1$ (Table 6.5) | $B_1$ (Table 6.5) | $p_c$ (kips/ft) | $p_m = B_1 P_c$ | $p_u = A_1 P_c$ | $m = \frac{p_u - p_m}{y_u - y_m}$ (see note 1) | $n = \frac{p_m}{m y_m}$ | $C = \frac{p_m}{y_m^{1/n}}$ | $y_k = \left(\frac{C}{n_h x}\right)^{n/n-1}$ (see note 2) (ft) | $p = cy^{1/n}$ | $p_k = C y_k^{1/n}$ (kips/ft) |
|---|---|---|---|---|---|---|---|---|---|---|---|---|
| 2 | 0.67 | 1 | 0.8 | 4.144 | 3.3 | 4.1 | 12.8 | 5.2 | 5.9 | 0.0144 | $5.9y^{0.19}$ | 2.6 |
| 4 | 1.34 | 1.1 | 0.86 | 11.400 | 9.8 | 12.5 | 43.2 | 4.5 | 19.2 | 0.0227 | $19.2y^{0.22}$ | 8.3 |
| 10 | 3.34 | 0.93 | 0.65 | 51.840 | 33.7 | 48.2 | 232 | 2.9 | 93.6 | 0.0313 | $93.6y^{0.34}$ | 28.3 |
| 15 | 5 | 0.88 | 0.55 | 106.935 | 58.8 | 94.1 | 564.8 | 2.1 | 245 | 0.0384 | $245y^{0.48}$ | 51.9 |

1. $y_m = B/60 = 3/60 = 0.05$ ft, $y_u = 3B/80 = 0.1125$ ft, $y_u - y_m = 0.0625$ ft.
2. $n_h = 90$ kips/ft$^3$.

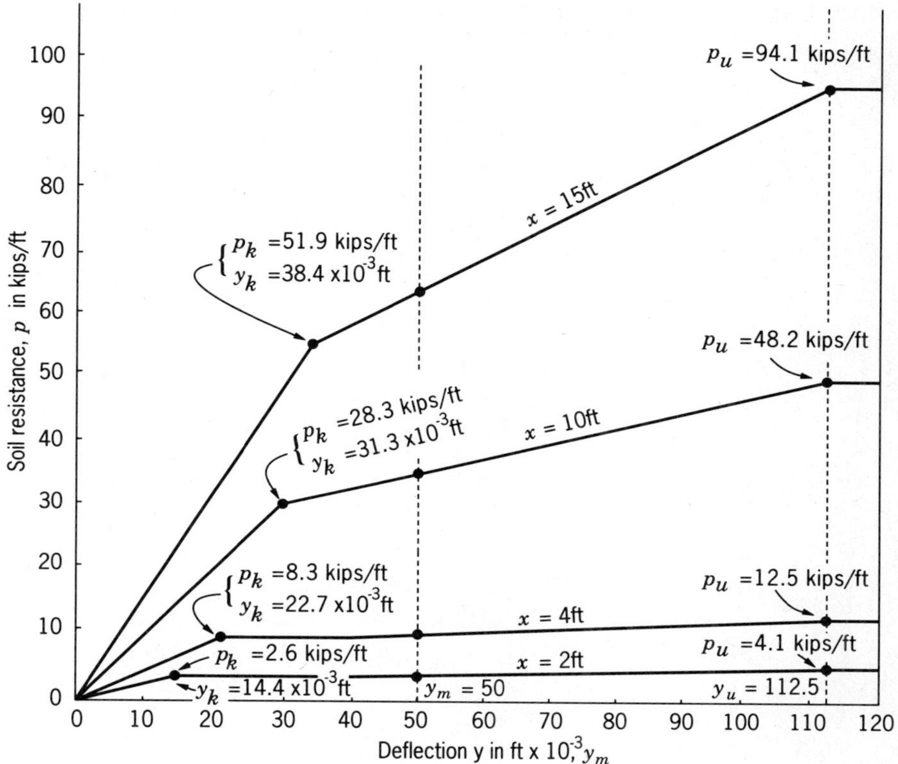


**Figure 6.25** *p–y* curves for different depths (Example 6.6).

lateral load, only the values of $P_{cr}$ will be applicable to the $p$–$y$ curves. The $p$–$y$ curves for various depths are then calculated in Table 6.7 and are plotted in Figure 6.25.

6(b) *Calculation of lateral deflections with depth*

$$n_h = 90 \text{ kips/ft}^3$$

$$EI = 3.38 \times 10^6 \text{ kips-ft}^2 \text{ from step 3(a)}$$

$$T = 8.2 \text{ ft for single pile, step 3(a) } (T)_{\text{trial}} = 8.2 \text{ ft}$$

$$\frac{L}{T} = \frac{60}{8.2} = 7.3$$

Therefore, coefficients $A_y$ and $B_y$ can be obtained from Tables 6.3 and 6.4

$$Q_g = 50 \text{ kips}, \quad M_g = 90 \text{ kips-ft}$$

**TABLE 6.8 Calculations for Lateral Deflection**

| $X$ (ft) | $Z=\frac{X}{T}$ | $A_y$ | $B_y$ | $y_A$ (ft) | $y_B$ (ft) | $y=y_A+y_B$ (ft) |
|---|---|---|---|---|---|---|
| 2 | 0.24 | 2.064 | 1.248 | $16.9\times10^{-3}$ | $2.3\times10^{-3}$ | $19.2\times10^{-3}$ |
| 4 | 0.48 | 1.705 | 0.925 | $14.0\times10^{-3}$ | $1.7\times10^{-3}$ | $15.7\times10^{-3}$ |
| 10 | 1.2 | 0.738 | 0.223 | $6.0\times10^{-3}$ | $0.4\times10^{-3}$ | $6.4\times10^{-3}$ |
| 15 | 1.8 | 0.247 | $-0.03$ | $2.03\times10^{-3}$ | $-0.05\times10^{-3}$ | $2.0\times10^{-3}$ |

**TABLE 6.9 Calculation of $k_h = E_s$ with depth, $x$**

| $x$ (ft) | $y$ (ft) | $p^a$ (kips/ft) | $k_h = E_s = \frac{p}{y}$ |
|---|---|---|---|
| 2 | $19.2\times10^{-3}$ | 2.7 | 140.6 |
| 4 | $15.7\times10^{-3}$ | 6.0 | 382 |
| 10 | $6.4\times10^{-3}$ | 5.0 | 781 |
| 15 | $2.0\times10^{-3}$ | 3.0 | 1500 |

[a]Values of $p$ are obtained from $p$–$y$ curve corresponding to above $y$ values from Figure 6.25.

$$y = y_A + y_B = A_y\frac{Q_g T^3}{EI} + B_y\frac{M_g T^2}{EI} \tag{6.34}$$

$$y = A_y\frac{50(8.2)^3}{3.38\times10^6} + B_y\frac{90(8.2)^2}{3.38\times10^6}$$

$$y = 8.2\times10^{-3}A_y + 1.8\times10^{-3}B_y$$

These values are given in Table 6.8.

6(c) *Determination of $E_s$, $(k_h)$* The value of $E_s$ is as calculated in the Table 6.9 and plotted in Figure 6.26.

6(d) Determination of $T$.

$$n_h = 100\ \text{kips/ft}^3 \text{ from first trial (Figure 6.26)}$$

$$(T)_{\text{obtained}} = \left(\frac{3.38\times10^6}{100}\right)^{1/5} = 8.05\ \text{ft}$$

The value of $T$ in the first trial was 8.2 ft

6(e) Assume $T = 8.1$ ft (i.e., tried $T = 8.1$ ft)
Determination of $y$ based on assumed values

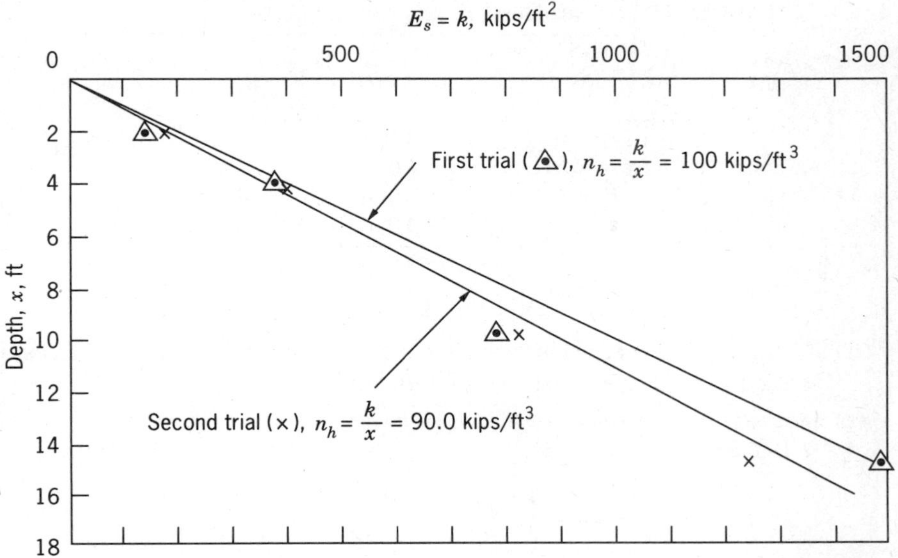


**Figure 6.26** Variation of $E_s$ with depth (Example 6.6).

$$y = y_A + y_B = A_y \frac{50(8.1)^3}{3.38 \times 10^6} + B_y \frac{90(8.1)^2}{3.38 \times 10^6}$$

$$= 7.86 \times 10^{-3} A_y + 1.75 \times 10^{-3} B_y$$

These values are tabulated in Table 6.10.

Values of $E_s$ from Table 6.11 are plotted with depth in Figure 6.26.

$$n_h = 90 \text{ kips/ft}^3$$

$$(T)_{\text{obtained}} = \left( \frac{3.38 \times 10^6}{90} \right)^{1/5} = 8.22$$

**TABLE 6.10 Calculations for Lateral Deflection**

| $X$ (ft) | $Z = \frac{X}{T}$ | $A_y$ | $B_y$ | $y_A$ (ft) | $y_B$ (ft) | $y = y_A + y_B$ (ft) |
|---|---|---|---|---|---|---|
| 2 | 0.21 | 2.096 | 1.278 | $16.7 \times 10^{-3}$ | $2.24 \times 10^{-3}$ | $18.71 \times 10^{-3}$ |
| 4 | 0.42 | 1.766 | 0.977 | $13.88 \times 10^{-3}$ | $1.71 \times 10^{-3}$ | $15.59 \times 10^{-3}$ |
| 10 | 1.05 | 0.850 | 0.328 | $6.68 \times 10^{-3}$ | $0.57 \times 10^{-3}$ | $7.25 \times 10^{-3}$ |
| 15 | 1.57 | 0.405 | 0.041 | $3.18 \times 10^{-3}$ | $0.07 \times 10^{-3}$ | $3.25 \times 10^{-3}$ |

**TABLE 6.11 Calculation of $E_s = k_h$ with Depth, X**

| $X$ (ft) | $y$ (ft) | $p$ (kips/ft$^3$) | $E_s = \frac{P}{y}$ |
|---|---|---|---|
| 2 | $18.71 \times 10^{-3}$ | 3.0 | 160.0 |
| 4 | $15.59 \times 10^{-3}$ | 6.0 | 384.8 |
| 10 | $7.25 \times 10^{-3}$ | 6.0 | 827.5 |
| 15 | $3.25 \times 10^{-3}$ | 4.0 | 1230.8 |

6(f) All these $T$ values are close to each other. Therefore, $T = 8.15$ ft can be used in further analysis without any error.

6(g) *Determination of Deflections and Moments.*
*Deflections*:

$$y = y_A + y_B = A_y \frac{Q_g T^3}{EI} + B_y \frac{M_g T^2}{EI} \qquad \frac{L}{T} = 7.36 \geqslant 5$$

therefore Table 6.3 and 6.4 can still be used for $A_y$ and $B_y$.

$$y = A_y \frac{50(8.15)^3}{3.38 \times 10^6} + B_y \frac{90(8.15)^2}{3.38 \times 10^6}$$

$$= 8 \times 10^{-3} A_y + 1.77 \times 10^{-3} B_y$$

From these equations, the values of deflection $y$ are obtained for various depths as given in Table 6.12.

*Moments*:

$$M = A_m Q_g T + B_m M_g$$

$$= 407.5 A_m + 90 B_m$$

where $Q_g = 50$ kips, $M_g = 90$ kips-ft, $T = 8.15$ ft at $X/T = 0$, $A_m = 0$, $B_m = 1$.

**TABLE 6.12 Calculation of Deflections with Depth**

| $X$ (ft) | $Z = \frac{X}{T}$ | $A_y$ | $B_y$ | $y_A$ (ft) | $y_B$ (ft) | $y$ (ft) |
|---|---|---|---|---|---|---|
| 0 | 0.00 | 2.435 | 1.623 | $19.5 \times 10^{-3}$ | $2.9 \times 10^{-3}$ | $22.4 \times 10^{-3}$ |
| 2 | 0.25 | 2.032 | 1.218 | $16.3 \times 10^{-3}$ | $1.2 \times 10^{-3}$ | $17.5 \times 10^{-3}$ |
| 4 | 0.50 | 1.644 | 0.873 | $13.2 \times 10^{-3}$ | $1.5 \times 10^{-3}$ | $14.7 \times 10^{-3}$ |
| 6 | 0.75 | 1.285 | 0.591 | $10.3 \times 10^{-3}$ | $1.0 \times 10^{-3}$ | $11.3 \times 10^{-3}$ |
| 12 | 1.50 | 0.463 | 0.071 | $3.7 \times 10^{-3}$ | $0.1 \times 10^{-3}$ | $3.8 \times 10^{-3}$ |
| 20 | 2.50 | 0.034 | $-0.079$ | $0.3 \times 10^{-3}$ | $-0.1 \times 10^{-3}$ | $0.2 \times 10^{-3}$ |

**TABLE 6.13 Calculation of Moments with Depth**

| $X$ (ft) | $Z=\frac{X}{T}$ | $A_m$ | $B_m$ | $407.5A_m$ | $90B_m$ | $M$ (kips-ft) |
|---|---|---|---|---|---|---|
| 2 | 0.25 | 0.245 | 0.997 | 99.8 | 89.7 | 189.50 |
| 4 | 0.50 | 0.459 | 0.976 | 187.0 | 87.8 | 274.80 |
| 6 | 0.75 | 0.622 | 0.927 | 253.5 | 83.4 | 336.90 |
| 13 | 1.60 | 0.746 | 0.594 | 304.0 | 53.5 | 357.50 |
| 18 | 2.25 | 0.527 | 0.318 | 214.8 | 28.6 | 243.40 |
| 26 | 3.20 | 0.168 | 0.034 | 68.5 | 3.1 | 71.60 |
| 32.6 | 4.00 | 0.000 | −0.042 | 0.0 | −3.78 | −3.78 |

So $M = 90$ kips-ft at ground level. Values of moments with depth are given in Table 6.13. Values of deflections and moments with depth are plotted in Figure 6.27. From this figure the following are obtained.

$$y_{\max} = 22.4 \times 10^{-3}\,\text{ft} = 0.27\,\text{in.}$$
$$M_{\max} = 380\,\text{kips-ft}$$

(a) Allowable lateral load on each pile = 34 kips.

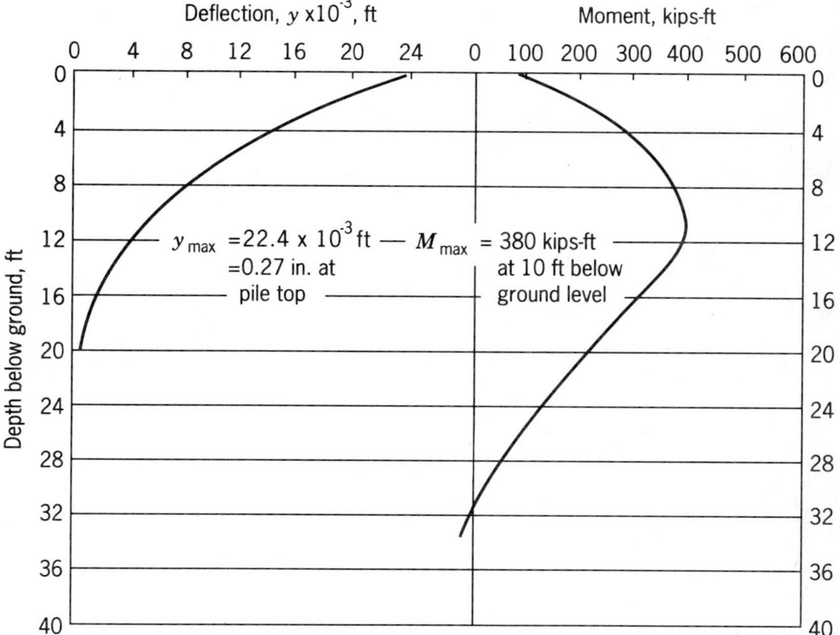


**Figure 6.27** Deflections and moments with depth (Example 6.6).

(b) Maximum bending moment along pile length for the applied lateral load = 178 kips-ft. Maximum deflection of the pile = 0.20 in.
(c) If the pile is allowed to yield beyond elastic limit, then for 50 kip lateral load and 90 kip-ft moment:
 (i) Maximum pile deflection = 0.27 in.
 (ii) Maximum bending moment along pile length = 380 kips-ft.

## 6.4 ULTIMATE LATERAL LOAD RESISTANCE OF A SINGLE PILE IN COHESIVE SOILS

Ultimate lateral load resistance of a single pile in cohesive soils can be determined by using Brinch Hansen's (1961) and Broms' (1964a) methods. Basic theory and assumptions for these methods have been discussed in Section 6.1. In this section, application of these methods for piles in cohesive soils are described.

### Brinch Hansen's Method

Equation (6.3) presents the relationship for ultimate soil reaction at any depth. For cohesive soils $\phi = 0$ and $c = c_u$. Also as shown in Figure 6.3 for $\phi = 0$, $k_q = 0$. Then the ultimate soil resistance $p_{xu}$ can be expressed by the following relationship:

$$p_{xu} = c_u K_c \tag{6.69}$$

where $K_c$ can be obtained from Figure 6.3. The procedure for calculating ultimate lateral resistance then consists of the following steps:

1. Divide the soil profile into a number of layers.
2. Determine $K_c$ for each layer and calculate $p_{xu}$ for these layers. Plot $p_{xu}$ with depth.
3. Assume a point of rotation at a depth $x_r$ below ground and take moment about the point of application of lateral load $Q_u$.
4. If this moment is small or near zero then $x_r$ is the right value. If not, repeat steps (1) through (3), until the moment is near zero.
5. Once $x_r$, the depth to the point of rotation, is determined, take moment about the point (center) of rotation and calculate $Q_u$.

The overall procedure is similar as presented in Examples 6.1 and 6.13. The only difference will be that equation (6.69) will be used to calculate $p_{xu}$ for cohesive soils.

### Broms' Method

Broms (1964a) analyzed free-head and fixed-head piles separately and also considered short and long piles separately. Basic theory and assumptions made in this theory were discussed in section 6.1. In the following paragraphs the application of this method are presented.

*Free-Head (Unrestrained) Piles*

1. **Short Piles ($L/R \leqslant 2$)** Figure 6.4 (a) and (c) present the possible failure mode and the distribution of ultimate soil reaction and moment for short free-head piles in cohesive soils, respectively. Unlike long piles whose lateral capacity is primarily dependent on the yield moment of the pile, the lateral capacity of short piles is solely dependent on the soil resistance. Taking moments about the point of maximum moment (a distance $1.5B + x_0$ below ground), we get (Broms 1964a):

$$M_{\max} = Q_u(e + 1.5B + 0.5x_0) \tag{6.70}$$

The length $(L - x_0)$ of the pile resists maximum bending moment, which is given by:

$$M_{\max} = 2.25Bc_u(L - x_0)^2 \tag{6.71}$$

where

$$x_0 = (Q_u/9c_uB) \tag{6.72}$$

Equations 6.71 and 6.72 can be solved to obtain $Q_u$. The solution is provided in Figure 6.28a where if $L/B$ and $e/B$ ratios are known then $(Q_u/c_uB^2)$ can be obtained. Thus the $Q_u$ value can be calculated.

2. **Long Piles ($L/R \geqslant 3.5$)** Figure 6.5a shows possible failure mode, the distribution of ultimate soil reaction, and moment for long free-head pile in cohesive soils. Equations (6.70) and (6.72) apply also for this case (Broms 1964a). The solutions are plotted in Figure 6.28b. Thus, for a known $(M_u/c_uB^3)$, one can obtain $(Q_u/c_uB^2)$ and finally $Q_u$ can be obtained.

*Fixed-Head (Restrained) Piles*

1. **Short Piles** Failure mechanism, distribution of ultimate soil reaction, and the distribution of bending moment are shown in Figure 6.4 (b) and (d), respectively. The following relationships are applicable for these piles (Broms 1964a):

$$Q_u = 9c_uB(L - 1.5B) \tag{6.73}$$

$$M_{\max} = 4.5c_uB(L^2 - 2.25B^2) \tag{6.74}$$

These relationships are plotted in Figure 6.28a.

2. **Long Piles** Again, the failure mechanism and distribution of ultimate soil reaction and moment for these piles are shown in Figure 6.5b. The $Q_u$ value

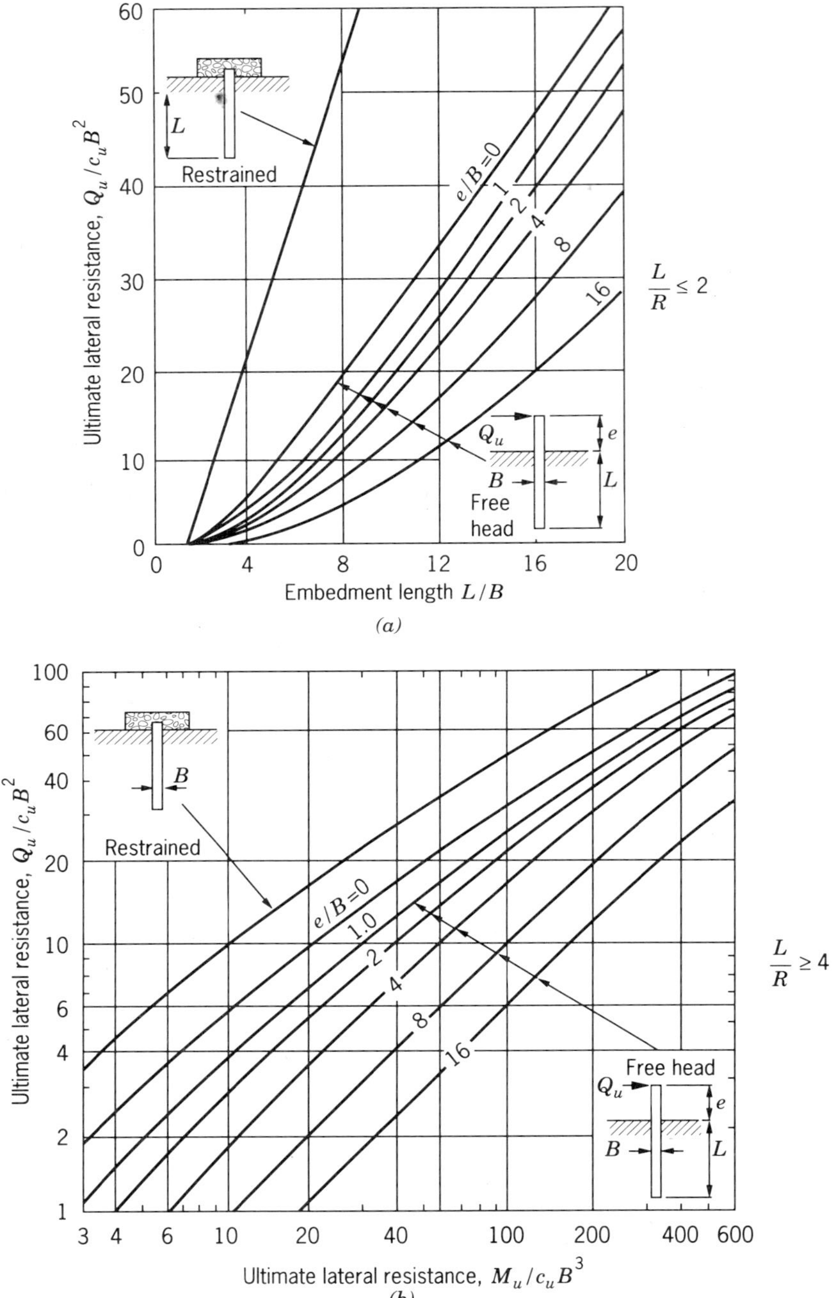


**Figure 6.28** Ultimate lateral load capacity of short and long piles in cohesive soils (a) Short piles, (b) long piles (Broms, 1964a).

for these piles can be obtained from the following relationship:

$$Q_u = \frac{2M_u}{(1.5B + 0.5x_0)} \tag{6.75}$$

This relationship is plotted in Figure 6.28b where

$$x_0 = \frac{Q_u}{9c_u B} \tag{6.72}$$

***Example 6.7*** Soil conditions at a site consist of very stiff clay to 45 ft (13.7 m) below ground. Laboratory tests on undisturbed samples of this clay showed that unconfined compressive strength ($q_u$) is 4 kips/ft$^2$ (200 kN/m$^2$) and unit weight ($\gamma$) is 125 lb/ft$^3$. Steel pipe piles 12.75-in. (273 mm) outside diameter and 0.25-in. (6.4 mm) wall thickness 35 ft (10.7 m) long are to be driven at this site. Assume that Young's modulus $E$ for steel is 29,000 ksi (20 MN/m$^2$) and yield strength $f_y$ is 35 ksi (241 M Pa). Calculate the allowable lateral load ($Q_{\text{all}}$) for a factor of safety of 2.5 for the following cases:

(a) Free-head pile.
(b) Fixed-head pile.

$k$ for soil pile system is 141.74 kips/ft$^3$.

SOLUTION

(a) *Free-head pile*

$$E = 29{,}000 \times 144\,\text{ksf} = 4176 \times 10^3\,\text{ksf}$$

$$I = \frac{\pi}{64}(12.75^4 - 12.25^4) = 192\,\text{in.}^4 = 0.0093\,\text{ft}^4$$

$$EI = 4176 \times 10^3 \times 0.0093 = 38.84 \times 10^3\,\text{kips-ft}^2$$

$$Z = \frac{I}{B/2} = \frac{192 \times 2}{12.75} = 30.1\,\text{in.}^3 = 0.0175\,\text{ft}^3$$

$$M_u = Zf_b = Z(0.6f_y) = 0.0175 \times 0.6 \times 35 \times 144 = 52.92\,\text{kips-ft}$$

$$R = \left(\frac{EI}{K}\right)^{1/4} = \left(\frac{38.84 \times 10^3}{141.74}\right)^{1/4} = 4.07\,\text{ft}$$

$$\frac{L}{R} = 35/4.07 = 8.6 > 3.5 \text{ long piles}$$

$$\frac{M_u}{c_u B^3} = \frac{52.92}{2(1.063)^3} = 22.03,\ c_u = \frac{q_u}{2} = 2\,\text{kips/ft}^2$$

From Figure 6.28b, for $M_u/c_u B^3 = 22.03$ and $e/B = 0$, $Q_u/c_u B^2 = 11$.

$$Q_u = 11 \times 2\,(1.063)^2 = 24.8\,\text{kips}$$

Using a factor of safety of 2.5,

$$Q_{\text{all}} = \frac{24.8}{2.5} = 9.9 \text{ kips}$$

(b) *Fixed-head pile*

From Figure 6.28b, for $M_u/C_uB^3 = 22.03$, $Q_u/c_uB^2 = 18$.

$$Q_u = 18 \times 2\,(1.063)^2 = 40.7 \text{ kips}$$

$$Q_{\text{all}} = \frac{40.7}{2.5} = 16.3 \text{ kips}$$

## 6.5 ULTIMATE LATERAL LOAD RESISTANCE OF PILE GROUPS IN COHESIVE SOIL

As mentioned in Section 6.1.2, only limited data are available on ultimate lateral load resistance of pile groups. The concept of group efficiency, $G_e$, as discussed for cohesionless soils can also be applied for cohesive soils. Thus ultimate lateral load resistance $(Q_u)_G$ of a group can be calculated from the following relationship:

$$(Q_u)_G = G_e n Q_u \tag{6.76}$$

where $n$ is the number of piles in a group, and $Q_u$ is the ultimate lateral resistance of a single pile.

**TABLE 6.14 Group Efficiency $G_e$, for Piles in Cohesive Soils**[a]

| | $G_e$ | | |
|---|---|---|---|
| $S/B$ | 2 × 2 group | 3 × 3 group | Recommended |
| 1 | 2 | 3 | 4 |
| 3 | 0.42 | 0.39 | 0.40 |
| 3.5 | 0.50 | 0.42 | 0.45 |
| 4.0 | 0.57 | 0.44 | 0.50 |
| 4.5 | 0.61 | 0.47 | 0.55 |
| 5.0 | 0.63 | 0.48 | 0.55 |
| 6.0[b] | — | — | 0.65 |
| 8.0[b] | — | — | 1.00 |

$S$ = center-to-center pile spacing.
$B$ = Pile diameter or width.
[a] These values have been obtained from curves provided by Prakash and Saran (1967).
[b] = Extrapolated values.

A series of model pile groups had been tested for lateral loads in clay by Prakash and Saran (1967). The group efficiency, $G_e$, from these tests can be used in equation (6.76). These values are presented in Table 6.14. The piles tested had $L/B = 32$, and the two groups tested consisted of a $2 \times 2$ and $3 \times 3$ set of piles. The group efficiency for the $3 \times 3$ set was found to be lower than the values for the $2 \times 2$ group. Also, when compared with cohesionless soils (Table 6.2) the $G_e$ values for cohesive soils are lower.

Table 6.14 shows $G_e$ values that were interpolated from the graph. The ultimate lateral load resistance of pile group can then be calculated by using equation (6.76) and Table 6.14. There is a need to carry out further laboratory and full-scale tests on pile groups.

Since $G_e$ values beyond $S/B = 5$ are not available, a value of 0.65 can be used for $S/B = 6$ by extrapolation and $G_e$ can be taken as unity for $S/B \geqslant 8$.

## 6.6 LATERAL DEFLECTION OF A SINGLE PILE IN COHESIVE SOILS

The two methods that can be used to calculate lateral deflection of a single pile in cohesive soils are the subgrade reaction approach and the elastic approach. Theoretical aspects of these two approaches were discussed in the beginning of this chapter. Application of these two approaches are presented in the following paragraphs.

### 6.6.1 Subgrade Reaction Approach

For normally consolidated clays, the modulus of subgrade reaction increases linearly with depth. Therefore, for such clays the analysis and method of calculating deflection for lateral load presented in Section 6.1.3 shall apply.

For overconsolidated clays, subgrade modulus is constant with depth. For such clays, deflection coefficients $A$ and $B$ are defined as

$$\frac{y_A EI}{Q_g R^3} = A_{yc} \tag{6.77a}$$

$$\frac{y_B EI}{M_g R^2} = B_{yc} \tag{6.77b}$$

where

$A_{yc}, B_{yc}$ = Deflection coefficients in clay for $Q_g$ and $M_g$.

Letting $y = y_A + y_B$ as in equation (6.34), we get deflection $y$ at any depth.

$$y = A_{yc}\frac{Q_g R^3}{EI} + B_{yc}\frac{M_g R^2}{EI} \tag{6.78a}$$

Similarly, moment $M$ at any depth is

$$M = A_{mc} Q_g R + B_{mc} M_g \tag{6.78b}$$

Solutions for $A$ and $B$ coefficients similar to those presented in section 6.1.3 had

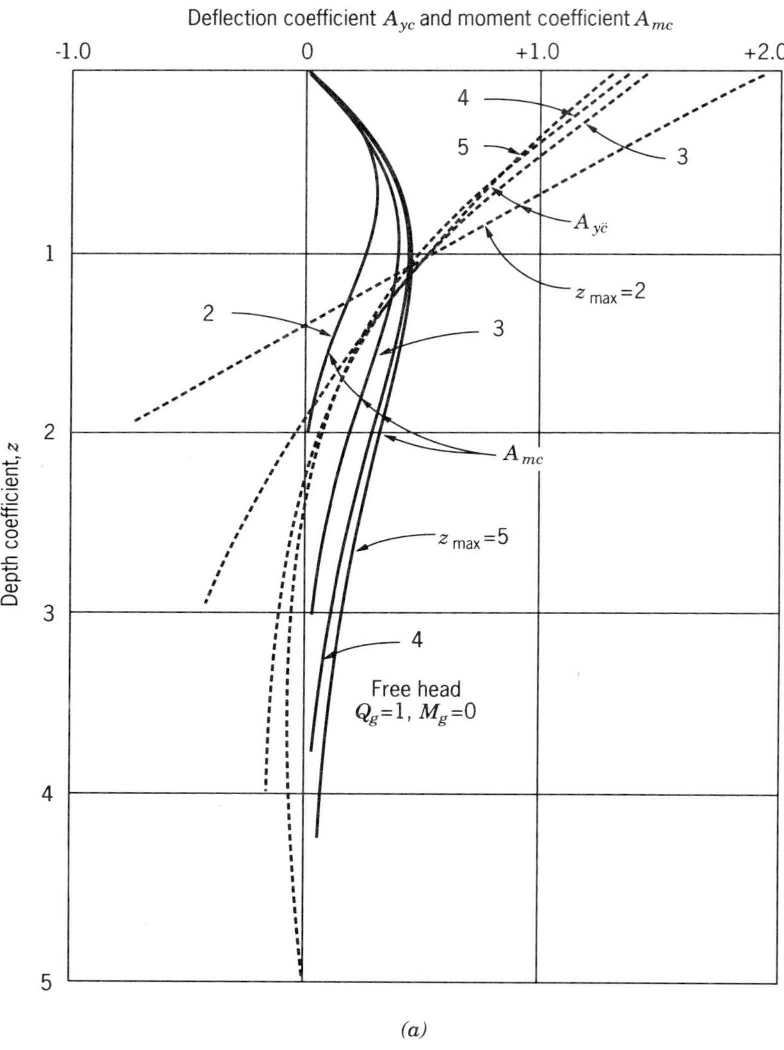


(a)

**Figure 6.29** Coefficients of moments and deflections for free-head pile in soils with constant soil modulus (a) Coefficients of deflections and bending moment for free-head pile carrying horizontal load at head and zero moment, (b) coefficients of deflections and bending moment for piles carrying moment at head and zero lateral load (Davisson and Gill, 1963).

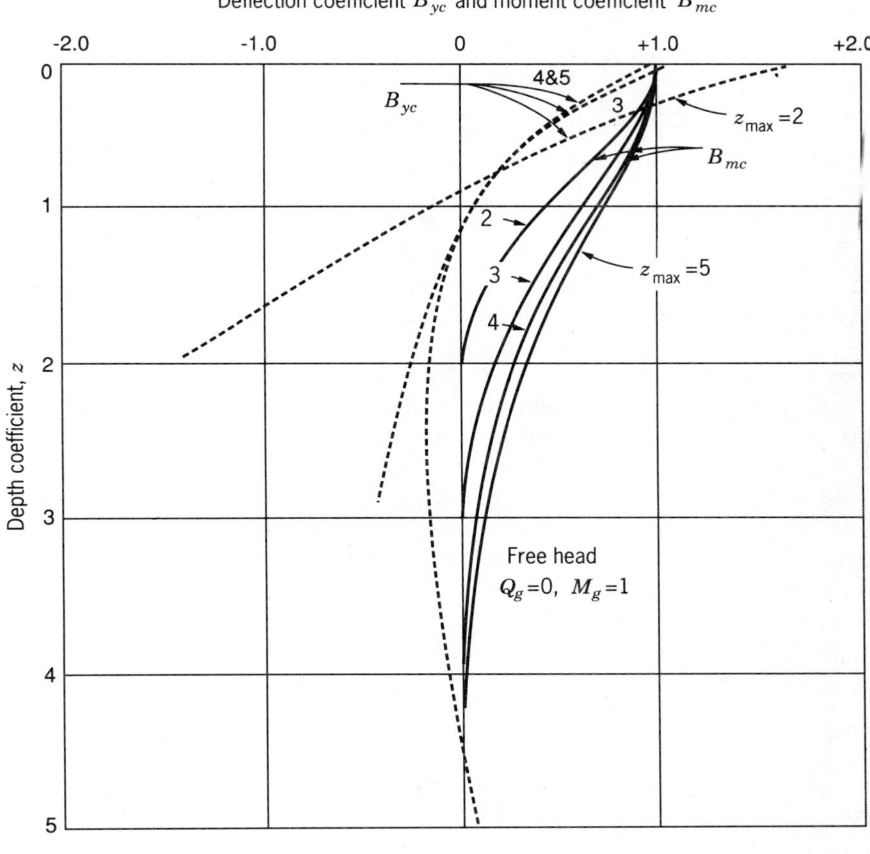


*(b)*

been obtained by Davisson and Gill (1963). In equation (6.41), by replacing $A_y$ with $A_{yc}$, we get

$$\frac{d^4 A_{yc}}{dz^4} + \phi(x) A_{yc} = 0 \tag{6.79}$$

Now putting $\phi\,(x) = 1$, $k_h = k$ and replacing $T$ with $R$, equation (6.33) becomes:

$$\frac{kR^4}{EI} = 1 \tag{6.80a}$$

$$R = \left(\frac{EI}{k}\right)^{1/4} \tag{6.80b}$$

and

$$Z = \frac{x}{R} \tag{6.80c}$$

Substituting the above equations in equation (6.79), the solutions for $A$ and $B$ coefficients can be obtained in a similar manner as for cohesionless soils (section 6.1.3)

The solutions for $A_{yc}$ and $A_{mc}$ have been plotted with nondimensional depth coefficient $z$ in Figure 6.29a and $B_{yc}$ and $B_{mc}$ in Figure 6.29b. It will be seen in Figure 6.29a that if $z_{\max}(= L/R) \leqslant 2$, the pile behaves as a rigid pile or a pole. And for $z_{\max}$ $(= L/R) \geqslant 4$, the pile behaves as an infinitely long pile.

***Example 6.8*** A 40-ft (12.2 m) long 10.75 in. (273 mm) outside diameter steel pile is driven into a clay with undrained strength ($S_u = c_u$) of 1.8 kips/ft$^2$ (85 k N/m$^2$). The pile has 0.25-in. (6.4 mm) wall thickness and Young's modulus for steel is $4176 \times 10^3$ ksf (20 MN/m$^2$). The pile head is free and is to be subjected to a lateral load. Due to superstructure requirements, the pile head cannot deflect more than 0.25 in. (6.4 mm). Calculate the maximum allowable lateral load and the corresponding maximum bending moment on the pile. There is no moment on the head.

SOLUTION From Table 4.16, the value of $k = 67c_u = 67 \times 1.8 = 120.6$ kips/ft$^2$

$$I = \frac{\pi}{64}(10.75^4 - 10.25^4) = 114 \text{ in.}^4 = 0.0055 \text{ ft}^4$$

$$EI = 4176 \times 10^3 \times 0.0055 = 22{,}968 \text{ kips-ft}^2$$

From equation (6.80b):

$$R = \left(\frac{22{,}968}{120.6}\right)^{1/4} = 3.73 \text{ ft}$$

$$\frac{L}{R} = \frac{40}{3.73} = 10.7$$

Deflection:

$$y_x = A_{yc}\frac{Q_g R^3}{EI} + B_{yc}\frac{M_g R^2}{EI} \tag{6.78a}$$

$$M = 0 \qquad y = \frac{0.25}{12} = 0.02 \text{ ft} \qquad EI = 22{,}968 \text{ kips-ft}^2 \qquad R = 3.73 \text{ ft}$$

$A_{yc} = 1.4$ from Figure 6.29a

Substituting these values in above equation, we get:

$$0.02 = 1.4 \frac{Q_g(3.73)^3}{22{,}968}$$

$$Q_g = 6.3 \text{ kips}$$

Moment:

$$M_g = A_{mc} Q_g R + B_{mc} M_g \tag{6.78b}$$

$$M_g = 0 \qquad Q_g = 6.3 \text{ kips} \qquad R = 3.73$$

From Figure 6.29a, the maximum $A_{mc} = 0.4$. Then substituting these values in above equation, we get:

$$M_{\max} = 0.4 \times 6.3 \times 3.73 = 9.4 \text{ kips-ft}$$

### 6.6.2 Application of *p–y* Curves to Cohesive Soils

As in the case of cohesionless soils (Section 6.1.4), the procedure for determination of *p–y* curves in cohesive soils has been proposed by Matlock (1970). The basis of these procedures is (1) field tests with an instrumented pile and (2) laboratory model testing.

Three loading conditions were considered for the design of laterally loaded piles in soft normally consolidated marine clay. These are (1) short-time static loading, (2) cyclic loading such as would occur during the progressive buildup of a storm, and (3) subsequent reloading with forces less than previous maxima.

In the field test, the steel test pile was 12.75 in. in diameter, and 35 pairs of electric resistance strain gauges were installed in the 42-ft embedded portion. The pile was calibrated to provide extremely accurate determinations of bending moment. Gauge spacings varied from 6 in. near the top to 4 ft in the lowest section.

The bending moment diagram with depth was differentiated and integrated twice to obtain the *p* and *y*, respectively, at any particular depth at different loads, and *p–y* curves were then drawn.

Basic theoretical aspects and the general concepts of *p–y* curves have been presented in detail in Section 6.1.4. In the following paragraphs, the procedures to establish *p–y* curves based on the above test program for soft and firm clays, for stiff clays, and for stiff overconsolidated clays are presented. With the help of these *p–y* curves, deflection and bending moment of a laterally loaded pile can be determined as a function of depth.

***Procedure for Establishing p–y Curves for Laterally Loaded Piles in Soft to Firm Clays*** The procedure for establishing *p–y* curves for soft to firm clays as described by Matlock (1970) consists of the following steps:

**Step 1** Carry out the field or laboratory testing to estimate the undrained strength $c_u$ and the unit weight $\gamma$ for the soil at the site.

**Step 2** Calculate the following factors:

$$x_r = \frac{6B}{\frac{\gamma B}{c_u} + J} \tag{6.81}$$

where

$x_r$ = critical depth below ground level
$B$ = pile width
$\gamma$ = unit weight of overburden soil
$c_u$ = undrained strength of clay
$J$ = an empirical factor

Based on experimental work, Matlock (1970) recommended $J = 0.5$ for soft clay and $J = 0.25$ for a stiff clay.

$$N_c = 3 + \frac{\gamma x}{c_u} + \frac{Jx}{B} \qquad \text{for } x < x_r \tag{6.82}$$

$$N_c = 9 \qquad \text{for } x \geqslant x_r \tag{6.83}$$

where

$N_c$ = the bearing capacity factor
$x$ = the depth below ground level

Other factors have been defined earlier.

**Step 3** First select a particular depth at which the $p$–$y$ curve will be drawn. Compare this depth with the critical depth $x_r$ and determine if equation (6.82) or (6.83) applies for $N_c$. Then calculate values for the $p$–$y$ curve (Figure 6.30) using the following steps.

**Step 4** Calculate the following:

$$p_u = c_u N_c B \tag{6.84}$$

where $p_u$ = the ultimate soil resistance per unit length of pile. Now, the deflection $y_c$ at soil resistance $p_u$ is the deflection corresponding to the strain $\varepsilon_c$ at the maximum stress resulting from the laboratory stress–strain curve in undrained triaxial compression and is expressed as

$$y_c = 2.5\varepsilon_c B \tag{6.85}$$

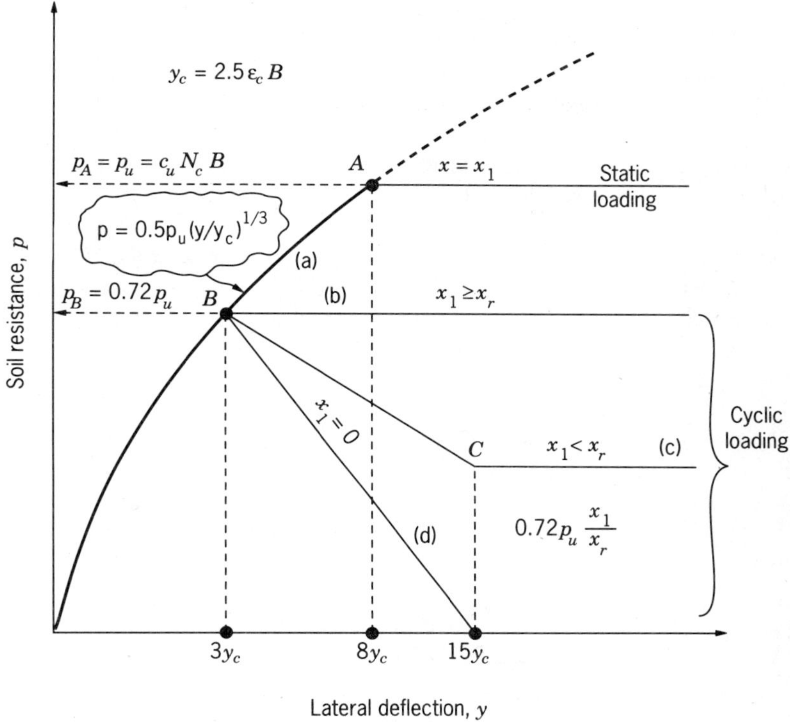


**Figure 6.30** Establishing the $p$–$y$ curve for soft to firm clay.

In the absence of laboratory test results, Matlock (1970) recommends the following $\varepsilon_c$ values: (1) 0.005 for brittle and sensitive clays, (2) 0.02 for disturbed or remolded clays or for unconsolidated sediments, and (3) 0.01 for normally consolidated clays.

**Step 5** The shape of the $p$–$y$ curve (curve $a$ in Figure 6.30) is defined by the following equation:

$$p = 0.5p_u(y/y_c)^{1/3} \tag{6.86}$$

If laboratory undrained triaxial compression test results are available, then the shape of the $p$–$y$ can be derived from that of the stress–strain curve. Alternatively, the shape of the curve can also be derived from the load-settlement curve in a plate load test.

The effect of cyclic loading on $p$–$y$ curve can be incorporated in Figure 6.30 in the following way:

1. $x_1 \geqslant x_r$: Cut off the $p$–$y$ curve at $p_B = 0.72p_u$ (curve $b$).

2. $x_1 < x_r$: Join point $B$ to $C$ and then draw a horizontal line (curve $c$). Point $C$ corresponds to lateral deflection $y = 15y_c$ and $p = 0.72p_u x_1/x_r$.
3. $x_1 = 0$: Join point $B$ to a point where $y = 15y_c$ (curve $d$). It is assumed that at $y \geqslant 15y_c$, soil resistance tends to zero.

**Step 6** Repeat the foregoing procedure for various depths to obtain $p$–$y$ curves with depth below ground surface.

***Procedure for Establishing p–y Curves for Laterally Loaded Piles in Stiff Clay*** According to Reese and Welch (1975), the procedure for establishing $p$–$y$ curves for laterally loaded piles in stiff clay consists of the following steps:

*Short-term Static Loading*

**Step 1.** Obtain the best possible estimates of the variation of the following: undrained shear strength $c_u$ with depth, effective unit weight $\gamma'$ with depth, and strain corresponding to one-half the maximum principal stress difference, $\varepsilon_{50}$. If value of $\varepsilon_{50}$ is not available, use a value of 0.005 or 0.010; the larger value is more conservative.

**Step 2** The ultimate soil resistance per unit length of the pile shaft $p_u$ is the smaller of the values obtained from following two equations:

$$p_u = \left(3 + \frac{\gamma' x}{c_u} + 0.5\frac{x}{B}\right)c_u B \tag{6.87}$$

$$p_u = 9C_u B \tag{6.88}$$

where

$\gamma'$ = average effective unit weight of soil from ground surface to depth $x$
$c_u$ = average undrained shear strength from ground surface to depth $x$
$B$ = width of the pile

**Step 3** Compute the deflection $y_{50}$ at one-half the ultimate soil resistance from the following:

$$y_{50} = 2.5B\varepsilon_{50} \tag{6.89}$$

**Step 4** Points describing the $p$–$y$ curve may then be obtained by the following equations:

$$p = 0.5p_u(y/y_{50})^{1/4} \tag{6.90}$$

$$p = p_u \text{ for } y \text{ greater than } 16y_{50}$$

*Cyclic Loading*

**Step 1** Obtain $p$–$y$ curves for short-term static loading by the procedure described previously.

**Step 2** Make an estimate of the number of times the design lateral load will be applied to the foundation.

**Step 3** For several values of $p/p_u$, obtain the values of $C_1$ and $C_2$ by using data from laboratory tests and the following equation:

$$C_2 = \frac{(\varepsilon_c - \varepsilon_i)}{(\varepsilon_{50} \log N)} \qquad \text{from laboratory data}$$

where

$\varepsilon_c$ = strain after $N$ cycles of repeated loading
$\varepsilon_i$ = strain on initial loading
$C_1$ is given by equation (6.91)

In absence of the laboratory tests, use the following equation to determine the value of $C$:

$$C_1 = C_2 = C = 9.6R^4 \tag{6.91}$$

where

$R = (p/p_u) = (\sigma_1 - \sigma_3)/(\sigma_1 - \sigma_3)_{max}$ = stress ratio

**Step 4** At the value of $p$ corresponding to the values of $(p/p_u)$ selected in step (3), compute new values of $y$ for cyclic loading from the following:

$$y_c = y_s + y_{50} C_1 \log N \tag{6.92}$$

where

$y_c$ = deflection after $N$ cycles of repeated loading
$y_s$ = deflection upon initial loading
$C_1$ = a parameter describing the effect of repeated loading on deflection and is equal to $9.6R^4$ as discussed previously.
$N$ = number of cycles of repeated loading

**Step 5** The "$p$–$y_c$" curves define the soil response after $N$ cycles of load.

***Procedure for Establishing p–y Curves for Laterally Loaded Piles in Stiff, Overconsolidated Clays*** Methods previously described apply to soils that have modulus linearly increasing with depth. These soils are either granular materials or normally consolidated clays. For stiff, overconsolidated clays, the soil modulus is constant with depth. Only limited experimental data are available for establishing $p$–$y$ curves for such soils.

Based on the load-test data and analysis, Bhushan et al. (1979) conclude that for short, rigid piers in stiff, overconsolidated clays, procedures proposed by Reese and Welch (1975) for piles in stiff clays will apply with the following modifications:

$$p_u = \left(3 + \frac{\gamma' x}{c_u} + 2\frac{x}{B}\right) c_u B \tag{6.93}$$

$$y_{50} = 2B\varepsilon_{50} \tag{6.94}$$

$$p = 0.5/p_u(y/y_{50})^{1/2} \tag{6.95}$$

All the terms are explained in the earlier section for stiff clay, and procedures described there for establishing the $p$–$y$ curve shall also apply here if equations (6.87), (6.89) and (6.90) are replaced with equations (6.93), (6.94), and (6.95), respectively.

These procedures are applicable to single pile only. Their application to pile groups requires further research.

***Example 6.9*** A 40-in. (1004 mm) diameter steel pipe pile is to be driven into a medium consistency (firm) clay with undrained shear strength, $c_u = 1\ \text{kip/ft}^2$ (47.9 kN/m²) and the unit weight of 120 lb/ft³ (1920 kg/m³). Draw the $p$–$y$ curve at depths $x = 0, x = 5$ ft (1.5 m), and $x = 10$ ft (3.0 m) below the ground surface for cyclic loadings.

SOLUTION As given above $c_u = 1\ \text{kip/ft}^2$, $\gamma = 120\ \text{lb/ft}^3$.

$$x_r = \frac{6 \times 3.33}{\dfrac{(120 \times 3.33)}{1000} + 0.25} = 30.757\ \text{ft} \tag{6.81}$$

where

$$B = \frac{40}{12} = 3.33\ \text{ft}$$

and

$J = 0.25$ for firm clay

$$N_c = 3 + \frac{120x}{1000} + \frac{0.25x}{3.33} \tag{6.82}$$

for $x < x_r(= 30.757\ \text{ft})$

$$N_c = 3 + 0.195x$$

Select $x = 0, N_c = 3.$

$$p_u = 1 \times 3 \times 3.33$$
$$= 9.99 \text{ kips/ft} \tag{6.84}$$
$$y_c = 2.5 \times 0.01 \times 3.33$$
$$= 0.083 \text{ ft} \tag{6.85}$$

where $\varepsilon_c = 0.01$.

The shape of the $p$–$y$ curve can be defined by $p = 0.5x\ 9.99(y/0.083)^{1/3}$. (6.86)

or

$p = 11.357\, y^{1/3}$, $p$ in kips/feet, and $y$ in feet.

For the effect of cyclic loading, $p_B = 0.72 p_u$, where $p_B$ will define point $B$ on the $p$–$y$ curve.

$p_B = 0.72 \times 9.99 = 7.19$ kips/ft

Therefore, the curve will have following points for $x = 0$ depth:

| | |
|---|---|
| $y = 8y_c = 0.664$ ft | $p_u = 9.99$ kips/ft |
| $= 0$ | $p = 0$ |
| $= 1/12$ ft | $= 5$ kips/ft |
| $= 2/12$ ft | $= 6.287$ kips/ft |
| $= 3y_c = 0.249$ ft | $= 7.178$ kips/ft |
| $= 15y_c = 1.245$ ft | $= 0$ for cyclic loading |

These values are plotted in curve (a) in Figure 6.31.

For $x = 5$ ft.

$$N_c = 3.975 \qquad p_u = 1 \times 3.975 \times 3.33 = 13.236 \text{ kips/ft} \qquad y_c = 0.083 \text{ ft}$$
$$p = 0.5 \times 13.236(y/0.083)^{1/3} = 15.047 y^{1/3}$$
$$p_B = 0.72 \times 13.236 = 9.529 \text{ kips/ft at } y = 3y_c = 0.249 \text{ ft}$$

$$p_B \text{ at } y = 15y_c = 1.245 \text{ ft is } p = p_B \frac{x}{x_r} = 9.529 \times \frac{5}{30.757} = 1.549 \text{ kips/ft}$$

| | |
|---|---|
| $y = 8y_c = 0.664$ ft | $p_u = 13.236$ kips/ft |
| $= 0$ | $p = 0$ |
| $y = 1/12$ ft | $p = 6.627$ kips/ft |
| $= 2/12$ ft | $= 8.330$ kips/ft |

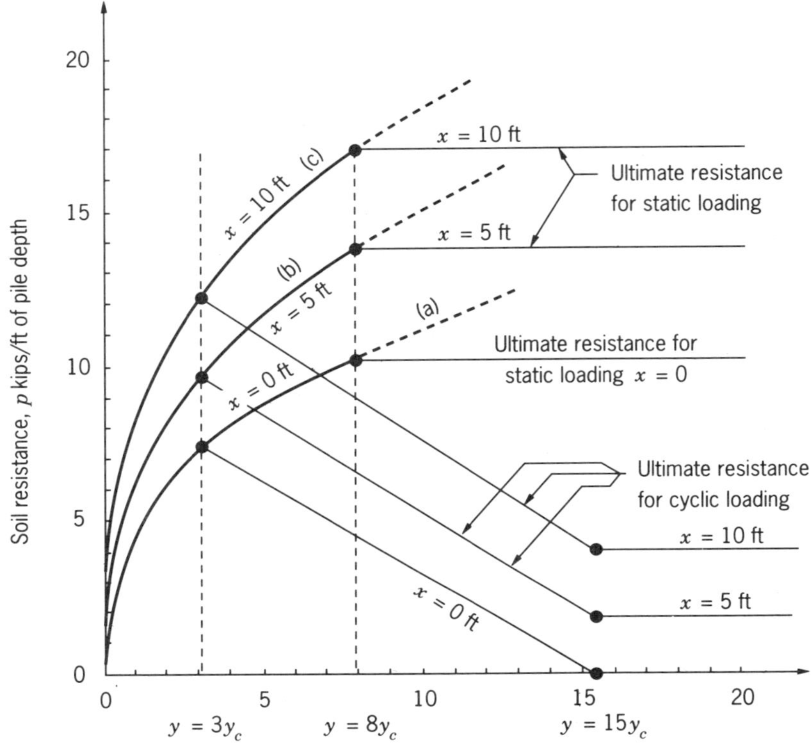


**Figure 6.31** $p$–$y$ curves at different depths (Example 6.9).

$$= 3y_c = 0.249 \text{ ft} \qquad p_b = 9.529 \text{ kips/ft}$$

$$= 15y_c = 1.245 \text{ ft} \qquad p = 1.549 \text{ kips/ft}$$

For $x = 10$ ft:

$$N_c = 4.95 \qquad p_u = 1 \times 4.95 \times 3.33 = 16.483 \text{ kips/ft} \qquad y_c = 0.083 \text{ ft}$$

$$p = 0.5 \times 16.483(y/0.083)^{1/3} = 18.739y^{1/3}$$

$$p_B = 0.72 \times 16.483 = 11.867 \text{ kips/ft at } y = 3y_c = 0.249 \text{ ft}$$

$$y = 15y_c = 1.245 \text{ ft} \qquad p = 11.867 \times \frac{10}{30.757} = 3.858 \text{ kips/ft}$$

$$= 8y_c = 0.664 \text{ ft} \qquad p_u = 16.483 \text{ kips/ft}$$

$$= 0 \qquad p = 0$$

$= 1/12\,\text{ft}$ $\quad\quad = 8.253\,\text{kips/ft}$

$= 2/12\,\text{ft}$ $\quad\quad = 10.374\,\text{kips/ft}$

$= 3y_c = 0.249\,\text{ft}$ $\quad\quad = p_b = 11.867\,\text{kips/ft}$

$= 15y_c = 1.245\,\text{ft}$ $\quad\quad = 3.858\,\text{kips/ft}$

Figure 6.31 shows the $p$–$y$ curve for these depths $x = 0$ (curve (a)) $x = 5\,\text{ft}$ (curve (b)), and $x = 10\,\text{ft}$ (curve (c)).

### 6.6.3 Application of the Elastic Approach

As discussed earlier, the soil displacements in this approach have been evaluated from the Mindlin equation for horizontal loads within a semiinfinite elastic mass, and the pile displacements have been determined by using the equation of flexure of a thin strip.

Solutions for lateral deflections and maximum moments for normally consolidated clays whose soil modulus increases with depth are similar to those for cohesionless soils. Therefore, solutions provided in Section 6.1.5 are also applicable in this case.

Poulos (1971) obtained solutions for deflection and maximum moments for laterally loaded piles in soils whose modulus is constant with depth (e.g., overconsolidated clays). Solutions described below for free-head and fixed-head piles may be used for such soils.

***Free-Head Piles*** The lateral displacement $y$ for a free-head pile can be expressed by the following relationship:

$$y = I_{\rho H}\frac{Q_g}{E_s L} + I_{\rho m}\frac{M_g}{E_s L^2} \tag{6.96}$$

where $I_{\rho H}$ and $I_{\rho m}$ are given in Figure 6.32. Similarly, the rotation $\theta$ of a free-head pile at the ground surface is given by:

$$\theta = I_{\theta H}\frac{Q_g}{E_s L^2} + I_{\theta M}\frac{M_g}{E_s L^2} \tag{6.97}$$

where $I_{\theta H}$ and $I_{\theta M}$ are influence factors as given in Figures 6.32 and 6.33 respectively. As shown in these figures, the influence factors are a function of pile flexibility factor $K_R$, which is expressed as follows:

$$K_R = \frac{E_p I_p}{E_s L^4} \tag{6.98}$$

Typical values of $K_R$ for various types of piles and soils are given in Table 6.15.

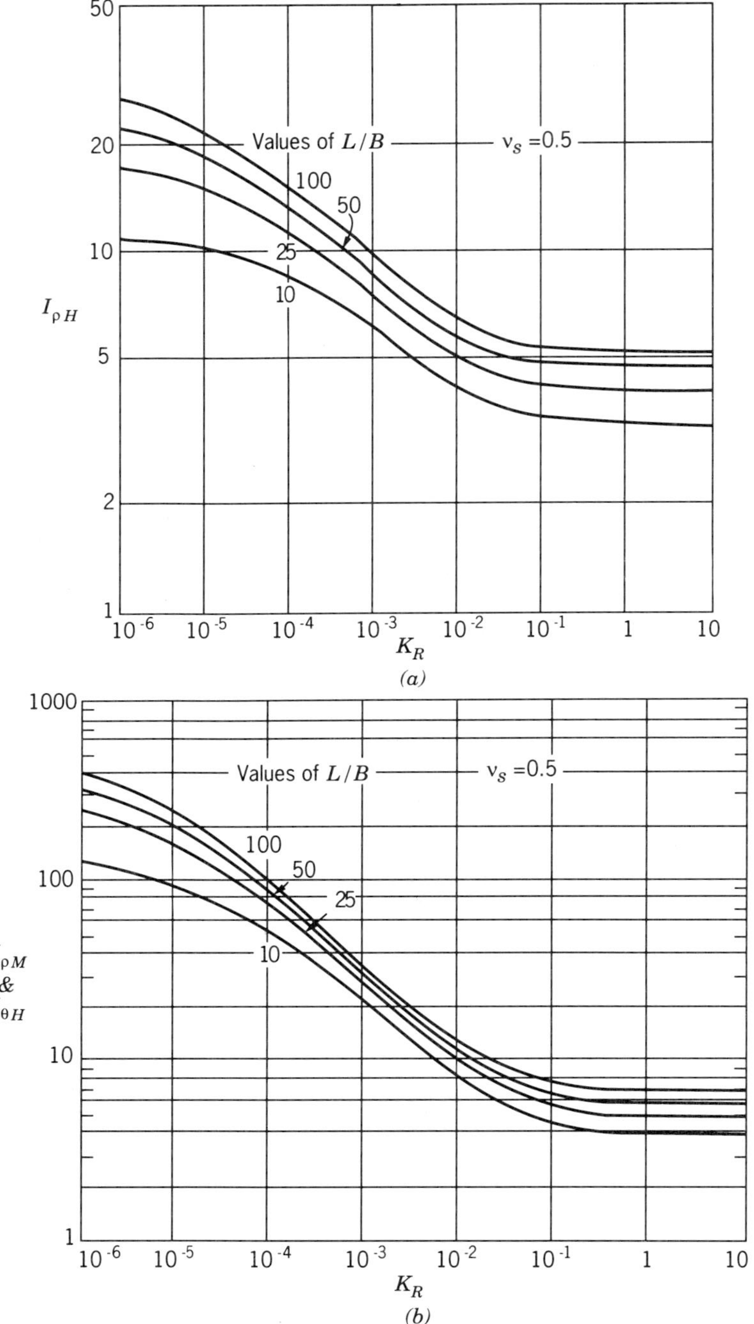


**Figure 6.32** Influence factor $I_{\rho H}$, $I_{\rho M}$, and $I_{\theta M}$ for constant modulus (a) $I_{\rho H}$ versus $K_R$ for free-head piles, (b) $I_{\rho M}$ and $I_{\theta H}$ versus $K_R$ for free-head piles (Poulos, 1971a).

**TABLE 6.15 Typical Values of Pile Flexibility Factor $K_R$ for Various Soils (Poulos, and Davis 1980)**

| Pile Type | Soft Clay | | Medium Clay | | Stiff Clay | | Loose Sand | | Dense Sand | |
|---|---|---|---|---|---|---|---|---|---|---|
| | Pile length, in feet | | | | | | | | | |
| | 20 | 50 | 20 | 50 | 20 | 50 | 20 | 50 | 20 | 50 |
| 1-ft diameter concrete | $6.2 \times 10^{-3}$ | $1.6 \times 10^{-4}$ | $3.1 \times 10^{-3}$ | $8.0 \times 10^{-5}$ | $1.2 \times 10^{-3}$ | $3.1 \times 10^{-5}$ | $3.7 \times 10^{-3}$ | $9.5 \times 10^{-5}$ | $9.2 \times 10^{-4}$ | $2.4 \times 10^{-5}$ |
| 3-ft diameter concrete | $5.0 \times 10^{-1}$ | $1.3 \times 10^{-2}$ | $2.5 \times 10^{-1}$ | $6.4 \times 10^{-3}$ | $9.4 \times 10^{-2}$ | $2.4 \times 10^{-3}$ | $3.0 \times 10^{-1}$ | $7.7 \times 10^{-3}$ | $7.5 \times 10^{-2}$ | $1.9 \times 10^{-3}$ |
| 1-ft (average) diameter timber | $3.1 \times 10^{-3}$ | $7.9 \times 10^{-5}$ | $1.5 \times 10^{-3}$ | $3.8 \times 10^{-5}$ | $6.0 \times 10^{-4}$ | $1.5 \times 10^{-5}$ | $1.8 \times 10^{-3}$ | $4.6 \times 10^{-5}$ | $4.7 \times 10^{-4}$ | $1.2 \times 10^{-5}$ |
| 14-in. × 14-in. × 117-1b steel H-pile | $2.7 \times 10^{-2}$ | $6.9 \times 10^{-4}$ | $1.3 \times 10^{-2}$ | $3.4 \times 10^{-4}$ | $5.0 \times 10^{-3}$ | $1.3 \times 10^{-4}$ | $1.6 \times 10^{-2}$ | $4.1 \times 10^{-4}$ | $4.0 \times 10^{-3}$ | $1.0 \times 10^{-4}$ |

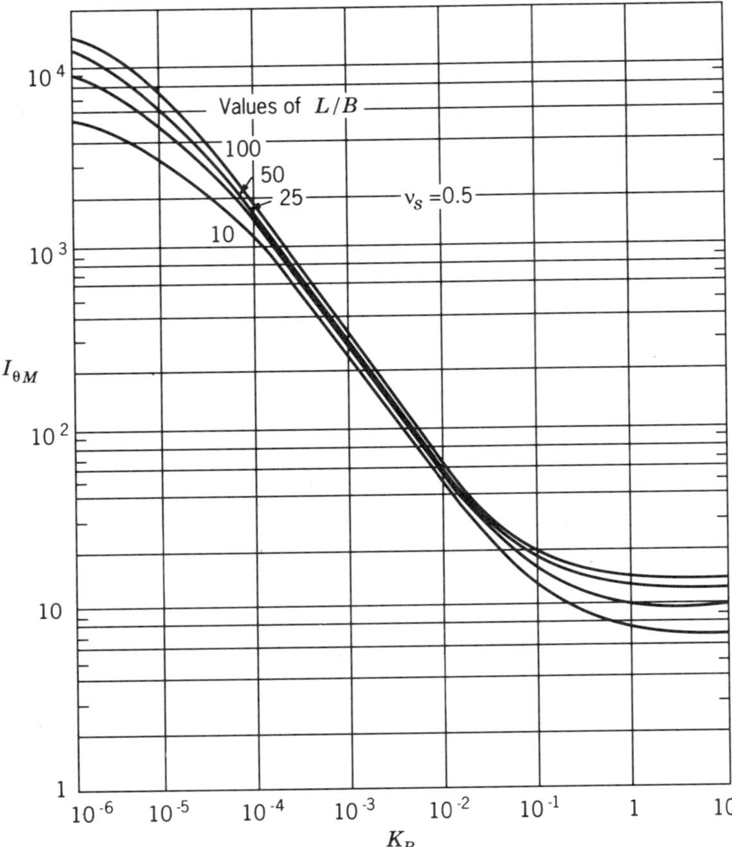


**Figure 6.33** Influence factors $I_{\theta M}$ versus $K_R$ for free-head pile with constant soil modulus (Poulos, 1971a).

The values of $E_s$ from which the $K_R$ values are derived have been obtained from the empirical correlation, $E_s = 40c_u$, where $c_u$ is undrained shear strength of soil (Poulos, 1971). The maximum moment in a free-head pile subjected to horizontal load can be obtained from Figure 6.34.

***Fixed-Head Pile*** The horizontal displacement $y$ of a fixed-head pile is obtained from the following relationship:

$$y = I_{\rho F} \frac{Q_q}{E_s L} \tag{6.99}$$

where $I_{\rho F}$ is an influence factor and can be obtained from Figure 6.35.

For a fixed-head pile the maximum moment at the pile head ($M_f$) can be obtained from Figure 6.36.

***Example 6.10*** A 12-in. (305 mm) diameter concrete pile is installed in a clay with unconfined compressive strength of 3 kips/ft$^2$ (144 kN/m$^2$). The pile is 20 ft (6 m) long and is subjected to a lateral load of 20 kips (89 kN) and a moment of 30 kip-ft (40.7 kN-m) at its free head. Calculate the deflection of pile head and the maximum moment in the pile.

SOLUTION The unconfined compressive strength of 3 kips/ft$^2$ indicates that the soil is stiff clay. From Table 6.15 for a 12-in. diameter concrete pile in stiff clay

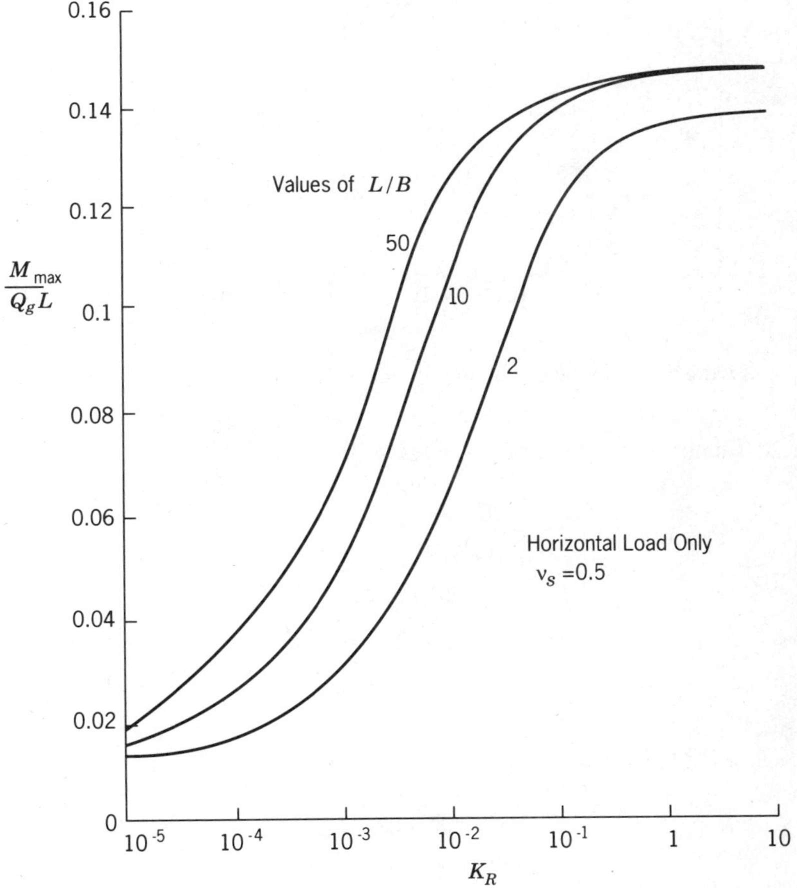


**Figure 6.34** Maximum moment in free-head pile (Poulos, 1971a).

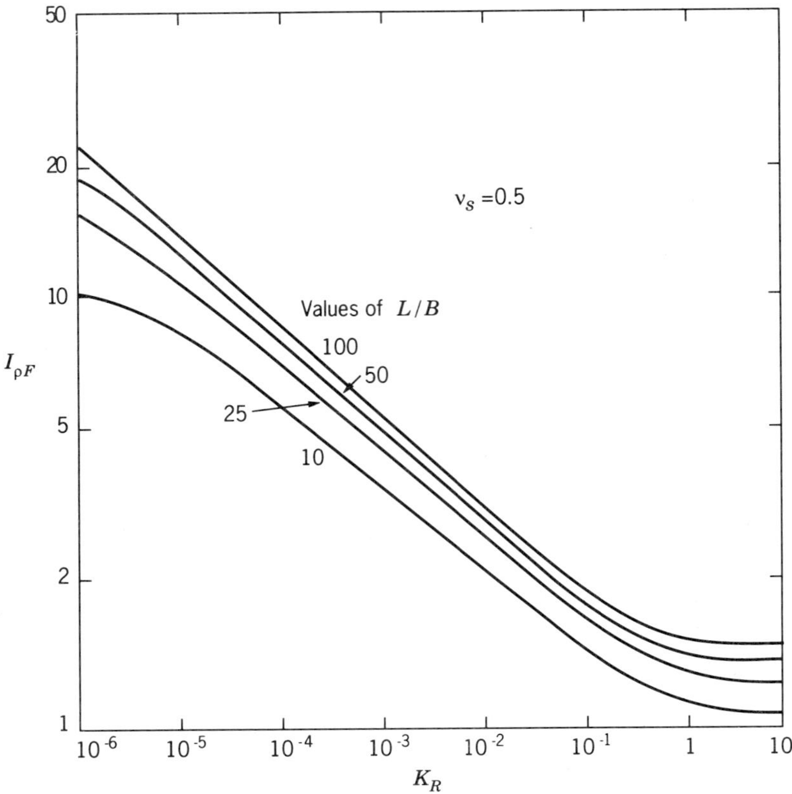


**Figure 6.35** Influence factors $I_{\rho F}$ for fixed-head pile (Poulos, 1971a).

with 20 ft length, the value of $K_R = 1.2 \times 10^{-3}$.

$$\frac{L}{B} = \frac{20}{1}$$

$$I_{\rho H} = 8 \text{ from Figure 6.32a}$$

$$I_{\rho M} = 45 \text{ from Figure 6.32b}$$

$$E_s = 40 \times \frac{3}{2} = 60 \text{ kips/ft}^2$$

where $c_u = q_u/2 = 3/2 = 1.5 \text{ kips/ft}^2$ and $E_s = 40c_u$

$$Q_g = 20 \text{ kips}$$

$$M_g = 30 \text{ kips-ft}$$

$$L = 20 \text{ ft}$$

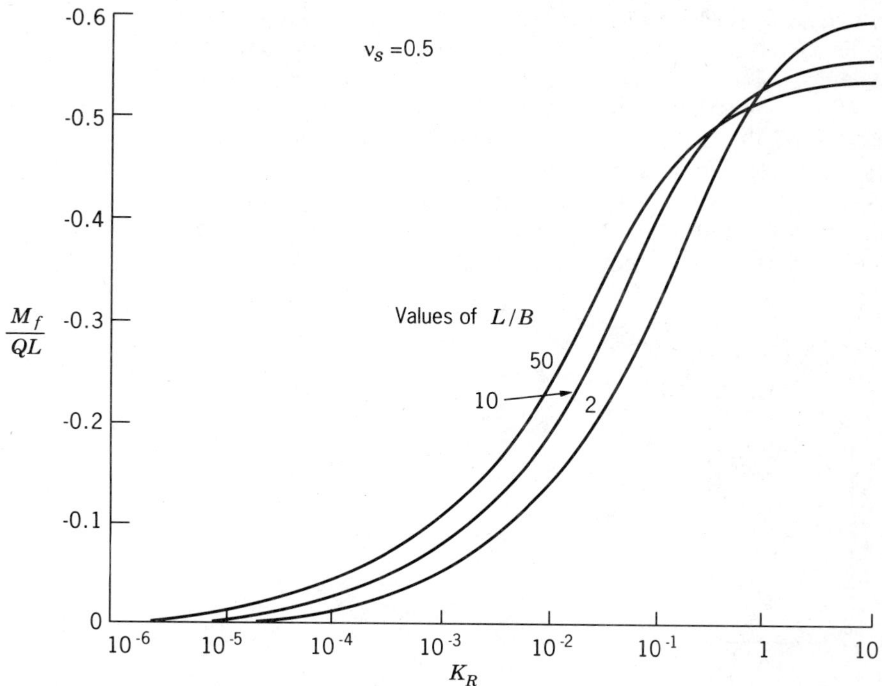


**Figure 6.36** Fixing moment at head of fixed-head pile (Poulos, 1971a).

Then from equation (6.96)

$$y = I_{\rho H}\frac{Q_g}{E_s L} + I_{\rho M}\frac{M_g}{E_s L^2} = \frac{8 \times 20}{60 \times 20} + \frac{45 \times 30}{60(20)^2} = 0.189\,\text{ft} = 2.3\,\text{in.}$$

The maximum moment can be obtained from Figure 6.34.

$$\text{For } K_R = 1.2 \times 10^{-3} \qquad \text{and } \frac{L}{B} = 20,$$

$$\frac{M_{\max}}{Q_g L} = 0.083$$

$$M_{\max} = 20 \times 20 \times 0.083 = 33.2\,\text{kips-ft}$$

## 6.7 LATERAL DEFLECTION OF PILE GROUPS IN COHESIVE SOIL

As discussed in Section 6.2 if piles in a group are spaced at less than $8B$ parallel to the direction of lateral load, individual pile capacity needs to be reduced. The reduction factor is applied to the modulus of horizontal subgrade reaction. Table 6.6 gives the values of these group reduction factors for various pile spacings.

With appropriately reduced modulus of subgrade reaction for pile spacing parallel to the direction of the lateral load, the individual pile capacity for allowable deflection can then be calculated by the procedure discussed in Section 6.6 (subgrade reaction approach). Pile group capacity will then be the sum of individual pile capacities calculated on the basis of reduced $k_h$. Poulos interaction factors are equally applicable to cohesive soils.

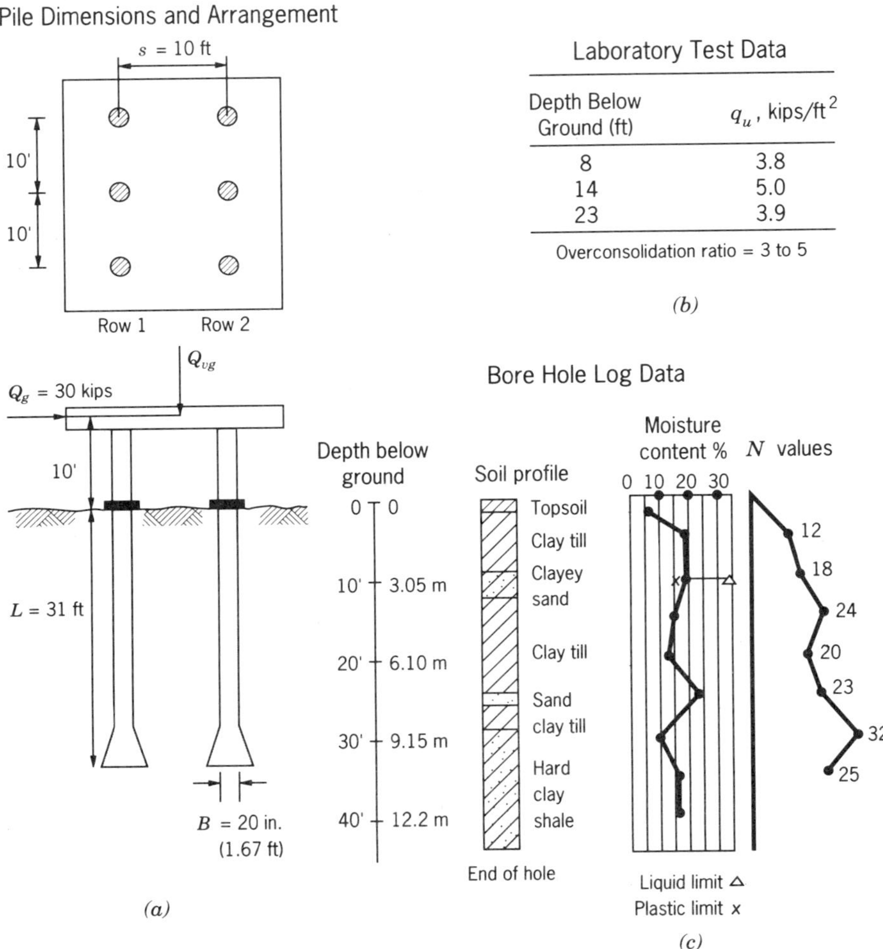


**Figure 6.37** (a) Pile dimensions and arrangement (b) soil properties and (c) soil profile along the pile depth. $S/B = 6$, $N$ = standard penetration values, $q_u$ = unconfined compressive strength on undisturbed sample (Example 6.11).

***Example 6.11*** As shown in Figure 6.37, a group of six piles is supporting a chemical storage tank above ground. The piles are installed through stiff to very stiff consistency clay till and support the truss above ground. Axial downward loads required that the pile tips be located at least 2 ft into the clay shale. The piles selected were drilled cast-in-place concrete piles having a 20-in. (500 mm) shaft diameter and 31 ft (9.5 m) length. The modulus of elasticity $E$ for concrete is $3.6 \times 10^6$ psi ($25 \times 10^6$ kN/$m^2$) and the moment resistance $M_u$ for pile shaft section is 75 kips-ft (102 kN-m). The pile heads are assumed to be free. Check if the pile group can resist a total lateral load, $Q_g = 30$ kips. The maximum deflection that the pile head at ground surface can undergo is 0.25 in. (6.35 mm).

SOLUTION

(a) *Soil Profile* Figure 6.37 shows the soil profile and the results of laboratory test data. From this information:

$$\text{Average } q_u = 4.24\,\text{kips/ft}^2$$

(b) *Pile Dimensions and Arrangement* As shown in Figure 6.37 piles are 31 ft long and have an $S/B$ ratio of 6. The arrangement and pile dimensions were established to meet axial compression loading requirements.

(c) *Calculation of Ultimate Lateral Resistance* Since the clay is overconsolidated and $q_u$ is constant with depth, the soil modulus will be constant with depth.

(i) *Single piles*

$$I = \frac{\pi B^4}{64} = \frac{\pi(1.67)^4}{64} = 0.382\,\text{ft}^4$$

$$EI = \frac{3.6 \times 10^6}{1000} \times 144 \times 0.382 = 198.028 \times 10^3\,\text{kips-ft}^2$$

$$R = \left(\frac{EI}{K}\right)^{1/4} = \left(\frac{198{,}028}{142}\right)^{1/4} = 6.1\,\text{ft}$$

$$K \simeq 67S_u\ (\text{Table 4.16}); \quad c_u = S_u$$

$$L/R = \frac{31}{6.1} = 5.0 > 3.5\ (\text{pile behavior as a long pile})$$

Ultimate lateral resistance of a free-head long pile:

$$\frac{M_u}{c_u B^3} = \frac{75}{2.12(1.67)^3} = 7.59 \qquad c_u = \frac{q_u}{2} = 2.12\,\text{kips/ft}^2$$

$$\frac{e}{B} = 10/1.67 = 6$$

From Figure 6.28b, for $e/B = 6$ and $M_u/c_u B^3 = 7.59$:

$$\frac{Q_u}{c_u B^2} = 1.57 \qquad Q_u = 1.57 \times 2.12(1.67)^2 = 9.3\,\text{kips}$$

with a factor of safety = 2.5, $Q_{\text{all}} = 9.3/2.5 = 3.7$ kips.

(ii) *Pile group* Since pile spacing is 6B, there will be interference between piles. From equation 6.76, using $Q_{\text{all}}$ for $Q_u$, we get

$$(Q_{\text{all}})_g = 6 \times 0.65 \times 3.7 = 14.43\,\text{kips where } G_e = 0.65 \text{ from Table 6.14}$$

(d) *Calculation of Lateral Resistance for Allowable Lateral Deflection*

(i) *Single piles*

$$y_{x=0} = A_{yc}\frac{Q_g R^3}{EI} + B_{yc}\frac{M_g R^2}{EI} \quad \text{from equation (6.78a)}$$

$$y_{x=0} = 0.25\,\text{in.} = 0.02\,\text{ft} \qquad R = 6.1 \qquad EI = 198{,}028\,\text{kips-ft}^2$$

$M = Q(10)$ since lateral load to applied at 10 ft above the pile head.

From Figure 6.29, for $Z_{\max} = L/R = 5$ at $x/R = 0$, $A_{yc} = 1.4$, $B_{yc} = 1$. Substituting these values in above equation, we get:

$$0.02 = 1.4\frac{Q_g(6.1)^3}{198{,}028} + 1\frac{10Q_g(6.1)^2}{198{,}028}$$

$$= 0.0016Q_g + 0.0018Q_g \qquad Q_g = 5.9\,\text{kips}$$

(ii) *Pile group* For $S/B = 6$, group reduction factor = 0.7 (from Table 6.6).

$$K = 0.7 \times 142 = 99.4\,\text{kips/ft}^3$$

$$R = \left(\frac{198{,}028}{99.4}\right)^{1/4} = 6.6 \qquad \frac{L}{R} = 4.6\,\text{ft}$$

From Figure 6.29, $A_{yc}$ and $B_{yc}$ values will be the same in step d(i), since $Z_{\max} = L/R$ remains close to 5. All the values to be substituted in equation. (6.78a) will be the same as in step d(i) except that the $R$ will be 6.6 instead of 6.1. Substituting this new $R$ values gives:

$$0.02 = 1.4\frac{Q_g(6.6)^3}{198{,}028} + 1\frac{10Q_g(6.6)^2}{198{,}028}$$

$$= 0.002Q + 0.002Q$$

$$Q = 5\,\text{kips}$$

$$(Q_{all})_G = 6 \times 5 = 30\,\text{kips}.$$

Lateral load of 5.9 kips was pile capacity without group reduction factor and 5 kips was pile capacity with the group reduction effect.

(e) *Allowable Lateral Load* From steps c(ii) and d(ii), the allowable lateral load on the group is the lower value.

$$(Q_G)_{all} = 14.43\,\text{kips} < 30\,\text{kips}$$

Therefore, the pile group cannot resist the imposed lateral load of 30 kips.

## 6.8 DESIGN PROCEDURE FOR PILES IN COHESIVE SOILS

Based on the study of behavior and analysis of piles in cohesive soils, the following design procedure is proposed.

1. **Soil Profile** From proper soils investigations establish the soil profile, ground water levels and note soil properties on the profile based on the field and laboratory tests.
2. **Pile Dimensions and Arrangement** Normally, pile dimensions and arrangements are established from axial compression loading requirements. The ability of these pile dimensions and their arrangement to resist imposed lateral loads and moments is then checked by following procedure.
3. **Calculation of Ultimate Lateral Resistance and Maximum Bending Moment**
   (A) Single Piles
      (i) (a) For normally consolidated clays whose soil modulus increase linearly with depth, determine the $n_h$ value from Table 4.16.
         (b) Calculate the relative stiffness, $T = (EI/n_h)^{1/5}$.
         (c) Determine the $L/T$ ratio and check if the pile behavior is as a short ($L/T \leqslant 2$) or long ($L/T \geqslant 4$) pile.
      (ii) (a) For overconsolidated clays whose soil modulus is constant with depth, determine $k$ from Table 4.16.
         (b) Calculate the relative stiffness $R = (EI/k)^{1/4}$.
         (c) Determine the $L/R$ ratio and check to see whether the pile behavior is as a short ($L/R \leqslant 2$) or long ($L/R \geqslant 3.5$) pile.
      (iii) Calculate the ultimate lateral resistance $Q_u$, the allowable lateral resistance $Q_{all}$, and the maximum bending moment $M_{max}$ for the applied loads by Brom's method outlined in Section 6.4.

   (B) Pile Group
   From Table 6.14, determine $G_e$ for $(S/B)$ ratio of the group. The allowable lateral resistance of the group $(Q_{all})_G$ is then the $G_e$ times $(nQ_{all})$ where $n$ is the number of piles in the group, and $Q_{all}$ is obtained from step 3(A).

4. **Calculation of Lateral Resistance and Maximum Moment for Allowable Lateral Deflection**

   (A) Single Piles

   (i) (a) For normally consolidated clays whose soil modulus increases linearly with depth, determine the $n_h$ value from Table 4.16.

   (b) Calculate the relative stiffness, $T = (EI/n_h)^{1/5}$. Determine the $L/T$ ratio.

   (c) Calculate the allowable lateral load for the specified lateral deflection and maximum bending moment for the design loading conditions by the subgrade reaction approach as outlined in Section 6.1.3.

   (ii) (a) For overconsolidated clays whose soil modulus is constant with depth, determine $k$ from Table 4.16.

   (b) Calculate the relative stiffness $R = (EI/k)^{1/4}$. Determine the $L/R$ ratio.

   (c) Calculate the allowable lateral load for the specified lateral deflection and maximum bending moment for the design loading conditions by the subgrade reaction approach as outlined in Section 6.6.1.

   (B) Pile Group

   (i) From Table 6.6, determine the group reduction factor for $n_h$ or $k$, as applicable, for the $S/B$ ratio of the group. Then the new $n_h$ or $k$ will be obained by multiplying $n_h$ or $k$ values, obtained in step 4(A) (i) (a) or (ii) (a), by group reduction factor. Finally, as outlined in step 4(A), calculate the allowable lateral load capacity of single pile based on this new $n_h$ or $k$ as applicable.

   (ii) The pile group capacity is the allowable lateral load capacity of a single pile, obtained in B (i), multiplied by the number of piles. The maximum bending moment will be for the new allowable lateral load for the group.

5. **Allowable Lateral Load and Maximum Bending Moment** Allowable lateral load is the lower of the values obtained in Steps 3 and 4. The maximum bending moment corresponds to the allowable lateral load

6. **Special Design Feature** Calculate the deflection and moment beyond the elastic range (where soil is allowed to yield plastically) for given lateral load and moment.

   (A) Establish the $p$–$y$ curve, for the type of soil encountered by the procedure outlined in Section 6.6.2.

   (B) Use $T$ or $R$, as applicable, obtained from step 3 and determine deflections along pile depth for the imposed lateral load and moment as follows:

   (i) For soils with modulus linearly increasing with depth, use method outlined in Section 6.1.3.

(ii) For soils with modulus constant with depth use the method outlined in Section 6.6.1.

The $T$ or $R$ value used here will be the first trial value and will be referred as $(T)_{\text{trial}}$ or $(R)_{\text{trial}}$ as follows.

(C) For the deflections determined in step 6(B), obtain the corresponding pressures from the $p$–$y$ curves established in step 6(A). Then obtain the soil modulus $k = (p/y)$, where $p$ is soil resistance, and $y$ is pile deflection. This is the *first trial* value for $k$. Plot the value of $k$ with depth. The $k$ may either increase with depth or be constant with depth depending on the type of clay.

(D) From $k$ obtained in step 6(C), calculate $n_h$ or $k$ as follows:

(i) $n_h = k/x$, where $x$ is the depth below ground

(ii) $k$ for modulus constant with depth

Then compute $T$ or $R$, as applicable, and call it the obtained value. Compare this $(T)_{\text{obtained}}$ or $(R)_{\text{obtained}}$ with the $(T)_{\text{trial}}$ or $(R)_{\text{trial}}$ value calculated in step 6(B). If these values do not match, proceed with the second trial as follows:

(E) Assume a $T$ or $R$ value closer to the value obtained in step 6(D). Repeat steps 6(B), 6(C), and 6(D) and obtain a new $T$ or $R$.

(F) Plot $(T)_{\text{obtained}}$ or $(R)_{\text{obtained}}$ values on the ordinate and $(T)_{\text{trial}}$ or $(R)_{\text{rial}}$ on the abscissa and join the points. Draw a line at 45° from the origin. The intersection of this line with the trial line will give the actual $T$ or $R$.

(G) With the finally obtained $T$ or $R$ value, calculate deflections $y$, soil resistance $p$, and moments $M$, along the pile length by the methods outlined in Sections 6.1.3 or 6.6.1, as applicable.

Design example 6.11 outlines steps 1 through 5. Example of design feature ($p$–$y$ curve) incorporated in step 6 will be similar to Example 6.9.

## 6.9 LATERAL RESISTANCE AND DEFLECTION OF PILES IN A LAYERED SYSTEM

Most soil deposits occur in layers. If some of these layers are too thin, they can be neglected. In cases where all the layers are of comparable thickness, but their properties do not vary significantly, soil properties can be averaged. In both cases, the soil can be considered as a homogeneous material and can be classified either as a cohesionless or a cohesive soil. The methods of analyzing laterally loaded piles discussed in the previous articles are applicable for such cases. In situations where thick layers of soils with differing soil properties exist, the analyses presented need modifications as discussed in the following paragraphs.

### 6.9.1 Ultimate Resistance in Layered Systems

Brinch Hansen's (1961) method is applicable for short, rigid piles installed in layered systems. The basic theory for this method has been discussed in the beginning of Section 6.1. The method consists of dividing the soil profile into a number of layers and then determining the ultimate soil resistance $p_{xu}$ for each layer by equation (6.3). The point of rotation $x_r$ is then determined by a trial-and-error method. Once $x_r$ is determined, the ultimate lateral resistance $Q_u$ is calculated by taking the moment about the point of rotation. Design procedure and an example (6.13) for this method are described later.

### 6.9.2 Lateral Deflection of Laterally Loaded Piles in Layered Systems

Davisson and Gill (1963) provide solutions for a two-layer soil system by using the modulus of subgrade reaction approach. The variation of modulus of subgrade reaction of the two layers used in this analysis are provided in Figure 6.38.

The total thickness of soil along the pile of length $L$ is divided into two layers, the top layer of thickness certain percent of $L$ and the bottom layer of the balance. The stiffness of top layer is characterized by soil modulus $K_T$ and that of the bottom layer by $K_b$.

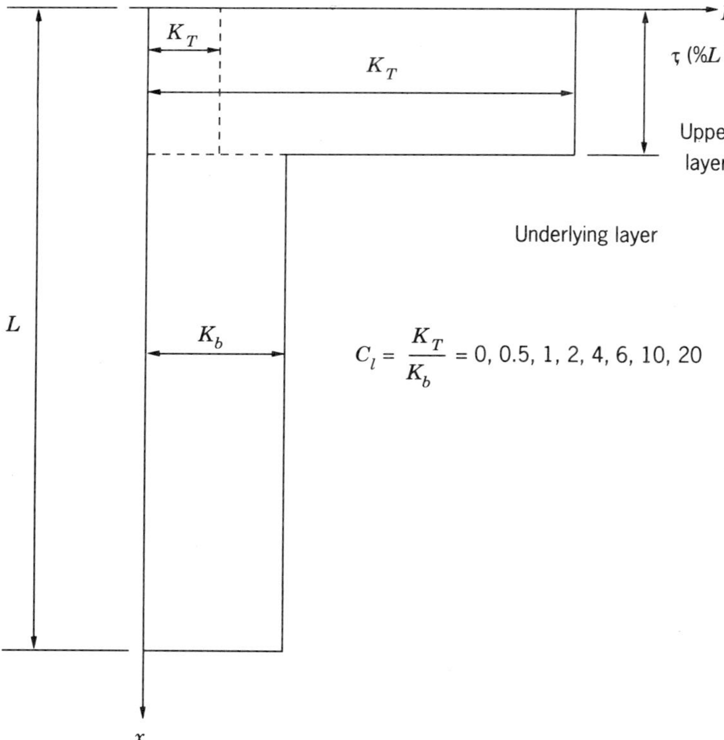


**Figure 6.38** Variations of the modulus of subgrade reaction in layered cohesive soils (Davisson and Gill, 1963).

The ratio $C_l = K_T/K_b$ has been varied from 0 to 20 in the foregoing solution (see Figure 6.38).

As discussed earlier, the differential equation for a beam on elastic foundation, assuming independent closely spaced elastic springs, can be expressed by equation (6.9). This equation can be rewritten in the following form:

$$\frac{d^4y}{dx^4} + \left(\frac{K_T}{K_b}\right)y = 0 \tag{6.100}$$

where

$$Z = \frac{x}{R} \quad \text{and} \quad R = \left(\frac{EI}{K_b}\right)^{1/4}$$

From Figure 6.38, $K_T/K_b = C_l$ for the upper layer and $K_T/K_b = 1$ for the lower layer and the governing differential equations become:

$$\frac{d^4y}{dx^4} + C_l y = 0 \quad \left(\text{for } 0 < \frac{x}{R} < \tau\frac{L}{R}\right) \tag{6.101}$$

$$\frac{d^4y}{dx^4} + y = 0 \quad \left(\text{for } \tau\frac{L}{R} < \frac{x}{R} < \frac{L}{R}\right) \tag{6.102}$$

Deflections and moments for free-head and fixed-head piles can then be obtained from the following relationship, derived by Davisson and Gill (1963).

***Free-Head Piles***

$$y_x = A_{yc}\frac{Q_gR^3}{EI} + B_{yc}\frac{M_gR^2}{EI} \tag{6.103}$$

$$M_x = A_{mc}Q_gR + B_{mc}M_g \tag{6.104}$$

***Fixed-Head Piles***

$$y_x = C_{yc}\frac{Q_gR^3}{EI} \tag{6.105a}$$

$$M_x = C_{mc}Q_gR \tag{6.105b}$$

Davisson and Gill (1963) obtained these solutions by analog computer and Figures 6.39 through 6.47 provide the values of above nondimensional factors $A_{yc}$, $A_{mc}$, $B_{yc}$, and $B_{mc}$ for free-head piles and $C_{yc}$ and $C_{mc}$ for fixed-head piles. An inspection of these figures shows that the soil from the ground surface to depths of $0.2R$ to $0.4R$ are important for surface deflections and maximum moments of laterally loaded piles.

In using these solutions, the stiffness of surface layer is defined in terms of the

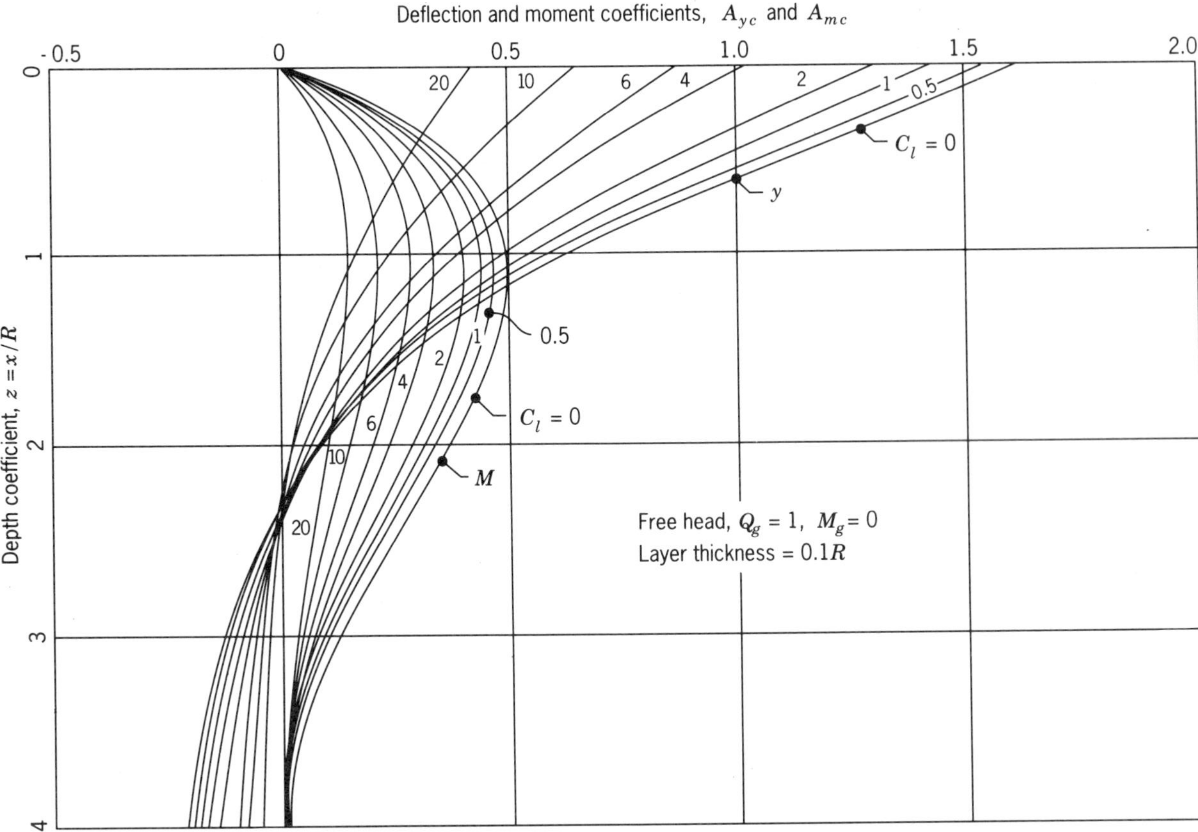


**Figure 6.39** Deflection and moment coefficients ($A_{yc}$ and $A_{mc}$): Free-head piles–layer thickness = 0.1$R$ (Davisson and Gill, 1963).

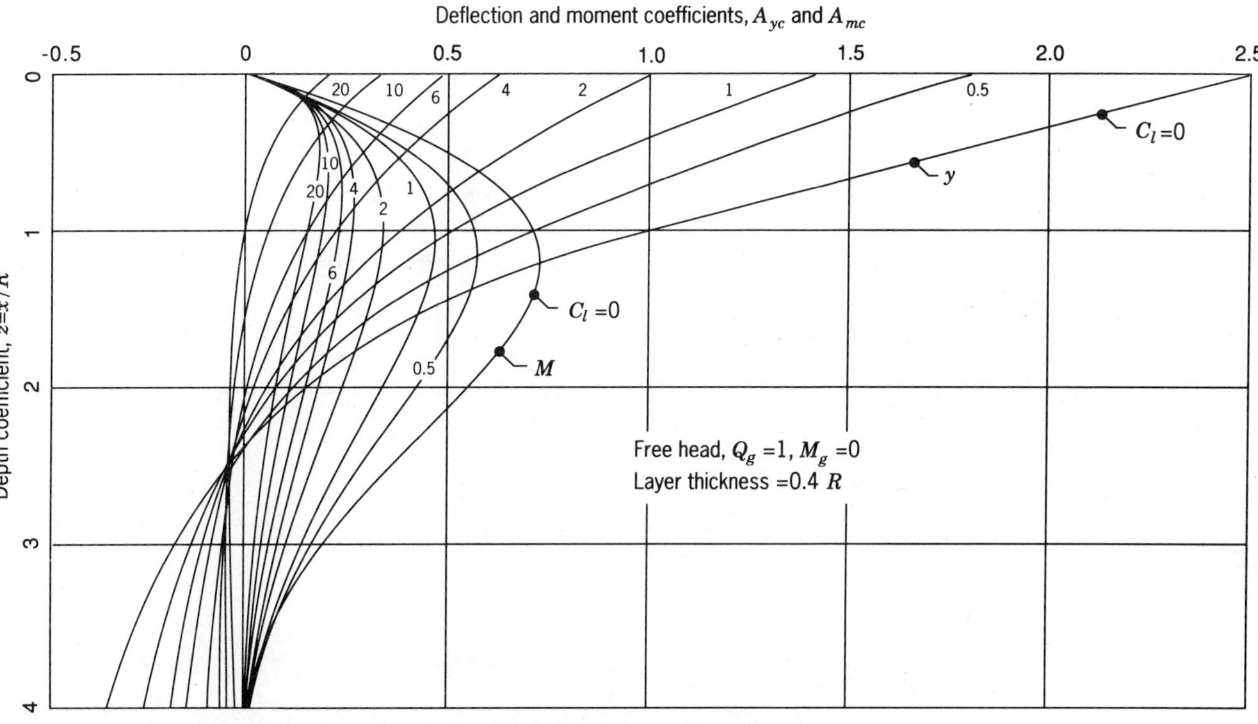


**Figure 6.40** Deflection and moment coefficients ($A_{yc}$ and $A_{mc}$): Free-head piles–layer thickness – 0.4$R$ (Davisson and Gill, 1963).

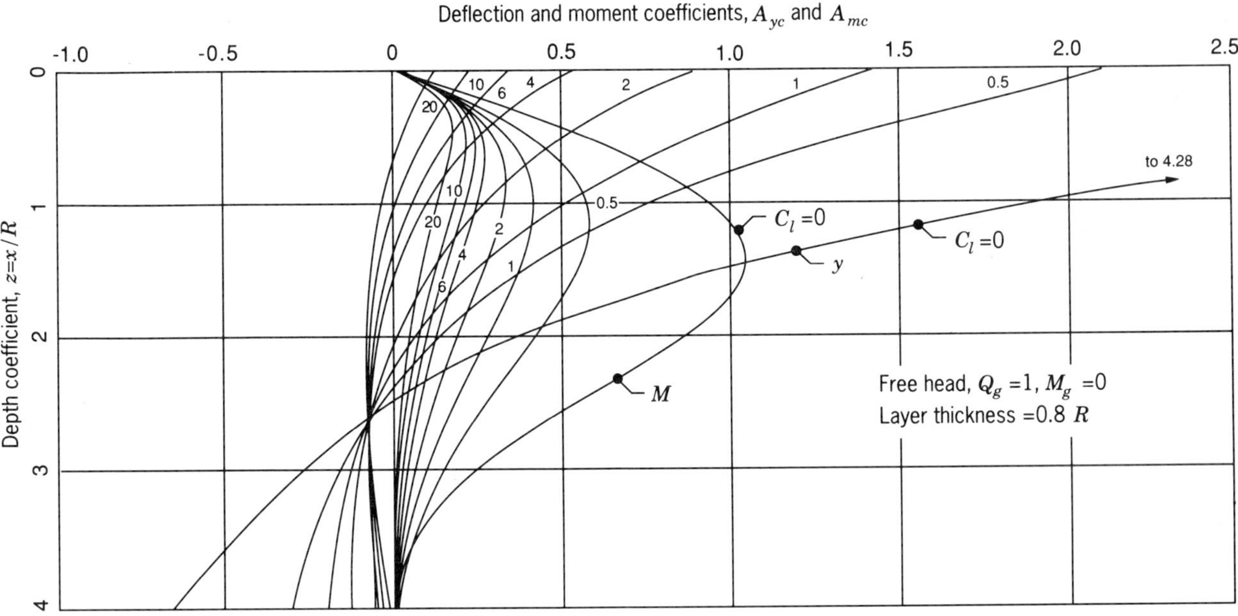


**Figure 6.41** Deflection and moment coefficients ($A_{yc}$ and $A_{mc}$): Free-head piles – layer thickness = $0.8R$ (Davisson and Gill, 1963).

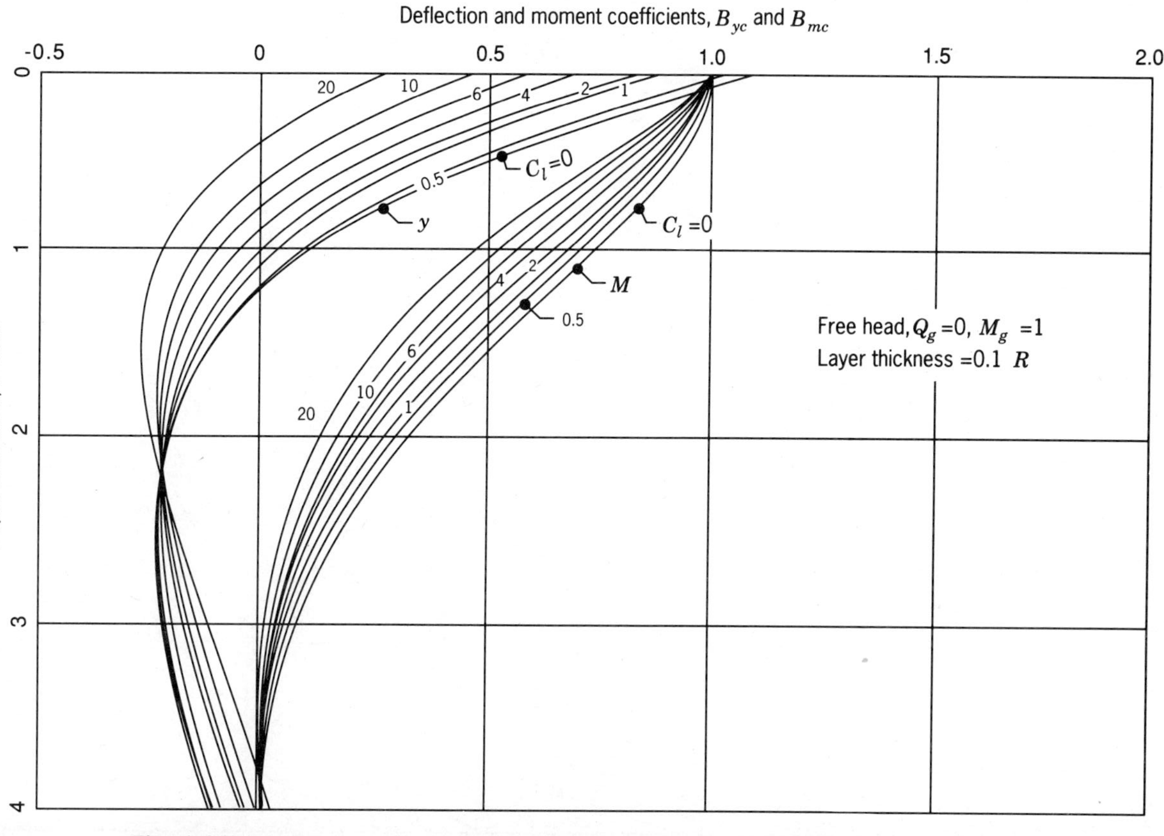


**Figure 6.42** Deflection and moment coefficients ($B_{yc}$ and $B_{mc}$): Free-head piles – layer thickness $= 0.1R$ (Davisson and Gill, 1963).

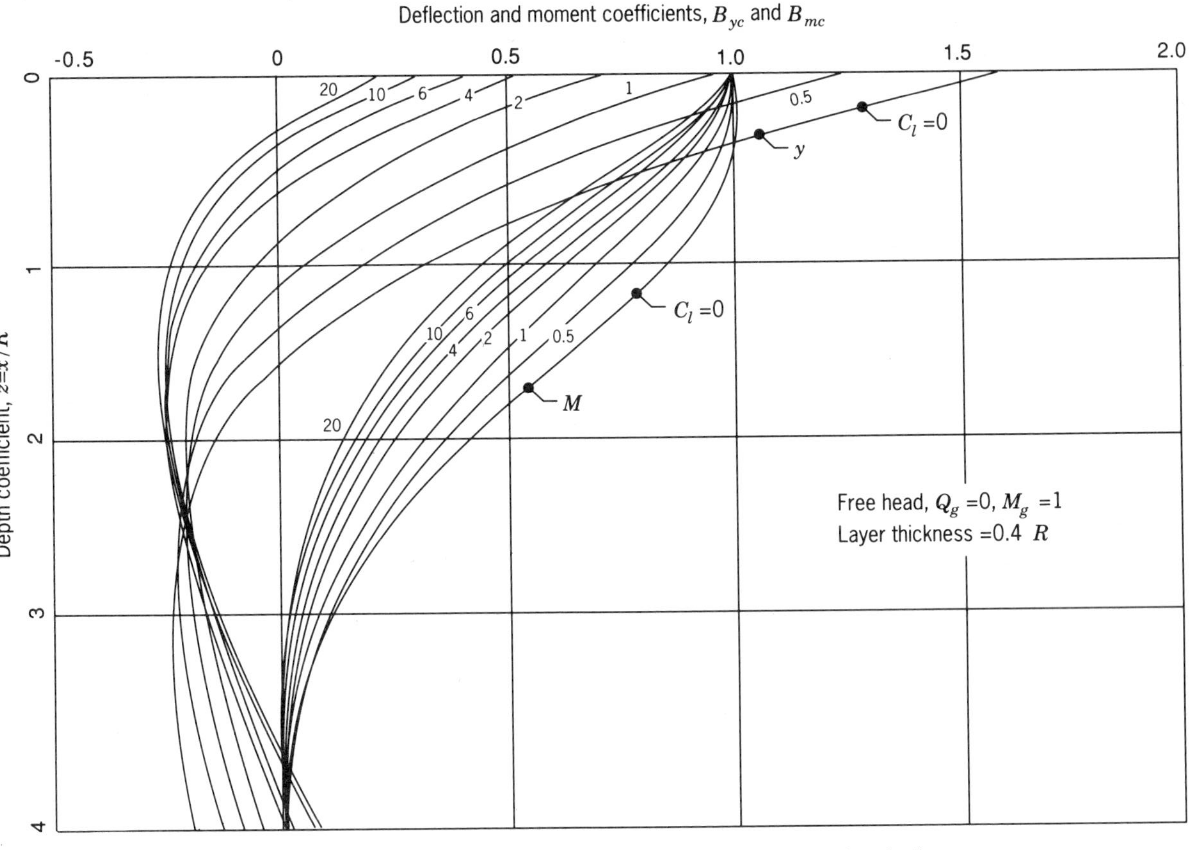


**Figure 6.43** Deflection and moment coefficients ($B_{yc}$ and $B_{mc}$): Free-head pile moment load – layer thickness = $0.4R$ (Davisson and Gill, 1963).

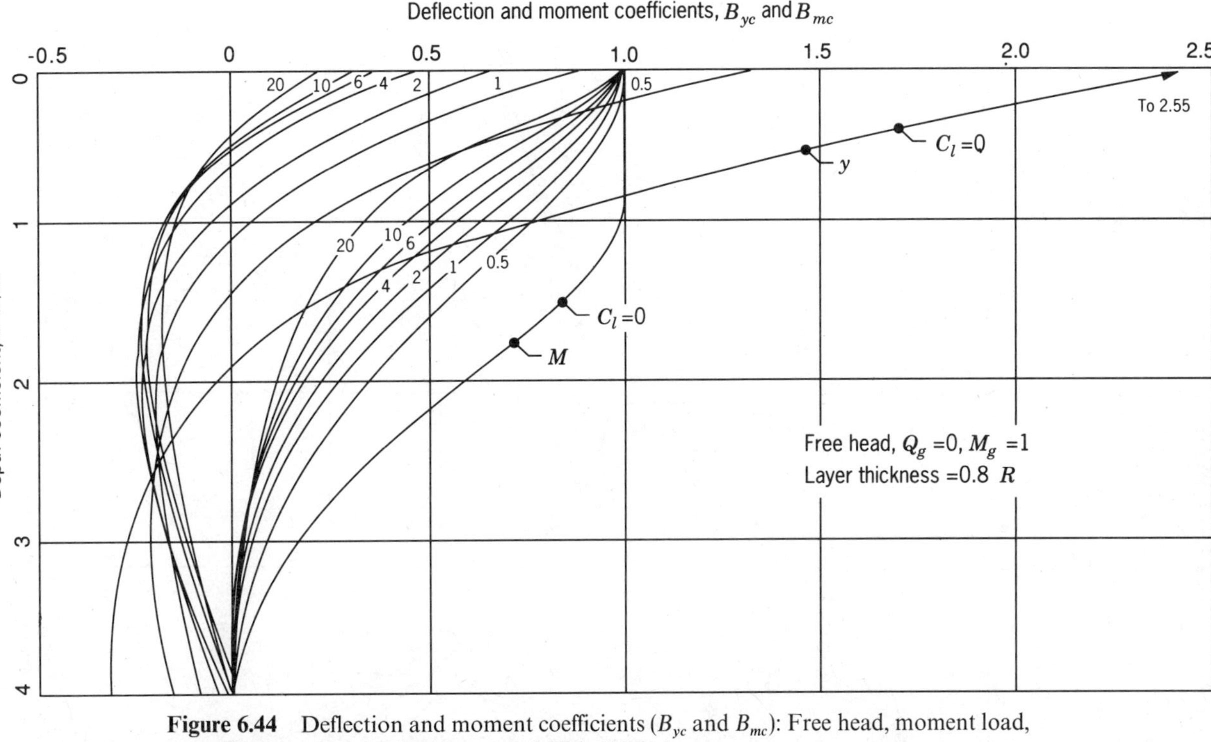


**Figure 6.44** Deflection and moment coefficients ($B_{yc}$ and $B_{mc}$): Free head, moment load, layer thickness = $0.8R$ (Davisson and Gill, 1963).

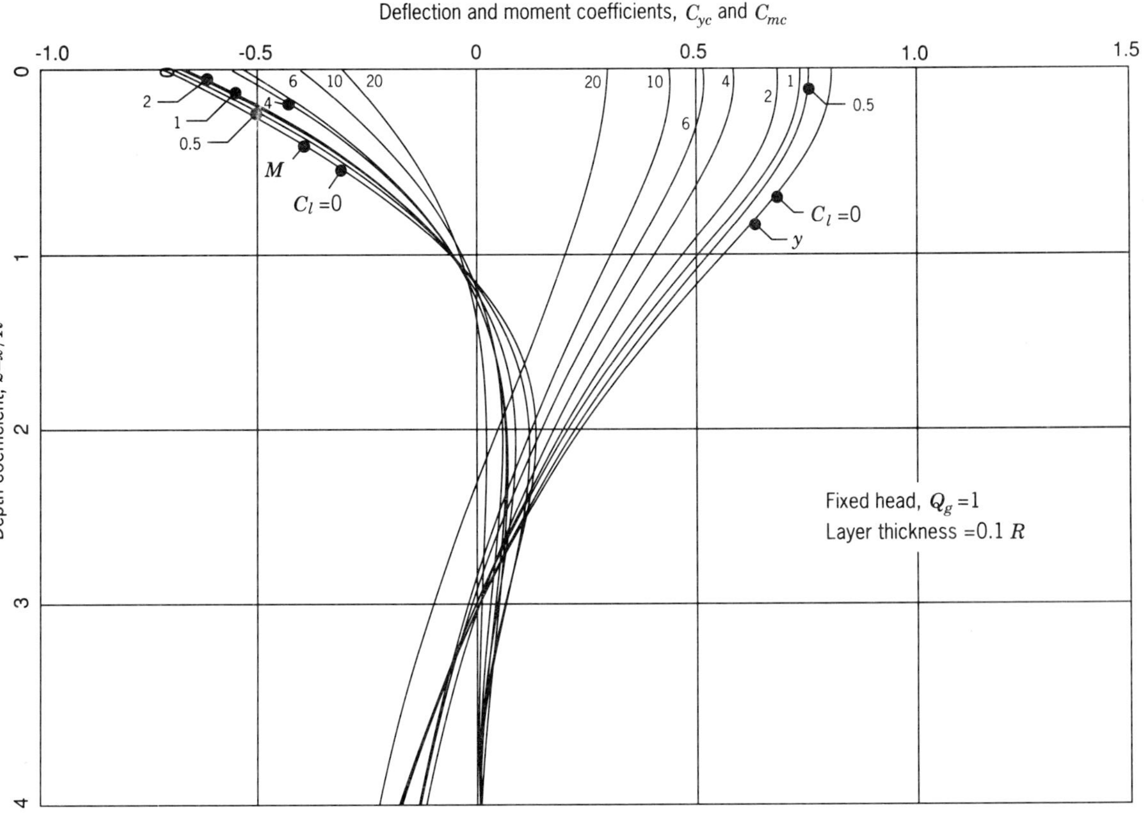


**Figure 6.45** Deflection and moment coefficients ($C_{yc}$ and $C_{mc}$): Fixed-head, Layer thickness = 0.1 $R$ (Davisson and Gill, 1963).

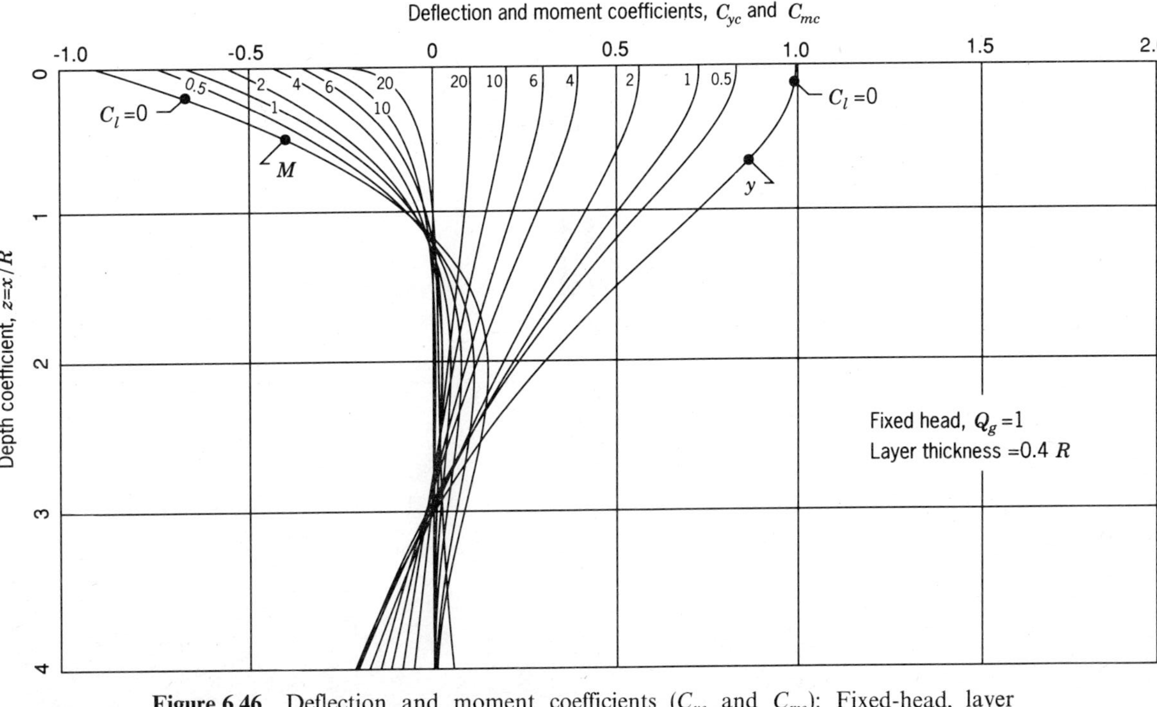


**Figure 6.46** Deflection and moment coefficients ($C_{yc}$ and $C_{mc}$): Fixed-head, layer thickness = $0.4R$ (Davisson and Gill, 1963).

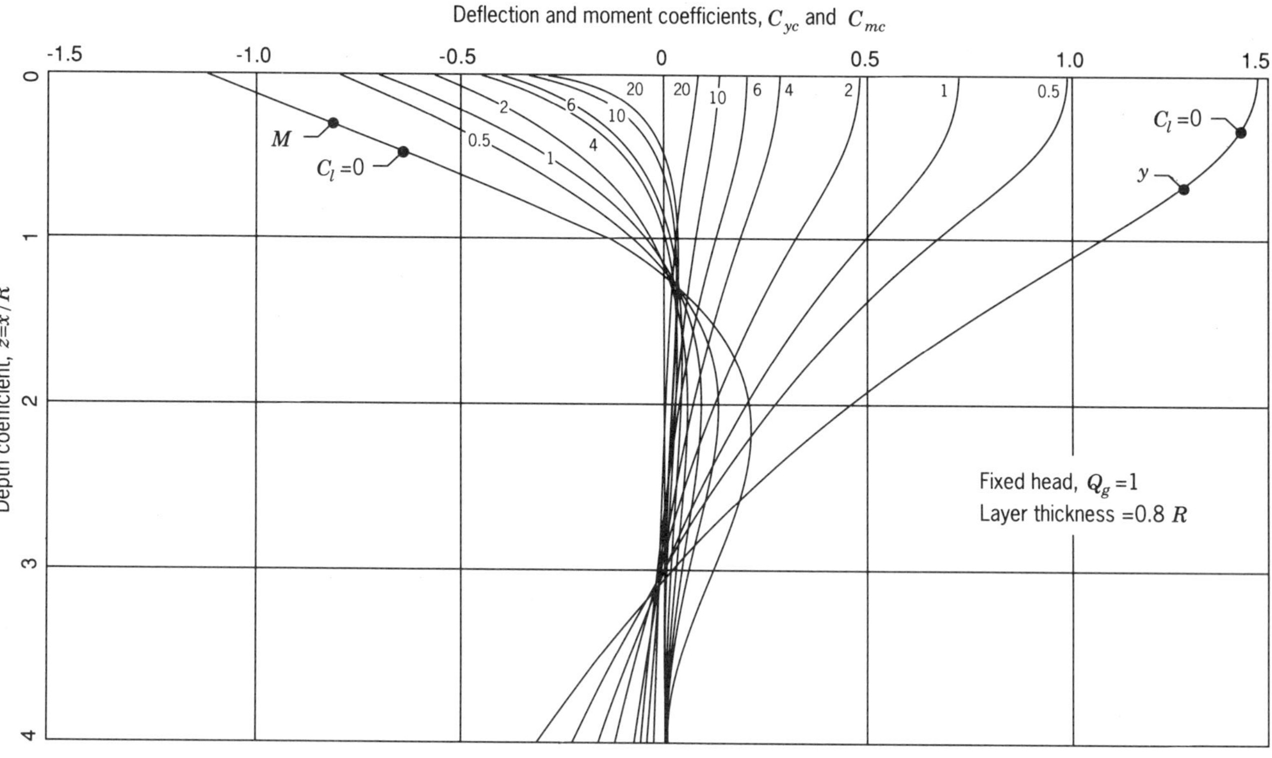


**Figure 6.47** Deflection and moment coefficients ($C_{yc}$ and $C_{mc}$): Fixed-head, layer thickness = 0.8 $R$ (Davisson and Gill 1963).

underlying layer. Thus, the stiffness factor $R$ is determined for the lower layer having soil modulus of $k_b$. These solutions are applicable only for long flexible piles where $L/R \geqslant 4$. Since most piles, in practice, are usually embedded at least $4R$, these solutions apply.

***Example 6.12*** For a single pile in cohesive soil in Example 6.8, illustrate the effect of an upper stiff layer of clay on the load carrying capacity for 0.25-in. deflection for the following cases.

(a) $C_1 = \dfrac{K_T}{K_b} = 2$, thickness of stiff layer = 1.5 ft

(b) $C_1 = 2$, the thickness of stiff layer = 3 ft

(c) $C_1 = 4$, the thickness of stiff layer = 1.5 ft

(d) $C_1 = 4$, the thickness of stiff layer = 3 ft

SOLUTION The following data has been taken from Example 6.8:

$$S_u = 1.8\ \text{kip/ft}^2$$

$$EI = 22{,}968\ \text{kips-ft}^2$$

$$R = 3.73\ \text{ft}$$

$$L = 40\ \text{ft}$$

$$Z_{\max} = 10.72$$

$$y_g = 0.25\ \text{in.}$$

$$Q_g = 6.3\ \text{kips (computed value)}$$

The deflection for free-head piles is computed from equation (6.103) for $M_g = 0$. The values of $A_{yc}$ for the foregoing four cases are 0.99, 0.9, 0.53, and 0.52 from Figures 6.40 and 6.41. The substitution of these values in equation (6.103) for a ground deflection of 0.25 inch gives the loads $Q_g$ listed in the table below.

| Case | $C_l = K_T/K_b$ | Thickness of Stiff Layer (ft) | Thickness of Stiff Layer % $R$ | $Q_g$(kips) | Percent Increase from Uniform Case |
|---|---|---|---|---|---|
| Uniform layer | 1 | — | — | 6.3 | — |
| (a) | 2 | 1.5 | 0.4 | 8.9 | 41 |
| (b) | 2 | 3.0 | 0.8 | 9.8 | 55 |
| (c) | 4 | 1.5 | 0.4 | 14.0 | 122 |
| (d) | 4 | 3.0 | 0.8 | 16.96 | 169 |

Observe from the above computations that stiff layer of even a small thickness near the ground surface increases the load carrying capacity of a single pile appreciably. Alternatively, the deflection will decrease appreciably at a given applied load. Therefore, advantage should be taken of the existence of any stiff layer in computing the lateral load capacity of pile in such situations.

## 6.10 DESIGN PROCEDURE FOR PILES IN LAYERED SYSTEM

1. **Soil Profile** From proper investigations establish the soil profile, ground water levels and note soil properties on the soil profile based on the field and laboratory tests.
2. **Pile Dimensions and Arrangement** Normally pile dimensions and arrangements are established from axial compression loading requirements. The ability of these piles to resist imposed lateral loads and moments is then checked by the following procedure.
3. **Calculation of Ultimate Lateral Resistance**

   3.1 *Single Piles*

   (i) Estimate $n_h$ or $k$ from Table 4.16, as applicable. Calculate $T = (EI/n_h)^{1/5}$ or $R = (EI/K_b)^{1/4}$, as applicable.

   (ii) If $L/R$ or $L/T \leqslant 2$ then the pile will behave as short rigid pile and ultimate lateral resistance can be calculated by Brinch Hansen's method.

   (iii) If $L/R \geqslant 4$ the piles will behave as flexible piles and lateral resistance shall be calculated for allowable lateral deflections as detailed in step 4.

   3.2 *Pile Group* From Table 6.2 or 6.14, as applicable, determine $G_e$ value for $(S/B)$ ratio. The allowable lateral resistance of the group is then the product of (1) number of piles $n$, (2) group efficiency, $G_e$ and (3) the allowable capacity of a single pile $Q_{\text{all}}$.

4. **Calculation of Lateral Resistance and Maximum Moment for Allowable Lateral Deflection** This method is only applicable for long flexible piles ($L/R \geqslant 4$) in cohesive soils.

   4.1 *Single Piles*

   (i) Calculate $R$ as mentioned in step 3.1 and check if $L/R \geqslant 4$.

   (ii) Determine $C_l = K_T/K_b$ and the thickness of surface layer in terms of $R$. Then calculate deflections and moments from equations (6.103) to (6.105), as applicable.

   (iii) For allowable lateral deflection at ground surface, the allowable lateral loads can be calculated by using equations (6.103) or (6.105a) for free-head or fixed-head condition, respectively.

4.2 *Pile Group*

(i) From Table 6.6 determine group reduction factors for $K_b$. and $K_T$. Calculate new $R$ and follow steps of 4.1 to calculate allowable lateral load of a single pile based on reduced $K_b$.

(ii) Pile group capacity is then the allowable lateral capacity from (i) above times the number of piles. Maximum bending moment on a

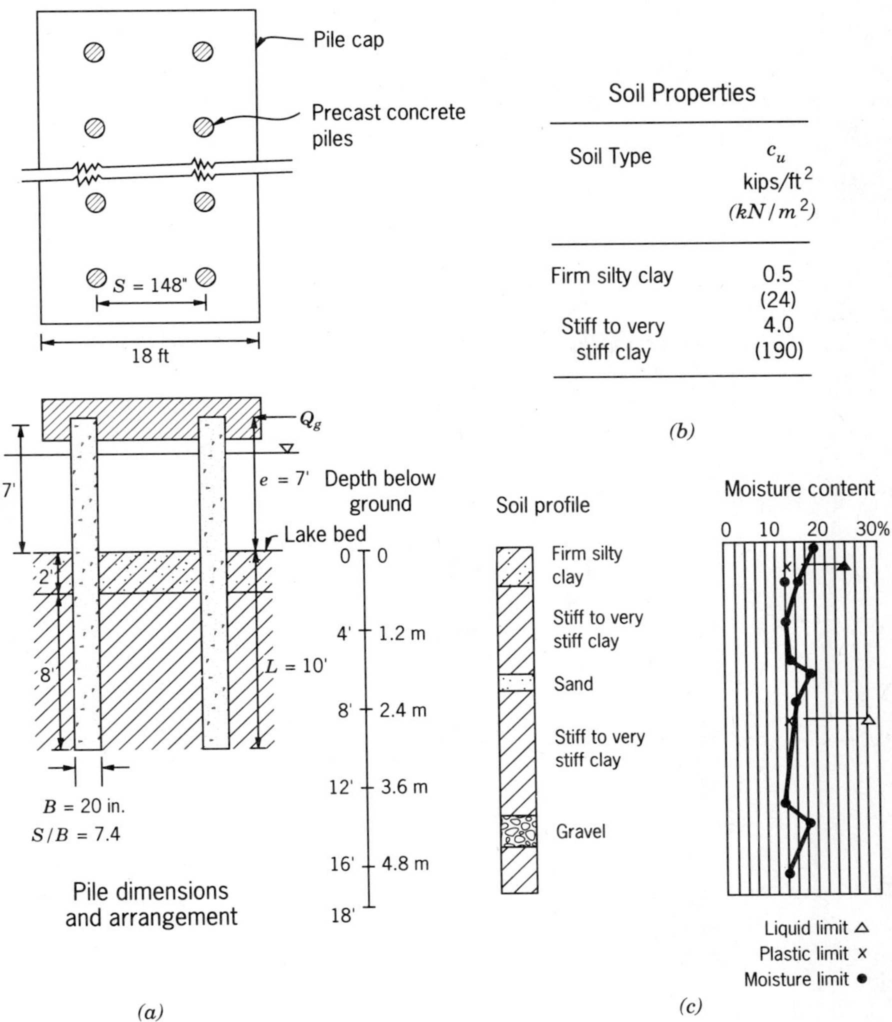


**Figure 6.48** (a) Pile dimensions, arrangements, (b) soil properties and (c) soil profile along the pile depth (example 6.13).

pile will correspond to this reduced allowable lateral load of a single pile.

***Example 6.13*** A servicing platform for ferryboats in a large lake consists of drilled precast concrete piles supporting the platform. The piles are 20 in. (500 mm) diameter, free headed, and embedded to 10 ft into the soil. The top 2 ft of the soil is firm silty clay and lower 8 ft is stiff to very stiff clay. Figure 6.48 shows the soil stratigraphy and parameters. Modulus of elasticity for concrete is $5 \times 10^5$ kips/ft$^2$ ($24 \times 10^6$ kN/m$^2$). Total number of piles in the group is 24.
(a) Calculate the ultimate and allowable lateral load that can be applied to the pile cap.
(b) If the pile embedment is increased to 20 ft calculate the allowable lateral load that can be applied to the pile cap for an allowable lateral pile deflection of 0.25 inch (6.35 mm) at the lake bed.

SOLUTION FOR (a)

1. *Soil profile* Figure 6.48b and c show soil properties and profile.
2. *Pile dimensions and arrangement* Figure 6.48a shows the pile dimensions and arrangements.
3. *Calculation of ultimate lateral resistance*
(i) An average $c_u$ for the soil $= (2 \times 0.5 + 8 \times 4)/(10) = 3.3$ kips/ft$^2$

$$q_u = 6.6\ \text{kips/ft}^2 = 3.3\ \text{tons/ft}^2$$

From Table 4.16, $k = 67\ c_u = 67 \times 3.3 = 221$ kips/ft$^2$

$$B = \frac{20}{12} = 1.67\ \text{ft}$$

$$I = \frac{\pi B^4}{64} = \frac{\pi(1.67)^4}{64} = 0.38\ \text{ft}^4$$

$$R = \left(\frac{EI}{K}\right)^{1/4} = \left(\frac{5 \times 10^5 \times 0.38}{221}\right)^{1/4} = 5.4\ \text{ft}$$

$$\frac{L}{R} = \frac{10}{5.4} \cong 2$$

Therefore, the piles will behave as short rigid piles and Brinch Hansen's method will be applicable. Also, $S/B = 148/20 = 7.4$ (from Figure 6.48). Since this ratio is close to 8, pile will behave as single piles and group effect will not be considered.
(ii) Values of $p_{xu}$ at various depths are calculated with the help of equation (6.3) and presented in the last column of Table 6.16. Average values of $p_{xu}$ at the

**TABLE 6.16 Values of Ultimate Soil Resistance with Depth (Example 6.13)**

| $X$ (ft) Below Lake Bed | $X/B$ | $K_c$ | $c_uK_c$ (kips/ft$^2$) | $p_{xu} = c_uK_c$ |
|---|---|---|---|---|
| 0 | 0 | 2 | 1.0 | 0 |
| 2 | 1.19 | 5 | 2.5<br>20.0 | 2′ |
| 4 | 2.39 | 6 | 24.0 | 4′ |
| 6 | 3.59 | 6.5 | 26.0 | 6′ |
| 8 | 4.79 | 7 | 28.0 | 8′ |
| 10 | 5.99 | 7 | 28.0 | 10′ |

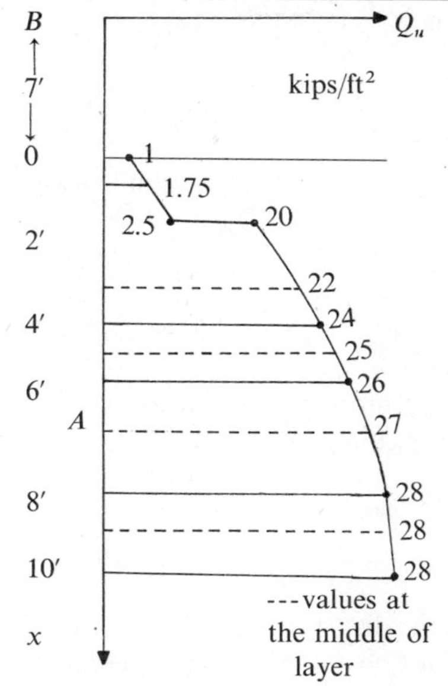

1. from Figure 6.3b.

middle of each 2 ft thick layer is shown by dotted lines in the last column of this table.

(a) Assume a trial point of rotation at 7.0′ below the lake bottom. This trial point is shown by point $A$ in Table 6.16.

Taking moment about the point of application of load, $Q_u$. This point is point $B$.

$$\sum M = 1.75 \times 2 \times 8 + 22 \times 2 \times 10 + 25 \times 2 \times 12 + 26.5 \times 1 \times 13$$
$$- 27.5 \times 2 \times 15 - 28 \times 1 \times 16.5 = 125.5 \text{ kips-ft/ft width}$$

(b) Assume a second trial point of rotation at 6.75′ below the lake bottom and then take moment about $B$.

$$\sum M = 1.75 \times 2 \times 8 + 22 \times 2 \times 10 + 25 \times 2 \times 12 + 26.5 \times 0.75$$
$$\times 13.375 - 27.25 \times 1.25 \times 14.375 - 28 \times 2 \times 16 = 48.18 \text{ kips-ft/ft}$$

The remainder is small and is significantly close to zero.

Therefore, the center of rotation at 6.75 ft below the lake bottom will yield $\sum M$ close to zero.

(c) Then, taking moment about the point of rotation:

$$Q_u(7 + 6.75) = 1.75 \times 2 \times 5.75 + 22 \times 2 \times 3.75 + 25 \times 2 \times 1.75 + 26.5 \times 0.75 \times 0.75/2 - 27 \times 1.25 \times \frac{1.25}{2} - 28 \times 2 \times 1 = 202.98 \text{ kips-ft/ft}$$

$$Q_u = \frac{202.98}{13.75} = 14.76 \text{ kips/ft width}$$

$$= 14.76 \times 1.67 = 24.6 \text{ kips}$$

(d) Ultimate moment resistance of pile section, $M_u$:
For a rectangular section:

$$M_u = \frac{A_s d}{\alpha_f}$$

where $d$ is the distance of extreme compression end to the center of tension bar of area $A_s$. For $f_y = 400$ MPa, and $f'_c = 25$ MPa, $\alpha_f = 3100$ from Canadian Portland Cement Association (1978). For a circular section, the foregoing relationship becomes:

$$M_u = 0.78\frac{A_s d}{\alpha_f} = 0.78 \quad \frac{A_s d}{3100} = \frac{A_s d}{3974}$$

Using No. 8 (25 mm), 6 bars with No. 3 (10 mm) ties at 12 in. (300 mm) center to center:

$$d = \text{Pile diameter} - \text{cover} - \text{tie diameter} - \text{bar radius} = 500 - 50 - 10 - 25/2 = 427.5 \text{ mm}$$

$A_s = \frac{\pi}{4}(25)^2 \times 3 = 1472.62 \text{ mm}^2$, three bars will be on the compression side and three will be on the tension side.

$$M_u = \frac{1472.62 \times 427.5}{3974} = 158.23 \text{ kN-m} = 116.6 \text{ kips-ft}$$

(e) Point of zero shear (maximum moment). Now assume that the point of zero shear is at 2.75 ft below the lake bottom. Then equating all horizontal forces above this point, we get:

$$\sum H = 0$$

$$14.76 - 1.75 \times 2 - 21 \times 0.75 = -4.5, \text{ which is close to zero}$$

Hence this is the point of maximum moment.
(f) Now equating maximum bending moment with the ultimate resistance, we get:

$$Q_u(7 + 2.75) = M_u = 116.6$$
$$= 12\,\text{kips}$$

Using a factor of safety of 2.5, the allowable lateral capacity will be:

$$Q_{\text{all}} = \frac{12}{2.5} = 4.8\,\text{kips (say 5 kips)}$$

$$(Q_G)_{\text{all}} = 24 \times 5 = 120\,\text{kips}$$

SOLUTION FOR (b)

4. *Calculation of lateral resistance for allowable deflection*

$$L = 20\,\text{ft}$$

$$c_u = 4\,\text{kips/ft}^2 \qquad q_u = 8\,\text{kips/ft}^2 \text{ for the bottom layer}$$

$$K = 67\,c_u = 268\,\text{kips/ft}^2$$
$$B = 1.67\,\text{ft} \qquad I = 0.38\,\text{ft}^4$$

$$R = \left(\frac{EI}{K}\right)^{1/4} = \left(\frac{5 \times 10^5 \times 0.38}{268}\right)^{1/4} = 5.1\,\text{ft}$$

$$\frac{L}{R} = \frac{20}{5.1} = 3.9 \cong 4$$

Therefore the pile will behave as a long flexible one and the Davisson and Gill (1963) method can be used.
Also, $S/B = 7.4 \cong 8$, Therefore, the piles will behave as single piles and group effect will not be considered.

$$\text{Top layer thickness} = 2\,\text{ft} = \frac{2}{5.1}R = 0.39\,R(\text{take} = 0.4R)$$

For $c_u = 0.5\,\text{kips/ft}^2$ $\quad k = k_T = 67\,c_u = 33.5\,\text{kips/ft}^2$

For $c_u = 4\,\text{kips/ft}^2$ $\quad k = k_b = 268\,\text{kips/ft}^2$

$$C_l = \frac{k_T}{k_b} = \frac{33.5}{268} = 0.125$$

For free-head piles from equation (6.103), we get

$$y_x = A_{yc}\frac{Q_g R^3}{EI} + B_{yc}\frac{M_g R^2}{EI}$$

From Figure 6.40, $A_{yc} = 2.25$ at $x = 0$ and for $C_l = 0.125$
From Figure 6.43, $B_{yc} = 1.5$ at $x = 0$ and for $C_l = 0.125$

$$EI = 5 \times 10^5 \times 0.38 \text{ kips-ft}^2$$

$$R = 5.1 \text{ ft}$$

$$M_g = 7\,Q_g$$

$$y_{x=0} = 0.25/12 = 0.02 \text{ ft}$$

Substituting these values in the above equation:

$$0.02 = 2.25\frac{Q_g(5.1)^3}{5 \times 10^5 \times 0.38} + 1.5\frac{7Q_g(5.1)^2}{5 \times 10^5 \times 0.38}$$

$$= 157 \times 10^{-5} \quad Q_g + 143.7 \times 10^{-5} \quad Q_g = 300.7 \times 10^{-5} Q_g$$

$$Q_{\text{all}} = \frac{0.02}{300.7 \times 10^{-5}} = 6.7 \text{ kips}$$

Total number of piles = 24

$$(Q_G)_{\text{all}} = 24 \times 6.7 = 160.8 \text{ kips (say 160 kips)}$$

The allowable lateral load that can be applied to pile cap = 160 kips.

## 6.11 PILES SUBJECTED TO ECCENTRIC AND INCLINED LOADS

In the previous sections, the behavior of a single vertical or groups of vertical piles subjected to central (or axial) vertical loads or lateral loads were discussed. In many situations such as under bridges and offshore structures, the pile groups may be subjected to simultaneous central vertical loads, lateral loads and moments. As shown in Figure 6.49a, such loads may either be resisted by a group of vertical piles or a pile group containing both the vertical and batter piles. Combination of such loads on the pile group may result into a system that is subjected to an eccentric and inclined load (Figure 6.49b).

In general, the following four methods are available to analyze this problem:

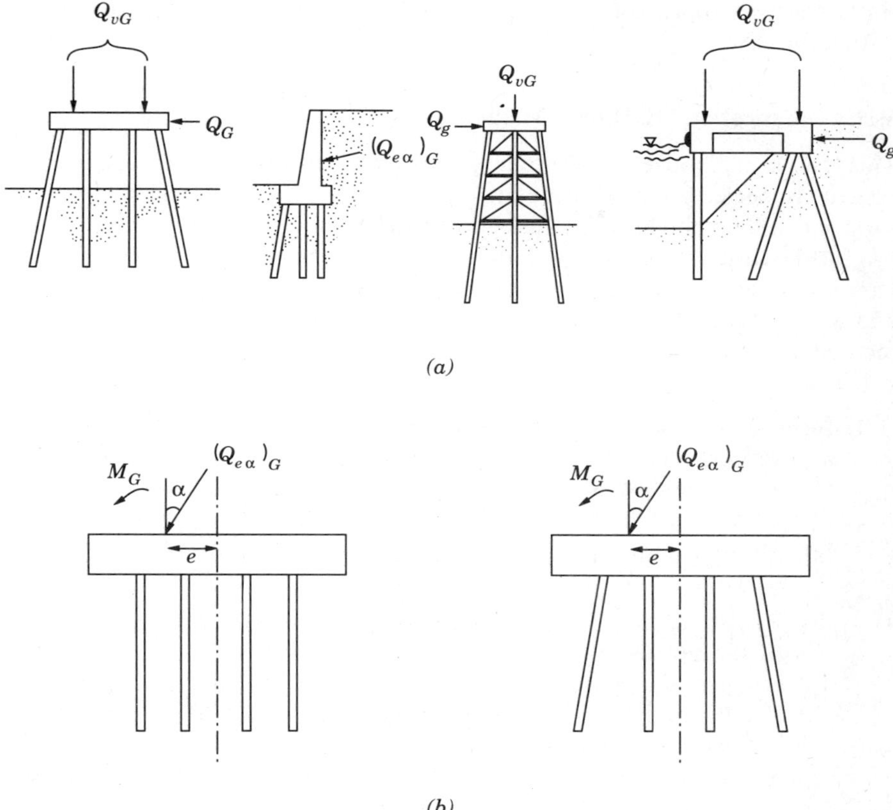


**Figure 6.49** Piles subjected to eccentric and inclined loads. (a) Examples of structural systems with vertical and batter piles (Vesic, 1977), (b) pile foundations subjected to eccentric and inclined loads.

1. Statical or traditional methods: This consists of analyzing the pile group as a simple, statically determinate system but ignoring the effect of the soil
2. Considering pile group as a structural system utilizing the theory of subgrade reaction for soil support
3. Consider interaction between piles and the soil by assuming soil to be an elastic continuum
4. Interaction relationships between soil and pile by determining bearing capacity of piles under eccentric inclined loads

In the following paragraphs, the first three methods will be briefly outlined. Following this, the fourth method will be presented in detail. This method is simple, easy to apply to practical problems, and supported by extensive model testing on piles.

### 6.11.1 Statical or Traditional Method

This simple method considers pile group as a simple statically determinate system. It neglects the contribution of soil to support the load. Due to its simplicity, this method is widely used in design but should only be limited to small projects because little is known of the reliability of this method. In the following paragraphs, two general cases—first, the inclined load on vertical and batter piles and second, eccentric vertical load on vertical piles—are analyzed by this method.

1. **Inclined Load on Vertical and Batter Piles** The simplified analysis of batter and vertical piles assumes that all piles are subjected to axial loads. The method of analysis described below is based on Culman's method as described by Chellis (1961) and consists of the following steps:
   (a) As shown in Figure 6.50, case (*A*) represents the resultant force by *R*.
   (b) Replace each group of similar piles by an imaginary pile at the center of the group. For example, in Figure 6.50, case (A) item (a), it is assumed that group A, group B, and group C offer the axial forces $R_A$, $R_B$, and $R_C$, respectively. Values of $R_A$, $R_B$, and $R_C$ can then be obtained by following procedure:
   (i) As shown in (b), draw pile cap and lines parallel to $R_A$, $R_B$, and $R_C$.
   (ii) Extend *R* to intersect $R_A$ at point *a*.
   (iii) Extend $R_B$ and $R_C$ to intersect at point *b*. Join points *a* and *b*.
   (iv) As shown in (c), first draw line *ac* parallel to and equal to *R* by selecting an appropriate scale. From *a* draw *ab* parallel to *ab* shown in item (b). Then from point *c* draw *cb* parallel to $R_A$ to intersect *ab* at point *b*. From *b* draw a line parallel to $R_B$ and from point *a* draw a line parallel to $R_C$ to obtain point *d*.

   Then $R_A$ will equal *cb*, $R_B$ will equal *bd* and $R_C$ will equal *ad*. Figure 6.50, case (A), item (c), shows these forces drawn to scale: The force direction (e.g., tension and compression) are also shown on this force diagram. Similarly, when the piles are subjected to a resultant pullout force $(P_{pull})_{\alpha,G}$, then the force polygon can be drawn as shown in Figure 6.50, case (B).
2. **Eccentric Vertical Load on Vertical Piles** Load on an individual vertical pile ($R_n$) from an eccentric vertical load can be obtained from the following relationship (Figure 6.51):

$$R_n = \frac{(Q_e)_G}{n} \pm \frac{(Q_e)_G \bar{x} x_n}{I_{yy}} \pm \frac{(Q_e)_G \bar{y} y_n}{I_{xx}} \tag{6.106}$$

Example: Case (A)

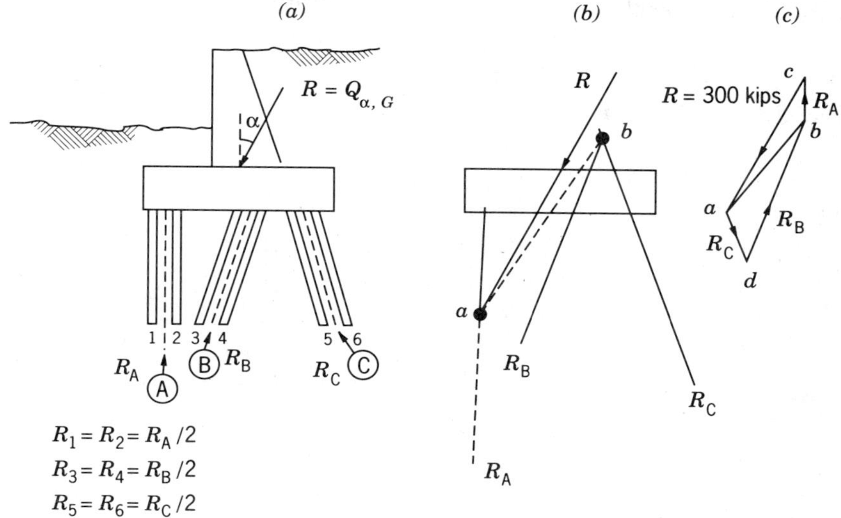


Example: Case (B)

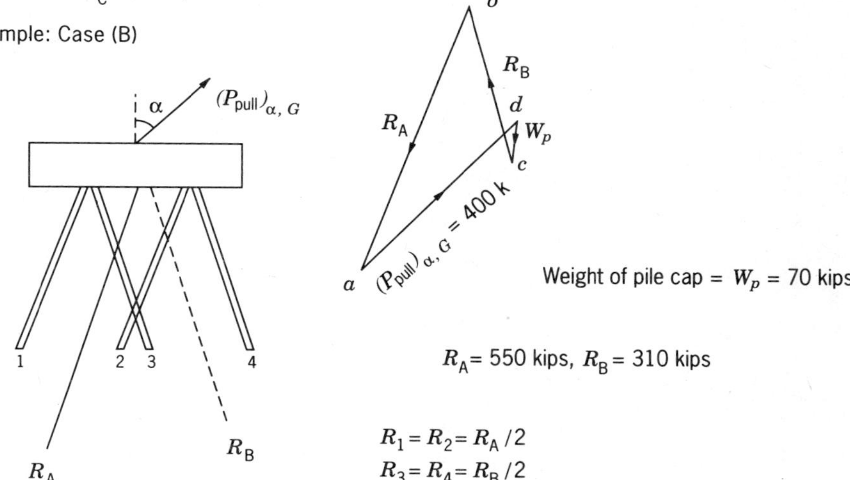


**Figure 6.50** Analysis of load distribution for vertical and batter piles.

where

$R_n$ = load or reaction on any pile
$(Q_e)_G$ = total eccentric vertical load on pile group
$n$ = number of piles in the group

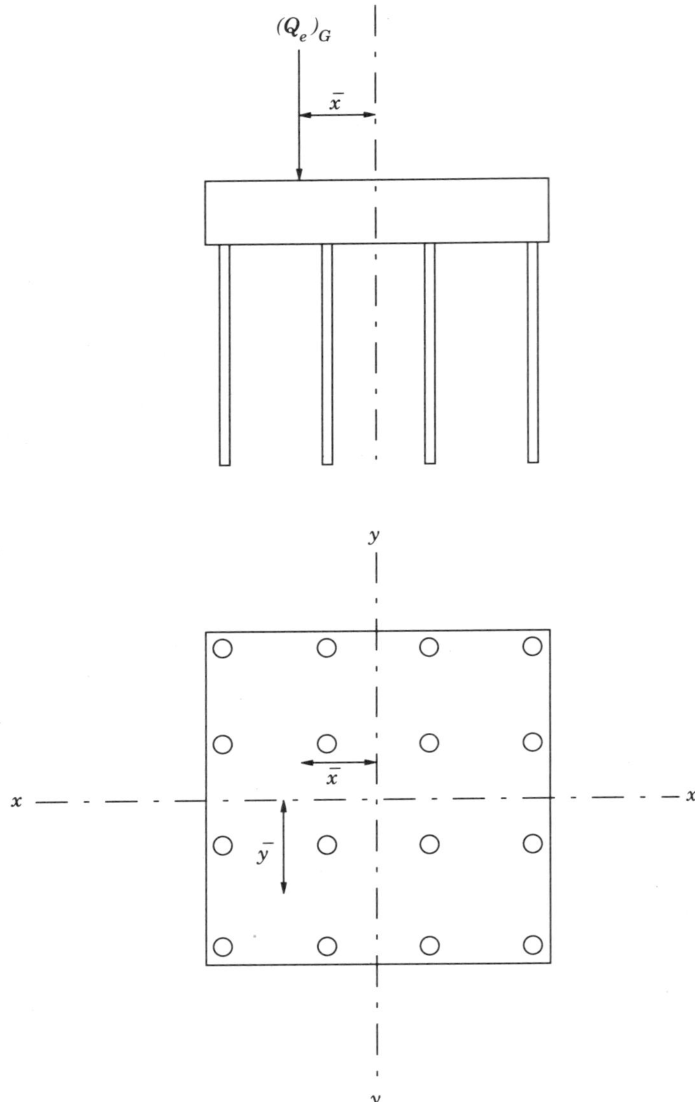


**Figure 6.51** Eccentric vertical load on vertical piles.

$\bar{x}, \bar{y}$ = eccentricities in the $xx$ and $yy$ directions, respectively (i.e., distances from the center of gravity of applied loads to the center of gravity of the pile group in the $x$ and $y$ directions)

$x_n, y_n$ = distances from center of gravity of pile group for each pile in the $x$ and $y$ directions, respectively

$I_{xx}, I_{yy}$ = moment of inertia of pile group about the $xx$ and $yy$ axes, respectively.

$A$ = pile cross section area

$$I_{xx} = Ay_1^2 + Ay_2^2 + \cdots + Ay_n^2 \tag{6.107}$$

$$I_{yy} = Ax_1^2 + Ax_2^2 + \cdots + Ax_n^2 \tag{6.108}$$

Figure 6.51 exhibits all these dimensions. Once $R_n$ is calculated, it should then be compared with the allowable axial compression (or downward load) capacity of the piles. Methods to calculate allowable axial downward load capacity are discussed in Chapter 5.

### 6.11.2 Theory of Subgrade Reaction Solution for a Pile Group

In situations where pile groups are subjected to eccentric and inclined loads, the problem formulation consisting of a group of vertical and batter piles rigidly connected by a pile cap as shown in Figure 6.52a consists of the following:

1. The externally applied pile group load in axial direction $P_G$, lateral direction $Q_G$, and the moment $M_G$ having a resultant $R$, will displace the foundation in the following three ways:
   (a) in axial downward or vertical direction, $x$
   (b) in horizontal direction $y$
   (c) tilting, $\theta$
2. The piles will resist above displacements by normal forces $P$, shear forces $Q$, and moments $M$.
3. In order to calculate bending moments and shears in individual piles, the aforementioned pile reactions ($P$, $Q$, and $M$) are to be determined.

Thus the above will require a structural analysis of the system.

Following assumptions are made to solve this problem (Vesic, 1977). Saul (1968), and Reese et al. (1970) present soil–pile interactions on similar assumptions.

1. The passive pressure and friction along the sides and on the pile cap base are neglected. This assumption is justified in situations where the supporting soil can be eroded by scour or is either weak or compressible. In other cases, it may lead to results on the safe side.
2. The pile spacing is such that they do not influence each other through the soil mass. According to Prakash (1962), this may be justified if pile spacings are eight times pile diameters in the direction of lateral load and three times pile diameters in the perpendicular direction. However, this influence can

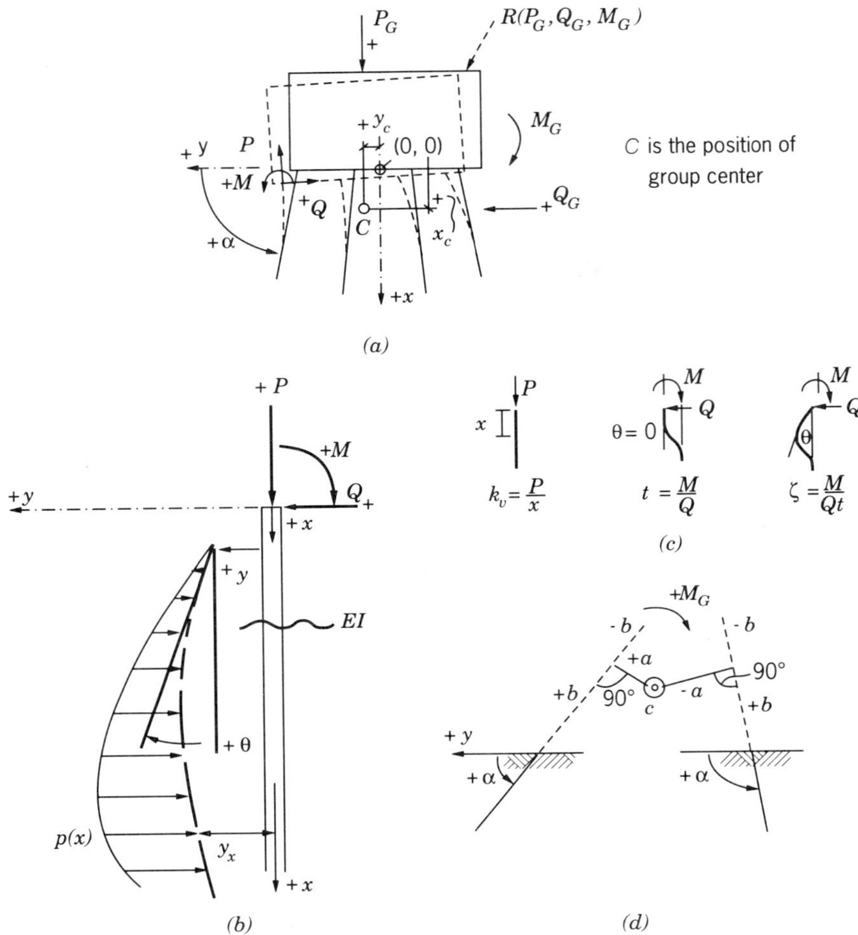


**Figure 6.52** Problem of pile foundations subjected to eccentric and inclined loads (a) Pile group under eccentric and inclined loads, (b) single pile under lateral load, (c) definition of pile coefficients, (d) sign convention for (a) and (b). (Vesic, 1977).

be accounted for by multiplying $n_h$ value with an appropriate reduction factor depending on the pile spacing (Table 6.6).

3. The components of displacement ($x$, $y$, and $\theta$) of a pile head are linear functions of reactions $P, Q$, and $M$, respectively. Also, the components are independent of reactions of other piles (Figure 6.52b).

The assumption (3) is generally not justified because of the nonlinearity in a load-displacement relationship. The effect of nonlinearity can, however, be accounted for by selecting secant values of $n_h$. This assumption leads to the following relationship:

$$P = k_v x \tag{6.109}$$

$$Q = k_t(y + t\theta) \tag{6.110}$$

$$M = k_t t(y + \xi t\theta) \tag{6.111}$$

$k_v$ and $k_t$ have the dimensions of (force/length).

Figure 6.52c provides definitions for these terms.

where

$k_v$ = axial load/axial settlement $= \dfrac{p}{x}$

$k_t$ = (lateral load/lateral deflection in pure translation) under fixed-head conditions

$t$ = (moment/lateral load) for fixed-head conditions

$\xi$ = moment/(lateral load for pure rotation × $t$)

$k_v$ can either be obtained from an axial pile load test or can be calculated by computing a settlement under working axial load. $k_t$, $t$, and $\zeta$ can be determined from lateral load tests. These values can also be obtained from soil–pile properties. For the case where soil modulus increases linearly with depth, the following relationships can be used (Vesic, 1977):

$$k_t = 1.075 n_h T^2 \tag{6.112}$$

$$t = 0.926T \tag{6.113}$$

$$\zeta = 1.62 \tag{6.114}$$

where $T = (EI/n_h)^{0.2}$ in which $E$ is the modulus of elasticity for the pile, $I$ is the moment of inertia of pile of width $B$, and $n_h$ is the constant of modulus of subgrade reaction and can be obtained from Table 4.16.

Based on these assumptions Vesic (1977) gives a relatively simple solution to this problem for a two-dimensional case and a rigid cap. For such a system (Figure 6.52a) if there is a point $C$ called center of the pile group such that the system experiences no rotation when resultant of the external forces $R$ passes through it, the position of $C$ can be found from geometrical and statical requirements. If, for a system that has $n$ equal size piles, the axis of each pile makes an angle $\alpha$ with the $y$ axis, located on the plane of pile head and assuming a coordinate center (0, 0) anywhere on the $y$ axis, with the $x$ axis as shown in Figure 6.52a then the coordinates of center $C$ are given by:

$$y_c = \frac{M_2 S_1 - M_1 S_2}{S_1 S_3 - S_2^2} \tag{6.115}$$

$$x_c = \frac{M_2 S_2 - M_1 S_3}{S_1 S_3 - S_2^2} \tag{6.116}$$

where

$$S_1 = \Sigma(\cos^2 \alpha + \sin^2 \alpha) \tag{6.117}$$
$$S_2 = (1 - \lambda)\Sigma \sin \alpha \cos \alpha$$
$$S_3 = \Sigma(\sin^2 \alpha + \lambda \cos^2 \alpha)$$
$$M_1 = (1 - \lambda)\Sigma y \sin \alpha \cos \alpha$$
$$M_2 = \Sigma y(\sin^2 \alpha + \lambda \cos^2 \alpha)$$

$\lambda = k_t/k_v$, $\alpha$ is the angle of each pile axis with the $y$ axis (Figure 6.52a) and $\Sigma y$ is the summation of distances of each pile head from point $C$

$P$, $Q$, and $M$ can then be obtained for each pile by the following:

$$P = C_{ph} Q_G + C_{pv} P_G + C_{pm} M_G \tag{6.118}$$

$$Q = C_{qh} Q_G + C_{qv} P_G + C_{qm} M_G \tag{6.119}$$

$$M = (C_{mh} Q_G + C_{mv} P_G + C_{mm} M_G)t \tag{6.120}$$

where the coefficients in equations (6.118), (6.119), and (6.120) are defined by the following:

$$C_{ph} = \frac{S_3 \cos \alpha - S_2 \sin \alpha}{S_1 S_3 - S_2^2} \tag{6.121a}$$

$$C_{pv} = \frac{S_1 \sin \alpha - S_2 \cos \alpha}{S_1 S_3 - S_2^2} \tag{6.121b}$$

$$C_{qh} = \lambda \frac{S_3 \sin \alpha + S_2 \cos \alpha}{S_1 S_3 - S_2^2} \tag{6.121c}$$

$$C_{qv} = -\lambda \frac{S_1 \cos \alpha + S_2 \sin \alpha}{S_1 S_3 - S_2^2} \tag{6.121d}$$

$$C_{pm} = -\frac{k_v a}{\Sigma_m} \tag{6.121e}$$

$$C_{qm} = \frac{k_t(b + t)}{\Sigma_m} \tag{6.121f}$$

$$C_{mm} = \frac{k_t(b + \xi t)}{\Sigma_m} \tag{6.121g}$$

$$\Sigma_m = \Sigma[k_v a^2 + k_t(b + t)^2 + k_t(\xi - 1)t^2] \tag{6.121h}$$

in which $a$ is the shortest distance between the pile axis and the center of the group and $b$ is the distance between the pile head and the point on the pile axis closer to the center of the group (Figure 6.52d).

With this formulation, the values of $P$, $Q$, $M$, $x$, $y$, and $\theta$ can be obtained. In spite of many assumptions made, this will still require elaborate computations and solutions by computer will be needed.

### 6.11.3 Pile Group Solution with Soil as an Elastic Medium

The elastic continuum approach discussed in Section 6.1 for lateral loads can be extended to cover piles and pile groups subjected to eccentric and inclined loads. Poulos and Madhav (1971) present a method to analyze single batter piles subjected to axial, lateral, loads and moments acting simultaneously. Analytical approach presented by Poulos (1974) can be used to analyze pile groups under eccentric and inclined loads. As discussed in Section 6.1, this approach still has to overcome the obstacle of applying the theory to the practical problems and field varifications. Therefore, this approach will not be discussed further here.

### 6.11.4 Bearing Capacity of Piles Under Eccentric and Inclined Loads: Interaction Relationship

When a pile is subjected to a horizontal load or pure moment its ultimate capacity can be obtained by Brinch Hansen's method (1961). For cases where piles are subjected to eccentric and inclined loads as under bridges and offshore structures, the ultimate bearing capacity can be obtained by the methods proposed by Meyerhof and Sastry (1985). These methods are based on tests carried out on fully instrumented rigid model piles jacked into homogeneous sand and clay. Methods are also available to make adjustments to take into account the flexibility of the pile (Meyerhof and Yalcin, 1984).

In the following sections, first the vertical piles subjected to eccentric and inclined loads in cohesionless soils will be presented. Then the behavior of vertical piles in cohesive soils will be discussed. Finally, this approach is extended to a group consisting of both vertical and batter piles.

## 6.12 VERTICAL PILES SUBJECTED TO ECCENTRIC AND INCLINED LOADS IN COHESIONLESS SOIL

Figure 6.53a shows a single vertical pile subjected to eccentric inclined load $Q_{e\alpha}$. The load is applied at angle $\alpha$ and eccentricity $e$ from the axis of the pile. The

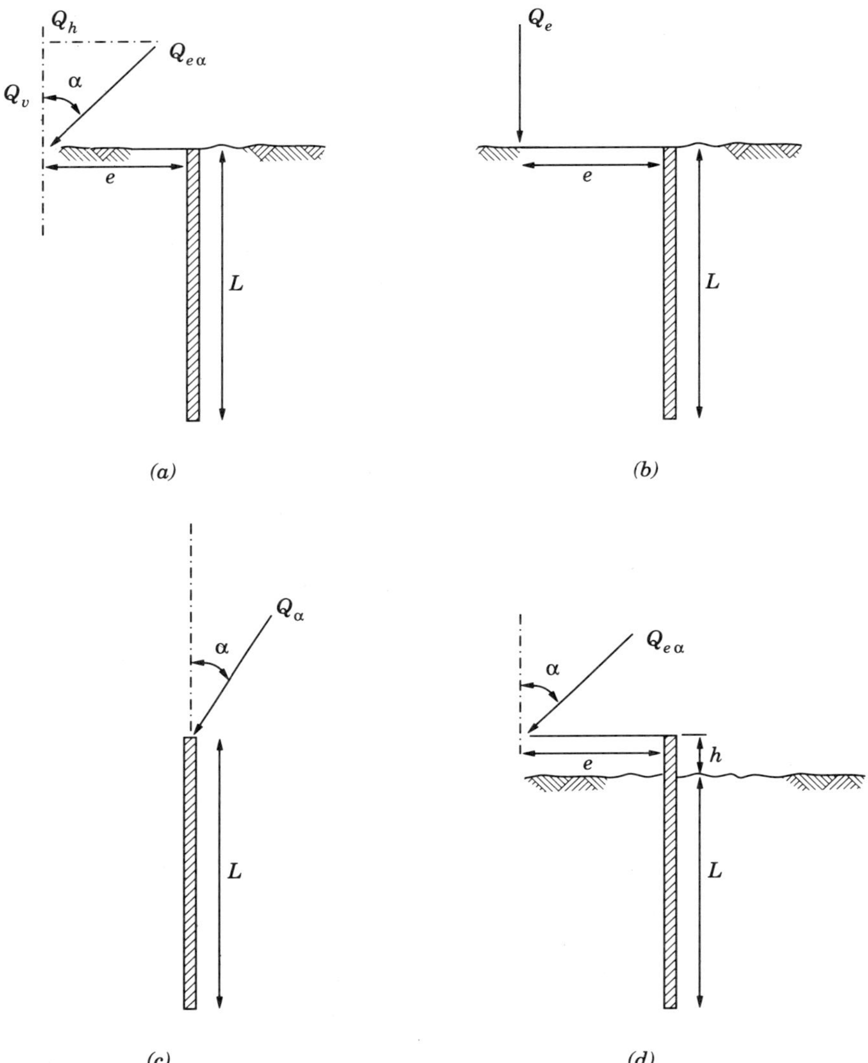


**Figure 6.53** Single vertical pile under eccentric inclined load. (a) Eccentric inclined load, (b) eccentric vertical load, (c) central inclined load, (d) a partially embedded single vertical pile under eccentric inclined load.

eccentric inclined load may be considered as composed of (1) $Q_e$, eccentric vertical load, Figure 6.53b, and (2) $Q_\alpha$, inclined load acting on the pile head. First, the ultimate eccentric vertical load capacity ($Q_{eu}$) of a rigid pile will be calculated. Following this, the ultimate central inclined load capacity ($Q_{\alpha u}$) will be obtained. Then the combined effect of eccentricity and inclined load on ultimate capacity will be presented. Moment on pile head due to projection above ground surface will also be discussed (Figure 6.53d). Finally, corrections for pile flexibility and group effects will be presented. A numerical example will be given at the end of this article to explain the concepts presented here.

As shown in Figure 6.54a, if a rigid pile of length $L$ is subjected to an ultimate load, $Q_{e\alpha,u}$, at an inclination $\alpha$ and eccentricity $e$ with the axis of the pile, then the lateral forces $Q_1$ and $Q_2$ can be approximated by triangular distribution. It should be seen that $Q_1$ and $Q_2$ are inclined at angles $+\delta_1$ and $-\delta_2$ with the horizontal where $\delta$ is anlge of friction between pile and soil. Plus sign has been taken in one case and negative sign in the other. This type of pressure distribution has been supported by measurements on instrumented piles and compared with theoretical relationships proposed by Krey (1936). These comparisons are shown in Figure 6.54b. The load eccentricities had eccentricity to depth $(e/D)$ ratio of 0.16, 0.38, and $\infty$ (pure moment) and load inclinations were $\alpha = 30°$, $60°$, and $90°$. The angle of internal friction $\phi$ in plane strain was $\phi_p = 35°$ for these tests. As expected, the figure shows that the observed lateral pressures decrease with smaller eccentricity ($e$) and smaller load inclination ($\alpha$).

Based on the results of model tests on fully instrumented piles and their analysis, Meyerhof and Sastry (1985) recommend the following semiempirical relationships for calculating ultimate capacity of vertical piles under eccentric and inclined load.

### 6.12.1 Ultimate Capacity Under Eccentric Vertical Loads

The ultimate capacity under eccentric vertical load $Q_{eu}$ at an eccentricity $e$ can be obtained from the following semiempirical interaction relationship proposed by Meyerhof et al. (1983).

$$\left(\frac{Q_{eu}}{P_u}\right)^2 + \frac{Q_{eu}e}{M_0} = 1 \tag{6.122}$$

where

$M_0$ = ultimate moment for a pile under pure moment without any axial load
$P_u$ = ultimate axial vertical load of pile

$M_0$ can be theoretically obtained by considering a smooth pile surface ($\delta_1 = \delta_2 = 0$) and setting $Q_1 = Q_2$. Thus, according to Meyerhof and Sastry (1985), $M_0$ can be given by the following expression:

$$M_0 = 0.09\gamma BL^3 K_b \tag{6.123}$$

where
$\gamma$ = unit weight of soil
$B$ = pile diameter
$L$ = pile length
$K_b$ = lateral earth pressure coefficient

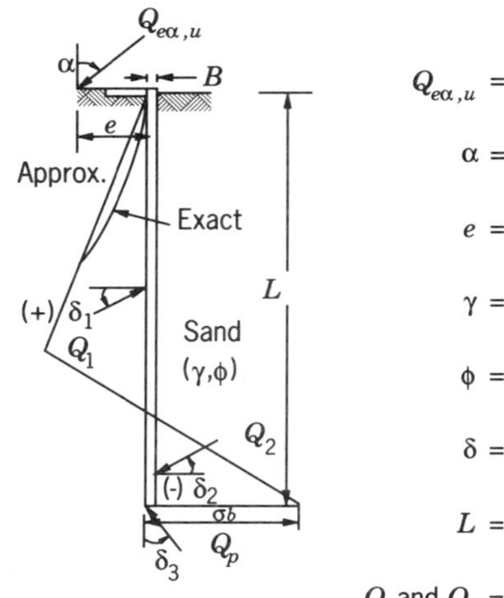


$Q_{e\alpha,u}$ = eccentric inclined ultimate load

$\alpha$ = inclination of load from vertical

$e$ = eccentricity of the load

$\gamma$ = unit weight of soil

$\phi$ = angle of internal friction of soil

$\delta$ = angle of skin friction

$L$ = pile length

$Q_1$ and $Q_2$ = lateral forces

$Q_p$ = pile base resistance inclined at angle $\delta_3$ with the vertical

$\sigma_b$ = lateral pressure at pile base level

*(a)*

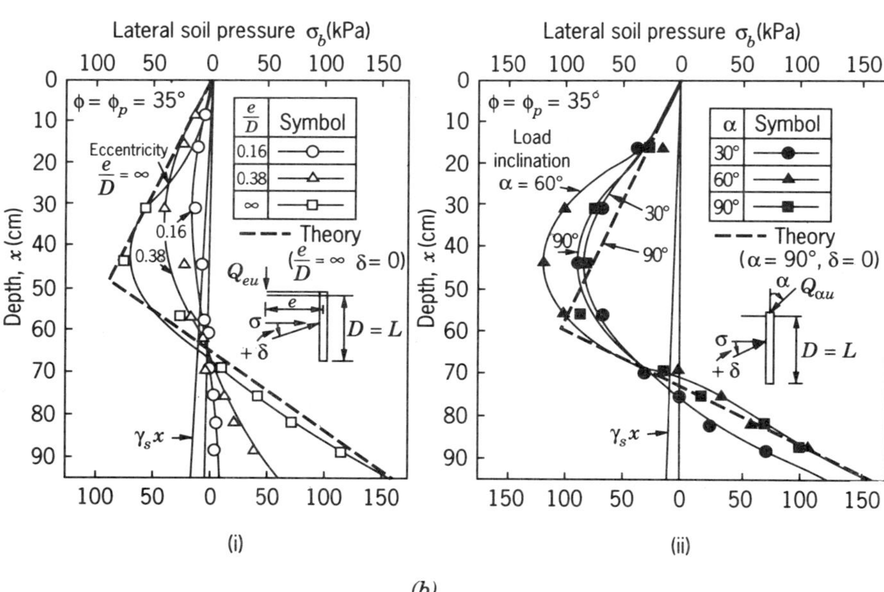


*(b)*

**Figure 6.54** Forces and soil pressures for a pile under eccentric and inclined loads in sand (a) Forces at failure of pile under eccentric inclined load, (b) distribution of lateral soil pressure on pile shaft at failure, (i) eccentric load, (ii) inclined load. (Meyerhof and Sastry, 1985)

For a free-head rigid vertical wall, the lateral earth pressure coefficient $K_b$ is equal to $[\tan^2(45+\phi/2)-\tan^2(45-\phi/2)]$ in cohesionless soils having an angle of internal friction equal to $\phi$. However, due to the existence of shearing resistance on vertical sides of the failure wedges in the soil around a pile, the ultimate lateral resistance per unit width of a rigid pile is greater than that of a corresponding wall. In order to account for this three-dimensional effect of a pile, the earth pressure on a pile can be obtained by multiplying the net earth pressure on a wall with a shape factor $S_b$ that varies from unity at the ground surface to that corresponding to the ultimate bearing capacity of a vertical strip footing at great depth ($x/B$ greater than 10 for loose sand to $x/B$ greater than 30 for dense sand). Based on the analytical data supported by model tests on piles, Meyerhof et al. (1981) show that an overall shape factor $S_{bu}$ can be used to estimate the total ultimate lateral resistance of a free-head rigid vertical pile in homogeneous sand. The values of $S_{bu}$ for various $(L/B)$ ratios of pile embedment are shown in Figure 6.55. Then $K_b$ can then be obtained from the following relationship:

$$K_b = [\tan^2(45+\phi/2)-\tan^2(45-\phi/2)]S_{bu} \tag{6.124}$$

$S_{bu}$ is the shape factor and can be obtained from Figure 6.55. The theoretical ultimate axial vertical load $P_u$ can be obtained from the following relationship (Meyerhof, 1976):

$$P_u = \gamma L N_q A_p + K_s \gamma L \tan\delta\,(A_s/2) \tag{6.125}$$

where

$\gamma$ = unit weight of soil
$L$ = pile length
$N_q$ = bearing capacity factor (Chapter 5)
$A_p$ = area of pile tip
$K_s$ = average coefficient of earch pressure on pile shaft (Chapter 5)
$\delta$ = angle of skin friction
$A_s$ = the area of pile shaft

This relationship has already been discussed in Chapter 5.

### 6.12.2 Ultimate Capacity Under Central Inclined Loads

According to Meyerhof and Ranjan (1972), the ultimate central inclined load, $Q_{\alpha u}$, on the pile cap at an inclination $\alpha$ can be obtained from the following semiempirical relationship:

$$\left(\frac{Q_{\alpha u}\cos\alpha}{P_u}\right)^2+\left(\frac{Q_{\alpha u}\sin\alpha}{Q_u}\right)^2=1 \tag{6.126}$$

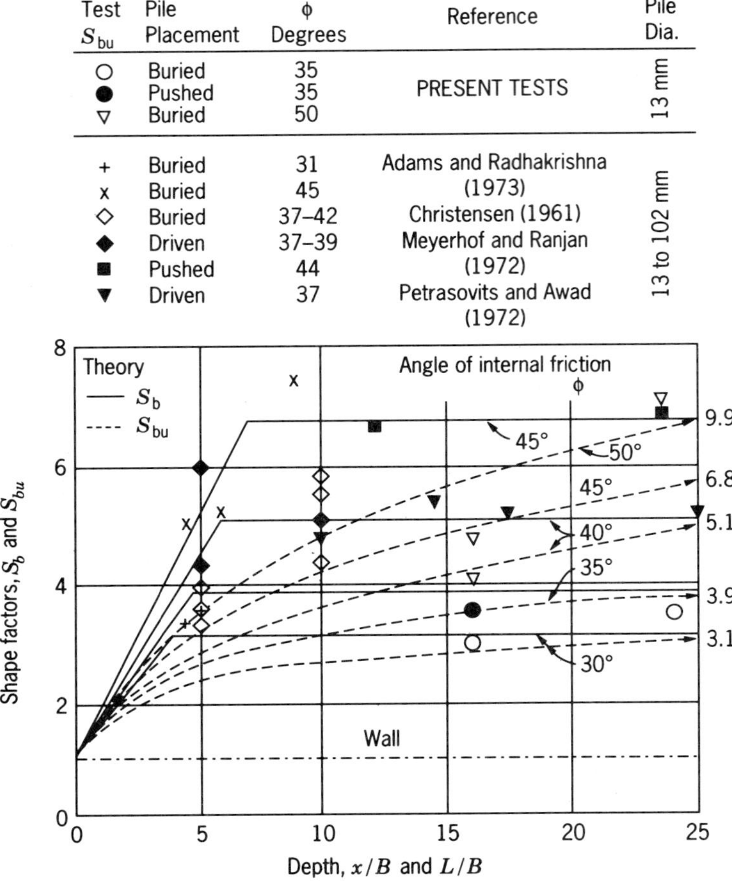


**Figure 6.55** Shape factors for laterally loaded piles in sand (Meyerhof et al., 1981).

where

$P_u$ = ultimate axial vertical load capacity
$Q_u$ = ultimate pile capacity under horizontal load ($\alpha = 90°$)

This can be theoretically obtained from the triangular pressure distribution (Figure 6.54b(ii)) for $\delta_1 = \delta_2 = 0$. The value of $Q_u$ can be obtained from the following relationship:

$$Q_u = 0.125\gamma BL^2 k_b \tag{6.127}$$

All the terms have been defined earlier.

### 6.12.3 Ultimate Capacity Under Eccentric Inclined Loads

Meyerhof et al. (1983) and Meyerhof and Yalcin (1984) suggest that by combining equations (6.122) and (6.126), the semiempirical interaction relationship for a rigid fully embedded pile with an eccentric inclined load becomes

$$\frac{(Q_{e\alpha,u}\cos\alpha)^2}{\left(1-\dfrac{Q_{eu}e}{M_0}\right)P_u^2}+\left(\frac{Q_{e\alpha,u}\sin\alpha}{Q_u}\right)^2=1 \tag{6.128}$$

where

$Q_{e\alpha,u}$ = ultimate eccentric inclined load
$Q_{eu}$ = ultimate eccentric vertical load
$P_u$ = ultimate axial vertical load
$M_0$ = ultimate moment for a pile under pure moment without any axial load
$L$ = pile length

From equation (6.128) ($Q_{e\alpha,u}$) can be determined for a given case. All other terms have been defined earlier. $Q_{eu}$ is equal to $Q_{e\alpha,u}\cos\alpha = Q_{vu}$.

In practice, it is often more convenient to use the vertical component ($Q_{vu} = Q_{e\alpha,u}\cos\alpha$) of the ultimate eccentric inclined load with an approximate overall reduction factor. $Q_{vu}$ can be obtained from the following expression (Meyerhof et al., 1983).

$$Q_{vu}=\frac{(1-\alpha/90)^2}{(1+(e/mL)^2)^{0.5}}P_u \tag{6.129}$$

where $m$ is obtained from the following relationship:

$$m=\frac{M_0}{P_uL} \tag{6.130}$$

All other terms have been discussed earlier.

### 6.12.4 Ultimate Load Capacity due to Partial Embedment

Meyerhof et al. (1983) suggest the following interaction relationship for partially embedded pile subjected to an eccentric inclined load at the free head at a distance $h$ above ground level (Figure 6.53d).

$$\frac{(Q_{e\alpha,u}\cos\alpha)^2}{\left(1-\dfrac{Q_{eu}e}{M_0}\right)P_u^2}+\frac{(Q_{e\alpha,u}\sin\alpha)^2}{\left(1-\dfrac{Q_{hu}h}{M_0}\right)^2Q_u^2}=1 \tag{6.131}$$

where $Q_{hu} = Q_u/(1 + 1.4\,h/L)$ All other terms have been defined earlier. From this equation, $Q_{e\alpha,u}$ can be obtained.

### 6.12.5 Pile Stiffness

According to Meyerhof (1976), a free-head pile in homogeneous elastic soil may be considered rigid for all practical purposes if its relative stiffness, $K_r \geqslant 0.01$, where $K_r$ is given by the following expression:

$$K_r = \frac{E_p I_p}{E_h L^4} \tag{6.132}$$

where

$E_p$ = modulus of elasticity of pile
$I_p$ = moment of inertia of pile
$E_h$ = average horizontal soil modulus along pile = $k_h$
$L$ = pile length

Meyerhof and Yalcin (1984) suggest that in case of flexible piles ($K_r < 0.01$) under eccentric or inclined loads, an effective embedment length $L$ can be approximated from the following relationship:

$$L_e = 3(K_r L)^{0.2} \leqslant L \tag{6.133}$$

The value of $K_r$ is to be calculated from equation (6.132). This value of $L_e$ should be used instead of $L$ in equations (6.123) and (6.127) to obtain pile capacities (Meyerhof and Sastry, 1985).

### 6.12.6 Pile Groups

The analysis of single piles subjected to eccentric inclined loads can be extended to pile groups with customary pile spacing of about three times the pile diameters (Meyerhof et al., 1983). Accordingly, the ultimate bearing capacity will be the smaller of the following two:

1. The sum of individual pile capacities. Support provided by pile cap is neglected because of the uncertainty in mobilizing this support. This assumption will be on the safe side.
2. The ultimate capacity of an individual pier consisting of the piles and the enclosed soil mass having a width $b$ as shown in Figure 6.56.

The ultimate capacities for a single pile have been discussed earlier. The ultimate capacity of a pier consisting of individual piles and the enclosed soil is presented below.

Based on the results of tests on model piles Meyerhof et al. (1983) recommend following semiempirical interaction relationships for ultimate bearing capacity of a pier (Figure 6.56) consisting of piles and the enclosed soil mass.

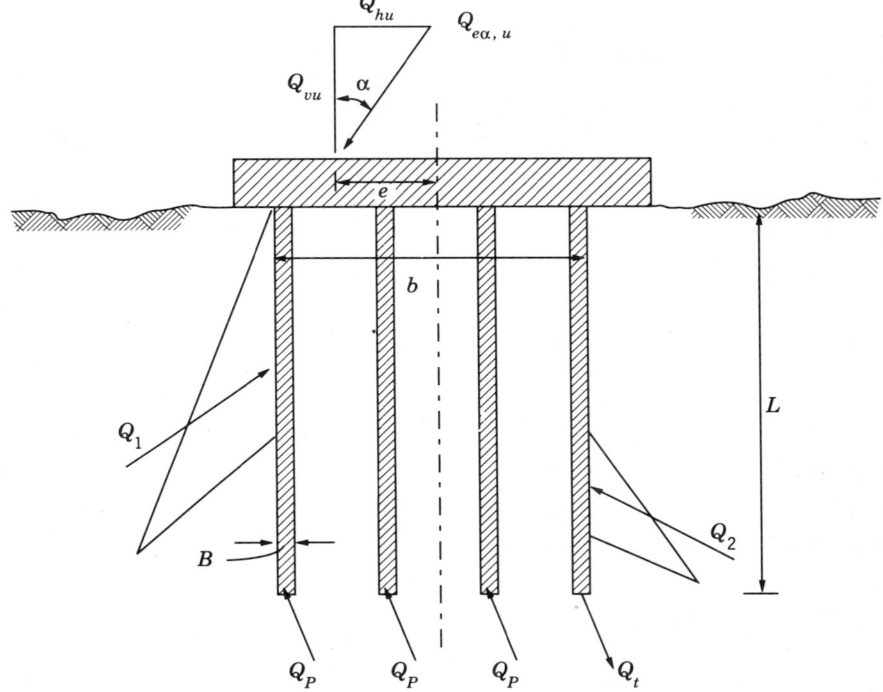


**Figure 6.56** Forces at failure of rigid pile group under eccentric inclined load in cohesionless soil.

### 6.12.7 Ultimate Eccentric Vertical Load

The model test results indicate that the normalized interaction relationship for ultimate loads and moments at the head of single pile, equation (6.122), should be modified for pile groups as given by the following:

$$\left(\frac{Q_{eu}}{P_u} - 0.4\right)^2 + \frac{Q_{eu}e}{M_0} = 1 \qquad \text{for } \frac{Q_{eu}}{P_u} \geqslant 0.4 \tag{6.134a}$$

$$\frac{Q_{eu}}{P_u} = \frac{2}{3}\left(\frac{Q_{eu}e}{M_0} - 0.4\right) \qquad \text{for } \frac{Q_{eu}}{P_u} \leqslant 0.4 \tag{6.134b}$$

$M_0$ and $P_u$ are obtained from equations (6.123) and (6.125) by using overall group width $b$ (Figure 6.56) instead of individual pile width $B$ and area of the group instead of area of the single pile.

### 6.12.8 Ultimate Central Inclined Load

Test results on pile group support that the semiempirical interaction relationship of equation (6.126) can be used for central inclined load on pier foundation (pile group consisting of piles and the enclosed soil). The values of $P_u$ and $Q_u$ can be determined from equations (6.125) and (6.127) by substituting $b$ instead of $B$.

***Eccentric Inclined Loads*** For combined eccentricity and inclination of the load the following approximate relationship has been found to support experimental results for obtaining the vertical component $Q_{vu}$ of the ultimate load $Q_u$.

$$\frac{Q_{vu}}{P_u} = \frac{(1 - \alpha/90)^4}{(1 + (e/mL)^2)^{0.5}} \qquad \text{for } \frac{Q_{e\alpha}}{P_u} \geqslant 0.4 \tag{6.135a}$$

$$\frac{Q_{vu}}{P_u} = \frac{0.4(1 - \alpha/90)^4}{((e/mL) - 1.5)} \qquad \text{for } \frac{Q_{e\alpha}}{P_u} \leqslant 0.4 \tag{6.135b}$$

values of $P_u$, $\alpha$, $e$, $m$, and $L$ have already been defined. $Q_v$ and $Q_{e\alpha}$ are shown in Figure 6.53a for single pile and Figure 6.56 for a pile group.

### 6.12.9 Ultimate Load due to Partial Embedment

Meyerhof et al. (1983) suggest that if an eccentric inclined load is applied to a rigid pile group at a height $h$ above ground level, then an approximate overall interaction relationship can be given by the following:

$$\frac{Q_{vu}}{P_u} = \frac{(1 - \alpha/90)^4}{(1 + (e/mL)^2)^{0.5}(1 + 1.4\,h/L)} \qquad \text{for } \frac{Q_{e\alpha,u}}{P_u} \geqslant 0.4 \tag{6.136a}$$

and

$$\frac{Q_{vu}}{P_u} = \frac{0.4(1 - \alpha/90)^4}{(1 + (e/mL - 1.5)(1 + 1.4h/L)} \qquad \text{for } \frac{Q_{e\alpha,u}}{P_u} \leqslant 0.4 \tag{6.136b}$$

All the terms have previously been defined.

***Example 6.14*** A group of nine vertical piles driven 25 ft (7.6 m) into dense sand with average $N = 38$, $\phi = 36^\circ$ and $\gamma = 120\,\text{lb/ft}^3$ ($1920\,\text{kg/m}^3$) is subjected to an eccentric inclined load $Q_{e\alpha}$ at an inclination of 30° and eccentricity 6 ft. The piles are 12 in. (304.8 mm) outside diameter and 0.5 in. (12.7 mm) wall thickness steel pile having modulus of elasticity $E_p = 29{,}000$ ksi ($200 \times 10^3$ MPa). The average horizontal soil modulus $E_h$ is $12\,\text{kips/ft}^2$. Calculate the allowable eccentric inclined load that can be applied on the pile group if the factor of safety against bearing capacity failure is taken as 2.5. The pile cap is resting on the ground and piles are fully embedded into the soil and are closed at their tips with a steel plate. Assume water table is near ground surface. This is shown in Figure 6.57.

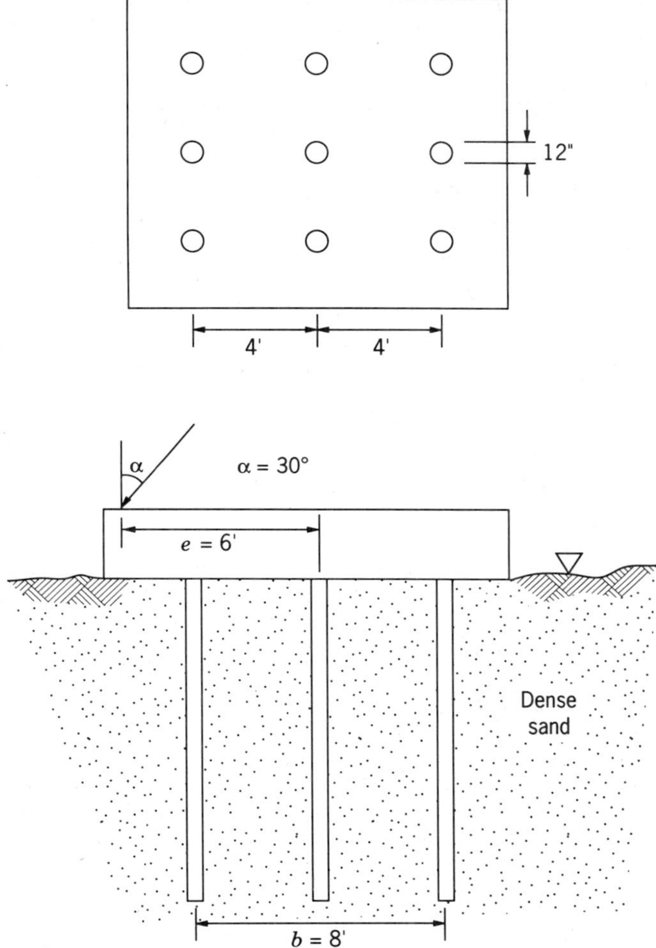


**Figure 6.57** Pile group arrangement and the eccentric inclined load (Example 6.14).

SOLUTION

**1. Pile Stiffness**

$$K_r = \frac{E_p I_p}{E_h L^4} \tag{6.132}$$

$$= \frac{(29 \times 144 \times 10^3)\,0.015}{12\,(25)^4} \qquad \text{where } I_p = \frac{\pi}{64}\left[\left(\frac{12}{12}\right)^4 - \left(\frac{11}{12}\right)^4\right] = 0.015\,\text{ft}^4$$

$$= 0.0134 \geqslant 0.01$$

Therefore, the piles will behave as rigid piles subjected to eccentric inclined loads, and no correction is required in pile length to account for pile flexibility.

**2. Individual Pile Capacities**

$$P_u = \gamma' L N_q A_p + k_s \gamma' L \tan\delta (A_s/2) \tag{6.125}$$

$$\gamma = 120\,\text{lb/ft}^2 \qquad \gamma' = 120 - 62.5 = 57.5\,\text{lb/ft}^3$$

$$L = 25\,\text{ft}$$

$$N_q = 60 \text{ for } \phi = 36^\circ \text{ from Table 5.2}$$

$$A_p = \pi/4(1)^2 = 0.785\,\text{ft}^2$$

$$k_s = 1 \text{ from Table 5.3}$$

$$\delta = 2/3\phi = (2/3)36^\circ = 24^\circ$$

$$\tan\delta = 0.445$$

$$A_s = \pi BL = \pi \times 1 \times 25 = 78.5\,\text{ft}^2$$

$$P_u = 57.5 \times 25 \times 60 \times 0.78 + 1 \times 57.5 \times 25 \times 0.445 \times \frac{78.5}{2}$$

$$P_u = 90.28 + 25.1 = 115.38\,\text{kips}$$

$$M_0 = 0.09\gamma' BL^3 k_b \tag{6.123}$$

where

$$k_b = [\tan^2(45 + \phi/2) - \tan^2(45 - \phi/2)]S_{bu} \tag{6.124}$$

$$S_{bu} = 4 \text{ for } \phi = 36^\circ \text{ and } L/B = 25/1 = 25 \text{ from Figure 6.55}$$

$$k_b = [\tan^2(45 + 18) - \tan^2(45 - 18)]4 = [3.85 - 0.53]4 = 3.32 \times 4$$
$$= 13.28$$

$$M_0 = 0.09 \times 57.5 \times 1(25)^3 \times 13.28 = 1073.8\,\text{kips-ft}$$

$$m = \frac{M_0}{P_u L} \quad \text{from equation (6.130)}$$

$$= \frac{1073.8}{115.38 \times 25} = 0.372$$

$$\frac{Q_{vu}}{P_u} = \frac{(1 - \alpha/90)^2}{(1 + (e/mL)^2)^{0.5}} \tag{6.129}$$

From Figure 6.57, three piles have $e = 2\,\text{ft}$, middle three piles have $e = 6\,\text{ft}$

and farthest three piles have $e = 10\,\text{ft}$.

$$e = 2\,\text{ft:}\quad \frac{Q_{vu}}{P_u} = \frac{(1 - 30/90)^2}{\sqrt{1 + (2/0.372 \times 25)^2}} = \frac{0.445}{1.023} = 0.434$$

$$= 6\,\text{ft:}\quad \frac{Q_{vu}}{P_u} = \frac{(1 - 30/90)^2}{\sqrt{1 + (6/0.372 \times 25)^2}} = \frac{0.445}{1.19} = 0.373$$

$$= 10\,\text{ft}\quad \frac{Q_{vu}}{P_u} = \frac{(1 - 30/90)^2}{\sqrt{1 + (10/0.372 \times 25)^2}} = \frac{0.445}{1.48} = 0.3$$

$$\frac{Q_{vu}}{P_u} = 3 \times 0.434 + 3 \times 0.373 + 3 \times 0.3 = 3.3$$

$$Q_{vu} = 3.3P_u = 3.3 \times 115.39 = 380\,\text{kip}$$

$$Q_{e\alpha,u} = Q_{vu}/\cos\alpha = 380/\cos 30 = 438\,\text{kips}$$

**3. Pile Capacity of the Pier Consisting of the Piles and the Enclosed Soil Mass**

$A_s$ is shown in Figure 6.57, $b = 8\,\text{ft}$

$$A_p = 8 \times 8 = 64\,\text{ft}^2$$

$$A_s = 4 \times 8 \times 25 = 800\,\text{ft}^2$$

Then from equation (6.125):

$$P_u = 57.5 \times 25 \times 60 \times 64 + 1 \times 57.5 \times 25 \times 0.445 \times \frac{800}{2}$$

$$= 7360 + 255.8 = 7615.8\,\text{kips}$$

For $x/B = L/B = 25/8 = 3.125$, $S_{bu} = 2.5$ for $\phi = 36°$ from Figure 6.55, then from equation (6.123):

$$M_0 = 0.09 \times 57.5 \times 8 \times (25)^3 \times 3.32 \times 2.5 = 5369.0\,\text{kips-ft}$$

$$m = \frac{M_0}{P_u} = \frac{5369}{7615.8 \times 25} = 0.028$$

From equation (6.135a):

$$\frac{Q_{vu}}{P_u} = \frac{Q_{e\alpha,u}\cos\alpha}{P_u} = \frac{(1 - \alpha/90)^4}{(1 + (e/mL)^2)^{0.5}} \qquad \text{for } \frac{Q_{e\alpha}}{P_u} \geqslant 0.4$$

$$\frac{Q_{e\alpha,u}\cos 30}{P_u} = \frac{(1 - 30/90)^4}{(1 + (6/0.028 \times 25)^2)^{0.5}} = \frac{0.189}{8.63} = 0.022$$

$$\frac{Q_{e\alpha,u}\cos 30}{P_u} = 0.022 \qquad \text{or} \qquad \frac{Q_{e\alpha,u}}{P_u} = 0.025$$

From equation (6.135b) for $Q_{e\alpha,u}/P_u \leqslant 0.4$

$$\frac{Q_{vu}}{P_u} = \frac{Q_{e\alpha,u}\cos\alpha}{P_u} = \frac{0.4(1-\alpha/90)^4}{(e/mL - 1.5)} = \frac{0.4 \times 0.189}{8.6 - 1.5} = 0.01$$

$$\frac{Q_{e\alpha,u}\cos 30}{P_u} = 0.01$$

$$\frac{Q_{e\alpha,u}}{P_u} = \frac{0.01}{\cos 30} = \frac{0.01}{0.866} = 0.012$$

From both the equation (6.135a) and (135b)

$$\frac{Q_{e\alpha,u}}{P_u} < 0.4$$

therefore equation (6.135b) applies.

$$\frac{Q_{e\alpha,u}}{P_u} = 0.012$$

$$Q_{e\alpha,u} = 0.012\,(7615.8) \cong 92\,\text{kips}$$

From step 2.0 and 3.0 above the smaller $Q_{e\alpha,u}$ is 92 kips
Therefore an allowable eccentric inclined load $Q_{e\alpha,u}$ for group $= Q_{e\alpha,u}/FS = 92/2.5 = 37$ kips.

## 6.13 VERTICAL PILES SUBJECTED TO ECCENTRIC AND INCLINED LOADS IN COHESIVE SOIL

As shown in Figure 6.58a, if a rigid pile of length $L$ is subjected to an ultimate load $Q_{e\alpha,u}$ at an inclination $\alpha$ and eccentricity $e$ with the axis of the pile then the interactive soil resistance forces are (1) lateral forces $Q_1$ and $Q_2$, (2) soil adhesion forces $C_1$ and $C_2$, and (3) point resistance $Q_p$.

Figure 6.58b (i) and (ii) show the distribution of lateral soil pressures. The theoretical lateral pressure distribution on the pile shaft was obtained from the earth pressure coefficients suggested by Meyerhof (1972). As shown in Figure 6.58b, a reasonable agreement was obtained between the observed lateral soil pressures from pile load tests and the predicted values of lateral pressures (Meyerhof and Sastry, 1985). The observed lateral pressures decrease rapidly with

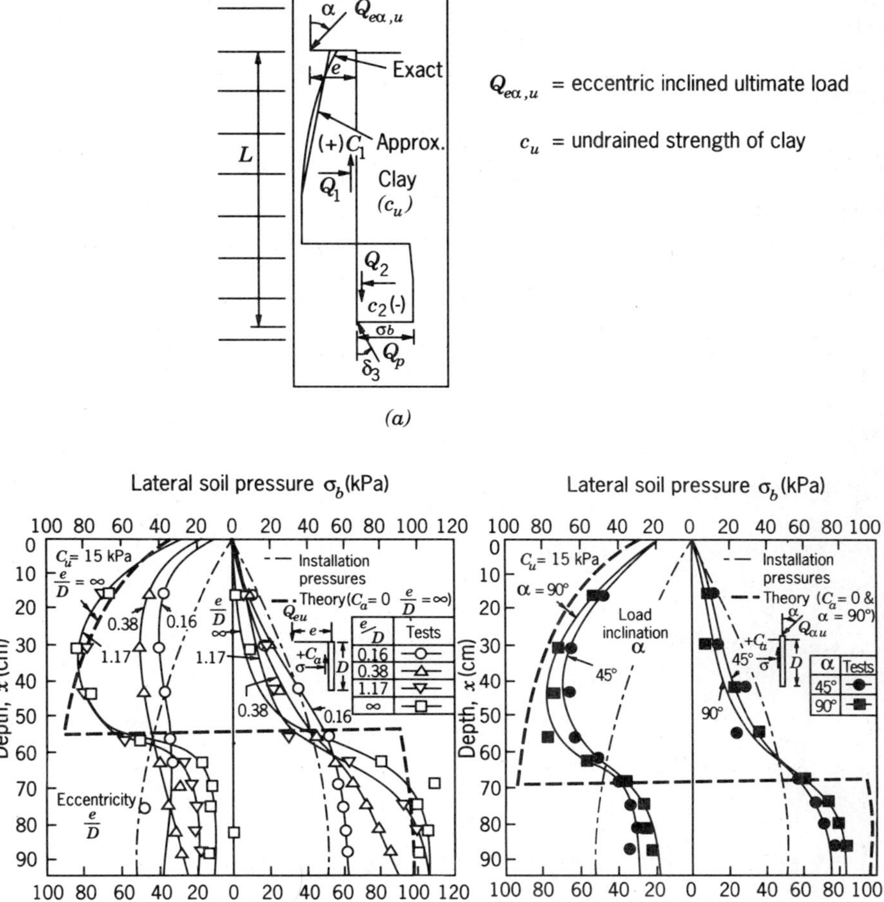


**Figure 6.58** Forces and soil pressures for a pile under eccentric and inclined loads in clay (a) Forces at failure of pile under eccentric inclined load, (b) distribution of lateral soil pressure on pile shaft due to installation and loading to failure, (i) eccentric load, (ii) inclined load. (Meyerhof and Sastry, 1985).

smaller eccentricity and load inclination. Figure 6.58 also shows that the roughly rectangular pressure distribution suggested by Brinch Hansen (1961) was found to be applicable in estimating ultimate pile capacities.

Meyerhof and Sastry (1985) recommend following semiempirical relationships for calculating ultimate capacities of vertical piles under eccentric and inclined load. This method is based on the results of model tests on fully instrumented piles and their analysis. Various terms used here are as defined in Figure 6.53.

### 6.13.1 Ultimate Capacity Under Eccentric Vertical Load

The ultimate eccentric vertical load $Q_{eu}$ at an eccentricity $e$ can be obtained from the following semiempirical interaction relationship (Meyerhof and Yalcin, 1984):

$$\left(\frac{Q_{eu}}{P_u}\right)^2 + \left(\frac{Q_{eu}e}{M_o}\right) = 1 \tag{6.137}$$

where

$M_0$ = the ultimate moment for a pile under pure moment without any axial load
$P_u$ = ultimate axial vertical load
$e$ = eccentricity

$M_0$ can be obtained by setting $C_a = 0$, as deduced from the measured lateral pressures (Meyerhof et al., 1981) and is given by following.

$$M_0 = 0.2C_{us}BL^2K_c \tag{6.138}$$

where

$C_{us}$ = average undrained shear strength of clay along the shaft
$K_c = 2\tan(45 + \phi/2)S_{cu}$

The shape factor $S_{cu}$ can be obtained from Figure 6.59. The theoretical ultimate axial vertical load $P_u$ can be obtained from the following relationship (Meyerhof, 1976).

$$P_u = 9C_{up}A_p + rC_{us}A_s \tag{6.139}$$

where

$C_{up}$ = average undrained shear strength of clay near the pile tip
$A_p$ = area of the pile at its tip
$C_{us}$ = average undrained shear strength of clay along pile shaft
$A_s$ = area of the pile shaft
$r$ = adhesion factor ($= C_a/C_{us}$) whose value depends on the magnitude of $C_{us}$ and method of pile installation.

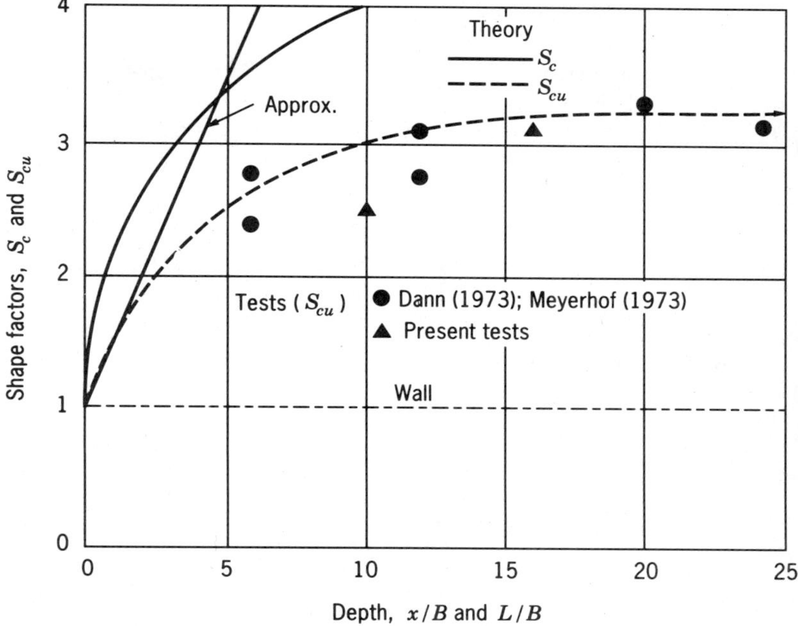


**Figure 6.59** Shape factors for laterally loaded piles in clay (Meyerhof et al., 1981).

### 6.13.2 Ultimate Capacity Under Central Inclined Load

For any given load inclination $\alpha$, the ultimate central inclined load, $Q_{\alpha u}$, can be obtained from the following relationship.

$$\left(\frac{Q_{\alpha u}\cos\alpha}{P_u}\right)^2 + \left(\frac{Q_{\alpha u}\sin\alpha}{Q_u}\right)^2 = 1 \tag{6.140}$$

where $Q_u$ = ultimate pile capacity under horizontal load ($\alpha = 90$) and can theoretically be estimated using zero adhesion as back calculated from the observed lateral soil pressures. For a fully embedded pile, the following expression was obtained by Meyerhof et al. (1981) to calculate the value of $Q_u$:

$$Q_u = 0.4C_{us}BLK_c \tag{6.141}$$

where $C_{us}$, $B$, $L$, and $K_c$ have been defined earlier.

### 6.13.3 Ultimate Capacity Under Eccentric Inclined Load

The semiempirical interaction relationship given by equation (6.128) is also applicable here to determine the ultimate eccentric inclined load $Q_{e\alpha,u}$. The values of $Q_{eu}$, $P_u$ and $Q_u$ shall, however, be determined from equations (6.137), (6.139), and (6.141), respectively.

In practice, it is more convenient to use the vertical component ($Q_{vu} = Q_{e\alpha} \cos\alpha$) of the ultimate eccentric inclined load with an approximate overall reduction factor. Meyerhof and Yalcin (1984) provide following expression to calculate $Q_{vu}$ for cohesive soils

$$Q_{vu} = \frac{P_u \cos\alpha}{\left(1 + \left(\frac{2e}{L}\right)^2\right)^{0.5}} \tag{6.142}$$

### 6.13.4 Ultimate Load Capacity due to Partial Embedment

Meyerhof and Yalcin (1984) suggest the following interaction relationship for partially embedded pile subjected to an eccentric load at the free head at a distance $h$ above the ground level (Figure 6.53d).

$$\frac{Q_{vu}^2}{\left(1 - \frac{M_{vu}}{M_o}\right) P_u^2} + \frac{Q_{hu}^2}{\left(1 - \frac{M_{hu}}{M_o}\right)^2 Q_u^2} = 1 \tag{6.143}$$

where

$Q_{vu} = Q_{e\alpha,u} \cos\alpha$
$Q_{hu} = Q_{e\alpha,u} \sin\alpha$
$M_{vu} = Q_{vu}\, e$
$M_{hu} = Q_{hu}\, h$

***Pile Stiffness*** Discussion on pile stiffness presented in section 6.12.5 is also applicable in this case. Therefore, for flexible piles, the $L_e$ value given in equation (6.133) may be used in equations (6.138) and (6.141).

***Pile Groups*** The above analysis of single piles subjected to eccentric inclined loads can be extended to pile groups with customary pile spacing of about three times the pile diameters both for cohesionless and cohesive soils (Meyerhof and Yalcin, 1984). Accordingly, the ultimate bearing capacity will be the smaller of the pile capacities determined in the following two ways.

1. The sum of the individual pile capacities and of the pile cap, if resting on the soil. Normally, support due to pile cap is neglected because of the uncertainty in mobilizing pile cap soil support.
2. The ultimate capacity of an individual pier consisting of the piles and enclosed soil mass having a width b, as shown in Figure 6.60.

The individual pile capacities under eccentric inclined loads have been discussed

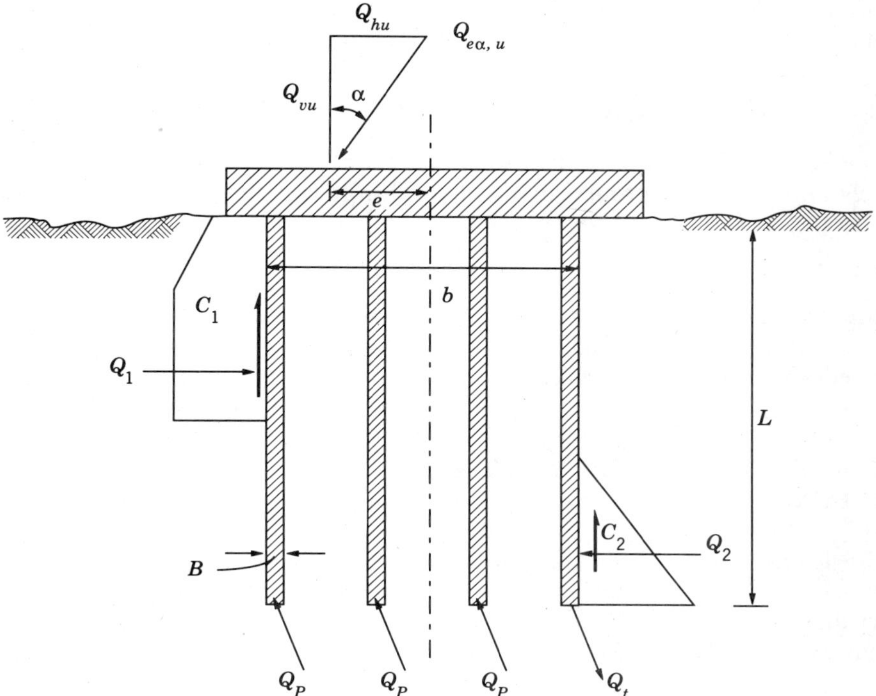


**Figure 6.60** Forces at failure of rigid pile group under eccentric inclined load in cohesive soils.

earlier. The ultimate capacity of an individual pier consisting of the piles and enclosed soil mass of width $b$ is presented as follows.

Based on the results of tests on model piles and their analyses, Meyerhof and Yalcin (1984) recommend following semiempirical interaction relationships for ultimate bearing capacity of a pier (Figure 6.60) consisting of piles and the enclosed soil mass.

### 6.13.5 Ultimate Eccentric Vertical Loads

The test results indicate that the normalized interaction relationship for ultimate loads and moments at the head of single piles as given by equation (6.137) can also be used for pile groups. $M_o$ and $P_u$ will be obtained from equations (6.138) and (6.139) respectively by using overall group width $b$ and area instead of individual pile width $B$ and individual pile area.

### 6.13.6 Ultimate Central Inclined Loads

Equation (6.140) can be used to obtain ultimate central inclined load for a pile group. This has been supported by test results. The values of $P_u$ and $Q_u$ can be determined from equations (6.139) and (6.141) by substituting pile group dimensions instead of individual pile dimensions.

### 6.13.7 Eccentric Inclined Loads

For combined eccentricity and inclination of the load semiempirical interaction relationship for pile group can be obtained from equations (6.128) and (6.142).

### 6.13.8 Ultimate Load due to Partial Embedment

Meyerhof and Yalcin (1984) suggest that if an eccentric inclined load is applied to a rigid pile group at a height $h$ above the ground level, an approximate overall interaction relationship given by equation (6.143) can be used here.

## 6.14 BATTER PILES SUBJECTED TO ECCENTRIC AND INCLINED LOADS

***Single Pile*** Figure 6.61 shows the comparison between a single vertical pile and the equivalent batter pile. This figure indicated that a vertical pile subjected to an inclined load at an angle $\alpha$ is equivalent in behavior to a batter pile inclined at an angle $\beta$ and subjected to vertical load. This equivalent behavior is apparent from

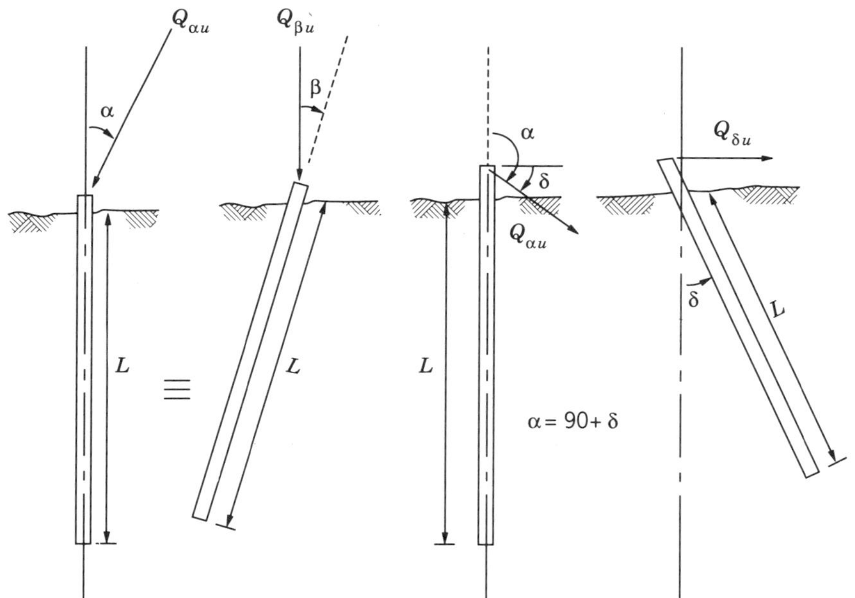


**Figure 6.61** Single vertical pile and the equivalent batter pile.

**TABLE 6.17 Comparison of Load Capacity of Vertical and Batter Piles**[1]

| $\alpha$ or $\beta$ (deg) | $\frac{Q_{\alpha u}}{(Q_{vu})}$ | $\frac{Q_{\beta u}}{(Q_{vu})}$ |
|---|---|---|
| 0 | 1 | 1 |
| 7.5 | 1.14 | 1.04 |
| 15 | 1.25 | 1.20 |
| 22.5 | 1.35 | 1.34 |
| 30 | 1.28 | 1.31 |

1. see Figure 6.61.
$\alpha$ = inclination of load on vertical pile.
$\beta$ = inclination of batter pile.
$Q_{\alpha u}$ = ultimate inclined load on vertical pile.
$Q_{vu}$ = ultimate vertical load on vertical pile.
$Q_{\beta u}$ = ultimate vertical load on battered pile.

the data presented in Table 6.17 (Awad and Petrasovits, 1968). This table exhibits that ultimate inclined load capacity of a vertical pile ($Q_{\alpha u}$) is equivalent to the ultimate vertical load capacity ($Q_{\beta u}$) of a batter pile inclined at an angle $\beta = \alpha$. Thus, the behavior of a batter pile may be analyzed by considering an equivalent vertical pile subjected to inclined loads. Figure 6.61 presents equivalent vertical and batter piles.

The ultimate capacity $Q_{\beta u}$ of a single batter pile can therefore be calculated if the ultimate inclined load of an equivalent vertical pile is determined by the methods outlined in Sections 6.12 and 6.13.

***Pile Group*** Once the equivalence between a single vertical pile having inclined load and a batter pile with vertical or lateral load is established, the pile group may be analyzed in the similar manner as described in Section 6.12 and 6.13. Accordingly, the ultimate bearing capacity will be the smaller of the following two capacities:

1. The sum of the individual pile capacities: Both vertical and batter pile capacities are to be determined if they form a part of the group. The support provided by pile cap is normally neglected because of the uncertainty in mobilizing soil support at its base.
2. The load capacity of a pier consisting of the piles and the enclosed soil mass.

Some load test results on the performance of batter pile groups are available (Tschebotarioff, 1953; Prakash and Subramanyam, 1964; and Simek, 1966). Further information on the behavior of batter pile groups under eccentric and inclined loading conditions is required. Until such data are obtained the foregoing design criteria can be used to analyze batter pile groups subjected to inclined and eccentric loads.

***Example 6.15*** A 12-in. (304.8 mm) outside diameter and 0.5-in. (12.7 mm) wall thickness 25 ft long steel pile is driven in clay. The pile is installed at an angle $\beta = 15°$ from the vertical. The modulus of elasticity $E_p$ for the pile is 29,000 ksi ($200 \times 10^3$ MPa) and average horizontal soil modulus $E_h = 12\,\text{kips/ft}^2$. The undrained strength $c_u = S_u$ of soil is $2\,\text{kips/ft}^2$ and adhesion $c_a = 1\,\text{kips/ft}^2$. Calculate the ultimate vertical capacity of the batter pile that is closed at its tip with a steel plate.

(a) *Equivalent vertical pile* From Figure 6.61, $\alpha = \beta = 15°$. Therefore, if $Q_{\alpha u}$ of the vertical pile is determined, then it will be equal to $Q_{\beta u}$ for the batter pile.

(b) *Pile stiffness* For equation (6.132):

$$K_r = \frac{E_p I_p}{E_h L^4} = \frac{(29 \times 144 \times 10^3)(0.015)}{12\,(25)^4} = 0.0134 \geqslant 0.01$$

where $I = \pi/64\,((12/12)^4 - (11/12)^4) = 0.015\,\text{ft}^4$. Therefore, the pile will have a rigid pile behavior and no correction in pile length is required.

(c) *Ultimate inclined load capacity of vertical pile* From equation (6.139), we get:

$$P_u = 9C_{up}A_p + rC_{us}A_s$$

where

$C_{up} = 2\,\text{kips/ft}^2 \qquad r = C_a/C_{us} = 1/2 = 0.5 \qquad B = 1\,\text{ft} \qquad L = 25\,\text{ft}$
$C_{us} = 2\,\text{kips/ft}^2 \qquad A_p = \pi/4\,(1)^2 = 0.785\,\text{ft}^2 \qquad A_s = \pi BL = 78.5\,\text{ft}^2$
$P_u = 9 \times 2 \times 0.785 + 0.5 \times 2 \times 78.5 = 14.13 + 78.5 = 92.63\,\text{kips}$

From equation (6.141), we get:

$$Q_u = 0.4C_{us}BLk_c$$

$$k_c = 2\tan(45 + \phi/2)S_{cu} = 2 \times 3.25$$

$$S_{cu} = 3.25 \text{ for } \frac{x}{B} = \frac{L}{B} = 25 \quad \text{and} \quad \text{for } \phi = 0 \quad \text{(see Figure 6.59)}$$

$$Q_u = 0.4 \times 2 \times 1 \times 25 \times 2 \times 3.25 = 40 \times 3.25 = 130\,\text{kips}$$

From equation (6.140), we get:

$$\left(\frac{Q_{\alpha u}\cos\alpha}{P_u}\right)^2 + \left(\frac{Q_{\alpha u}\sin\alpha}{Q_u}\right)^2 = 1$$

where

$$\alpha = \beta = 15°, P_u = 92.63\,\text{kips}, Q_u = 130\,\text{kips}$$

$$\left(\frac{Q_{\alpha u}\cos 15}{92.63}\right)^2+\left(\frac{Q_{\alpha u}\sin 15}{130}\right)^2=1$$

$$Q_{\alpha u}^2(0.000108+0.0000039)=1$$

$$Q_{\alpha u}^2=\frac{1}{0.0001119}=8931.46$$

$$Q_{\alpha u}=94.5\,\text{kips}=Q_{\beta u}$$

Therefore, the ultimate vertical load capacity of the batter pile is 94.5 kips.

## 6.15 LIMIT STATE ANALYSIS FOR PILE FOUNDATION DESIGN

Conventional geotechnical analysis and structural design discussed up to this point are usually based on the following two requirements: (1) allowable loads and (2) allowable settlements or deformations.

Allowable loads are obtained by dividing the failure or ultimate loads with a safety factor. These ultimate or failure loads in soil are called *ultimate limit states.* Allowable settlements are the limits beyond which the structure would fail due to excessive deformation or settlement. The onset of excessive deformation is called the *serviceability limit state.*

Meyerhof (1984) discusses the details of these limit states as applicable to geotechnical engineering analyses. Following main items, directly applicable to pile foundation design, are based on this recommendation.

### 6.15.1 Ultimate Limit States

A total safety factor (= ultimate load/applied load) normally ranges between 2 to 3 for pile foundations as presented in Table 6.18. The upper value of safety factors apply to normal loads and service conditions, and lower values are used for maximum loads and worst environmental conditions. The lower values have been used in conjunction with performance observations, large field tests, and temporary work.

A more consistent approach leading to a more uniform margin of safety for different types and components of foundations under different loading conditions

**TABLE 6.18 Values of Minimum Total Safety Factors for Foundations**

| Conditions | Safety Factors (FS) |
|---|---|
| Values to be used in conjunction with performance observations, large field tests, analyses of failures of similar structures at the end of the service life, and for temporary works | 2 |
| Normal loads and service conditions | 3 |

**TABLE 6.19 Values of Minimum Partial Factors (Meyerhof, 1984)**

| Category | Item | Load Factor | Resistance Factor |
|---|---|---|---|
| Loads | Dead loads | $(f_d)$ 1.25(0.85) | |
| | Live loads, wind, or earthquake | $(f_l)$ 1.5 | |
| | Water pressures | $(f_u)$ 1.25(0.85) | |
| Shear strength | Cohesion ($c$) (stability:earth pressures) | | $(f_c)$ 0.65 |
| | Cohesion ($c$) (foundations) | | $(f_c)$ 0.5 |
| | Friction ($\tan\phi$) | | $(f_\phi)$ 0.8 |

1. $f_c$, $f_\phi$ = resistance factors applied to actual (representative) shear strength parameters of cohesion, $c$, and friction, $\tan\phi$, respectively, to obtain corresponding (reduced) factored shear strength parameters of soil, as used in ultimate limit states analysis to estimate the factored soil resistance, factored earth pressure, or factored foundation capacity.

$f_d$, $f_l$, $f_u$ = load factors applied to actual (characteristic) dead loads, live or environmental loads, and water pressures respectively, to obtain corresponding (generally increased) factored loads and forces, as used in ultimate limit states analysis.

2. When live and environmental loads both act together, a load combination factor of 0.7 may be applied to both loads, but the total effect must not be less than that for full live or environmental load acting alone.

may be obtained by using partial safety factors. These factors are used in limit state design. In summary limit state design consists of the following:

1. Specified loads and forces are multiplied by load factors that generally are greater than unity
2. Resistance of material such as shear strength parameters of soil are multiplied by resistance factors that are less than unity

Thus, the limit state analysis is based on factored loads and factored resistance using partial factors. The values of minimum partial factors are given in Table 6.19. The ultimate limit states of foundations are governed by bearing capacity failure under the applied loads using the load factors of Table 6.20. These load and resistance factors agree with those specified in the National Building Code of Canada (1985) and in the Ontario Highway Bridge Design Code (1983), respectively. Ontario Bridge Design Code also recommends these values for pile foundations. Load factors given in the parentheses of this table apply to dead loads and water pressures when their effect is beneficial as for the dead loads resisting instability by sliding, overturning or uplift.

Additional load modification factor ($f_q$) and resistance modification factor ($f_r$) are required to ensure that the same margin of safety as obtained in the conventional total safety factor is achieved. A performance factor ($f_p$) is

**TABLE 6.20 Modification and Performance Factors for Deep Foundations (Meyerhof, 1984)**

| Item | Load Modification Factor, $f_q$ | Performance Factor, $f_p$ | Resistance Modification Factor $f_r$ |
|---|---|---|---|
| Down-drag loads (negative skin friction) | 1.25 | | |
| Bearing capacity (static cone test) | | 0.5 | |
| Bearing capacity (sand) (standard penetration test) | | 0.3 | |
| Load test, dynamic analysis | | 0.5 | |
| Shaft resistance (effective stress) | | | 0.6 |
| Horizontal capacity (sand) | | | 0.5 |
| Horizontal capacity (clay) | | | 0.8 |

1. $f_p$ = performance factor used to modify (reduce) the foundation capacity determined by load tests, or estimated from in-situ tests such as penetrometer testing.
2. $f_q$ = load modification factor used to modify (increase) factored loads or factored forces.
3. $f_r$ = resistance modification factor used to modify (reduce) the factored soil resistance or factored foundation capacity.

recommended if the ultimate capacity of pile foundations has been determined from load tests or estimated from dynamic methods of analysis with field measurements. These factors can be improved if more full scale field pile load tests are carried out (Jaeger and Bakht, 1983). Table 6.20 provides modification and performance factors for pile foundations.

### 6.15.2 Serviceability Limit States

The allowable settlements of structures can only be determined for each particular case separately. The serviceability limit states are checked by using a load factor of unity on all loads. However, when both live and environmental loads act together, a load combination factor of 0.7 may be used. The partial factors for deformation and compressibility properties of soils should be taken as unity. For settlement estimates based on load tests or penetration tests, a performance factor of unity and partial factor of 0.7 should be used.

## 6.16 OVERVIEW

This chapter presented the analysis and design of vertical piles under lateral loads both in cohesionless and cohesive soils. This includes the determination of allowable lateral loads based on ultimate lateral resistance, elastic lateral deformation, and the lateral deformation beyond the elastic range. Design information for piles subjected to lateral loads in layered system has also been presented in this chapter. Pile foundations under bridges and offshore structures are often subjected to eccentric inclined loads. Therefore, the latter part of the chapter discussed the analysis and design of vertical and batter piles subjected to eccentric and inclined loads.

The allowable lateral loads on piles is smaller of the lateral load obtained by dividing the ultimate lateral resistance by an adequate factor of safety and the load corresponding to an acceptable lateral deflection. Therefore, the methods for calculating allowable lateral loads of vertical piles can be divided into methods of calculating the ultimate lateral resistance and acceptable deflection at working lateral load.

The two methods of calculating ultimate lateral resistance of vertical piles are Brinch Hansen's method (1961) and Broms' method (1964). Both methods are based on the earth pressure theory. Brinch Hansen's method is applicable both for $c - \phi$ soils and layered soil system. The major disadvantage of this method is that it is applicable only for short piles. Broms' method, on the other hand, is applicable both for short and long piles. Simplifying assumptions made in this method have resulted in the development of design graphs and simple equations. Although this method is not applicable for $c - \phi$ soils and layered system, it is widely used because of its easy application in the design both for cohesive and cohesionless soils, separately.

Methods of calculating acceptable deflection at working load are the modulus of subgrade reaction approach (Reese and Matlock, 1956) and the elastic approach (Poulos, 1971). The modulus of subgrade reaction approach treats laterally loaded piles supported on a series of equivalent, independent linearly elastic springs. The method is relatively simple and can incorporate factors such as nonlinearity, variation of subgrade reaction with depth, and the layered system. Corrections to the modulus can be applied for group action, cyclic and dynamic loadings (Davisson, 1970; Prakash, 1981). This method is widely used because a considerable amount of experience has been gained in applying this theory to practical problems.

The modulus of subgrade reaction approach can be extended beyond the elastic range where soil yields plastically. This can be done by employing $p$–$y$ curves (Matlock, 1970; Reese et al., 1974, Reese and Welch, 1975; Bhushan et al., 1979). Empirical procedures for establishing $p$–$y$ curves around a single pile for cohesionless as well as cohesive soils have been explained. Although empirical relationships provided in $p$–$y$ curves have been supported by field tests on single piles, further testing should be carried out to supplement these relationships, particularly to justify their application to a pile group.

The elastic approach for calculating lateral deflection assumes that the laterally loaded pile is supported by an ideal elastic continuum and gives solutions for varying soil modulus. Although the approach is theoretically more realistic, its major problem is the application of the theory to the practical problem specially in the determination of soil modulus, $E_s$. Furthermore, this approach needs more field verification by applying the theoretical concept to practical problems.

The ultimate lateral resistance of pile groups in cohesionless soils can be obtained by using the concept of group efficiency. Such factors have been determined by Oteo (1972) for cohesionless soils and by Prakash and Saran (1967) for cohesive soils. Lateral deflection of pile groups can be calculated by applying the reduction factors in the coefficient of subgrade reaction. Ultimate resistance and deflections of pile groups are based on limited model pile group tests. There is a need to carry out confirmatory field and laboratory tests on laterally loaded pile groups.

In many field conditions, such as under bridges and offshore structures, the pile groups may be subjected to simultaneous vertical loads, lateral loads and moments. Combination of such loads on the pile group may result into a system that is subjected to an eccentric and inclined load. In general, such a system can be analyzed by four methods: (1) statical method, (2) considering pile group as a structural system supported on springs (3) considering soil–pile interaction assuming soil as an elastic continuum, and (4) interaction relationships between the soil–pile system. The statical or traditional method neglects the contribution of soil support and should be used only on small projects. The method that considers pile group as a structural system supported on springs makes various assumptions regarding soil-pile interaction (Vesic, 1977; Saul, 1968; Reese, et al., 1970). Based on these assumptions, mathematical formulation for solving pile loads can be made. In spite of these simplifying assumptions, these mathematical formulas require elaborate computations. The method of solving by soil–pile interaction assuming soil as an elastic continuum as proposed by Poulos (1974) needs further field verification and therefore has not been discussed here. The use of interaction relationships between soil–pile system to obtain bearing capacity of piles under eccentric and inclined loads has been studied in detail by Meyerhof and Ranjan (1972), Meyerhof et al. (1981 and 1983), Meyerhof and Yalcin (1984), and Meyerhof and Sastry (1985). Various interaction solutions have been provided based on the extensive tests carried out on fully instrumented model piles jacked into homogeneous sand and clay. These interaction relations are simple to use for design. However, confirmatory field tests are required to gain further confidence on this method.

A method of analysis of batter piles subjected to eccentric and inclined loads has also been included. This method is based on establishing an equivalence between a batter pile inclined at an angle $\beta$ that is subjected to vertical load and a vertical pile subjected to an inclined load at an angle $\alpha$ (Awad and Petrasovits, 1968). Only limited test data are available on single and batter pile groups. Further tests and analyses on the behavior of batter piles under eccentric and inclined loads are required. Until such data are obtained design methods proposed in this chapter can be used to analyze batter piles.

All the above methods have become available in the previous three decades and represent a state of the art on behavior, analysis, and design of piles under lateral loads. The design of pile foundations is still more an art than a science despite all these advances. Therefore, considerable judgment is needed in selection of a design value of soil modulus, group reduction factor and group efficiency values. The information in this chapter will serve as an aid in developing this judgment along with study of case histories on the subject (see Chapter 11).

## REFERENCES

American Petroleum Institute Recommended Practice for Planning, Designing, and Constructing Fixed Offshore Platforms, API RP 2A, Washington DC, January 1982.

Awad, A. and Petrasovits, G., "Considerations on the Bearing Capacity of Vertical and Batter Piles Subjected to Forces Acting in Different Directions", *Proceedings of the 3rd Budapest Conference, Soil Mechanics and Foundation Engineering*, Budapest, 1968, pp. 484–497.

Bhushan, K., Haley, S. C., and Fong, P. T., "Lateral Load Tests on Drilled Piers in Stiff Clays," *J. Geotech. Eng. Div.* ASCE, Vol. 105, No. GT 8, August 1979, pp. 969–985.

Brinch Hansen, J., "The Ultimate Resistance of Rigid Piles Against Transversal Forces", Danish Geotechnical Institute (Geoteknisk Institut) Bull. No. 12, Copenhagen, 1961, p. 5–9.

Broms, B., "The Lateral Resistance of Piles in Cohesive Soils," *J. Soil Mech. Found. Div.*, ASCE, Vol. 90, No. SM2, March 1964a, pp. 27–63.

Broms, B., "The Lateral Resistance of Piles in Cohesionless Soils," *J. Soil Mech. Found. Div.*, ASCE, Vol. 90, No. SM3, May 1964b, pp. 123–156.

Building Code of the City of Boston, 1964.

Canadian Foundation Engineering Manual, Canadian Geot. Soc. Vancouver B.C. (Canada) 1985 2nd Ed. 456 p.

Canadian Portland Cement Association, *Metric Design Handbook for Reinforced Concrete Elements*, 1978, pp. 1–27.

Chellis, R. D., "Pile Foundations," McGraw HillBook Co NY, 1961.

Cox, W. R., Reese, L. C., and Grubbs, B. R., "Field Testing of Laterally Loaded Piles in Sand," Offshore Technology conference, Houston, TX, 1971, pp. 459–472.

Davisson, M. T., "Lateral Load Capacity of Piles," *Highway Research Record*, Washington, DC, 1970, pp. 104–112.

Davisson, M. T. and Gill, H. L., "Laterally-Loaded Piles In a Layered Soil System," *J. Soil Mech. Found. Div.*, ASCE, Vol. 89, No. SM3, May 1963, pp. 63–94.

Davisson, M. T. and Prakash, S., "A Review of Soil Pile Behavior," *Highway Research Record*, No. 39, 1963, pp. 25–48.

Jaeger, L. G. and Bakht, B. "Number of Tests Versus Design Pile Capacity," J. *Geotech. Eng. Div.* ASCE, Vol. 109, No 6 June 1983 pp. 821–831.

Kim, J. B., Singh, L. P. and Brungraber, R. J., "Pile Cap Soil Interaction From Full-Scale Lateral Load Tests," *J. Geotech. Eng. Div.* ASCE, Vol. 105, No. GT 5, May 1979, pp. 643–653.

Krey, H., *Erddruck, Erdwiderstand und Tragfahigkeit des Baugrunder*. W. Ernest and Sohn, Belin, W. Germany, 1936.

Matlock, H., "Correlation for Design of Laterally Loaded Piles in Soft Clay," *Proceeding Offshore Technology Conference*, Houston, TX, Paper OTC 1204, 1970.

Matlock, H. and Reese, L. C., "Foundation Analysis of Offshore Pile Supported Structures," *Proceedings Fifth International Conference on Soil Mechanics and Foundation Engineering*, Paris, Vol. 2, 1961, pp. 91–97.

Matlock, H. and Reese, L. C., "Generalized Solutions for Laterally Loaded Piles," *Transactions of the American Society of Civil Engineers*, Vol. 127, 1962, part 1, pp. 1220–1247.

Meyerhof, G. G., "Stability of Slurry Trench Cuts in Saturated Clay," *Proceedings Speciality Conference on the Performance of Earth and Earth Supported Structures*, Purdue University, Lafayette, IN, Vol. 1, 1972, pp. 1451–1466.

Meyerhof, G. G., "Bearing Capacity and Settlement of Pile Foundations," *J. Geotech. Div.* ASCE, Vol. 102, No. GT 3, March 1976, pp. 197–228.

Meyerhof, G. G. and Ranjan, G., "The Bearing Capacity of Rigid Piles Under Inclined Loads in Sand, I: Vertical Piles," *Canadian Geotechnical Journal*, Vol. 9, 1972, pp. 430–446.

Meyerhof, G. G., "Safety Factors and Limit States Analysis in Geotechnical Engineering," *Canadian Geotechnical Journal*, Vol. 21, No. 1, February 1984, pp. 1–7.

Meyerhof, G. G., Mathur, S. K., and Valsangkar A. J., "Lateral Resistance and Deflection of Rigid Walls and Piles in Layered Soils," *Canadian Geotechnical Journal*, Vol. 18, No. 2 May 1981, pp. 159–170.

Meyerhof, G. G., Yalcin, A. S., and Mathur, S. K., "Ultimate Pile Capacity for Eccentric Inclined Load," *J. Geotech. Eng. Div.* ASCE, Vol. 109, No. GT 3, March 1983, pp. 408–423.

Meyerhof, G. G. and Yalcin, A. S., "Pile Capacity for Eccentric Inclined Load in Clay," *Canadian Geotechnical Journal*, Vol. 21, No. 3, August 1984, pp. 389–396.

Meyerhoff, G. G. and Sastry, V. V. R. N., "Bearing Capacity of Rigid Piles Under Eccentric and Inclined Loads," *Canadian Geotechnical Journal*, Vol. 22, No. 3, August 1985, pp. 267–276.

*NAVFAC Foundations and Earth Structures, Design Manual 7.2*, Department of the Navy, Alexandria, VA, May 1982.

Ontario Highway Bridge Design Code, Ministry of Transportation and Communication, Toronto, 1983.

Oteo, C. S., "Displacements of a Vertical Pile Group Subjected to Lateral Loads," *Proceedings 5th European Conference of Soil Mechanics and Foundation Engineering*, Madrid, Vol. 1, 1972, pp. 397–405.

Palmer, L. A. and Thompson, J. B., "The Earth Pressure and Deflection Along the Embedded Lengths of Piles Subjected to Lateral Thrust," *Proceedings Second International Conference on Soil Mechanics and Foundation Engineering*, Rotterdam, Holland, Vol. V, 1948, pp. 156–161.

Poulos, H. G., "Behavior of Laterally Loaded Piles: I-Single Piles," *J. Soil Mech. Found. Div.*, ASCE, Vol. 97, No. SM 5, 1971a, pp. 711–731.

Poulos, H. G., "Behavior of Laterally Loaded Piles: II-Pile Groups," *J. Soil Mech. Found. Div.* ASCE, Vol. 97, No. SM5, 1971b, pp. 733–751.

Poulos, H. G., "Analysis of Pile Groups Subjected to Vertical and Horizontal Loads," *Aust. Geomechanics J.*, Vol., G4, No. 1, 1974, pp. 26–32.

Poulos, H. G. and Davis, E. H., *Pile Foundation Analysis and Design.* Wiley, New York, 1980.

Poulos, H. G. and Madhav, M. R., "Analysis of the Movement of Battered Piles," *Proceedings 1st Australion–New Zealand Conference on Geomechanics*, Melbourne, Australia, 1971, pp. 268–275.

Prakash, S., "Behavior of Pile Groups Subjected to Lateral Loads," Ph.D. Thesis, University of Illinois, Urbana, 1962, p. 397.

Prakash, S., *Soil Dynamics.* McGraw-Hill, Inc., New York, 1981.

Prakash, S. and Saran D., "Behavior of Laterally Loaded Piles in Cohesive Soils," *Proceedings 3rd Asian Regional Conference on Soil Mechanics and Foundation Engineering*, Haifa (Israel), 1967, pp. 235–238.

Prakash, S. and Subramanayam, G., "Load Carrying Capacity of Battered Piles," *Roorkee University Research Journal*, Vol. VII, No. 1 and 2, September 1964, pp. 29–46.

Reese, L. C., "Laterally Loaded Piles: Program Documentation," *J. Ceotech. Eng. Div.*, ASCE, Vol. 103, No. GT 4, April 1977, pp. 287–305.

Reese, L. C., Cox, W. R., and Koop, F. D., "Analysis of Laterally Loaded Piles in Sand," *Proceedings Offshore Technology Conference*, Houston, TX, Paper No. OTC 2080, 1974, pp. 473–483.

Reese, L. C. and Matlock, H., "Non-dimensional Solutions for Laterally Loaded Piles with Soil Modulus Assumed Proportional to Depth," *Proceedings 8th Texas Conference on Soil Mechanics and Foundation Engineering*, Austin, TX, 1956, pp. 1–41.

Reese, L. C., O'Neill, M. W., and Smith, E., "Generalized Analysis of Pile Foundations," Proceedings *J. Soil Mech. Found. Div.*, Vol. 96, No. SM1, 1970, pp. 235–250.

Reese, L. C. and Welch, R. C., "Lateral Loading of Deep Foundations in Stiff Clay," *J. Geotech. Eng. Div.*, ASCE, Vol. 101, No. GT 7, July 1975, pp. 633–649.

Saul, W. E., "Static and Dynamic Analysis of Pile Foundations," *J. Struct. Div.* ASCE Vol. 94, No. ST 5, 1968, pp. 1077–1100.

Simek, J., "Resultats d'observations de l'Influence d'une Force Horizontale sur des Groupes de Pieux," *Sols-Soils*, No. 18–19, 1966, pp. 11–18.

Tschebotarioff, G. P., "The Resistance to Lateral Forces of Single Piles and Pile Groups," American Society of Testing and Materials, Special Technical Publication No. 154: 1953.

Vesic, A. S., "Design of Pile Foundations," Transportation Research Board, National Research Council Washington, D.C., 1977.

# 7

# PILE FOUNDATIONS UNDER DYNAMIC LOADS

A sand mass under vibrations tends to increase in density with a corresponding decrease in voids. In a mass of saturated sand below groundwater level, soils may be subjected to liquefaction resulting in increases in density. The movement of soil grains is associated with the decrease of effective stresses. If the soil is under a certain initial shear stress, the effect of vibrations is felt to a different degree (Prakash, 1981).

A pile introduces additional shear stresses in the soil mass. Excessive settlements are likely to occur under vibrations. In order to study the effect of vibrations on piles, Swiger (1948) reported tests on piles in sand. A static load was first applied on a pile. This was then vibrated under this static load. The vibrator consisted of a plate 12 in. (30 cm) in diameter and 1 in. (2.5 cm) thick that was mounted with an eccentricity of 1 in. The speed of the vibrator could be varied from about 400 revolutions per minute to 3000 revolutions per minute. The pile was vibrated at its natural frequency of 500 revolutions per minute, which had been determined experimentally. The static loads on the pile were 61 and 121 kips. The rate of settlement with the higher static load was several times that with the smaller load.

Agarwal (1967) and Prakash and Agarwal (1971) reported tests on vertical model piles embedded in sand at 33 percent relative density. The piles were loaded with a predetermined fraction of upward static pullout resistance. The tank containing piles was subjected to vertical vibrations at 2.3 and 5.2 Hz. It was found that the number of cycles of motion needed to pull out the pile a predetermined distance of 0.8 in. (2 cm) decreased with an increase in the static vertical upward load and the vertical peak acceleration.

Ghumman (1985) conducted a comprehensive series of model tests on penetration testing of piles under vertical vibrations. A model pile 2.4 in. (6 cm) in

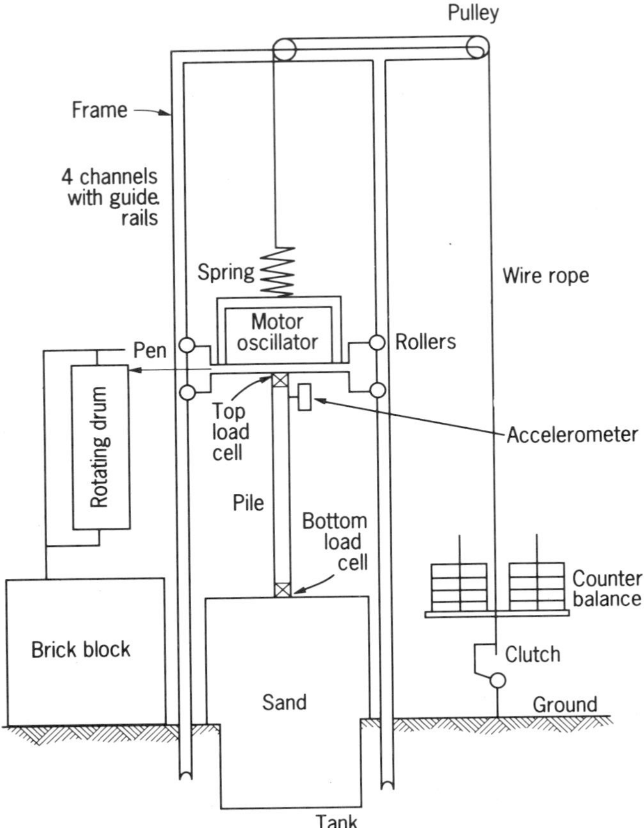


**Figure 7.1** A setup for study of penetration of piles under axial vibrations (after Ghumman, 1985).

diameter and 64 in. (160 cm) long was subjected to a predetermined static load. The vertical vibrations were then imparted to the pile by a fully counterbalanced mechanical oscillator, which could be excited to different frequencies (Figurg 7.1). A typical penetration record with time at a frequency of oscillations of 10 Hz is shown in Figure 7.2. A static load of 165 lb (75 kg) had been applied on the pile head and the dynamic force level had been varied from 99 lb (45 kg) in test no. 1.5 to 132 lb (60 kg in test no. 1.6) and 198 lb (90 kg in test no. 1.8). Both the rate of penetration and total penetration increased with dynamic force. The foregoing experimental behavior highlights the importance of vibrations in inducing the settlement of piles.

Earthquakes introduce lateral forces on piles. The energy supplied to a structure may be absorbed in the elastic and plastic deformations of both the

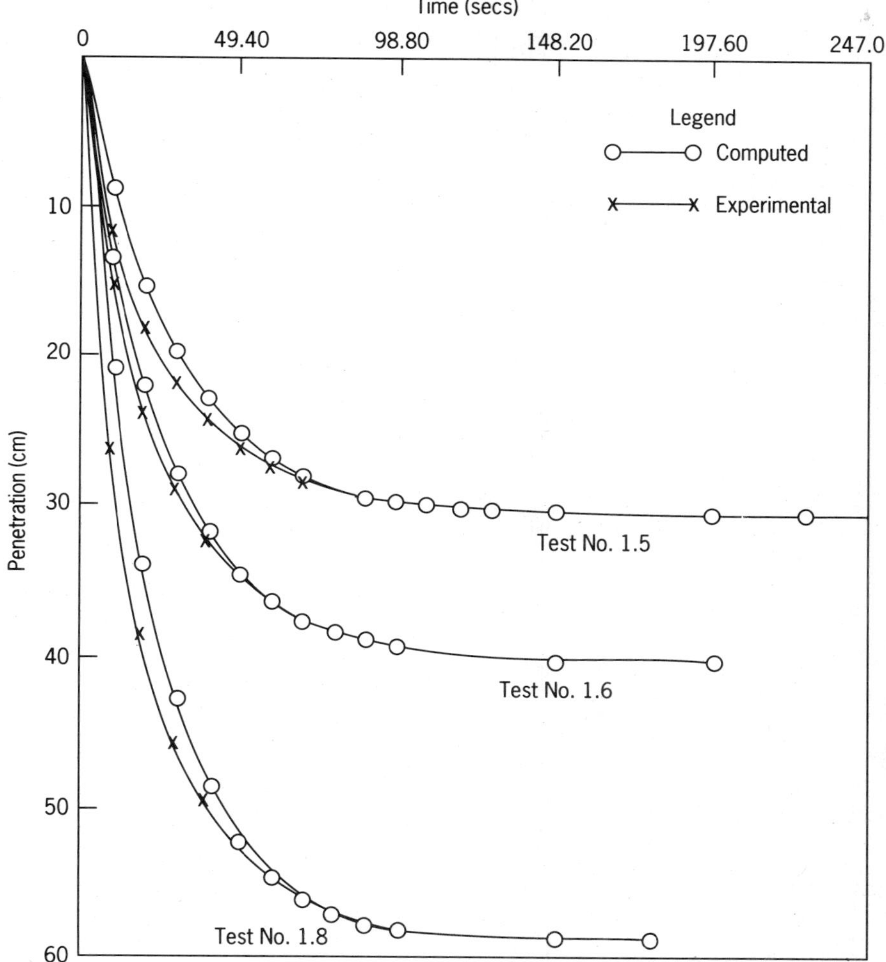


**Figure 7.2** Penetration versus time graph for test no. 1.5, 1.6, and 1.8 (after Ghumman, 1985).

superstructure and substructure. Eccentric and inclined loads and moments may be introduced on the pile heads and pile caps.

Lateral forces on the superstructure are assumed to be transferred to the ground through the pile cap as lateral loads and moments, and the stability of the piles is checked against these loads. Vertical loads are always present. These may cause buckling of the piles, particularly if freestanding lengths are large, or they may increase the deflections. Therefore, buckling of the piles and the beam–column action become important (Prakash, 1985, 1987). The pile caps of individual columns are interconnected by grade beams.

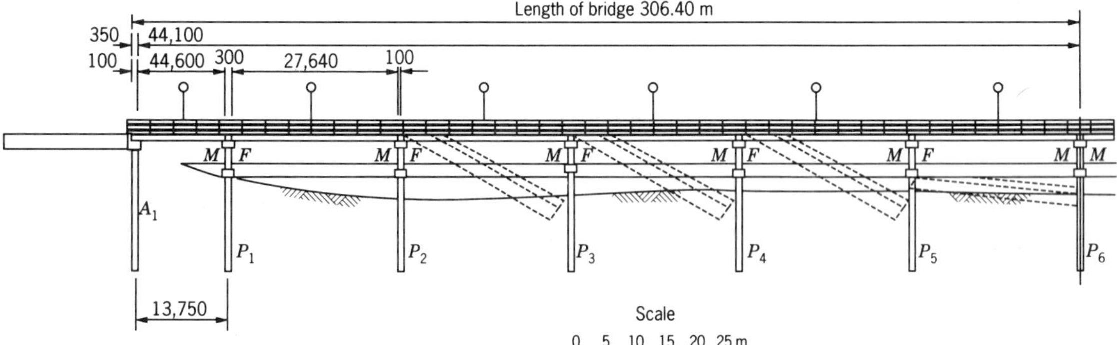


**Figure 7.3** Profile of Showa Bridge showing damage to deck slabs due to out-of-phase motions of piers (Fukuoka, 1966).

Fukuoka (1966) reported classical damage to the Showa Bridge in the Niigata, Japan, earthquake of 1964 due to vibrations of piles and pile-supported piers. Showa Bridge was completed about a month before the earthquake. This bridge has 12 composite girders, its width is about 80 ft (24 m), and its total length is about 1023.3 ft (307 m). Its main span length is about 93.3 ft (28 m), and its side span length is about 50 ft (15 m). A typical pier is composed of steel pipe piles of 2 ft (60 cm) in diameter and wall thickness of 0.64 to 0.36 in. (16 to 9 mm). Its profile is shown in Figure 7.3. Five main spans out of 10 fell down. A girder between Pier No. 5 ($P_5$) and Pier No. 6 ($P_6$) fell down completely and submerged into the river. Girders between $P_2 - P_3$, $P_3 - P_4$, $P_5 - P_6$, $P_6 - P_7$ inclined; fixed ends of the girders remained on the top of pile piers; and movable ends dropped into the river. Upper parts of $P_5$ and $P_6$ above the riverbed bent by about 90°. $P_5$ bent toward the left, and $P_6$ bent toward the right. This damage occurred due to out-of-phase motions of the piers $P_2 - P_3$, $P_3 - P_4$, and $P_4 - P_5$.

A pile of pier No. 4 ($P_4$) was taken out after the earthquake (Figure 7.4). The maximum deflection of the pile at the mud line is approximately 40 in. (1000 mm) Bending and buckling of the pile shows important soil–pile–soil interaction effects.

Piles may be used to support the foundations in buildings, machines, and offshore structures. In buildings, the soils near the ground surface will be of poor quality, necessitating the transfer of loads to deeper depths. In machine foundations in addition to the above consideration, it may be necessary to increase the natural frequency of foundation soil system and decrease their amplitudes. In offshore structures, piles may be of very large lengths (up to 1000 ft or so) always with considerable freestanding lengths. The introduction of piles makes the system stiff, and both the natural frequency and the amplitudes of motion are effected. In all vibration problems, resonance needs to be avoided. Hence, the natural frequency of the soil–pile system is necessarily evaluated.

In the following sections, the natural frequency of the soil–pile system, dynamic analysis, and the design of piles against earthquakes and under machine foundations are discussed.

## 7.1 PILES UNDER VERTICAL VIBRATIONS

Barkan (1962) proposed determination of soil pile stiffness from a cyclic vertical pile load test similar to a cyclic plate load test (Prakash and Puri, 1988). A plot of load $P$ and elastic settlement $z_1$ may be represented by a straight line up to the working load in many situations. The constant of proportionality ($k$), *the coefficient of elastic resistance of the pile* is then:

$$P = kz_1 \tag{7.1}$$

where $z_1$ is elastic deflection of pile.
It represents the load required to induce a unit elastic settlement of the pile. The

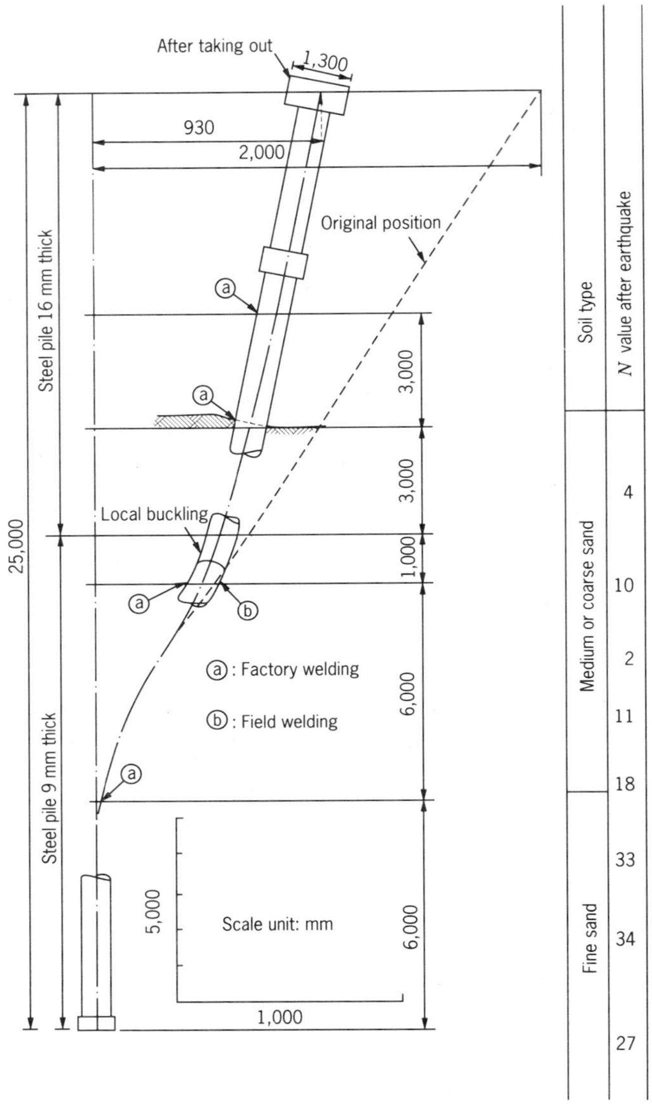


**Figure 7.4** Pipe pier no. 4, taken out from the ground after the Niigata earthquake (Fukuoka, 1966).

coefficient of elastic resistance of the pile depends on soil properties, pile characteristics (e.g., length), and the length of time the pile has been in the soil. For example, the elastic resistance of a pile may have different values during driving and some time later, particularly in soft clays.

The natural frequency of the pile in vertical vibration is then given by;

$$f_{nz} = \frac{1}{2\pi}\sqrt{\frac{k}{m}} \tag{7.2}$$

where $m$ = mass of the pile and static load on pile.

Based on the above simple concepts, Barkan (1962) described test data and typical values of elastic constants of piles and pile groups under both vertical and horizontanl vibrations. This analysis does not consider damping in the system and the dynamics of the problem.

It has been shown in Chapter 4, that the soil modulus depends upon the strain in the soil. Therefore, $k$ in equation (7.2) will have different values for machine foundation problems and for earthquake loading. No simple and direct relationship between strains in the soil along a pile, particularly in horizontal vibrations and soil deformations around the pile are available.

Since the elastic soil constants $E$, $G$, and $k$ are strain or displacement dependent, the values of the elastic constant $k$ determined from a lateral deflection of the order of 3.4 mm in Barkan's test are not applicable to machine foundation problems.

### 7.1.1 End-Bearing Piles

If piles are driven in soft soil and are embedded in sound rock or a hard stratum at their tip, the piles may be considered as end bearing piles. Deformations of the pile tip will not occur when dynamic loads are transferred to the pile. The pile may then be considered as an elastic rod fixed at its tip (base) and free at the top, with a mass $m$ resting on the top (Figure 7.5).

If no mass rests on top, we then have a solid resonant column with the fixed-free condition, which has a resonant frequency given by (Prakash and Puri, 1988).

$$\omega_n = \frac{(2n-1)\pi v_r}{2l} \tag{7.3}$$

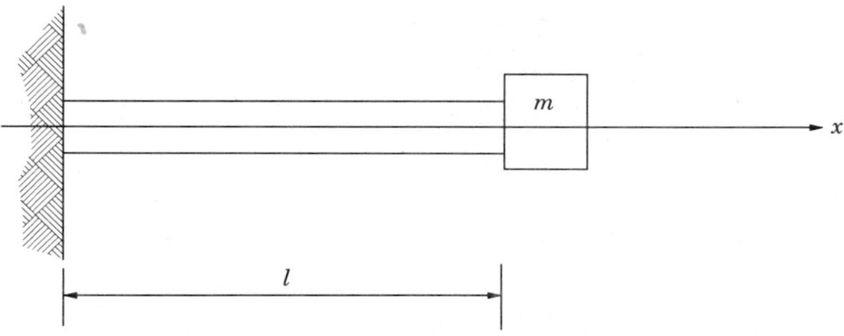


**Figure 7.5** Fixed-free rod pile with mass attached to free end.

where

$\omega_n$ = natural circular frequency (rad/sec)
$v_r$ = velocity of longitudinal wave propogation in the rod (ft/sec)
$l$ = length of the rod (ft)

for $n = 1$

$$\omega_n = \frac{\pi v_r}{2l}$$

or

$$f_n = \frac{v_r}{4l} = \frac{1}{4l}\sqrt{\frac{E}{\rho}} \tag{7.4}$$

where

$f_n$ = natural frequency of rod in cycles per sec

$E$ = Young's modulus of elasticity of the pile

$\rho = \frac{\gamma}{g}$ = mass density of the pile material

$l$ = length of the pile

Displacement $u(= f(x, t))$ of a vibrating rod is given by

$$u = U(A \cos \omega_n t + B \sin \omega_n t) \tag{7.5}$$

For the case in which the weight of the pile is negligible as compared to the supported mass, the natural frequency may be obtained by applying the end condition

$$U = 0 \qquad \text{at } x = 0$$

to a vibrating rod with zero weight. (Prakash and Puri, 1988). The displacement function of such a vibrating rod is given by:

$$U = D \sin \frac{\omega_n x}{v_r} \tag{7.6}$$

where

$U$ = a function of $x$, which defines displacement (Figure 7.5)
$\omega_n$ = natural circular frequency of rod

$A, B, D$ = constants that depend on initial conditions
$t$ = time

Differentiating equation (7.5) with respect to $x$ and $t$ gives:

$$\frac{\partial u}{\partial x} = \frac{\partial U}{\partial x}(A \cos \omega_n t + B \sin \omega_n t) \tag{7.7a}$$

and

$$\frac{\partial^2 u}{\partial t^2} = -\omega_n^2 U(A \cos \omega_n t + B \sin \omega_n t) \tag{7.7b}$$

For longitudinal excitation of the rod in Figure 7.5, displacement is zero at the fixed end. At the free end, a force that is equal to the inertia force of the concentrated mass is exerted on the rod. The equation of dynamic equilibrium may be written as:

$$F = \frac{\partial u}{\partial x} AE = -m\frac{\partial^2 u}{\partial t^2} \tag{7.8}$$

Substituting equation (7.7) into equation (7.8), we get,

$$AE\frac{\partial U}{\partial x} = m\omega_n^2 U \tag{7.9}$$

Finally, substituting $U$ from equation (7.6) into equation (7.9), we get

$$AE\frac{\omega_n}{v_r}\cos\frac{\omega_n l}{v_r} = \omega_n^2 m \sin\frac{\omega_n l}{v_r} \tag{7.10a}$$

which can be reduced to:

$$\frac{Al\gamma}{W} = \frac{\omega_n l}{v_r}\tan\frac{\omega_n l}{v_r} \tag{7.10b}$$

where

$Al\gamma$ = weight of rod
$W$ = weight of added mass

The solution of equation (7.10) is plotted in Figure 7.6, from which the natural frequency in vertical vibrations $f_n$ may be determined.

In order to illustrate the influence of axial loading on the resonant frequency of end-bearing piles on rock, Richart (1962) included the effect of axial load, pile length, and pile material (Figure 7.7). The three curves in the upper part of the

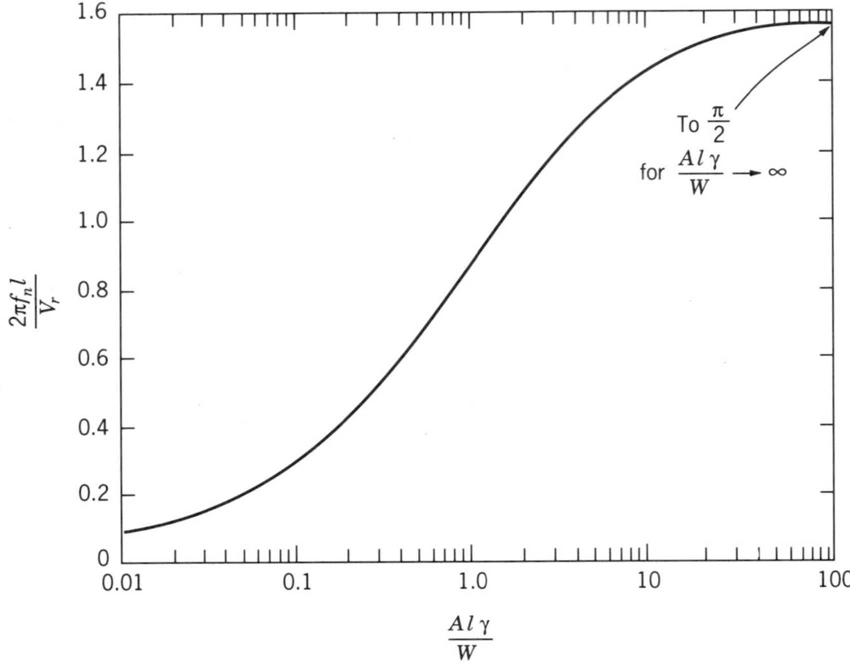


**Figure 7.6** Graphical solution for equation (7.10) (after Richart et al., 1970).

diagram illustrate the resonant frequencies of unloaded steel, concrete, and wooden piles, computed from equation (7.10). As the axial load is increased on a pile of given length, the resonant frequency is reduced (Figure 7.7).

### 7.1.2 Friction Piles

In floating piles, unlike end-bearing piles, the load is transferred from the shaft to the soil, and their analysis under vertical vibrations is quite different than that for end-bearing piles. Some of the methods employed to determine the response of floating piles to vertical dynamic loads are as follows:

1. A three-dimensional analysis (e.g., using the finite element method) considering the propagation of waves through the pile and soil
2. Solution of the one-dimensional wave equation, for example, in a manner similar to the solution of this equation to analyze the pile-driving process
3. An analysis of the response of a lumped mass–spring–dashpot system representing the pile and soil
4. An elastic analysis in which it is assumed that the elastic waves propagate only horizontally

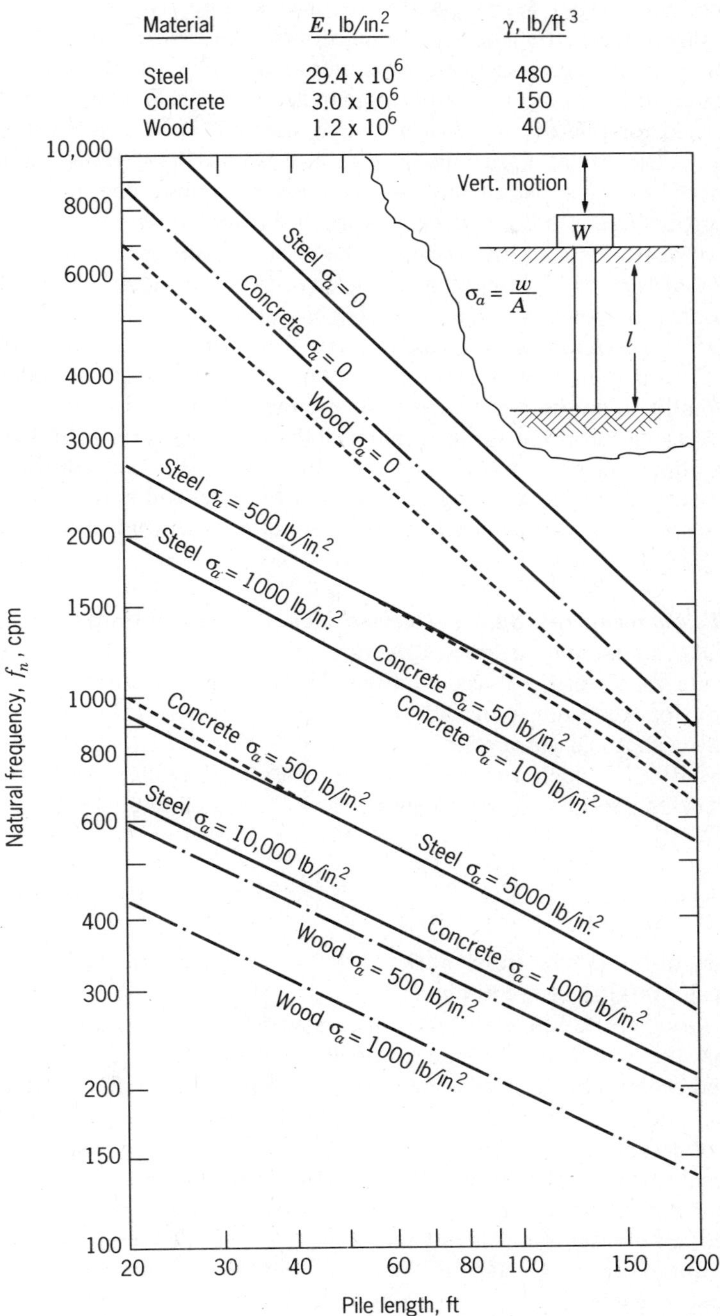


**Figure 7.7** Resonant frequency of vertical oscillation for a point-bearing pile resting on a rigid stratum and carrying a static load $W$ (after Richart, 1962).

A three-dimensional analysis is too expensive and involved for every day use. For pile-supported turbogenerator foundations in nuclear power plants where tolerance limits are very critical, such methods are in use. Solution of one-dimensional wave equations, involving extension of the numerical method of analysis used for pile driving, has not been used for solving problems of pile response under vertical vibrations (Poulos and Davis, 1980). However, detailed treatment of bearing capacity determination has been described in Chapter 5. A single degree of freedom lumped-mass–spring–dashpot system has been used for solution of vertical vibrations of piles by Barkan (1962) and Maxwell et al. (1969). Madhav and Rao (1971) used a two-degree-of-freedom model.

The fourth approach has been used by Novak (1974, 1977b) and Sheta and Novak (1982) to obtain an approximate solution for pile response to vertical loading. The soil has been assumed as composed of a set of independent infinitesimally thin horizontal layers of infinite extent. This model could be thought of as a generalized Winkler material that possesses inertia and dissipates energy. By applying small harmonic excitations, Novak derived solutions for the equivalent stiffness and damping constants of the pile–soil system. This model predicts response of vertically vibrating piles better than that of Maxwell et al. (1969).

***Maxwell's Lumped-Mass Spring–Dashpot Model*** The vibrating pile is shown in Figure 7.8a and its single-degree-of-freedom model is shown in Figure 7.8b. With appropriate values of the mass, damping, and spring constant selected for the system, the foundation response can be determined from solutions of elementary theory of mechanical vibrations.

The solution for such a system is given by Prakash (1981) and Prakash and Puri (1988). In Figure 7.8, various terms have been defined.

$$Z_0 = \frac{F_0}{\sqrt{(k - m\omega^2)^2 + (c\omega)^2}} \tag{7.11}$$

This solution differs from Barkan's solution since Maxwell et al. (1969) considered damping in the system.

In this model, the equivalent mass $m$ has been considered as the mass of the oscillator, the pile cap, and the static load above the ground. Tests were performed on steel H piles and concrete-filled pipe piles in silty sand, and clay overlying sand. The values of equivalent stiffness $k$ and damping ratio $\xi$ (defined as ratio of actual damping to critical damping of a vibrating system) had been back-calculated from the test results. At resonance, the dynamic value of $k_n$ was found to be greater than the static stiffness for comparable piles.

The computed damping ratio $\xi$ for single piles was of the order of 0 to 0.04. A significant finding was that both the stiffness and the damping ratio varied with frequency. In particular, the response at resonance was not reliably predicted from data on stiffness and damping computed at nonresonant frequencies. The variation of stiffness, expressed in terms of a stiffness ratio $k/k_n$ (where $k_n$

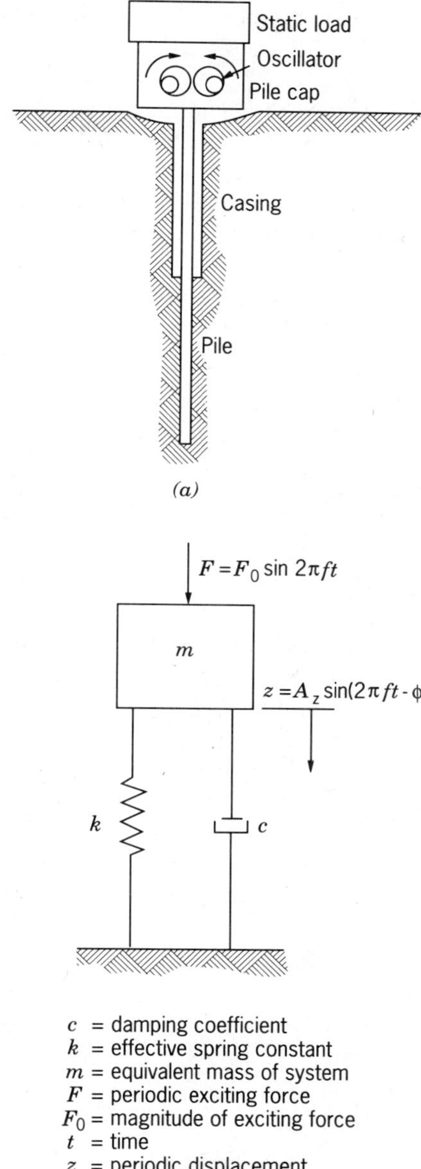


**Figure 7.8** Analytical model for floating pile. (a) Pile and soil system, (b) mechanical model system (Maxwell et al., 1969). (Reprinted by permission of the American Society for Testing and Materials.)

= stiffness at resonant frequency), and damping ratio $\xi$ with frequency ratio $f/f_n$, for pipe pile D-1 are plotted in Figure 7.9 where $f$ and $f_n$ are operating and natural frequencies, respectively.

Contact of the cap with the ground is an important factor that may affect both the natural frequency and amplitude of vibrations of the system in all modes of vibrations. In this case, typical test results (1) with the cap in contact with the soil and (2) after excavating beneath the cap showed that the dynamic displacements of the pile cap were approximately 0.0385 in. and 0.145 in., respectively under excitation by a force of constant amplitude 4*t* (ton). Since the stiffness of a pile foundation is generally greater than that of a corresponding surface foundation, the natural frequency of the foundation–soil system will be increased by the use of piles.

## 7.2 PILES UNDER LATERAL VIBRATIONS

The response of a single pile subjected to a time-dependent horizontal force and moment has been studied by several methods, including the following:

1. The pile is considered to be an equivalent cantilever and the effect of the soil is neglected.

   Hayashi (1973), Prakash and Sharma (1969), and Prakash and Gupta (1970) determined the natural frequencies of the soil–pile system in this manner. The soil–pile system is idealized as a massless equivalent cantilever with a single concentrated mass at the top. Its natural frequency is determined by using Rayleigh's method. The exciting frequency is used to check the frequency of the system for resonance. This is not a realistic approach and no frequency dependence on the vibration parameter and damping are considered. Also, no information can be obtained on the moments, stresses, and displacements along the length of the pile for dynamic loads.
2. The pile is considered as a beam on an elastic foundation subjected to time-dependent loading and analyzed by finite differences. Moments, stresses, and displacements along the length of the pile may be analyzed, and impact loads as well as harmonic loads can be considered (Tucker, 1964).
3. The approximate analytical technique developed by Novak (1974) derives stiffness and damping constants for piles and pile groups, with the help of which lateral response is determined.

   Complete solutions for vertical, lateral and torsional vibrations are presented in Section 7.4.
4. The fourth approach is in which the soil–pile system has been modeled by a set of discrete (lumped) masses, springs, and dashpots. This approach can be used to incorporate the depth and nonlinearity variations of the soil properties that depend on the definition of the local soil stiffness and geometric damping (Penzien, 1970; Prakash and Chandrasekaran, 1973,

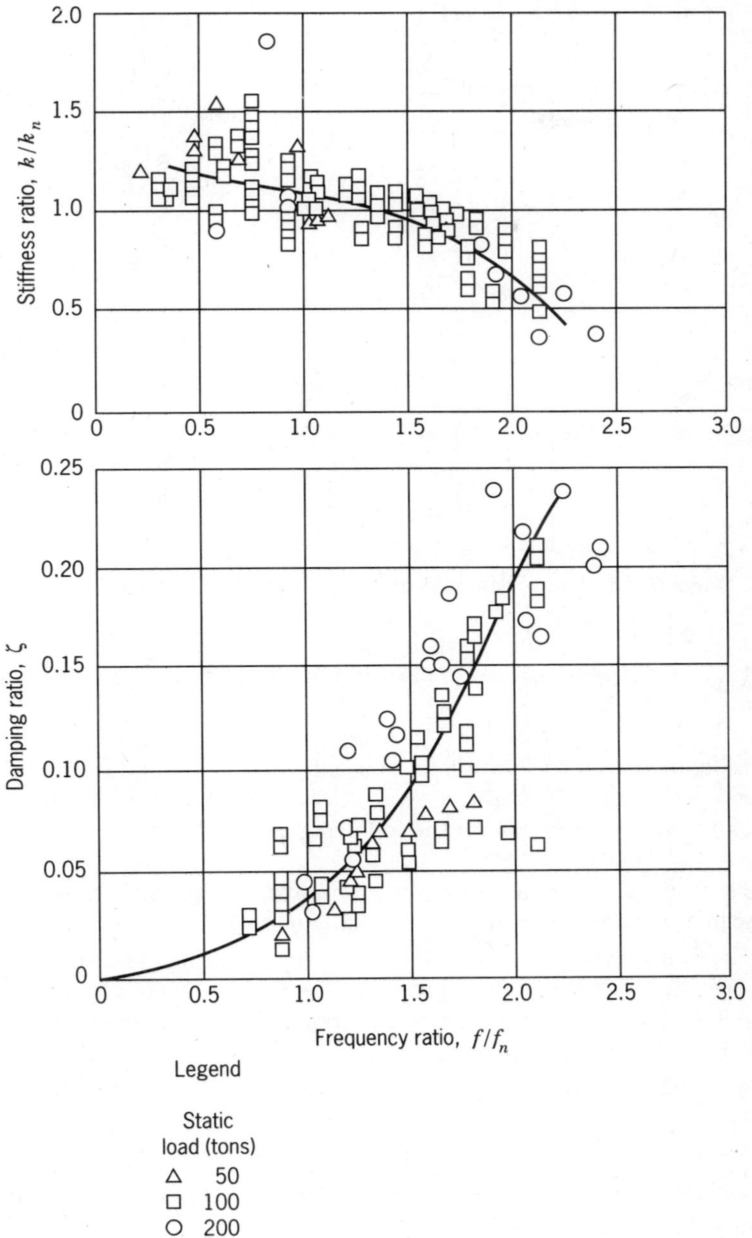


**Figure 7.9** Stiffness and damping ratio vs. frequency ratio, pipe pile D-1 (Maxwell et al., 1969). (Reprinted by permission of the American Society for Testing and Materials.)

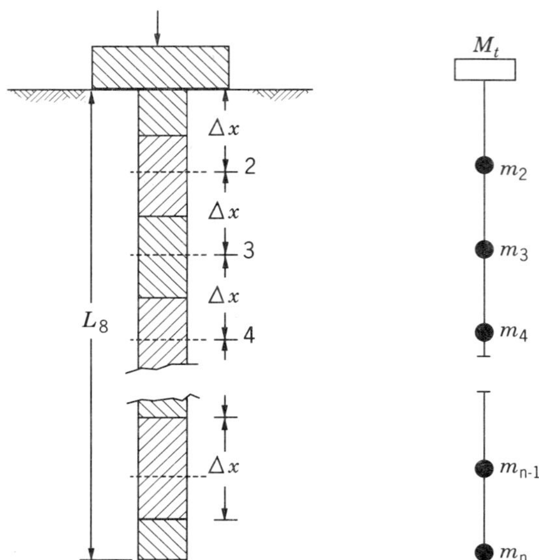


**Figure 7.10** Pile structure idealization (after Chandrasekaran, 1974).

1977). This is an extension of the solution for the static case based on the soil modulus concept.

A resonably practical solution for soil–pile interaction under dynamic loads has been proposed by Chandrasekaran (1974) (Prakash and Chandrasekaran, 1980). This analysis is based on the following assumptions:

1. The pile is divided into a convenient number of segments and mass of each segment is concentrated at its center (Figure 7.10).
2. The soil is considered as a linear Winkler's spring. The soil reaction is separated into discrete parts at the center of the masses in Figure 7.10. The soil modulus variation is considered both linearly varying with depth and constant with depth (Figure 7.11).
3. A fraction of the mass of the superstructure is concentrated at the pile top as $M_t$.
4. The system is one dimensional.
5. The pile top conditions are either completely free to undergo translation and rotation $F$ or completely restrained against rotation but free to undergo translation $F_t$. Partial fixity at the top can be solved by interpolation. The pile tip is free.

For determination of the free-vibration characteristics, modal analysis was performed by using successive approximations of the natural frequencies of the

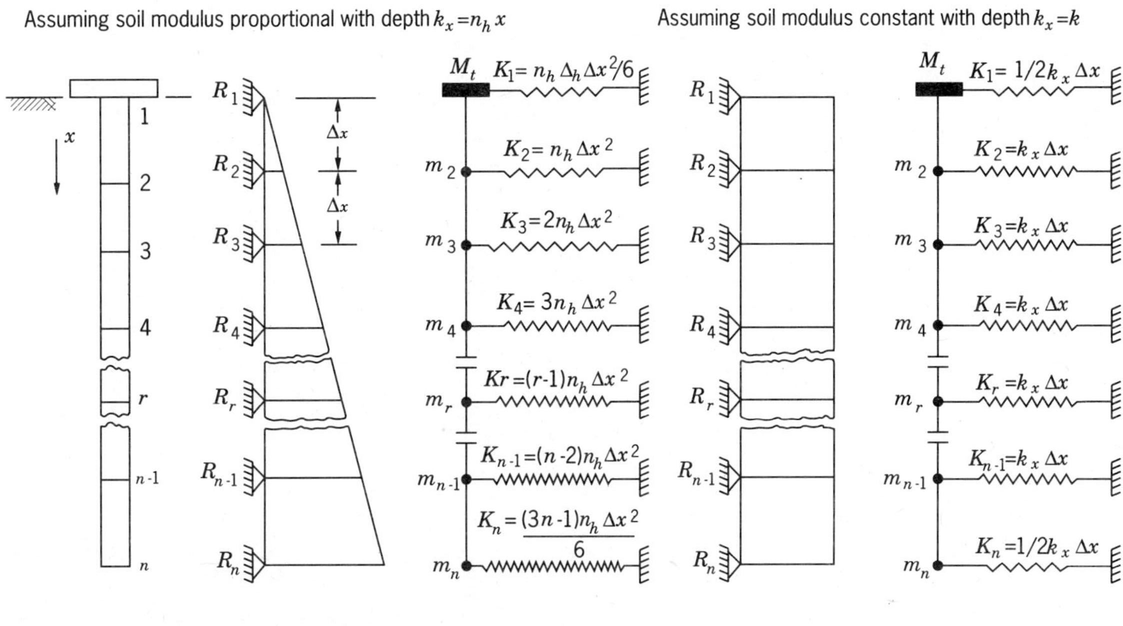


**Figure 7.11** Discretization of soil–pile interaction effects. (a) Soil modulus linearly varying with depth, (b) soil modulus constant with depth (after Chandrasekaran, 1974).

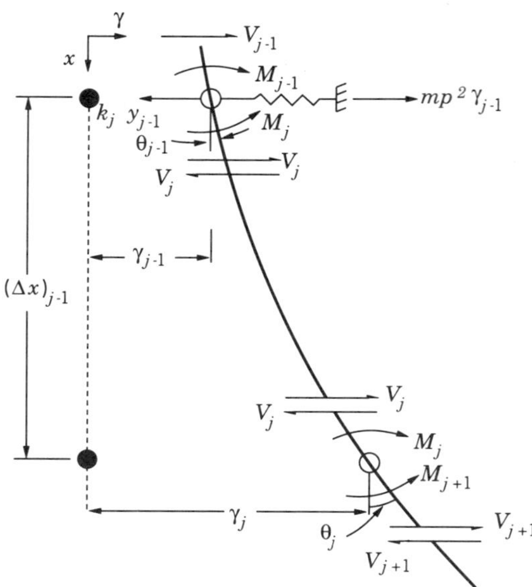


**Figure 7.12** Deflections and forces at two adjoining nodal points (after Chandrasekaran, 1974).

system with an initially assumed value and related end conditions. The assumed end conditions are also utilized to generate the transfer equations and to evaluate the unknown quantities in terms of the known quantities, either at the pile top or the pile tip. These modal quantity values at different station points define the mode shapes. Values at the bottom or top of the piles assist in determining the natural frequencies of vibrations in different modes. The forces and displacements in two different station points are illustrated in Figure 7.12 (Prakash and Chandrasekaran, 1977). For details refer to Chandrasekaran (1974).

The soil stiffness has been defined by a modulus of horizontal reaction $k_h$ $(FL^{-2})$. This has been considered to vary (1) linearly with depth and (2) remain constant with depth. In both of these cases, solutions have been obtained for natural frequency, modal displacements, slopes, bending moments, shear forces, and soil reactions along the lengths of the piles in the first three modes of vibrations (Chandrasekaran, 1974; Prakash and Chandrasekaran, 1980). Only typical solutions for handling a practical problem shall be presented in the following paragraphs.

### 7.2.1 Range of Variables

A large number of pile sizes, soil moduli values, pile stiffness and relative stiffness factors (R or T) were selected for parametric study. In soils for which the soil modulus may be assumed constant with depth, the range of values are listed in

**TABLE 7.1 Range of Variables After Prakash and Chandrasekaran (1980)**

| Quantity | Units | Range |
|---|---|---|
| Diameter of pile | m | 0.3, 0.4, 0.5, 0.6, and 0.7 |
| $k$ | ton-m$^{-2}$ | 94.25 to 368.55 |
| $EI$ | ton-m$^2$ | $4.77 \times 10^2 \cdots 141 \times 10^2$ |
| $R$ | m | 1 to 3 |
| $Z_{max}$ | — | 1 to 15 |

Case – k = constant with depth.

**TABLE 7.2 Range of Variables After Chandrasekaran and Prakash (1980)**

| Quantity | Units | Range |
|---|---|---|
| Diameter of pile | m | 0.3, 0.4, 0.5, 0.6, and 0.7 |
| $n_h$ | ton-m$^{-3}$ | 58.2 to 4634.397 |
| $EI$ | ton-m$^2$ | $4.77 \times 10^2 \cdots 141 \times 10^2$ |
| $R$ | m | 1 to 3 |
| $Z_{max}$ | — | 1 to 15 |

Case $k = n_h \cdot x$.

Table 7.1. In soils for which soil modulus may be assumed to vary linearly with depth, the range of variables are listed in Table 7.2.

### 7.2.2 Natural Frequencies

Based on the foregoing analysis, nondimensional frequency factors have been obtained with respect to the basic soil parameters.

The variables constituting $F_{cL}$, *the nondimensional frequency factor* for piles embedded in soils in which the soil modulus remains constant with depth, is given as:

$$\omega_{n1} = F_{CL_1} \text{ (or } F'_{CL_1}) \div \sqrt{\frac{W}{gkR}} \tag{7.12}$$

where

$\omega_{n1}$ = the first natural angular frequency in radians per second

$\frac{W}{g}$ = lumped mass at the top of the pile

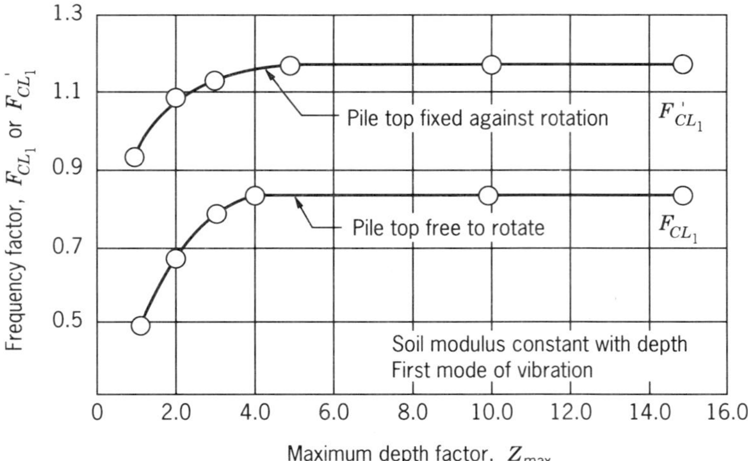


**Figure 7.13** Nondimensional frequency factors in first mode of vibrations. Soil modulus constant with depth (Prakash and Chandrasekaran, 1977).

$k$ = soil modulus

$R$ = relative stiffness factor, defined as follows:

$$R = \sqrt[4]{\frac{EI}{k}} \tag{6.80b}$$

In Figure 7.13, the variation of frequency factor $F_{CL_1}$ or $F'_{CL_1}$ with $Z_{\max}$ has been plotted, in which $Z_{\max} = L/R$. $F_{CL}$ and $F'_{CL}$ refer to cases with the pile top free to rotate and the pile top restrained against rotation, respectively.

Similarly in Figure 7.14a and b, frequency factors $F_{SL1}$ and $F'_{SL1}$ for soils whose moduli vary linearly with depth have also been plotted for cases where the pile top is free to rotate and the pile top is restrained against rotation, respectively. The definitions of $F_{SL1}$ and $F'_{SL1}$ for the pile tops free to rotate and the pile top restrained against rotation are identical and given as:

$$\omega_{n_1} = (F_{SL_1} \text{ or}) F'_{SL_1} \div \sqrt{\frac{W}{g} \frac{1}{n_h T^2}} \tag{7.13}$$

in which $n_h$ is the constant of horizontal subgrade reaction $k_x (= n_h x)$ and

$$T = \sqrt[5]{\frac{EI}{n_h}} \tag{6.44}$$

It will be seen from Figures 7.13 and 7.14 that the natural frequency attains a

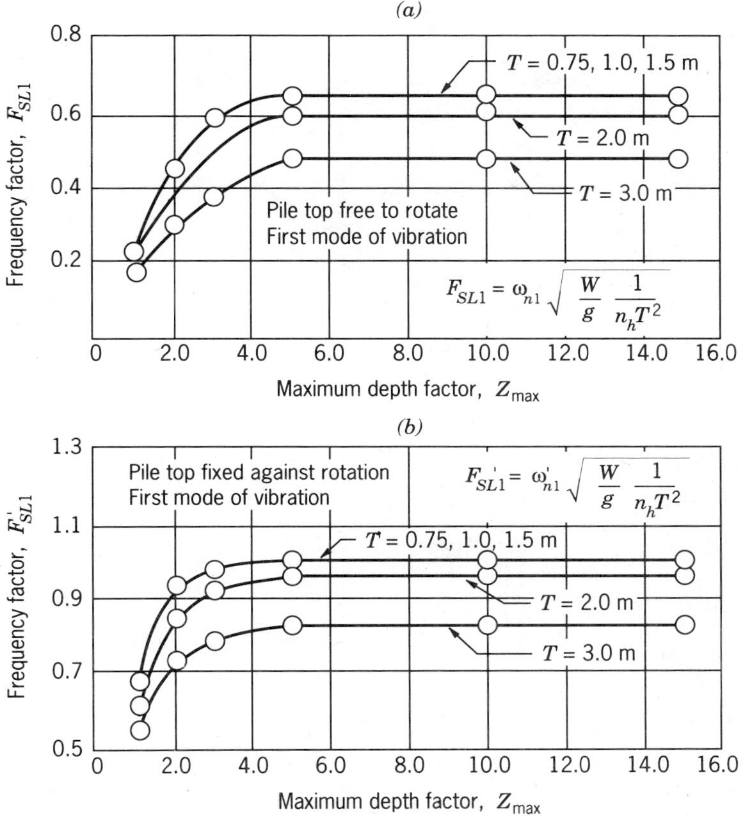


**Figure 7.14** Nondimensional frequency factors in first mode of vibrations. (a) Soil modulus linearly varying with depth and pile top free, (b) soil modulus linearly varying with depth and pile top restrained against rotation (Prakash and Chandrasekaran, 1977).

constant value for $Z_{max} \geqslant 4.5$ in all cases. Therefore, piles with embedded depths ($L_s \geqslant 5R$ or $5T$ behave as "*long*" piles as under static loading (Chapter 6).

With these two sets of curves, the natural frequency of the soil pile system and hence the time period for the first mode of vibrations maybe determined, if the soil–pile characteristics, length, and fixity conditions are known. Similar frequency factors and mode shapes parameters for determining natural frequencies and mode shapes in the second and third modes of vibrations have been plotted by Chandrasekaran (1974). From the mode shapes and frequencies of the system, the overall response can be computed by principle of modes superposition Here, only the solutions for the first modes of vibrations have been presented and a design procedure based on these solutions formulated.

Figure 7.15 shows a plot of nondimensional displacement with depth factor $z(=x/T)$ in the first mode of vibration when the pile top is restrained against

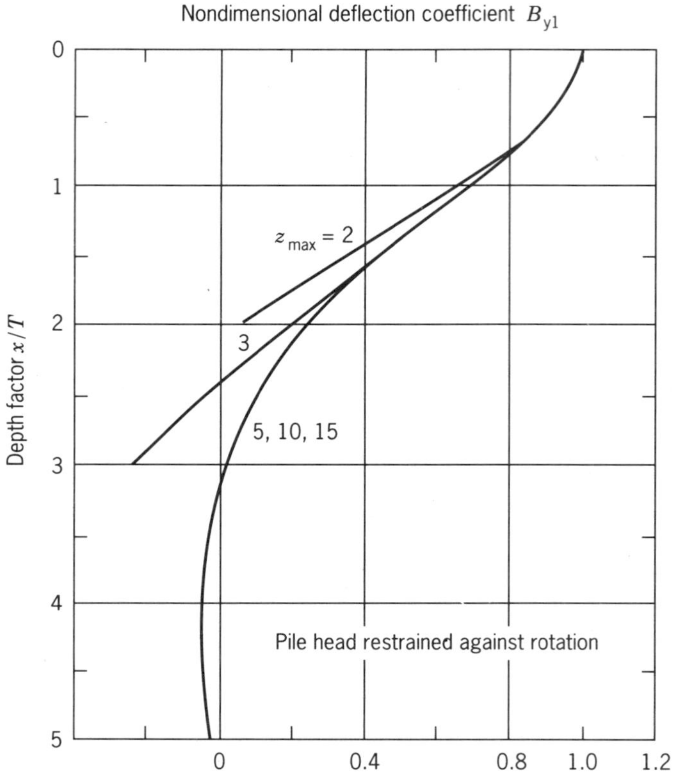


**Figure 7.15** Nondimensional deflection coefficients assuming soil modulus proportional to depth (Chandrasekaran, 1974)

rotation but is free to translate and is embedded in a soil with modulus proportional with depth. Similarly, the variation of the nondimensional bending moment coefficient with the depth of the pile in the first mode of vibrations is shown in Figure 7.16.

It can be seen from Figure 7.16 that for a pile with $Z_{max} \geqslant 5$, the maximum negative bending moment coefficient $B_{me1} = 0.90$ and the maximum value occurs at $x/T = 0$ (i.e., at the connection of the pile with the pile cap). These data are sufficient for the design of piles, and the entire curve is not needed (See Section 7.3).

## 7.3 ASEISMIC DESIGN OF PILES

Based on the foregoing analysis and the concept of the response spectrum, the following method of analysis and design of piles against earthquakes may be used. For this analysis, the following data must be obtained first:

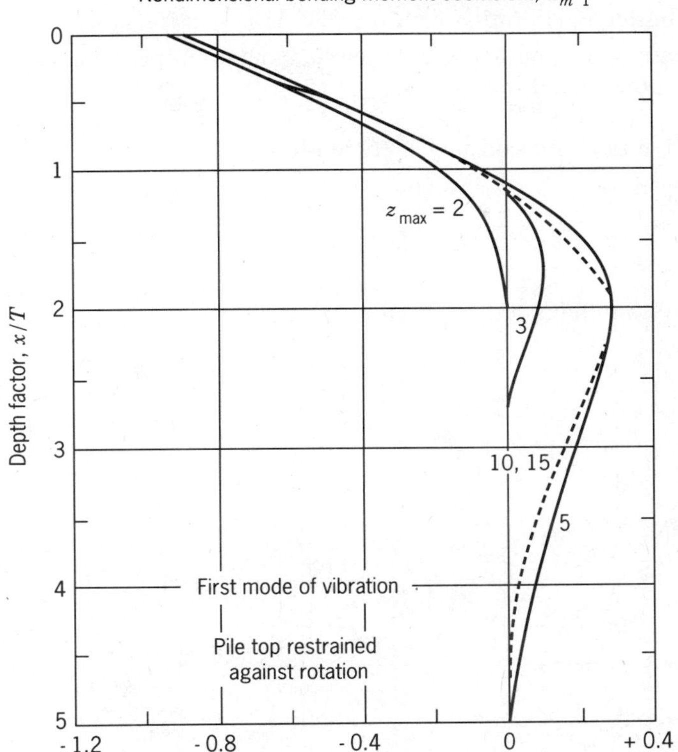


**Figure 7.16** Nondimensional bending moment coefficient assuming soil modulus proportional to depth (Chandrasekaran, 1974).

1. Soil characteristics, and boring logs of the site
2. Pile characteristics, size, $EI$, length, and type of pile
3. Lateral load deflection of the pile under static conditions for estimation of $k$ or $n_h$

**Design Steps**

1. Estimate the dynamic soil modulus $k$ or $n_h$ based on the principles discussed in Chapter 4. In the absence of realistic data, the values from a static lateral load test may be modified based on engineering judgment.
2. Compute the relative stiffness factor $R$ or $T$.
3. Calculate the maximum depth factor $Z_{max}$ for a pile; $Z_{max}$ in most practical cases will be greater than 5.
4. For the computed value of the maximum depth factor and the pile end condition, read the frequency factor (Figures 7.13 or 7.14).

5. Estimate the dead load on the pile. The mass at the pile top which may be considered vibrating with the piles is only a fraction of this load.
6. Determine the natural frequency $\omega_{n_1}$ and time period in first mode of vibrations as follows:

   (a) Soil modulus constant with depth:

$$\omega_{n_1} = F_{CL_1} \text{ (or } F'_{CL_1}) \div \sqrt{\frac{W}{gkR}} \tag{7.12}$$

   (b) Soil modulus proportional to depth:

$$\omega_{n_1} = (F_{SL^1} \text{ or}) F_{SL_1} \div \sqrt{\frac{W}{g}\frac{1}{n_h T^2}} \tag{7.13}$$

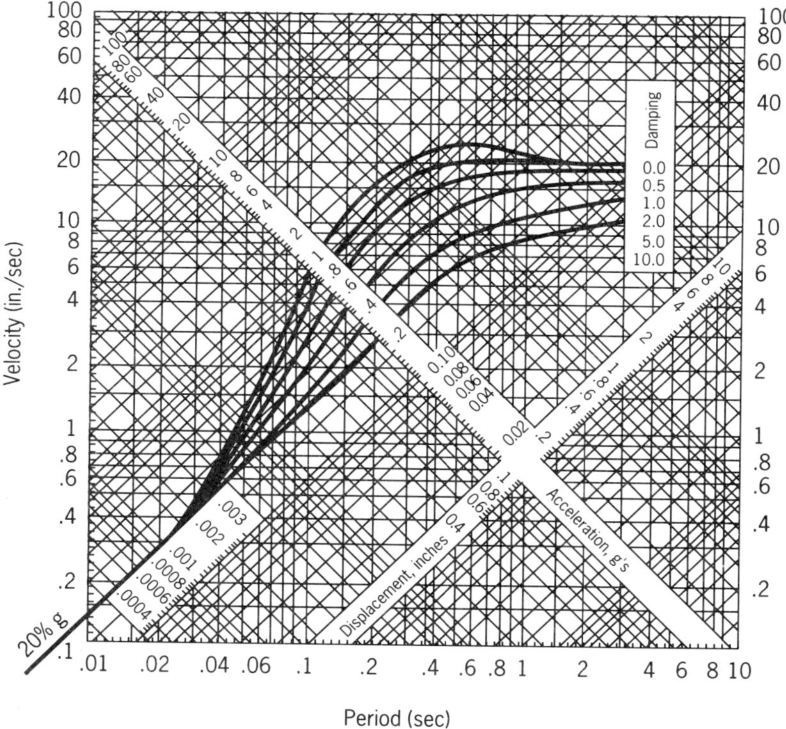


**Figure 7.17** Combined plot of design spectrum giving $S_\alpha$, $S_v$, and $S_d$ as a function of period and damping, scaled to 20 percent of acceleration at zero period. (Reproduced with permission of Prentice-Hall, Inc., Englewood Cliffs, N.J., from G. W. Housner, "Design Spectrum", *Earthquake Engineering*, R. L. Wiegel, ed.).

**TABLE 7.3 Maximum Values of Coefficient $A_{me}$[a]**

| Maximum Depth Factor, $Z_{max}$ | Coefficient $A_{me}$ | | |
|---|---|---|---|
| | Pile Top Free to Rotate | Pile Fixed at Top Against Rotation | |
| | | Negative | Positive |
| 2 | 0.13 | 0.9 | 0 |
| 3 | 0.24 | 0.9 | 0.04 |
| 5–15 | 0.32 | 0.9 | 0.18 |

[a]After Chandrasekaran (1974).

Then

$$T_{n_1} = 2\pi/\omega_{n_1} \tag{7.14}$$

7. For the foregoing time period, determine the spectral displacement $S_d$ for assumed damping from Figure 7.17. This is the maximum displacement of the pile head. If an accelogram for a site has been selected, spectral response is determined for this ground motion. For the soil pile system, 5 to 10 percent damping may be assumed (Prakash, 1981).
8. Estimate the maximum bending moment in the pile section.

   (a) Soil modulus constant with depth:

   $$\text{Bending moment} = A_{me} \times kR^2 \times S_d \tag{7.14a}$$

   The maximum values of $A_{me}$ are given in Table 7.3.

   (b) Soil modulus increasing linearly with depth:

   $$\text{Bending moment} = B_{me} \times n_h T^3 \times S_d \tag{7.14b}$$

   The maximum values of $B_{me}$ are given in Table 7.4. The pile section should be able to stand the foregoing moments.
9. For the computed maximum ground displacement, the displacement all along the length of the pile may be determined by assuming that the deflected shape in vibrations in similar to one under static conditions (See Chapter 6). For soil modulus constant with depth or soil modulus linearly varying with depth, the solutions of Davisson and Gill (1963) and Reese and Matlock (1956) may be used for two cases of soil modulus. The soil reaction is then computed all along the pile lengths as follows:

**TABLE 7.4 Maximum Values of Coefficient $B_{me}$[a]**

| Maximum Depth Factor, $Z_{max}$ | Coefficient $B_{me}$: Pile Top Free to Rotate | Pile Fixed at Top Against Rotation: Negative | Pile Fixed at Top Against Rotation: Positive |
|---|---|---|---|
| 2 | 0.100 | 0.93 | 0 |
| 3 | 0.255 | 0.93 | 0.10 |
| 5–15 | 0.315 | 0.90 | 0.28 |

[a] After Chandrasekaran (1974).

(a) For soil modulus constant with depth:

$$p_x = k \cdot y_x \tag{7.16a}$$

(b) For soil modulus linearly varying with depth:

$$p_x = n_h \cdot x \cdot y_x \tag{7.16b}$$

The allowable soil reaction may be taken as that corresponding to the Rankine passive pressure at all depths (Prakash et al., 1979).

The solution of pile deflection, bending moments, and soil reactions will be obtained for the two cases of pile restraint: the pile top free to rotate and the pile top restrained against rotation but free to translate. Fixity conditions of the actual piles must be estimated and the solution obtained for this case by linear interpolation.

The deflections, bending moments, and soil reactions under static loading are added to the corresponding values under dynamic loading to arrive at the final values.

For this analysis, the soil modulus values recommended in Chapter 4 and modified for appropriate dynamic conditions may be used.

***Group Action*** The value of $k$ needs to be corrected for group action. The following guidelines are recommended.

1. In cases where the center-to-center spacing of piles is $8d$ in the direction of loading where $d$ is the diameter of the pile, and the center-to-center spacing is at least $2.5d$ in the direction perpendicular to the load, group action is neglected. The piles may be arranged to behave as *individual* piles. If the spacing in the direction of the load is $3d$, the effective value of $k$ ($k_{eff}$) is $0.25k$. For other spacing values, a linear interpolation may be made. This

recommendation is based on model tests on piles in sands under static loads (Prakash, 1962).

2. If a *cyclic* load is applied, the deflections increase and $k_{eff}$ decreases. It has been observed that the deflections after 50 cycles of load application are double the deflections under the first cycle (Prakash, 1962). The soil modulus decreases to 0.30 times and 0.4 times for soils with linearly increasing and constant modulus with depth, respectively.

If group action and oscillatory loads are considered, the soil modulus is decreased on two counts, and the final value may be less than 10 percent of $k$ for a single pile. These recommendations may be regarded as tentative. When more data become available, these recommendations may need to be revised.

## 7.4 NOVAK'S DYNAMIC ANALYSIS OF PILES

In this section, soil pile analyses developed by Novak for vertical vibrations of piles and piles under lateral and rocking motion are presented. In these procedures, the soil pile stiffness and damping have been evaluated for the system. A complete dynamic analysis can then be performed.

### 7.4.1 Vertical Vibrations

The main assumptions in Novak's analysis are (Novak, 1974, 1977a):

1. The pile is vertical and of circular cross section.
2. The pile material is linearly elastic.
3. The pile is perfectly connected to the soil (i.e., there is no separation between soil and pile under vibrations).
4. The pile is a floating pile.
5. The soil above the tip is modeled as a linear elastic layer composed of infinitesimally thin independent layers, which means that the elastic waves propagate only horizontally. The soil reaction acting on the tip is assumed to be equal to that of an elastic halfspace.
6. The motion is small and excitation is harmonic, which yields the impedance functions and the equivalent stiffness and damping constants of the soil–pile system that can be used in structural analysis.

In Figure 7.18, an elastic vertical pile is shown undergoing complex vertical vibration $w(z, t)$ (Novak, 1977) such that:

$$w(z, t) = w(z)e^{i\omega t} \tag{7.17}$$

where

$w(z)$ = complex amplitude at depth $z$

$i = \sqrt{-1}$

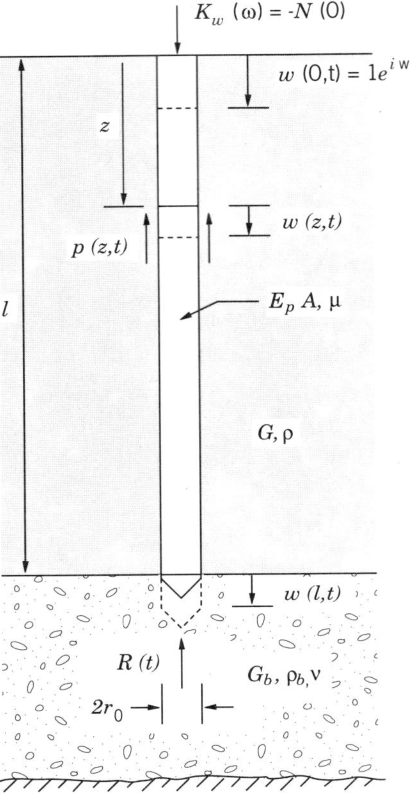


**Figure 7.18** Vertical pile and notations.

$\omega$ = circular frequency
$t$ = time

The motion of the pile is resisted by soil reaction along the pile length and a concentrated reaction at the tip. Therefore, the soil reaction appears in the equation of motion for an element $dz$, and the concentrated reaction represents the boundary conditions at the tip.

The distributed soil reaction acting on pile element $dz$ at depth $z$ is written as (Baranov, 1967; Prakash and Puri, 1988)

$$p(z,t)dz = G(S_{\omega 1} + iS_{\omega 2})w(z,t)dz \tag{7.18}$$

where

$G$ = shear modulus of soil surrounding the pile; and

$$S_{\omega 1} = 2\pi a_0 \frac{J_1(a_0)J_0(a_0) + Y_1(a_0)Y_0(a_0)}{J_0^2(a_0) + Y_0^2(a_0)} \tag{7.19}$$

$$S_{\omega 2} = \frac{4}{J_0^2(a_0) + Y_0^2(a_0)} \tag{7.20}$$

where

$J_0(a_0), J_1(a_0)$ = Bessel functions of the first kind of order zero and one
$Y_0(a_0), Y_1(a_0)$ = Bessel functions of the second kind of order zero and one
$S_{\omega_1}$ and $S_{\omega_2}$ = functions of the dimensionless frequency
$a_0 = r_0\omega/V_s$
$r_0$ = pile radius
$V_s = \sqrt{G/\rho}$
$\rho$ = mass density of soil

Parameters $S$ are shown in Figure 7.19.

With the soil reactions defined by equation (7.18), the differential equation of damped axial vibration of the pile is

$$m_1 \frac{\partial^2 w(z,t)}{\partial t^2} + c\frac{\partial w(z,t)}{\partial t} - E_p A \frac{\partial^2 w(z,t)}{\partial z^2} + G(S_{\omega_1} + iS_{\omega_2})w(z,t) = 0 \tag{7.21}$$

where

$m_1$ = mass of the pile per unit length
$c$ = coefficient of pile internal damping
$E$ = Young's modulus of the pile
$A$ = area of the pile cross section

Equation (7.21) reduces to an ordinary differential equation with the harmonic motion described by equation (7.17) as follows:

$$w(z)[-m_1\omega^2 + ic\omega + G(S_{\omega_1} + iS_{\omega_2})] - E_p A \frac{d^2 w(z)}{dz^2} = 0. \tag{7.22}$$

The solution to this equation is:

$$w(z) = B\cos\Lambda\frac{Z}{l} + C\sin\Lambda\frac{Z}{l} \tag{7.23}$$

where

$l$ = pile length
$B, C$ = integration constants

and the complex frequency parameter

$$\Lambda = l\sqrt{\frac{1}{E_p A}[m_1\omega^2 - GS_{\omega_1} - i(c\omega + GS_{\omega_2})]} \tag{7.24}$$

Note:

$$\Lambda_0 = l\sqrt{\frac{m_1\omega^2}{E_pA}} \qquad K = \frac{l^2G}{E_pA} \tag{7.25}$$

which, for a pile of circular cross section

$$\Lambda_0 = \frac{l}{r_0}\sqrt{\frac{\rho_p}{\rho}\frac{G}{E_p}}a_0 = \frac{l}{r_0}\frac{V_s}{V_c}a_0 \tag{7.26}$$

and

$$K = \frac{1}{\pi}\frac{G}{E_p}\left(\frac{l}{r_0}\right)^2 = \frac{1}{\pi}\frac{\rho}{\rho_p}\left(\frac{V_s}{V_c}\frac{l}{r_0}\right)^2 \tag{7.27}$$

where

$V_c = \sqrt{E_p/\rho_p}$ = longitudinal wave velocity in the pile
$\rho_p$ = mass density of the pile

Denote further:

$$a = \Lambda_0^2 - KS_{\omega 1} \qquad b = -K\left(c\omega\frac{1}{G} + S_{\omega 2}\right) \tag{7.28}$$

and

$$r = \sqrt{(a^2+b^2)} \qquad \tan\phi = \frac{b}{a} \tag{7.29}$$

Then the frequency parameter $\Lambda$ is more conveniently written as

$$\Lambda = \Lambda_1 + i\Lambda_2 \tag{7.30}$$

where

$$\Lambda_1 = \sqrt{r}\cos\frac{\phi}{2} \qquad \Lambda_2 = \sqrt{r}\sin\frac{\phi}{2} \tag{7.31}$$

The integration constants $B$ and $C$ are given by the boundary conditions.

Harmonic motion with a unit amplitude is assumed to be $w(0,t) = 1e^{i\omega t}$ at the head of the pile, since this form of excitation defines the stiffness and damping of the soil–pile system at the pile head. Therefore, the first boundary condition is

$$w(0) = 1 \tag{7.32}$$

The motion of the pile generates a concentrated reaction $R(t)$ of the soil at its tip. This can be described approximately as the reaction to the vertical motion of a rigid circular disk of an elastic halfspace and can be written as $R(t) = Re^{i\omega t}$, the amplitude of which is:

$$R = -G_b r_0 (C_{\omega_1} + iC_{\omega_2}) w(l) \tag{7.33}$$

where

$G_b$ = shear modulus of the soil below the tip
$w(l)$ = the complex amplitude of the tip
$C_{\omega_1}$, and $C_{\omega_2}$ = dimensionless parameters depending on the dimensionless frequency ($a_0$) and Poisson's ratio ($\nu$)

The shear wave velocity of the soil below the tip is

$$V_b = \sqrt{G_b/\rho_b}$$

where

$G_b, \rho_b$ = shear modulus and mass density of the soil near the tip, respectively.

As $G_b \to \infty$ the motion of the tip vanishes corresponding to an end-bearing pile. With $G_b \to G$, the pile becomes floating. The distributed soil reaction, $p(z, t)$, contributes to the total stiffness and damping of the system in both the end bearing and the floating pile but to different degrees.

Using Bycroft's (1956) solution, the polynomial expressions for the parameters $C_\omega$ for $\nu$ of 0.25 are:

$$C_{\omega_1} = 5.33 + 0.364a_0 - 1.41a_0^2 \tag{7.34a}$$

$$C_{\omega_2} = 5.06a_0 \tag{7.34b}$$

and for $\nu$ of 0.5

$$C_{\omega_1} = 8.00 + 2.18a_0 - 12.63a_0^2 + 20.73a_0^3 - 16.47a_0^4 + 4.458a_0^5 \tag{7.34c}$$

$$C_{\omega_2} = 7.414a_0 - 2.98a_0^2 + 4.324a_0^3 - 1.782a_0^4 \tag{7.34d}$$

The parameters $C_\omega$ described by equations (7.34) have been plotted against dimensionless frequency in Figure 7.19.

The axial force in the pile, positive for tension, is

$$N(Z) = E_p A \frac{dw(z)}{dz} = E_p A \frac{\Lambda}{l}\left(-B \sin \Lambda \frac{z}{l} + C \cos \Lambda \frac{z}{l}\right) \tag{7.35}$$

The end force of the pile must be equal to the soil reaction given by equation (7.33). Thus, the boundary condition for the tip, $z = l$, is

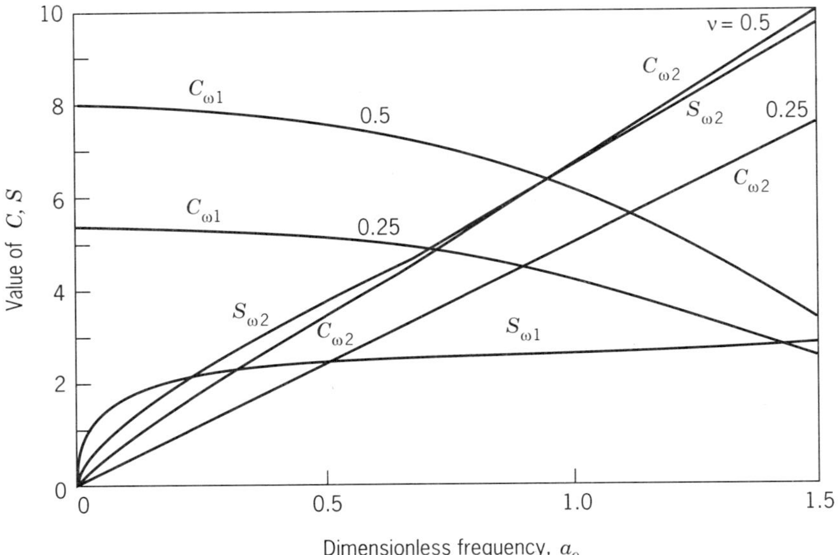


**Figure 7.19** Parameters $S_{\omega 1}$, $S_{\omega 2}$, $C_{\omega 1}$, and $C_{\omega 2}$ (Novak, 1977).

$$E_p A \frac{\Lambda}{l}(-B\sin\Lambda + C\cos\Lambda) = -G_b r_0 (C_{\omega_1} + iC_{\omega_2})(B\cos\Lambda + C\sin\Lambda) \tag{7.36}$$

Equations 7.32 and 7.33 give:

$$B = 1 \tag{7.37}$$

The second integrational constant from equation (7.36) is:

$$C(\Lambda) = \frac{K'\Lambda\sin\Lambda - (C_{\omega_1} + iC_{\omega_2})\cos\Lambda}{K'\Lambda\cos\Lambda + (C_{\omega_1} + iC_{\omega_2})\sin\Lambda} \times 1 \tag{7.38}$$

where $C_{\omega_1}$ and $C_{\omega_2}$ are evaluated for frequency $a_0 = r_0\omega/V_b$ and

$$K^1 = \frac{E_p A}{G_b l r_0} \tag{7.39}$$

For a circular pile, $K^1$ becomes

$$K^1 = \pi \frac{r_0}{l}\frac{E_p}{G_b} = \pi \frac{r_0}{l}\frac{\rho_p}{\rho_b}\left(\frac{v_c}{V_b}\right)^2 \tag{7.40}$$

From the integration constants, the amplitude of the pile displacement becomes:

$$w(z) = 1\cos\Lambda\frac{z}{l} + C(\Lambda)\sin\Lambda\frac{z}{l} = w_1 + iw_2 \tag{7.41}$$

where $C(\Lambda)$ is obtained from equation (7.38).

The unit appearing in equations 7.37, 7.38, and 7.41 is actually the amplitude of the head and thus has the dimension of length.

The real amplitude of motion is:

$$w(z) = \sqrt{w_1^2 + w_2^2} \tag{7.42}$$

and the phase angle is given by:

$$\phi(z) = a\tan\frac{w_2}{w_1} \tag{7.43}$$

Novak (1977a) determined the variation of the amplitude and phase with (1) relative depth $z/l$, (2) slenderness ratio $l/r_0$, (3) wave velocity ratio $V_s/v_c$, (4) frequency ratio $a_0$ for $v = 0.5$, (5) density ratio $\rho/\rho_p = 0.7$, which is typical of reinforced concrete piles, and (6) shear wave velocity ratios $V_b/V_s = 1$ and 10,000 that characterize floating and end-bearing piles, respectively. Internal damping of the pile has been neglected.

These plots indicated that:

1. The tip condition is particularly important in weak soils (small $V_s/v_c$) in which even a very long pile can vibrate almost as a rigid body.
2. It is only the upper part of a pile that undergoes significant displacement in stiff soils.

The increase in the phase shift where visible is indicative of increased damping. In the design of pile-supported footings and structures, the stiffness and damping constants of the soil–pile system at the level of the pile head are needed. Having determined these quantities, the remaining procedure is the same as that for end-bearing piles.

The complex stiffness is equal to the force that produces a unit dynamic displacement of the pile head at a certain frequency. Thus it is:

$$K_w = -N(0)$$

where

$$N(z) = E_p A\, dw(z)/dz$$

Differentiating equation (7.41) and substituting $z = 0$, we obtain the complex stiffness as:

$$K_w = \frac{E_p A}{l} F_w(\Lambda) \tag{7.44}$$

where

$$F_w(\Lambda) = -\Lambda C(\Lambda) = F_w(\Lambda)_1 + iF_w(\Lambda)_2 \tag{7.45}$$

and $C(\Lambda)$ is given by equation (7.38). In equation 7.45, subscript 1 denotes the real part of $F_w$, which defines the real stiffness and subscript 2 indicates the imaginary (out of phase) part that relates to damping.

The stiffness constant $k_w$ of one pile can be rewritten as

$$k_w^1 = \frac{E_p A}{r_0} f_{w_1} \tag{7.46}$$

where

$$f_{w_1} = \frac{F_w(\Lambda)_1}{l/r_0} \tag{7.47}$$

The constant of equivalent viscous damping of one pile is $E_p A F_w(\Lambda)_2/(l\omega)$, which can be written as:

$$c_w^1 = \frac{E_p A}{V_s} f_{w_2} \tag{7.48}$$

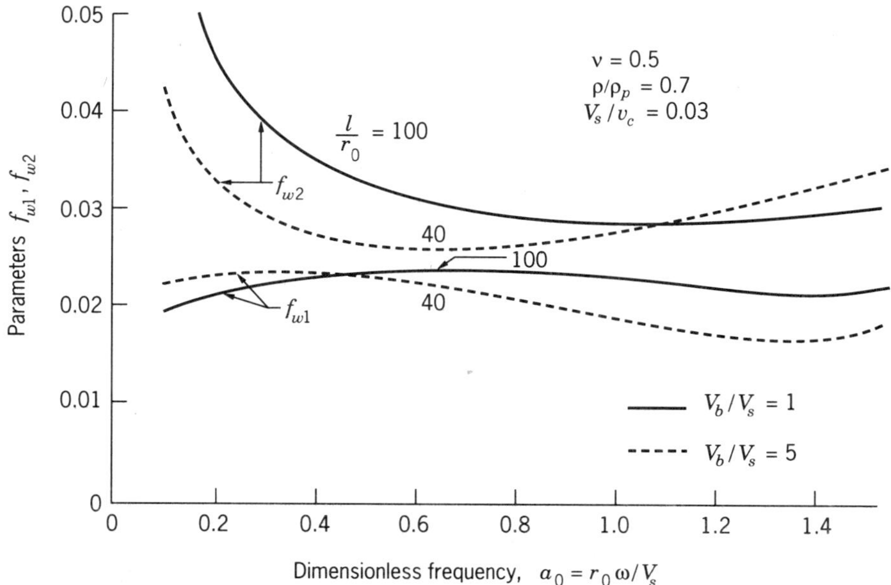


**Figure 7.20** Variations of stiffness and damping parameters of pile with frequency (Novak, 1977).

where

$$f_{w_2} = \frac{F_w(\Lambda)_2}{a_0 l / r_0} \tag{7.49}$$

The stiffness and damping of piles vary with frequency, as shown in Figure 7.20. In this figure, parameter $f_{w_1}$ characterizes stiffness, and parameter $f_{w_2}$ characterizes damping. These parameters have been plotted for a few typical cases. This figure shows that:

1. The dynamic stiffness of the soil–pile system varies only moderately with frequency both for slender as well as rigid piles.
2. The damping decreases rapidly with increasing frequency but levels off in the range of moderate frequencies.

Since stiffness and damping do not depend much on frequency, Novak (1977a) has recommended parameters $f_{w_1}$ and $f_{w_2}$ for design purposes which are independent of frequency. Figure 7.21 shows the variation of the stiffness and damping parameters of the pile with the shear wave velocity ratio, $V_b/V_s$, of the

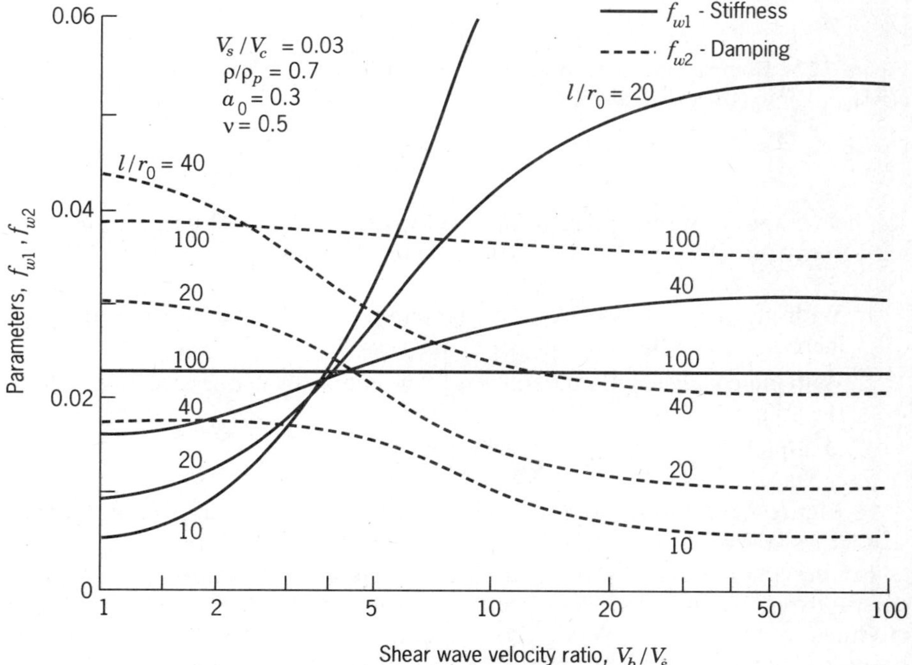


**Figure 7.21** Variations of stiffness and damping parameters of pile with ratio of shear wave velocities of soil below and above tip (after Novak, 1977).

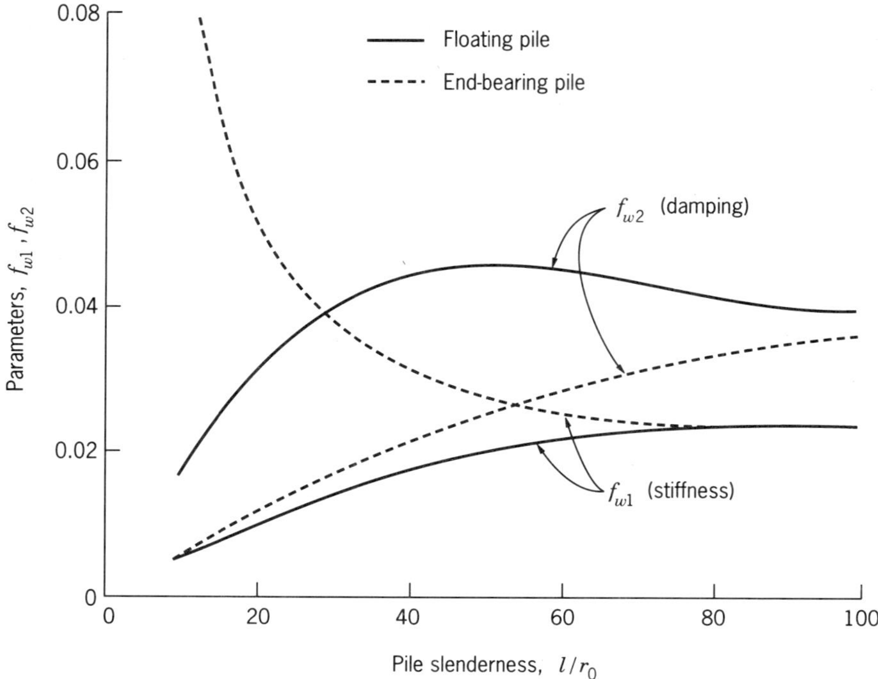


**Figure 7.22** Comparison of floating piles with end bearing piles ($\rho/\rho_p = 0.7$, $v = 0.5$, $a_0 = 0.3$, $V_s/V_c = 0.03$) (Novak, 1977).

soil below and above the pile tip. The slenderness ratios ($l/r_0$) used in this plot vary from 10 to 100, and $V_s/V_c = 0.03$. It is seen from this figure that:

1. With increasing stiffness of the soil below the tip, the stiffness of the pile increases while the damping decreases.
2. With increasing length, the stiffness of the end-bearing piles decreases while the stiffness of floating piles increases.
3. Damping increases with pile length in most cases.

In Figure 7.22, stiffness and damping parameters have been plotted against slenderness ratio ($l/r_0$) for floating as well as end-bearing piles.

For design of both end-bearing and floating piles, the constants $f_{w_1}$ and $f_{w_2}$ in equations (7.46) and (7.48) had been solved by Novak (1974, 1977), for soil modulus constant with depth. Novak and El-Sharnouby (1983) included solutions for shear modulus decreasing upward in a quadratic parabola for end bearing piles (Figure 7.23a) and floating piles (Figure 7.23b).

The geometric damping ratio for a single pile may be determined from

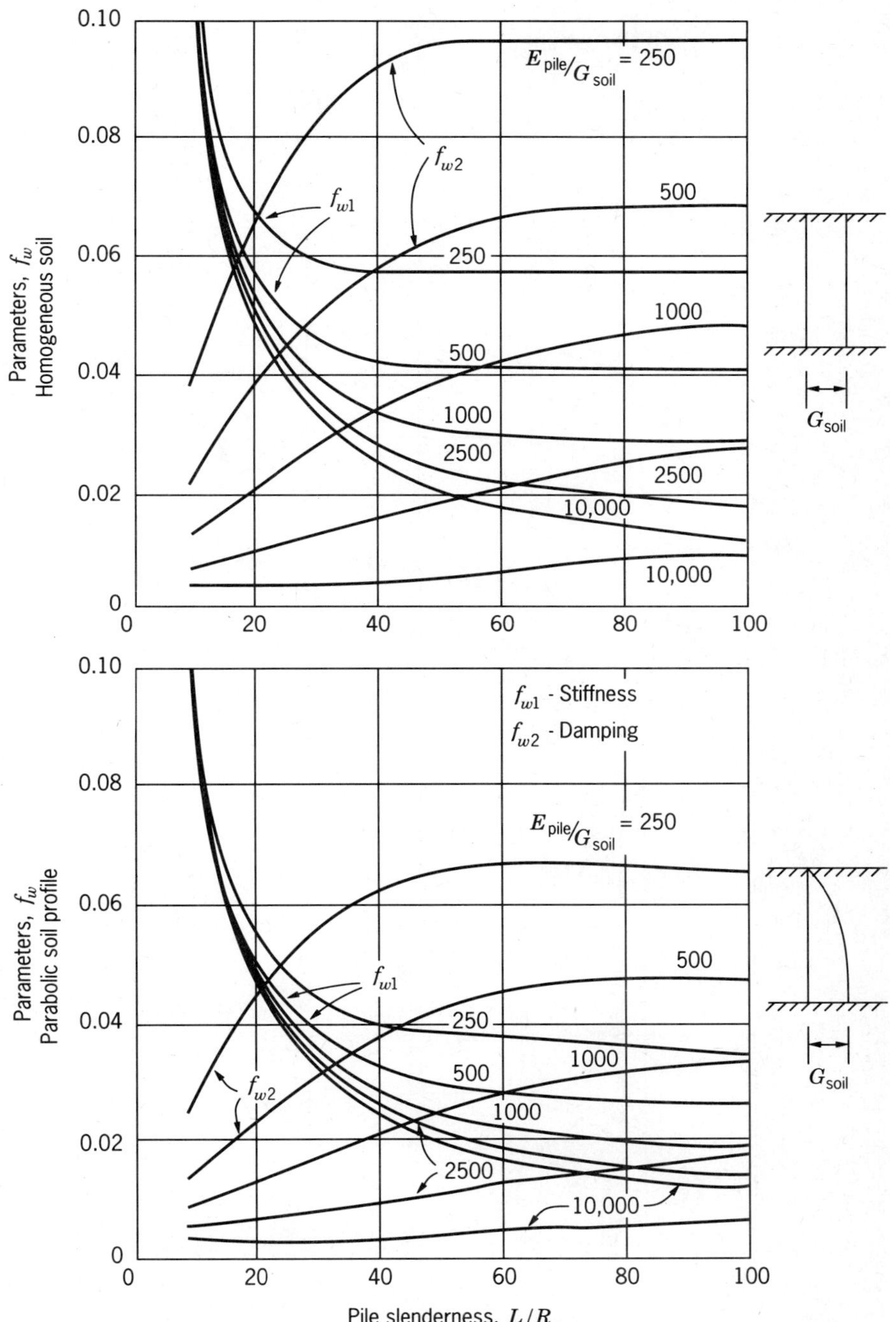


**Figure 7.23a** Stiffness and damping factors for fixed tip vertically vibrating piles (Novak and El-Sharnouby, 1983).

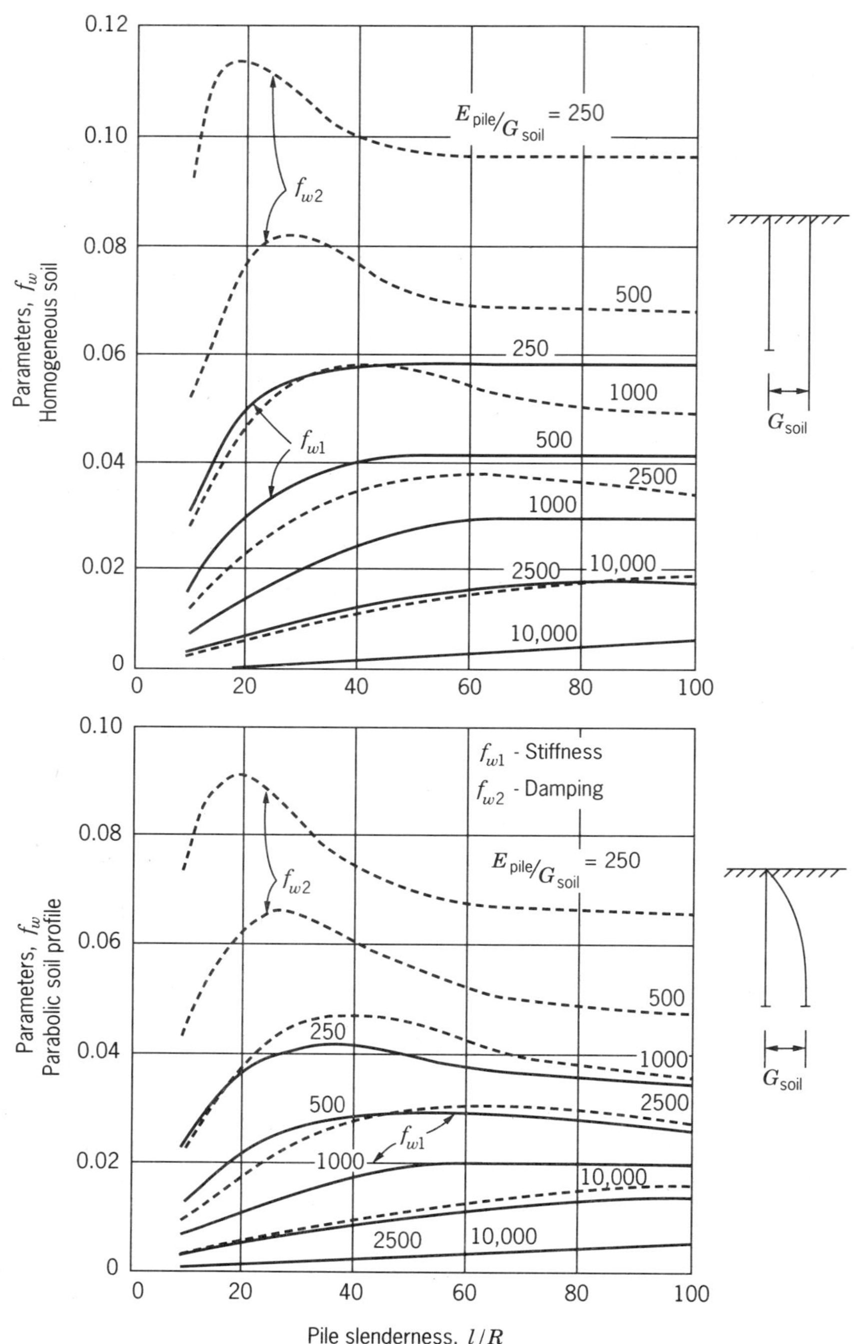


**Figure 7.23b** Stiffness and damping parameters of vertical response of floating piles (Novak and El-Sharnouby, 1983).

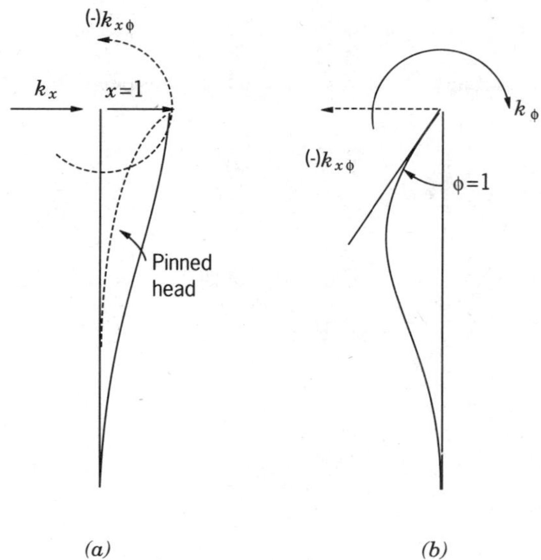


**Figure 7.24** Generation of lateral pile stiffness in individual directions: (a)Horizontal, (b) rotation (Novak and El-Sharnouby, 1983).

equation (7.50):

$$D_w^1 = \frac{c_w^1}{2\sqrt{k_w^1 m_c}} \tag{7.50}$$

where $m_c$ is the mass of the cap plus the portion of the structure load vibrating in phase with the cap.

### 7.4.2 Lateral Vibrations

Novak (1974) had derived lateral stiffness and damping constants for single piles with soil modulus constant with depth. He considered (1) translation alone, Figure 7.24a, (2) rotation alone, Figure 7.24b, and (3) coupled rotation and translation. Novak and El-Sharnouby (1983) extended these solutions to include parabolic variation of soil-shear modulus also. Equations (7.51) to (7.56) summarize the stiffness and damping coefficients and Table 7.5 lists values of constants used.

Translation stiffness constant,

$$k_x^1 = \frac{E_p I_p}{r_0^3}(f_{x_1}) \tag{7.51}$$

**TABLE 7.5 Stiffness and Damping Parameters of Horizontal Response for Piles with $L/r_0 > 25$ for Homogeneous Soil Profile and $L/r_0 > 30$ for Parabolic Soil Profile**

| | | Stiffness Parameters | | | | Damping Parameters | | | |
|---|---|---|---|---|---|---|---|---|---|
| $\nu$ (1) | $E_{pile}/G_{soil}$ (2) | $(f_{\phi_1})$ (3) | $f_{(x\phi_1)}$ (4) | $(f^*_{x_1})$ (5) | $(f^p_{x_1})$ (6) | $(f_{\phi_2})$ (7) | $f_{(x\phi_2)}$ (8) | $(f^*_{x_2})$ (9) | $(f^p_{x_2})$ (10) |
| Homogeneous Soil Profile | | | | | | | | | |
| 0.25 | 10,000 | 0.2135 | −0.0217 | 0.0042 | 0.0021 | 0.1577 | −0.0333 | 0.0107 | 0.0054 |
| | 2,500 | 0.2998 | −0.0429 | 0.0119 | 0.0061 | 0.2152 | −0.0646 | 0.0297 | 0.0154 |
| | 1,000 | 0.3741 | −0.0668 | 0.0236 | 0.0123 | 0.2598 | −0.0985 | 0.0579 | 0.0306 |
| | 500 | 0.4411 | −0.0929 | 0.0395 | 0.0210 | 0.2953 | −0.1337 | 0.0953 | 0.0514 |
| | 250 | 0.5186 | −0.1281 | 0.0659 | 0.0358 | 0.3299 | −0.1786 | 0.1556 | 0.0864 |
| 0.40 | 10,000 | 0.2207 | −0.0232 | 0.0047 | 0.0024 | 0.1634 | −0.0358 | 0.0119 | 0.0060 |
| | 2,500 | 0.3097 | −0.0459 | 0.0132 | 0.0068 | 0.2224 | −0.0692 | 0.0329 | 0.0171 |
| | 1,000 | 0.3860 | −0.0714 | 0.0261 | 0.0136 | 0.2677 | −0.1052 | 0.0641 | 0.0339 |
| | 500 | 0.4547 | −0.0991 | 0.0436 | 0.0231 | 0.3034 | −0.1425 | 0.1054 | 0.0570 |
| | 250 | 0.5336 | −0.1365 | 0.0726 | 0.0394 | 0.3377 | −0.1896 | 0.1717 | 0.0957 |
| Parabolic Soil Profile | | | | | | | | | |
| 0.25 | 10,000 | 0.1800 | −0.0144 | 0.0019 | 0.0008 | 0.1450 | −0.0252 | 0.0060 | 0.0028 |
| | 2,500 | 0.2452 | −0.0267 | 0.0047 | 0.0020 | 0.2025 | −0.0484 | 0.0159 | 0.0076 |
| | 1,000 | 0.3000 | −0.0400 | 0.0086 | 0.0037 | 0.2499 | −0.0737 | 0.0303 | 0.0147 |
| | 500 | 0.3489 | −0.0543 | 0.0136 | 0.0059 | 0.2910 | −0.1008 | 0.0491 | 0.0241 |
| | 250 | 0.4049 | −0.0734 | 0.0215 | 0.0094 | 0.3361 | −0.1370 | 0.0793 | 0.0398 |
| 0.40 | 10,000 | 0.1857 | −0.0153 | 0.0020 | 0.0009 | 0.1508 | −0.0271 | 0.0067 | 0.0031 |
| | 2,500 | 0.2529 | −0.0284 | 0.0051 | 0.0022 | 0.2101 | −0.0519 | 0.0177 | 0.0084 |
| | 1,000 | 0.3094 | −0.0426 | 0.0094 | 0.0041 | 0.2589 | −0.0790 | 0.0336 | 0.0163 |
| | 500 | 0.3596 | −0.0577 | 0.0149 | 0.0065 | 0.3009 | −0.1079 | 0.0544 | 0.0269 |
| | 250 | 0.4170 | −0.0780 | 0.0236 | 0.0103 | 0.3468 | −0.1461 | 0.0880 | 0.0443 |

*Source*: Novak and El-Sharnouby (1983).
$f^p_{x_1}$ and $f^p_{x_2}$ are parameters for pinned head.
*Fixed-translating head.

Translation damping constant,

$$c_x^1 = \frac{E_p I_p}{r_0^2 V_s}(f_{x_2}) \tag{7.52}$$

Rotation stiffness constant,

$$k_\phi^1 = \frac{E_p I_p}{r_0}(f_{\phi_1}) \tag{7.53}$$

Rotation damping constant,

$$c_\phi^1 = \frac{E_p I_p}{V_s}(f_{\phi_2}) \tag{7.54}$$

Cross-stiffness constant,

$$k_{x\phi}^1 = \frac{E_p I_p}{r_0^2} f_{(x\phi_1)} \tag{7.55}$$

Cross-damping constant,

$$c_{x\phi}^1 = \frac{E_p I_p}{r_0 V_s} f_{(x\phi_2)} \tag{7.56}$$

in which

$I_p$ = moment of inertia of pile cross-section
$E_p$ = Young's modulus of pile
$V_s$ = shear wave velocity in soil
$V_c$ = longitudinal wave velocity in pile
$r_0$ = pile diameter
$f$ = constants in Table 7.5

It was found, as in case of vertical vibrations, that the frequency dependence of stiffness and damping can generally be ignored, and that the important parameters are the ratio of Young's modulus of the pile and shear modulus of the soil and the slenderness ratio $L/r_0$.

Also in Table 7.5, coefficients have been included for both pin-headed and fixed-translating headed piles. For a pin-headed pile, $f_{x_1}^p$ gives translation stiffness and $f_{\phi_1} = 0$ (i.e., $k_{\phi_1}^1 = 0$). The stiffness and damping of pin-headed piles are much less than for fixed (translating) head piles.

The soils very near the surface control the load deformation properties of the pile. In addition, a gap may be formed behind a pile under lateral vibrations. Therefore, the value of $G$ or $V_s$ to be used for such a case is smaller than the value used for vertical analysis. This holds both for static as well as dynamic analysis.

The effect of vertical static load may be significant only with extremely soft or loose soils. Most stiffness and damping parameters were reduced by the presence of axial load, but the damping caused by rotation is increased.

### 7.4.3 Torsional Vibrations

Novak and Howell (1977) developed an analysis for torsional vibrations of piles. The main assumptions in this analysis are:

1. The pile has a circular cross section, and is vertical and elastic. It is perfectly connected to the soil.
2. The pile is end bearing.
3. The soil is modeled as a linear viscoelastic medium with frequency independent material damping of the hysteretic type.
4. The soil reaction per unit length of the pile is assumed to be equal to that derived for plane strain conditions (i.e., for uniform rotation of an infinitely long pile).
5. The excitation is harmonic and the motion of the pile is small.

In Figure 7.25 the vertical pile undergoes a complex harmonic rotation (equation (7.57)) about its vertical axis.

$$\psi(z, t) = \psi(z)e^{i\omega t} \tag{7.57}$$

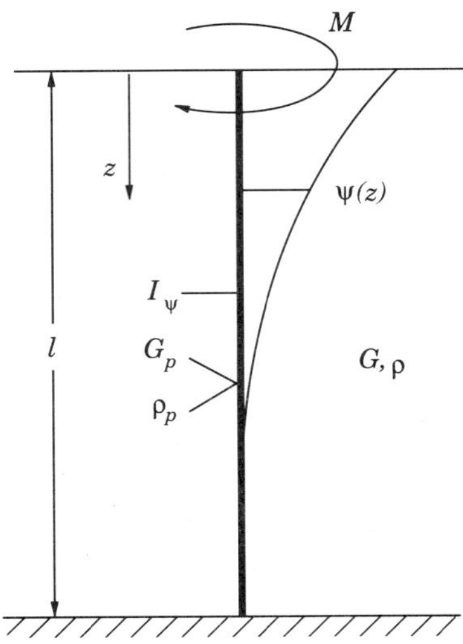


**Figure 7.25** Torsion of vertical pile and rotation.

in which

$\psi(z)$ = complex amplitude of the pile rotation at depth $z$
$i = \sqrt{-1}$
$\omega$ = circular frequency of excitation
$t$ = time.

The resistance to motion of the pile is provided by torsional soil reaction acting on pile element $dz$ and may be written as (Novak and Sach, 1973):

$$Gr_0^2(S_{\psi 1} + iS_{\psi 2})(\psi(z,t))dz \tag{7.58}$$

where the stiffness parameter

$$S_{\psi 1}(a_0) = 2\pi\left(2 - a_0\frac{J_0J_1 + Y_0Y_1}{J_1^2 + Y_1^2}\right) \tag{7.59}$$

and the damping parameter

$$S_{\psi 2}(a_0) = \frac{4}{J_1^2 + Y_1^2} \tag{7.60}$$

Here, dimensionless frequency $a_0 = \dfrac{r_0\omega}{V_s}$

where

$r_0$ = pile radius
$V_s = \sqrt{G/\rho}$ = shear wave velocity
$G$ = shear modulus of soil
$\rho$ = mass density of soil
$J_0(a_0), J_1(a_0)$ = Bessel functions of the first kind or order zero and one, respectively
$Y_0(a_0), Y_1(a_0)$ = Bessel functions of the second kind of order zero and one, respectively

For noncircular piles, $r_0$ is the equivalent radius of the possible slip circle around the pile.

The material damping may be included by the addition of an out of phase component to the soil shear modulus, which then becomes

$$G^* = G_1 + iG_2, \qquad G^* = G_1(1 + i\tan\delta) \tag{7.61}$$

in which

$\tan\delta = G_2/G_1$
$G_1, G_2$ = real and imaginary parts, respectively, of the complex soil shear

modulus, $G^*$
$\delta$ = loss angle.

Thus, $G^*$ replaces $G$ in equation (7.58) and enters equations (7.59 and 7.60) through $a_0$.

The hysteretic material damping significantly increases the damping, $S_{\psi 2}$, by an almost constant amount, equal to $4\pi \tan \delta$ at low frequencies, and reduces the stiffness, $S_{\psi 1}$, slightly at higher frequencies. Experiments by Novak and Howell (1977) have shown that material damping may be neglected for other vibration modes but is very important for torsion. Also, the displacement of slender piles quickly diminishes with increasing depth and varies with frequency to a lesser degree. In addition, the effect of the tip conditions is less significant for the more slender pile, in which case the tip is fixed by the soil. The degree of this fixity depends on pile slenderness and the stiffness of soil (wave velocity ratio, $V_s/V_p$).

Stiffness and damping constants $k_\psi^1$ and $c_\psi^1$ for fixed-tip single piles are given by

$$k_\psi^1 = \frac{G_p J}{r_0} f_{T,1} \tag{7.62}$$

and

$$c_\psi^1 = \frac{G_p J}{V_s} f_{T,2} \tag{7.63}$$

in which

$G_p$ = shear modulus of pile material
$J$ = polar moment of inertia of the cross section
$r_0$ = effective radius of one pile
$V_s$ = shear wave velocity of soil
$f_{T,1}, f_{T,2}$ = parameters in Figures 7.26 and 7.27 that have been plotted for dimensionless input parameters for timber piles and reinforced concrete piles, respectively.

These figures show that damping parameter $f_{T,2}$ varies with frequency much more than the stiffness parameter $f_{T,1}$.

The marked effect of material damping may be seen from the broken lines in Figures 7.26 and 7.27, which were calculated with $\tan \delta = 0.1$, a representative value for soils. The material damping of the soil increases significantly the total torsional damping of the pile, particularly at low frequencies, and makes the equivalent viscous damping constant somewhat less frequency dependent than it is with $\tan \delta = 0$ (for higher frequencies). The effect of material damping on the torsional stiffness of the pile at higher frequencies is negligible.

***Stiffness and Damping Constants of Group*** The torsional stiffness and damping constants of a pile have been obtained in the above analysis as moments

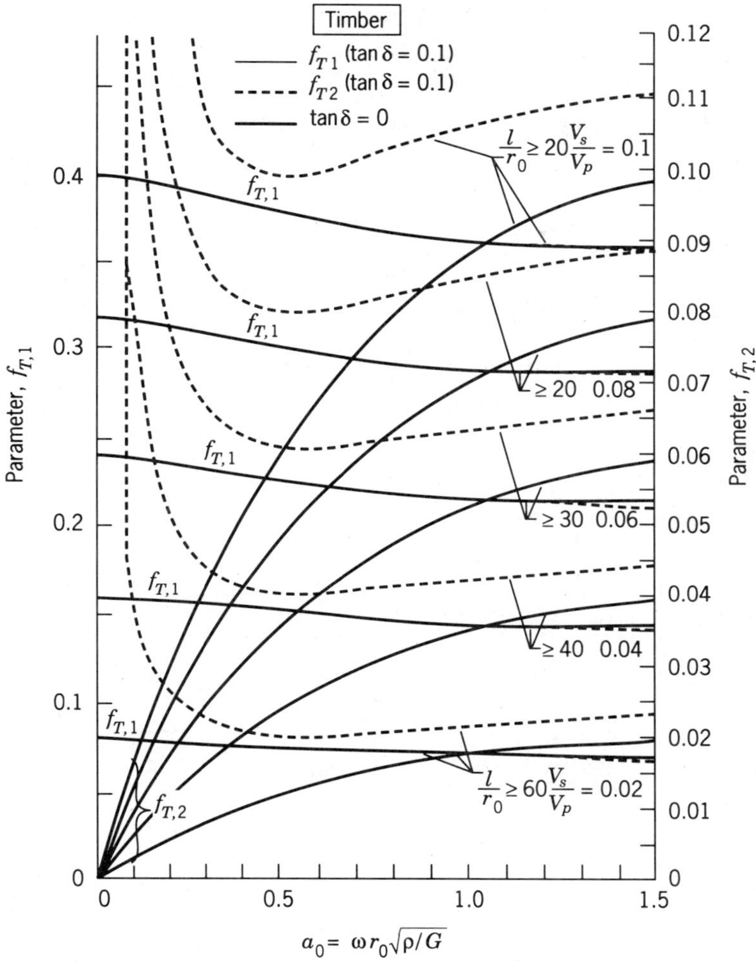


**Figure 7.26** Torsional stiffness and damping parameters of timber piles ($\rho/\rho_p = 2$) (Novak and Howell, 1977).

that correspond to unit rotational displacement and velocity. For a pile located beyond the reference point, these moments are composed of two parts: (1) that which twists the pile and (2) that which translates it. In Figure 7.28, $x_r$ and $y_r$ are distances of any pile from the C.G. of the group. Then the torsional stiffness constant of a pile group is

$$k_\psi = \sum [k_\psi^1 + k_x^1(x_r^2 + y_r^2)] \tag{7.64}$$

and the torsional damping constant is

$$c_\psi = \sum [c_\psi^1 + c_x^1(x_r^2 + y_r^2)] \tag{7.65}$$

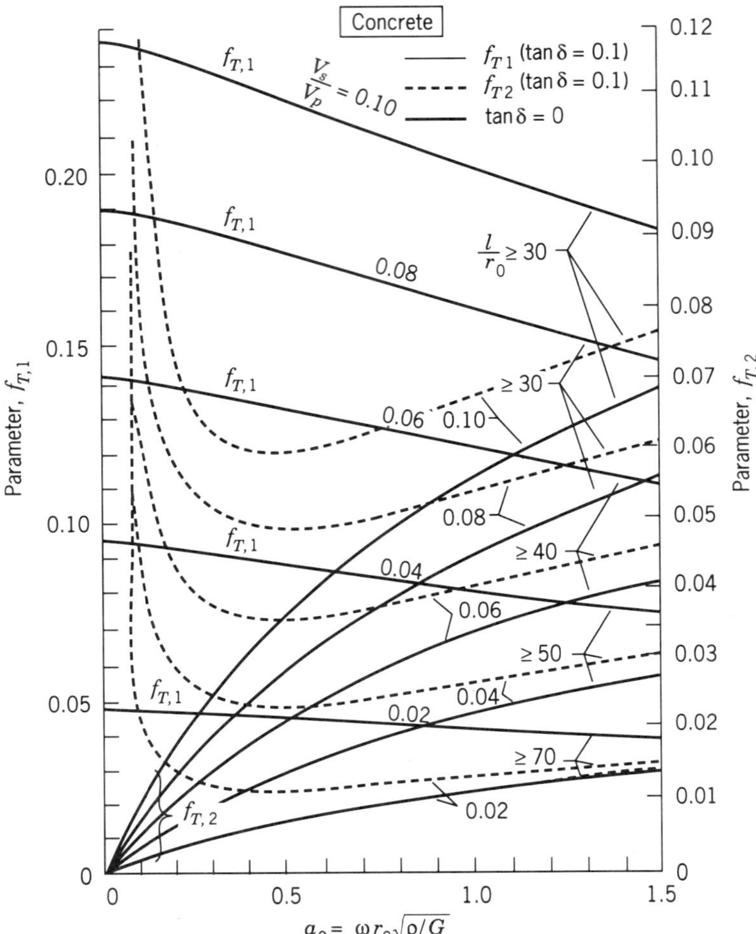


**Figure 7.27** Torsional stiffness and damping parameters of reinforced concrete piles ($\rho/\rho_p = 0.7$) (Novak and Howell, 1977).

The summation is extended over all the piles. In equations (7.64) and (7.65), $k_\psi^1$ and $c_\psi^1$ are stiffness and damping constants, respectively, of a pile subjected to torsion (equations (7.62) and (7.63), and $k_x^1$ and $c_x^1$ are stiffness and damping constants, respectively, of a pile subjected to horizontal translation (equations (7.51) and (7.52)), respectively.

Equations (7.64) and (7.65) show clearly that the contribution of the translation components increases with the square of the distance from the reference point, $R = \sqrt{x_r^2 + y_r^2}$. Therefore, in practice, the contribution to torsion of each pile depends on the ratio of the torsional stiffness to the stiffness caused by horizontal translation.

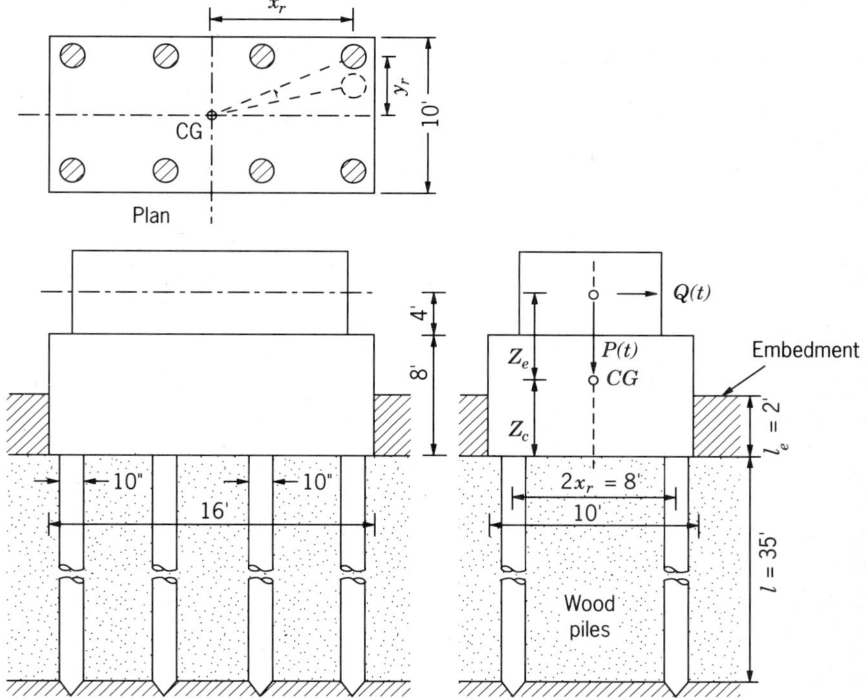


**Figure 7.28** Dimensions of pile foundation.

It has been shown by Novak and Howell (1977) that the contribution of the pile torsion decreases quickly with the ratio $R/r_0$. Therefore, the torsion of the piles may be more important for footings supported by a small number of large piles than for footings supported by a large number of slender piles spread far from the reference point. The maximum effect of twisting will become important if the foundation is a caisson, which may behave as one large diameter pile. If the centroid of the footing coincides with the elastic center of the piles in plan, the excitation moment $M_0 \cos \omega t$ produces pure torsional response of the footing $\psi_0$, given by

$$\psi_0 = \frac{M_0}{\sqrt{(k_\psi - I_p\omega^2)^2 + (c_\psi\omega^2)^2}} \tag{7.66}$$

where $I_p$ = polar mass moment of inertia of the pile group including the pile cap about the vertical axis.

## 7.5 GROUP ACTION UNDER DYNAMIC LOADING

Piles are generally used in groups. The stiffness and damping of pile groups need be evaluated from considerations of group action. It is not correct to assume that group stiffness and damping are the simple sum of the stiffness and damping of individual piles. The extent of group action depends on the ratio of spacing to diameter of piles. The smaller the spacing, the larger the group action and vice versa. In Section 7.3, the group action under lateral vibrations was discussed based on results of model piles. Here, the stiffness and damping coefficients of the pile groups will be presented based on analytical solutions.

### 7.5.1 Vertical Vibrations

Novak and Grigg (1976) proposed that the deflection factors of Poulos for group action of statically loaded piles based on elastic analysis may also be applied to a pile group undergoing steady-state vibration. Therefore, stiffness of pile group $k_w^g$ may be obtained from equation (7.67):

$$k_w^g = \frac{\sum_1^n k_w^1}{\sum_1^n \alpha_A} \tag{7.67}$$

where

$n$ = number of piles
$\alpha_A$ = axial displacement interaction factor for a typical reference pile in the group relative to itself and to all other piles in the group, assuming the reference pile and all other piles carry the same load

The factor $\alpha_A$ is obtained from Figure 7.29.

The equivalent geometric damping ratio for the group is given by

$$c_w^g = \frac{\sum_1^n c_w^1}{\sum_1^n \alpha_A} \tag{7.68}$$

If the pile cap is not in contact with the ground, equations (7.67) and (7.68) can be used directly to compute the response of the pile group in vertical vibrations. Embedment of the pile cap results in increase of the stiffness and damping values of the pile group.

However, it may be assumed that, in practice, embedment is provided only in the development of side friction between the cap and soil and only when dense granular backfill is used. The soil beneath the base of the cap is likely to be of poor quality and may settle away from the cap both in cohesive and noncohesive soils. Also, cohesive backfill may shrink away from the sides and become ineffective.

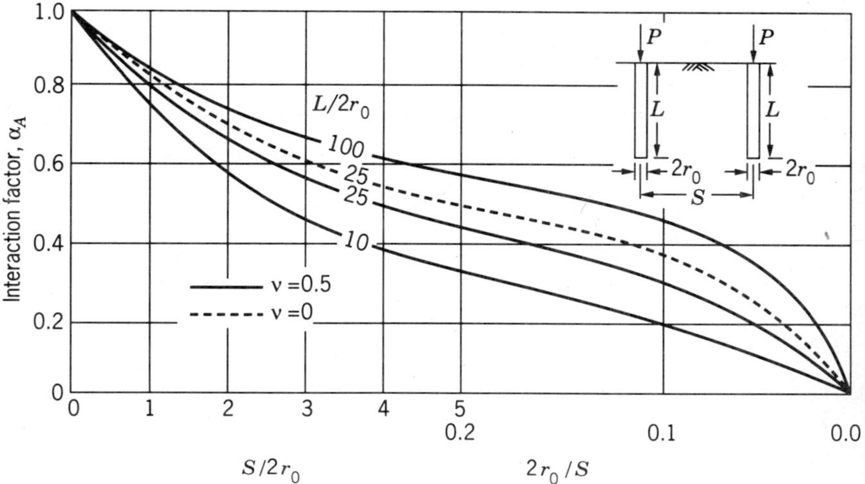


**Figure 7.29** $\alpha_A$ as a function of pile length and spacing (Poulos, 1968).

Novak and Beredugo (1972) have developed expressions for calculating stiffness and geometric damping constants for embedded footings that can be applied to pile caps. These are added to the stiffness and damping values obtained in equations (7.67) and (7.68). The sum of the two stiffness and damping values give the total system stiffness and damping for a group of piles. The stiffness ($k_w^f$) and damping ($c_w^f$) values due to side friction of the pile cap is expressed as (Prakash and Puri, 1988):

$$k_w^f = G_s h \bar{S}_1. \tag{7.69}$$

$$c_w^f = h r_0 \bar{S}_2 \sqrt{G_s \rho_s} \tag{7.70}$$

where

$h$ = depth of embedment of the cap
$r_0$ = equivalent radius of the cap

$G_s$ and $\rho_s$ are the shear modulus and total mass density of the backfill and $\bar{S}_1$ and $\bar{S}_2$ are constants and are 2.70 and 6.70, respectively.

Novak (1974) computed vertical response of a machine and its foundations, Figure 7.28. The foundation consisted of a rectangular block of concrete 16 ft long × 10 ft wide × 8 ft high (4.8 m × 3 m × 2.4 m high). It was considered both embedded 2 ft into the soil and having no embedment. It was supported on 35-ft-long, fixed-top timber piles in a medium stiff clay. The machine weight was 10 tons. The response of the pile foundation with varying frequency is shown in Figure 7.30 for four cases:

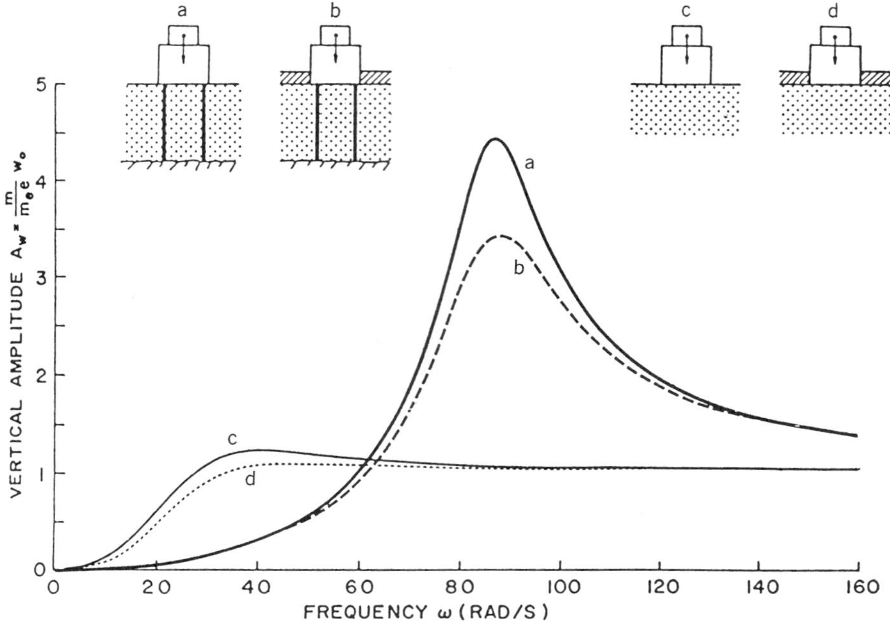


**Figure 7.30** Vertical response of (a) Pile foundation, (b) embedded pile foundation, (c) shallow foundation, and (d) embedded shallow foundation ($B = mr_{0z}^3 = 5.81$) (Novak, 1974. Reproduced by permission of National Research Countil of Canada).

1. The foundation is supported on pile with no pile cap embedment.
2. The pile cap is embedded in a soil layer.
3. The foundation is supported on elastic half space.
4. The foundation is embedded in a soil layer.

It can be seen from this figure that, in pile-supported structures,

1. The piles increase the resonant frequency, and in this case, increased displacement amplitude at resonance (curves *a* and *b*).
2. Damping can be increased by embedding the pile cap in a soil layer (curve *b*).
3. Damping in a pile-supported foundation is very low as compared to soil supported footings.

Material damping was not considered in this particular analysis.

It may, however, be seen that if the operating frequency of the machine is less than 60 rad/sec, the amplitude of vertical vibrations is reduced by use of piles. If the operating frequency is less than 40 rad/sec, the amplitudes are reduced to less than one-third of their corresponding values without piles.

Sheta and Novak (1982) presented an approximate theory for vertical vibrations of pile groups that accounts for dynamic interactions of piles in a group, weakening of soil around the pile because of high strain, soil layering, and arbitrary tip conditions. The effect of pile interaction on damping and stiffness of pile groups, distribution of internal forces in the piles, and response of pile-supported foundations to harmonic excitation have been studied. It was further found that (1) dynamic group effects differ considerably from static group effects and (2) dynamic stiffness and damping of pile groups are much more frequency dependent than those of single piles.

### 7.5.2 Lateral Vibrations

In lateral vibrations, the stiffness and damping for groups of piles is given by

$$k_x^g = \frac{\sum_1^n k_x^1}{\sum_1^n \alpha_L} \tag{7.71}$$

and

$$c_x^g = \frac{\sum_1^n c_x^1}{(\sum_1^n \alpha_L)} \tag{7.72}$$

where $\alpha_L$ = displacement factor for lateral translation and may be adopted from Figure 7.31.

Again, as for vertical vibrations, the spring costant $k_x^f$ and damping $c_x^f$ due to pile cap translation are, respectively:

$$k_x^f = G_s h \bar{S}_{x1} \tag{7.73}$$

and

$$c_x^f = h r_0 \sqrt{G_s \rho_s} . \bar{S}_{x2} \tag{7.74}$$

where

$h$ = depth of embedment
$r_0$ = equivalent radius of the cap
$G_s$ and $\rho_s$ = the shear modulus and total mass density of the backfill and
$\bar{S}_{x1}$ and $\bar{S}_{x2}$ = constants in Table 7.6

The total stiffness and total damping values are sums of equations (7.71) and (7.73) and (7.72) and (7.74), respectively, as:

$$\text{Total } k_x^g = k_x^g + k_x^f \tag{7.75}$$

**TABLE 7.6 Stiffness and Damping Constants for Half-Space and Side Layers for Sliding Vibrations**

| Poisson's Ratio $\nu$ | Validity Range | Constant Parameter |
|---|---|---|
| 0.0 | $0 < a_0 < 1.5$<br>$0 < a_0 < 1.5$ | $\bar{S}_{x1} = 3.6$<br>$\bar{S}_{x2} = 8.2$ |
| 0.25 | $0 < a_0 < 2$<br>$0 < a_0 < 1.5$ | $\bar{S}_{x1} = 4.0$<br>$\bar{S}_{x2} = 9.1$ |
| 0.4 | $0 < a_0 < 2.0$<br>$0 < a_0 < 1.5$ | $\bar{S}_{x1} = 4.1$<br>$\bar{S}_{x2} = 10.6$ |

After Beredugo and Novak (1972).

and

$$\text{Total } c_x^g = c_x^g + c_x^f \tag{7.76}$$

For rocking vibrations, the effect of pile groups and the pile cap is accounted for as for sliding and equations have been written in Section 7.6 in the section on design procedure. The use of these equations has been illustrated in the design example.

A comparison of the observed and predicted response of pile groups has been presented in Section 7.9, where it has been shown that there are several deficiencies in the analysis of pile groups at this time (1990). Therefore, it is recommended that approximate methods described in this chapter may be used in practice until better and simpler methods of analysis are developed.

## 7.6 DESIGN PROCEDURE OF PILES UNDER DYNAMIC LOADS

The design procedure essentially consists of the computation of the stiffness of the pile group considering group action and the damping of the pile group considering group action. This procedure has been developed based on the analytical formulation of stiffness and damping in different modes of vibrations in the preceding section. The response of the foundations may then be computed either by the spectral response technique described in Section 7.3 for earthquake loading or by response equations for machine foundation loads as illustrated below.

### Design Procedure for Pile-Supported Machine Foundations

Based on the analysis presented in the previous sections, a design procedure of piles under vertical vibrations, horizontal vibrations, and torsion will now be described. The following soil and pile properties and dimensions must be determined.

***Soil Properties*** Shear modulus $G_s$, Poisson's ratio $\nu_s$, and unit weight $\gamma_s$ for the soil both around the pile and below its tip.

***Pile Properties and Geometry*** Pile length, cross section, and spacing in the group, unit weight $\gamma$ of pile and pile cap, and Young's modulus of pile material.

Based on the above information: $V_s$ = shear wave velocity in soil and $V_c$ = compression wave velocity in pile are computed.

**Vertical Vibrations**

1. Compute spring stiffness and damping of single pile.

$$k_w^1 = \frac{E_p A}{r_0} f_{w_1} \tag{7.46}$$

$$c_w^1 = \frac{E_p A}{V_s} f_{w_2} \tag{7.48}$$

The values of functions $f_{w_1}$ and $f_{w_2}$ are obtained from Figures 7.23.

2. Compute spring stiffness and damping of pile group, $k_w^g$ (piles only).

$$k_w^g = \frac{\sum_1^n k_w^1}{\sum_1^n \alpha_A} \tag{7.67}$$

and

$$c_w^g = \frac{\sum_1^n k_w^1}{\sum_1^n \alpha_A} \tag{7.68}$$

in which $\alpha_A$ is taken from Figure 7.29.

3. Determine spring stiffness and damping due to side friction $k_w^f$.

$$k_w^f = G_s h \bar{S}_1 \tag{7.69}$$

$$c_w^f = h r_0 \cdot \bar{S}_2 \sqrt{G_s \rho_s} \tag{7.70}$$

Values of $\bar{S}_1$ and $\bar{S}_2$ are listed previously. (See Section 7.5.1).

4. Compute total spring stiffness and total damping.

$$\text{total} \quad k_w^g = k_w^g + k_w^f \tag{7.69a}$$

$$\text{total} \quad c_w^g = c_w^g + c_w^f \tag{7.70a}$$

**Translation**

1. Compute stiffness and damping of a single pile.

$$k_x^1 = \frac{E_p I_p}{r_0^3}(f_{x1}) \tag{7.51}$$

$$c_x^1 = \frac{E_p I_p}{r_0^2 V_s}(f_{x2}) \tag{7.52}$$

in which $f_{x1}$ and $f_{x2}$ are given in Table 7.5.

2. Compute stiffness and damping of the pile group (of piles only).

$$k_x^g = \frac{\sum_1^n k_x^1}{\sum_1^n \alpha_L} \tag{7.71}$$

$$c_x^g = \frac{\sum_1^n c_x^1}{\sum_1^n \alpha_L} \tag{7.72}$$

in which $\alpha_L$ is taken from Figure 7.31.

3. Compute stiffness and damping due to pile cap.

$$k_x^f = G_s h \bar{S}_{x1} \tag{7.73}$$

$$c_x^f = h r_0 \sqrt{G_s \rho_s}\, \bar{S}_{x2} \tag{7.74}$$

values of $\bar{S}_{x1}$ and $\bar{S}_{x2}$ are listed in Table 7.6.

4. Total stiffness and total damping are then the sum of the stiffness and damping values computed in steps 2 and 3, respectively and given by equations 7.75 and 7.76.

**Rocking**

1. Compute stiffness and damping of a single pile in both rocking alone as well as in coupled motion.

$$k_\phi^1 = \frac{E_p I_p}{r_0} \times f_{\phi 1} \tag{7.53}$$

$$c_\phi^1 = \frac{E_p I_p}{V_s} \times f_{\phi 2} \tag{7.54}$$

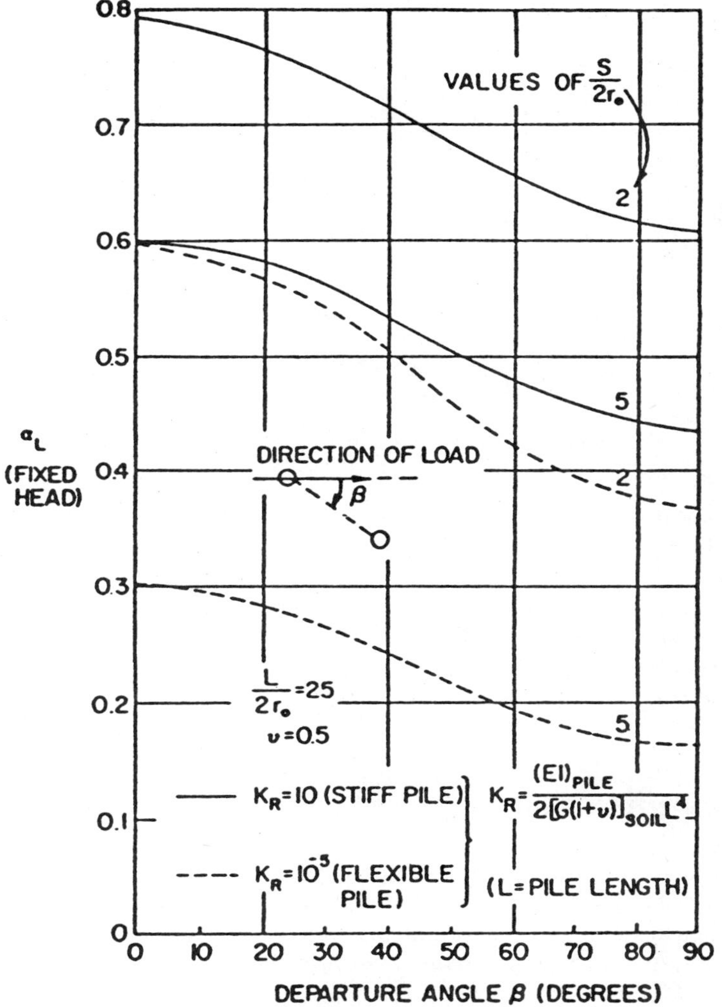


**Figure 7.31** Graphical solution for $\alpha_L$ (Poulos, 1972).

$$k^1_{x\phi} = \frac{E_p I_p}{r_0^2} \times f_{(x\phi 1)} \tag{7.55}$$

$$c^1_{x\phi} = \frac{E_p I_p}{r_0 V_s} \times f_{(x\phi 2)} \tag{7.56}$$

Values of $f$ parameters are listed in Table 7.5.

2. Compute stiffness and damping of the pile group (Novak, 1974).

$$k^g_\phi = \sum_1^n (k^1_\phi + k^1_w x_r^2 + k^1_x z_c^2 - 2z_c k^1_{x\phi}) \tag{7.77}$$

$$c_\phi^g = \Sigma_1^n (c_\phi^1 + c_w^1 x_r^2 + c_x^1 z_c^2 - 2z_c c_{x\phi}^1) \tag{7.78}$$

in which

$x_r$ = distance of each pile from the C.G. (Figure 7.28)
$z_c$ = height of center of gravity of the pile cap above its base (Figure 7.28).

$$\delta = \frac{h}{r_0}$$

3. Compute stiffness and damping of the pile cap.

$$k_\phi^f = G_s r_0^2 h \bar{S}_{\phi 1} + G_s r_0^2 h[(\delta^2/3) + (z_c/r_0)^2 - \delta(z_c/r_0)]\bar{S}_{x1} \tag{7.79}$$

$$c_\phi^f = \delta r_0^4 \sqrt{G_s \gamma_s / g} \left\{ \bar{S}_{\phi 2} + \left[ \frac{\delta^2}{3} + \left(\frac{z_c}{r_0}\right)^2 - \delta\left(\frac{z_c}{r_0}\right)\right] \bar{S}_{x2} \right\} \tag{7.80}$$

4. Total stiffness and total damping are then the sum of stiffness and damping values computed in steps 2 and 3.

Once the stiffness and damping of the system are computed, its response can be determined from principles of elementary mechanical vibrations (Prakash and Puri, 1988). See also Section 7.8 for detailed procedure. Analysis of pile groups in torsional vibrations was presented in Section 7.4.3.

## 7.7 CENTRIFUGE MODEL TESTS ON PILES

In order to check the various methods of analysis of piles under dynamic conditions, it is desirable to carry out field dynamic testing of full-scale piles. Only a few such tests have been conducted. Novak and Grigg (1976) carried out vibration tests on large model (or small prototype) piles. Prakash and Sharma (1969) and Woods (1984) report lateral dynamic tests on full size piles. Hassini and Woods (1989) report model tests under lateral vibrations on single pile and pile groups in sands.

In many fields of engineering, scaled models of large structures are used to study physical phenomena. Scaled models of geotechnical structures under earth's gravity, however, do not satisfy similitude conditions because the stress levels in the model do not match those in the prototype. By placing the model in an appropriately increased gravitational field, the model material is made heavier, and prototype stress levels in the model are achieved. Such a gravitational field is created by spinning the centrifuge arm at an appropriate angular speed such that the centrifugal acceleration at the location of the model on the arm is $ng$ where $g$ is the acceleration due to gravity and $n$ is the model scale. The scaling relationships used in centrifugal modeling studies are summarized in Table 7.7.

**TABLE 7.7 Scaling Relations Between Prototype and Centrifuge Model**

| Quantity | Full-Scale (prototype) | Model at *ng*'s |
|---|---|---|
| Linear dimension, displacement | 1 | $1/n$ |
| Area | 1 | $1/n^2$ |
| Volume | 1 | $1/n^3$ |
| Stress | 1 | 1 |
| Strain | 1 | 1 |
| Force | 1 | $1/n^2$ |
| Acceleration | 1 | $n$ |
| Velocity | 1 | 1 |
| Time | | |
| In dynamic terms | 1 | $1/n$ |
| in diffusion cases | 1 | $1/n^2$ |
| Frequency in dynamic problems | 1 | $n$ |

After Scott (1979).

Since each model is of finite size, different parts of the model are at different radii from the rotational axis of the centrifuge. Therefore, different parts of the model will be subjected to different gravitational intensities. The greater is the radial distance of the model compared with the dimension of the model in the direction of the centrifuge arm, the more uniform the acceleration field across the model will result.

### 7.7.1 Studies of a Model and a Prototype

Centrifuge studies on models of geotechnical structures and pile foundations under dynamic conditions have been used more recently. A few such important studies will now be described. Scott et al. (1982) performed both prototype tests on piles and model tests in centrifuge, and compared the results.

#### *Prototype Tests*

*Pile, Soil, and Tests* Two piles, of 24 in. (0.60 m) outside diameter and 0.5 in. (13 mm) thick wall, were driven to a depth of 32 ft (9.6 m). One of these piles was instrumented with strain gauges to indicate the bending moment in the pile. A steel platform was welded to one of the piles and loaded with 24 tons of lead weights, approximately equaling the working axial load on the pile. Two shaking machines were mounted on the platform. By adding weights to the rotating baskets of these machines, and by changing the frequency, the acting dynamic force on the pile was varied from a few hundred to a few thousand pounds (1 to 10 KN). At the higher level, this force was a substantial fraction of the lateral dynamic force that the pile might encounter in an earthquake as part of a structure.

The load and displacement of pile and bending moment in the piles were monitored. The platform displacement and acceleration were recorded during the dynamic tests, and the ground movements were observed adjacent to the piles through the use of an array of seismometers. Also, pore pressure gauges were installed in the soil at varying distances from the pile.

The soil at site consists of 18 to 20 ft (5.4 to 6 m) of medium-dense uniform silty sand overlying strata of silty clay, silty sand, and silt-stone. The upper layer of sand is of most interest for the deflections of piles since effects of the pile on the soil below about 10 pile diameters (20 ft or 6.0 m) are generally negligible for 2 ft (0.6 m) diameter piles. The silty sand exhibited a cone penetration resistance in the range 30 to 60 tsf (2.9 to 5.8 $MN/m^2$) up to a depth of 20 ft (6.0 m) below the ground surface. The in-place dry density was about 105 pcf (1.68 $T/m^3$). At 20 ft depth, there was a thin layer of clayey silt 6 in. to 1 foot thick (0.15 to 0.3 m) underlain by a much denser layer of sand with a cone penetration between 100 and 200 tsf (9 to 18 $MN/m^2$). The bottom of the pile penetrated this sand layer. The standard penetration test $N$ value in the upper layer of silty sand was approximately 15. The dense sand layer between 25 and 30 ft (7.5 to 9.0 m) depth had an $N$ value of 25 to 30.

In dynamic tests, the frequency of excitation was gradually increased to beyond the first natural frequency in horizontal vibrations. The rotation speed of the vibrator was again gradually reduced to zero. Typical peak amplitudes of displacement monitored at first mode resonance at 1 ft (0.3 m) above ground surface ranged from 0.025 in. (0.6 mm) in a test at 364 lb force (1.62 kN), to greater than 0.43 in. (10.9 mm) at 1762 lb peak force (7.84 kN). Typical damping factor ($\xi$) ranged from 4 to 6 percent. Computed peak pile accelerations at first mode resonance at 1 ft (0.3 m) above ground surface varied from 0.02 to 0.17 g while peak moments in the pile at resonance reached maximum values of $5.8 \times 10^5$ lb-in. (67 kN-m). Scott et al. (1982) reported only one set of test data at a medium force level that is discussed in this chapter for comparison with the model tests in the centrifuge.

Figure 7.32 shows the response curve for displacement near ground surface during the medium force level. In Figure 7.33 the bending moment and displacement in the pile are plotted for test condition of Figure 7.32. It will be seen from this figure that the maximum moment induced in the pile at resonance in this test occurred at about 15 ft (4.5 m) from the top of the pile, or 7 ft (2.1 m) below ground surface. This indicates that the pile was fairly flexible relative to the soil system.

### *Centrifuge Model Tests*

*Testing Arrangement* The model pile consisted of a stainless steel tube 1/2 in. (13 mm) in diameter with a 0.010-in. (0.25 mm) thick wall and was tested at centrifuge accelerations of about 50 g, which corresponded almost exactly to the correct scaling of the prototype pile (1:50).

The soil in the centrifuge tests was obtained from the site of the prototype tests.

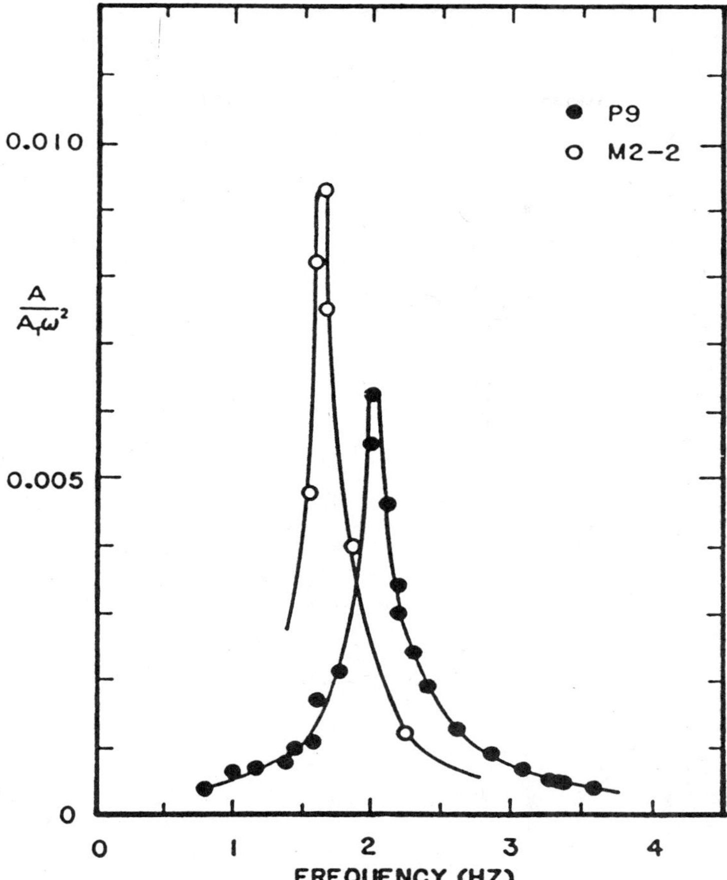


**Figure 7.32** Displacement response versus frequency for prototype and model piles (Scott et al., 1982).

The dry density of soil in these tests was 103 pcf (1.65 T/m$^3$). Two sets of tests were carried out on the centrifuge: one with the silty sand from the test site prepared dry, the other with the soil saturated. The model pile was instrumented with 10 pairs of strain gauges mounted on the inside of the tube at opposite ends of a diameter.

The model was placed by pushing it into the soil at 1 g as usual. This may raise a question regarding the similitude of the stress fields around the model and prototype piles. However, in the case of dynamic tests, the frequency and duration of the dynamic shaking tends to eliminate the detailed structure of stress in the vicinity of the pile over the depth range that contributes most of the soil's resistance to pile deflections. This is in contrast to the differences that exist in

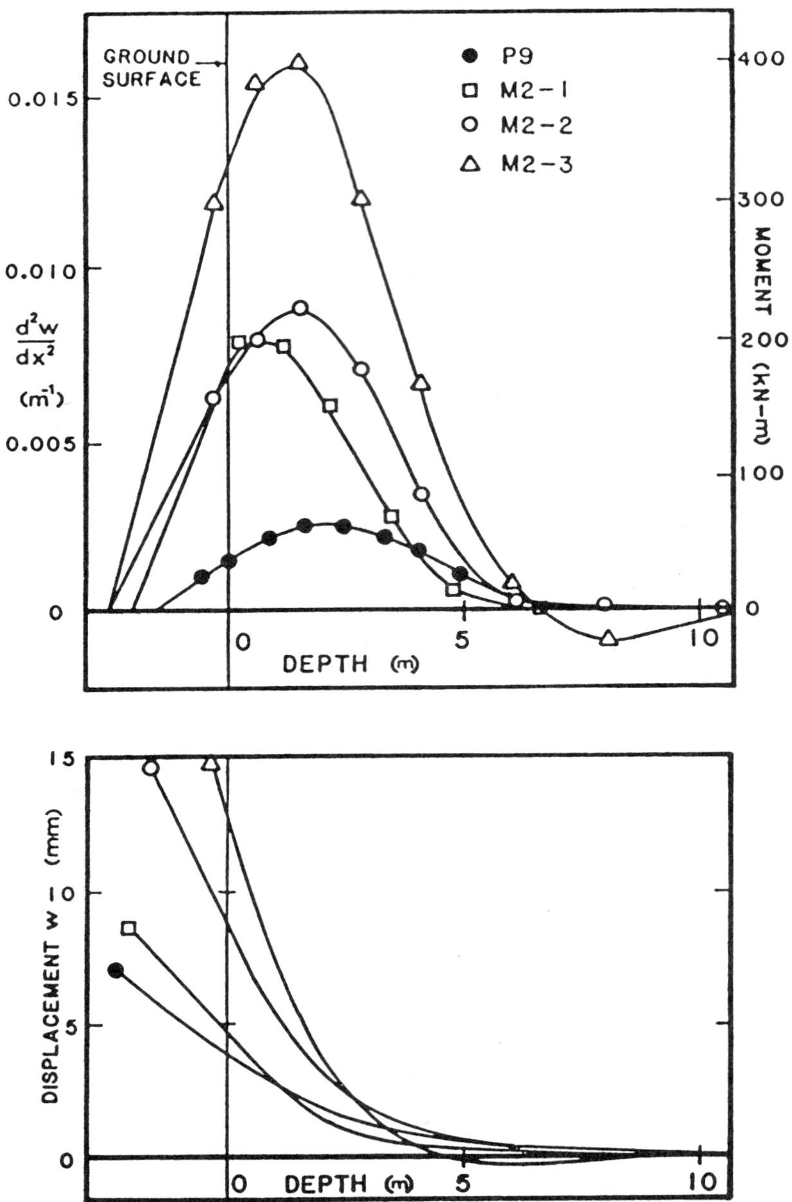


Figure 733 Moment and displacement versus depth for prototype and model piles (Scott et al., 1982).

static pile tests and may be important. The pile was excited by a miniature compressed air-driven shaking device. The speed of rotation could be varied up to 500 Hz (30,000 rpm).

A model test was conducted by running the shaking machine through a range of frequencies while the pile motions were observed on one of the strain gauge pair outputs. When the first mode frequency had been determined in this way, a series of tests was run at frequencies around the resonant frequency.

The signals were recorded by converting the analog to digital output. For each individual frequency sampling it was only necessary to record about 0.2 sec of test data. Depending on the frequency, this would consist of 10 to 20 cycles of the model pile.

The centrifuge test most nearly similar to prototype test P9 in terms of test conditions was M2-2 (Table 7.8). All the results on the model have been reported in terms of prototype dimensions. The modal frequency, $f_{n1}$, and peak amplitude of displacement at ground surface were fairly close for model and prototype tests (Table 7.8 and Figure 7.32). The test differences might be due to the different heights of the line of action of the dynamic force above ground surface, which was 8.8 ft for the model and 5 ft for the prototype. But for this difference, the model and prototype responses may be considered quite close (Scott et al., 1982). In addition, the damping ratios of Table 7.8, expressed as a percentage of the critically damped value, are very similar in model and prototype and are quite low compared to what might be expected for a system involving a yielding soil.

These centrifuge model tests on piles may be regarded as a good starting point.

**TABLE 7.8 Comparison of Prototype and Model Performance**

| Test number | M2-2 | P9 |
|---|---|---|
| Soil | Wet | Wet |
| Shaking level | Medium | Medium |
| First mode frequency $F_{n1}$, Hz | 1.65 | 2.01 |
| Peak displacement, $A_1$, in. | 0.34 | 0.24 |
| Damping, % critical | 4.2 | 4 |
| Shaker force, lb at 1 Hz | 369 | 367 |
| Maximum moment, $10^5$ lb-in. | 20.1 | 4.8 |
| Maximum moment depth, ft | 5 | 7 |
| Height of force above ground, ft | 8.8 | 5 |
| Uniform Winkler $k$, psi | 940 | 645 |

After Scott et al. (1982).

All data are given in prototype dimensions. M—model; P—prototype

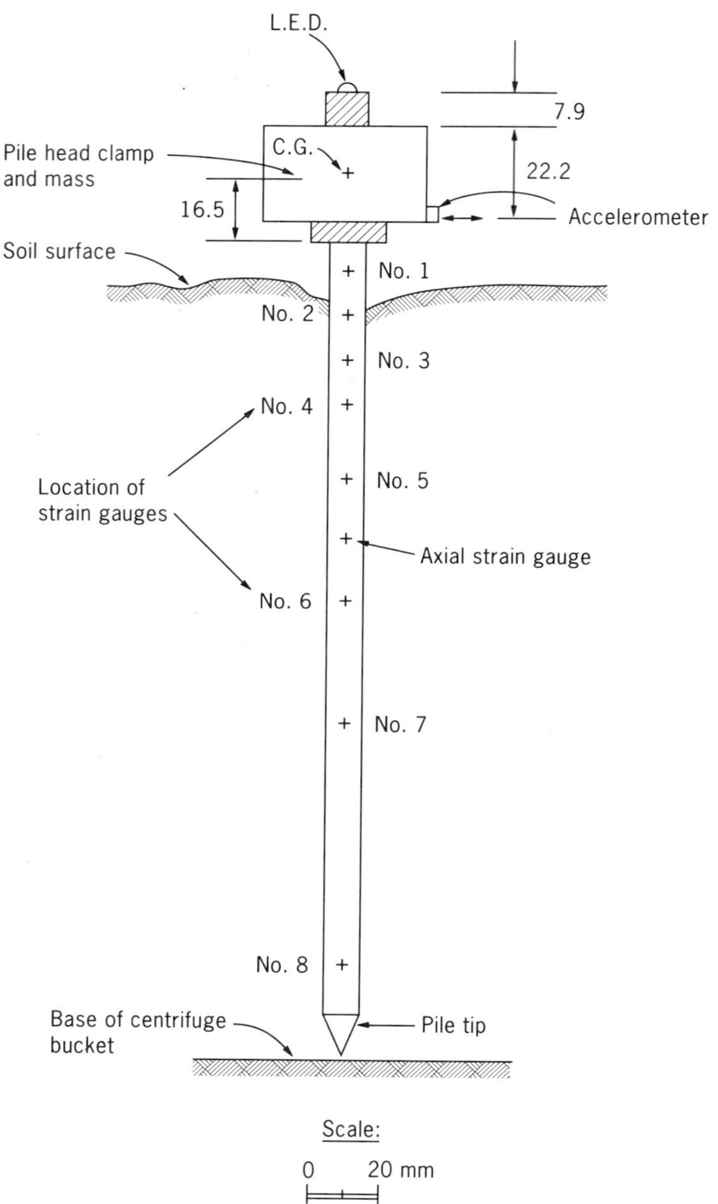


**Figure 7.34** Single pile showing instrumentation layout (Finn and Gohl, 1987).

All the answers to the response of piles under dynamic condition have not been obtained. However, these tests showed that the centrifuge model testing holds promise in understanding the dynamic pile behavior. A more comprehensive series of centrifuge model tests have been reported by Finn and Gohl (1987) which will now be described.

### 7.7.2 Studies of Model Piles and Pile Groups

The tests by Finn and Gohl (1987) represent very carefully conducted tests on piles in the centrifuge and provide a data base against which currently available analytical models used to predict the lateral response of piles to earthquake loading could be checked. Several tests on single piles and pile groups were performed but data on a single-pile and two-pile groups embedded in dry sand under lateral loading were presented. The single pile was subjected to both sinusoidal and random earthquake motion while the pile groups were subjected to sinusoidal wave motion only.

The excitation levels for the pile groups were kept low enough to ensure approximately linear elastic response so that the accuracy of elastic solutions could be checked. The distribution of shear moduli in the foundation layer were measured while the centrifuge was in flight using piezoceramic bender elements to measure the distribution of shear wave velocities from which the shear moduli were computed. This has been achieved in centrifuge tests for the first time and should make predictions and checking of data against analytical methods more reasonable and accurate (Finn and Gohl, 1987).

In these tests, the acceleration varied from 55 g at the surface of the model to 68 g at the base. An average centrifuge scale factor, $n$ equal to 60 was used in converting model test quantities to prototype scale.

The model pile in the single tests was stainless steel tubing 0.375 in. (9.52 mm) outside diameter having a 0.010 in. (0.25 mm) wall thickness (Figure 7.34). Eight pairs of foil type strain gauges were mounted on the outside of the pile to measure bending strains. The $EI$ of the instrumented pile was determined to be 13.98 N-m$^2$.

A mass was screwed to a clamp attached to the head of the pile to simulate the influence of superstructure inertia forces acting on the pile during excitation Table 7.9. The center of gravity of the pile head mass was calculated to be 16.5 mm above the base of the pile head clamp. The pile head mass was instrumented using a noncontact photovoltaic displacement transducer. The locations of the accelerometer and light-emitting diode (LED) used by the displacement sensor are shown in Figure 7.34. Pile head displacements were measured with respect to the moving base of the soil container. Pile tests were carried out in both "loose" and "dense" sands at void ratios of 0.83 and 0.57, respectively. Instrumented piles were pushed into the soil by hand in loose sand. In dense sands, a low level vibration of the sand foundation was used to assist penetration.

Tests on two-pile groups were conducted at various spacings to evaluate

**TABLE 7.9 Summary of Model Pile and Pile Head Masses—Single Pile**

| Item | Dimensions (mm) | Weight (N) |
|---|---|---|
| Pile head mass | Diameter = 43.7<br>Height = 23.1 | 2.356 |
| Pile head insert | Height = 9.5<br>Diameter = 9.3 | 0.016 |
| Pile head clamp | Area = 19.0 × 19.0<br>Height = 5.08 | 0.044 |
| Conical pile tip | Diameter = 9.6 (nominal)<br>Height = 10.9 | 0.014 |
| Weight of steel tube, including strain gauges, glue, and lead wires from base of pile head mass to tip of pile | Length = 209.5<br>Diameter = 9.52 | 0.114 |
| Weight of steel tube | Length = 209.5<br>Diameter = 9.52 | 0.109 |

After Finn and Gohl (1987).

interaction effects (Figure 7.35). Both piles were instrumented to measure bending strains. In addition, one pile was instrumented to record axial strains caused by rocking of the pile foundation during shaking. The piles in the group were rigidly attached to a pile cap and an additional mass was bolted to the pile cap to simulate the effects of a superstructure as in the case of a single pile (Table 7.10). The center of gravity of the pile cap assembly was 17.0 mm above the base of the pile cap. The pile cap mass assembly was instrumented with an accelerometer and displacement LED (Figure 7.35) as for the single pile.

After model pile installation, four lightweight settlement plates were placed at a minimum of eight pile diameters from the center of any pile to measure surface settlement. The settlements result from two causes: (1) settlement due to the increase in self-weight of the soil during spin-up of the centrifuge and (2) settlement due to the cyclic shear strains generated by the base motion.

The two types of settlements results were monitored independently. The void ratio of the foundation layer was decreased to 0.78, due to increase in self-weight in loose sands, a reduction of 0.05 from the void ratio in the 1-g environment. The corresponding void ratio changes in dense sand were negligible.

The soil shear modulus was measured at several locations with piezoceramic bender elements Figure 7.36.

***Single-Pile Response*** All data are presented at prototype scale. In test 12, the pile was subjected to a moderate level of shaking (peak base acceleration 0.15 g),

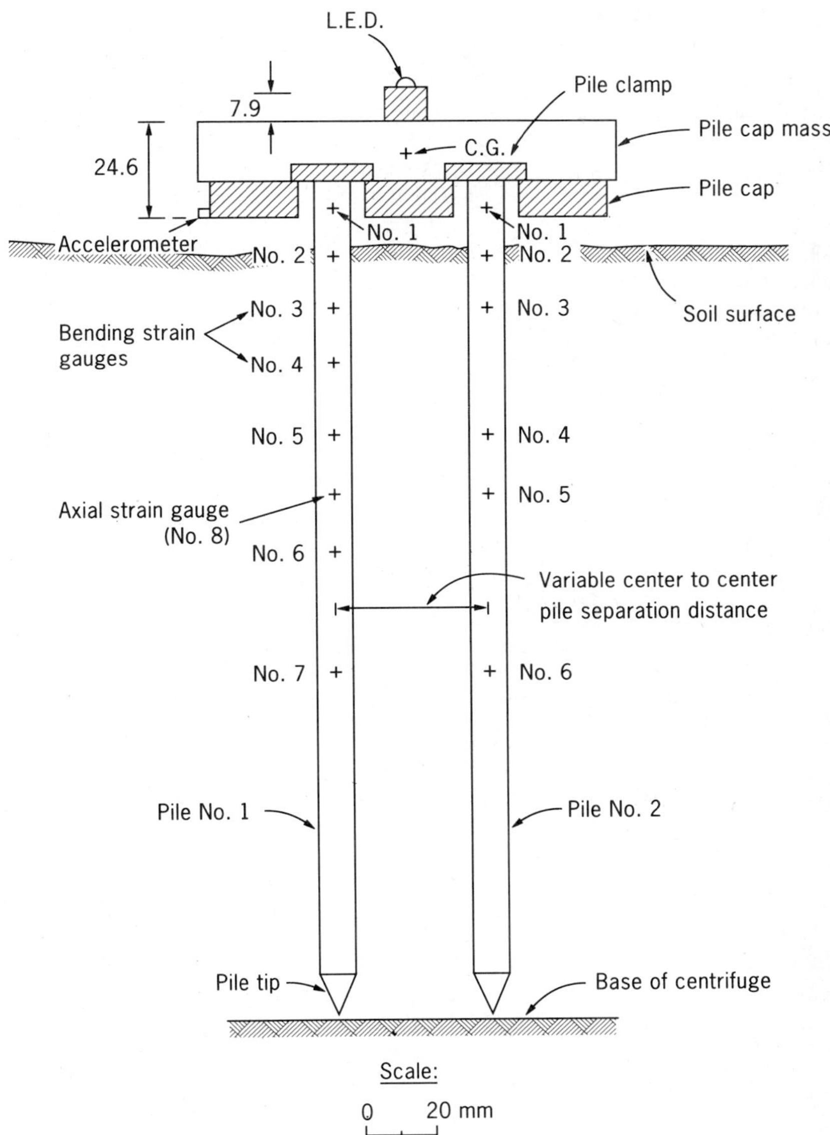


**Figure 7.35** Pile group showing instrumentation layout (Finn and Gohl, 1987).

**TABLE 7.10 Summary of Pile and Pile Cap Dimensions and Weights—Pile Group Tests**

| Item | Dimensions (mm) | Weight (N) |
|---|---|---|
| Pile cap mass | 108.0 × 47.8 × 14.9 (L × W × H) | 4.123 |
| Pile cap | 101.6 × 37.8 × 9.7 (L × W × H) | 0.728 |
| Pile head inserts (2) | Height = 9.5<br>Diameter = 9.3 | 0.016 ea. |
| Pile head clamps (2) | Area = 19.0 × 19.0<br>Height = 5.08 | 0.044 ea. |
| Piles Nos. 1 and 2, including strain gauges, glue and lead wires from base of pile head mass to tip of pile | Length = 209.5<br>Diameter = 9.52 | 0.114 ea. |

After Finn and Gohl (1987).

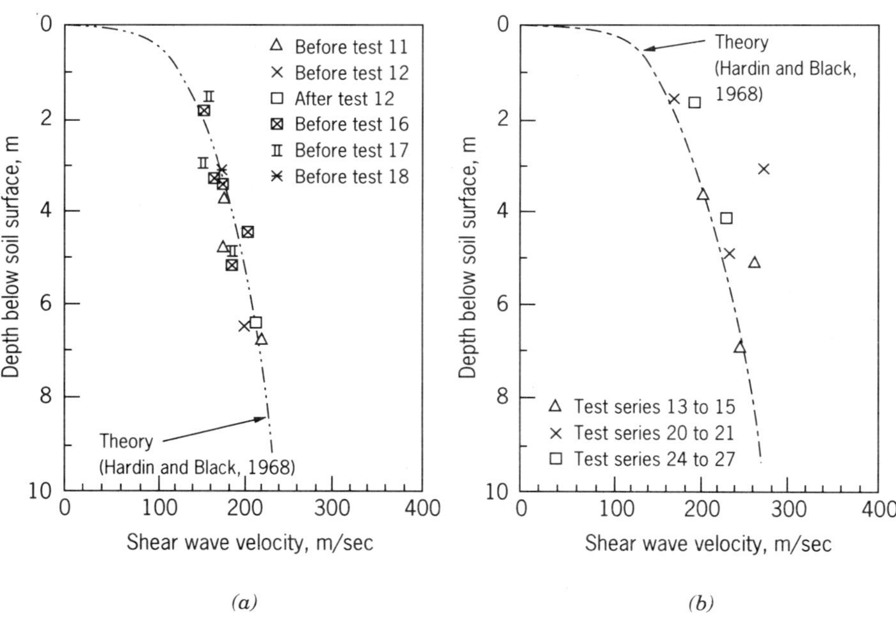


**Figure 7.36** Shear wave velocities during centrifuge flight (a) Loose sand, (b) dense sand (Finn and Gohl, 1987).

**TABLE 7.11 Single-Pile Test Characteristics**

| Test | Soil Type | $Z_{cg}$[a] | Base Motion Type | Peak Base Accel. (g) | Peak Pile Head Accel. |
|---|---|---|---|---|---|
| 12 | Loose (avg. Void ratio after consolidation = 0.78) | 1.95 | Earthquake (30-sec duration) | 0.15 | 0.18 |
| 41 | Very dense (avg. void ratio = 0.57) | 1.89 | Sine wave (20 cycles) | 0.04[b] | 0.041 |

After Finn and Gohl (1987).
[a] Distance of center of gravity of pile head mass above ground surface.
[b] Averaged over steady state portion of base input motion.

while in test 41, twenty cycles of a sine wave base motion with a peak steady state acceleration of 0.04 g was applied (Table 7.11).

The acceleration input at the base of the model and accelerations recorded in the free field at the surface of the soil layer and at the pile head are shown in Figure 7.37a, b, and c. Pile head displacements are shown in Figure 7.38a and b. The time histories of pile bending moment at various points along the pile are shown in Figure 7.39a, b, and c for strain gauge stations, 1, 4, and 7 (see Figure 7.34). The bending moment distribution along the pile at a time when maximum pile head deflection occurs ($t = 12.0$ sec) is shown in Figure 7.40.

From the data in Figures 7.38 to 7.40, the following observations may be made:

1. In Figure 7.37, the maximum input base acceleration was 0.15 g. The peak free-field acceleration was 0.26 g, and the peak pile head acceleration was 0.18 g. Thus, both the pile head and free-field peak accelerations were magnified relative to the input base acceleration.
2. The predominant period of the pile head response was longer than that of the free-field ground surface response. Therefore, strong interaction takes place between them.
3. A comparison of Figures 7.38a and 7.39a, b, and c shows that pile displacements at the top of the pile head mass in the direction of shaking (X direction) and bending moments along the pile have the same general frequency content as the pile head accelerations.
4. In Figure 7.40, the bending moments increase to a maximum near strain gauge 4 and then decrease to approximately zero at greater depths. This variation is typical of a long pile in the sense that the lower parts of the pile

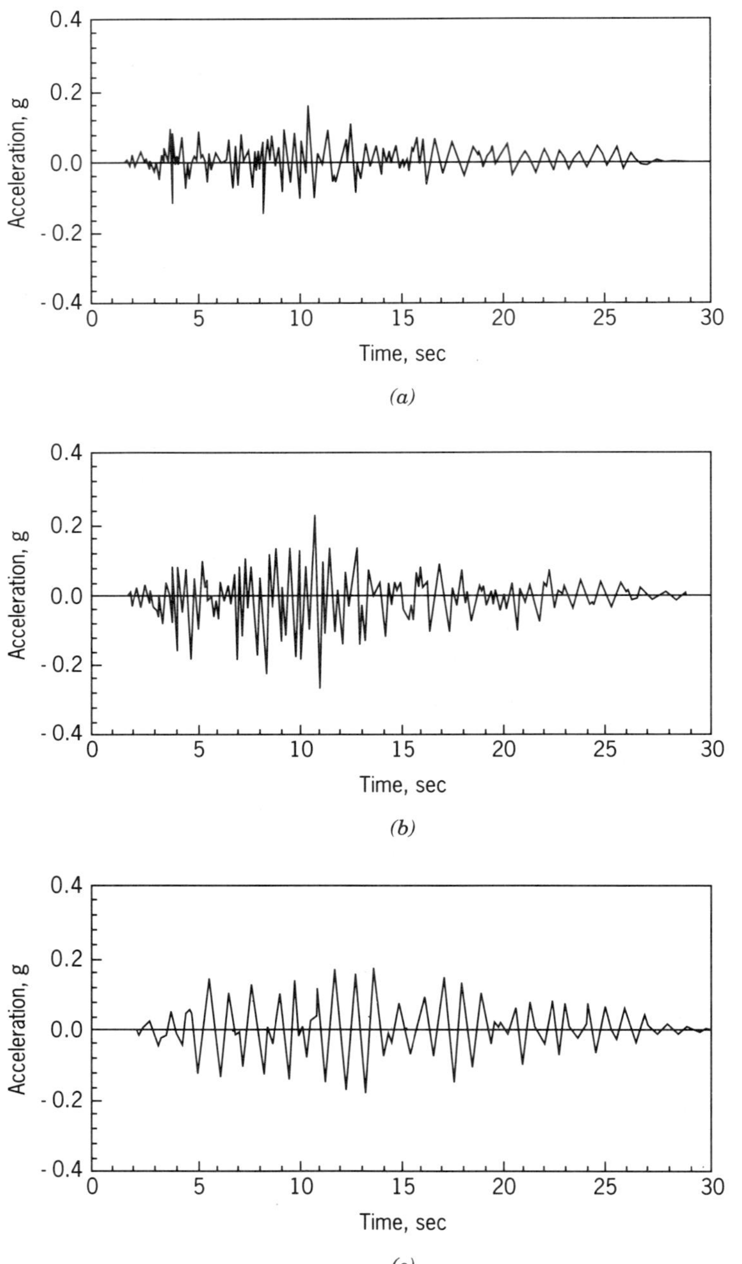


**Figure 7.37** Acceleration time histories—single pile test no. 12 (a) Input base motion, (b) free field acceleration, (c) lateral pile head acceleration (Finn and Gohl, 1987).

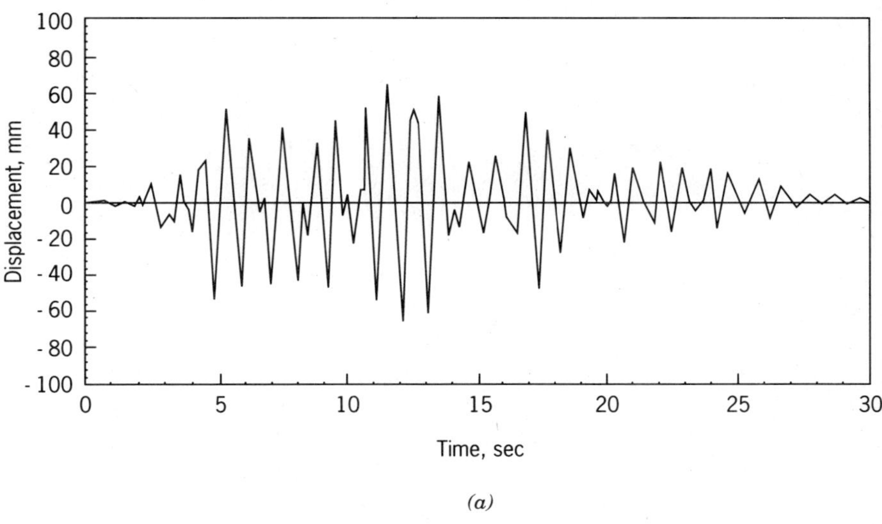


*(a)*

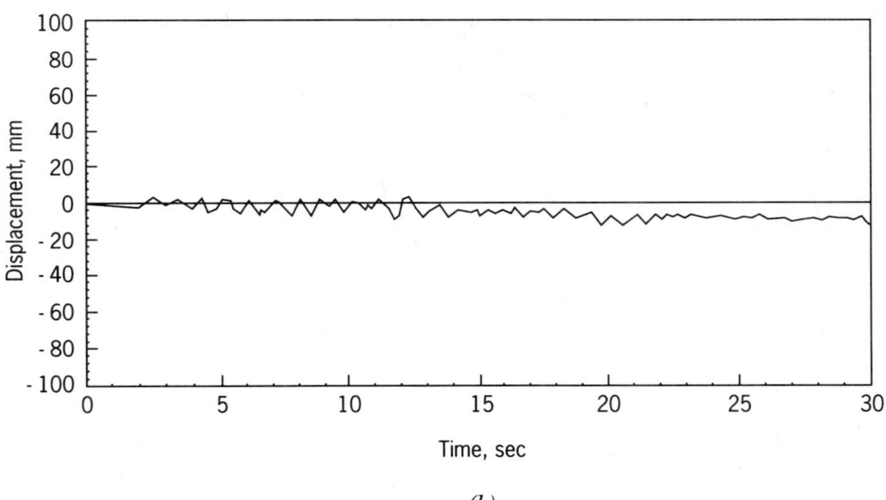


*(b)*

**Figure 7.38** Pile head displacement time histories—single pile test no. 12 (a) Displacement in the direction of load (X-direction), (b) displacement in the direction perpendicular to load (Y-direction). (Finn and Gohl, 1987).

do not influence the pile head response to the inertia forces applied at the pile head (Finn and Gohl, 1987).

5. The spatial variation of bending moments along the pile (Figure 7.40) shows that all points along the pile experience the same sign of bending moment at any instant in time. Thus, all points are vibrating in phase, suggesting that the free-headed pile is vibrating in its first mode.

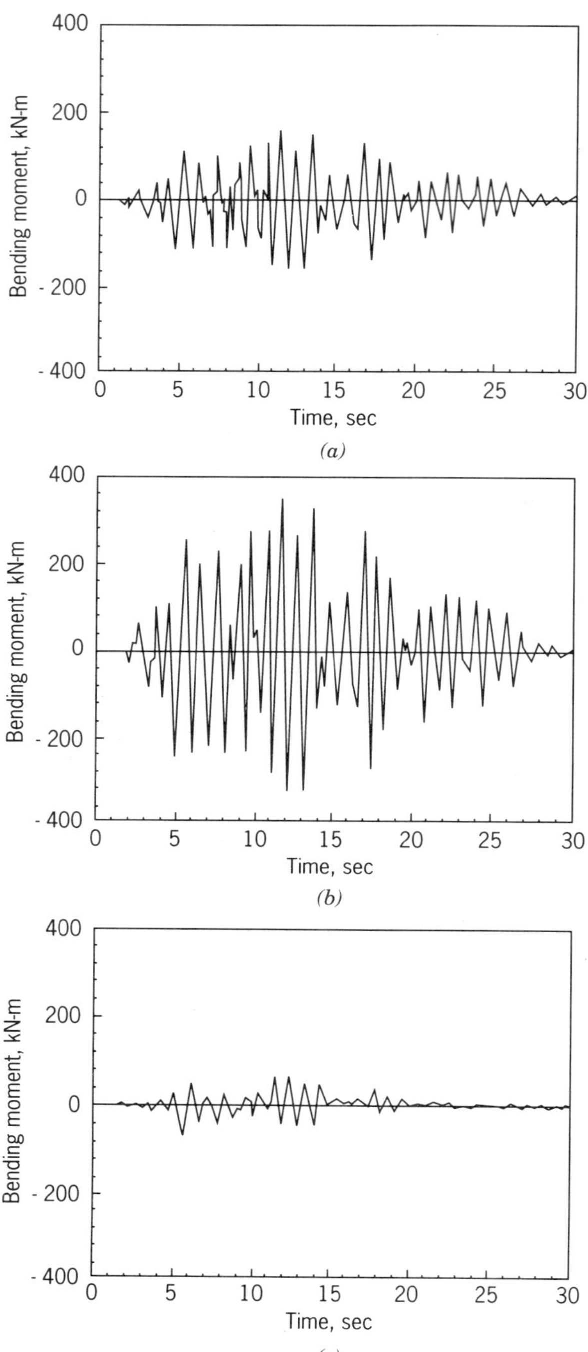


**Figure 7.39** Bending moment time histories—single pile test no. 12 (a) Strain gauge no. 1, (b) strain gauge no. 4, (c) strain gauge no. 7. (Finn and Gohl, 1987).

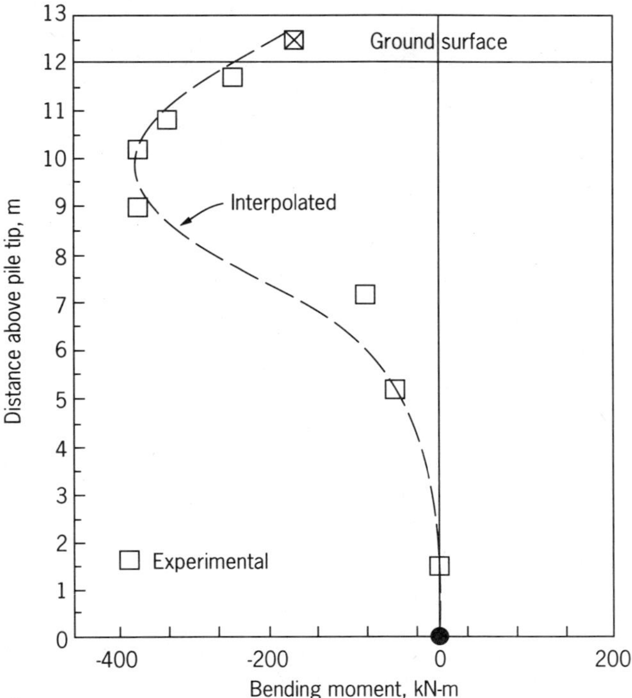


**Figure 7.40** Bending moment versus depth at peak pile head displacement (Finn and Gohl, 1987).

Test 41 was conducted in dense sand in order to provide data on single pile response required for the analysis of pile group tests in dense sand under excitation levels low enough to allow the assumption of linear elastic response. Therefore, the peak input base acceleration of pile test 41 was only 0.04 g.

The response of the single pile in test 41 was analyzed using a Winkler foundation model with a lateral stiffness $k$ proportional to the square root of the depth (i.e., $k = \alpha z^{1/2}$).

The value of $\alpha$ was determined as 20,000 $\mathrm{kN/m^{5/2}}$ by analyzing the soil property data in Figure 7.36. The displacements at ground surface were computed by integrating twice the measured moment distribution in the pile. As a check on the validity of the Winkler model assumed above, the moment distribution in the pile was computed and compared with the measured distribution. The comparison is shown in Figure 7.41, which appears satisfactory.

***Pile Group Response*** Two-pile groups were tested at various spacings at low levels of excitation using an approximately harmonic base motion (Figure 7.42). Bending moment distributions in piles with a center-to-center spacing equal to

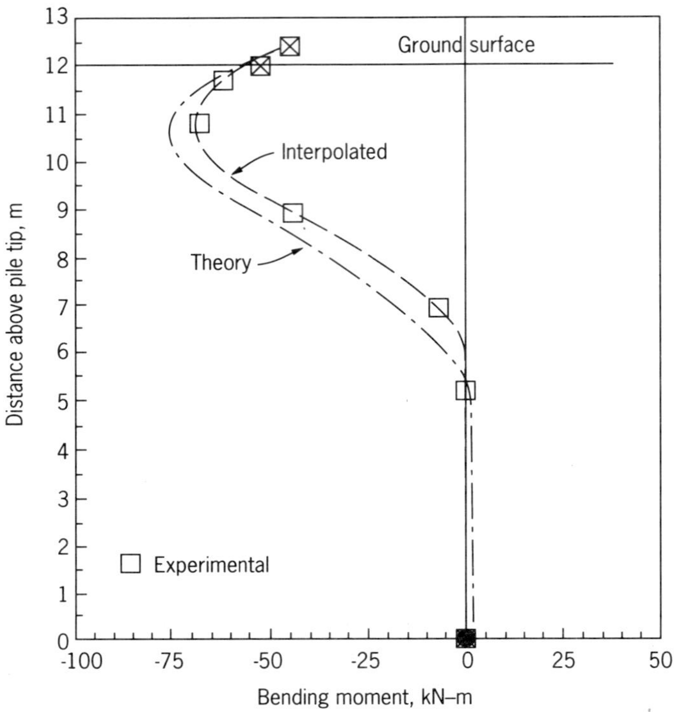


**Figure 7.41** Bending moment versus depth at peak displacement in tenth cycle—single pile test no. 41—dense sand (Finn and Gohl, 1987).

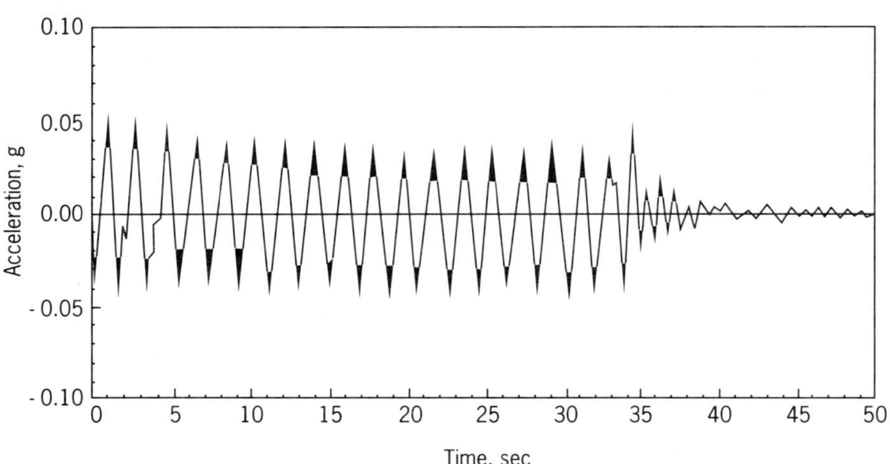


**Figure 7.42** Input base acceleration time history—pile group test no. 25 (Finn and Gohl, 1987).

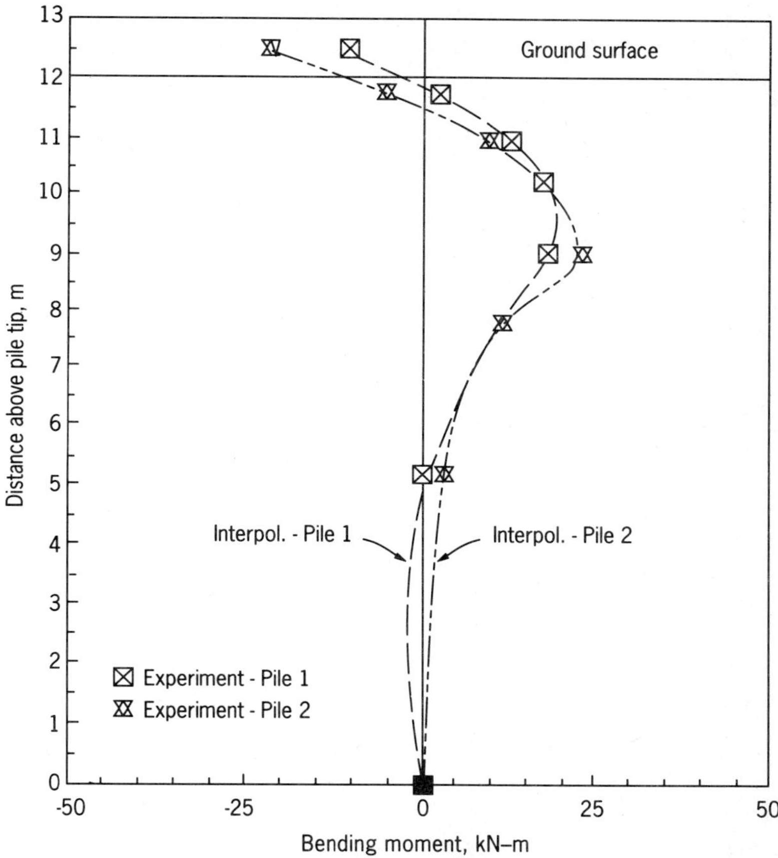


**Figure 7.43** Bending moment versus depth at maximum pile deflection loading in the direction of pile spacing ($s/b = 2$) dense sand—group test no. 21 (Finn and Gohl, 1987).

two pile diameters for the tenth load cycle at a time when pile cap deflection is a maximum during the cycle are shown in Figure 7.43.

In this figure, at peak displacement (inline loading), the bending moment changes sign indicating the restraint of the pile cap against rotation. The moment distributions in the two piles are sufficiently different to suggest significant interaction (Finn and Gohl, 1987).

The steady-state peak pile cap displacement is plotted against the pile spacing ratio, $s/b$, for ratios between 2 and 6 for inline shaking in Figure 7.44. This figure suggests that the pile cap displacements at the same level of excitation depend very strongly on pile spacing for inline shaking. This indicates strong interaction between piles in the group. Computed value of displacements for single pile are also shown in Figure 7.44. The results suggest that interaction

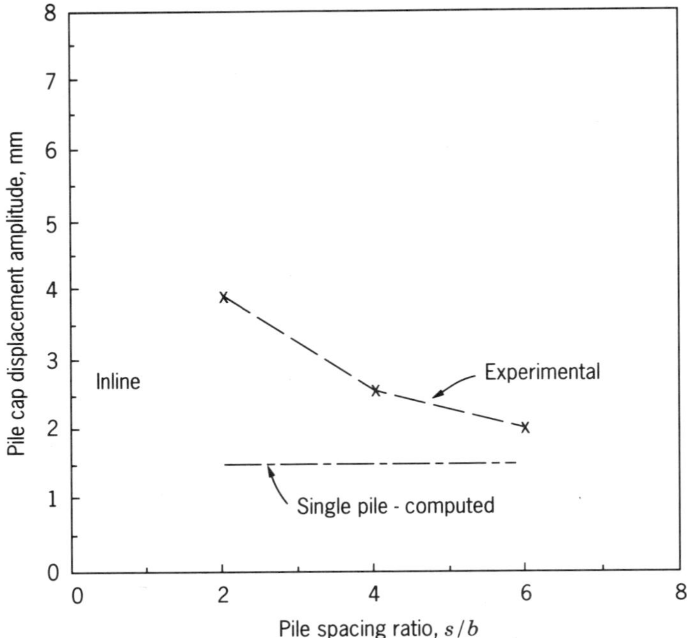


**Figure 7.44** Influence of pile interaction on pile cap displacement in inline loading (Finn and Gohl, 1987).

effects at spacings beyond about 6 pile diameters are insignificant (Finn and Gohl, 1987).

The tests of Finn and Gohl had been conducted with maximum spacing($s$) of six times the diameter of the pile. Their data in Figure 7.44, however, suggests that probably the experimental curve will become asymptotic to the value of the single pile displacement if the $s/b$ ratio approaches 8, as previously shown by Prakash (1962) in his model tests in piles in sand.

The tests data reported above had a unique feature. In these centrifuge tests, the in-situ distribution of shear moduli in the soil was measured during flight using piezoceramic bender elements. This gives data to check the measurements of the pile behavior with the predicted response.

Sufficiently more data are needed to check the validity of various analytical formulations with the measured response both in case of single pile and pile groups. A comprehensive model study on single piles and pile groups is underway at the University of Missouri, Rolla (Sreerama, 1990). It appears that data are being collected by various investigators, and in the not too distant future, better comparisons of the predicted and measured responses will be available.

## 7.8 EXAMPLES

***Example 7.1*** A four-pile group carries 75 tons ($t$) vertical load per pile. $EI$ of the pile is $1.2 \times 10^{10}$ lb-in$^2$, diameter of the piles is 12 in., and length of the piles is 45 ft. The soil is noncohesive with $\phi = 30°$ $\gamma_t = 110$ pcf and $n_h = 50$ lb/in$^3$. If this pile group is subjected to ground motion similar to that of EL Centro 1940 NS-component, determine (assuming appropriate group action):

1. Maximum displacement of the pile head
2. Maximum bending moment in the pile under dynamic condition
3. Soil reaction along the pile

SOLUTION

*(a) Free pile top*

$n_h = 50$ lb/in$^3$—Assume spacing of piles in the group of $3 \times d$ where $d$ = diameter of pile the $k = k_{\text{eff}} = 0.25\,k = 0.25 \times 50 = 12.5$ lb/in.$^3$ (From Table 6.6).

$$T = 5\sqrt{\frac{EI}{n_h}} = 5\sqrt{\frac{1.2 \times 10^{10}}{12.5}} = 62.58 \text{ in.} = 1.6 \text{ m}$$

$$Z_{\max} = \frac{L_s}{T} = \frac{45 \times 12}{62.58} = 8.6286$$

$Z_{\max} > 5$. As usual, it is a "long" pile
From Figure 7.14a, b for $Z_{\max} > 5$ and $T = 1.6$ m.

$F_{SL1} = 0.64$ for free pile head
$F'_{SL1} = 0.99$ for pile head restrained against rotation
Dead load on pile $= 75\,t$

The mass participating in vibrations is only a fraction of this load (choose 80 percent).
Using 80 percent of dead load, the mass at the pile top is:

$$M_t = 75 \times \frac{1}{32.2} \times \frac{80}{100} = \frac{60\,t \sec^2}{32.2 \times 12 \text{ in.}}$$

Using Equation (7.13).

$$\omega_{n1} = F_{SL1} \div \sqrt{\frac{W}{g}\frac{1}{n_h T^2}}, \text{ pile head free to rotate}$$

$$= 0.64 \times \sqrt{\frac{32.2 \times 12 \times 12.5}{60 \times 2000}} \times 62.58$$

$$= 8.0352 \text{ rad/sec}$$

$$f_{n1} = \frac{\omega_{n1}}{2\pi} = \frac{8.0352}{2 \times \pi} = 1.27885 \text{ (cycles/sec)}$$

$$T_{n1} = \frac{1}{f_{n1}} = \frac{2\pi}{\omega_{n1}} = \frac{1}{1.27885} = 0.7820 \text{ sec}$$

From Figure 7.17, in which combined earthquake response spectra for *EL* centro May 1940 N.S. has been plotted, assuming 5% damping spectral displacement $S_d = 1.2$ in. = maximum displacement

*Maximum bending moment*

$$M_g = B_{me} \times n_h \times T^3 \times S_d \tag{7.14b}$$

Using Table 7.4, the maximum value of the coefficient $B_{me}$ for a pile top free to rotate $Z_{\max} = 5$ to 15, $B_{me} = 0.315$.

$$\therefore \quad M_g = 0.315 \times \frac{12.5}{2000} \times (62.58)^3 \times 1.2 = 579.00 \text{ in } t = 48.25 \text{ t-ft}$$

Because the pile head is free to rotate:

$$Y_g = A_y \frac{Q_g T^3}{EI} = A_y B \text{ (constant)}$$

For soils with modulus increasing linearly with depth (from Table 6.3)

$$A_y = 2.435$$

(constant $B$) in the above equation is $\dfrac{S_d}{A_y} = \dfrac{1.2}{2.435} = 0.4928$

$$y_x = 0.4928 \times A_y$$

Soil reaction $p = n_h \cdot x \cdot y_x$ has been computed in Table 7.12a.

*(b) Restrained pile head*

$$F'_{SL1} = 0.99$$

Using Equation 7.13(a)

$$\omega_{n1} = F'_{SL1} \div \sqrt{\frac{W}{g} \frac{1}{n_h T^2}}$$

**TABLE 7.12a Computation of $y_x$ and $p_x$ Along the Pile Length for a Pile that is Free to Rotate (Example 7.1)**

| $X$ (ft) | $Z$ $(x/T)$ | $A_y$ | $y_x = 0.4928A_y$ (in.) | $k_x = \hat{n_h x}$ (lb/in.$^2$) | $p_x = k_x y_{(x)}$ (lb/in.) |
|---|---|---|---|---|---|
| 1 | 2 | 3 | 4 | 5 | 6 |
| 0 | 0 | 2.435 | 1.2 | 0 | 0 |
| 1 | 0.19 | 2.096 | 1.0329 | 150 | 155 |
| 2 | 0.38 | 1.827 | 0.9003 | 300 | 270 |
| 3 | 0.58 | 1.526 | 0.7520 | 450 | 338 |
| 4 | 0.77 | 1.257 | 0.6194 | 600 | 372 |
| 5 | 0.96 | 1.012 | 0.4987 | 750 | 374 |
| 7.5 | 1.44 | 0.511 | 0.2518 | 1125 | 283 |
| 10.0 | 1.92 | 0.184 | 0.0906 | 1500 | 136 |
| 12.5 | 2.40 | 0.055 | 0.0271 | 1875 | 51 |
| 15.0 | 2.88 | −0.049 | −0.0241 | 2250 | −54 |
| 17.5 | 3.36 | −0.066 | −0.0325 | 2625 | −85 |
| 20.0 | 3.83 | −0.054 | −0.0266 | 3000 | −80 |
| 22.5 | 4.31 | −0.037 | −0.0182 | 3375 | −61 |
| 25.0 | 4.79 | −0.018 | −0.0089 | 3750 | −33 |
| 30.0 | 5.75 | −0.009 | −0.0044 | 4500 | −20 |

Soil reaction $p$ may be plotted from column 6 with depth of pile $X$ in column 1.

$$\omega_{n1} = 0.99 \times \sqrt{\frac{32.2 \times 12 \times 12.5}{60 \times 2000}} \times 62.58$$

$$= 12.43 \text{ rad/sec}$$

$$f_{n1} = \frac{\omega_{n1}}{2\pi} = 1.9782 \text{ Hz}$$

$$T_{n1} = \frac{2\pi}{\omega_{n1}}$$

$$= 0.51 \text{ sec.}$$

From Figure 7.17 for a period $T_{n1}, = 0.51$ and damping of 5%, spectral displacement $S_d = 0.67$ in. Maximum bending moment $M_g = B_{me} \times n_h \times T^3 \times S_d$ from equation (7.14b).
For a restrained pile with $Z_{max}$ 5 to 15:

$$B_{me} = -0.90 \text{ (from Table 7.4)}$$

$$\therefore \quad M_g = -0.90 \times \frac{12.5}{2000} \times (62.58)^3 \times 0.67 = -923.53 \text{ t-in.}$$

$$= -76.96 \text{ t-ft}$$

$$y_g = (A_y - 0.93\, B_y)$$

$$\text{At } Z = 0 \qquad y_g = S_d = 0.67 \text{ in.}$$

Then $y_g = C_y B$ (constant $B$).
The constant $B$ can be calculated for $Z = 0$:

$$A_y = 2.435$$

$$B_y = 1.623$$

$$S_d = (2.435 - 0.93(1.623))\, B$$

$$= C_y B$$

$$B = \frac{S_d}{C_y} = \frac{0.67}{2.435 - 0.93(1.623)} = 0.7238$$

**TABLE 7.12b Computation of $y_x$ and $p_x$ Along the Pile Length for a Pile Fixed Against Rotation (Example 7.1)**

| $X$ (ft) | $Z$ $(x/T)$ | $A_y$ | $B_y$ | $C_y = A_y - 0.93B_y$ | $y_x = C_y \times 0.7238$ | $k_x = n_h \times x$ (lb/in.$^2$) | $p_x = k_x y_x$ (lb/in.) |
|---|---|---|---|---|---|---|---|
| 1 | 2 | 3 | 4 | 5 | 6 | 7 | 8 |
| 0 | 0 | 2.435 | 1.623 | 0.9256 | 0.6700 | 0 | 0 |
| 1 | 0.19 | 2.096 | 1.309 | 0.8786 | 0.6359 | 150 | 95 |
| 2 | 0.38 | 1.827 | 1.031 | 0.8682 | 0.6284 | 300 | 188 |
| 3 | 0.58 | 1.526 | 0.776 | 0.8043 | 0.5822 | 450 | 262 |
| 4 | 0.77 | 1.257 | 0.571 | 0.7260 | 0.5255 | 600 | 315 |
| 5 | 0.96 | 1.012 | 0.398 | 0.6419 | 0.4646 | 750 | 348 |
| 7.5 | 1.44 | 0.511 | 0.095 | 0.4227 | 0.3060 | 1125 | 344 |
| 10.0 | 1.92 | 0.184 | −0.062 | 0.2417 | 0.1749 | 1500 | 262 |
| 12.5 | 2.40 | 0.055 | −0.078 | 0.1275 | 0.0923 | 1875 | 173 |
| 15.0 | 2.88 | −0.049 | −0.087 | 0.0319 | 0.0231 | 2250 | 52 |
| 17.5 | 3.36 | −0.066 | −0.067 | −0.0037 | −0.0027 | 2625 | −7 |
| 20.0 | 3.83 | −0.054 | −0.038 | −0.0187 | −0.0135 | 3000 | −41 |
| 22.5 | 4.31 | −0.037 | −0.019 | −0.0193 | −0.0140 | 3375 | −47 |
| 25.0 | 4.79 | −0.018 | −0.006 | −0.0124 | −0.0090 | 3750 | −34 |
| 30.0 | 5.75 | −0.009 | 0.000 | −0.009 | −0.0065 | 4500 | −29 |

The soil reaction may now be plotted with depth.

$$y_x = B(C_y)_x$$

$$y_x = 0.7238(A_y - 0.938\, B_y)_z$$

See Table 7.12b, for computation of soil reaction along the pile.

*(c) Partial fixity*

Fixity conditions of the actual piles in the group must be estimated and the solution obtained for that fixity value by linear interpolation. In this case, let us assume 50 percent fixity. Compute displacement for 50 percent fixity. The displacement under dynamic condition is $S_d = (1.2 + 0.67)/2 = 0.935$ in. $S_d$ = maximum displacement for 50 percent fixity = 0.935 in.

Computation of the maximum bending moment for 50 percent fixity is:

$$B_{me} = \frac{0 - 0.90}{2} = -0.45$$

$$M_g = \frac{-0.45 \times 12.5}{2000} \times (62.58)^3 \times 0.935 = -644.48 \text{ ton-in.}$$

Soil reaction $p$:

$$y_g = (A_y - 0.465\, B_y)(\text{constant } B)$$

$$\text{At } Z = 0 \qquad y_g = S_d = (2.435 - 0.463(1.623))\,(B)$$

$$B \text{ (constant)} = \frac{0.935}{2.435 - 0.463(1.623)} = 0.5554$$

Thus, $y_x = B(C_y)_x = 0.5554(C_y)_x$.

See Table 7.12c for computation of soil reaction along the pile.

***Example 7.2*** Several groups of piles are to be proportioned for different column loads. The concrete piles are 12 in. in diameter and 60 ft long.

$$EI = 1.2 \times 10^{10} \text{ lb-in.}^2$$

The following soil and pile properties may be assumed:

$$\text{Soil:} \quad G_s = 400 \text{ tsf}$$

$$\gamma_{sat} = 110 \text{ pcf}$$

$$\text{Pile:} \quad \gamma_p = 150 \text{ pcf}$$

$$E_p = 2.5 \times 10^5 \text{ tsf}$$

Assume that $G$ is constant with depth and the piles are end bearing.

**TABLE 7.12c Computation of Soil Reaction with 50 percent Fixity of Pile Top (Example 7.1)**

| $X$ (ft) | $Z$ $(x/T)$ | $A_y$ | $B_y$ | $C_y = A_y - 0.465B_y$ | $y_x = 0.554C_y$ | $k_x = n_h x$ (lb/in.$^3$) | $p = k_x y_x$ (lb/in.) |
|---|---|---|---|---|---|---|---|
| 1 | 2 | 3 | 4 | 5 | 6 | 7 | 8 |
| 0 | 0 | 2.435 | 1.623 | 1.6835 | 0.9350 | 0 | 0 |
| 1 | 0.19 | 2.096 | 1.309 | 1.4900 | 0.8275 | 150 | 124 |
| 2 | 0.38 | 1.827 | 1.031 | 1.3506 | 0.7501 | 300 | 225 |
| 3 | 0.58 | 1.526 | 0.776 | 1.1667 | 0.6480 | 450 | 292 |
| 4 | 0.77 | 1.257 | 0.571 | 0.9926 | 0.5513 | 600 | 331 |
| 5 | 0.96 | 1.012 | 0.398 | 0.8277 | 0.4597 | 750 | 345 |
| 7.5 | 1.44 | 0.511 | 0.095 | 0.4670 | 0.2594 | 1125 | 292 |
| 10.0 | 1.92 | 0.184 | −0.062 | 0.2127 | 0.1181 | 1500 | 177 |
| 12.5 | 2.40 | 0.055 | −0.078 | 0.0911 | 0.0506 | 1875 | 95 |
| 15.0 | 2.88 | −0.049 | −0.087 | −0.9987 | −0.0048 | 2250 | −11 |
| 17.5 | 3.36 | −0.066 | −0.067 | −0.0350 | −0.0194 | 2625 | −51 |
| 20.0 | 3.83 | −0.054 | −0.038 | −0.0304 | −0.0202 | 3000 | −61 |
| 22.5 | 4.31 | −0.037 | −0.019 | −0.0282 | −0.0157 | 3375 | −53 |
| 25.0 | 4.79 | −0.018 | −0.006 | −0.0152 | −0.0085 | 3750 | −32 |
| 30.0 | 5.75 | −0.009 | 0.000 | −0.009 | −0.0050 | 4500 | −22 |

The soil reaction may now be plotted with depth.

(a) Estimate the stiffness and damping values of the single pile and pile group in vertical vibrations for pile spacing of 3.3 ft center to center for the following groups: 2 × 2, pile cap thickness 3 ft, 3 × 3, pile cap thickness 4 ft, 4 × 4, pile cap thickness 5 ft.
The pile cap projection may be assumed 6 in. beyond the pile edge.
Show also if the selection of a particular reference pile will affect your result for 4 × 4 group. Neglect contribution of stiffness and damping due to base reaction of the pile cap. The load per pile is 55 *t*. Compute natural frequency and amplitudes of motions for the 4 × 4 group if vertical unbalanced load per pile is $P_{(z)} = (2 \sin 2\pi f)$ tons and $f = 3\,H_z$.
(b) For a 4 × 4 group, estimate the damping and stiffness in rocking and horizontal vibrations for the single as well as the pile group including contribution of pile cap from side reactions. Assume reduced soil properties around the pile cap by an appropriate factor.

SOLUTION

*(a) Vertical vibrations: Single pile*

Diameter $B = 1$ ft, length $l = 60$ ft
$EI = 1.2 \times 10^{10}$ lb.in$^2$
Soil $G_s$ = constant with depth = 400 tsf

$\gamma_{sat} = 110$ pcf
Pile, $\gamma_p = 150$ pcf, $E_p = 2.5 \times 10^5$ tsf

Assumption: Piles are end bearing. Let $\nu_{soil} = 0.5$
*Estimation of stiffness and damping values:*
Single Pile:

$$k_w^1 = \frac{E_p \cdot A}{r_0} f_{w1} \tag{7.46}$$

where $r_0$ = equivalent radius = 0.5 ft.

$$c_{w^1} = \frac{E_p \cdot A}{V_s} f_{w_2} \tag{7.48}$$

and $f_{w_1}$ and $f_{w_2}$ are obtained from Figure 7.23

$$V_s = \sqrt{\frac{G_s \cdot g}{\gamma_s}} = \sqrt{\frac{400 \times 2000 \times 32.2}{110}} = 483.92 \text{ ft/sec}$$

$$E_p/G_s = \frac{2.5 \times 10^5}{400} = 625$$

$$\frac{l}{r_0} = \frac{60}{0.5} = 120 > 100$$

Use $l/r = 100$ and $E_p/G_s = 625$.
From Figure 7.23*a*

$$\frac{E_p}{G_s} = 500 \qquad f_{w_1} = 0.041 \qquad f_{w_2} = 0.068$$

$$\frac{E_p}{G_s} = 1000 \qquad f_{w_1} = 0.029 \qquad f_{w_2} = 0.048$$

For

$$\frac{E_p}{G_s} = 625 \quad f_{w_1} = 0.041 - \left(\frac{0.041 - 0.029}{500}\right) 125 = 0.038$$

and

$$f_{w_2} = 0.068 - \left(\frac{0.068 - 0.048}{500}\right) 125 = 0.063$$

$$k_w^1 = \frac{2.5 \times 10^5}{0.5} \times \frac{\pi \times 1^2}{4} \times 0.038 = 14923\,\text{t/ft} = 1244\,\text{t/in.}$$

and

$$c_{w^1} = \frac{2.5 \times 10^5}{483.92} \times \frac{\pi \times 1^2}{4} \times 0.063 = 25.56\,\text{t-sec/ft} = 2.13\,\text{t-sec/in.}$$

*2 × 2 Pile Group*

To consider group effect, assume that any pile in the group is a reference pile (see Figure 7.45a). With $r_0 = 6$ in., the values of $S/2r_0$ are calculated for other piles. For adjacent piles:

$$S/2r_0 = \frac{3.3}{1} = 3.3$$

and for the diagonal pile:

$$S/2r_0 = \frac{\sqrt{3.3^2 + 3.3^2}}{1} = 4.67$$

$$L/2r_0 = \frac{60}{1} = 60 \quad \text{let } v = 0.5$$

$$\alpha_A = 1 \text{ for reference pile (pile no. 1)}$$

For piles 2 and 3, $\alpha_A$, interpolate for $S/2r_0 = 3.3$ and $L/2r_0 = 60$ (From Figure 7.29).
Interpolating

$$L/2r_0 = 25 \qquad \alpha_A = 0.54$$

$$L/r_0 = 100 \qquad \alpha_A = 0.65$$

for

$$L/2r_0 = 60 \qquad \alpha_A = 0.54 + \left(\frac{0.65 - 0.54}{75}\right) \times 35 = 0.59$$

Similarly for diagonal pile no. 4:

$$\alpha_A = 0.52$$

$$\sum \alpha_A = 1 + 2(0.59) + 0.52 = 2.70$$

Combined stiffness of piles:

$$k_w^g = \Sigma k_w^1/\Sigma\alpha_A \tag{7.67}$$

$$= \frac{4 \times 1244}{2.7} = 1842\ \text{t/in.}$$

$$c_w^g = \Sigma c_w^1/(\Sigma_A) \tag{7.68}$$

$$= \frac{4 \times 2.13}{2.7} = 3.15\ \text{t-sec/in.}$$

*Determination of spring stiffness and damping due to side friction of pile cap* $k_w^f$

Assume pile cap is embedded 2.5 ft. (Figure 7.45b).

$$\bar{S}_1 = 2.7$$

Assume $G_s$ of backfilled soil as 400 tsf.

$k_w^f = 400 \times 2.5 \times 2.7 = 2780$ t/ft $= 225$ t/in. (From equation 7.69).

$$\text{Cap size} = s + 2r_0 + 1$$

$$= 3.3 + 1 + 1 = 5.3\ \text{ft}$$

$$r_0(\text{cap}) = \left(\frac{5.3 \times 5.3}{\pi}\right)^{1/2} = 2.99\ \text{ft}$$

$$c_w^f = hr_0 \cdot \bar{S}_2 \cdot \sqrt{G_s \rho_s}$$

$$\bar{S}_2 = 6.70$$

$$\therefore\ c_w^f = 2.5 \times 2.99 \times 6.70 \sqrt{\frac{400 \times 2000 \times 110}{32.2}}$$

$$= 82799\ \text{lb sec/ft} = 3.45\ \text{t-sec/in.}$$

$$\text{Total } k_w^g = 1842 + 225 = 2066.84\ \text{t/in.}$$

$$\text{Total } c_w^g = 3.15 + 3.45 = 6.6047\ \text{t-sec/in.}$$

It will be seen that the pile cap contributes significantly to the damping of the whole system in a small group.

*3 × 3 Pile Group*

Cap thickness = 4 ft

*Step 1:* Select the center pile (no. 5) as the reference pile (Figure 7.45c).

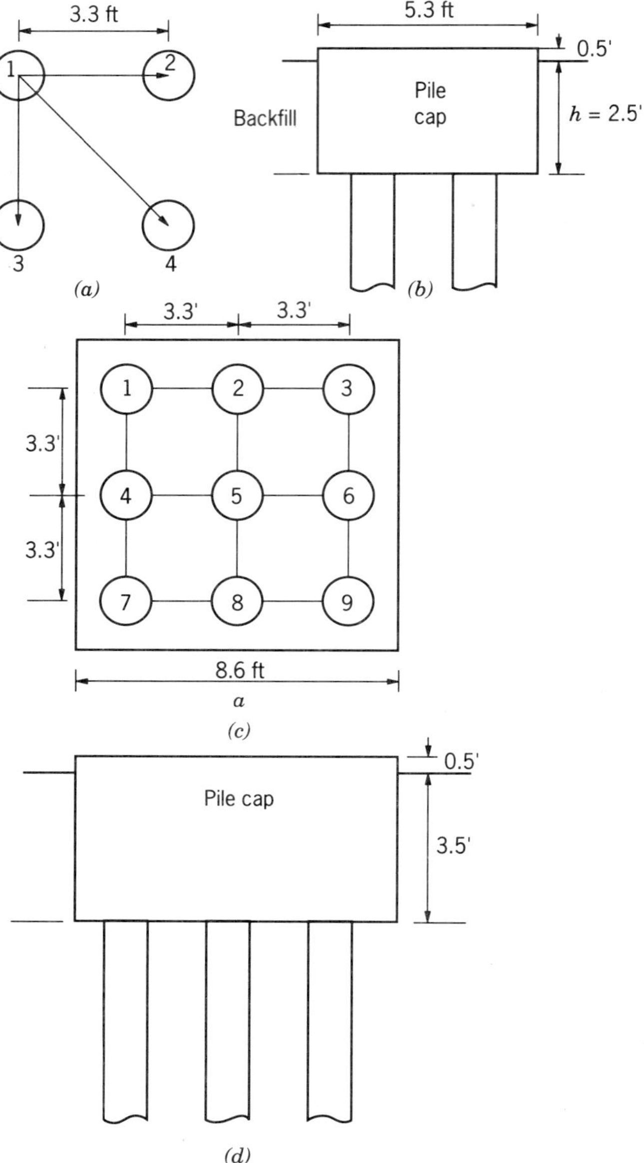


**Figure 7.45** Pile groups. (a) Plan (four-pile group), (b) section (four-pile group), (c) plan (nine-pile group), (d) section nine-pile group, (e) plan 16 piles group, (f) section (16 pile group) (Example 7.2).

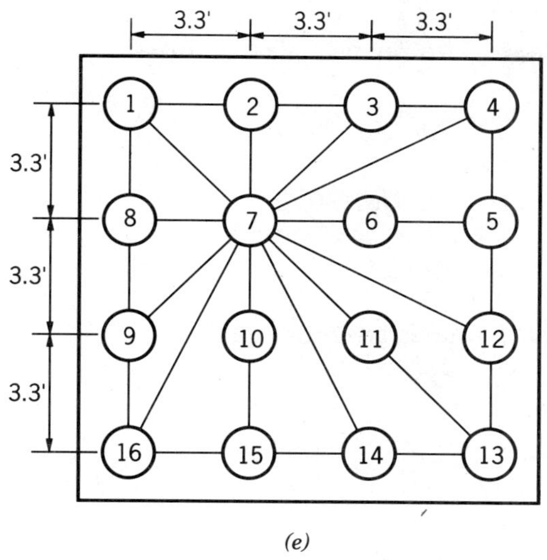


*(e)*

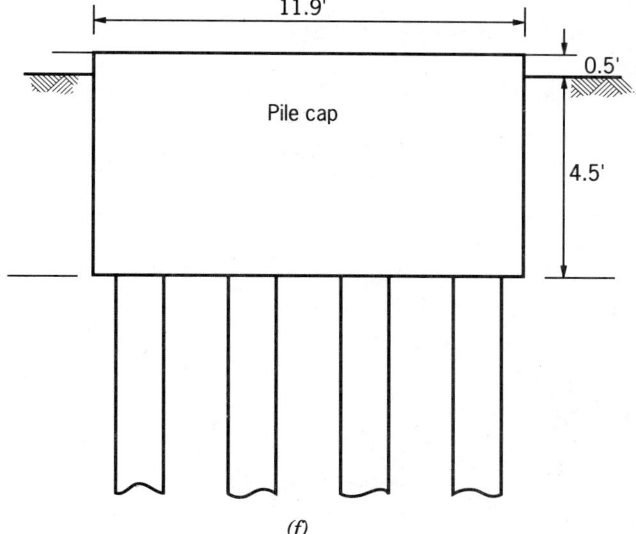


*(f)*

*Calculations of* $\alpha_A$

For adjacent piles 2, 4, 6, and 8:

$S/2r_0 = 3.3$

For diagonal piles 1, 3, 7, and 9:

$S/2r_0 = 2 \times (3.3)^2 = 4.67$

From Figure 7.29:

$$\alpha_A \text{ for reference pile (no. 5)} = 1$$

$$2r_0 = 3.3, \ \alpha_A \text{ for adjacent piles} = 0.59$$

$$\frac{S}{2r_0} = 4.67, \ \alpha_A \text{ for diagonal piles} = 0.52$$

$$\sum \alpha_A = 1 + 4(0.59) + 4(0.52) = 5.44$$

*Step 2:* Stiffness for pile group is given by equation 7.67

$$k_w^g = \Sigma\, k_w^1/\Sigma\alpha_a$$

$$= \frac{9 \times 1244}{5.44} = 2056 \text{ t/in.}$$

Damping for pile group is given by equation 7.68.

$$c_w^g = \frac{\Sigma c_w^1}{\Sigma \alpha_A} = \frac{9 \times 2.13}{5.44} = 3.52 \text{ t-sec/in.}$$

*Step 3:* Determine spring stiffness and damping due to side friction on pile cap, $k_w^f$.

Assume pile cap is embedded 3.5 ft in the ground (Figure 7.45d)

$$k_w^f = G_s \cdot h \cdot \bar{S}_1 \tag{7.69}$$

$$\bar{S}_1 = 2.7$$

$$k_w^f = 400 \times 3.5 \times 2.7 = 3780 \text{ t/ft} = 315 \text{ t/in.}$$

Dimensions of cap $= 2 \times 3.3 + 1 + 1 = 8.6$ ft.
Equivalent radius for pile cap:

$$r_0(\text{cap}) = \left(\frac{8.6 \times 8.6}{\pi}\right)^{1/2} = 4.85 \text{ ft.}$$

$$c_w^f = h r_0 \bar{S}_2 \sqrt{G_s \rho_s}$$

$$\bar{S}_2 = 6.7$$

$$c_w^f = 3.5 \times 4.85 \times 6.7 \times \sqrt{400 \times 2000 \times \frac{110}{32.2}} = 188095 \text{ lb-sec/ft.}$$

$$c_w^f = 7.84 \text{ t-sec/in.}$$

*Step 4.* Total stiffness

$$k_w^g = 2056 + 315\,\text{t/in.} = 2371\,\text{t/in.}$$

Total Damping:

$$c_w^g = 3.52 + 7.84$$

$$c_w^g = 11.36\,\text{t-sec/in.}$$

A larger pile cap contributes to damping in a larger measure as compared to a smaller pile cap as in a 2 × 2 pile group. Similar remarks apply to a 4 × 4 pile group.

*4 × 4 Pile Group*

Cap thickness = 5 ft

*Step 1:* Select pile 1 (Figure 7.45e) as reference pile and calculate $\alpha_A$ for pile group.

| Pile No. | $S/2r_0$ | Interpolation | | $\alpha_a$ |
|---|---|---|---|---|
| 1 | Reference Pile | | | 1 |
| 2, 8 | 3.3 | From 2 × 2 group | = | 0.59 |
| 3, 9 | 6.6 | $\left[0.4 + \frac{0.55 - 0.4}{75} 35\right]$ | = | 0.47 |
| 4, 16 | 9.9 | $\left[0.3 + \frac{0.46 - 0.3}{75} 35\right]$ | = | 0.375 |
| 12, 14 | 11.898 | $\left[0.285 + \left(\frac{0.45 - 0.285}{75}\right)35\right]$ | = | 0.362 |
| 5, 15 | 10.436 | $\left[0.29 + \left(\frac{0.455 - 0.29}{75}\right)35\right]$ | = | 0.367 |
| 6, 10 | 7.38 | $\left[0.38 + \left(\frac{0.53 - 0.38}{75}\right)35\right]$ | = | 0.450 |
| 13 | 14 | $\left[0.28 + \left(\frac{0.4 - 0.28}{75}\right)35\right]$ | = | 0.336 |
| 7 | 4.667 | From 2 × 2 group | = | 0.52 |
| 11 | 9.334 | $\left[0.33 + \left(\frac{0.49 - 0.33}{75}\right)35\right]$ | = | 0.405 |

$$\Sigma\alpha_A = 1 + 2(0.59) + 2(0.47) + 2(0.375) + 2(0.362) + 2(0.367)$$

$$+ 2(0.45) + 0.336 + 0.52 + 0.405$$
$$= 7.489$$

*Step 2*: Group stiffness

$$k_w^g = \Sigma\, k_w^1 / \Sigma\, \alpha_A = \frac{16 \times 1244}{7.489}$$
$$= 2657 \text{ t/in.}$$

Group Damping

$$c_w^g = \Sigma\, c_w^1 / \Sigma\, \alpha_A = \frac{16 \times 2.13}{7.489}$$
$$= 4.55 \text{ t-sec/in.}$$

Now let us select pile 7 as reference pile and calculate $\alpha_A$:

| Pile No. | $S/2r_0$ | $\alpha_A$ |
|---|---|---|
| 7 | Reference | 1 |
| 2, 6, 8, 10 | 3.3 | 0.59 |
| 1, 3, 9, 11 | 4.667 | 0.52 |
| 4, 12, 14, 16 | 7.38 | 0.45 |
| 5, 15 | 6.6 | 0.47 |
| 13 | 9.334 | 0.405 |

$$\alpha_A = 1 + 4(0.59) + 4(0.52) + 4(0.45) + 2(0.47) + 0.405 = 8.59$$

The combined stiffness and damping were calculated as above and are listed below for pile group:

| Ref. Pile | $k_w^g$ | $c_w^g$ |
|---|---|---|
| 1 | 2657 t/in. | 4.55 t-sec/in. |
| 7 | 2317 t/in. | 3.97 t-sec/in. |

*Step 3*: Determine spring stiffness and damping due to side friction on pile cap, $k_w^f$.

Assume that pile cap is embedded 4.5 ft in the ground $k_w^f$ (Figure 7.45f):

$$= G_s h \cdot \bar{S}_1$$
$$= 400 \times 4.5 \times 2.7 = 4860 \text{ t/ft} = 405 \text{ t/in.}$$

$$\text{where } \bar{S}_1 = 2.7 \tag{7.69}$$

$$\text{Dimensions of cap} = 3 \times 3.3 + 1 + 1 = 11.9$$

$$\text{Size of pile cap} = 11.9\,\text{ft} \times 11.9\,\text{ft}$$

Equivalent radius for pile cap:

$$r_0(\text{cap}) = \left(\frac{11.9 \times 11.9}{\pi}\right)^{1/2} = 6.72\,\text{ft}$$

Damping:

$$c_w^f = h.r_0\bar{S}_2\sqrt{G_S\rho_s} \tag{7.70}$$

$$= 4.5 \times 6.72 \times 6.7 \times 1653.15 = 334643\,\text{lb-sec/ft}$$

$$c_w^f = 13.94\,\text{t-sec/in.}$$

$$\text{where } \bar{S}_2 = 6.7$$

*Step 4*:

$$\text{Total stiffness} = k_w^g + k_w^f$$

$$\text{Total } k_w = 2657 + 405 = 3062\,\text{t/in.}$$

or

$$k_w = 2317 + 405 = 2722\,\text{t/in.}$$

Total damping:

$$c_w = 4.55 + 13.94 = 18.49\,\text{t-sec/in.}$$

or

$$= 3.97 + 13.94 = 17.91\,\text{t-sec/in.}$$

It will be seen that appropriate selection of a reference pile affects the computed stiffness by about 10 to 15 percent and damping by about 20 percent. Novak (1974) does not provide guidelines for selection of the reference pile.

*Step 5*: Determination of natural frequency and amplitude of vertical vibration. For the $4 \times 4$ pile group with pile cap thickness 5 ft:

1. Dimensions of pile cap. $11.9 \times 11.9 \times 5.0$ in feet.
2. Assume pile cap is made of the same material as piles.

$$\gamma_p = 150\,\text{pcf}$$

$$E_p = 2.5 \times 10^5\,\text{tsf}$$

3. Unbalanced forces:

$$P_z = 2 \times 16t$$

$$f = 3\text{ cycles/sec} \qquad \omega = 2\pi f = 18.84\,\text{rad/sec}$$

$$P(t) = P_z \sin \omega \text{t} = 32 \sin 2\pi(3)t = 32 \sin 6\pi t$$

4. Total mass—assumed to be equal to the mass of pile cap and the superstructure load.
5. The effective vertical load of the static column load vibrating with the pile cap may be assumed to be 80 percent of the superstructure load

$$0.8 \times 55 = 44t \text{ per pile}$$

6. Effective superstructure load for pile group

$$= 4 \times 4 \times 44 = 704t$$

With the stiffness, damping, and masses established, the response of the pile group may now be determined from principles of mechanical vibration (Prakash and Puri, 1988) as below:
Natural frequency:

$$\omega_{nz}^g = \sqrt{\frac{k_w^g}{m^g}}$$

where

$\omega_{nz}^g$ = natural frequency of pile group in vertical vibrations
$k_w^g$ = total stiffness of pile group and cap
$m^g$ = effective vibrating mass

The lowest values of $k_w^g$ and $c_w^g$ for 16-pile group are used.
Effective mass for the pile group:

$$m^g = (4 \times 4 \times 44 \times 2000 + 150 \times 11.9 \times 11.9 \times 5)/32.2$$

$$= 47025 \text{ lb sec}^2/\text{ft}$$

$$\omega_n = \sqrt{\frac{2722 \times 2000 \times 12}{47025}} = 37.27 \text{ rad/sec}$$

Maximum amplitude of vibration in vertical mode:

$$(Z_0)_{\max} = \frac{P_z}{\sqrt{(k - m\omega^2)^2 + (c\omega)^2}}$$

$$= \frac{32 \times 2000}{\sqrt{(2722 \times 2000 \times 12 - 47025(18.84)^2)^2 + (17.91 \times 2000 \times 12 \times 18.84)^2}}$$

$$= 0.00129 \text{ ft}$$

$$= 0.0156 \text{ in.}$$

(b) *Translation and rocking*
Single Piles:

$$EI = 1.2 \times 10^{10}\ \text{lb/in.}^2$$

$$E_p = 2.5 \times 10^5\ \text{tsf} = 2.5 \times 2000 \times 10^5 \times \frac{1}{144}\ \text{lb/in.}^2 = 3472222\ \text{lb/in.}^2$$

$$\therefore I = 3456\ \text{in.}^4$$

Let the reduced values of $G_s$ be 60 percent of original.

$G_s$ around pile cap = 240 tsf.
$G_s$ around pile (itself) reduced to 75 percent of original.

$$G_s \text{ around pile} = \frac{75 \times 400}{150} = 300\ \text{tsf}$$

$$V_s = \sqrt{G_s/\rho} = \sqrt{\frac{300 \times 2000 \times 32.2}{110}} = 419\ \text{ft/sec}$$

Assuming $\nu = 0.4$ and a homogeneous soil profile ($G$ = constant with depth).

$$\frac{E_p}{G_s} = \frac{2.5 \times 10^5}{300} = 833$$

*Sliding*
Interpolation of stiffness and damping
Parameters of horizontal response for piles with $l/R > 25$ for homogenous soil profile from Table 7.5.

$$\frac{E_p}{G_s} = 1000, \qquad f_{x1} = 0.0261$$

$$= 500, \qquad f_{x1} = 0.0436$$

$$= 833, \qquad f_{x1} = 0.0261 + \frac{(0.0436 - 0.0261)}{500}\, 333$$

$$= 0.0378$$

$$\frac{E_p}{G_s} = 1000, \qquad f_{x2} = 0.0641$$

$$= 500, \qquad f_{x2} = 0.1054$$

$$= 833, \qquad f_{x2} = 0.0641 + \frac{(0.1054 - 0.0641)}{500}\, 333 = 0.0916$$

$$f_{x2} = 0.0916$$

Horizontal stiffness constant:

$$k_x^1 = \frac{E_p I_p}{r_0^3}(f_{x1}) \tag{7.51}$$

$$= \frac{2.5 \times 10^5 \times 3456 \times 0.0378}{6^3 \times 144}$$

$$= 1050\,\text{t/in}$$

$$c_x^1 = \frac{E_p I_p}{r_0^2 V_s}(f_{x2}) \tag{7.52}$$

$$= \frac{2.5 \times 10^5 \times 3456 \times 0.0916}{144 \times 6^2 \times 419 \times 12}$$

$$= 3.0362\,\text{t-sec/in.}$$

*Rocking*

For $\nu = 0.4$, $\dfrac{E_p}{G_s} = 833$. Find $f_{\phi 1}$ and $f_{\phi 2}$ from Table 7.5 as above.

$$\frac{E_p}{G_s} = 1000, \qquad (f_{\phi 1}) = 0.3860$$

$$= 500, \qquad (f_{\phi 1}) = 0.4547$$

$$= 833, \qquad (f_{\phi 1}) = 0.3860 + \frac{(0.4547 - 0.3860)}{500} \times 333$$

Stiffness coefficient $f_{\phi 1} = 0.4318$:

$$\frac{E_p}{G_s} = 1000 \qquad (f_{\phi 2}) = 0.2677$$

$$= 500 \qquad (f_{\phi 2}) = 0.3034$$

$$= 833.33 \qquad (f_{\phi 2}) = 0.2677 + \frac{(0.3034 - 0.2677)}{500}\, 333$$

$$(f_{\phi 2}) = 0.2915$$

Rotational stiffness and geometric damping constants for single pile:

$$k_\phi^1 = \frac{E_p I_p}{r_0}(f_{\phi 1}) \tag{7.53}$$

$$= \frac{2.5 \times 10^5 \times 3456 \times 0.4318}{144 \times 6}$$

$$= 4.318 \times 10^5\,\text{in. t/rad}$$

$$c_\phi^1 = \frac{E_p I_p}{V_s}(f_{\phi 2}) \tag{7.54}$$

$$= \frac{2.5 \times 10^5 \times 3456 \times 0.2915}{144 \times 419 \times 12}$$

$$= 347.835 \text{ t-sec/rad.}$$

Stiffness and damping parameters of pile group for piles only

*Translation*

Letting the departure angle $\beta = 0$ and using Figure 7.31, obtain $\alpha_L$

$$K_R = \frac{(EI)_{\text{Pile}}}{2G(1+v)_{\text{soil}} L^4} = \frac{1.2 \times 10^{10}}{2 \times 300 \times 2000(1+0.4) \times (60^4 \times 144)}$$

$$= 3.827 \times 10^6$$

$\therefore$ Pile is flexible pile. Hence, use dotted lines on Figure 7.31. For the $4 \times 4$ pile group, calculate $S/2r_0$ using pile 7 as reference pile (Figure 7.45e)

| Pile No. | $\frac{S}{2r_0}$ | $\alpha_L$ |
|---|---|---|
| 7 | 0 | 1.0 |
| 2,6,8,10 | 3.3 | 0.47 |
| 1,3,9,11 | 4.6667 | 0.33 |
| 4,12,14,16 | 7.38 | 0.08 |
| 5,15 | 6.6 | 0.12 |
| 13 | 9.334 | 0 |

Figure 7.31 is for $L/2r_0 = 25$ and $v = 0.5$. In the above solution, it is assumed that although $v = 0.5$, the plot may be used for $v = 0.4$ as well.

$$\therefore \Sigma\alpha_L = 1 + 4(0.47) + 4(0.33) + 4(0.08) + 2(0.12) = 4.76$$

$$k_x^g = \frac{\Sigma k_x^1}{\Sigma \alpha_L} = 16 \times \frac{1050}{4.76} = 3529 \text{ t/in.}$$

$$c_x^g = \frac{\Sigma c_x^1}{\Sigma \alpha_L} = \frac{16 \times 3.0362}{4.76} = 10.21 \text{ t-sec/in.}$$

For pile caps:

$$k_x^f = G_s h S_{x1} \tag{7.73}$$

$$= 240 \times 4.5 \times 4.1$$

$$= 4428\,\text{t/ft} = 369\,\text{t/in.}$$

$$\bar{S}_{x1} = 4.1\text{—Table 7.6}$$

$$c_x^f = hr_0\sqrt{(G_s\rho_s)}\cdot\overline{S_{x2}} \tag{7.74}$$

$$= 4.5 \times 6.72\frac{(240 \times 2000 \times 110)^{1/2}}{32.2}10.6 = 410{,}099\,\text{lb-sec/ft.}$$

$$= 17.0875\,\text{t-sec/in.}$$

$$\overline{S_{x2}} = 10.6\text{—Table 7.6}$$

Total stiffness:

$$k_x = 3529.00 + 369$$

$$= 3898.00\,\text{t/in.}$$

Total damping:

$$c_x = 10.21 + 17.0875 = 27.297\,\text{t-sec/in.}$$

It will be seen that the pile cap contributes about 10 percent to the stiffness, while its contribution to damping is more than the damping due to piles alone.
*Cross-coupling constants*

$$k_{x\phi}^1 = \frac{E_pI_p}{r_0^2}(f_{x\phi1}) \tag{7.55}$$

$$c_{x\phi}^1 = \frac{E_pI_p}{r_0V_s}(f_{x\phi2}) \tag{7.56}$$

$$\frac{E_p}{G_s} = 833$$

Interpolate between 1000 and 500 from Table 7.5.

$$\frac{E_p}{G_s} = 1000 \qquad f_{x\phi1} = -0.0714$$

$$= 500 \qquad f_{x\phi1} = -0.0991$$

$$= 833 \qquad f_{x\phi1} = -0.0991 + \frac{(-0.0714)-(-0.0991)333}{500}$$

$$= -0.0806$$

$$= 1000 \qquad f_{x\phi 2} = -0.1052$$
$$= 500 \qquad f_{x\phi 2} = -0.1425$$
$$= 833 \qquad f_{x\phi 2} = -0.1425 + \frac{(-0.1052) - (-0.1425)333}{500} = -0.1176$$

$$\therefore \; k^1_{x\phi} = \frac{1.2 \times 10^{10}}{6^2} \times (-0.0806) = 26.8667 \times 10^6 \,\text{lb/in.}$$
$$= -13433.33 \,\text{t}$$
$$c^1_{x\phi} = \frac{1.2 \times 10^{10}}{6 \times 419 \times 12}(-0.1176)$$
$$= -23.3879 \,\text{t-sec/in}$$

Rocking stiffness and damping due to pile group

$$k^g_\phi = \Sigma^n_1[k^1_\phi + k^1_w x^2_r + k^1_x z^2_c - 2z_c k^1_{x\phi}] \tag{7.77}$$

$x_r = 1.65'$ for piles no. 2, 3, 6, 7, 10, 11, 14, 15

$x_r = 4.95'$ for piles no. 1, 4, 5, 8, 9, 12, 16, 13

$$k^g_\phi = 8\left[2 \times 431{,}800 + 1244\{(1.65 \times 12)^2 + (4.95 \times 12)^2\} + 1050 \times 2 \times \frac{(5 \times 12)^2}{2} - 4\left(\frac{5 \times 12}{2}\right)(-13433.3)\right]$$
$$= 73.94 \times 10^6 \,\text{ton/in}$$

$$c^g_\phi = \Sigma[c^1_\phi + c^1_w x^2_r + c^1_x z^2_c - 2z_c c^1_{x\phi}] \tag{7.78}$$
$$= 8\left[2 \times 347.835 + 2 \times 3.0362\left(\frac{5 \times 12}{2}\right)^2 - \frac{4 \times 5 \times 12}{2}(-23.3879) + 2.13\{(1.65 \times 12)^2 + (4.95 \times 12)^2\}\right]$$
$$= 164.949 \times 10^3 \text{t-sec/in.}$$

*Rocking stiffness and damping due to pile cap*

$$\delta = \frac{h}{r_0} = \frac{4.5}{6.714} = 0.67$$

From Table 7.6, frequency independent constants for embedded pile cap with side resistance.

$$\overline{S_{x1}} = 4.1 \quad \overline{S_{x2}} = 10.6 \quad \overline{S_{\phi 2}} = 2.5 \quad \text{and} \quad \overline{S_{\phi 2}} = 1.8$$

$$k^f_\phi = G_s r^2_0 h \bar{S}_{\phi 1} + G_s r^2_0 h\left[(\delta^2/3) + \left(\frac{z_c}{r_0}\right)^2 - \delta\left(\frac{z_c}{r_0}\right)\right]\bar{S}_{x1} \tag{7.79}$$

$$\therefore k_\phi^f = 240 \times 6.714^2 \times 4.5 \times 2.5 + 240 \times 6.174^2 \times 4.5\left[\frac{0.67^2}{3} + \left(\frac{2.5}{6.7}\right)^2 - 0.67\left(\frac{2.5}{6.7}\right)\right] \times 4.1$$

$$= 129{,}455 \text{ t/ft} = 10788 \text{ t/in.}$$

$$c_\phi^f = \delta r_0^4 \sqrt{G_s \gamma_s / g}\left\{\bar{S}_{\phi 2} + \left[\frac{\delta^2}{3} + \left(\frac{z_c}{r_0}\right)^2 - \delta\left(\frac{z_c}{r_0}\right)\right]\bar{S}_{x2}\right\} \tag{7.80}$$

$$= 0.67 \times 6.7^4 \sqrt{\frac{240 \times 110}{32.2}}\left\{1.8 + \left[\frac{0.67^2}{3} + \left(\frac{2.5}{6.7}\right)^2 - \frac{0.67 \times 2.5}{6.7}\right]10.6\right\}$$

$$= 15738 \text{ t-sec/ft} = 1311 \text{ t-sec/in.}$$

*Total stiffness and damping values*
Total stiffness $k^g = 73.94 \times 10^6 + 10788$

$$k^g = 74.02 \times 10^6 \text{ t/in.}$$

Total $c^g = 164.949 \times 10^3 + 1311$

$$c^g = 166.26 \times 10^3 \text{ t-sec/in.}$$

Total sliding and rocking stiffness and damping have been worked with foregoing computations. Response of a systems can then be determined from theory of mechanical vibrations (Prakash and Puri 1988).

## 7.9 COMPARISON OF PREDICTED RESPONSE WITH OBSERVED RESPONSE OF SINGLE PILES AND PILE GROUPS

Several lateral dynamic load tests on full-sized single piles were performed to check if the predicted response tallied with the measured response, (Gle, 1981; Woods, 1984). No tests have been performed on pile groups. Also, Novak and El-Sharnouby (1984) performed tests on a group of model piles to compare predictions with performance. No single pile tests were performed. The predicted response did not tally with the measured response in either case.

### 7.9.1 Tests of Full-Size Single Piles

Fifty-five steady-state lateral vibration tests were performed on 11 pipe piles 14 in. in outside diameter with wall thickness of 0.188 in. to 0.375 in. (0.47 cm to 0.94 cm) at three sites in southeast Michigan (Woods, 1984). The end-bearing piles were 50 to 160 ft (15 to 48 m) long.

Figure 7.46a shows response curves for the pile GP 13-7, 157 ft (47.1 m) long in soft clay. The pile was excited in steady-state oscillation by attaching an eccentric

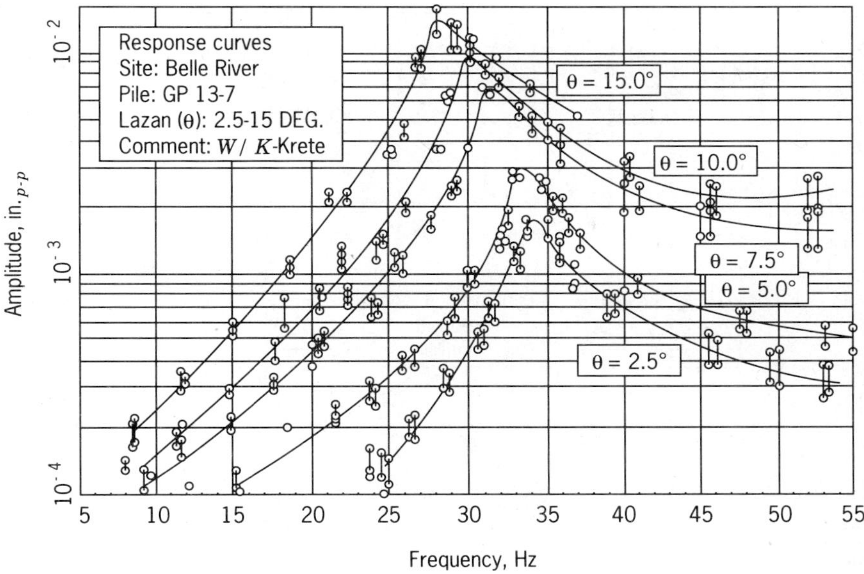


**Figure 7.46a** Response curves; a decrease in resonant frequency with increasing amplitudes (Gle, 1981).

weight vibrator (Lazan oscillator) to the head of the pile, whose response was monitored by two velocity transducers. At the conclusion of the first steady-state test, the eccentricity of the Lazan oscillation was increased to increase the oscillating force and the test was repeated. To cover the range of lateral displacements covered by most machine foundations, four or five increasing eccentricities were used. It was observed that the frequency of maximum response decreased as the force level increased, indicating non-linear response. A PILAY computer program was used by Woods (1984) to determine stiffness and damping of the pile (Novak and Aboul-Ella 1977). PILAY is a continuum model accommodating a multilayered soil based on the elastic soil layer approach of Baranov (1967). However, PILAY assumed that the soil surrounding the pile in a given layer is the same at all distances from the pile.

A dynamic response curve with this solution is shown in Figure 7.46b along with the field data. The correlation between predicted and measured response is very poor. In all tests, computed response based on stiffness and damping from PILAY and measured response showed that the amplitudes of motion were greater than predicted and the frequency of maximum response was lower than predicted.

In an attempt to match the measured response with the computed response the following two approaches were adopted.

1. For predicting the response, only a fraction of the rocking and translation

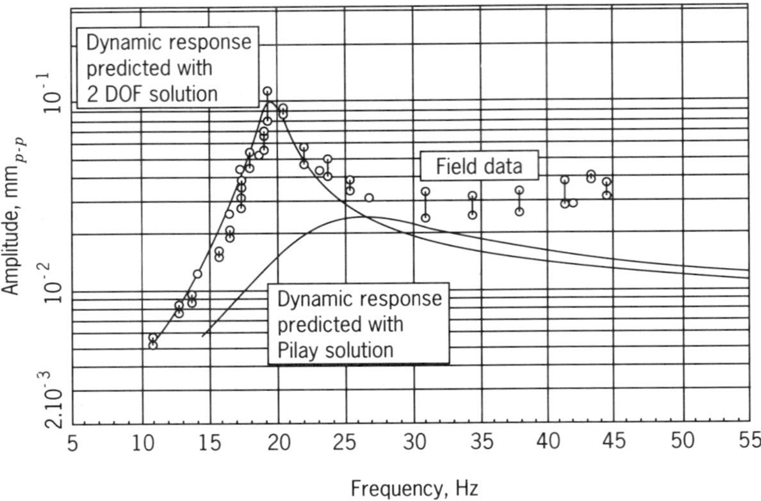


**Figure 7.46b** Typical response curves predicted by PILAY superimposed on measured pile response (Woods, 1984).

stiffness computed by PILAY was used. It was found that even with a wide variation in rocking stiffness, the observed amplitudes in the frequency range just above the horizontal translation peak was still higher than the predicted amplitude. The observed increase is more likely due to change in soil parameters caused by pile driving.

A better representation of the disturbed soil zone had been obtained by replacing the original soil with sand fill around the pile up to four feet depth.

2. Because of the poor correlation achieved in the initial attempt, a second correlation with the analytical procedure—PILAY 2—was attempted. PILAY 2 permits an inclusion of a "*softened*" or "*weakened*" zone surrounding the pile, simulating the disturbance to the soil caused by pile installation.

A good match of the measured and predicted response could be obtained by a considerably reduced soil moduls in the softened zone (one-tenth to two-tenths of the original value) and the extent of the softened zone (one-half to one times the pile radius). A loss of contact of the soil with pile for a short length close to the ground surface also improved the predicted response. No tests on pile groups were performed at any of these sites.

### 7.9.2 Tests on Groups of Model Piles

El-Sharnouby and Novak (1984) performed dynamic tests on a 102 steel pipe piles group. The piles were 42.5 in. (106 cm) long with outside and inside diameters of 1.068 in. (26.7 mm) and 0.837 in. (20.93 mm), respectively. The slenderness ratio ($l/r_0$) of piles was greater than 40 and the pile spacing was about 3 diameters. The pile group was placed in a hole in the ground, which was backfilled with a specially prepared soil mixture. The pile cap was 2.4 in. (6 cm) above the ground level. The pile group was excited by a Lazan oscillator at frequencies of 6 to 60 Hz in the vertical and horizontal directions and in the torsional mode. Free vibration tests and static tests were also performed. The measured response curves were very linear for small amplitudes and indicated relatively small nonlinearity at amplitudes of 0.008 in. (0.2 mm). The test results of Gle (1981) and Woods (1984) show definitely nonlinear behavior of in situ piles.

Novak and El-Sharnouby (1984) analyzed the data as above by the following methods:

1. Using static interaction factors by Poulos (1971, 1975, 1979) and Poulos and Davis (1980)
2. Concept of equivalent piers
3. Using dynamic interaction factors by Kaynia and Kausel (1982)
4. Direct dynamic analysis of Waas and Hartmann (1981)

### 7.9.3 Horizontal Response

Horizontal, rocking, cross stiffness, and damping constants, $k_x$, $k_\phi$, $k_{x\phi}$, $c_x$, $c_\phi$, and $c_{x\phi}$ were calculated for a single pile using the computer program PILAY 2. A group interaction factor, $\sum\alpha_L$, of the group of 102 piles based on Poulos' charts (1975, 1979) was estimated approximately as 13. This interaction factor was applied only to the horizontal stiffness $k$. The theoretical horizontal component of coupled response to horizontal excitation, based on the static interaction factor, is shown together with the experimental one in Figure 7.47. Four theoretical response curves have been plotted against the experimental one. Curve $a$ represents the group response without any interaction effect, while curve $b$ was calculated using the static interaction factor for stiffness only. It can be seen that a much lower value of the interaction factor is needed for the stiffness if the resonant frequency is to be matched. Therefore, an interaction factor of 2.85 was introduced for stiffness of yield curve $c$. The best agreement between the theoretical and experimental curves was achieved by increasing the damping constant by 45 percent (curve $d$). Yet some discrepancy between the theoretical and experimental response curves occurs at frequencies other than the resonant frequency. This indicates the limits of the applicability of static interaction factors (Novak and El-Sharnouby, 1984).

The experimental curves approach unity as frequency increases, which suggests that no correction with regard to the apparent mass appears necessary in the case of horizontal response. An apparent mass was determined and introduced in computations for matching the predicted response with the measured response in vertical vibrations. The correction factors have been

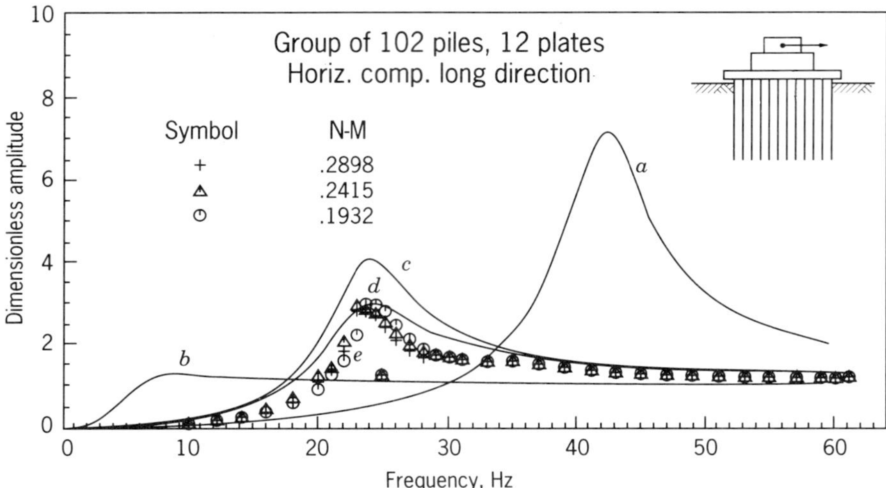


**Figure 7.47** Experimental horizontal response curves and theoretical curves calculated with static interaction factors. (a) No interaction, (b) static interaction factors applied to stiffness only (c) interaction factor of 2.85 applied to stiffness only; (d) with 2.85 and $-1.40$ interaction factors for stiffness and damping respectively; and (e) experimental data (Novak and El-Sharnouby, 1984).

applied arbitrarily to the theoretically computed stiffness and damping values, to match the predicted and experimental curves. Thus, there is a need to revise the interaction factors $\alpha$'s analytically.

### 7.9.4 Concept of Equivalent Pier

The equivalent radius, moment of inertia, and mass moment of inertia were calculated. The stiffness and damping constants of the equivalent pier were evaluated from the PILAY 2 computer program.

The behavior of the layers below the pier tip was considered in two ways: (1) the actual thickness and shear wave velocity of the layers up to a depth of 3 times the pier length were considered, and (2) an average of soil characteristics below the tip was taken to characterize a half-space lying just below the pier tip.

The latter approximation was considered since the first one may under estimate the stiffness for the coupled horizontal and rocking motion. The response was calculated assuming 2 degrees of freedom (i.e., sliding and rocking).

The dimensionless horizontal component of coupled response to horizontal excitation is plotted against the experimental response in Figure 7.48. The approach of taking the average shear wave velocity for the layers under the pier (curve *b*) yields results somewhat closer to the experimental data. It can be seen that the pier concept provides a very good estimate of stiffness (resonant

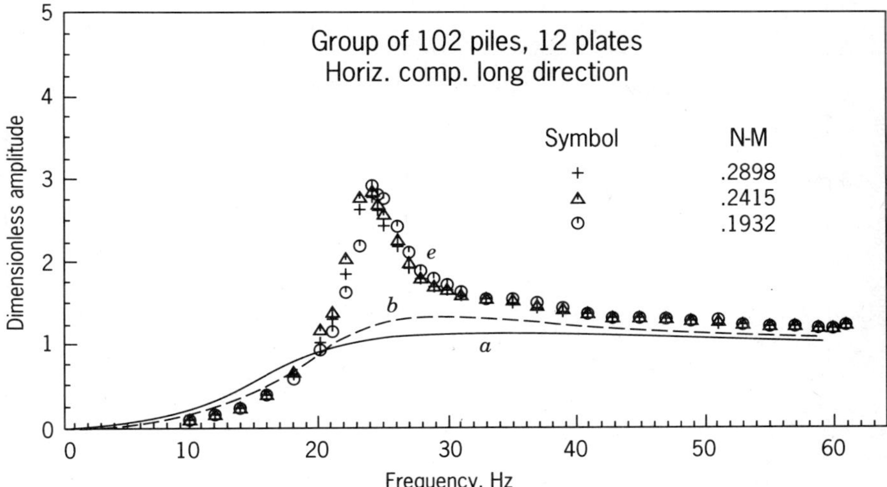


**Figure 7.48** Experimental horizontal response curve and theoretical curve based on equivalent pier concept (a) Considering soil layers under pier, (b) taking average of soil layer properties under pier tip as half space, and (e) experimental data (Novak and El-Sharnouby, 1984).

frequency) but overestimates damping. A far better match with experimental curve is achieved by considering only 40 percent of the theoretical damping constant (Figure 7.49).

Novak and El-Sharnouby compared the experimental data with the predictions by Kaynia and Kausel's (1982) method. For horizontal dynamic loading, interaction factors have been presented in the form of charts by Kaynia and Kausel. The dynamic interaction factors for the group of 102 piles were established as 4.2 and − 1.4 for the horizontal stiffness and damping constants, respectively. The computed response curve *a* (Figure 7.50) is plotted along with the experimental data. The stiffness is moderately underestimated and the damping ratio somewhat overestimated.

The horizontal dynamic impedances of Waas' equivalent axisymmetric model for the Novak and El-Sharnouby (1984) group were computed by Waas using the Waas and Hartmann analysis (1981), and the PILAY computer program for a frequency range of 0 to 50 Hz. The theoretical dimensionless response curve based on Waas' impedance is also shown in Figure 7.50 (curve *b*). It can be seen that the theoretical stiffness is somewhat underestimated and the damping considerably overestimated, but considering the complexity of the problem, the response prediction may be considered reasonably good. Waas used soil material damping ratio $\beta = 0.1$; his prediction would be even better for smaller damping (Novak and El-Sharnouby, 1984). Also, comparisons of the theoretical and measured response both in vertical as well as in torsional vibrations by several methods have been presented by the authors. The above discussion points to the fact that

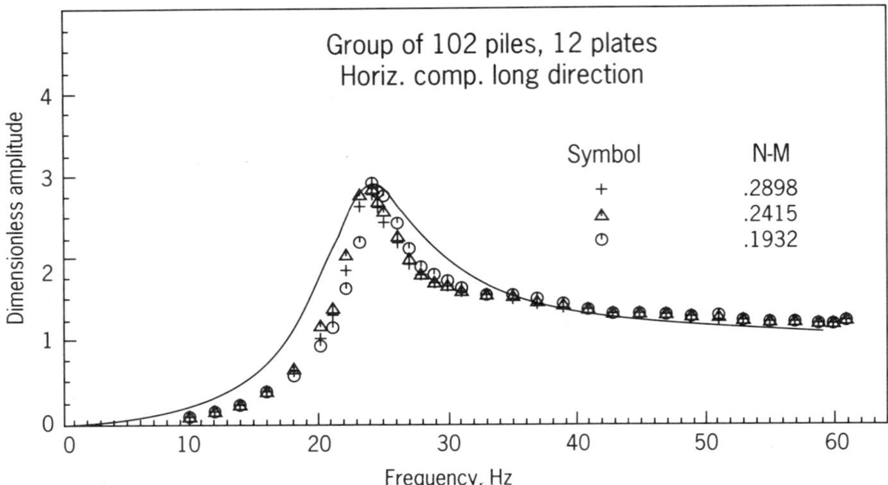


**Figure 7.49** Experimental horizontal response curve and theoretical curve based on equivalent pier concept considering 40 percent only of its damping constant (Novak and El-Sharnouby, 1984).

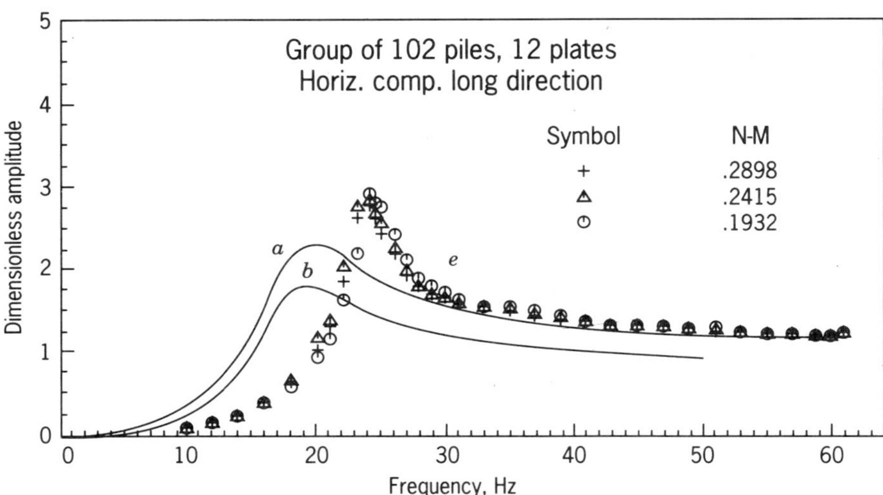


**Figure 7.50** Experimental horizontal response curve and theoretical curves (a) Calculated with Kaynia and Kausel dynamic interaction factors, (b) calculated with Waas and Hartmann impedences, (e) experimental (Novak and El-Sharnouby, 1984).

dynamic interaction is very complicated and further theoretical and experimental research is needed in dynamic behavior of piles groups.

## 7.10 PILES IN LIQUEFYING SANDS

As earthquake induced pore-water pressure rises in a saturated sand deposit, the effective stress decreases and, consequently, the bearing capacity of the soil also decreases. A piled foundation supported in such a deposit will thus experience additional settlement, which will be related to the increase in excess pore pressure.

According to Vesic (1964) the displacements required to reach ultimate pile tip loads are roughly proportional to pile diameter. In the studies of DeAlba (1983a, b) on a shake table, therefore, both static bearing capacity and dynamic behavior have been related to pile diameter.

In single-pile tests, static bearing capacity was determined as the load required to produce a pile deformation of 10 percent of its diameter, as suggested by Vesic, (1977b) for full-size driven piles. Dynamic pile settlement under increasing pore pressure is reported in fractions of diameter against pore pressure ratio $r$, defined as pore pressure increase, $\Delta u$, divided by initial vertical effective stress, $\sigma_v'$. It was considered that dimensionless results obtained in this way would be applicable to the prototype situation.

DeAlba (1983a) reported tests of a model study on a single pile embedded in a saturated sand placed in a tank that could be excited by a constant amplitude sinusoidal force. A commercially available washed and screened Holliston 00 sand was used in his tests. Holliston 00 is a clean, uniform medium sand with subangular grains composed basically of quartz, feldspars, ferromagnesiams, and mica with the following properties:

1. $D_{50} = 0.40$ mm; $D_{60} = 0.43$ mm; $D_{10} = 0.24$ mm
2. Coefficient of uniformity: $C_u = 1.8$
3. Percent passing 200 mesh: 1 percent
   Maximum density (ASTM-D2049-64): 107.4 lb/ft$^3$ (16.9 kN/m$^3$)
   Minimum density: 88.2 lb/ft$^3$ (13.9 kN/m$^3$)
   Mean specific gravity of grains: 2.69
   Friction angle, $\phi$, at $D_r = 50$ percent: 37°

The degree of saturation of the sand was between 99.5 percent and 99.9 per cent. Cyclic loading was applied in undrained conditions to induce liquefaction and subsequent cyclic mobility. Basic parameters measured in each test included (1) pore pressure, (2) vertical pile displacement, (3) horizontal base displacement, and (4) applied load.

Altogether, 35 successful tests were performed in the program. Besides pile diameter and static safety factor, basic variables considered were (1) relative density, (2) effective confining stress level, (3) stress history, and (4) length of pile inserted.

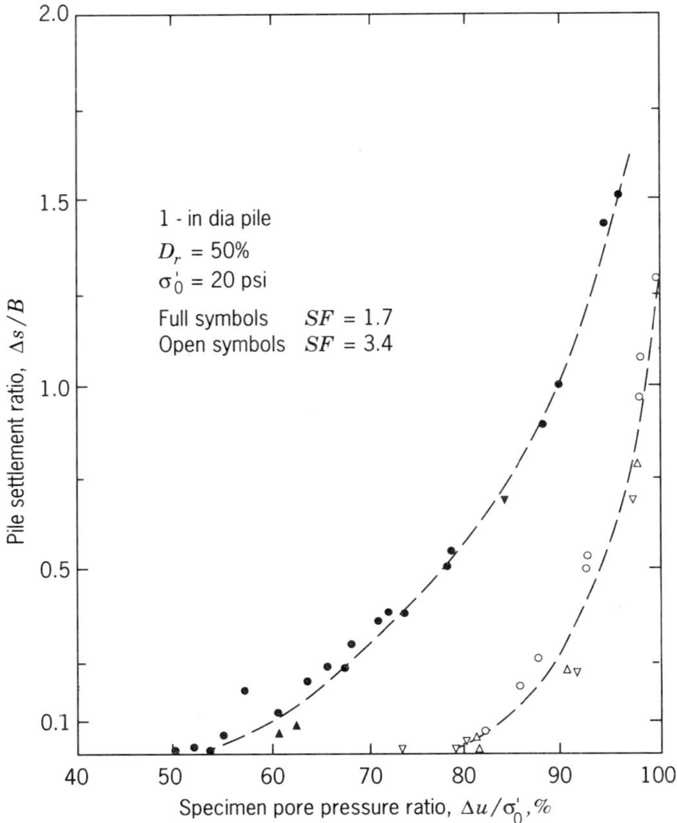


**Figure 7.51** Settlement of 1-in (25-mm) diameter pile, $D_r = 50$ percent; $\sigma_0' = 20$ psi (140 k Pa) (DeAlba, 1983a).

Figure 7.51 summarizes the settlement ratio, $\Delta s/B$ in sand with $D_r = 50$ percent. Figure 7.52 shows results for $D_r = 68$ percent under the same conditions.

These results show that pile settlement is very sensitive to the level of earthquake-induced pore pressure, and that settlement $\Delta s/B$ amounting to pile failure will occur before liquefaction $\Delta u/\sigma_0' = 1$ is reached.

DeAlba (1983b) reported further tests with a group of four closely spaced piles. For details of the experimental setup see DeAlba (1983a, b).

The variables that were monitored during each test included the (1) applied cyclic load, (2) the dynamically induced pore water pressure, and (3) the displacement of the control pile. All the pile group tests were carried out with specimens at a relative density of 50 percent.

In Figure 7.53 the pile settlement ratio ($\Delta s/B$) and pore pressure ratio $\Delta u/\sigma_0'$ for

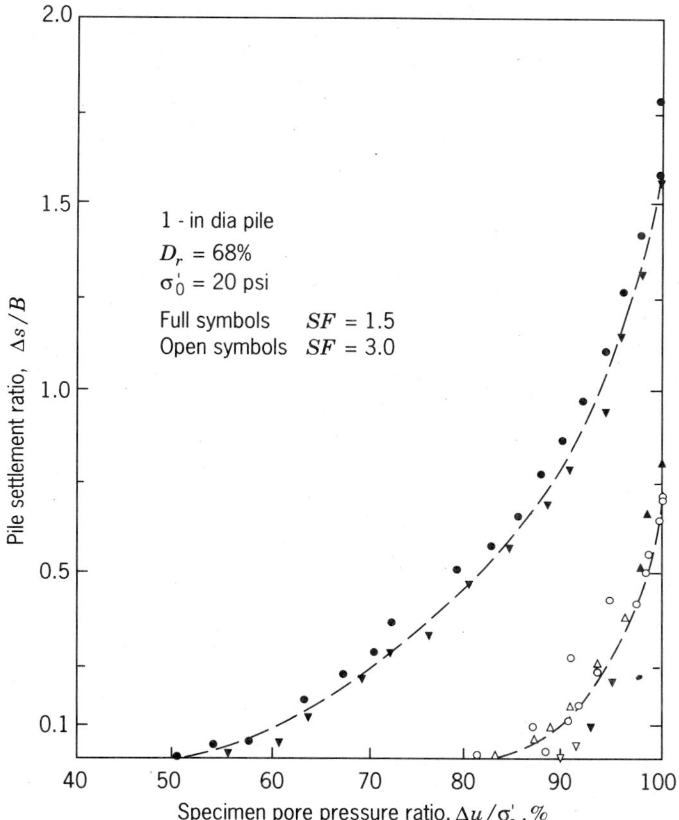


**Figure 7.52** Settlement of 1-in. (25-mm) diameter pile, $D_r = 68$ percent, $\sigma'_0 = 20$ psi (140 k Pa) (DeAlba, 1983a).

the control pile in the four-pile group are plotted. The figure shows that essentially no settlement was observed, within the range of safety factors studied, until $\Delta u/\sigma'_0$ rose beyond a threshold level. Pile settlement then developed relatively slowly until $\Delta s/B$ reached about 0.05 to 0.1, and increased rapidly thereafter as residual pore pressure built up, with pile movement generally exceeding one diameter before liquefaction was reached.

Failure pore pressure ratio under dynamic loading was defined as that value of $\Delta u/\sigma'_0$ for which a $\Delta s/B$ value of 0.1 was observed. This definition is consistent with the definition of failure settlement accepted for static loads (DeAlba, 1983a).

Figure 7.54 from the single pile study, shows that the failure $\Delta u/\sigma'_0$ have unique values at different relative densities if the static safety factors are the same.

Figure 7.55 shows $\Delta s/B$ versus $\Delta u/\sigma'_0$ for a single pile in a deep deposit at a

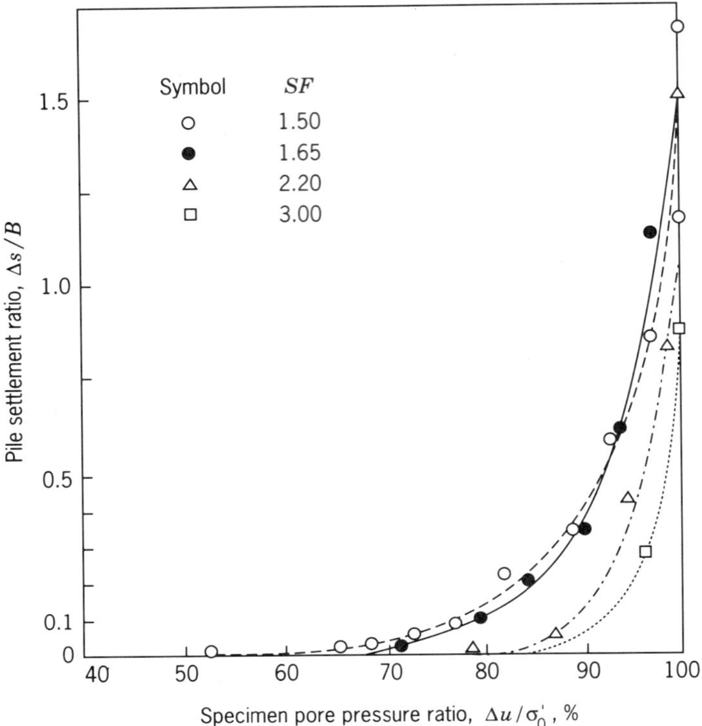


**Figure 7.53** Pile group: Control pile settlement (DeAlba, 1983b).

relative density of 50 percent. DeAlba (1983b) recommends that this figure may also represent earthquake-induced pore pressure at failure conditions for a pile group, failure being defined as an additional settlement of 0.1 diameters under dynamic loading. The individual piles in the group are expected to exhibit the settlement behavior shown in dimensionless form in Figure 7.55 (DeAlba, 1983b).

The model test data in the preceding paragraphs need verification with actual pile behavior from piles subjected to earthquakes. In the meantime, considerable engineering judgement may be needed to apply these results to an actual problem.

## 7.11 OVERVIEW

Piles are used extensively for supporting building foundations, in seismic zones for machine foundations and for offshore structures. The nature of pile response and pile interactions are quite different in all three cases. Earthquake loading for piles under buildings may cause large deformations and soil nonlinearity. On the

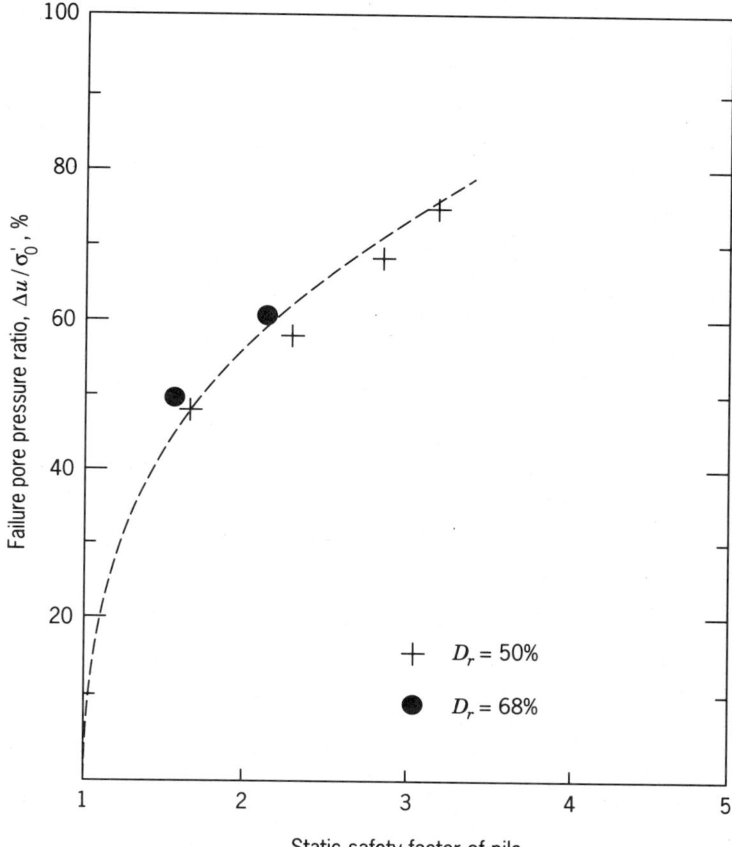


**Figure 7.54** Failure conditions for single pile. Tests at $D_r = 50$ and 68 percent (DeAlba, 1983b).

contrary, machines may cause only small amplitudes of vibrations, and soils may behave as elastic materials. In offshore structures, the piles are especially long and slender with considerable freestanding lengths.

In earthquake loading, only lateral vibrations may be important while in machine foundations, the piles may be subjected to vertical oscillations, horizontal translation and rocking, and torsion.

Solutions based on beam on elastic foundation for static loads has been extended for dynamic loading by Chandrasekaran (1974) Penzien (1970) and a design procedure has been proposed based on spectral response technique. For pile-supported machine foundations, simple solutions for single piles in all the modes of vibrations have been included in this chapter. Also, group action on the behavior of the total system as compared to that of the single pile has been

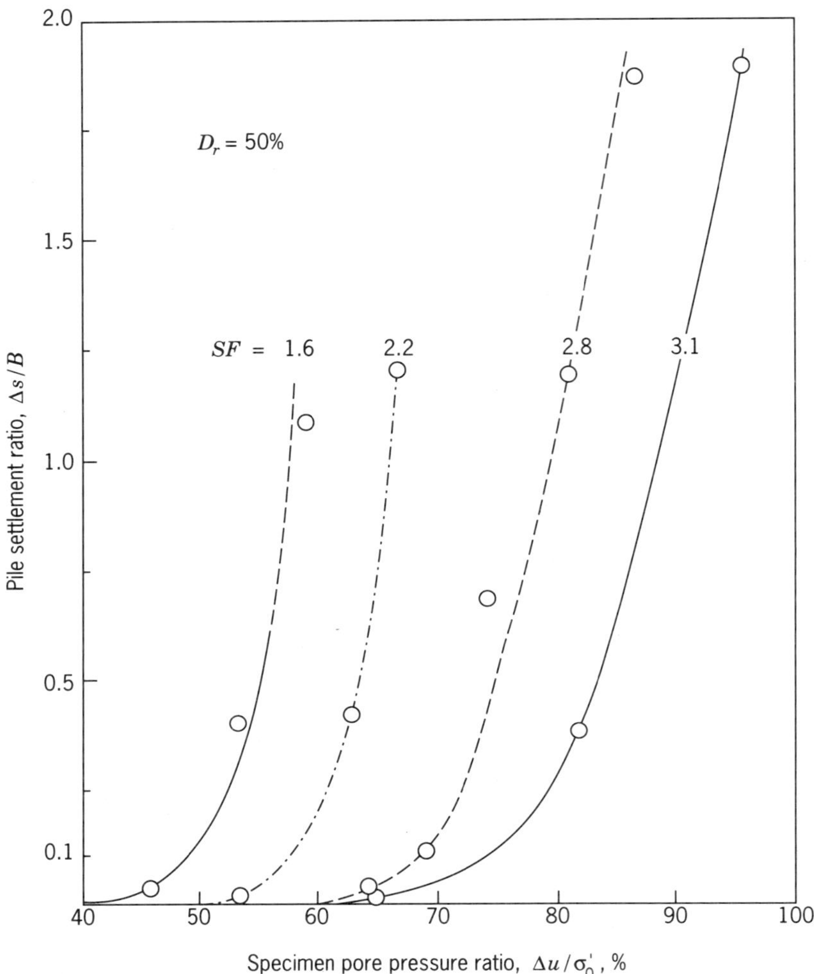


**Figure 7.55** Pile settlement in deep deposits (DeAlba, 1983b).

evaluated, and a complete analysis has been included. However, there are certain definite gaps in the present (1990) understanding of single pile and pile group action under dynamic loads.

Nogami (1983) and Nogami and Liang (1983) have also obtained solutions for pile groups and have shown that the concept of the Winkler soil model could be applicable to pile groups for the frequency range higher than the fundamental natural frequency of the soil deposit. It was further found that:

1. A dynamic group effect can be strongly frequency dependent and depends on the ratio between the pile spacing and the wavelength propagating in the soil, due to the phase shifts between the directly induced pile motion and the

transmitted motions. Thus, the type of predominant waves induced in the soil, frequency, and pile spacing control the frequency dependent behavior of pile groups. The effect of material damping of the soil is primarily a reduction of the amplitude of the motion.

2. A dynamic group effect is more pronounced in larger groups with stiffer piles.
3. Under the dynamic load, the group effect may increase or decrease the values of the stiffness and damping parameters per pile in a group from the values of a single pile, although it always decreases the stiffness value under a static load.

Initial analyses by Barkan (1962) and Maxwell et al. (1969) have been shown to have only limited application. For vertically vibrating piles, Novak's (1974) analysis for single piles is reasonable and uses rational soil and pile properties. However, in case of groups, static interaction factors have been used (Novak and Grigg, 1976). Sheta and Novak (1982) developed an approximate theory for vertical vibrations of pile groups.

On the basis of comparison of predicted and measured response of 102 closely spaced pile groups in vertical vibrations, Novak and El-Sharnouby (1984) have shown that:

1. Correction for the apparent mass in vertical vibrations may be necessary, particularly for rigid floating closely spaced piles.
2. The static interaction factor provided quite a good estimate of the group stiffness, but the group damping could not be predicted.
3. For closely spaced piles, the equivalent pier concept provided a reasonable agreement with the experimental data if the theoretical damping constant was reduced to 40 percent.

The solutions are by no means simple in their present form. Therefore, more research is needed to solve the problem completely and put it in a form which can be easily used by the practicing engineer.

Horizontal vibrations of piles have been investigated by considering the pile as

1. an equivalent cantilever,
2. a beam on elastic foundation (Tucker 1964 and Prakash 1981), and
3. installed in a continuum (Novak 1974).

The equivalent cantilever method does not consider realistic behavior of soil pile system. Solutions for beams on elastic foundations need to be developed further to put them in readily usable forms. Novak's solution for a single pile and for pile groups for horizontal vibrations is subject to the same limitation as that for vertical vibrations.

The tests of Novak and El-Sharnouby (1984) showed that the static group interaction effects differed considerably from dynamic group effects in horizontal vibrations. However, the equivalent pier concept predicted the stiffness well but not the damping for that particular case. For single piles also, Woods (1984) found that softened zone around the pile in clay alters the behavior and needs to be considered in a realistic analysis.

For torsional vibrations of vertical piles, Novak and Howell (1977) developed solutions for the dynamic stiffness and damping, which depend on soil–pile interaction in terms of such dimensionless parameters as (1) shear wave velocity ratio (ratio of soil shear wave velocity to pile shear wave velocity), (2) slenderness ratio (ratio of pile length to effective radius), (3) mass ratio (ratio of specific mass of the soil to specific mass of the pile), (4) dimensionless frequency, and (5) material damping ratio.

For a group of piles, the contribution from torsion to the total stiffness and damping decreases with the relative distance of the pile from the centroid of the footing. Pile foundations can have smaller natural frequencies in torsion than shallow footings, but the increased damping of the system gives lower resonant amplitudes. This contrasts with other modes of vibration. Since pile slippage and other effects such as method of installing the piles are not accounted for in this theory, comparison with experiments is desirable.

The interaction of pile cap with soil affects the dynamic response of the system, which can be accounted for in all modes of vibrations on the basis of principles of embedded foundations (Prakash and Puri, 1988).

Based on the approximate solutions in the preceding sections, a step-by-step design procedure has been developed. Solved problems have been included to illustrate the developed design procedure.

The soil properties used in defining the stiffness and damping parameters are (1) shear wave velocity $V_s$ and shear modulus $G$ and (2) Poissons ratio $\nu$, which may be determined as described in Chapter 4.

Aubry and Postel (1985) considered the soil–pile system as a fiber-reinforced composite material and the technique of homogenization of composite materials was used to compute equivalent modulus that were used to compute the seismic response of the equivalent foundation at the soil surface. This method has been shown to be useful particularly for very large number of piles beneath a foundation. This method may be regarded as a complimentary solution to Novak's equivalent pier concept for closely spaced piles.

Gazetas and Dobry (1984) proposed a method to compute the response of a single, fixed head pile under horizontal excitation at its head. In this method, the solution is based on realistic estimation of (1) deflections of the pile under static lateral load, (2) dashpots attached to the pile at every elevation, (3) a dashpot at its head, and (4) a variation of spring coefficient and damping ratio with frequency.

The applicability of the proposed method has been illustrated in three linearly hysteretic soil deposits: homogeneous deposit with modulus constant with depth, in-homogeneous deposit with modulus increasing linearly with depth, and layered deposit. Hassini and Woods (1989) have studied stiffness change in model pile groups in sands with spacing of piles in both vertical and horizontal vibrations.

Centrifuge studies on models of pile foundations have been performed. More recently (1987), the tests of Finn and Gohl have shown quantitatively the extent of group action in a four-pile group under lateral vibrations. The soil shear velocity was measured with piezoceramic bender elements in the soil with depth. This data have shown that the carefully conducted centrifuge tests, short of full-scale tests, hold promise for validation of theoretical formulations. The question of settlement and failure of single pile and pile group under vibrations in liquefying sand has been studied by DeAlba (1983a, b). The results of the model study are promising, although their field verification is needed.

The philosophy and methods described in this chapter will change as the understanding of the dynamic pile behavior improves. There is an urgent need for more theoretical and experimental research. (See Sreerama, 1990).

## REFERENCES

Agarwal, H. P. "Effect of Vibrations on Skin Friction of Piles," Master of Engineering Dissertation, University of Roorkee, Roorkee, India. 1967.

Aubry, D. and Postel, M., "Dynamic Response of a Large Number of Piles by Homogenization" *Proceedings Second International Conference on Soil Dynamics and Earthquake Engineering, aboard the Queen Elizabeth II*, New York–Southampton, June 1985, pp. 4–105 to 4–119.

Barkan, D. D., *Dynamics of Bases and Foundations* McGraw-Hill Book Co. New York 1962.

Baranov, V. A., "On the Calculation of Excited Vibrations of an Embedded Foundation," (in Russian) *Voprosy Dynamiki Prochnocti*, No. 14 Polytechnical Institute of Riga, Latvia. 1967, pp. 195–209.

Beredugo, Y. O. and Novak, M., "Coupled Horizontal and Rocking Vibrations of Embedded Footings," *Can. Geotech. J.*, Vol. 9, No. 4, November 1972, pp. 477–497.

Bycroft, G. N., "Forced Vibration of a Rigid Circular Plate on a Semi-Infinite Elastic Half Space on Elastic Stratum," *Philosophical Transactions of the Royal Society*, London, U.K., Series A, Vol. 248, No. 948, 1956, pp. 327–368.

Chandrasekaran, V., "Analysis of Pile Foundations Under Static and Dynamic Loads," Ph.D. Thesis, University of Roorkee, Roorkee, India 1974.

Davisson, M. T. and Gill, H. L., "Laterally Loaded Piles in a Layered Soil System," *J. Soil Mech. Found. Div.*, ASCE, Vol. 89, No. SM 3, 1963, pp. 63–94.

DeAlba, P. A., "Pile Settlement in Liquefying Sand Deposit," *J. Geot. Eng. Dn.* ASCE Vol. 109, No. 9, September, 1983a, pp. 1165–1180.

DeAlba, P. A., "Group Effect on Piles in a Liquefying Sand Deposit," Proceedings of the Conference of Geotechnical Practice in Off-Shore Piling, University of Texas, Austin, TX, April, 1983b, pp. 300–314.

Finn, W. D. and Gohl, W. B., "Centrifuge Model Studies of Piles Under Simulated Earthquake Loading," Dynamic Response of Pile Foundation, Geotechnical Special Publication, No. 11, ASCE, Atlantic City, NJ, 1987, pp. 21–28.

Fukoka, M. "Damage to Civil Engineering Structures" Soils and Foundations Tokyo, Japan, Vol. 6, No. 2, March 1966, pp. 45–52.

Gazetas, G. and Dobry, R., "Horizontal Response of Piles in Layered Soils," *J. Geot. Eng. Div.*, ASCE, Vol. 110, No. GT1, January, 1984, pp. 20–40.

Gazetas, G. and Roesset, J. M., "Vertical Vibrations of Machine Foundations," *J. Geot. Eng. Dn.*, ASCE, Vol. 105, No. GT12, 1979, pp. 1435–1454.

Ghumman, M. S., "Effect of Vertical Vibrations on the Penetration Resistance of Piles," Ph.D. Thesis, University of Roorkee, Roorkee, India 1985.

Gle, D. R., "The Dynamic Lateral Response of Deep Foundations," Ph.D. Dissertation, The University of Michigan, Ann Arbor 1981.

Hassini, S., and Woods, R. D., "Dynamic Experiments with Model Pile Foundations," Proc. 12th International Conference on Soil Mechanics and Foundation Engineering, Vol. II, pp. 1135–1138, Rio de Janeiro (Brazil) 1989.

Hayashi, S. C., "A New Method of Evaluating Seismic Stability of Steel Structures, *Proceedings Fifth World Conference on Earthquake Engineering*, Rome, Italy, Vol. 2, 1973, pp. 2602–2605.

Housner, G. W., "Design Spectrum," in R. L. Wiegel (ed.), *Earthquake Engineering* Prentice-Hall, Englewood Cliffs, NJ, 1970.

Kaynia, A. M. and Kausel, F., "Dynamic Behavior of Pile Groups," Proceedings, Second International Conference on Numerical Methods in Offshore Piling, Austin, TX, 1982, pp. 509–532.

Madhav, M. R. and Rao, N. S. V. K., "Model for Machine Pile Foundation Soil System," *J. Soil Mech. and Found. Div.*, ASCE, Vol. 97, No. SMI, 1971, pp. 295–299.

Maxwell, A. A., Fry, Z. B., and Poplin, J. K., "Vibratory Loading of Pile Foundations," ASTM, *Special Technical Publication No. 444*, 1969, pp. 338–361.

Nogami, T., "Dynamic Group Effect in Axial Responses of Grouped Piles," *J. Geotech. Eng.*, ASCE, Vol. 109, No. GT2, 1983, pp. 220–223.

Nogami, T. and Liang, H., "Behavior of Pile Groups Subjected to Dynamic Loads," *Proceedings 4th Canadian Conference on Earthquake.* Engineering, 1983, pp. 414–420.

Novak, M., "Dynamic Stiffness and Damping of Piles," *Can Geotech. J.*, Vol. 11, No. 4, 1974, pp. 574–598.

Novak, M., "Vertical Vibration of Floating Piles," *J. Eng. Mech. Div.*, ASCE, Vol. 103, No. EMI, 1977a, pp. 153–168.

Novak, M., "Foundations and Soil Structure Interaction," Theme Report, Topic 4, *Proceedings VI World Conference on Earthquake Engineering*, Vol. 2, New Delhi, 1977b, pp. 1421–1448.

Novak, M. and Aboul-Ella, E., "PILAY—A Computer Program for Calculation of Stiffness and Damping of Piles in Layered Media," Report No. SACDA 77–30, University of Western Ontario, London, Ontario, Canada, 1977.

Novak, M. and Beredugo, Y. O., "Vertical Vibration of Embedded Footings," *J. Soil Mech. Found. Div.*, ASCE, Vol. 98, No. SM12, 1972, pp. 1291–1310.

Novak, M. and El-Sharnouby, B., "Stiffness and Damping Constants of Single Piles," *J. Geotech. Eng. Div.*, ASCE, July, Vol. 109, No. 7, 1983, pp. 961–974.

Novak, M. and El-Sharnouby, B., "Evaluation of Dynamic Experiments on Pile Group," *J. Geotech. Eng. Div.*, ASCE, Vol. 110, No. 6, 1984, pp. 738–756.

Novak, M. and Grigg, R. F., "Dynamic Experiments with Small Pile Foundation," *Can. Geot. J.*, Vol. 13, No. 4, 1976, pp. 372–395.

Novak, M. and Howell, J. F., "Torsional Vibrations of Pile Foundations," *J. Geot. Eng. Div.*, ASCE Vol. 103, No. GT4, 1977, 271–285.

Novak, M. and Sach, K., "Torsional and Coupled Vibrations of Embedded Footings," *Int. J. Earthquake Eng. Structural Dynamics*, Vol. 2, No. 1, 1973, pp. 11–33.

Penzien, J., "Soil–Pile Foundation Interaction," in R. L. Wiegel (ed.), *Earthquake Engineering*, Prentice-Hall, Inc., Englewood Cliffs, NJ. 1970.

Poulos, H. G., "Analysis of the Settlement of the Pile Groups," *Geotechnique*, Vol. XVIII, No. 4, 1968, pp. 449–471.

Poulos, H. G., "Behavior of Laterally Loaded Piles. II—Pile Groups," *J. Soil Mech. and Found Div.*, ASCE, Vol. 97, No. SM5, 1971, pp. 733–751.

Poulos, H. G., Lateral Load Deflection Prediction for Pile Groups," *J. Geotechn. Eng. Div.*, ASCE, Vol, No. GT1, 1975, pp. 19–34.

Poulos, H. G., "Groups Factors for Pile-Deflection Estimation," *J. Geotech. Eng. Div.*, ASCE, Vol. 105, No. GT12, 1979, pp. 1489–1509.

Poulos, H. G. and Davis, E. H., *Pile Foundation Analysis and Design*, Wiley, New York, 1980.

Prakash, S., "Behavior of Pile Groups Subjected to Lateral Loads," Ph.D. Thesis University of Illinois, Urbana, 1962.

Prakash, S., *Soil Dynamics*, McGraw-Hill Book Co., New York, 1981.

Prakash, Sally "Buckling Loads of Fully Embedded Piles," M.S. Thesis, University of Missouri-Rolla, 1985.

Prakash, Sally, "Buckling Loads of Fully Embedded Piles," *Int. J. Comp. Geotech.*, Vol. 4 (1987) pp. 61–83.

Prakash, S. and Agarwal, H. P., "Effect of Vibrations on Skin Friction of Piles *Proceedings Fourth Asian Regional Conference on Soil Engineering*, Bangkok, Thailand, Vol. 1, 1971.

Prakash, S. and Agarwal, S. L., "Effect of Pile Embedment on Natural Frequency of Foundations," *Proceedings South East Asian Regional Conference on Soil Mechanics and Foundation Engineering*, Bangkok, Thailand, 1967, pp. 333–336.

Prakash, S. and Chandrasekaran, V., "Pile Foundations Under Lateral Dynamic Loads," *Proceedings Eighth International Conference on Soil Mechanics and Foundation Engineering*, Moscow, Vol. 2, 1973, pp. 199–203.

Prakash, S. and Chandrasekaran, V., "Free Vibration Characteristics of Piles," *Proceedings Ninth International Conference on Soil Mechanics and Foundation Engineering*, Tokyo, Vol. 2, 1977, pp. 333–336.

Prakash, S. and Chandrasekaran, V., "Analysis of Piles in Clay Against Earthquakes," Preprint no. 80–109, ASCE Convention and Exposition, Portland, OR, April 14–18. 1980.

Prakash, S. and Gupta, L. P., "A Study of Natural Frequency of Pile Groups," *Proceedings Second South East Asian Regional Conference on Soil Engineering*, Singapore, Vol. 1, 1970, pp. 401–410.

Prakash, S. and Puri, V. K., *Foundation for Machines, Analysis and Design*, Wiley, New York, 1988.

Prakash, S., Ranjan, G., and Saran, S. *Analysis and Design of Foundations and Retaining Structures*, Sarita Prakashan Meerut, UP, India. 1979.

Prakash, S. and Sharma, H. D., "Analysis of Pile Foundations Against Earthquakes," *Indian Concr. J.*, 1969, pp. 205–220.

Reese, L. C., and Matlock, H., "Non-dimensional Solutions for Laterally Loaded Piles with Soil Modulus Assumed Proportional to Depth," *Proceedings 8th Texas Conf. on Soil Mechs. and Found Eng.* Austin TX 1956, pp. 1–41.

Richart, F. E., Jr., "Foundation Vibrations," *Transactions*, ASCE, Vol. 127, Part 1, 1962, pp. 863–898.

Richart, F. E., Hall, J. R., and Woods, R. D., *Vibrations of Soils and Foundations*, Prentice-Hall, Inc., Englewood Cliffs, NJ. 1970.

Scott, R. F., Ting, J. M., and Lee, J., "Comparison of Centrifuge and Full-Scale Dynamic Pile Tests," *Proceedings Soil Dynamics and Earthquake Engineering Conference*, Southampton, UK, Vol. I, 1982, pp. 281–301.

Sheta, M. and Novak, M., "Vertical Vibrations of Pile Groups," *J. Geot. Eng.*, ASCE, Vol. 108, No. GT4, April, 1982, pp. 570–590.

Sreerama, K., "Dynamic Soil Pile Interactions," Ph.D. thesis, Civil Engineering Departments, University of Missouri-Rolla (1990) In preparation.

Swiger, W. F., "Effect of Vibration of Piles in Loose Sand," *Proceedings Second International Conference on Soil Mechanics and Foundation Engineering*, Rotterdam, Vol. 2, 1948, p. 19.

Tucker, R. L., "Lateral Analysis of Piles with Dynamic Behavior," *Proceedings North American Conference on Deep Foundations*, Mexico City, Vol. I, 1964, pp. 157–171.

Vesic, A., "Model Testing of Deep Foundations in Sand and Scaling Laws," Panel discussion, Session II, *Proceedings of the North American Conference on Deep Foundation*, Mexico Cit, Vol. II, 1964, pp. 525–533.

Vesic, A., "Design of Pile Foundations," National Cooperative Highway Research Program, Synthesis of Highway Practice No. 42, TRB, NRC, Washington DC 1977.

Waas, G. and Hartmann, H. G., "Pile Foundations Subjected to Dynamic Horizontal loads," European Simulation Meeting, "Modelling and Simulating of Large Scale Structural Systems," Capri, Italy, 1981, p. 17. Also SMIRT, Paris.

Wiegel, R. L. *Earthquake Engineering*, Prentice Hall, Englewood Cliffs, NJ 1970.

Winkler, E., *Die Lehre von Elastizitat und Festigkeit* Prague, 1867, p. 182.

Woods, R. D., "Lateral Interaction between Soil and Pile," *Proceedings International Symposium on Dynamic Soil Structure Interaction*, Minneapolis, MN, September, 1984, pp. 47–54.

# 8

# ANALYSIS AND DESIGN OF PILE FOUNDATION IN PERMAFROST ENVIRONMENTS

In permafrost, there are many additional considerations that control the behavior of piles. In Section 4.3, the mechanical and deformation (creep) behavior of frozen soils having high ice and unfrozen water contents was shown to be greatly affected by changes in temperature. Also, the foundations are frequently subjected to large uplift forces caused by frost action in the active layer. This causes vertical forces on the grade beams or pile caps and *frost grip* or *adfreeze forces* along the sides of piles.

Two distinct soil condition—thaw stable and thaw unstable—necessitate altogether different types of pile analysis. In this chapter, basic definitions peculiar to permafrost are presented first, followed by general pile design consideration and piles subjected to vertical and lateral loads. Based on the behaviour of piles in permafrost, a design procedure has also been developed.

## 8.1 DEFINITIONS

The following definitions and terms applicable to permafrost have been used in this chapter.

**Active Layer** The top layer of ground above the permafrost table that is subject to annual freezing and thawing. This is also termed as *annual frost zone* that thaws each summer and refreezes each fall.

**Afreeze Bond Strength** The bond or the adhesive strength that is developed between the pile surface and the surrounding frozen soil.

**Cold Regions** Regions where frost penetrates the ground to a depth of about 0.3 m or more at least once in 10 years.

**Creep** The time-dependent shear strain or shear deformation behavior under undrained conditions. Frozen soils exhibit substantial deformation under sustained loading due to a complex phenomena of melting of ice and movement of water accompanied by a breakdown of the ice and their bonds with soil particles. This may result in plastic deformation of pore ice and soil particle readjustment.

**Freezing Index** The accumulated freezing days below 0°C during a single freezing season. It actually is the area between the 0°C line and the curve of mean daily temperatures below 0°C and is represented by the number of degree-days. It is generally used as a measure of potential frost penetration below ground.

**Freezeback Time** The time required for freezing the artificially thawed ground or warm soil-slurry backfilled in an annular space around a pile placed into the augered hole. Freezeback occurs due to natural conduction from the surrounding permafrost.

**Frost-susceptible Soils** Soils in which significant ice segregation occurs due to the growth of ice lenses when water flows to the freezing plane. Most widely accepted criteria for frost-susceptible soils is due to Casagrande (1932), which states that nonuniform soils containing more that 3 percent of particles smaller than 0.02 mm and uniform soils containing more that 10 percent smaller than 0.02 mm particle size can be classified as frost-susceptible soils.

**Frost Depth** The depth below ground surface to which the soil or rock freezes in winter. This depth depends on the ground temperature during winter, the soil type through which frost penetrates, and the ground surface conditions (e.g., depth of snow cover, vegetation, and exposure to sunlight).

**Frost Heave** It is the upward movement of ground due to the formation and growth of ice lenses in frost-susceptible soils.

**Frost Table** The frozen surface to which thawing of the seasonal frozen ground has occurred at any time in spring and summer. This table is in the active layer.

**Frozen Ground** It is the zone in ground where soil or rock is at the temperature below 0°C.

**Ice-poor Frozen Soils** Soils that do not have enough ice content to fill the pore spaces completely. Normally, these soils have bulk density greater than 1700 kg/m$^3$ and also exhibit some intergranular contact. Therefore, interparticle friction can be mobilized in these soils.

**Ice-rich Frozen Soils** Soils that contain ice in excess to that required to fill pore spaces. Thus, there is no grain to grain contact in these soils. Bulk density—including segregated ice—for these soils is typically less than about 1700 kg/m$^3$.

**Latent Heat of Fusion** The amount of heat that is required to melt the ice or freeze the water in a unit volume of soil without changing the temperature.

**Non-frost-susceptible Soils** These soils do not display ice segregation during freezing. Most sands and gravels that do not contain silt fall in this category. These soils do not exhibit frost heave that is characteristic during freezing Basically, soils that are not frost susceptible are frost stable and therefore fall in this category.

**Permafrost** The thermal condition of the ground when the soils or rocks are permanently frozen or are at temperature below 0°C continuously for over at least two consecutive winters and the intervening summer. Moisture or ice may or may not be present in these materials.

**Permafrost Degradation (Thawing)** The process that results in a decrease in permafrost thickness or an increase in the active layer due to artificial (e.g., removal of an insulating vegetation layer or construction activity) or natural (e.g., climatic warming) causes resulting in thawing.

**Permafrost Table** The surface that represents the upper boundary of permafrost.

**Primary Creep** A characteristic exhibited by frozen soils when their creep rate or deformation rate under constant stress continuously decreases with time. It appears that at low stress levels, low ice content, or ice-poor frozen soils exhibit this behavior.

**Seasonal Frost** The freezing phenomena of the soils below ground caused by subzero surface temperatures. This keeps the earth materials frozen only during winter months.

**Secondary Creep** Characterized by frozen soils when their creep rate is constant with time. This is also called steady-state creep. Ice-rich frozen soils under moderate stress conditions exhibit this behavior.

**Tertiary Creep** Characterized by an accelerated creep rate. This would normally lead to soil failure and is exhibited at high stress levels.

**Thaw Settlement** Downward movement of ground due to the dissipation of water on melting of excess ice in the soil.

**Thermal Conductivity** A measure of the quantity of heat that will flow through a unit area of unit thickness in unit time under a unit temperature gradient.

**Thermal Diffusivity** The ratio of thermal conductivity and the volumetric heat capacity and is an index of the facility with which a substance will have temperature change.

**Volumetric Heat Capacity** The amount of heat that is required to change the temperature of a unit volume of material by one degree.

## 8.2 GENERAL DESIGN CONSIDERATIONS

Chapter 2 (Section 2.6.2) described various types of piles used in the permafrost environment, Chapter 3 (Section 3.4.4) provided information on installation methods for these piles, and Chapter 4 (Section 4.3) covered the mechanical and deformation (creep) behavior of frozen soils. This chapter discusses the analysis and design of piles in permafrost environment.

The design of pile foundation in permafrost area is influenced primarily by the following factors:

1. The type and use of the structure (e.g., if the structure imparts heat to the underlying permafrost then the depth of active layer—the depth below which soil is permanently frozen—increases). This causes thawing of frozen soils, resulting in the decrease of soil strength and the increase of settlements.
2. Strength and deformation characteristics of foundation soils (e.g., mechanical properties and deformation (creep) behavior of frozen soils are temperature dependent). Frequently, these soils have high ice content (ice in excess to that required to fill pore spaces), thus there is no grain to grain contact. Therefore, these soils exhibit temperature unstable behavior. This has been discussed in Chapter 4 (Section 4.3) and will be further discussed in Section 8.2.1.
3. Ground thermal regime (e.g., temperature profile below ground both in summer and winter). Assessment of ground temperatures should be made both prior to and after construction during the entire life of the structure because the behavior of frozen soils is greatly affected by temperatures.

4. Nature of foundation soils. If these soils are frost susceptible (their pore space is such that on freezing they encourage the growth of ice lenses if water is available), then these soils heave due to frost action resulting in the application of adfreeze forces (upward forces along pile surface in an active zone due to frost heave) on piles during winter. These soils also exhibit thaw consolidation (downward movement of ground due to escape of water on melting of excess ice) resulting in the application of downdrag forces on pile. Section 8.2.2 further discusses frost heave forces, and Section 8.3.3 discusses downdrag forces applied along the pile surface. If foundation soils are composed of competent material such as ice-free rock, dense glacial till, non-frost-susceptible soils (clean, well-drained sand or gravel), then pile foundation design can be carried out in a conventional manner and frozen conditions can be neglected. On the other hand, if the materials are frost susceptible and are thaw unstable, then the following design concepts and procedures should be used.

### 8.2.1 Load-Settlement Behavior of Foundation in Frozen Soils

The load-settlement behavior of frozen soils depends on interparticle friction, particle interlocking, cohesion, and the bonding of particles by ice. The bonding of particles by ice, however, is the dominant factor that controls the behavior of frozen soils. At very high ice content (ice-rich frozen soils), the behavior of frozen soil is similar to that of ice, and factors such as temperature, pressure, strain rate, grain size, crystal orientation, and density influence its behavior. However, at low ice content (ice-poor frozen soils), the presence of unfrozen water films that surround the soil particles begin to influence soil behavior because interparticle forces, especially in fine-grained frozen soils, become effective.

When a frozen soil is subjected to deviatoric stress, it develops stress concentration on the ice component between soil particles. Similar stress concentrations on the ice may develop due to hydrostatic pressures. These would result in pressure melting of ice in frozen soils causing an increase in the amount of unfrozen water with pressure. It has been reported by Tsytovich (1960), Low et al. (1968) and Chamberlain et al. (1972) that pressure melting due to the application of hydrostatic or deviatoric stresses result in water flows to regions of lower stresses where it freezes again. This movement of water under stress results in breaking of structural and ice-cementation bonds. The mineral particles may therefore slip. This is a time-dependent process and may result in strength reduction of the soil. On the other hand, an increase in soil strength may take place due to the formation of some new ice-cementation bonds and increase in intermolecular bonds caused by simultaneous consolidation effects. This time-dependent strength decrease or increase in low-ice-content soils is important in understanding the behavior of frozen soils under loads. Such soils can be tested in uniaxial compression creep tests on cylindrical frozen soils. As shown in Figure 8.1a, if a series of uniaxial compression creep tests are carried out at the same temperature, then creep curves for different constant stress levels are

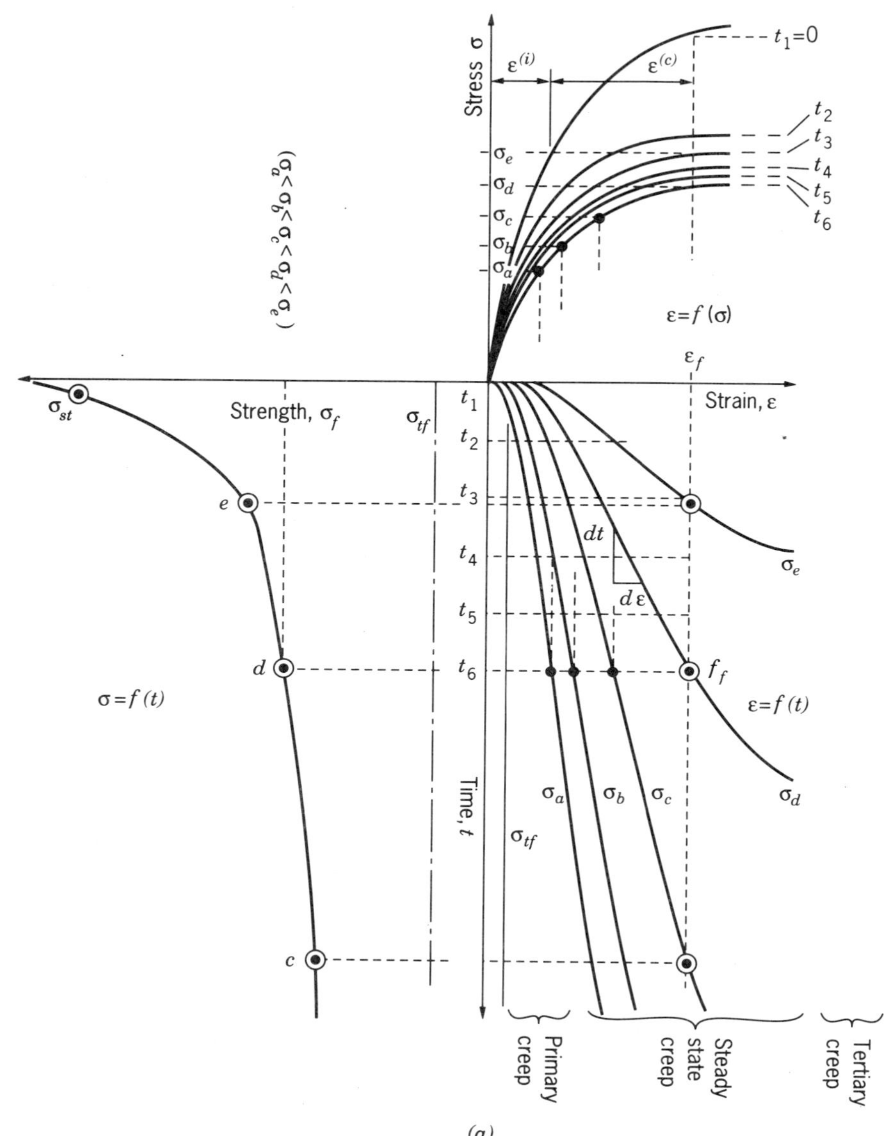


(a)

**Figure 8.1** Behavior of frozen soil under uniaxial creep test at constant temperature. (a) Typical plots of data from uniaxial compression creep tests conducted at a constant temperature and confining pressure (Ladanyi, 1972), (b) strain-time relationship in a constant stress creep test.

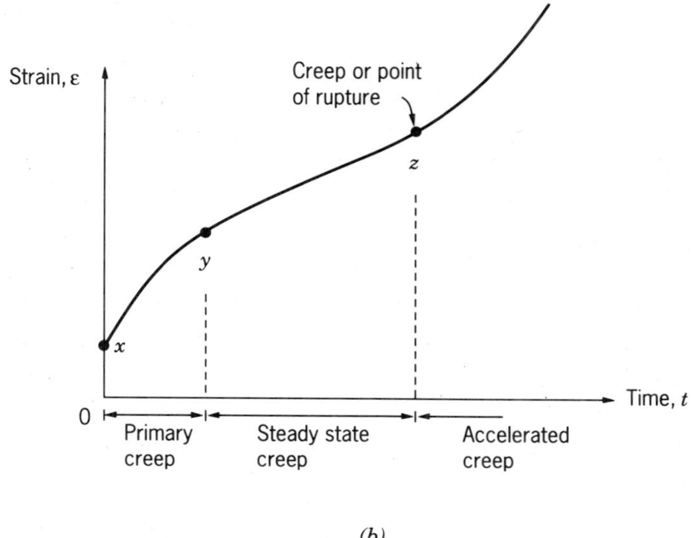


(b)

represented by plots shown on the lower right-hand corner of this figure. These curves show that three distinct stages of creep, as shown in Figure 8.1b, can be observed.

The first stage is called primary creep, which exhibits a continually decreasing creep rate or slope of the time ($t$) and strain ($\varepsilon$) plot. The second stage is called steady-state creep where the slope is constant and a minimum creep rate is reached. The third stage is when creep rate continually increases leading to failure and is termed as tertiary or accelerated creep. The portion of the total curve, each segment of curve represents, depends on the material and the stress level. Primary creep dominates at low stress level in low-ice (ice-poor) content frozen soils, while steady-state creep is exhibited by ice-rich soils under moderate stress levels. At high stress levels, accelerated creep occurs leading to specimen failure after a short period of time.

In Figure 8.1a, horizontal lines have been drawn at constant time $t_1, t_2 \ldots t_6$. At any one time, say $t_2$, the stress ($\sigma$) and the corresponding strain ($\varepsilon$) have been plotted on the top side of this figure. These are isochronous (equal time) stress–strain curves. These plots show that stress–strain plots for frozen soils are time dependent. Furthermore, if failure is defined by an arbitrary strain, $\varepsilon_f$, then failure stress, $\sigma_f$, or the strength for each time can be obtained from Figure 8.1a and can be plotted with time as shown on the lower left-hand section of the figure. This shows that the strength of frozen soils is time dependent and decreases with time. Furthermore, similar creep curves can be obtained for varying temperatures and confining pressures, which establishes that stress–strain and strength behavior of frozen soil depends on time, temperature of test, soil type, and confining pressure.

The strain-time or the deformation-time behavior of frozen soils results in

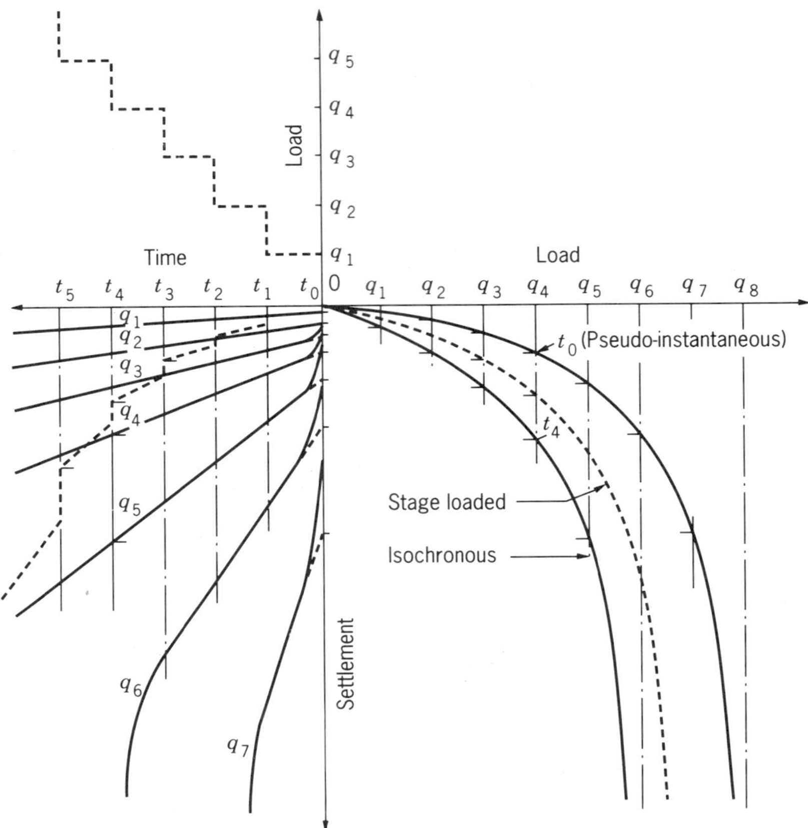


**Figure 8.2** Schematic presentation of the load-settlement–time relationship in loading tests (Ladanyi and Johnston, 1974).

creep and strength reduction with time. This phenomena must be considered when designing foundations on frozen soils. Figure 8.2 exhibits settlement-time relationships of frozen soils under loaded plates subjected to constant loads $q_1$ through $q_7$ where $q_1 < q_2 < q_3 \ldots < q_7$. For small loads on ice-poor soils that have some particle to particle contact, the deformation will gradually stop with time. As shown by curves $q_1$ through $q_5$, ice-rich soils may exhibit steady-state creep under low to medium loads. Heavier loads ($q_6$ and higher) will exhibit continuing deformations leading finally to failure. Actual values of these loads ($q$) producing different creep patterns will depend on soil type, soil temperature, and confining pressures.

Load-settlement behavior of a loaded plate obtained from settlement-time data is shown on right side of Figure 8.2. A pseudo-instantaneous load-settlement curve corresponds with settlement-time curves at $t = 0$ for various loads ($q$). An isochronous load-settlement curve is for same time (e.g., $t_4$ used in this case) obtained in the same manner as isochronous strength–strain curves in Figure 8.1a. The dotted lines show the stage loading in which case the stress is increased after same time from $q_1$ to $q_2 \ldots$ to $q_6$. Various load-settlement curves exhibit that settlement of a loaded plate increases with time. Also, ultimate bearing capacity, indicated by a vertical tangent to the load-settlement curve, decreases with time. Furthermore, these settlement-time relations are significantly influenced by temperature. For example, according to Womick and LeGoullon (1975) an increase in ground temperature from $-1.1°C$ to $-0.8°C$. resulted in a considerable increase in settlement rate beneath a piled structure in Fairbanks, Alaska. This clearly indicates that the load-settlement behavior of foundations on frozen ground is significantly influenced by time and ground temperatures and should be carefully evaluated.

In summary, unlike unfrozen soils whose bearing capacity is solely governed by shear strength and allowable settlements, the bearing capacity of foundations on frozen ground may be governed by additional factors such as time and the ground temperatures.

### 8.2.2 Frost Heave and Adfreeze Forces

Frost heave is generally characterized by the vertical upward movement of ground due to freezing. This can cause upward pressures at the base of foundations. The upward moving ground can also grab the surface of a pile shaft and apply an upward force on it. This application of upward force along pile shaft is called *adfreeze force* and the phenomena is called *frost jacking*. The following three conditions must be satisfied for frost heave to occur at a site:

1. The soil must be frost susceptible (i.e., the physical composition of the soil should be such that it promotes the migration of surrounding moisture to the freezing front and result in the formation of ice lenses). Generally silty soils exhibit this behavior. Chapter 4 (Section 4.3) describes the grain size criteria established by Casagrande (1932) for such soils.
2. There must be cold temperatures to propagate freezing.
3. There must be a source of water supply to feed the growth of ice lenses.

If all the above three conditions exist, frost heave would result first in 9 percent volume increase due to freezing of pore water in soil and second due to the migration of free moisture from below towards the colder surface. When this water reaches the frost line it freezes, forming ice lenses. As these lenses expand and grow, the ground surface moves upward resulting in increased heave. Thus, the largest part of the frost heave occurs due to the formation and the growth of ice lenses.

Frost-heaving ground can develop heaving forces on foundations in two ways. The first consists of the development of vertical upward forces acting on the underside of a foundation or grade beam. The second consists of the mobilization of vertical forces along side of pile shaft in the frozen zone and is called adfreeze force. Foundation design should therefore consider these vertical heaving forces in the following two ways:

1. Frost heave forces acting on the underside of the foundations and the grade beams: Prediction of these heave forces on horizontal surfaces such as underside of foundations, grade beams supported on ground, or spanning between two piles and underside pile caps is difficult because the forces depend on many variables, such as soil type, variation of soil temperature with depth, availability of water, and overburden pressures. Heave force pressures of as much as 110 psi (760 kPa) have been calculated by indirect methods, such as from the estimation of weight of buildings known to have been lifted by frost heave (Johnston, 1981). Heave pressures as high as 260 psi (1800 kPa) have also been measured on a 12-in. (300 mm) diameter anchored steel plate (Penner, 1970). It is recommended that rather than designing for such high potential uplift heave forces, foundations should be placed well below the depth of seasonal frost penetration (the depth below ground surface up to which the soil freezes during winter and thaws due to seasonal temperatures). Similarly, in permafrost areas, foundations should be placed below the active layer (the depth below ground surface where soil thaws in summer and refreezes each winter).
2. Adfreeze forces acting upward tangentially on the sides or perimeter of foundations of piles: Adfreeze forces develop when the foundation unit such as the pile is pulled upward by the surrounding frozen soil during the process of heaving. The magnitude of these forces should depend on the development of peak adfreeze bond strength between the pile surface and the frozen soil, a phenomena similar to the adhesive forces developed between pile and the surrounding clay in temperate zones. Adfreeze bond strengths have been measured in the field in various soil types and their range varies between 6 to 30 psi (40 to 210 kPa) (Kiselev, 1974). Also, according to the measurements made on uncoated steel piles, average adfreeze bond stresses greater than 40 psi (275 kPa) have been reported (Crory and Reed, 1965). The magnitude and the factors affecting these adfreeze bond stresses and hence the adfreeze forces on pile perimeter has not yet been fully investigated. Average adfreeze bond strength of 10 psi (70 kPa) for wood and concrete piles and 15 psi (100 kPa) for steel piles have been suggested by Davison et al. (1981). The magnitude of these upward adfreeze bond strength is a function of ground temperature, depth below ground, and ice content in the soil. Also, these values vary during the season (e.g., peak adfreeze bonds are developed early in the season when heave rates are high). However, the maximum uplift force along the pile surface will develop when the length under adfreeze forces is maximum which would occur when the maximum frost penetration has occurred. Also, it should be recognized that these adfreeze forces decrease if relaxation of

stress occurs due to small deformation of piles. This means maximum adfreeze forces will develop on rigidly anchored foundations while any yielding or movement would result in the reduction of these forces.

As discussed above, prediction of adfreeze forces on a pile perimeter is difficult. An indication of the maximum adfreeze forces ($F$) acting upward on a pile can be obtained by using Dalmatov's equation (Tsytovich, 1959; Davison et al., 1981).

$$F = pt_f(c - 0.5bT_m) \tag{8.1}$$

where

$F$ = total upward adfreeze force, kg
$p$ = Pile perimeter in contact with frozen soil, cm
$t_f$ = thickness of frozen soil zone, cm
$b$ = experimental parameter = 1.5 to 2.8 psi (10 to 19 kPa)
$c$ = experimental parameter = 6 to 10 psi (40 to 70 kPa)
$T_m$ = minimum soil temperature in freezing zone, °C

Andersland and Anderson (1978) suggest that tangential upward stresses generated on vertical surfaces should be measured directly from field tests. However, in absence of such field data the following design values as recommended by Vyalov and Porkhaev (1969) can be used:

1. For soil temperatures 27°F (−3°C) or higher, the upward adfreeze forces are 11.4 psi (78.5 kPa).
2. For soil temperatures below 27°F (−3°C) the upward adfreeze forces are 8.5 psi (58.8 kPa).

In summary, the magnitude of actual frost heave and adfreeze forces depends on soil type, moisture content, and permafrost temperatures and should be measured in the field. In absence of field data, the foregoing values can be used as a guide.

### 8.2.3 Frost Heave Control Methods

Estimation of frost heave and adfreeze forces, as discussed above, might serve only as rough guide for design purposes. Exact determination of the values of these forces is difficult. Therefore, methods to prevent or reduce the effects of these forces on foundations should be considered. In most situations, it is a common practice to locate the foundation below the permafrost table or below the depth of seasonal frost depth, as applicable. This technique may prevent the upward frost heave forces at the base of the foundation, but the upward adfreeze forces along the pile perimeter in freezing zone will still be effective. The following frost heave control methods should therefore be considered, where possible.

1. **Excavation and Replacement of Frost-susceptible Soil** In order to eliminate the formation and growth of ice lenses within the seasonal frost depth or the active layer in permafrost area, the frost-susceptible soil that promotes the formation of ice lenses within the frost depth or active zone should be excavated. This can then be replaced with compacted granular material that does not display ice segregation during freezing and thus eliminates adfreeze force development.
2. **Thermal Insulation** Thermal insulation such as styrofoam can be installed along and around the pile foundation that would prevent or reduce the depth of frost penetration (Robinsky and Bespflug, 1973). Similarly the insulation can also reduce the depth of permafrost thawing by preventing the surface heat conduction below ground. This technique will significantly reduce, if not eliminate, the frost heave and adfreeze forces by reducing the depth of seasonal frost and the active layer.
3. **Foundation Anchoring** The basic principle behind this technique is to provide uplift or pullout resistance by providing sufficient anchorage below seasonal frost or the active layer depth. Theoretically, this can be done by calculating the upward adfreeze forces and comparing it with the mobilized resisting forces below active layer. A rule of thumb, based on field experience, is that a pile should be embedded below the permafrost table to at least twice the thickness of the active layer. Similar guidelines are used for embedment depths below seasonal frost depths. This guideline needs further evaluation and field testing.
4. **Foundation Isolation from Heaving Forces** This method of eliminating or significantly reducing adfreeze forces consists of drilling a hole larger than the outside diameter of the pile in the active layer where the upward adfreeze forces are developed. A steel casing or sleeve is placed around the pile. The annular space between the steel sleeve and the bore hole is filled with sand slurry. The annular space between the sleeve and the pile is usually filled with a mixture of 70 percent oil and 30 percent wax. This prevents the entry of water and soil in between the pile and the sleeve. This technique ensures that the surrounding heaving soil in frost zone is completely isolated from the pile shaft. This method had been successfully used in permafrost areas to eliminate the development of adfreeze forces along pile perimeter in active zone. The only disadvantage of this method appears to be that the lateral load capacity of a pile is significantly reduced because of the unsupported lateral pile length in active layer.

### 8.2.4 Freezeback of Piles

Piles in permafrost areas are designed to carry loads by end bearing when there is a competent material such as rock or nonfrost susceptible dense sand or gravel at reasonable depths. Alternatively, a pile carries the loads by adfreeze bond strength in frozen soils or by skin friction in unfrozen soils. The bond strength is the bond or the adhesive force developed between the pile and the surrounding frozen soil and will be further discussed in Section 8.3. A combination of adfreeze bond and the end bearing can also be mobilized to carry pile loads in permafrost areas. When adfreeze bond is utilized to carry loads, piles must be well anchored in permafrost before the loads are applied.

As discussed in Chapter 3 (Section 3.4.4) the most common method of installing piles in permafrost areas is to auger a hole about 4 to 8 in. (100 to 200 mm) larger in size than the required pile diameter. The pile is then dropped into the hole, and the annular space between the pile and the surrounding soil is filled with soil slurry. On freezing, an adfreeze bond is developed between the slurry and the pile. This would require that the freezeback time, the time required for the backfilled slurry to freeze in the annular areas surrounding the pile, be estimated to ensure that adequate adfreeze bond has been mobilized before the design load can be applied after pile installation.

The main factors that govern the freezeback time are:

1. The permafrost ground temperatures should be low.
2. The volume of slurry surrounding the pile should be minimum so that freezeback can occur fast.
3. The latent heat of fusion of the slurry should be as low as possible which is largely governed by its water content. The moisture (water) content of the backfilled slurry should therefore be kept as low as possible.
4. In order for the freezeback time to be minimum, the heat source from the slurry should be as far as possible. This means that the spacing of piles should be as far as possible.

The latent heat $Q$ of slurry per meter of pile in joules per meter, depends on the volume of slurry, the slurry water content $w$, and the dry density $\gamma_d$, of the slurry. If it is assumed that the heat is conducted radially away from the pile and the slurry, the latent heat of slurry is then given by the following expression (Crory, 1963):

for round pile section $$Q = \pi L(r_2^2 - r_1^2)w\gamma_d \tag{8.2a}$$

for H-pile section $$Q = L(\pi r_2^2 - A)w\gamma_d \tag{8.2b}$$

where

$L$ = latent heat of water, J/m$^3$, $= 334 \times 10^3$ J/kg
$r_2$ = radius of drilled hole, m
$r_1$ = radius of circular pile section, m
$A$ = area of cross-section of H-pile section, m$^2$
$w$ = water content in percent of dry slurry weight
$\gamma_d$ = dry unit weight of slurry
$L_{\text{slurry}}$ = latent heat of slurry $= 334\gamma_d\left(\dfrac{w}{100}\right) \times 10^3$ J/m$^3$

The general solution of pile slurry freezeback time $t$ provided by Crory (1963) and modified by Davison et al. (1981) is given as follows:

$$t \simeq \frac{r_2^2}{\alpha}\left(\frac{Q}{9.3Cr_2^2\Delta T}\right)^{1.34} \tag{8.3}$$

where

$t$ = freezeback time, s
$C$ = volumetric heat capacity of permafrost, J/m$^3$°C
$\alpha$ = thermal diffusivity of permafrost, m$^2$/s
$Q$ = latent heat of slurry per m length of pile, J/m
$\Delta T$ = initial temperature of permafrost, °C below freezing
$r_2$ = radius of drilled hole for the pile, m

When the freezeback time $t$ calculated from equation (8.3) by natural dissipation of the heat to the surrounding frozen soil is longer than allowed by construction period, and the load has to be applied on the pile sooner than the period $t$, calculated above, then artificial refrigeration methods may be required. These methods are discussed in Chapter 2 (Section 2.6.2). It is a good practice to check freezeback time after pile installation by monitoring field installed thermocouples before the pile can be fully loaded (Crory, 1963).

***Example 8.1*** A 200-mm (8-in.) diameter ($2r_1$) steel pile was installed in a 300-mm (12-in.) diameter ($2r_2$) drilled hole. The annular space around the pile is backfilled with sand slurry having a maximum temperature of 4°C. Estimate the natural freezeback time without allowing permafrost temperature to exceed − 1°C. Based on the field investigations and laboratory tests it was found that the average permafrost temperature $\Delta T$ is − 2°C, volumetric heat capacity $C$ of sand slurry is $2.75 \times 10^6$ J/m$^3$°C, the latent heat of slurry $L_{\text{slurry}}$ is $155 \times 10^6$ J/m$^3$, the thermal diffusivity of permafrost $\alpha$ is $15 \times 10^{-7}$ m$^2$/s, and heat capacity $C$ of permafrost is $1.88 \times 10^6$ J/m$^3$.

SOLUTION

1. *Volumetric Latent Heat (Q)* Volumetric latent heat of slurry per meter of pile length will be a sum of slurry latent heat plus the sensible (temperature) heat conducted into the permafrost surrounding a drilled hole.

$$\begin{aligned} Q &= \pi(r_2^2 - r_1^2)L_{\text{slurry}} + \pi(r_2^2 - r_1^2)C(T)\ \text{Jm}^2/\text{m}^3 \\ &= \pi(0.15^2 - 0.1^2)155 \times 10^6 + \pi(0.15^2 - 0.1^2)2.75 \times 10^6(4+1) \\ &= 0.039(155 \times 10^6 + 13.75 \times 10^6) = 6.58 \times 10^6\ \text{J/m} \end{aligned}$$

where equation (8.2) was used for latent heat of slurry, and the second term $\pi(r_2^2 - r_1^2)CT$ represents heat due to temperature change from 4°C to − 1°C.

2. *Freezeback Time* $(t)$ From equation (8.3):

$$t = \frac{r_2^2}{\alpha}\left(\frac{Q}{9.3Cr_2^2\Delta T}\right)^{1.34}$$

$r_2 = 0.15\,\text{m}$ $\alpha = 15 \times 10^{-7}\,\text{m}^2/\text{s}$ $Q = 6.58 \times 10^6\,\text{J/m}$

$C = 1.88 \times 10^6\,\text{J/m}^3$ $\Delta T = 2°\text{C}$

$$t = \frac{(0.15)^2}{15 \times 10^{-7}}\left(\frac{6.58 \times 10^6}{9.3 \times 1.88 \times 10^6 \times 0.15^2 \times 2}\right)^{1.34}$$

$$= 297,754.5\,\text{sec}$$

$$\simeq 3.5\,\text{days}$$

## 8.3 PILES SUBJECTED TO AXIAL COMPRESSION LOADS

In this section, the allowable axial compression load capacity, settlement, frost action in active layers, and permafrost thawing effects on vertical piles in permafrost are discussed. Permafrost may contain soils that are ice rich. These soils contain ice in quantities that are significantly more than the volume of water present in the same soil in the unfrozen state. Because of this, an engineer requires the knowledge of thermal regime (i.e., the effects of changes in the ground thermal profile due to seasonal temperature changes and the heat introduced into the ground from pile foundations). Once the ground thermal aspects have been established, the response of the frozen ground to the loading has to be assessed. This section, therefore, first discusses the permafrost area thermal aspects followed by a brief description of mechanical properties of frozen soils. Finally, the load carrying capacity of vertical piles to axial loads are presented.

***Permafrost Area Thermal Aspects*** Figure 8.3a exhibits the typical ground temperature profile responding to the annual cycle of temperature change at the ground surface. This figure also shows that in winter the ground is completely frozen as exhibited by temperature profile on the left hand side. During summer, the ground temperature to a depth, $t_{al}$, is above 0°C, and the ground thaws to this depth. This layer, which thaws in each summer and then refreezes in winter, is called the *active layer*. The summer temperature profile below ground in this case is shown by the broken line in Figure 8.3a.

This thermal regime can be manipulated, for better or worse, by construction activities. For example, a well-ventilated air space below a structure supported on piles provides a shaded, cooler area in the summer. Also, this area is relatively snow free in winter, resulting in lower ground temperatures due to the lack of insulation from smaller or no snow cover. This results in a lower depth of active layer and thus improves foundation temperature conditions. On the other hand, activities such as right-of-way clearing that could remove the surface cover of organic layers and expose the surface to summer heat or introduction of heat to

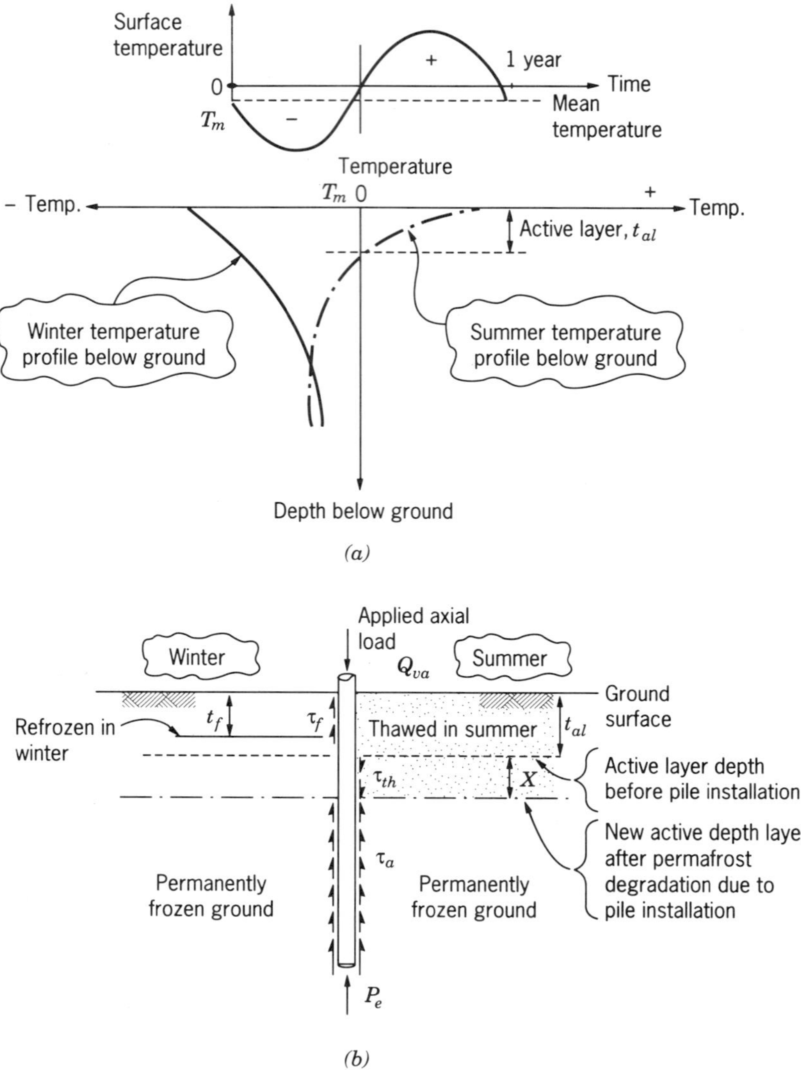


**Figure 8.3** Ground thermal aspect and typical upward and downward forces on a pile. (a) Annual ground temperature variation, (b) schematic presentation of forces acting on pile in permafrost during summer and winter including effects of permafrost degradation.

the ground from pile foundations will deepen the active layer locally. This is called *permafrost degradation*. The excess water in this thawed layer would then dissipate due to consolidation resulting in a downward movement of the soil. This applies downward pressures, $\tau_{th}$, on the pile similar to the negative skin friction discussed in Chapter 5 (Section 5.1.12). In winter, when the ground refreezes, the near surface soil applies upward adfreeze stresses, $\tau_f$, along the pile perimeter. This concept has already been discussed in Section 8.2.2.

***Forces Acting on Axially Loaded Pile*** Figure 8.3b shows a schematic presentation of forces acting on a pile that has been subjected to an imposed applied downward axial load $Q_{va}$. The right-hand side of this figure shows that in summer, the soil used to thaw to a depth of $t_{al}$, before pile installation. After pile installation and due to the introduction of heat into the ground from construction activity, the permafrost has thawed to an additional depth $X$, causing downdrag stresses $\tau_{th}$, on pile surface for this thawed depth $X$. The left-hand side of Figure 8.3b exhibits the application of upward adfreeze stresses $\tau_f$, due to frost action or heave when ground refreezes in winter. Below the depth of new active layer, the ground remains permanently frozen both in summer and winter months. In this area the pile is bonded with the surrounding frozen soil by adfreeze bond strength $\tau_a$. The adfreeze bond strength, $\tau_a$ is discussed in Section 8.3.1. Thus a pile foundation subjected to axial downward load $Q_{va}$ in permafrost must be designed for the following conditions:

1. **To Support Axial Downward Load $Q_{va}$ without Causing Failure** This requires estimating ultimate capacity that depends on the development of adfreeze bond strength $\tau_a$, as will be shown later. This is discussed in Section 8.3.1.
2. **Total Differential Settlements Are within an Allowable Range** This requires understanding the behavior and estimating the magnitude of time-dependent settlements of piles in permafrost. This is discussed in Section 8.3.2.
3. **Adfreeze Forces in Winter** In winter, the pile is subjected to adfreeze stresses $\tau_f$, due to frost action in the frozen zone $t_f$. These upward acting heave forces are to be resisted by the adequately developed adfreeze bond strength in the permafrost zone. These forces were discussed in Section 8.2.2.
4. **Downward Drag Forces** Finally, downdrag forces due to stresses $\tau_{th}$, developed by the settlement of soil in degraded permafrost depth $X$, should also be considered in the pile design. This is presented in Section 8.3.3.

It is important to note that the depth of active layer that may develop during the service life of a strucure must be estimated based on field probes during site investigations and past local experience. This depth should be increased by about 2 ft (0.6 m) to account for local variations and increased thaw around pile due to construction activity and the introduction of heat into the ground from the pile.

### 8.3.1 Axial Compression Pile Load Capacity

The development of ultimate capacity of a pile in frozen soil is related to the rupture of adfreeze bond that is developed due to the mobilization of adfreeze

**TABLE 8.1 Summary of Coefficient *m***

| Pile type | $m$ |
|---|---|
| Steel | 0.6 |
| Concrete | 0.6 |
| Timber (uncreosoted) | 0.7 |
| Corrugated steel pile | 1.0 |

After Weaver and Morgenstern (1981).

bond strength $\tau_a$ along pile perimeter in permanently frozen soil. The concept of adfreeze bond strength is similar to the adhesive strength between pile and the unfrozen soil. Like adhesive strength, the adfreeze bond strength is directly related to the roughness of the pile. However, adfreeze bond strength is also inversely related to soil ice content and the ground temperature. Long-term adfreeze bond strengths are determined from long-term laboratory and field tests. Based on these tests it has been proposed that adfreeze bond strength $\tau_a$ can be related to the long-term shear strength $S_{lt}$ by the following relationship (Weaver and Morgenstern, 1981):

$$\tau_a = mS_{lt} \tag{8.4}$$

where

$S_{lt} = C_{lt} + \sigma_n \tan \phi_{lt}$
$C_{lt}$ = long-term cohesion of permafrost
$\phi_{lt}$ = long-term angle of internal friction of permafrost

Typical values for $m$ are given in Table 8.1 for various pile materials and their surface features embedded in permafrost. The long-term shear strength $S_{lt}$ of a frozen soil is similar to that of unfrozen soil (i.e., it depends both on frictional $\phi_{lt}$ and cohesive $C_{lt}$ components. However, since the normal stress on the adfreeze plane between soil and pile is small (typically 100 kPa), and thus long-term frictional strength is generally insignificant. This term therefore may be neglected in Equation (8.4) which can then be rewritten in the following form:

$$\tau_a = mC_{lt} \tag{8.5}$$

Based on the review of long-term cohesive strengths $C_{lt}$ of frozen soils by Vialov (1959) and of polycrystalline ice by Voitkovskii (1960), we can conclude that $C_{lt}$ is primarily temperature dependent. These data are shown on Figure 8.4.

Thus, the long-term adfreeze strength $\tau_a$ equation (8.5), can be directly related to the long-term cohesive strength $C_{lt}$ and a constant $m$ that is dependent on pile material. The cohesive strength $C_{lt}$ should be determined for the frozen soil immediately at the pile-frozen soil interface. As an example, if a steel pile is driven into ice-rich varved clay, the soil structure remains intact and $C_{lt}$ of ice-rich

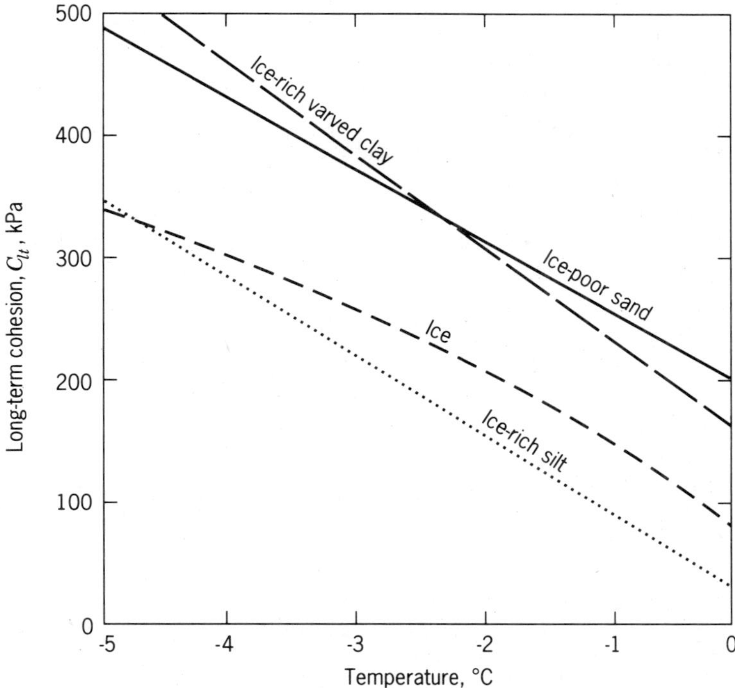


**Figure 8.4** Long-term cohesive strength, $C_{lt}$, of frozen soils (Weaver and Morgenstern, 1981).

varved clay should be used from Figure 8.4. If the pile is placed in a predrilled hole and the annular space is backfilled with a soil slurry mixed with the original excavated soil, the varved structure of the original soil is destroyed and $C_{lt}$ for ice-rich silt should be used from Figure 8.4. Furthermore, if this steel pile was installed in winter by slurry method, a thin layer of ice will coat the surface of pile due to moisture migration to the pile surface. In this situation, $C_{lt}$ for ice should be used from Figure 8.4 (Weaver and Morgenstern, 1981).

The ultimate capacity of piles in permafrost can therefore be assessed by consideration of adfreeze strength along the pile shaft. The contribution of end bearing in most situations is small and can be neglected, which is on the safe side. End bearing should only be considered when a dense, non-frost-susceptible ice-free stratum is encountered below the pile (Davison et al., 1981; Weaver and Morgenstern, 1981). The ultimate pile load capacity $(Q_v)_{\text{ult}}$ can, therefore, be determined from the following relationship:

$$(Q_v)_{\text{ult}} = 2\pi R \int_0^L \tau_a(z)\, dz \tag{8.6}$$

where

$R$ = pile radius
$L$ = pile embedment length in permafrost below the active layer
$\tau_a(z)$ = the adfreeze bond strength of the frozen soil layer for the maximum expected temperature along the depth

Parmeswaran (1981) and Frederking and Kerri (1983) also discuss the effect of temperature on adfreeze strength of frozen soils. Allowable pile load capacity $(Q_v)_{all}$ can then be calculated by using a safety factor of 2. After, $(Q_v)_{all}$ has been estimated based on adfreeze bond strength, a design based on time-dependent settlement (creep) has to be carried out to ensure that pile displacements under structural load are tolerable.

### 8.3.2 Pile Settlement

In this section, methods to calculate the allowable load on a pile to maintain the settlements within tolerable limits over the lifetime of the structure are discussed. Weaver and Morgenstern (1981) show that for piles in frozen soils, the load carried by end bearing is negligible. For example, for a 0.2-m diameter pile that is 25 m long installed in ice, the fraction of load supported in end bearing at $-1$°C is 0.5% and at $-10$°C is 0.65 percent of the total load. Similarly, end bearing supported 1.1 percent of the total load if this pile was installed in frozen Ottawa sand. Therefore, for all practical purposes, piles, installed in frozen soils can be treated as friction piles unless they are bearing on ice-free rock or dense, ice-free sand and gravel.

***Friction Piles in Ice-Rich Frozen Soils*** Consider the problem of a pile in frozen ground (Figure 8.5). The following simplifying assumptions are made to solve for settlement of a pile:

1. At a constant temperature, the pile material is considerably more rigid in the long-term loading than the surrounding frozen soil.
2. The permafrost is in fresh water.
3. The shear stress is uniformly distributed along the pile shaft.
4. The end-bearing stresses are zero.

The shear stress $\tau_a$ can then be expressed as follows:

$$\tau_a = \frac{Q_{va}}{2\pi RL} \tag{8.7}$$

where

$Q_{va}$ = axial downward load on pile = friction capacity = $Q_f$
$R$ = pile radius
$L$ = the embedded pile length in permafrost

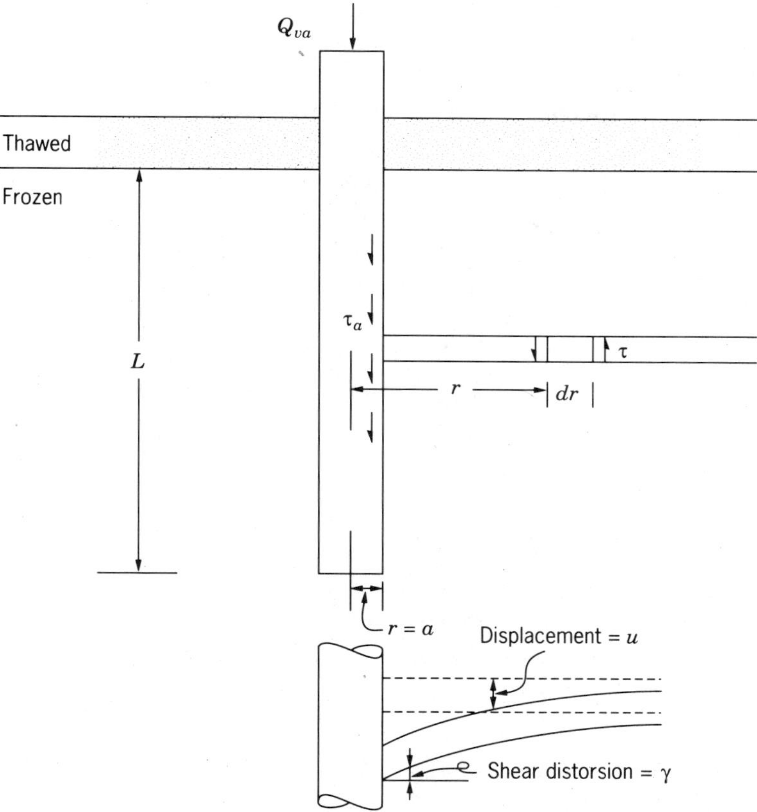


**Figure 8.5** Shear stresses and strains around pile in frozen soil (Nixon and McRoberts, 1976).

For a weightless soil, the applied shear stress $\tau_a$ at $r = a$ is related to the shear stress $\tau$ at any other radius $r(> R)$ by the following expression (Figure 8.5):

$$\tau = \tau_a \left( \frac{R}{r} \right) \tag{8.8}$$

Johnston and Ladanyi (1972) analyze the problem for a frozen soil subjected to simple shear under plane strain conditions and show that the flow law for a general state can be written as follows:

$$\dot{\gamma} = 3^{(n+1)/2} B \tau^n \tag{8.9}$$

where

$\dot{\gamma}$ = shear strain rate induced in frozen ground around a pile due to shear stress $\tau$

**TABLE 8.2 Secondary Creep Parameters: Ice-Rich Soils**

| Temperature | $B$ (kPa$^{-n}$·year$^{-1}$) | $n$ |
|---|---|---|
| $-1$°C | $4.5 \times 10^{-8}$ | 3.0 |
| $-2$°C | $2.0 \times 10^{-8}$ | 3.0 |
| $-5$°C | $1.0 \times 10^{-8}$ | 3.0 |
| $-10$°C | $5.6 \times 10^{-9}$ | 3.0 |

After Weaver and Morgenstern (1981).

$n$ and $B$ = constants obtained from a series of creep tests carried out on frozen soil at a constant temperature (Table 8.2)

$\tau$ = induced shear stress in frozen soil due to applied load $Q_{va}$ on a cylindrical pile in frozen ground

For the problem of a vertically loaded pile in frozen soil, the strain around the pile in the tangential $\theta$ direction is zero. Therefore, each element of the frozen soil deforms under plain strain conditions (Nixon and McRoberts, 1976). Equation (8.9) obtained for plain strain conditions can therefore be applied for this problem. This is done by substituting equation (8.8) into equation 8.9, which yields following equation:

$$\dot{\gamma} = 3^{(n+1)/2} B \left( \frac{\tau_a R}{r} \right)^n \tag{8.10}$$

The shear strain $\gamma$ can be related to displacement, as shown in Figure 8.5, and is given by following expression:

$$\gamma = -\frac{du}{dr} \tag{8.11}$$

where

$u$ = displacement at any radius $r$.

Similarly, shear strain rate $\dot{\gamma}$ will be:

$$\dot{\gamma} = -\frac{d\dot{u}}{dr} \tag{8.12}$$

Substituting $\dot{\gamma}$ of equation (8.12) in equation (8.10), we get:

$$\frac{d\dot{u}}{dr} = -3^{(n+1)/2} B \left( \frac{\tau_a R}{r} \right)^n \tag{8.13a}$$

On integrating above equation and using boundary condition that (1) at $r = R$, $\dot{u} = \dot{u}_R$ (i.e., at pile radius $R$ the soil displacement rate equals the pile displacement rate $\dot{u}_R$) and (2) at $r = \infty$, $\dot{u} = 0$ (i.e., at an infinite radius the displacement is zero), we obtain the following relation for the displacement rate of the pile under a load $Q_{va}$ (or shear stress $\tau_a$):

$$\frac{\dot{u}}{R} = \frac{3^{(n+1)/2} B \tau_a^m}{(n-1)} \tag{8.13b}$$

This relationship gives the settlement rate $\dot{u}$(mm/yr) for ice and ice-rich frozen soils. The load-carrying capacity $Q_f$ of a friction pile can then be obtained as follows by combining equations (8.7) and (8.13b):

$$Q_f = 2\pi RL\left(\frac{\dot{u}}{R}\right)^{1/n}\left(\frac{n-1}{3^{(n+1)/2}}\right)^{1/n}\left(\frac{1}{B}\right)^{1/n} = 2\pi RL\tau_{av} \tag{8.14}$$

All the terms of this equation have been defined earlier. For a known temperature, values of $n$ and $B$ have been taken from Table 8.2 and the deformation rate $\dot{u}/R$ and $\tau_a$ have been plotted for different temperatures in Figure 8.6.

From this figure, for an acceptable deformation rate $\dot{u}/R$ and at a known ground temperature, the average allowable shaft stress $\tau_{av}$, and hence the load-carrying capacity $Q_f$ of pile shaft, embedded in ice-rich soils, can be determined.

***Friction Piles in Ice-Poor Frozen Soils*** Unlike ice-rich soils whose time-dependent load settlement is governed by steady-state secondary creep, the behavior of ice-poor soils is governed by primary creep. In Section 8.2.1 this creep behavior of frozen soils was discussed. Ladanyi (1972) utilized the Mohr–Coulomb failure theory to model the effect of hydrostatic pressure on the steady-state (secondary creep) rates. This approach may be extended to nonsteady creep rates. Based on this, the primary creep for ice-poor soils can be expressed by following relationship (Weaver and Morgenstern, 1981):

$$\varepsilon_e = D\left[\frac{(j+2)}{3}\sigma_e + (1-j)\sigma_m\right]^C t^b \tag{8.15}$$

where

$$\sigma_m = \text{mean normal pressure (kP}_\text{a}) = \frac{(\sigma_1 + \sigma_2 + \sigma_3)}{3}$$

$$D = \left[\frac{1}{w(\theta+1)^k}\right]^C$$

$t =$ time elapsed after the application of load, h

$\theta =$ temperature below freezing point of water, 0°C

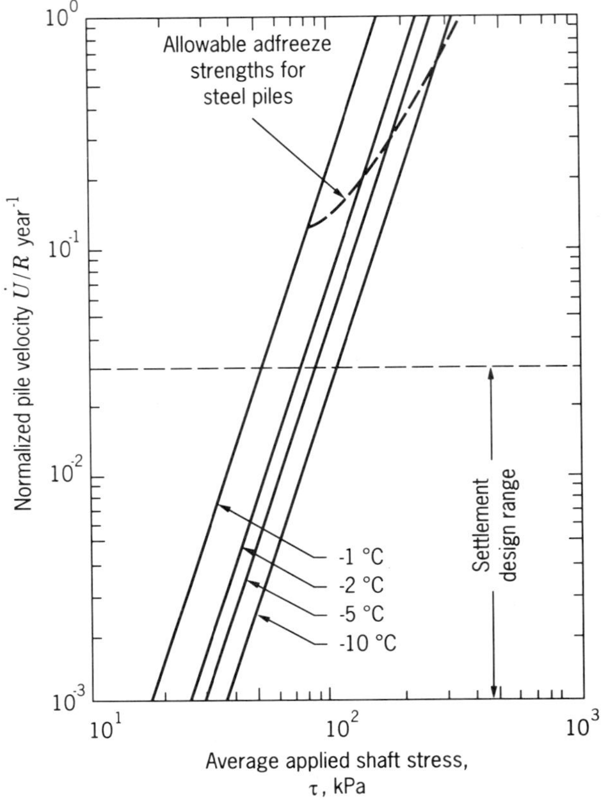


**Figure 8.6** Design chart for friction piles in ice or ice-rich soils (Weaver and Morgenstern, 1981).

$b, c, w, k =$ material constants given in Table 8.3

$$j = \frac{(1 + \sin\phi)}{(1 - \sin\phi)}$$

$\phi =$ angle of internal friction

$\sigma_e =$ applied constant stress, $\mathrm{kP_a}$

$\varepsilon_e =$ strain due to $\sigma_e$

In summary, ice-poor soils do not contain a continuous network of segregated ice. Therefore, creep in these soils is damped and can be approximated by equation (8.15). Weaver and Morgenstern (1981) have further suggested that for piles in ice-poor soils the ratio of $(\sigma_1/j\sigma_3)$ is typically less than 1.25 and conclude that a better accuracy may, therefore, be achieved by substituting $j = 1$ in equation (8.15). Further rearranging this equation and expressing it in terms of

**TABLE 8.3 Primary Creep Parameters: Ice-Poor Soils**

| Soil | $c$ | $b$ | $w$ ($MP_a\, h^{b/c}/°C^k$) | $k$ |
|---|---|---|---|---|
| Suffield clay | 2.38 | 0.333 | 0.73 | 1.2 |
| Bat-Baioss clay | 2.50 | 0.45 | 1.25 | 0.97 |
| Hanover silt | 2.04 | 0.151 | 4.58 | 0.87 |
| Callovian sandy loam | 3.70 | 0.370 | 0.88 | 0.89 |
| Ottawa sand[a] | 1.28 | 0.449 | 44.7 | 1.0 |
| Manchester fine sand | 2.63 | 0.631 | 2.29 | 1.0 |
| Ottawa sand[b] | 1.32 | 0.263 | 21.0 | 1.0 |

After Weaver and Morgenstern (1981).
[a]Study by others.
[b]Study by Weaver and Morgenstern (1981).

deformation $u_R$, the following expression has been suggested by Weaver and Morgenstern (1981) for creep rate in ice-poor frozen soils:

$$\frac{u_R}{Rt^b} = \frac{3^{(c+1)/2} D\tau_{av}^c}{(c-1)} \tag{8.16}$$

The load-carrying capacity of friction piles in ice-poor soils can then be expressed as follows:

$$Q_f = 2\pi RL\left(\frac{u}{Rt^b}\right)^{1/c}\left(\frac{c-1}{3^{(c+1)/2}}\right)^{1/c}\left(\frac{1}{D}\right)^{1/c} = 2\pi RL\tau_{av} \tag{8.17}$$

All the terms of this equation have been defined earlier.

This equation has been summarized in Figure 8.7. From this figure for an acceptable deformation rate $u/t^b$ and at a known ground temperature, the average allowable shaft stress, $\tau_{av}$, and hence the load-carrying capacity $Q_f$ of pile shaft, embedded in ice-poor soils, can be determined.

***Creep Settlement in Saline Permafrost*** Creep settlement data presented are for permafrost soils in fresh water. However, in coastal areas, both fine and coarse grained soils can be expected to contain some salt in pore water. Based on creep test on saline frozen fine-grained soils, Nixon and Lem (1984) reported that there is 10- to 100-fold increase in uniaxial creep rates when compared to freshwater frozen soil tests. These results suggest that there will be an increase in displacements as soil salinity increases. Nixon and Neukirchner (1984) applied the results of creep tests on piles in saline frozen soils and produced a chart similar to Figure 8.6 for pile displacement rates and the shaft stresses in saline soils as

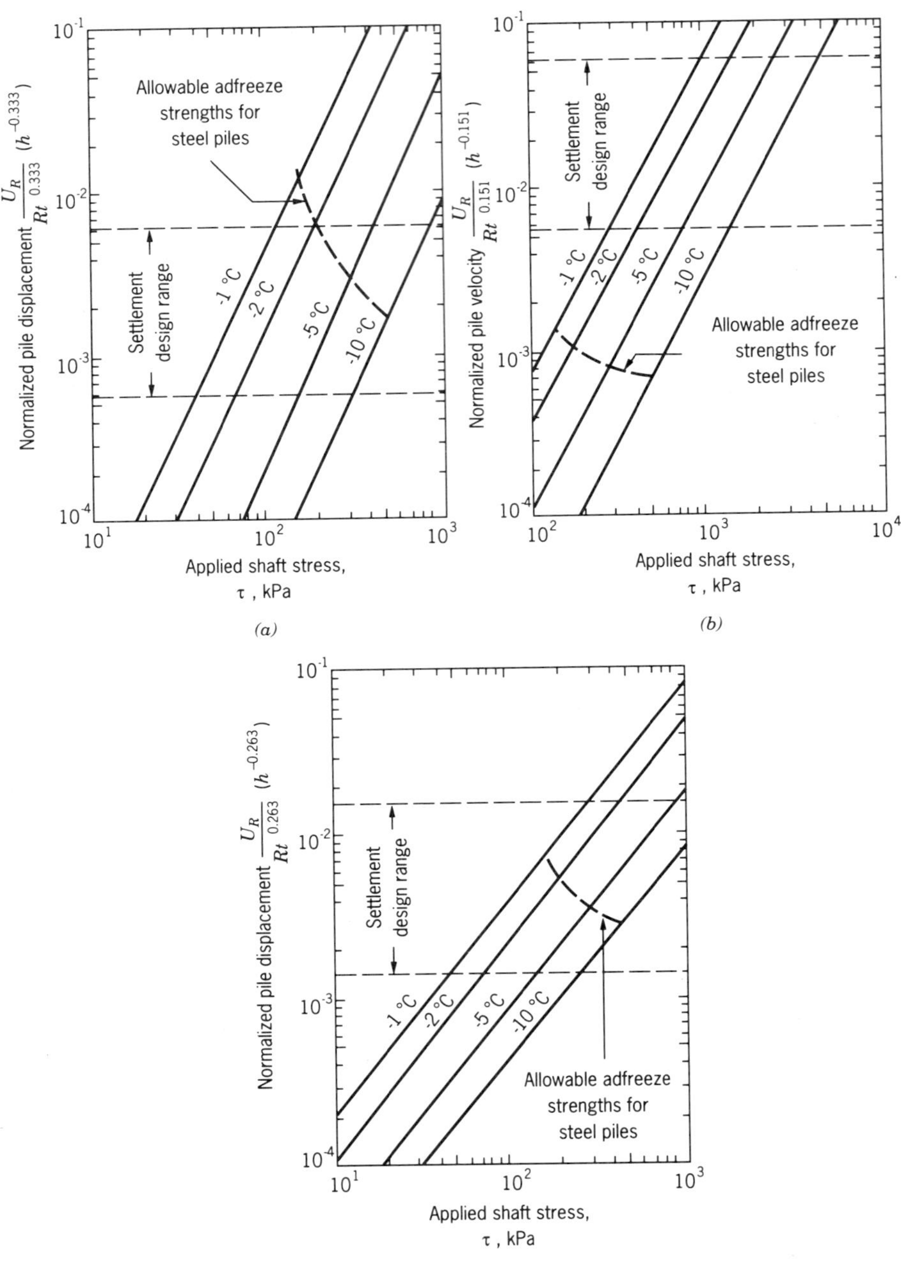


**Figure 8.7** Design charts for friction piles in ice-poor soils (a) Suffield Clay ($\gamma_f = 1.76\ \text{Mg/m}^3$), (b) Hanover soil ($\gamma_f = 1.78\ \text{Mg/m}^3$), (c) Ottawa sand ($\gamma_f = 2.00\ \text{Mg/m}^3$). (Weaver and Morgenstern, 1981).

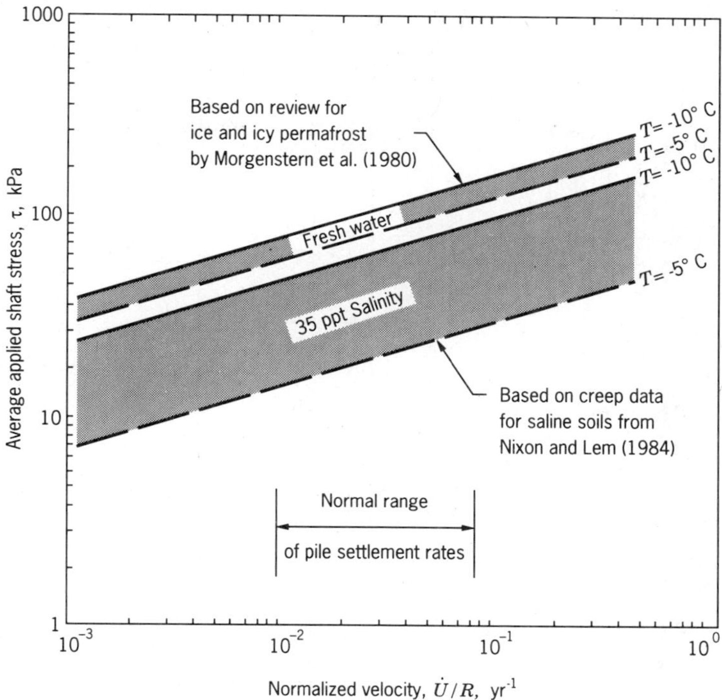


**Figure 8.8** Prediction of creep settlement for fresh water and saline ice-rich soils (Nixon and Neukirchner, 1984).

shown in Figure 8.8. Curves both for fresh water and saline water are shown on this figure. This figure clearly shows that creep displacements of vertically loaded piles in saline water are higher than for equivalent fresh water soils and can be used for design purposes.

***Example 8.2*** A 200-mm (8 in.) diameter steel pipe pile was driven 16 m in ice-rich silty soil. The life of the structure that is supported on this pile foundation is 20 years, and the maximum tolerable settlement throughout its life is 0.04 m. It is predicted that the maximum active layer will be 2.0 m thick, and the warmest ground temperature profile is expected to vary linearly from 0°C at 2 m depth to −5°C at 16 m depth. The downdrag on the pile due to permafrost degradation is neglected and frozen soils are in fresh water.

(a) Calculate the allowable axial load on pile.
(b) Calculate the allowable pile load if this pile is placed in augered hole filled with sand slurry. The soil conditions consist of 10 m of ice-rich silt over ice-poor sand.

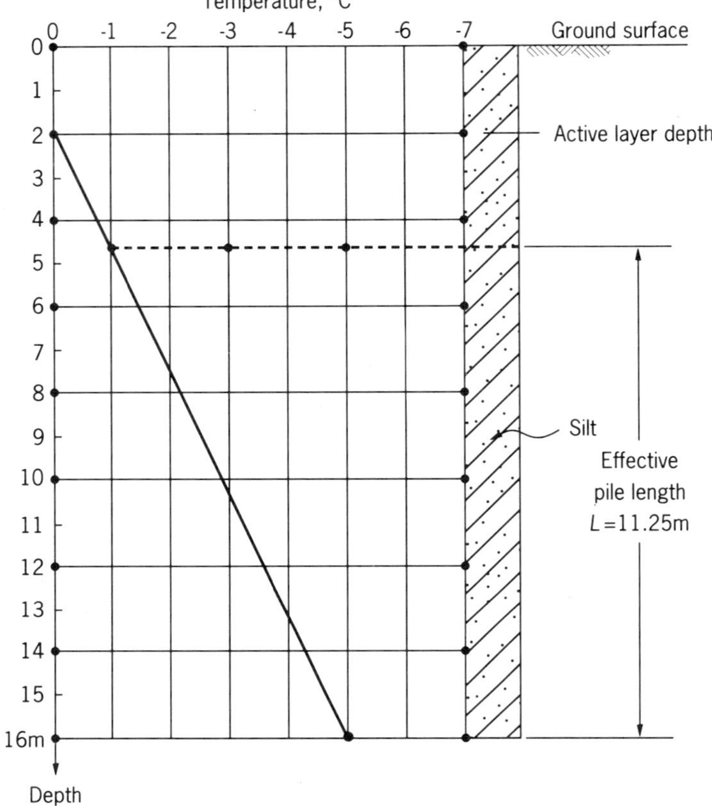


**Figure 8.9** Soil profile and ground temperature profile (for Example 8.2).

## SOLUTION

(a) Based on permafrost colder than $-1°$C and from Figure 8.9, the effective pile embedment depth $L = 11.25$ m. The average temperature over this depth is $-3°$C.

(i) *Pile Capacity based on Adfreeze Strength*
From Table 8.1, $m = 0.6$ for steel pile
From Figure 8.4, for ice-rich silt at $-3°$C

$$C_{lt} = 225\,\text{kP}_a$$

Therefore, from equation (8.5),

$$\tau_a = 0.6 \times 225 = 135\,\text{kP}_a$$

Then, from equation (8.6), the ultimate pile capacity $(Q_v)_{ult}$ is as follows:

$$(Q_v)_{ult} = 2\pi(0.1)(11.25)(135) = 954.26\,\text{kN}$$

(ii) *Pile Capacity Based on Pile Settlement*

$$\text{Allowable pile settlement rate} = \frac{0.04}{20} = 0.002\,\text{m/yr}$$

$$\frac{\dot{u}}{R} = \frac{0.002}{0.1} = 0.02/\text{yr}$$

From Figure 8.6, the allowable shaft stress $\tau$ is:

$$\tau = 70\,\text{kPa}$$
$$(Q_v)_{ult} = 2\pi(0.1)(11.25)(70) = 494.8\,\text{kN}$$

From (i) and (ii) the allowable pile load will be governed by settlement consideration.

$$(Q_v)_{all} = \frac{(Q_v)_{ult}}{FS} = \frac{494.8}{2} = 247\,\text{kN}$$

(b) As in (a), $L = 11.25$ m, and the average temperature for this depth is $-3°$C.

(i) *Pile Capacity Based on Adfreeze Strength*

From Table 8.1, $m = 0.6$

Now, in this case, the adfreeze strength will be applicable between pile and frozen sand slurry interface. Therefore, from Figure 8.4, long-term cohesion for ice-poor sand at $-3°$C will be:

$$C_{lt} = 375\,\text{kPa}$$

Then, from equation (8.5),

$$\tau_a = 0.6 \times 375 = 225\,\text{kPa}$$

From equation (8.6), $(Q_v)_{ult} = 2\pi(0.1)(11.25)(225) = 1590\,\text{kPa}$

(ii) *Pile Capacity Based on Pile Settlement*

Since the lower (16–10) = 6 m of the pile is located in ice-poor sand, the friction support provided by ice-rich silt may be ignored because, in the long run, most of the load will be transferred to the ice-poor soil. This soil is stiffer than the overlying ice-rich silt.

From Table 8.3 for ice-poor soil, the following are creep parameters: $c = 1.32$, $b = 0.263$, $w = 21.0\,\mathrm{MP}_a h^{b/c}/^\circ\mathrm{C}^k$, $k = 1.0$, then:

$$\frac{u}{Rt^{0.263}} = \frac{0.04}{0.1(20 \times 365 \times 24)^{0.263}} = 0.016h^{-0.263}$$

From Figure 8.7c for frozen Ottawa sand for $u/Rt^{0.263}h^{0.263} = 0.016$, and at $-3^\circ$C, we get $\tau = 500\,$kPa. Then, $(Q_v)_{\mathrm{ult}} = 2\pi(0.1)(6)(500) = 1885\,$k Pa. From (i) and (ii), the allowable pile load will be governed by adfreeze consideration.

$$(Q_v)_{\mathrm{all}} = \frac{(Q_v)_{\mathrm{ult}}}{FS} = \frac{1590}{2} = 795\,\mathrm{kN}$$

### 8.3.3 Downdrag due to Permafrost Thawing

As was discussed in the beginning of this chapter, construction activities such as right-of-way clearing could remove the surface cover of organic insulating layer and expose the soil surface to summer heat. Similarly, in some cases, above ground pipeline supporting piles may also change the existing ground thermal regime by introducing additional heat into the ground. This results in long-term thawing or degradation of the permafrost and increases the depth of active layer. The excess water in this thawed layer would then dissipate due to consolidation resulting in the downward movement of soil. This will apply additional downward pressure on the pile shaft in this zone. The depth of permafrost degradation is shown by $X$ and the downward pressures are shown by $\tau_{th}$ in Figure 8.3.

The depth $X$ in meters can be estimated from the data reported by Linell (1973) and is given by the following expression:

$$X = C_{th}(t)^{0.5} \tag{8.18}$$

where

$C_{th}$ = a constant varying between 0.9 to 1.3 m/yr$^{0.5}$
$t$ = time in years

Davison et al. (1981) also recommended this approach for estimating the depth of permafrost thawing ($X$).

The downward pressures $\tau_{th}$ can then be calculated by using the concepts of earth pressures at rest within the zone $X$ as follows:

$$\tau_{th} = K_0 \gamma' Z \tan\phi' \tag{8.19}$$

where

$\tau_{th}$ = downward drag pressure along pile surface

$K_0$ = coefficient of earth pressures at rest
$\gamma'$ = submerged weight of thawed soil
$Z$ = depth under consideration
$\phi'$ = effective angle of internal friction of thawed soil
Using $\phi' \simeq 36°$
$K_0 = 1 - \sin\phi' = 0.41$, then $K_0 \tan\phi' = 0.41\ (0.726) \simeq 0.3$

Also assuming a pile of radius $R$ being subjected to a downward drag force $D$ in thaw zone $X$, we can express $D$ as follows:

$$D = \left( K_0 \tan\phi' \gamma' \frac{X^2}{2} \right) \pi(2R) \tag{8.20}$$

Substituting $K_0 \tan\phi' = 0.3$, equation (8.20) can be rewritten as follows:

$$D = 0.3\pi R \gamma' X^2 \tag{8.21}$$

All the terms have been defined.

When these downdrag forces $D$ become excessive due to permafrost thawing or their effect unduly influences the length of pile embedment in permafrost, then methods to control permafrost thawing should be considered. These include the use of insulation or installing thermal piles. Thermal piles have already been described in Chapter 2 (Section 2.6.2). Also, a well-ventilated air space under a structure that provides shades in summer and reduces snow cover in winter (thus reducing ground temperatures in winter due to the lack of insulation by snow cover) is more likely to maintain permafrost conditions.

## 8.4 PILES SUBJECTED TO LATERAL LOADS

Pile-supported structures such as elevated buildings, above-grade pipelines, and pretensioned loads in an anchorage system can apply sustained lateral loads on vertical piles by horizontal components of inclined loads. These would cause horizontal creep displacements that are governed by the creep mechanism as discussed in Section 8.3. Nixon (1984) examined the relationship between horizontal load, applied moment, and the resulting horizontal pile displacement rate for short rigid piles. These results were then compared with available field and laboratory tests which showed good agreement. Basic theory and design charts obtained from Nixon (1984) are presented as follows:

### 8.4.1 Free-headed Short Rigid Piles

Figure 8.10 shows the configuration of a laterally loaded free-head rigid pile. The pile is assumed to rotate about some point at a depth $x = \beta L$ when a lateral load $Q$ is applied at a height $H$ above the ground surface. If the pile displacement

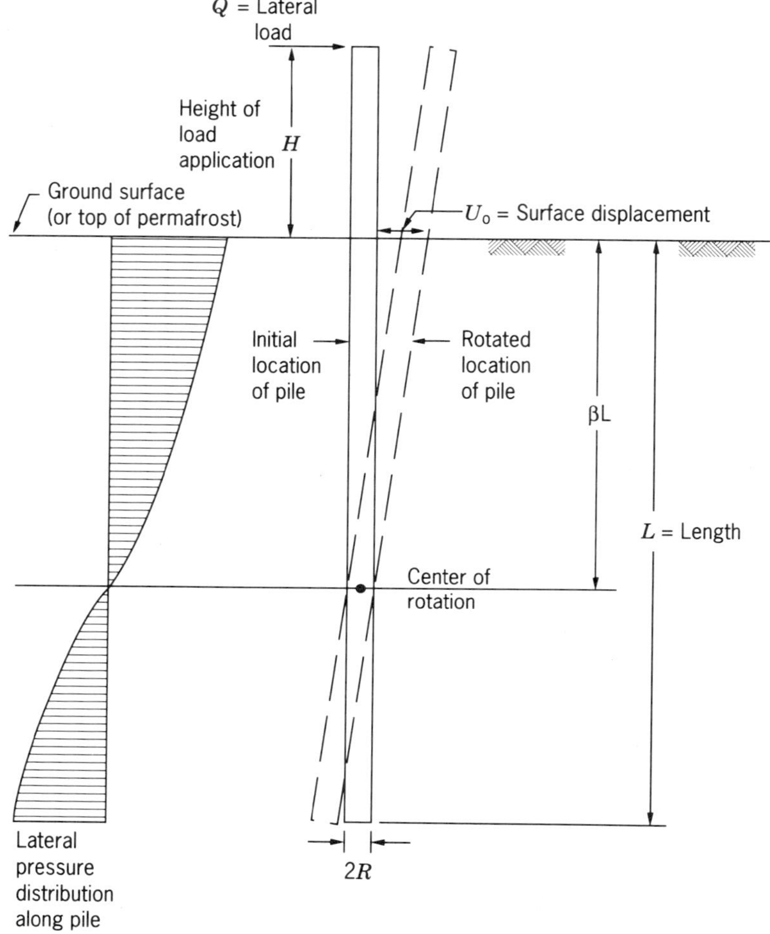


**Figure 8.10** Configuration of a laterally loaded free-headed rigid pile (Nixon, 1984).

at the ground surface is $u_0$, then displacement rate ($\dot{u} = \partial u/\partial t$) at any depth can be given as follows:

$$u = \dot{u}_0(1 - x/\beta L) \tag{8.22}$$

The creep rate $\dot{u}$ for a laterally loaded cylindrical tube or strip loaded area can be presented by the following expression (Nixon, 1978):

$$\dot{u} = IRB(\Delta p)^n \tag{8.23}$$

where

$I$ = the influence factor and is equal to $[(2/n)^n(\sqrt{3}/2)^{n+1}]$ (Ladanyi, 1975)
$n$ and $B$ = creep parameters (Table 8.2)
$\Delta p$ = horizontal stress on the loaded area
$R$ = pile radius or half width of loaded area

From equations (8.22) and (8.23), an expression for $\Delta p$ can be obtained as follows:

$$\Delta p = \left(\frac{\dot{u}_0}{IRB}\right)^{1/n}\left(1 - \frac{x}{\beta L}\right)^{1/n} \tag{8.24}$$

The two unknowns $\dot{u}_0$ and $\beta$ can be obtained by solving the following two equations obtained from horizontal force and moment equilibrium, respectively:

$$Q = 2R\int_0^L \Delta p\, dx - 2R\int_{\beta L}^{L} \Delta p\, dx \tag{8.25}$$

$$2R\int_{\mathrm{H}}^{H+\beta L} \Delta px\, dx - 2R\int_{H+\beta L}^{H+L} \Delta px\, dx = 0 \tag{8.26}$$

Equation (8.26) is obtained by taking moments about the point of load application.

On solving these equations, Nixon (1984) obtained the following general solution for lateral ground surface displacement rate ($\dot{u}_0$) for a rigid pile:

$$\frac{\dot{u}_0}{IRB\left(\dfrac{Q}{2RL}\right)^n} = \left[\frac{1 + (1/n)}{\beta(1 - E)}\right]^n \tag{8.27}$$

where $E = (1/\beta - 1)^{(1+1/n)}$ and other terms were defined earlier. It should be noted here that the coefficient $B$ is temperature dependent, and therefore this equation includes temperature. Figure 8.11a provides plots for dimensionless load ($Q/2RL$) against the surface displacement rate ($\dot{u}/R$) on a double logarithmic scale for various temperatures for $n = 3$, which is the case for all practical purposes as shown by Table 8.2. Figure 8.11b presents design charts for ($\dot{u}_0/R$) = 0.02 per year and 0.04 per year. Figure 8.11a and b can be used for calculating lateral load for specified lateral movement rate at the ground surface. Nixon compared the results of above analysis with field pile load tests carried out by Rowley et al. (1973) and Rowley et al. (1975) and laboratory model tests on laterally loaded piles. Comparisons between predictions made by above theoretical approach and the load tests show good agreement. Nixon (1984) suggests that further long-term testing should be carried out on piles that will provide more confidence on this approach.

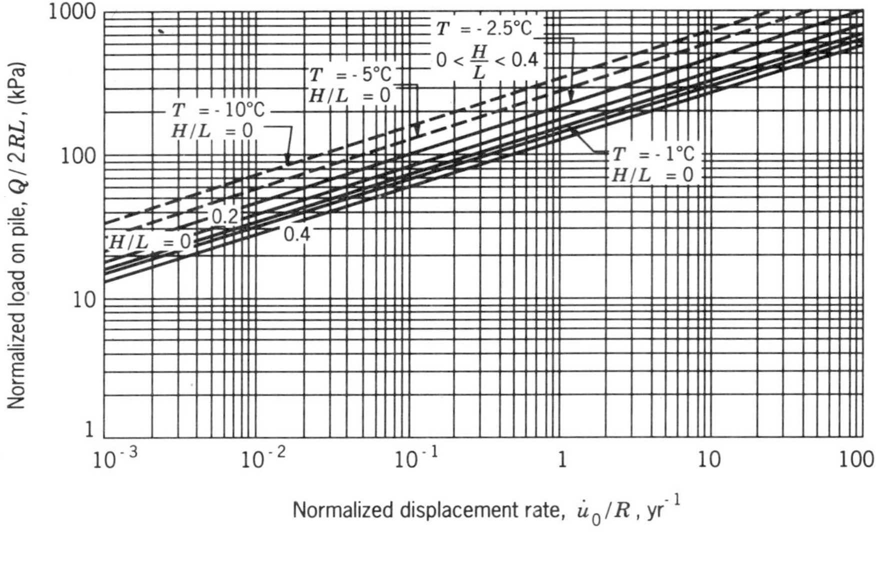


*(a)*

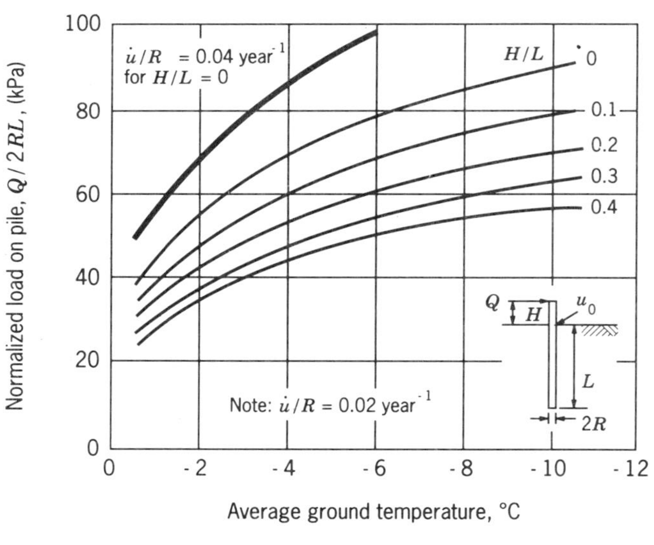


*(b)*

**Figure 8.11** Lateral load and displacement rates for rigid free-headed piles (Nixon, 1984). (a) Solution for lateral load versus displacement rate for free-headed rigid pile for $n = 3$, (b) design curves for creep data on ice and a specified lateral displacement rate.

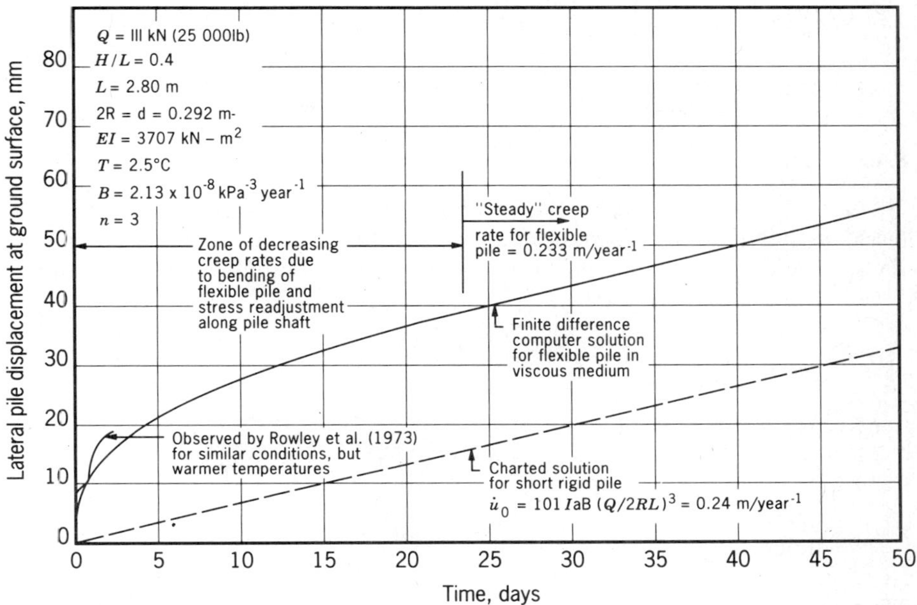


**Figure 8.12** Comparison of solutions for flexible piles and short piles (Nixon, 1984).

***Example 8.3*** A 250-mm (10 in.) diameter 6 m (19.7 ft) long steel pile is embedded in ice-rich silty frozen soil which is at $-3°C$. Calculate the allowable lateral load $Q$ that can be applied at 1.8 m (5.9 ft) above ground. The allowable lateral deformation over 20 year period at ground surface is 50 mm (2 inch).

SOLUTION Using terms of Figure 8.10:

$$L = 6\,\text{m} \qquad R = \frac{250}{2} = 125\,\text{mm}$$

$$H = 1.8\,\text{m}$$

$$H/L = 1.8/6 = 0.3$$

$$\dot{u}_0 = \frac{50}{20} = 2.5\,\text{mm/year}$$

$$\frac{\dot{u}_0}{R} = \frac{2.5}{125} = 0.02/\text{year}$$

From Figure 8.11b, for $\dot{u}_0/R = 0.02/\text{year}$, $H/L = 0.3$ and at $-3°C$, we get the following:

$$\frac{Q}{2RL} = 43\,\text{kPa}$$

$$Q = 43 \times 2 \times 0.125 \times 6 = 64.5\,\text{kN} \equiv 14.5\,\text{kips}$$

### 8.4.2 Laterally Loaded Flexible Piles

Nixon (1984) presented the following theoretical formulation for flexible piles based on the equation for bending of an elastic beam. The lateral deformation $u$ of such a pile under variable reaction $\Delta p$ can be given by following relationship:

$$EI\frac{\partial^4 u}{\partial x^4} = -2R\Delta p \tag{8.28}$$

where

$EI$ = pile material flexibility
$\Delta p = (\dot{u}/IRB)^{1/n}$ as given by equation (8.23)
$\dot{u} = \partial u/\partial t$
$x$ = depth
$t$ = time

The boundary conditions at the top and base of embedded pile for bending moment $M$ and the horizontal shear load $Q$ are as follows:

$$M = -EI\frac{\partial^2 u}{\partial x^2} = QH \qquad \text{at } x = 0 \tag{8.29}$$

$$\text{Shear} = EI\frac{\partial^3 u}{\partial x^3} = Q \qquad \text{at } x = 0 \tag{8.30}$$

$$M = -EI\frac{\partial^2 u}{\partial x^2} = 0 \qquad \text{at } x = L \tag{8.31}$$

$$\text{Shear} = EI\frac{\partial^3 u}{\partial x^3} = 0 \qquad \text{at } x = L \tag{8.32}$$

Nixon (1984) solved this problem by finite difference approximation by using a short program written for the digital computer. The program outputs the displacement, bending moment, shear load, and lateral soil reaction along pile depth at the nodes of pile that was divided into $m$ equal segments of length $\Delta x$. Figure 8.12 presents the lateral deformation versus time obtained for a 2.8 m long, 292 mm diameter pile. Other soil and pile details are provided in this figure. Results plotted in this figure show that a relatively long period must elapse before the flexible pile achieves its equilibrium-deformed shape. For example, it took about 20 days before the surface lateral displacement rates of flexible pile became equal to that predicted by the simplified analysis for a rigid pile. It appears that higher apparent creep rates for flexible piles can be anticipated for a significant time period due to the flexibility of the pile and stress readjustment along the pile shaft. This means that the time to achieve a near-steady pile creep velocity can be greatly reduced if a more rigid pile is used. Field lateral load pile tests and further analysis based on above formulation are required before design curves relating the pile rigidity to the creep properties of surrounding soils for a given applied lateral load and tolerable deformation can be established for design.

The foregoing analysis and design methods are based on limited field tests and therefore need further pile load tests in the field to verify and improve the theoretical treatment of the piles subjected to lateral loads in frozen soils. These methods can, however, be used for estimating allowable lateral load on pile and then the estimated load should be confirmed by field tests. When field tests are not practical following limiting guidelines may be used:

1. For short-term loadings, such as construction, wind and seismic loads for piles greater than or equal to 300 mm (12 in.) diameter and embedment length greater than or equal to 6 m (20 ft) the maximum allowable lateral loads should be limited to the following:
   (a) Steel pile: Maximum allowable lateral load = 77 kN (17.3 kips)
   (b) Timber pile: Maximum allowable lateral load = 58 kN (13 kips)
   These recommendations are based on full-scale pile load tests carried out on 300 mm (12 in.) diameter steel piles (Rowley et al., 1973, 1975).
2. For sustained long-term loads such as seasonal temperature loads that may last for 6 months and for pipeline anchor forces, the allowable lateral loads should not exceed 6 kips (Davison et al., 1981).

## 8.5 RECOMMENDATIONS FOR DESIGN

For pile design in permafrost area, the following steps are recommended:

**Step I** Follow temperate (warm) climate end-bearing pile design procedures if competent bedrock is within reasonable distance below ground surface. If this approach is not practical, permafrost pile design techniques are to be used.

**Step II** Since settlement and strength properties of warm frozen soils (temperature greater than $-1°C$) are still poorly defined, the following three design alternatives are used for such cases:

(a) If permafrost is thaw stable, carry out the design as if the soils are unfrozen. This case applies to clean coarse-grained frozen soils.
(b) If permafrost is thaw unstable, then prethaw and compact these soils.
(c) Lower the permafrost temperature by using artificial refrigeration as discussed in Chapter 2 (Section 2.6.2) and Chapter 3 (Section 3.4.4). Pile

design is then identical to that for cold permafrost (temperature less than $-1^\circ$C discussed as follows.

**Step III** If the ground temperature is colder than $-1^\circ$C, then the following procedure is recommended:

(a) Determine the soil profile at the site and obtain highest measured permafrost temperature. This can either be based on past experience or records or thermocouple measurements for the project. Based on this, determine the depth of permafrost table.

(b) Based on the nature of structure determine if permafrost thawing is likely to occur. If there is a possibility of permafrost thawing, use equations (8.18) and (8.21) to estimate the total downdrag force.

$$X = C_{th}(t)^{0.5} \tag{8.18}$$

$$D = 0.3\pi R\gamma' X^2 \tag{8.21}$$

(c) Based on the highest measured temperature profile, as mentioned in (a), estimate the ultimate adfreeze bond, strength from equation (8.5). Then determine $(Q_v)_{\text{ult}}$ from equation (8.6).

$$\tau_a = mC_{lt} \tag{8.5}$$

$$(Q_v)_{\text{ult}} = 2\pi R_0 \int_0^L \tau_a(z)dz \tag{8.6}$$

Use a proper factor of safety, say 2, to obtain $(Q_v)_{\text{all}}$. Then equate $(Q_v)_{\text{all}}$ with the following:

$$(Q_v)_{\text{all}} = \text{structure load} + \text{downdrag force} \tag{8.33}$$

Determine the total embedment length ($L$) required to satisfy above equation.

(d) If the soils are ice rich (i.e., $\gamma_f$, the frozen bulk density including segregated ice is less than about 1700 kg/m$^3$), calculate pile load capacity $Q_f$ for a given limiting deformation rate $\dot{u}$ and embedment length $L$ by using equation (8.14).

$$Q_f = 2\pi RL\left(\frac{\dot{u}}{R}\right)^{1/n}\left(\frac{n-1}{3^{(n+1)/2}}\right)^{1/n}\left(\frac{1}{B}\right)^{1/n} = 2\pi RL\tau_{av} \tag{8.14}$$

Use a safety factor of 2 to obtain allowable $Q_f$.

(e) If the frozen soils are ice poor ($\gamma_f > 1700$ kg/m$^3$) calculate $Q_f$ from equation (8.17). Use a safety factor of 2 to obtain allowable $Q_f$.

$$\text{if allowable } Q_f \geqslant (\text{structure load} + \text{downdrag force}) \text{ O.K.}$$

$$\text{if allowable } Q_f < (\text{structure load} + \text{downdrag force}) \text{ Not O.K.}$$

Redesign can be carried out by increasing pile embedment length $L$ until allowable $Q_f$ is greater than structural load plus the downdrag force.

(f) Estimate the frost heave force $F$ as discussed in Section 8.2.2. Then ensure that the following condition is satisfied.

$$F < (\text{structure load} + \text{downdrag force})$$

For most cases with sustained loads supported by embedment length in permafrost, seasonal frost heave forces should not be a problem.

## 8.6 DESIGN EXAMPLE

***Example 8.4*** Site investigations at a northern site indicated the following features:

(a) The maximum active layer is 2.0 m (6.6 ft).
(b) Based on measurements by thermocouples, the highest measured permafrost temperatures were 0°C at 2.2 m (6.6 ft) depth and then linearly decreasing to −4°C at 20 m (66 ft) depth below ground.
(c) The soils were ice-rich frozen silts.
(d) The seasonal frost depth was recorded to be 1.25 m (4 ft).

Design a pile to carry a maximum sustained vertical load of 200 kN (44.8 kips) when the maximum allowable pile settlement is 50 mm (2 in.) throughout the 20-year life of the structure. Available materials at the site are timber (spruce) with an allowable strength of 5000 kN/m$^2$ (104.5 kips/ft$^2$) and steel pipe and H pile with an allowable strength of 62,000 kN/m$^2$ (1295 kips/ft$^2$).

SOLUTION

1. Assume that a spruce pile is installed in a predrilled slurried hole and allowed to freeze back. The required pile diameter is:

$$\text{load} = \frac{\pi}{4}(2R)^2 \times \text{strength}$$

$$\text{or } 200 = \frac{\pi}{4}(2R)^2 \times 5000$$

$$\therefore (2R)^2 = \frac{200 \times 4}{\pi \times 5000}$$

$$\text{or } 2R = 225\,\text{mm} = \text{pile diameter}$$

Then use a 350-mm diameter predrilled hole and place the 225-mm diameter pile surrounded by slurry and then allowed to freeze back.

2. Estimate downdrag force due to permafrost thawing or degradation. From equation (8.18): $X = C_{th}(t)^{0.5}$.

$$X = 0.9(20)^{0.5} = 4\,\text{m}$$

when

$$C_{th} = 0.9 \quad \text{(see Section 8.3.3)}$$

$$t = 20 \text{ years}$$

From equation (8.21):

$$D = 0.3\pi R\gamma' X^2 \tag{8.21}$$

$$D = 0.3\pi\left(\frac{0.225}{2}\right)(5.9)(4)^2 = 10.0\,\text{kN}$$

where

$\gamma = 1.6\,\text{Mg/m}^3 = 15.7\,\text{kN/m}^3\ (100\,\text{lb/ft}^3)$
$\gamma' = \gamma - \gamma_w = 5.9\,\text{kN/m}^3\ (37.5\,\text{lb/ft}^3)$

3. Embedment length based on adfreeze strength. The temperature at 2 m depth is 0°C and at 20 m depth is −4°C. For simplification, the average temperature over this depth is assumed to be −2°C. For greater accuracy, the depth can be divided into layers and based on temperature variation with depth average temperature at middle of each layer can be used. Results based on above simplification will not be very much different from the one based on breaking the depth in layers. Then from equation (8.6):

$$(Q_v)_{\text{ult}} = \pi(2R)\int_0^L \tau_a(z)d_z. \tag{8.6}$$

If

$L = 20\,\text{m}$
$R = 0.225/2\,\text{m}$
$\tau_a = 0.7 \times 150 = 105\,\text{k Pa}$ (equation (8.5), Figure 8.4, and Table 8.1)
$(Q_v)_{\text{ult}} = 1484\,\text{kN} \qquad (Q_v)_{\text{all}} = 742\,\text{kN with } FS = 2$

If

$$L = 15\,\text{m} \qquad (Q_v)_{ult} = 1113\,\text{kN} \qquad (Q_v)_{all} = 556.5\,\text{kN}$$

If

$$L = 10\,\text{m} \qquad (Q_v)_{ult} = 742\,\text{kN} \qquad (Q_v)_{all} = 371\,\text{kN}$$

Sustained structure load + downdrag force = 200 + 10 = 210 kN. From above for $L = 10$ m, $(Q_v)_{all} = 371\text{ kN} > 210\text{ kN}$. Therefore use a pile embedment length of 10 m. Total pile length = embedment + active layer + degradation depth. Total pile length = 10 + 2 + 4 = 16 m (53 ft).

4. Embedment length based on settlement criteria. From equation (8.14):

$$Q_f = 2\pi RL\left(\frac{\dot{u}}{R}\right)^{1/n}\left(\frac{n-1}{3^{(n+1)/2}}\right)^{1/n}\left(\frac{1}{B}\right)^{1/n}$$

$$2R = 0.225\,\text{m} \qquad \dot{u} = 0.05/20\,\text{m/yr} \qquad n = 3 \qquad B = 2 \times 10^{-8} \text{ (Table 8.2)}$$

$$L = 10\,\text{m}$$

$$Q_f = \pi \times 0.225 \times 10\left(\frac{0.05 \times 2}{20 \times 0.225}\right)^{1/3}\left(\frac{2}{3^2}\right)^{1/3}\left(\frac{1}{2 \times 10^{-8}}\right)^{1/3} = 474\,\text{kN}$$

$$\text{Allowable load} = Q_f/FS = 237\,\text{kN} \qquad FS = 2$$

This is still greater than 210 kN (from step (3) above). Therefore, a 225-mm (9 in.) diameter pile with 33 ft (10 m) embedment length in permafrost and a total length of 53 ft (16 m) will have an allowable design load of 53.3 kips (237 kN). The design load is controlled by settlement criteria.

5. Heave force ($F$). From Section 8.2.2, $F = \pi(2R)$ depth of seasonal frost × adfreeze force. $F = \pi(0.225)(1.25)(78.5)\text{kN} = 69.4\,\text{kN}(15.6\,\text{kips})$. This force is less than sustained load plus downdrag force ⇒ O.K. The total pile length of 16 m (53 ft) can be reduced by using construction methods so that permafrost degradation is either reduced or avoided. One such method could be providing thermopiles. Another method is to provide shaded air space that would decrease ground temperatures. If this pile foundation supports a building with air space below the floor of the structure, the shade will inhibit the permafrost degradation. Thus, a 4.0-m depth of thaw degradation can be saved. The total required pile length will then be 10 + 2 = 12 m, instead of 16 m as shown above. This assumes that the pile does not introduce heat into ground by conduction, which is a reasonable assumption for timber piles.

## 8.7 OVERVIEW

If frozen soils in permafrost environment consist of competent materials such as ice-free rock, dense glacial till, or non-frost-susceptible soils, then pile design can be carried out in a conventional manner as discussed in Chapters 5 and 6. In such situations, frozen soil condition can be neglected. If, on the other hand, the materials are frost susceptible, procedures outlined in this chapter for permafrost environment are applicable.

The strength of frozen soils depends on ground temperatures, stress level, soil type, and the duration of test time. In general, the strength of these soils decreases with time and increases as soil temperatures decrease. Furthermore, these soils also exhibit creep phenomena under constant deviatoric stress. For example, ice-rich frozen soils exhibit steady-state creep under low to medium load levels. Based on the studies of load-deformation behavior of frozen soils under different confining pressures and temperatures it has been shown by Ladanyi (1972), Ladanyi and Johnston (1974), Nixon (1978), Nixon and Lem (1984) and other investigators that shear strength and load deformation behavior of frozen soils is significantly influenced by time and the ground thermal regime.

The upper part of the active layer thaws in summer and refreezes every winter. If frost-susceptible soils exist in this upper part, then frost heaving will occur, which can develop upward heaving forces on foundations in two ways. The first consists of the application of upward forces on underside of foundation surfaces, such as pile caps and grade beams. The second consists of the development of upward adfreeze forces along the pile perimeter in the frost zone. Magnitude of these heave forces depends on soil type, moisture content, ground thermal regime, and foundation flexibility. Heavy pressures as high as 260 psi (1800 kPa) have been reported in the literature (Penner, 1970). It is therefore recommended that rather than designing for such high potential uplift pressures, foundations should be placed well below the depth of active layer. The magnitude of adfreeze forces along pile surface in frost zone can also vary significantly. The magnitude and the factors affecting adfreeze bond stresses have not yet been fully investigated. Average values of 10 psi (70 kPa) for wood and concrete piles and 15 psi (100 kPa) for steel piles are generally used in practice for calculating adfreeze upward forces on piles. Further work is required in this area.

One of the most common methods of installing piles in permafrost areas is to auger a hole into the ground and then drop a steel or wood pile into it. The annular space around the pile is then filled with sand–water slurry. This slurry is allowed to freeze to develop an adfreeze bond between the pile and the frozen soil. The time required for this slurry to freeze by natural conduction is called *freezeback time*. This time can be estimated with reasonable accuracy by using equation (8.3). If natural freezeback time is too long and the load on pile has to be applied sooner than this, then artificial refrigeration methods may be required.

The two design criteria that must be satisfied for an axially loaded pile in permafrost are as follows:

1. Estimate the ultimate load capacity of a pile that is mobilized by adfreeze bond strength. This can be related to the long-term cohesion between the pile and the frozen soil for the warmest permafrost temperature that is

expected during its service life. The ultimate pile load capacity can then be calculated by using equation (8.6), Figure 8.4, and Table 8.1.

2. The load required to maintain the settlement within tolerable limits over the iife of a structure can be estimated by considering a friction pile subjected to simple shear under plain strain conditions. Equation (8.14) for ice-rich soils and equation (8.17) for ice-poor soils can be used to estimate this load. These equations require the use of experimentally determined creep parameters. Although a large database on these creep parameters has been collected, further site-specific data supported by long-term field pile load tests should be encouraged to gather more information on soil–pile creep behavior. Some information on creep settlement on saline permafrost is also available. Additional work is required in this area.

The allowable pile load capacity can then be calculated by dividing the lower value obtained from (1) or (2) above with a safety factor of 2. Analysis shows that for ice-rich frozen soils, load capacity calculated from tolerable settlement criteria governs while for ice-poor soils, the load capacity is generally governed by adfreeze bond strength.

Allowable lateral load capacity of a short free-headed vertical pile can be estimated from equation (8.27) and Figure 8.11. Nixon (1984) also gives a formulation for a long elastic laterally loaded pile, which can be solved by a computer program. These design methods are based on limited pile load tests. Further long-term pile load tests are required to provide more confidence in these design methods. In the meantime, design recommendations provided in Section 8.4 for estimating capacity of laterally loaded piles can be used.

## REFERENCES

Andersland, O. B. and Anderson, D. M. (eds.), *Geotechnical Engineering for Cold Regions.* McGraw-Hill Book Co., New York (1978).

Casagrande, A. "A New Theory of Frost Heaving: Discussion: *Proceedings U.S. Highway Research Board*, Vol. II, Part I, 1932, pp. 168–172.

Chamberlain, E., Groves, C. and Perham, R. "The Mechanical Behaviour of Frozen Earth Materials Under High Pressure Triaxial Test Conditions," *Geotechnique*, Vol. 22, No. 3, 1972, pp. 469–483.

Crory, F. E. "Pile Foundations in Permafrost," *Proceedings International Conference on Permafrost (1963)*, Lafayette, Indiana, U.S. National Academy of Sciences, Publ. 1287, 1966, pp. 467–476.

Crory, F. E. and Reed, R. E. "Measurement of Frost Heaving Forces on Piles," U.S. Army, CRREL, Technical Report 145, 1965, p. 27.

Davison, D. M., Harris, M. C., Hayley, D. W., Johnston, G. H., Ladanyi, B., McCormick, G., Nixon, J. F. and Penner, E. *Permafrost Engineering Design and Construction*, G. H. Johnston (ed.) Wiley, New York, 1981, pp. 247–343.

Frederking, R. and Kerri, J. "Effects of Pile Material and Loading State on Adhesive

Strength of Piles in Ice," *Can. Geotech. J.*, Vol. 20, No. 4, November 1983, pp. 673–680.

Johnston, G. H. (ed.), *Permafrost Engineering Design and Construction.* Wiley, 1981.

Johnston, G. H. and Ladanyi, B. "Field Tests of Grouted Rod Anchors in Permafrost," *Can. Geotech. J.*, Vol. 9, No. 2, 1972, pp. 176–194.

Kersten, M. S. *Thermal Properties of Soils*, University of Minnesota, Engineering Experiment Station Bulletin 28, 1949, p. 227.

Kiselev, M. F. "Standard Values of Specific Tangential Forces of Frost Heaving of Soils," *J. Soil Mech. Found. Eng.* (U.S.S.R), No. 3, 1974, pp. 41–43 (Translated by Consultants Bureau, New York).

Ladanyi, B., "An Engineering Theory of Creep of Frozen Soils," *Can. Geotech. J.*, Vol. 9, No. 1, 1972, pp. 63–80.

Ladanyi, B. "Bearing Capacity of Strip Footings in Frozen Soils," *Can. Geotech. J.*, Vol. 12, 1975, pp. 393–407.

Ladanyi, B. and Johnston, G. H. "Behaviour of Circular Footings and Plate Anchors Embedded in Permafrost," *Can. Geotech. J.*, Vol. 11, No. 4, 1974, pp. 531–553.

Linell, K. A., "Long-Term Effects of Vegetation Cover on Permafrost Stability in an Area of Discontinuous Permafrost," *Proceedings International Conference on Permafrost*, Yakutsk, U.S.S.R., North American Contribution, U.S. National Academy of Sciences, 1973, pp. 688–693.

Low, P. F., Anderson, D. M. and Hoekstra, P. "Some Thermodynamic Relationships for Soils at or Below the Freezing Point, 2. Effect of Temperature and Pressure on Unfrozen Soil Water," *Water Resources Research*, Vol. 4, No. 5, 1968, pp. 541–544.

Nixon, J. F., "First Canadian Geotechnical Colloquium: Foundation Design Approaches in Permafrost Areas," *Can. Geotech. J.*, Vol. 15, No. 1, 1978, pp. 96–112.

Nixon, J. F., "Laterally Loaded Piles in Permafrost," *Can. Geotech. J.*, Vol. 21, No. 3, 1984, pp. 431–438.

Nixon, J. F., and Lem, G. "Creep and Strength Testing of Frozen Saline Fine-Grained Soils," *Can. Geotech. J.*, Vol. 21, No. 3, 1984, pp. 518–529.

Nixon, J. F. and McRoberts, E. C. "A Design Approach for Pile Foundations in Permafrost," *Can. Geotech. J.*, Vol. 13, No. 1, 1976, pp. 40–57.

Nixon, J. F., and Neukirchner, R. J. "Design of Vertical and Laterally Loaded Piles in Saline Permafrost," *Proceedings, 2nd International Conference on Cold Regions Engineering*, Edmonton, Alberta, Canada, 1984, pp. 1–14.

Parameswaran, V. R., "Adfreeze Strength of Model Pile in Ice," *Can. Geotech. J.*, Vol. 18, No. 1, 1981, pp. 8–16.

Penner, E., "Frost Heaving Forces in Leda Clay," *Can. Geotech. J.*, Vol. 7, No. 1, 1970, pp. 8–16.

Robinsky, E. I., and Bespflug, K. E. "Design of Insulated Foundations," *J. Soil. Mech. Found. Div.*, ASCE, Vol. 99, No. SM9, 1973, pp. 649–667.

Rowley, R. K., Watson, G. H., and Ladanyi, B. "Vertical and Lateral Pile Load Tests in Permafrost," *Proceedings 2nd International Conference on Permafrost*, Yakutsk, U.S.S.R., North American Contribution, U.S. National Academy of Sciences, 1973, pp. 712–721.

Rowley, R. K., Watson, G. H. and Ladanyi, B. "Prediction of Pile Performance in Permafrost Under Lateral Load," *Can. Geotech. J.*, Vol. 12, No. 4, 1975, pp. 510–523.

Tsytovich, N. A. "Principles of Geocryology," Part II, Chapter III, Canada, National Research Council, technical translation TT 1239, 1959, pp. 28–79.

Tsytovich, N. A., "Bases and Foundations on Frozen Soils," U.S. Highways Research Board, Special Report 58, 1960, pp. 1–93.

Voitkovskii, K. F. "Mekharicheskiye Svoystva Idia" (The Mechanical Properties of Ice.) Issledovaniya Academii Nauk. (In Russian; English Translation by the Air Force Cambridge Research Laboratories, Bedford, MA, AFCRL-62-838, AMS-T-R-391) 1960, p. 92.

Vyalov, S. S. "Rheological Properties and Bearing Capacity of Frozen Soils, (Russian, translated in 1965) U.S. Army Corps of Engineers, Cold Regions Research and Engineering Laboratory, Army Translation No. 74, Hanover, NH, 1959, p. 219.

Vyalov, S. S., and Porkhaev, G. V. (eds.), "*Handbook for the Design of Bases and Foundations of Buildings and Other Structures on Permafrost*," National Research Council, Canada, Technical Translation 1865, 1976.

Weaver, J. S., and Morgenstern, N. R. "Pile Design in Permafrost," *Can. Geotech. J.*, Vol. 18, No. 3, 1981, pp. 357–370.

Womick, O., and LeGoullon, R. B. "Settling a Problem of Settling," *The Northern Engineer*, Vol. 7, No. 1, 1975, pp. 4–10.

# 9

# PILE LOAD TESTS

As discussed in Chapters 5 through 8, the estimation of pile load capacity and settlement under a load is based on the results of field investigations, laboratory testing and the empirical and semiempirical methods. These estimated values should then be confirmed by field pile load tests. Pile load tests, in practice, are normally executed in two alternative ways:

1. **Test Pile** Preliminary pile design is first carried out on the basis of site investigations, laboratory soil testing, and office study. Pile load tests are then carried out to refine and finalize the design. For these conditions, the test piles are generally tested to failure.
2. **Test on a Working Pile** In areas where previous experience is available, pile design is carried out based on the site investigations, laboratory soil testing, and office study. Pile load tests are then carried out on randomly selected actual piles to check the pile design capacities. In these situations, the piles are generally tested to two times the design capacity.

The equipment and test procedures for these two alternatives are essentially similar. The main difference is the level of final loading. Therefore, the details of tests presented below are applicable for both of the tests listed above. This chapter presents the details of pile load test for axial compression, pullout, and lateral and dynamic loads.

## 9.1 AXIAL COMPRESSION PILE LOAD TESTS

This section first discusses the test equipment and load and, movement measuring instruments required in an axial compression pile load test. Following this, the load test procedures and methods of interpreting test data are discussed. Finally, an example of pile load test and its interpretation are presented.

### 9.1.1 Test Equipment and Instruments

The main aspects of test equipment and instruments consist of load application arrangements and the instruments to measure the resulting movements or deformations. These two items are presented here separately.

***Load Application Arrangements*** As shown in Figure 9.1a, a typical example of axial compression load application arrangement consists of two anchor piles located on either side of a test pile. In order to minimize the interference between test and anchor piles, a minimum distance of five times the pile diameter is maintained between the piles. A reaction beam is placed on top of the anchor piles and the test pile is loaded by utilizing a hydraulic jack placed centrally on top of the test pile. This results in applying compressive load on the test pile and the tensile load on the anchor piles. A slightly different loading arrangement is shown in Figure 9.1b. As shown in Figure 9.2, an alternate loading arrangement such as a timber crib and weights can also be used in lieu of the anchor pile and reaction beam system. ASTM D 1143-81 (1989) cites other alternate loading arrangements.

Load applied by hydraulic jack is measured either by a calibrated load cell placed between the jack and the pile or by a calibrated pressure gauge located between the pump and the hydraulic jack. The load cell and the pressure gauge should be calibrated before each test program to an accuracy of not less than 5 percent of the applied load. Some engineers require that the ram and the gauge be calibrating as a unit. The advantage of calibrating as a unit is supposed to be that the effect of ram friction occurring along the sides of the ram (primarily at the location of the seal) can be taken into account. Davisson (1970), has shown that this can be an exercise in futility. Also if the ram is not perfectly aligned with the pile, eccentric loading may occur which may cause misalignment of the ram in the ram housing. This increases the potential for ram friction. Davisson (1970, 1989) recommends that a spherical bearing may be placed between the ram and the reaction bearing (Figure 9.1b). An introduction of a spherical bearing will also eliminate horizontal movements of the loading arrangements. These movements may be up to 1″ or more in a poor set up and be no more than 1/8″ in a good set up (Davisson, 1970). In order to provide a check and as a backup in case of one system malfunctioning, consideration should be given to employing both the load cell and the pressure gauge. At the time of load test planning, it should be ensured that the loading frame is designed for the maximum anticipated applied load and hydraulic jack rams have sufficient travel to provide for anticipated pile settlements, deflections of the reaction beam and elongation of connections of anchoring devices. It is also recommended that the loading frame should be conservatively designed so that at least 50 percent higher load can be applied on test piles in case the actual failure load is higher than the anticipated value. If this is not done, then valuable load movement behavior near failure may be missed (Nordlund, 1982; Sharma et al., 1986).

***Instruments for Measuring the Movements*** The two main types of movement measurements in a pile load test are pile butt axial movement measurement and incremental strain measurements along the pile length.

Pile butt axial movement measurements are required in all pile load tests. The incremental strain measurements are used to determine the distribution of load transfer from the pile to the soil and are generally considered as an optional measurement feature.

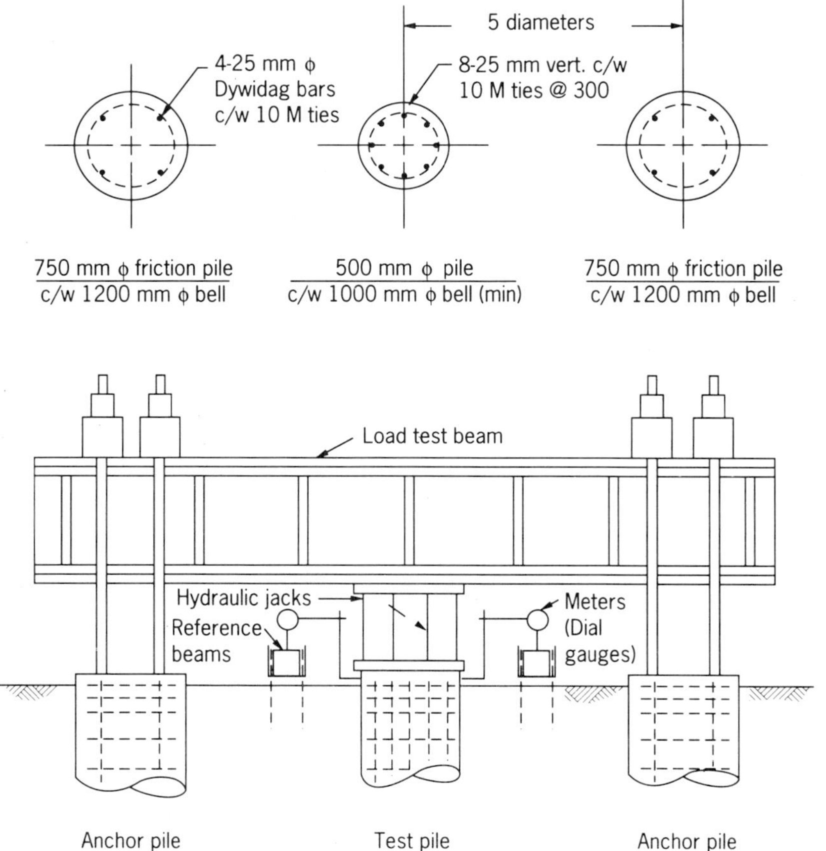


**Figure 9.1a** An example of a typical axial compression load application arrangement (Sharma et al., 1984).

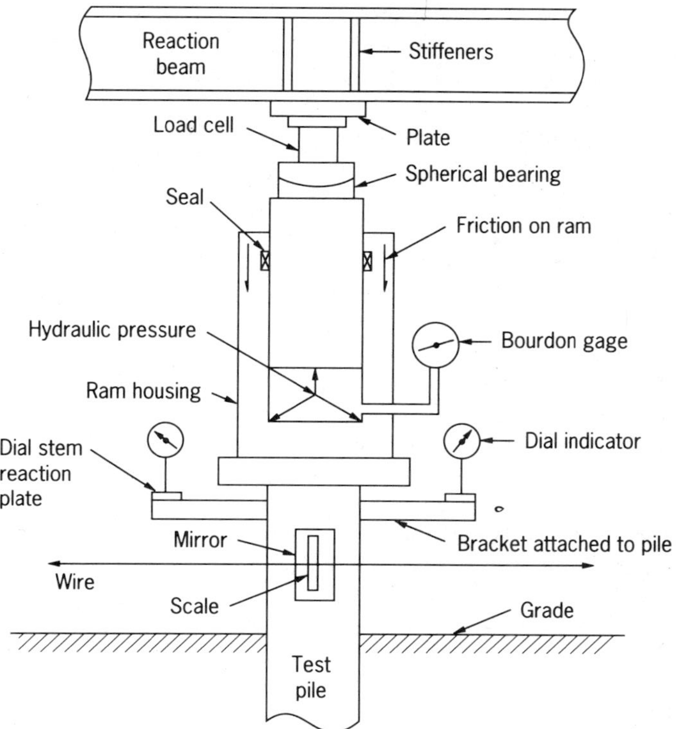


**Figure 9.1b** An example of hydraulic ram on a test pile with spherical bearing. (Davisson, 1970)

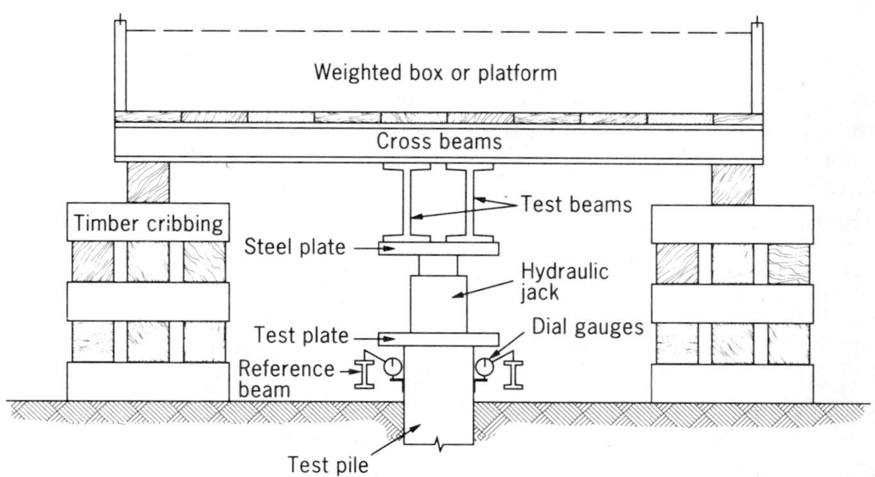


**Figure 9.2** Axial compression load application arrangement using timber cribbing and the weighted box (ASTM, 1986).

*Pile Butt Axial Movement* The most common types of instruments used to measure pile butt or head axial movement are the dial gauges, the wire, mirror and scale arrangement, and the surveyor's level system.

DIAL GAUGES In the dial gauge measuring system, two parallel reference beams, one on each side of the test pile, are independently supported on firm supports. These supports are located at least 8 ft (2.5 m) away from the test beam. Reference beams should be stiff to support the instruments and should be of such rigidity that excessive vibrations in readings do not occur. It is recommended that reference beams should be of timber so that temperature variations do not affect the readings. If steel beams are used then it should be ensured that one end of each beam is free to move horizontally to accommodate beam length changes with temperature variations. A minimum of two dial gauges, approximately equal distance and on opposite sides of the test pile, should be mounted on the reference beam. Davisson (1970) recommends that the two dial gauges *must* be located on a diameter of the pile cross section and placed at equal radial distances. In this manner, the two dial readings can be averaged to obtain the deflection of the center of the pile (Figure 9.1b). Dial gauges should have at least a 2 in. (50 mm) travel. Longer gauge stems should be provided where larger movement is anticipated. All gauges should also have a precision of at least 0.01 in. (0.25 mm). Figure 9.3(a) presents this measuring arrangement schematically.

THE WIRE, MIRROR, AND SCALE ARRANGEMENT As shown on Figure 9.3b, the wire, mirror, and scale system consists of two parallel wires, one on each side of the test pile and supported on both ends as far as practicable from anchor piles. Wires should be tightly anchored at the supports so that tension is maintained throughout the test. Piano wires or equivalent type are generally recommended. Each wire passes across the face of the test pile and is located at about 1 in. (25 mm) away from the face. A mirror and a scale are mounted on the face of the pile opposite to the wire. The pile butt axial movements can be recorded from the readings of the scale directly by lining up the wire and its image in the mirror. Davisson (1970) recommends the use of two dial gauges and two wire-scale-mirror systems at right angles to each other. For further details, the reader is referred to Davisson (1970).

SURVEYOR'S LEVEL SYSTEM A surveyor's level is generally used for measuring the axial movement as a check rather than as a primary means of movement measurement. The system consists of a surveyor's level stationed at least 10 pile butt diameters from both the test and the reaction piles (Davisson, 1970). Level shots are then taken on a reference scale fixed to the side of the test pile as shown on Figure 9.3b. These readings are referenced to a permanent bench mark located outside the load test area.

*Incremental Strain Measurements Along the Pile Length* Incremental strain

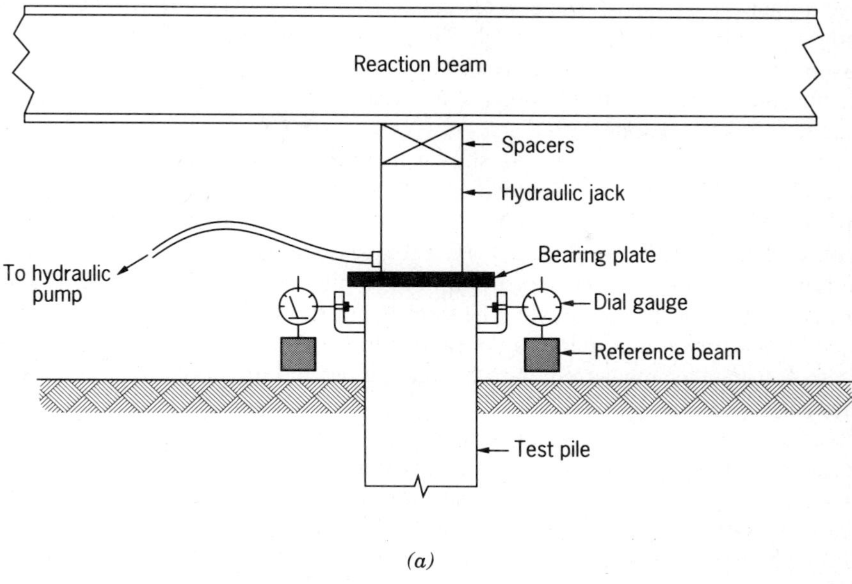


*(a)*

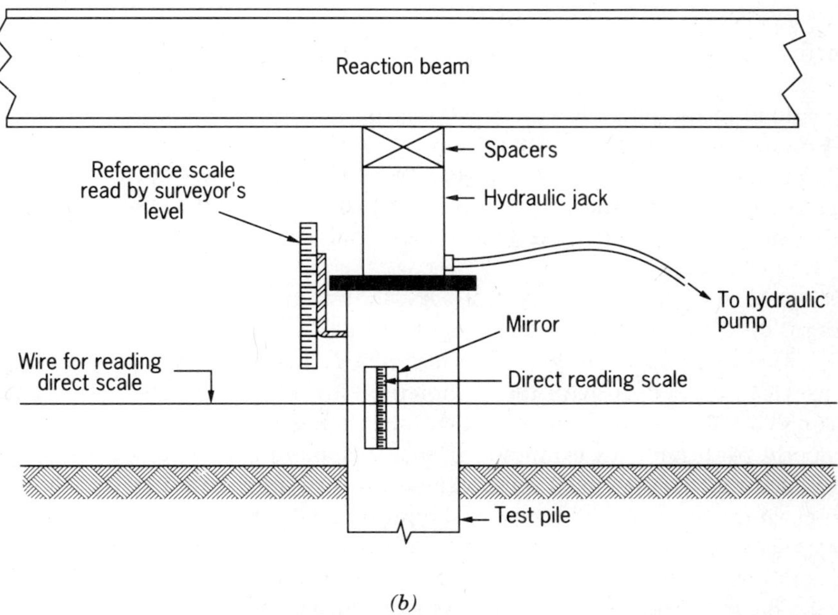


*(b)*

**Figure 9.3** Instruments for measuring pile butt axial vertical movement. (a) Dial gauges, (b) wire, mirror, and scale and the surveyor's level.

measurements along the pile length are carried out to determine the distribution of load transfer from the pile to the soil. These also provide information on pile tip movements or deflections along the pile. Instruments that can be used for such measurements are the strain rods (or telltales) and the electric strain gauges.

TELLTALE MEASURING SYSTEM As shown in Figure 9.4a telltales or strain rods normally consist of polyvinyl chloride (PVC) tubings extended to steel end plates embedded inside a concrete pile or welded on the steel pile at various locations along the pile length. Inside the PVC tubing, a stainless steel rod is installed extending from the end plate to the top of the pile. Both the PVC tube and the steel rod extend to the top of pile. The steel rod must be allowed to move freely in the tube. The movement of the top of each telltale or strain rod relative to the top of the test pile is measured with a dial gauge having 0.001 in. (0.025 mm) sensitivity.

Normally, telltale readings are referenced to the top of the pile. By noting the location of the specific telltale rod anchor plate and by measuring the relative movement of the individual rod, elastic shortening of pile at that location can be obtained. With this information the load in the pile at the midpoint between two telltale anchor plates separated by a distance $L$ can be obtained by the following relationship:

$$Q_{va} = A\frac{\Delta L}{L}E \tag{9.1}$$

where

$Q_{va}$ = load in the pile midway between two anchor plates
$A$ = cross section area of the pile
$\Delta L$ = difference in movement between two telltale rods
$L$ = distance along the pile between the two telltale anchor plates
$E$ = modulus of elasticity of the pile material

STRAIN GAUGE MEASURING SYSTEM As shown in Figure 9.4b electric strain gauges or vibrating wire strain gauges can be mounted along the pile length at various locations before the pile is installed. In cast-in-place drilled piles, these gauges can be tied up with the reinforcing bars and wires can be brought up through a PVC casing. In driven piles, the strain gauges and the wire should be properly protected. An example for such installation will be provided in the following paragraph. Since these gauges are temperature sensitive, additional temperature-compensating gauges should be used for each strain gauge. Long term measurements on concrete and timber piles may represent changes in the pile material itself as well as movement of the pile relative to the soil. By contrast steel pile cross sections are stable with respect to creep and changes in modulus and can provide information on long term pile support (Davisson, 1970). The strain $\varepsilon$ can be determined directly by noting the change in the strain gauge

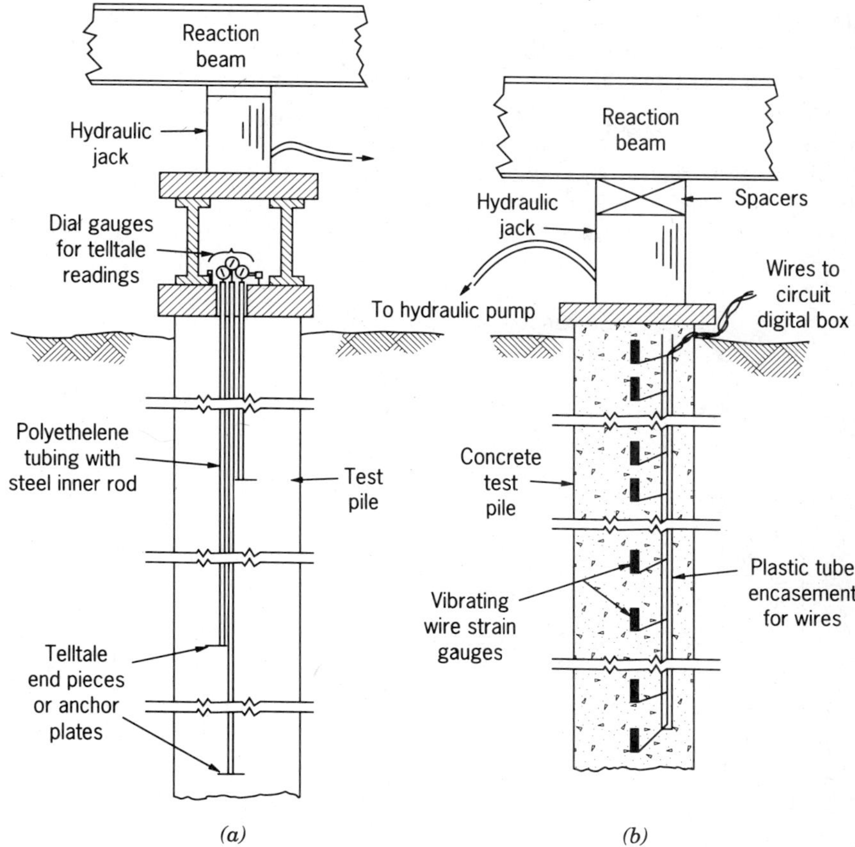


**Figure 9.4** Incremental strain-measuring systems along the pile length. (a) Telltale, (b) electric strain gauge measuring system.

reading from the unstrained to any desired load. The load at the point will then be calculated by following relationship:

$$Q_{va} = AE\varepsilon \tag{9.2}$$

where

$Q_{va}$ = load in the pile at the location of the strain gauge
$A$ = cross section area of the pile
$\varepsilon$ = strain gauge reading
$E$ = modulus of elasticity of the pile material

Rieke and Crowser (1986) cite a case where four instrumented W14 × 144 steel piles were successfully load tested. Two of these piles were tested under both axial compression and uplift and other two were tested in uplift only. The instrumentation consisted of installing four telltales and four vibrating wire strain gauges as shown in Figure 9.5. The telltale anchors were welded to the web of the pile at four locations. The telltale rods were approximately 1/4 in. in diameter and were attached to the pile by threading into short sections of No. 6 bars. The telltale rods were encased in PVC tubing having a nominal inside diameter of 1/4 in. and a nominal outside diameter of 1/2 in. Just below the telltale anchors, vibrating wire strain gauges were welded at locations shown in Figure 9.5. Steel channels

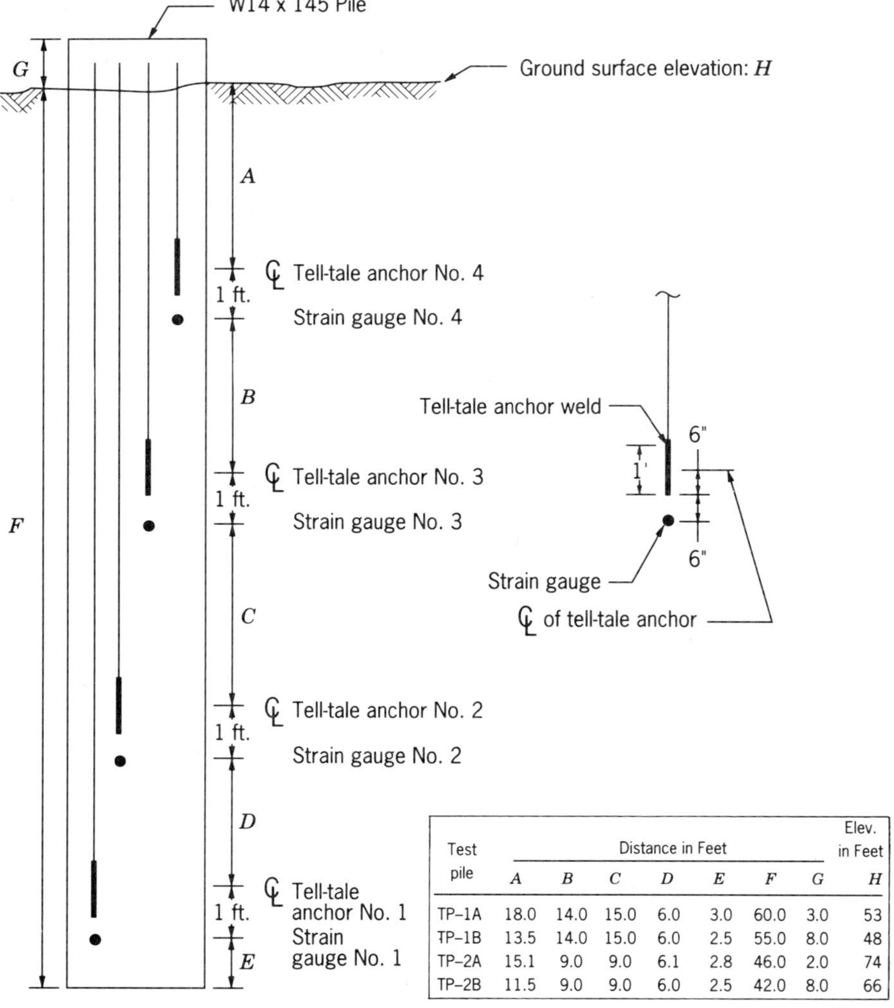


| Test pile | Distance in Feet A | B | C | D | E | F | G | Elev. in Feet H |
|---|---|---|---|---|---|---|---|---|
| TP–1A | 18.0 | 14.0 | 15.0 | 6.0 | 3.0 | 60.0 | 3.0 | 53 |
| TP–1B | 13.5 | 14.0 | 15.0 | 6.0 | 2.5 | 55.0 | 8.0 | 48 |
| TP–2A | 15.1 | 9.0 | 9.0 | 6.1 | 2.8 | 46.0 | 2.0 | 74 |
| TP–2B | 11.5 | 9.0 | 9.0 | 6.0 | 2.5 | 42.0 | 8.0 | 66 |

**Figure 9.5** A steel-driven test pile instrumentation at a site (Rieke and Crowser, 1986, courtesy of *The Slope Indicator Co.*)

$1 \times 1/2 \times 1/8$-in. were welded along the length of the pile on each side of the instrumentation. A steel strap about 18 in. long and 1 to 2 in. wide was welded across the channels every 2 ft along the pile length for additional strength. The space within the channel was then filled with a two-part epoxy that became hard overnight to a point about 5 ft (1.5 m) below the top of the instrumentation. This was to protect strain gauges and their wires and to keep PVC sections in place during driving. A steel plate was tack welded to the channel over the uppermost 5 ft (1.5 m) of the instrumentation. The space beneath this protective plate surrounding the instrumentation was filled with a two-part foam. Test results indicated that, although the telltale showed poor performance, all the vibrating wire strain gauges performed very well and did not appear to suffer any damage during the driving of the test pile.

Sharma et al. (1986) cite another case where two 26-in. (660 mm) shaft diameter and 36-in. (914 mm) bell diameter bored cast-in-place concrete test piles were instrumented with IRAD Model EM-5 vibrating wire strain gauges embedded at 5 ft (1.5 m) intervals down the pile shaft. The gauges were 6.63 in. (166 mm) long with $\pm 1\,\mu$ strain sensitivity and working temperature range of $-40$ to 160°F. To avoid damage to the gauges and also to ensure that large aggregates do not cause nonuniform strain fields near the gauge, the gauge were cast into cement topping mix with the same strength as the pile concrete. These gauges were then tied to the pile reinforcing cage. The wires from each gauge were brought to the pile top through a PVC tubing. The concrete was then poured into the pile borehole. Load test results indicated that most of these gauges performed satisfactorily during the axial compression pile load test. The foregoing examples clearly indicate that if properly protected and calibrated, the vibrating wire strain gauges can provide good data on incremental strains and load transfer along the pile.

### 9.1.2 Test Procedures

Practicing engineers and researchers have used many pile load test methods that have been reported in several publications (ASTM D1143-81, Butler and Hoy, 1977; Fellenius, 1975, 1980; Mohan et al., 1967; New York State DDT, 1974; Swedish Pile Commission, 1970; Weele, 1957; Whitaker, 1957, 1963; Whitaker and Cooke, 1961). From the available numerous load test methods the following four methods can be identified as the basic load test methods (Joshi and Sharma, 1987):

1. Slow Maintained Load Test Method (SM Test)
2. Quick Maintained Load Test Method (QM Test)
3. Constant Rate of Penetration Test Method (CRP Test)
4. Swedish Cyclic Test Method (SC Test)

1. **Slow Maintained Load Test Method (SM Test)** This test method, as recommended by ASTM D1143-81(1989), consists of the following steps:

(a) Load the pile in eight equal increments (i.e., 25 percent, 50 percent, 75 percent, 100 percent, 125 percent, 150 percent, 175 percent, and 200 percent) to 200 percent of the design load.

(b) Maintain each load increment until the rate of settlement has decreased to 0.01 in./h (0.25 mm/h) but not longer than 2 h.

(c) Maintain 200 percent load for 24 h.

(d) After the required holding time, remove the load in decrements of 25 percent with 1 h between decrements.

(e) After the load has been applied and removed, as above, reload the pile to the test load in increments of 50 percent of the design load, allowing 20 min between load increments.

(f) Then increase the load in increments of 10 percent of design load until failure, allowing 20 min between load increments.

This test method is commonly considered as the ASTM Standard Test method and is generally used for site investigation prior to installing contract piles and writing specifications. The main disadvantage of this test is that it is time consuming (e.g., a typical test period may last 40 to 70 h or more).

2. **Quick Maintained Load Test Method (QM Test)** This test method, as recommended by the New York State Department of Transportation, the Federal Highway Administration, and the ASTM 1143-81 (optional), consists of the following main steps:

(a) Load the pile in 20 increments to 300 percent of the design load (i.e., each increment is 15 percent of the design load).

(b) Maintain each load for a period of 5 min with readings taken every 2.5 min.

(c) Add load increments until continuous jacking is required to maintain the test load or test load has been reached.

(d) After a 5-min interval, remove the full load from the pile in four equal decrements with 5 min between decrements.

This test method is fast and economical. Typical time of test by this method is 3 to 5 h. This test method represents more nearly undrained conditions. This method cannot be used for settlement estimation because it is a quick method.

3. **Constant Rate of Penetration Test Method (CRP Test)** This method is recommended by Swedish Pile Commission, New York State Department of Transportation, and ASTM D1143-81 (optional). It consists of the following main steps:

(a) The pile head is forced to settle at 0.05 in/min (1.25 mm/min).

(b) The force required to achieve the penetration rate is recorded.

(c) The test is carried out to a total penetration of 2 to 3 in. (50 to 75 mm).

The main advantages of this method are that it is fast (2 to 3 h) and is economical. This method is of particular value for friction piles but may not be practical for end-bearing piles because of the high force requirements to cause penetration through hard-bearing stratum.

4. **Swedish Cyclic Test Method (SC Test)** This method as recommended by Swedish Pile Commission consists of the following main steps:

(a) Load the pile to one-third of the design load.

(b) Unload to one-sixth the design load. Repeat the loading and unloading cycles 20 times.

(c) Increase the load by 50 percent higher than the item (a) and then repeat as item (b).

(d) Continue until failure is reached.

This test method is time consuming, and cycling changes the pile behavior so the pile is different than the original pile. It is only recommended on special projects where cyclic loading may be of main importance.

As shown in Figure 9.6, the SM tests and SC tests are the slowest tests and the CRP test is the fastest. Figure 9.7 compares typical load-movement behavior for the four test types discussed. This figure shows that the shape of load–movement curve by the CRP test method is well defined and agrees well with the QM-test load-movement curve before the failure is reached. The SM test method is commonly used in North America because it is simple, most engineers are

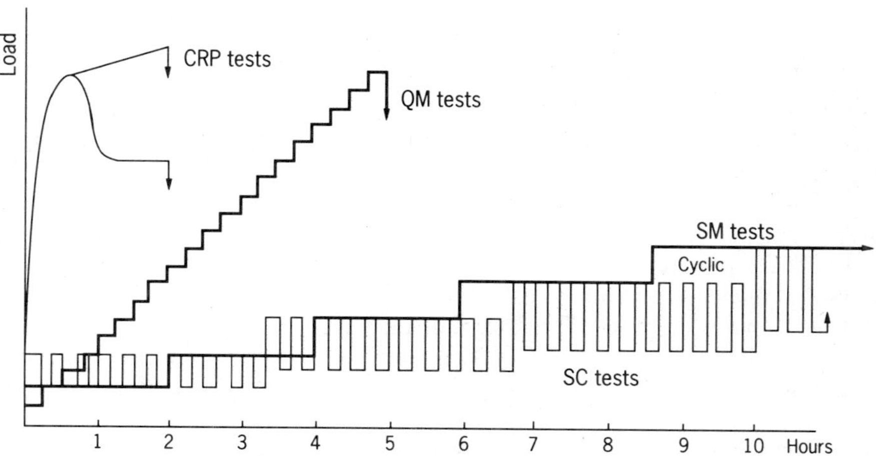


**Figure 9.6** Comparison of required time for various test methods (Fellenius, 1975).

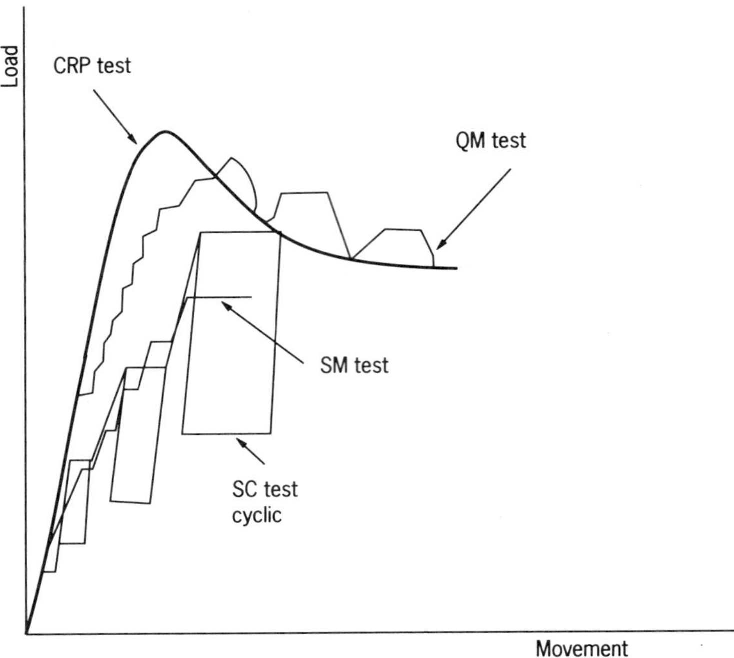


**Figure 9.7** Comparison of load–movement behaviour for test methods (Fellenius, 1975).

familiar with it, its interpretation based on gross and net settlements can be made easily, and it can furnish a rough estimate of the expected pile settlement under working load. Interpretation of the failure load from load-movement curves obtained from load tests will be discussed in the following section.

### 9.1.3 Interpretation of Test Data

Generally, load and settlement test data are plotted with load along the abscissa and settlement along ordinate. However, these coordinates can be interchanged depending on the engineer's preference. The plotted settlement could either be gross (the total movement of pile butt under full test load) or the net (the distance the pile has permanently moved after it has rebounded upon removal of the test load). These plotted data are then used to estimate the failure load so that allowable pile capacity can be calculated.

The ultimate failure load for a pile is defined as the load when the pile plunges or the settlements occur rapidly under sustained load. Plunging, however, may require large movements that may exceed the acceptable range of the soil–pile system. Other failure definitions consider arbitrary settlement limits such as the pile is considered to have failed when the pile head has moved 10 percent of the pile end diameter or the gross settlement of 1.5 in. (38 mm) and net settlement of

0.75 in. (19 mm) occurs under two times the design load. Many engineers define the failure load at the point of intersection of the initial tangent to the load–movement curve and the tangent to or the extension of the final portion of the curve. All these definitions for defining failure are judgemental. Ideally, a failure definition should be based on some mathematical rule and should result in repeatable values. Also, the value should be independent of scale effects and individual's personal opinion. The following interpretation methods have been used in the past for various load tests. First, these methods are reviewed and their applicability for different pile types discussed.

1. Davisson's method (1972)
2. Chin's method (1970, 1971)
3. De Beer's method (1967)
   or De Beer and Wallays' method (1972)
4. Brinch Hansen's 90 percent criterion (1963)
5. Brinch Hansen's 80 percent criterion (1963)
6. Mazurkiewicz's method (1972)
7. Fuller and Hoy's method (1970)
8. Butler and Hoy's method (1977)
9. Vander Veen's method (1953)

**1. Davisson's Method** The procedure for obtaining failure load by this method consists of the following steps:

(a) Draw the load-movement curve as shown in Figure 9.8a.
(b) Obtain elastic movement, $\Delta = (Q_{va})L/AE$ of the pile where $Q_{va}$ is the applied load, $L$ is pile length, $A$ is pile cross-sectional area, and $E$ is modulus of elasticity of the pile material.
(c) Draw a line $OA$ based on equation for elastic movement, $\Delta$, as identified in item (b) .
Draw a line $BC$ parallel to $OA$ at a distance of $x$ where $x = 0.15 + D/120$ in., ($D$ = diameter of pile in in.).
(e) The failure load is then at the intersection of $BC$ with load–movement curve (i.e., point $C$).

This method was originally recommended for driven piles, and its use is preferred for the QM test method. The main advantage of this method is that the limit line $BC$ can be drawn before starting the test. Therefore, it can be used as one of the acceptance criteria for proof-tested contract pile.

**2. Chin's Method** This method is shown in Figure 9.8b and consists of the following steps:

(a) Draw the $\Delta/Q_{va}$ versus $\Delta$ plot, where $\Delta$ is the movement and $Q_{va}$ is the corresponding applied load.

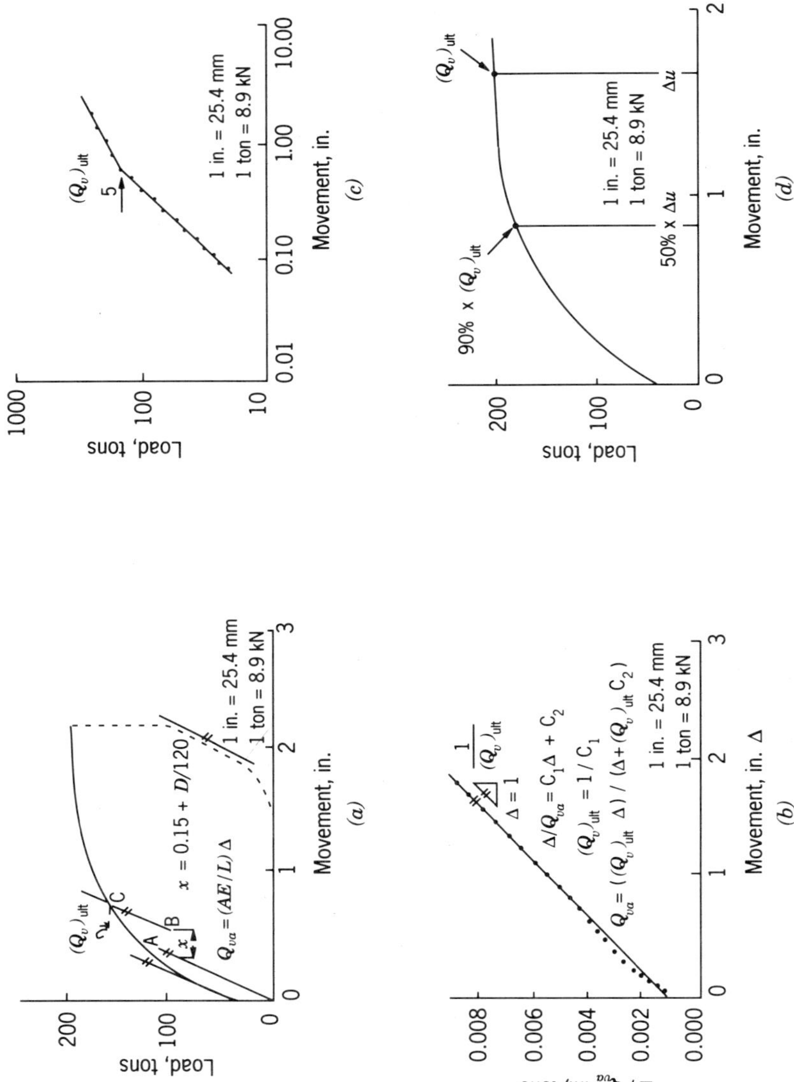


**Figure 9.8** Load test interpretation methods by Davisson, Chin, De Beer, and Brinch Hansen's 90 percent Criterion (Fellenius, 1980; Joshi and Sharma, 1987). (a) Davisson's method, (b) Chin's method, (c) De Beer's method, (d) Brinch Hansen's 90 percent criterion.

(b) The ultimate load $(Q_v)_{\text{ult}}$ is then equal to $1/C_1$. Figure 9.8b explains all the terms. The relationships given in this figure assume that the load–movement curve is approximately hyperbolic.

This method of ultimate load interpretation is applicable for both the QM and SM tests, provided constant time increments are used during the test. In selecting the straight line from the points, it should be understood that the data points do not appear to fall on the straight line until the test load has passed Davisson's limit value. This method may not provide realistic failure value for tests carried out as per ASTM Standard Method because it may not have constant time load increments.

3. **De Beer's Method** As seen in Figure 9.8c, this method consists of the following steps:

(a) Plot load and movement on logarithmic scales.
(b) These values then fall on two straight lines.
(c) The failure load is then defined as the load that falls at the intersection of these two straight lines.

This method was originally proposed for a slow test, such as SM tests.

4. **Brinch Hansen's 90 percent Criterion** This is a trial and error method and the method of interpretation is shown on Figure 9.8d and consists of the following steps:

(a) Plot the load–movement curve.
(b) Find the load $(Q_v)_{\text{ult}}$ and $\Delta_u$ that gives twice the movement of the pile head as obtained for 90 percent of the load $(Q_v)_{\text{ult}}$, where $(Q_v)_{\text{ult}}$ is the failure load.

This method is applicable to the CRP test method regardless of the soil type.

5. **Brinch Hansen's 80 percent Criterion** This method of interpretation is shown in Figure 9.9a and consists of the following steps:

(a) Plot $\dfrac{\sqrt{\Delta}}{Q_{va}}$ and $\Delta$ curve, where $\Delta$ is the movement and $Q_{va}$ is the load.
(b) Failure load $(Q_v)_{\text{ult}}$ and failure movement $\Delta_u$ are then given as follows:

$$(Q_v)_{\text{ult}} = \frac{1}{2\sqrt{C_1 C_2}} \tag{9.3a}$$

$$\Delta_u = \frac{C_2}{C_1} \tag{9.3b}$$

All the terms are defined in Figure 9.9a. This method assumes that the load–movement curve is approximately parabolic. The method is applic-

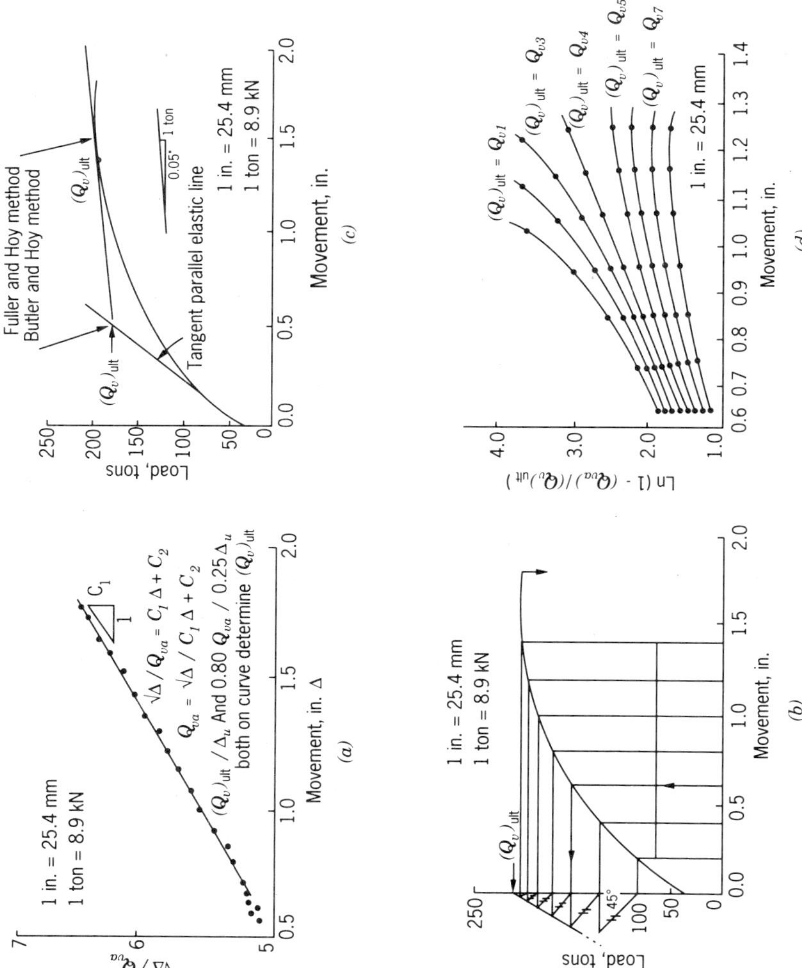


**Figure 9.9** Test load interpretation methods. (a) Brinch Hansens's 80 percent criterion method. (b) Mazurkiewicz's method. (c) Fuller and Hoy's and Butler and Hoy's methods. (d) Vander Veen's method (Fellenius, 1980; Joshi and Sharma, 1987).

able for both the quick and slow tests (e.g., QM and SM tests). The failure criteria agrees well with the plunging failure. However, the plot and calculations can not be performed in advance of the test loading. This method of interpretation is not suitable for test methods that include unloading cycles or where plunging failure is not achieved.

6. **Mazurkiewicz's Method** As shown in Figure 9.9b, this method consists of the following steps:
   (a) Plot the load–movement curve.
   (b) Choose a series of equal pile head movements and draw vertical lines that intersect on the curve. Then draw horizontal lines from these intersection points on curve to intersect the load axis.
   (c) From the intersection of each load, draw 45° line to intersect with the next load line.
   (d) These intersections fall approximately on a straight line. The point which is obtained by the intersection of the extension of this line on the vertical (load) axis is the failure load.

   This method assumes that load–movement curve is approximately parabolic. The failure load values obtained by this method should, therefore, be close to the 80 percent criterion. Furthermore, all the intersections of these lines do not always fall on a straight line. Therefore, some judgment may be required in drawing the straight line.

7. **Fuller and Hoy's Method** This consists of the following steps:
   (a) Plot a load–movement curve as shown in Figure 9.9c.
   (b) Find the failure load $(Q_v)_{\mathrm{ult}}$ on the curve where the tangent on the load–movement curve is sloping at 0.05 in./ton.

   This method is applicable for QM test. The main disadvantage with this method may be that it penalizes the long piles because they will have larger elastic movements and therefore 0.05 inch/ton slope will occur sooner.

8. **Butler and Hoy's Method** As shown on Figure 9.9c, this method consists of the following steps:
   (a) Plot the load–movement curve.
   (b) The failure load is then the intersection of the 0.05-in./ton slope line with either the initial straight portion of the curve (Figure 9.9c) or the line parallel to the rebound curve or the elastic line starting from the origin (not shown).

   This method is applicable for the QM test.

9. **Vander Veen's Method** This method consists of the following steps:
   (a) Choose a value of failure load, say $(Q_v)_{\mathrm{ult}}$.

(b) Plot $l_n(1 - Q_{va}/(Q_v)_{ult})$ for different values of $Q_{va}$ against the movement for various load, $Q_{va}$.

(c) When the plot becomes a straight line, then the corresponding $(Q_v)_{ult}$ represents the correct failure load as shown by $Q_{va}$ in Figure 9.9d.

The main disadvantage of this method is that time-consuming calculations are required to obtain the failure load.

Joshi and Sharma (1987) carried out failure load interpretations on five different load–movement curves obtained by using the SM test method. The length to diameter ratio for these piles varied between 12 to 32. Load–movement curves for all these piles indicated plunging failure. All nine failure load interpretations methods discussed above were used. Results obtained from this study provided the following conclusions:

(a) For bored and belled concrete piles, the Fuller and Hoy method provided a reasonable estimate for the failure load.

(b) For expanded-base-compacted (Franki) piles, the Davisson, Butler and Hoy; and Fuller and Hoy methods provide reasonable estimates for failure loads.

(c) For driven H piles, Brinch Hansen's 90 percent criterion and Fuller and Hoy's method predicted the failure load similar to the failure test load.

Fellenius (1980) carried out similar interpretations on a 12-in. (305mm) diameter concrete-driven pile that was tested by the CRP method. Interpretations indicated that Fuller and Hoy's, Brinch Hansen's 90 percent criterion, and Vander Veen's methods provided reasonable estimates of failure loads. The foregoing indicated that in call cases, Davisson's method predicted conservative values for failure loads, and Chin's method invariably yielded failure loads higher than the actual test failure loads. The Fuller and Hoy method appeared to yield failure loads that were reasonable approximations of the actual failure loads.

### 9.1.4 Example of a Pile Load Test

Sharma et al. (1984) reported two axial compression, a pullout, and two lateral pile load tests on cast-in-place bored and belled concrete piles. This section presents the data and the analysis on one of these axial compression pile load test. Section 9.2.4 presents the data and analysis for the pullout pile load test, and Section 9.3.4 presents similar data on a lateral pile load test.

As shown in Figure 9.10a, the general soil conditions at the site consisted of glacial clay till over bedrock that primarily consisted of clay shale and siltstone. Bedrock was layered, fissured, and slickensided and was under artesian water pressures.

The general test layout consisted of three reaction piles and three test piles as shown in Figure 9.10b. In order to have a minimum interference between two adjacent piles, the center-to-center pile spacing was kept five times the shaft diameter of the larger pile. The general arrangement for the axial compression

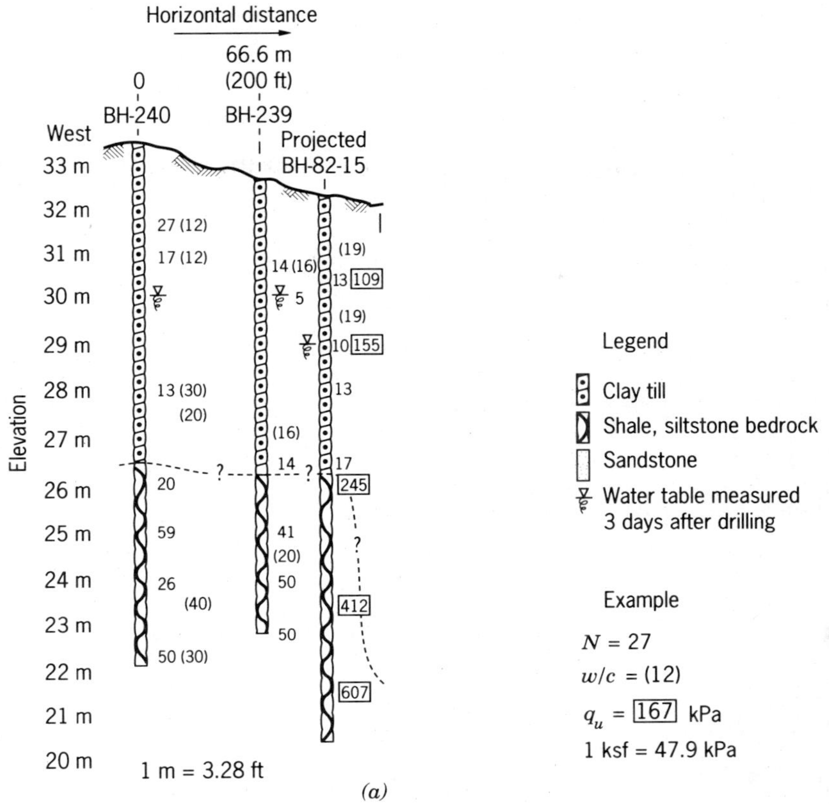

(a)

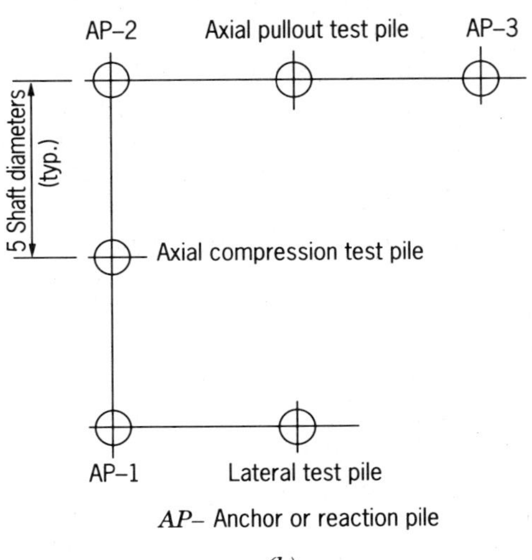


(b)

**Figure 9.10** (a) Soil stratigraphy and (b) layout for a pile load test program at a test site (Sharma et al., 1984).

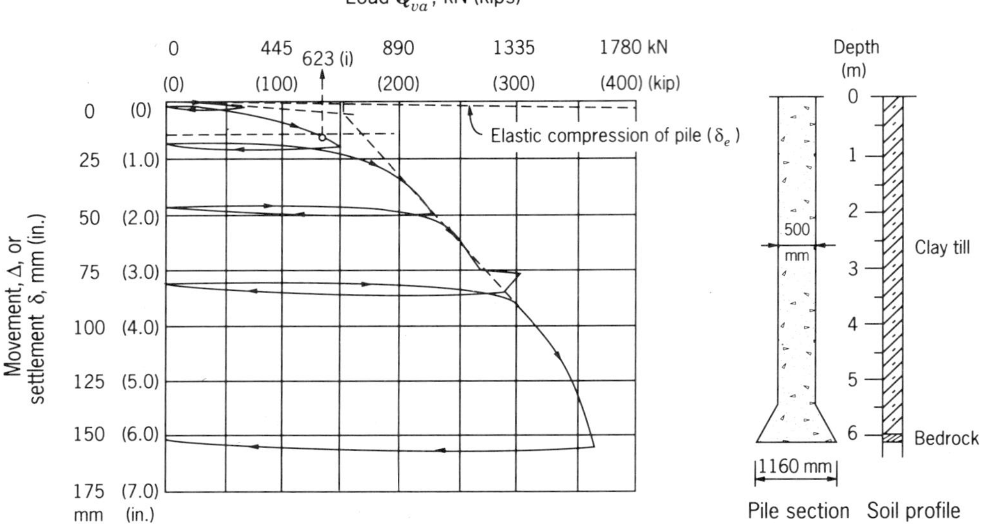


**Figure 9.11** Axial compression pile load test results (Sharma et al., 1984).

pile load test is shown in Figure 9.1. The test pile has a 20-in. (500 mm) shaft diameter and a 40-in. (1000 mm) bell diameter. Axial load was applied by a 600-kip (2670 kN) capacity hydraulic jack and by using two 30-in. (750 mm) shaft diameter anchor piles at each end of a reaction beam. Figure 9.1 also shows the pile head movement measurement arrangement. Axial compression load test was carried out as per ASTM 1143-81 and is detailed in Section 9.1.2 under the SM test method. All the gauges and the hydraulic jack were calibrated for the test.

Figure 9.11 exhibits the test pile dimensions, soil profile, and load–movement curve obtained from the load test. The test pile failed at an approximate axial compression load of 355 kips (1580 kN). The various interpretation methods presented in Section 9.1.3 were used for the load–movement data presented in Figure 9.11. Table 9.1 summarizes the failure loads interpreted by these methods. The results show that the De Beer method yields the lowest failure load followed by Davisson's and Butler and Hoy's methods. Brinch Hansen's 90 percent criterion could not be applied for this load–movement data because the shape of the curve was such that $(Q_v)_{\text{ult}}$ and $\Delta_u$ that gives twice the movement of the pile head as obtained for 90 percent of the load $(Q_v)_{\text{ult}}$ could not be found on the curve. Also, Mazurkiewicz's method could not be applied because the curve is not approximately parabolic, which is the assumption for this method. Vander Veen, Brinch Hansen's 80 percent criterion, and Chin's methods yielded failure loads higher than the test failure load. As discussed in Section 9.1.3, Fuller and Hoy's method gave interpreted failure load close to the test failure load for this bored and belled concrete pile.

**TABLE 9.1 Failure Loads Interpreted by Various Methods**

| Method | Failure Load, $(Q_v)_{ult}$ (kips) | (kN) |
|---|---|---|
| De Beer | 103 | (458) |
| Davisson | 138 | (614) |
| Butler and Hoy | 162 | (721) |
| Fuller and Hoy | 315 | (1402) |
| Vander Veen | 404 | (1798) |
| Brinch Hansen's 80 percent criterion | 448 | (1994) |
| Chin | 484 | (2154) |

Note: Test results showed that failure occurred at approximately 355 kips (1580 kN) (Sharma et al., 1984).

## 9.2 PULLOUT PILE LOAD TESTS

This section discusses the test equipment and instruments, test procedures, and the interpretation method for pullout pile load test data. Finally, an example of a pullout pile load test is presented.

### 9.2.1 Test Equipment and Instruments

The test equipment and instruments consist of the load application arrangement and instruments for measuring movements. In the following paragraphs, these will be presented separately.

***Load Application Arrangement*** Figures 9.12 and 9.13 show two typical setups for applying pullout loads on the test pile. Figure 9.12 shows an arrangement where the pullout load is applied to the pile by a hydraulic jack acting between supported test beam and a reaction frame anchored to the pile. Two reaction supports consist either of piles or cribbing installed on either side of the test pile. The clear distance between the test pile and the reaction piles or cribbing shall be at least five times the pile butt diameter but not less than 8 ft (2.5 m) so that there is no significant effect on the performance of test pile due to external loading. Figure 9.13 shows an alternate loading arrangement where the load is applied to the pile by hydraulic jacks acting at both ends of the test beam that is anchored to the pile. If this loading arrangement is used, then the load on the pile is twice the jacking load. ASTM (1989) provides details on other alternate loading arrangements. Any one of these loading arrangements can be used for a pile load test depending on their availability or the preference of design engineer.

Before a pile load test is started, it should be ensured that the complete jacking system including the hydraulic jack(s), hydraulic pump, and pressure gauge

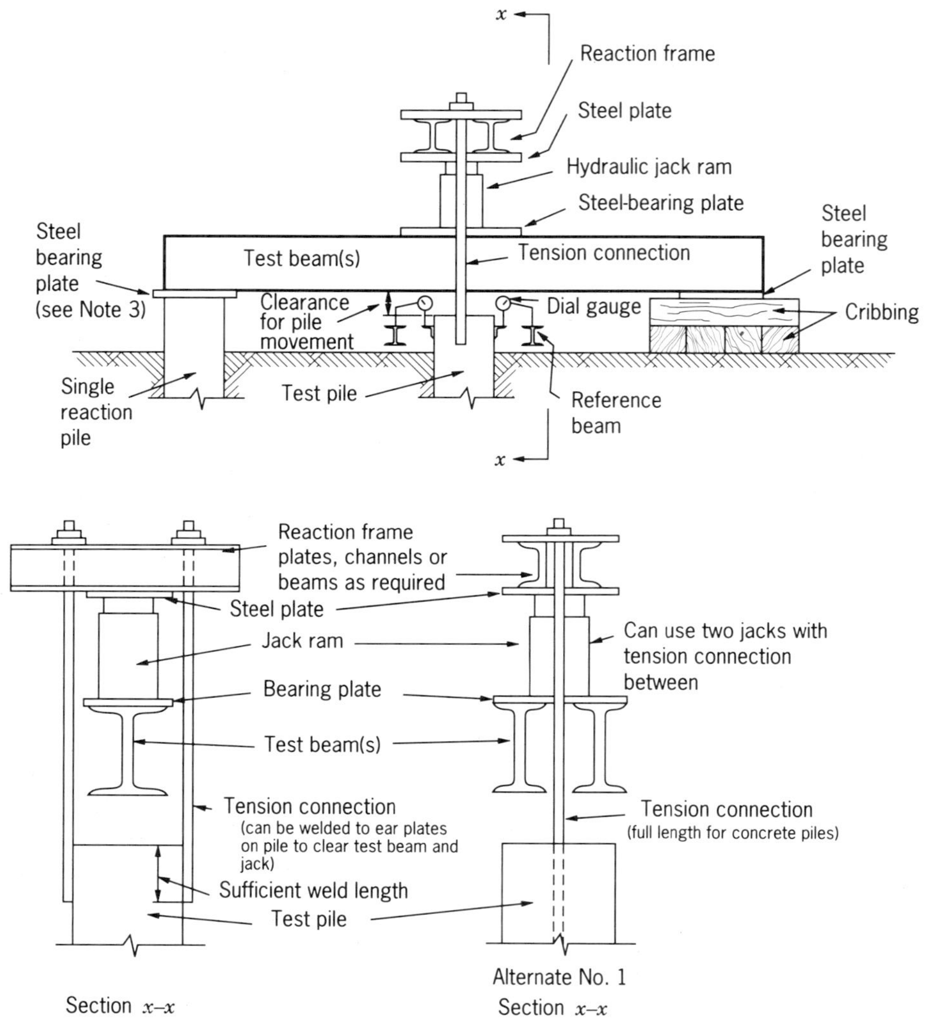


Note 1—Load on pile equals applied load.
Note 2—Use same type reaction (piles or cribbing) at both ends of test beam.
Note 3—Plate not required for steel reaction pile.
Note 4—Use stiffener plates between flanges of all beams where structurally required.

**Figure 9.12** Typical setup for applying pullout load by using hydraulic jack between beam and reaction frame (ASTM D 3689-83, 1989).

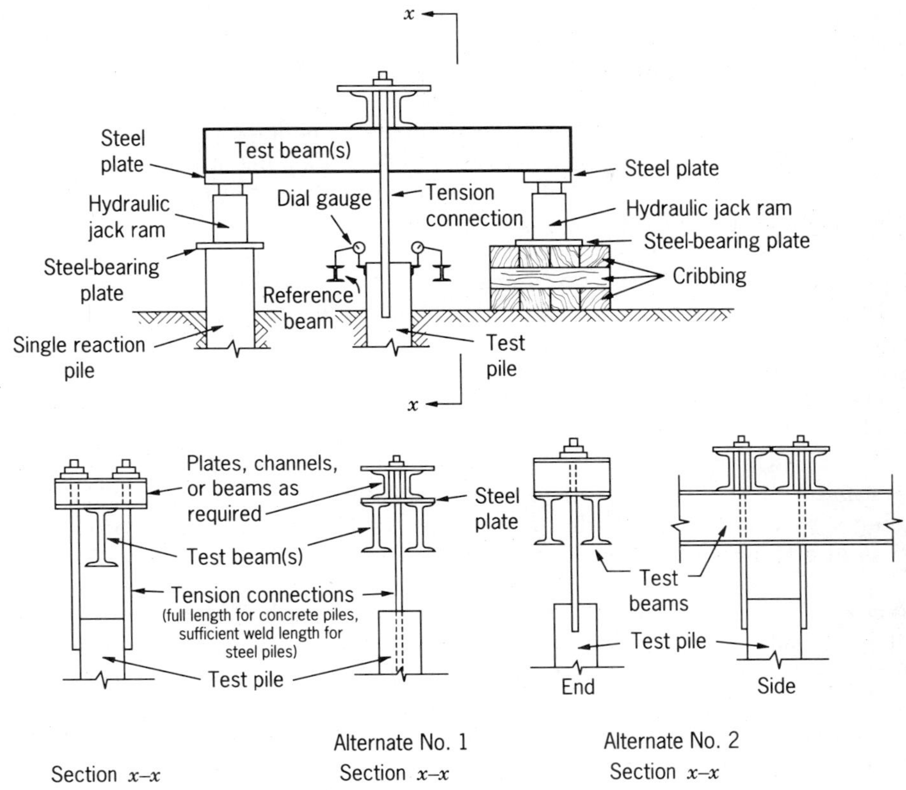


Note 1—Use same type reaction (piles or cribbing) at both ends of test beam.
Note 2—Load on pile is twice the jacking load unless the pressure gauge has been calibrated for the two-jack system.
Note 3—Use stiffener plates between flanges of all beams where structurally required.

**Figure 9.13** Typical setup for applying pullout load by using hydraulic jack, one at each end of the beam (ASTM D 3689-83, 1989).

should be calibrated as a unit. If more than one jack is used, both should be of the same ram diameter, connected to a common manifold and pressure gauge, and should be operated by a single hydraulic pump. It is a good practice to use a dual load measuring system (i.e., a pressure gauge and a load cell). This provides a check and acts as a backup in case one system malfunctions. The test beam and reaction piles should be designed so that they have enough strength for the test load. Finally, the hydraulic jack rams should have sufficient travel for the anticipated pile movements, anchor elongations, and the test beam deflections.

***Instruments for Measuring Movements*** As for axial compression pile load tests, the two main types of movement measurements in a pullout load test are pile butt axial movement measurement, and incremental strain measurements along the pile length.

The axial movement measurements are required in all the pile load tests. The incremental strain measurements are used to determine the distribution of load transfer along pile shaft and therefore are used on special projects and may be considered as an optional measurement.

*Pile Butt Axial Movement Measurement* Pile butt axial movement measurement consists of a primary and secondary system. Dial gauges and wire, mirror, and scale are used as the primary system while a surveyor's level is used as a secondary or a check system. All dial gauges should have at least 3-in. (75 mm) travel and 0.001-in. (0.025 mm) accuracy. As shown in Figures 9.12 and 9.13, dial gauges should be mounted on independent reference beams that are firmly supported in the ground at a clear distance of not less than 8 ft (2.5 m) from the test pile and from the reaction piles or cribbing. All other requirements of the dial gauges, wire, mirror and scale, and surveyor's level as detailed in Section 9.1.1 shall also be applicable here.

*Incremental Strain Measurements Along the Pile Length* Incremental strains along the pile length are measured to determine the distribution of load transfer from the pile to the soil. Instruments that are generally used to make such measurements are the strain rods (telltales) and the electric strain gauges. These instruments have been discussed in detail in Section 9.1.1. Similar instrumentation can be used for the pullout pile load test.

### 9.2.2 Test Procedures

The four basic load test methods identified for axial compression pile load test (Section 9.1.2) are also applicable for pullout tests with the difference that the load is applied in the upward direction on the test pile. ASTM D3689-83 (1989) provides details of these and other load testing procedures. The commonly used pullout test procedure in North America is the Slow Maintained Load Test Method (SM Test). Most engineers are familiar with this method and its interpretation.

### 9.2.3 Interpretation of Test Data

Methods of determining failure load from pullout pile load tests vary depending on the tolerable movement of the structure. In general, failure load for pullout test is more easily defined when compared with the axial compression load test data because the available pullout resistance generally decreases more distinctly after reaching failure. The generally accepted interpretation methods, in practice, for estimating ultimate pullout load is the lowest of the following three criteria (Sharma et al., 1984)

1. Failure load may be taken as the load value that produces a net upward pile butt movement of 0.25 in. (6.25 mm).
2. The upward failure load is at the point of intersection of tangents on the load–movement curve.
3. The upward failure load is the value at which upward movement suddenly increases disproportionately (i.e., the point of sharpest curvature on the load–movement curve).

These methods of interpretation are shown in Figure 9.15 and are discussed in the load test example, Section 9.2.4.

### 9.2.4 Example of a Pile Load Test

A pullout pile load test was carried out at the location shown in Figure 9.10b. Soil stratigraphy at the site is shown in Figure 9.10a. The pile was a 20-in. (500 mm)

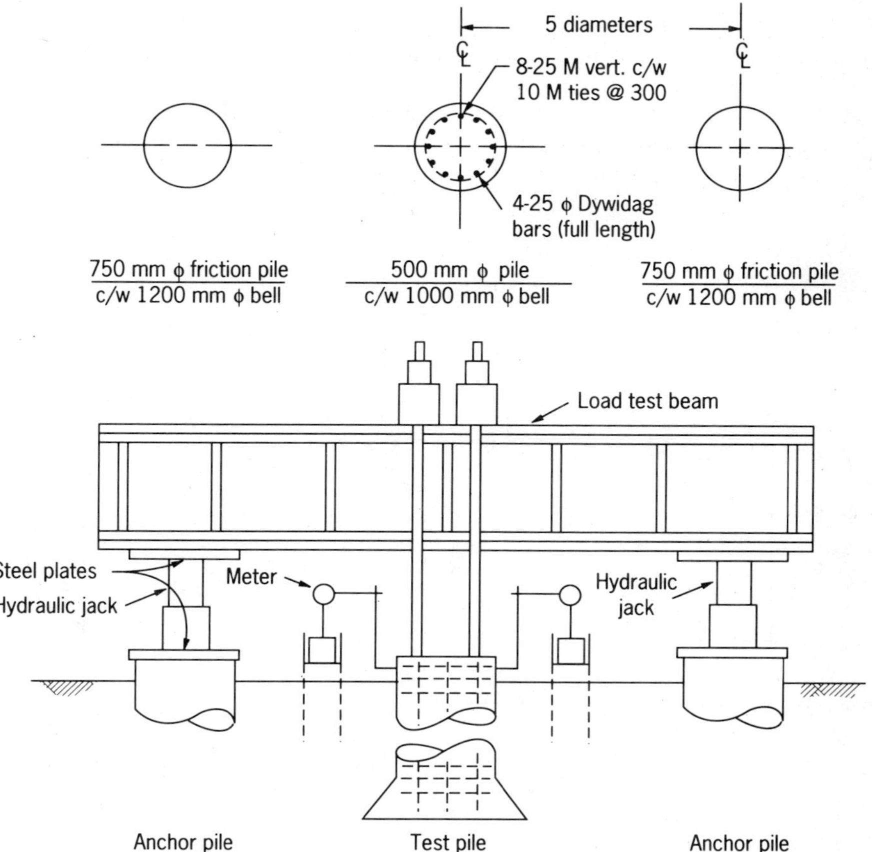


**Figure 9.14** Arrangement for axial pullout pile load test setup (Sharma et al., 1984).

shaft diameter, 40-in. (1000 mm) bell diameter, and 20 ft (6.1 m) long cast-in-placed concrete pile.

The general arrangement for the load test consisted of two 30-in. (750 mm) shaft diameter and 48-in. (1200 mm) bell diameter concrete reaction piles on either side of the test pile. Pull out load was applied by two hydraulic jacks as shown in Figure 9.14. Pile head upward movement was measured by two dial gauges, one on either side of the test pile. The load test was carried out as per Slow Maintenance Load Test Method (SM Test); this is also specified as Standard Loading Procedure and Loading in Excess of 200 percent Uplift Load as per ASTM D3689-83 (1989). This procedure is described in Section 9.1.2. The only difference being that the load was applied in pullout (tension) instead of the axial compression. All gauges and the hydraulic jacks were calibrated before the test was started.

Figure 9.15 shows the pile dimensions and the soil profile recorded during pile hole drilling operation. This figure also presents the load–movement curve obtained from the pullout load test. Load test exhibited that at about 190 kips (845 kN) the pile could not hold any pullout load. Ultimate loads as interpreted by the three methods identified in Section 9.2.3 are as follows:

1. **Failure Load Based on 0.25-in. (6.25 mm) Pile Head Upward Movement** As shown in Figure 9.15, the load corresponding to 0.25-in. (6.25 mm) pile head upward movement is 170 kips (758 kN). Therefore, failure load based on this criterion is 170 kips (758 kN).

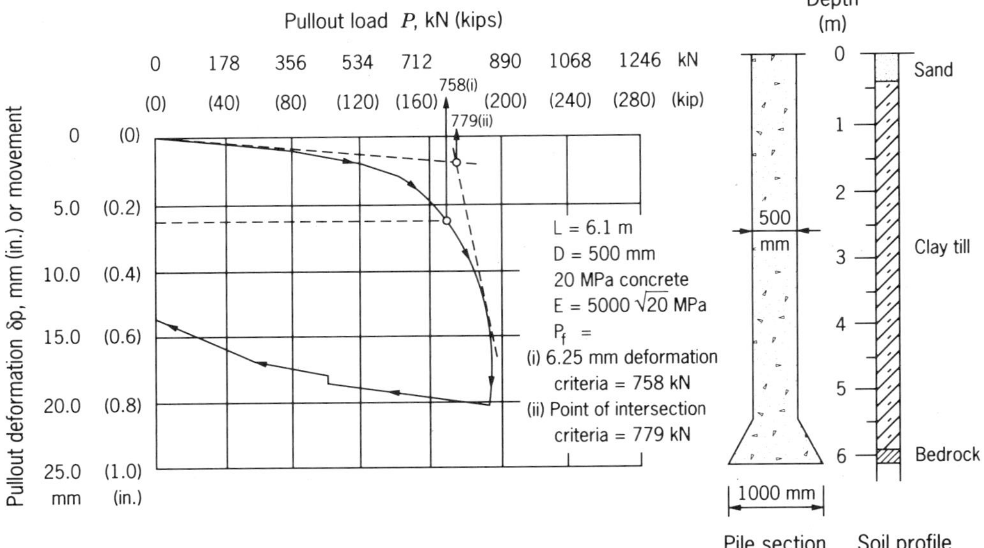


**Figure 9.15** Axial pullout pile load test data (Sharma et al., 1984).

2. **Failure Load At the Point of Intersection of Tangents** Again, as exhibited in Figure 9.15, the load corresponding to the point of intersection of tangents on the initial and final parts of the curve is 175 kips (779 kN). This will be the failure load based on this criterion.
3. **Failure Load at the Point of Sharpest Curvature** From Figure 9.15 it can be interpreted that the load at the point of least radius or the sharpest curvature is approximately 180 kips (800 kN). This interpretation method is very subjective and appears to depend a lot on the individual's judgment.

These three methods interpret failure load ranging from 170 kips (758 kN) to 180 kips (800 kN). These values are close to the test load of 190 kips (845 kN) at which the pile could not hold any pullout load and continued to move. The interpreted failure load is therefore 170 kips (758 kN), which is the lowest of the interpreted values discussed above.

## 9.3 LATERAL PILE LOAD TESTS

In this section, first the test equipment and load–movement measuring instruments for lateral load test are presented. Following this, the test procedures, interpretation of test data, and finally an example of the pile load test are discussed.

### 9.3.1 Test Equipment and Instruments

The test equipment and instruments consist mainly of the load application arrangement and the movement measuring instruments. These are presented separately.

***Test Equipment for Load Application*** As shown in Figure 9.16, the lateral load is applied to the test pile by using a hydraulic jack and a suitable reaction system. Examples of some of the reaction systems are one or more reaction piles(s), deadman, and weighted platform as shown in Figure 9.16(a), (b), and (c). A steel test plate of sufficient stiffness to prevent it from bending under lateral load and of sufficient size to accommodate the hydraulic jack cylinder is placed in full contact with the test pile. Blocking used between reaction system and the hydraulic cylinder should be of sufficient strength so that it can transfer applied lateral reaction without distortion. Davisson (1970) recommends the use of spherical bearing in lateral load test as a necessity because this test inherently involves rotation of the pile head.

A lateral load applied by hydraulic cylinder is either measured by a calibrated load cell or a pressure gauge. When a pressure gauge is used, it should be ensured that the complete system consisting of hydraulic cylinder, valves, pump, and pressure gauges are calibrated as one unit. Calibration of testing equipment should be done before each test.

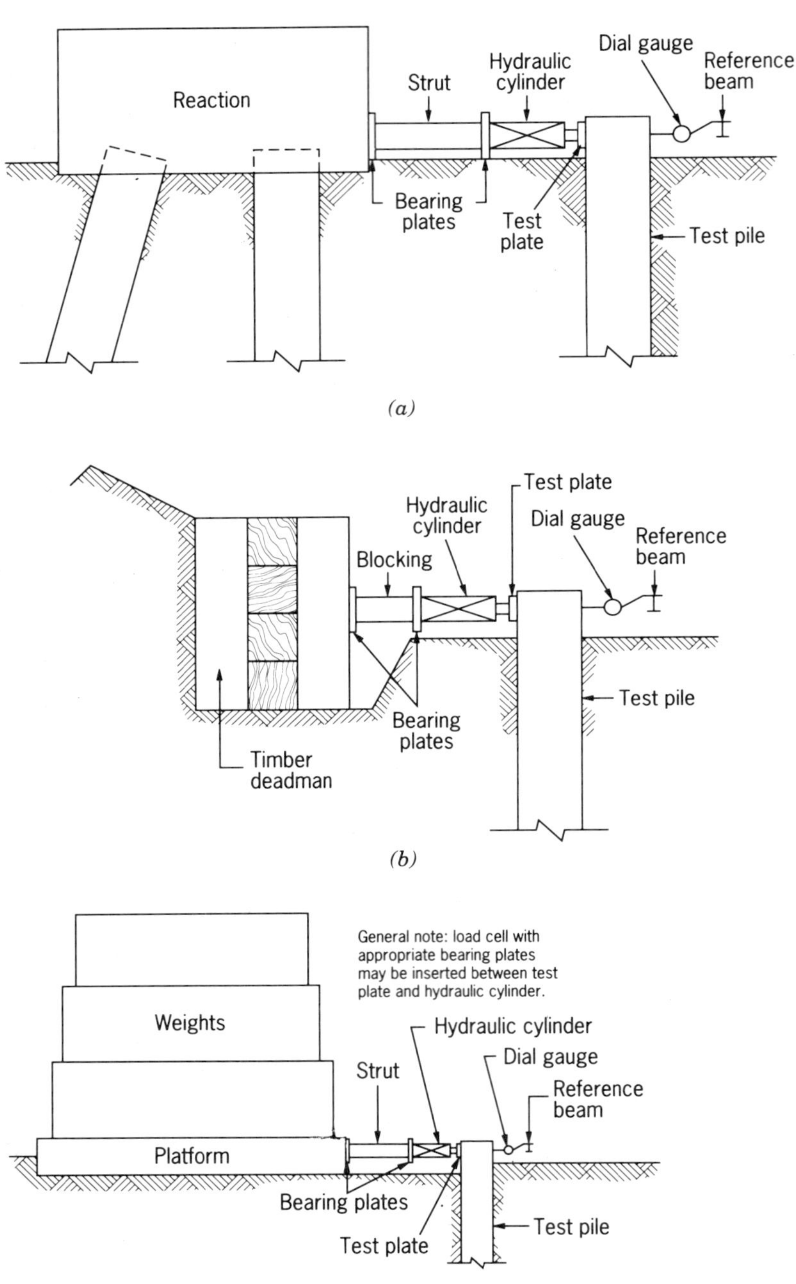


**Figure 9.16** Typical set-ups for applying lateral load. (a) Reaction piles, (b) deadman, (c) weighted platform. (ASTM D 3966-81, 1989).

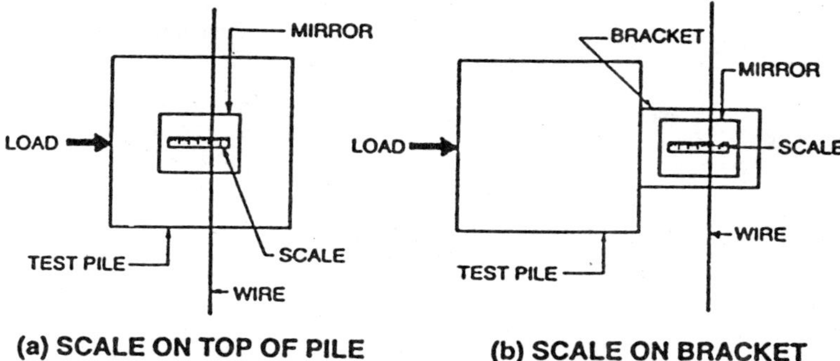


**Figure 9.17** Typical wire–scale arrangements to measure lateral movements: Top view (a) Scale on top of pill, (b) scale on bracket. (ASTM D 3966-81, 1989).

***Instruments for Measuring Movements*** The two common modes of movement measurement in lateral load tests are lateral movement of pile head along the line of load application and the lateral movement of pile axis with depth. The pile head movement is measured in all tests while the lateral movement with depth is measured in special projects only.

The lateral movement of pile head is commonly measured by dial gauges, wire–mirror and scale system, and the transit and the scale. It is a good practice to use two separate measuring systems in order to have a check on the data and to provide supporting data in case of accidental failure of one of the systems.

As shown in Figure 9.16, dial gauges are mounted on reference beams with a stem bearing against the side of the pile. The dial gauges should have at least 3 in. (75 mm) travel with a precision of at least 0.01 in. (0.25 mm). A typical wire–mirror and scale system is shown in Figure 9.17. This consists of mounting a mirror and a scale on the top center of the test pile or on a bracket mounted on the side of the pile. A piano wire is then stretched perpendicular to the line of load application and passing over the face of the scale. The scale should have 0.01 in. (0.25 mm) sensitivity. The mirror and the scale move with the pile and the piano wire is stationary. The difference of the final and the initial readings on the scale gives pile movement.

The lateral deflection or movement of the pile along its depth can also be measured by installing a tube or a duct along the axis of the pile at its center. This duct should be suitable to accommodate an inclinometer to measure lateral shift of the pile along its depth. Figure 9.18 shows a typical lateral load test set up, measuring devices and an inclinometer tube.

Rotation of pile head is usually of interest and can be measured by taking deflection measurements normal to the pile axis at two locations; the difference between the two readings is a measure of the change in slope (Davisson, 1970).

### 9.3.2 Test Procedures

The loading procedures that are frequently used in engineering practice are as follows:

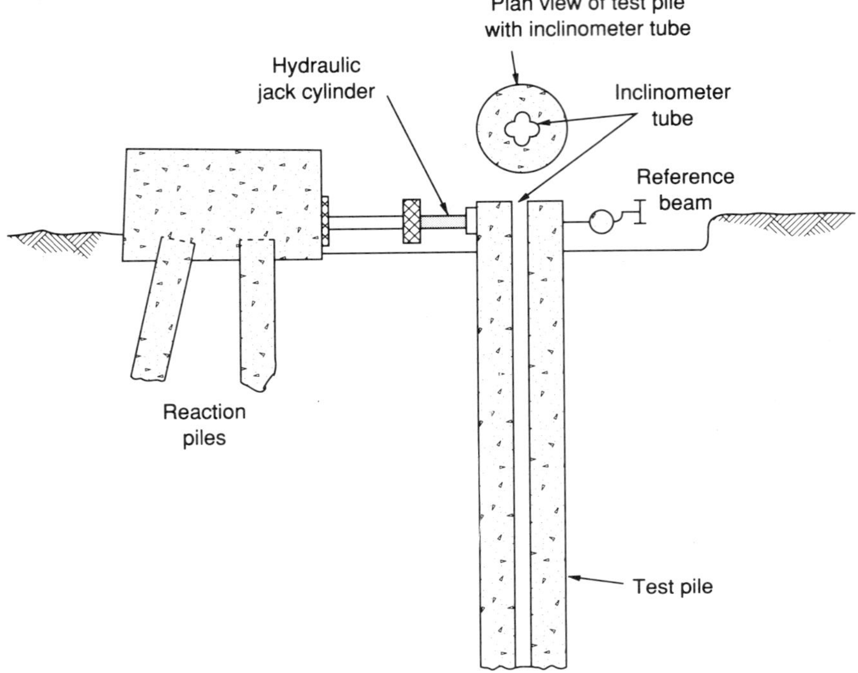


**Figure 9.18** A typical lateral load set-up and measuring devices with tube for inclinometer (Countesy: Thurber Consultants Ltd).

**1. Standard Loading Procedure**

(a) Apply the total load in 10 steps to 200 percent of design load (e.g., 25 percent, 50 percent, 75 percent, 100 percent 125 percent, 150 percent, 170 percent, 180 percent 190 percent, and 200 percent). The 25 percent and 50 percent of design load increments are applied for 10 min each and the 75 percent load increment is maintained for 15 min. Other load increments are maintained for 20 min each.

(b) After maintaining 200 percent design load for 60 min, unload the pile in steps of 50 percent of the design load (e.g., to 150 percent, 100 percent, 50 percent, and 0 percent, maintaining each load decrement for 10 min).

(c) For each step, record the load and movements. Finally, record the rebound after the full load has been removed.

**2. Loading in Excess of Standard Load**

(a) After the standard load has been applied as per procedure outlined in item (1), increase the load in steps of 50 percent of design load to 200 percent of design load maintaining each increment for 10 min (e.g., 50 percent, 100 percent, 150 percent, and 200 percent).

(b) Then increase the load in steps of 10 percent of the design load, each load level maintained for 15 min until failure, or to the maximum specified load (e.g., 210 percent, 220 percent, 230 percent, 240 percent, etc. to the maximum test specification).

(c) After the maximum load has been achieved, maintain the load for 30 min, then reduce to 75 percent, 50 percent, 25 percent, and 0 percent, maintaining each load level for 10 min.

On special projects, other load testing procedures such as cyclic loading, surge loading, reciprocal loading, and loading to maintain specified deflection may be utilized to suit project needs. Cyclic loading consists of applying and removing a percent of design loads in cycles and each level maintained for 10 to 20 min. Surge loading involves the application of any specified number of multiple loading cycles at any specified load level. Reverse loading consists of applying the lateral test load in either the push mode followed by the pull mode or vice versa. Load testing procedure, where load is applied to maintain specified deflection, is given in Section 9.3.4. These and other loading procedures are also detailed in ASTM D3966-81 (1989).

### 9.3.3 Interpretation of Test Data

Methods of determining failure load from lateral pile load tests vary depending on the tolerable movement of the structure supported by the piles. The generally accepted criteria for estimating the ultimate lateral load is the lower of the following two methods:

1. Failure load may be taken at 0.25 in. (6.25 mm) lateral movement or deformation.
2. Failure load may be considered at the point of intersection of tangents on the load–movement curve.

These two methods of interpretations are shown in Figure 9.20 and are discussed in the load test example Section 9.3.4.

### 9.3.4 Example of a Pile Load Test

A lateral load test was carried out at the location where a compression and a pullout test, discussed earlier and shown in Figure 9.10b, were also conducted. Soil stratigraphy at the site is shown in Figure 9.10a. The test pile was a 20-in. (500 mm) shaft diameter, 40-in. (1000 mm) bell diameter, and 20.5-ft (6.25 m) long cast-in-place concrete pile.

Figure 9.19 shows the general arrangement for this pile load test. As exhibited in this figure, the reaction pile was a 30-in. (750 mm) shaft diameter and 48-in. (1200 mm) bell diameter concrete pile. A system of steel bearing plates, a steel H-beam block, and a hydraulic jack was used for load application. Lateral movement was measured by two dial gauges supported independently on two reference beams. The load testing procedure applied the load to maintain specified deflection and consisted of the following steps:

1. Apply the load in the lateral direction until the deflection reaches 0.06 in. (1.59 mm). Maintain the load until the rate of movement from the previous load increment is less than 0.01 in./h (0.25 mm/h), or until 2 h have elapsed, whichever occurs first. Record the lateral load on the pile.
2. Continue the test by repeating step (1) above for the following deflections: 0.12 in. (3.18 mm), 0.19 in. (4.76 mm), 0.25 in. (6.35 mm), 0.38 in. (9.53 mm), 0.5 in. (12.7 mm), 0.64 in. (15.88 mm), and 1 in. (25.4 mm).
3. Remove the load by holding at deflections of 0.76 in. (19.05 mm), 0.5 in. (12.7 mm), and 0.25 in. (6.35 mm) for 2 h each (total time of 6 h). At this stage remove all the load to determine the residual deflections, if any.
4. Load piles and measure loads at deflections of 1 in. (25.4 mm), 1.5 in. (38.1 mm), 2 in. (50.5 mm), 2.5 in. (63.5 mm), and 3 in. (76.2 mm) etc., until

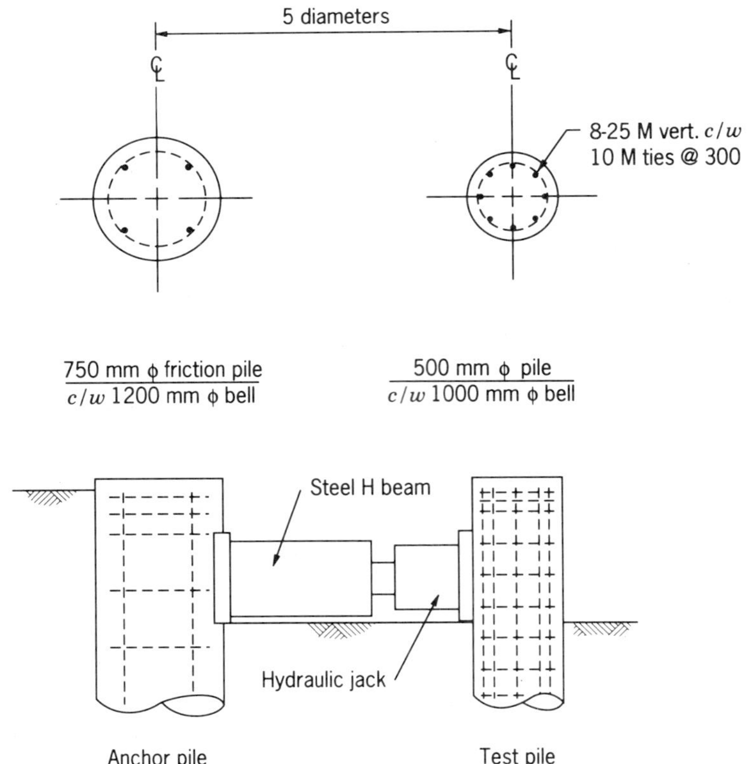


**Figure 9.19** Arrangement for lateral pile load test set-up (Sharma et al., 1984).

failure. At each interval, the rate of movement must not exceed 0.01 in./h (0.25 mm/h). If the rate cannot be maintained, the load should be reduced to the last interval capable of maintaining a movement of 0.01 in./h (0.25 mm/h).

Figure 9.20 presents the load-deflection curve for the tested pile. This figure also shows the pile dimensions and the soil profile recorded during pile installation. A load test exhibited that at about 47 kips (209 kN) the pile could not hold this amount of lateral load. Therefore, the load was removed back to zero. Ultimate loads as interpreted by the two methods were as follows:

1. Failure load for 0.25 in. (6.25 mm) deformation was 21.6 kips (96.5 kN).
2. Failure load exhibited by the intersection of tangent points was 22.5 kips (100 kN).

Failure loads interpreted by the two methods are also shown in Figure 9.20. The lower of these two values is 21.6 kips (96.5 kN) and was used as the failure load. It should be realized that actual instability at which the load could not be held was at about 47 kips (209 kN) when the pile head had deformed about 1 in. (25 mm). The definition of failure load should therefore be related to the acceptable or tolerate lateral deformation of the structure. Where no such criteria are available, 0.25 in. (6.25 mm) is considered as the criterion on which failure load is established.

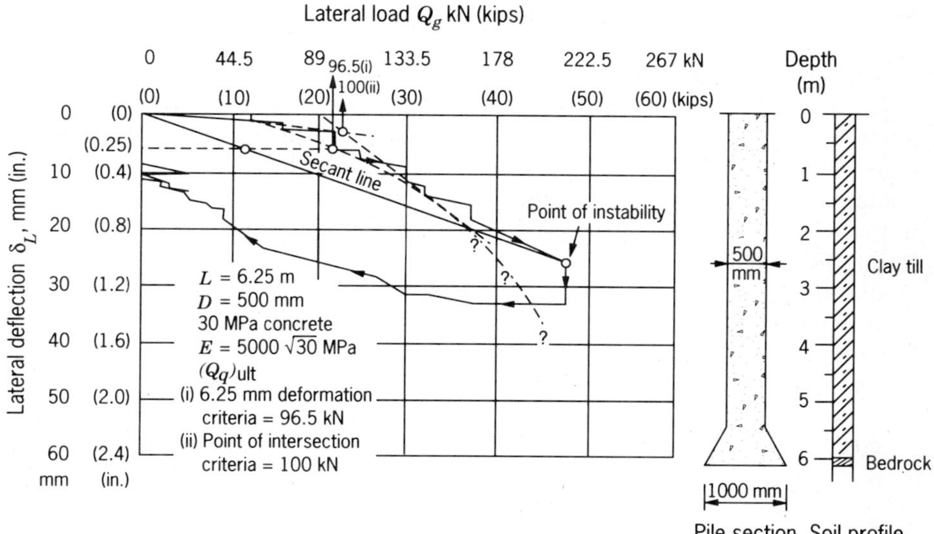


**Figure 9.20** Load–deflection curve for a lateral pile load test (Sharma et al., 1984).

## 9.4 DYNAMIC PILE LOAD TESTS

Pile installation methods used in conventional pile-driving operations can have a significant effect on the soil–pile boundary conditions and the dynamic behavior at the pile–foundation interface (see Chapter 1). Therefore, a dynamic pile load test is performed to study response parameter of a single pile or a pile group. The response of piles will be studied in (1) vertical vibrations, (2) horizontal translation and rocking (coupled modes), and (3) torsional vibrations depending on the nature of loading on the actual piles. Several full-scale pile tests have been performed (Prakash and Sharma, 1969; Gle, 1981; Gle and Woods, 1984; Prakash et al., 1985; Blaney et al., 1987). Different test methods have been adopted by each investigator.

There is no standard test method for dynamic tests on piles yet in the United States. One particular method for lateral vibratory load test on a single pile will be described. The procedure for other modes of vibrations will be only slightly different.

### 9.4.1 Test Equipment and Instruments

In a vibratory load, the dynamic stiffness and damping parameters for each degree of freedom of the foundation are evaluated. Gle and Woods (1984) describe a procedure for lateral vibratory test on a full-scale isolated steel pile soil system. A soil–pile–mass system will behave as a free-headed pile with coupled lateral translation and rocking degrees of freedom. Each degree of freedom has a resonance. The horizontal resonance can usually be defined reasonably well. However, depending on the power of the oscillator and the stiffness of the soil–pile–mass system, the rocking resonance may or may not be observed completely.

The pile was excited with a Lazan oscillator. It was generally necessary to add vertical load on the pile to reduce its natural frequency within the range of operation of the Lazen oscillator. This was accomplished by welding and bolting steel plates, 610 by 610 mm square and 19 mm thick, to the head of the pile. Each plate weighed 560 N. A hole slightly larger than the diameter of the pile to be tested was made in the center of one of the steel plates. This plate was slipped over the head of the pile and welded to the pile as shown in Figure 9.21. A few additional steel plates are then stacked on top of the base plate along with a housing that is used to mount the Lazan oscillator. The entire stack of steel plates is fastened together through each corner and at the center of each side with eight threaded steel rods. The steel rods are tightened enough to force the steel-plate mass to act as a rigid body.

A steady-state sinusoidal force is provided by a mechanical oscillator, which uses the centrifugal force of unbalanced masses mounted on two counterrotating shafts to generate a variable alternating force in a horizontal plane (see Figure 3.5a). The magnitude of this force is controlled by adjusting the phase angle between the masses. Speed of the oscillator is controlled by a variable speed

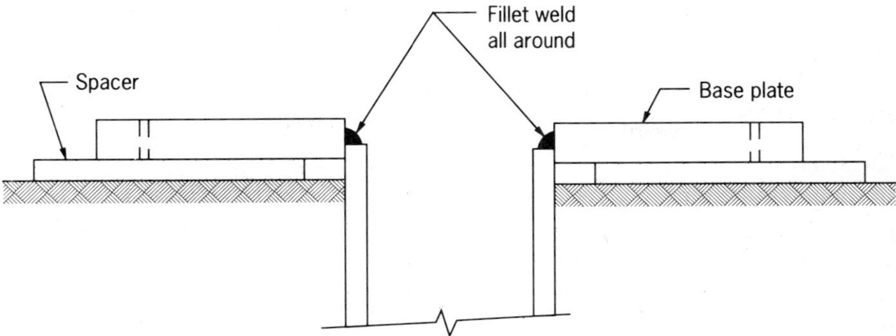


**Figure 9.21** Cross section showing attachment of base plate to pipe pile (Gle and Woods, 1984. Reprinted by permission of ASTM).

electric motor that is connected to the oscillator with a flexible shaft. This may generate a variable force within a frequency range of about 5 to 55 Hz to be applied to the pile. The displacement is measured with two velocity transducers mounted on each side of the mass as shown in Figure 9.22. Output signals from these transducers were recorded on a dual-channel, strip-chart recorder. Calibration of the velocity transducers helped accurate conversion of the recorded velocity to displacement. Mounting both transducers in a horizontal

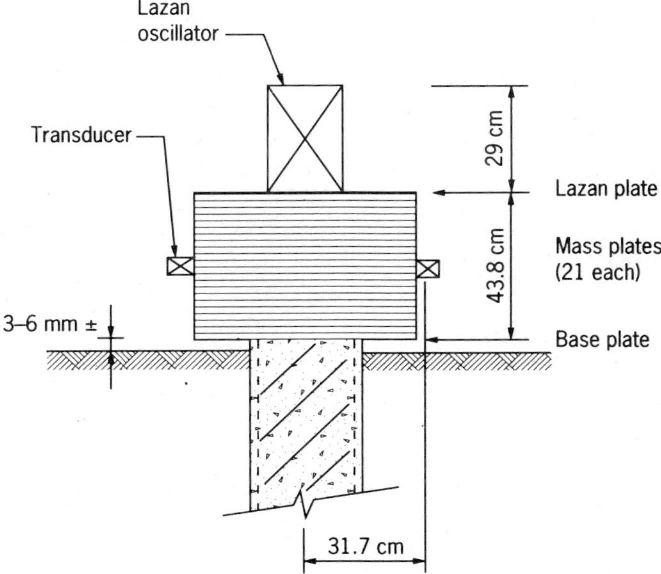


**Figure 9.22** Typical cross section of a dynamic lateral pile test (Gle and Woods, 1984. Reprinted by permission of ASTM).

plane on each side of the mass as shown provides an independent check on the calculated displacement amplitude (Gle and Woods, 1984).

### 9.4.2 Test Procedures

For determination of the dynamic lateral response of the mass caused by the soil pile interaction alone, it is recommended that the mass be located as close to the ground surface as possible without touching the soil. As the distance between the soil surface and the bottom of the mass is increased, the dynamic response becomes essentially the structural response of a mass on a cantilever. In this case, minimal information is obtained on the soil–pile interaction. In order to achieve this, thin removable spacers are used to facilitate positioning the base plate perpendicular to the centerline of the pile. The base plate is then leveled and tack welded in place until a fillet weld can be completed around the circumference of the pile. When the base plate is securely attached to the head of the pile, a selected number of mass plates and the Lazan oscillator housing plate are added and tightened in place. Threaded connections for the velocity transducers are welded to a few of the mass plates before the field testing. These steel plates are positioned as desired within the stack of mass plates.

The velocity transducers are located as close to the elevation of the center of mass as possible to minimize the contribution of the rocking mode to the recorded response unless the rocking response is of particular interest. The Lazan oscillator is connected to the drive motor through a flexible coupling. Appropriate weights in the form of steel plates may be added to the head of each pile for all dynamic lateral-load tests to bring the resonant frequency within the range of operation of the Lazan oscillator. The mass to be attached to the pile head will be determined in each case from the above considerations.

Initially, a relatively low-force level (mass eccentricity) is set on the Lazan oscillator. The Lazan oscillator is then used to drive the soil–pile–mass system through a frequency range from about 5 to 55 Hz. At each desired frequency, the oscillator is run for sufficient time to record the steady-state response. When the maximum output of the Lazan oscillator is reached, additional response data are obtained as the frequency is reduced, particularly around resonance.

Five to seven tests (using different Lazan force levels) are usually conducted on the same pile. Typical double-amplitude force levels ranged from about 20 to 4450 N in tests reported by Gle and Woods (1984). This will usually provide a broad spectrum of response curves sufficient to bracket the amplitude of vibration for most full-scale foundations unless stiff soils or high displacement amplitudes are expected.

It is recommended that a *plucking test* be necessarily performed on the single pile. In this test, the steel-plate mass is "plucked" by applying an impulse force to the mass and recording the free-vibration response of the soil–pile–mass system. The impulse force is applied by striking the mass horizontally with a wooden plank or a hammer. From the measured free-vibration response, damping and the damped natural frequency of the soil–pile–mass system are determined. This information supplements the values obtained by the steady-state tests.

Because the amplitude of vibration cannot be controlled and is usually much higher than the steady-state testing, this test must only be conducted at the conclusion of the steady-state dynamic testing. It is also desirable to conduct this test in a direction perpendicular to the steady-state testing direction to minimize the effect of any soil disturbance around the pile.

### 9.4.3 Interpretation of T st Data

At the frequency for a rotating mass excitation, the maximum amplitude of vibration is given by the equation as: (Gle and Woods, 1984; Prakash and Puri, 1988)

$$A_x/(m_e\bar{e}/M) = 1/[2\xi_x(1-\xi_x^2)^{1/2}] \tag{9.4}$$

where

$A_x$ = measured amplitude, m
$m_e\bar{e} = \sin(\theta/2)/2\pi^2$ (for double-amplitude Lazan force output)
$M$ = mass on the pile head, kg
$\xi_x$ = damping ratio in the horizontal translation direction

Equation (9.4) gives the value of damping in translation ($\xi_x$). The undamped natural frequency ($\omega_n$) of the soil–pile system is

$$\omega_n = \omega_{nd} \times \sqrt{1-2\xi_x^2} \tag{9.5}$$

where $\omega_{nd}$ = damped natural frequency.

An approximate value for the translation spring constant $k_x$ can also be backcalculated knowing the mass on the head of the pile. A similar procedure can be used for the rocking resonance if the peak response is well defined.

The dynamic response of the soil–pile–mass system is in fact correctly represented with a coupled sliding and rocking two-degree-of-freedom solution. Equations of motion can be written for each of the translation and rotation degrees of freedom and solved simultaneously for the dynamic response. Summing forces and moments about the center of mass in Figure 9.23 gives (Gle and Woods, 1984):

$$\begin{bmatrix} A & B & C & -D \\ -B & A & D & C \\ C & -D & E & F \\ D & C & -F & E \end{bmatrix} \begin{bmatrix} A_{x1} \\ A_{x2} \\ A_{\phi 1} \\ A_{\phi 2} \end{bmatrix} = \begin{bmatrix} F_x \\ 0 \\ F_x h_2 \\ 0 \end{bmatrix} \tag{9.6}$$

where $F_x$ is the steady-state, double-amplitude force generated by the Lazan

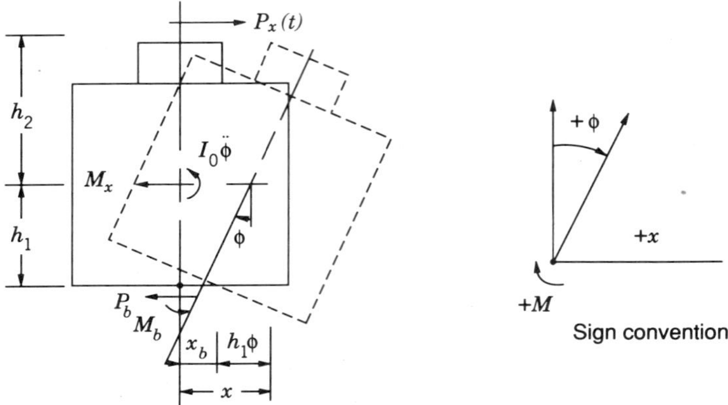


**Figure 9.23** Free-body diagram of forces and moments acting on the steel-plate mass (Gle and Woods, 1984. Reprinted by permission of ASTM).

oscillator, and the variables $A$ through $F$ are defined as:

$$
\begin{aligned}
A &= M\omega^2 - k_x & D &= C_x h_1 \omega \\
B &= C_x \omega & E &= I_0\omega^2 - k_\phi - h^2 k_x \\
C &= h_1 k_x & F &= C_\phi \omega + h_1 C_x \omega
\end{aligned}
\tag{9.7}
$$

where

$M$ = mass on the head of the pile, kg
$I_0$ = mass polar moment of inertia, N-m-s$^2$
$\omega$ = circular frequency, rad/s
$k_x$ = lateral-translation spring stiffness, N/m
$k_\phi$ = rotational spring stiffness N-m/rad
$C_x$ = lateral-translation damping value, N-s/m
$C_\phi$ = rotational damping value N-s/rad
$h_1$ = height of center of gravity of the pile cap above the base.

The horizontal translation of the mass is defined by

$$X = A_x \sin(\omega t - \theta_x) \tag{9.8}$$

and the rocking rotation of the mass by

$$\phi = A_\phi \sin(\omega t - \theta_\phi)$$

where

$$A_x = (A_{x1}^2 + A_{x2}^2)^{1/2}$$

$$A_\phi = (A_{\phi1}^2 + A_{\phi2}^2)^{1/2}$$

$$\theta_x = \tan^{-1}(A_{x2}/A_{x1})$$
$$\theta_\phi = \tan^{-1}(A_{\phi 2}/A_{\phi 1}) \tag{9.9}$$

### 9.4.4 Example of a Pile Load Test

Woods (1984) describes tests on a full-scale H pile (see Section 7.9). One of these test data has been selected for illustration here. Figure 7.46b compares the dynamic response data from a field test to that predicted analytically using the two degree-of-freedom solution. The translation stiffness and damping parameters were backcalculated from the single-degree-of-freedom equations (with viscous damping) and the rotational stiffness and damping parameters were as predicted by the PILAY program using soil modulus values measured in situ by the cross-hole method. The dynamic response curve predicted using the PILAY stiffness and damping values for both the translation and rotation parameters is also shown for comparison. Gle and Woods (1984) have not computed the stiffness parameters for the field test data. Rather, the stiffness was computed from in-situ properties and damping determined from both steady state vibratory and plucking tests and the computed response compared with the measured response. It was also found that the calculated values of the damping ratio in steady-state vibration tests was much smaller than that in plucking tests. The difference in the two values may be attributed to the response amplitudes.

## 9.5 OVERVIEW

This chapter discussed the test equipment and instrumentation, test procedures, data interpretation methods, and one example each for axial compression, axial pullout, lateral load and the dynamic pile load tests.

The four basic axial compression pile load test methods (SM test, QM tests, CRP tests, and SC tests) are well known to geotechnical engineers. However, the SM test method is generally used in North America. The pile is loaded to 200 percent of the design load when contract piles are load tested during construction. ASTM D1143-81 identifies this as "Standard-Loading Procedure." The piles are load tested to failure when the load test data are used to design the piles. Crowthers (1988) describes load testing methods of deep foundations in detail. Among the various load test interpretation methods, the Fuller and Hoy method always yields failure loads that are the best approximation of test failure loads. Davisson's method always predicts conservative values of failure loads, and Chin's method invariably yields failure loads that are higher than the actual failure loads. Further analysis is required to determine the applicability of these methods for length to diameter ratios of greater than 32.

Various load testing procedures for axial pullout and lateral load tests are available in the literature. Out of these methods standard loading procedures as specified in ASTM 3689-83 and ASTM 3966-81 appear to be widely used in

North America. The interpretation methods for load–movement data of pullout and lateral load tests are not yet well established. In absence of any specific criteria, the load corresponding to 0.25 in. (6.25 mm) pullout or lateral movement may be used as the failure load. Further work is required in this area.

Only a limited number of vibration tests have been performed on full scale piles under different modes of vibrations. Therefore, the test methods have not been standardized. However, guidance is available from description of several of these tests (Prakash and Sharma, 1969; Gle and Woods, 1984; Prakash et al., 1985; Blaney, et al., 1987). The test data have also not been interpreted uniformly.

There is an urgent need to generate more data on full-scale pile tests under vibrations and to develop a unified method of interpretation. Since the soil is disturbed due to pile driving and there may be loss of contact between the soil and the pile in lateral vibrations (see Chapter 1), it is unlikely that the response predicted on the basis of the soil–pile stiffness from in-situ soil properties would reasonably match the field test values. It is therefore recommended that the stiffness values be estimated from the pile-load test itself.

## REFERENCES

American Society for Testing and Materials ASTM D1143-81, "Standard Method of Testing Piles under Static Axial Compressive Load," Vol. 04.08, Philadelphia, 1989, pp. 179–189.

American Society for Testing and Materials ASTM D3689-83, "Standard Method of Testing Individual Piles Under Static Axial Tensile Load," Vol. 04.08, Philadelphia, 1989, pp. 474–484.

American Society for Testing and Materials ASTM, D3966-81, "Standard Method of Testing Piles under Lateral Loads," Vol. 04.08, Philadelphia, 1989, pp. 494–508

Blaney, G. W., Muster, G. L., and O'Neill, M. W. "Vertical Vibration Test of a Full-Scale Pile Group," Proceedings Dynamic Response of Pile Foundations, ASCE, *Geot. Special Publications* No. 11, Atlantic City, 1987, pp. 149–156.

Brinch Hansen, J. Discussion, "Hyperbolic Stress–Strain Response. Cohesive Soils," *J. Soil Mech. Found Div.* ASCE, Vol. 89, No. SM4, 1963, pp. 241–242.

Butler, H. D. and Hoy, H. E. "Users Manual for the Texas Quick-Load Method for Foundation Load Testing," *Federal Highway Administration*, Office of Development, Washington, DC, 1977, 59 pp.

Chin, F. K. "Estimation of the Ultimate Load of Piles not Carried to Failure," *Proceedings 2nd Southeast Asian Conference on Soil Engineering*, Singapore, 1970, pp. 81–90.

Chin, F. K. "Discussion, Pile Tests—Arkansas River Project," *J. Soil Mech. Found. Div.* ASCE, Vol. 97, No. SM6, 1971, pp. 930–932.

Crowthers, C. L. *Load Testing of Deep Foundations*, Wiley & Sons, New York, NY, 1988.

Davisson, M. T. "Static Measurement of Pile Behavior", *Proc. Conf. on Design and Installation of Pile Foundations and Cellular Structures*, Ed. H. Y. Fang and T. D. Dismuke, Bethlehem, (PA) 1970, pp. 159–164.

Davisson, M. T. "High Capacity Piles," *Proceedings, Lecture Series Innovations in Foundation Construction*, ASCE, Illinois Section, Chicago, 1972, 52 pp.

Davisson, M. T. "Foundations in Difficult Soils-State of the Practice Deep Foundations-Driven Piles", Seminar on Foundations in Difficult Soils, Metropolitan Section, ASCE, April 1989, New York.

De Beer, E. E. and Wallays, M. "Franki Piles with Overexpanded Bases," La Technique des Travaux, No. 333, 1972, 48 pp.

Fellenius, B. H. "Test Load of Piles and New Proof Testing Procedure," *J. Geotech. Eng. Div.*, ASCE, Vol. 101, No. GT9, 1975, pp. 855–869.

Fellenius, B. H. "The Analysis of Results from Routine Pile Load Tests," *Ground Engineering*, 1980, pp. 19–31.

Fuller, F. M. and Hoy, H. E. "Pile Load Tests Including Quick-load Test Method Conventional Methods and Interpretations," HRB 333, 1970, pp. 78–86.

Gle, D. R. "The Dynamic Lateral Response of Deep Foundations" Ph.D. Dissertation, Department of Civil Engineering, The University of Michigan, Ann Arbor 1981.

Gle, D. R. and Woods, R. D. "Suggested Procedure for Conducting Dynamic Lateral-Load Tests on Pile," *Symposium on Laterally Loaded Deep Foundation, Analysis and Performance*, ASTM STP835 Kansas City Missouri, 1984, pp. 157–171.

Joshi, R. C. and Sharma, H. D. "Prediction of Ultimate Pile Capacity From Load Tests on Bored and Belled, Expanded Base Compacted and Driven Piles," *Proceedings, International Symposium on Prediction and Performance in Geotechnical Engineering*, Calgary, Algebra, Canada, 1987, pp. 135–144.

Mazurkiewicz, B. K. "Test Loading of Piles According to Polish Regulations," Royal Swedish Academy of Engineering Sciences Commission on Pile Research. Report No. 35, Stockholm, 1972, 20 pp.

Mohan, D., Jain, G. S., and Jain, M. P. "A New Approach to Load Tests," *Geotechnique* Vol. 17, 1967, pp. 274–283.

New York State Department of Transportation, *Static Load Test Manual*, N.Y. DOT Soil Mechanics Bureau, Soil Control Procedure SCP4/74, 1974, 35 pp.

Nordlund, R. L. "Dynamic Formula for Pressure Injected Footings," *J. Geotech. Eng. Div.*, ASCE, Vol. 108, No. GT3, 1982, pp. 419–437.

Prakash, S. *Soil Dynamics*, McGraw-Hill Book Co., New York, 1981.

Prakash, S. and Puri, V. K. *Foundations for Machines*, Wiley, New York, 1988.

Prakash, S., Ranjan, G., and Kumar, K. "Dynamic Soil–Pile Constants for Turbo-Generator Foundations," Madras Refineries Report issued by Geotechnical Division, Central Building Research Institute, Roorkee, India, 1985.

Prakash, S., and Sharma, H. D., "Analysis of Pile Foundations Against Earthquakes," *Ind. Conc J.*, Vol. 43, No. 6 1969, pp. 205–220.

Rieke, R. D. and Crowser, J. C. "Instrumentation of Driven Piles," *The Indicator*, Slope Indicator Company, Seattle, Washington, 1986, pp. 2–5.

Sharma, H. D., Harris, M. C., Scott, J. D., and McAllister, K. W. "Bearing Capacity of Bored Cast-In-Place Concrete Piles in Oil Sand," *J. Geotech. Eng. Div.*, ASCE, Vol. 112, No. 12, 1986, pp. 1101–1116.

Sharma, H. D., Sengupta, S., and Harron, G. "Cast-In-Place Bored Piles on Soft Rock Under Arterian Pressures," *Canadian Geotech. J.* Vol. 21, No. 4, 1984, pp. 684–698.

Swedish Pile Commission "Recommendations for Pile Driving Test and Routine Test

Loading of Piles," Royal Swedish Academy of Engineering Sciences Commission on Pile Research, Report No. 11, Stockholm, 1970, 35 pp.

Vander Veen, C. "The Bearing Capacity of a Pile," *Proceedings, 3rd International Conference on Soil Mechanics and Foundation Engineering*, Vol. 2, Zurich, 1953, pp. 84–90.

Weele, A. F. A Method of Separating the Bearing Capacity of a Test Pile into Skin Friction and Point Resistance," *Proceedings, 4th International Conference on Soil Mechanics and Foundation Engineering*, Vol. 2, London, England, 1957, pp. 76–80.

Whitaker, T. "Experiments with Model Piles in Groups," *Geotechnique*, Vol. VII No. 4. 1957, pp. 147–167.

Whitaker, T. "The Constant Rate of Penetration Test for the Determination of the Ultimate Bearing Capacity of a Pile," *Proceedings, Institution of Civil Engineers*, Vol. 26, London, England, 1963, pp. 119–123.

Whitaker, T. and Cooke, R. W. "A New Approach to Pile Testing," *Proceedings, 5th International Conference on Soil Mechanics and Foundation Engineering*, Vol. 2, Paris, France, 1961, pp. 171–176.

Woods, R. D. "Lateral Interaction between Soil and Pile," *Proceedings International Symposium on Dynamic Soil Structure Interaction*, Minneapolis, MN, 1984, pp. 47–54.

# 10

# BUCKLING LOADS OF SLENDER PILES

Granholm (1929) showed that for piles of normal dimensions driven through soil, buckling should not take place except in extremely soft soil. However, very slender and long piles are increasingly used today (1990). In offshore structures, these piles also extend for a considerable distance mudline. Therefore, the possibility of buckling of such piles has received considerable attention. Research has been carried out to obtain more accurate estimates of buckling loads of piles. The majority of analytical methods proposed have employed the subgrade-reaction theory, described in this chapter. Both fully embedded and partially embedded piles are considered.

## 10.1 FULLY EMBEDDED PILES

Earlier solutions for the elastic buckling loads of embedded piles were based on a subgrade modulus for the soil which was assumed to be constant over the length of the pile. Hetenyi (1946) presented a survey of the work by Forssell (1918, 1926) and Grandholm (1929); the governing differential equation is

$$EI\frac{d^4y}{dx^4} + P\frac{d^2y}{dx^2} + ky = 0 \tag{10.1}$$

where

$EI$ = flexural stiffness of the pile
$P$ = axial load
$k$ = subgrade modulus

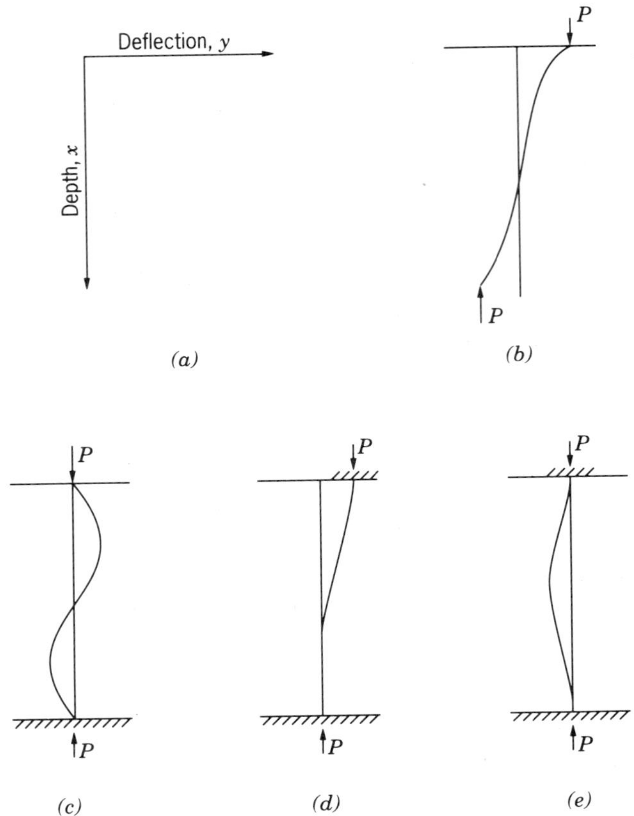


**Figure 10.1** Pile boundary conditions (a) Coordinate system (b) free (f), (c) pinned ($p$), (d) fixed translating ($Ft$), (e) fixed ($F$).

All the foregoing quantities were considered to be constants in those solutions. Figure 10.1 shows pile boundary conditions. The solutions of equation (10.1) have been obtained in the nondimensional form, letting

$$R = \sqrt[4]{\frac{EI}{k}} \quad \text{and} \quad Z = \frac{x}{R} \tag{10.2}$$

Then

$$Z_{\max} = \frac{L}{R} \tag{10.3}$$

where

$L$ = embedded length of the pile
$R$ = relative stiffness factor
$EI$ = flexural stiffness of pile
$Z$ = nondimensional depth coefficient

By substituting these definitions into equation (10.1) and rearranging, the following equation is obtained:

$$\frac{d^4y}{dz^4} + \frac{PR^2}{EI} \cdot \frac{d^2y}{dz^2} + y = 0 \tag{10.4}$$

Let $PR^2/EI$ be the axial load coefficient $U$; then

$$U_{cr} = \frac{P_{cr}R^2}{EI} \tag{10.5}$$

where subscript '*cr*' represents the critical values of $U$, and $P$.
By substitution, equation (10.6) is obtained:

$$\frac{d^4y}{dz^4} + U\frac{d^2y}{dz^2} + y = 0 \tag{10.6}$$

The critical values of the axial load coefficient, $U_{cr}$, are obtained by solving equation (10.6) for $U$ with due consideration to the pile boundary conditions and the pile length, $Z_{max}$. The boundary conditions are free ($f$), pinned ($p$), fixed-translating ($ft$), and fixed-non-translating, ($F$) (see Figure 10.1). An analog computer was used to obtain solutions for equation (10.6); the techniques and the computer program have been presented by Davisson and Gill (1963).

***Case 1: k = Constant*** In this solution, the axial load has been assumed to be constant in the pile, and no load transfer occurs. The pile is initially straight. The solutions are shown in Figure 10.2 in dimensionless form, as a plot of $U_{cr}$ versus $Z_{max}$ for several boundary conditions (e.g., *ft-p*, *p-p*, *ft-f*, *f-p*, *f-f*) (Davisson, 1963). Figure 10.2 shows that the boundary conditions exert a controlling influence on $U_{cr}$.

For pinned ends, the pile deforms into a number of sine half-waves, with the number of waves depending on the total length of the pile. $U_{cr}$ values were obtained for the first three modes (Davisson, 1963); for all modes, the $U_{cr}$ values are above 2, and at certain values of $Z_{max}$ become tangent to the line $U_{cr} = 2$. The lowest values of $U_{cr}$ for any given length $Z_{max}$ are the ones of interest; for practical purposes, $U_{cr}$ is considered equal to 2 (Davisson, 1963).

Another solution commonly referred to is the one for perfectly free ends ($f$-$f$). In this case, $U_{cr}$ is zero when $Z_{max}$ equals zero and increases with an increase in $Z_{max}$ until a maximum value of unity is reached. At this point, a mode change occurs and $U_{cr}$ dips below unity, but it returns to unity when the next mode change is about to occur. With increasing pile length, the magnitude of the deviation from unity becomes negligible. Because in most practical cases $Z_{max}$ is greater than 5, $U_{cr}$ can be considered equal to unity (Davisson, 1963). It will thus be seen that the boundary conditions exert a controlling influence on $U_{cr}$.

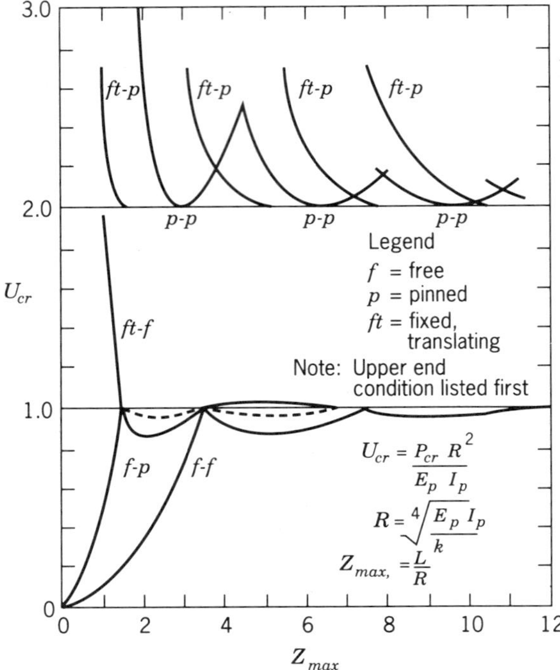


**Figure 10.2** Buckling load vs. length for $k_h$ = constant (Davisson, 1963).

For a free head and a pinned tip ($f$-$p$) pile (Figure 10.2), $U_{cr}$ increases rapidly with an increase in $Z_{max}$ up to the limiting value of unity. After first reaching unity, the higher modes indicated $U_{cr}$ values were little different from unity. The buckling appears to be controlled by the boundary offering the least restraint. It may be reasoned that a pile with a pinned head and a free tip would also have a limiting $U_{cr}$ value of unity (Davisson, 1963).

A pile with its head fixed against rotation but not translation ($ft$) represents a pile in a group. When combined with a free tip, the value of $U_{cr}$ oscillates slightly about 1. When combined with a pinned tip, the value of $U_{cr}$ becomes tangent to 2.

***Case 2: $k = n_h \cdot x$*** When a soil profile is considered for which $k = n_h \cdot x$, the boundary condition at the pile head becomes extremely important compared to the boundary condition at the pile tip. Because the pile tends to buckle where the subgrade modulus is the lowest, instability will tend to occur immediately adjacent to the pile head.

When $k = n_h x$, equation (10.1) becomes

$$EI\frac{d^4y}{dx^4} + P\frac{d^2y}{dx^2} + n_h xy = 0 \tag{10.7}$$

Let

$$T = \sqrt[5]{\frac{EI}{n_h}} \quad \text{and} \quad Z = \frac{x}{T} \tag{10.8}$$

then

$$Z_{\max} = \frac{L}{T} \tag{10.9}$$

where

$T$ = relative stiffness factor
$Z$ = nondimensional depth coefficient
$Z_{\max}$ = maximum value of the depth coefficient

By substituting the above into equation (10.7) and rearranging, we obtain:

$$\frac{d^4y}{dz^4} + \frac{PT^2}{EI} \cdot \frac{d^2y}{dz^2} + Zy = 0 \tag{10.10}$$

Let $V$ denote the axial load coefficient, $PT^2/EI$; then,

$$V_{cr} = P_{cr}\frac{T^2}{EI} \tag{10.11}$$

By substitution, equation (10.10) becomes

$$\frac{d^4y}{dz^4} + V\frac{d^2y}{dz^2} + Zy = 0 \tag{10.12}$$

Equation (10.12) was solved for $V_{cr}$ with the aid of an analog computer Davisson (1963). $V_{cr}$ versus $Z_{\max}$ for a pile with a free head and a free tip (*f-f*) is shown in Figure 10.3. $V_{cr}$ starts at zero and increases with an increase in $Z_{\max}$ up to a limiting value of approximately 0.71.

Other boundary conditions in Figure 10.3 are a pile with a free head and a pinned tip (*f-p*). Because of the increase in restraint that a pinned tip offers, compared to a free tip, $V_{cr}$ increases more rapidly with length than for the free-tip case. The maximum $V_{cr}$ was approximately 0.78, which is only slightly higher than that for the free-tip case. For a pinned-head, free-tip pile (*p-f*) a considerably higher value of $V_{cr}$ is observed at any given length $Z_{\max}$. This illustrates the effect of the restraint of a pinned-head pile when compared to a free-head pile. Two buckling modes were observed for this case, but for all practical pile lengths $V_{cr}$ exceeds 1.44. Generally, a pile will have a length exceeding a $Z_{\max}$ value of 3 to 4 (Davisson, 1963).

For a pile with its head fixed against rotation but not translation and a free-tip, (*ft-f*), the minimum value of $V_{cr}$ is approximately 0.88; it occurs at a very short pile length, namely, $Z_{\max} = 2.3$. $V_{cr}$ increases rapidly for pile lengths greater than

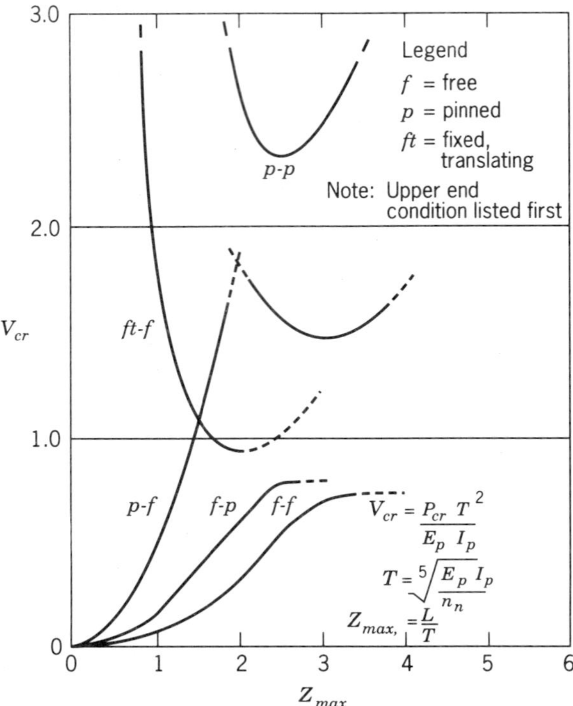


**Figure 10.3** Buckling load vs. length for $k_h = n_h x$ (Davisson, 1963).

$Z_{\max} = 2.3$. It can be reasoned that a pile with a fixed-translating head and a pinned tip would have higher $V_{cr}$ values, for any given pile length, than the free-tip case. By similar reasoning, it can be seen that a pile with a fixed-non-translating head and either a free tip or a pinned tip would also exhibit higher values. For a pile with both ends pinned (*p-p*), the minimum observed $V_{cr}$ value was 2.30 at $Z_{\max} = 2.60$.

Because most real piles are initially deformed, and because the theoretical elastic buckling load is an unconservative upper bound to the actual failure load, the computed buckling loads are often only an aid to the judgment of the engineer faced with the task of predicting the buckling load for a pile. The use of load tests is also unconservative. Most load tests are performed in a relatively short period of time during which a large part of the axial load in the pile is dissipated by skin friction (see Chapter 1). Under service conditions, the skin friction may be much less than that in short term tests and the tendency to buckle would be greater (Davisson, 1963).

Prakash (1987) obtained solutions for buckling loads in closed form by energy methods for fully embedded vertical piles for boundary conditions, pinned top-pinned tip (*p-p*), fixed top-fixed tip (*F-F*), and a linear variation of soil stiffness

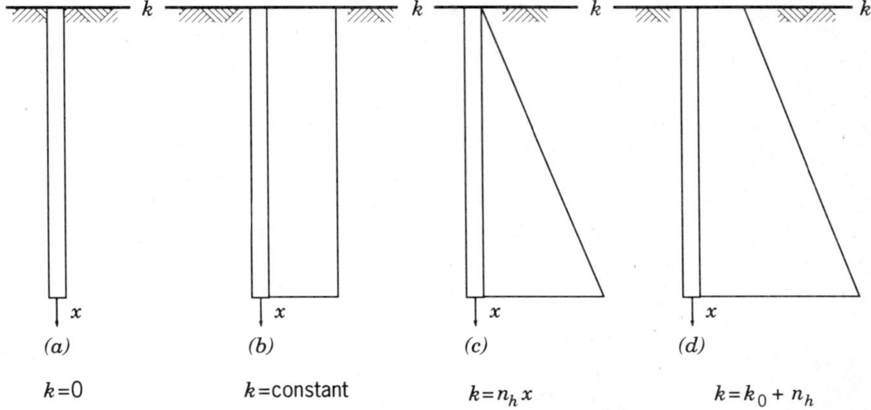


**Figure 10.4** Soil property variation along the depth of the pile.

with initial value $k_0 > 0$ (Figure 10.4). The effects of pile length, soil stiffness, and boundary conditions on buckling loads and mode of buckling have been studied for pile lengths up to 24 m with an $EI$ of 477 tm$^2$, $K_0$ from 0 to 2000 t/m$^2$ and $n_h$ from 0 to 2000 t/m$^3$.

Where $k_0 = k$ at the top of a fully embedded pile, and $k_L = k$ at the tip of a fully embedded pile, and $n_h$-constant of subgrade reaction, $n_h = (k_L - k_0)/L$. The variation of coefficient of subgrade reaction with depth has been shown in Figure 10.4. Four cases are shown:

1. Constant with depth $k_0 = k = 0$ (Figure 10.4a)
2. Constant with depth $k_0 = k_L =$ constant (Figure 10.4b)
3. Increasing linearly with depth with zero value at the surface, $k = n_h x$ (Figure 10.4c)
4. Increasing linearly with depth with nonzero value at the surface, $k = k_0 + n_h x$ in which $k_0 \neq 0$, as in Figure 10.4d

The critical load was determined by calculating the smallest eigenvalue of the leading principal submatrix.

The buckling loads were determined based on an energy method (i.e., the increment of the strain energy during the beam deflection will be equal to the work done by the external forces). The equations of the deflection curves satisfying different boundary conditions on the beam have been substituted into the work energy equation. In order to determine the buckling load $P_{cr}$, the derivative of the energy equation was set equal to zero and transformed into matrix notation with a standard eigenvalue form.

***Effect of Stiffness Linearly Increasing with Depth and k = Constant on the Buckling Load*** Figure 10.5 shows a plot of buckling load $P_{cr}$ and length $L$ of

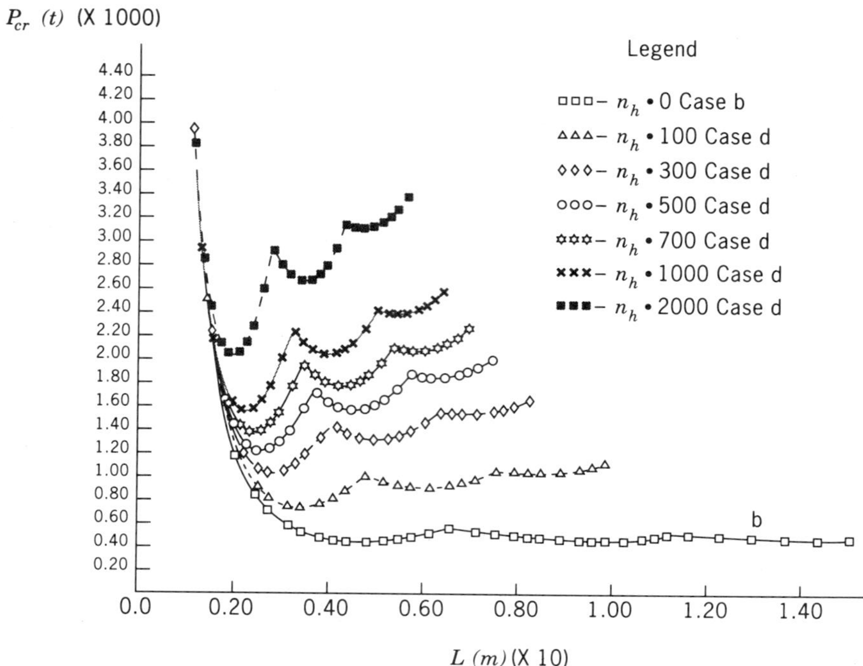


**Figure 10.5** Critical load of case b and d for a pinned-pinned end pile when $k_0 = 100\,\text{t/m}^2$ (Prakash, 1987).

the pile for a pile with $EI = 477\,\text{tm}^2$, (1) $k_0 = 100\,\text{t/m}^2$ (case b Figure 10.4), and (2) $k_0 = 100\,\text{t/m}^2$ $n_h$ increasing from zero to $2000\,\text{t/m}^3$ (case d). The buckling mode changes from the first mode to the second and then to the third as the length of the pile increases. The buckling load in general increases with increase in the value of $n_h$, which is obvious.

The minimum buckling load (in case d Figure 10.4) in a higher mode increases as compared to the corresponding value in the previous mode. This behavior is distinctly different from the situation in which $k$ was constant with depth, that is, in case b (Prakash, 1987). Similar behavior was observed with $k_0 = 500$, 1000, and $2000\,\text{t/m}^2$ (Prakash, 1985).

***Effect of Increasing $k_0$ Values when $n_h$ = Constant on the Buckling Loads*** Figure 10.6 shows a plot of buckling load $P_{cr}$ and length of pile $L$ with $EI = 477\,\text{t/m}^2$, $n_h = 100\,\text{t/m}^3$ and $k_0$ increasing from zero to $2000\,\text{t/m}^2$. As in the previous case, the buckling mode changes from the first mode to the second and then to the third as the length of the pile increases. The buckling load in general increases with the increase in the value of $k_0$ ($n_h$ = constant), which is to be expected.

The minimum buckling load in a higher mode increases as compared to the

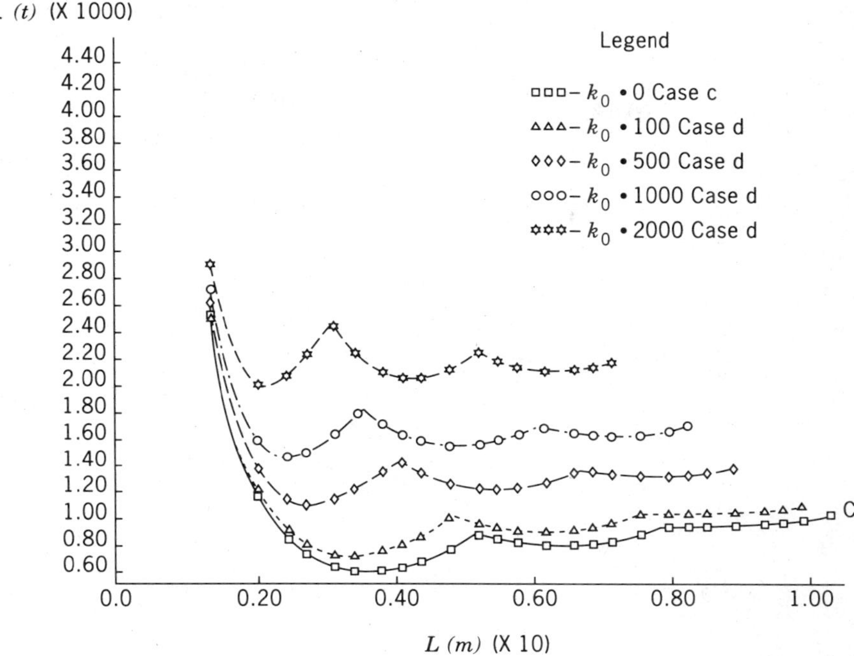


**Figure 10.6** Critical Load of case c and d for a pinned-pinned end Pile when $n_h = 100\,\text{t/m}^3$ (Prakash, 1987).

corresponding value in the previous mode. This is distinctly different than when $k$ was constant with depth and $n_h$ was zero that is, case b, Figure 10.5. Similar behavior was observed for $n_h = 500\,\text{t/m}^3$, $1000\,\text{t/m}^3$ and $2000\,\text{t/m}^3$ (Prakash, 1985).

***Effect of Boundary Conditions on the Buckling Load*** In Figure 10.7, $P_{cr}$ has been plotted against the length of the pile for $k_0 = 100\,\text{t/m}^2$ and $n_h = 100\,\text{t/m}^3$ (case d Figure 10.4) for two boundary conditions (i.e., pinned top-pinned tip ($p$-$p$) and fixed top-fixed tip ($F$-$F$)). It will be seen that the buckling load decreases sharply as the length of the pile increases and attains a minimum value of 724$t$ and 1413$t$ for $p$-$p$ and $F$-$F$ boundary conditions, respectively. The buckling loads in the higher modes are larger in both cases. The mode shape in both cases depends on the length of the pile (i.e., as the pile length increases, higher buckling modes appear). The buckling loads are highest for boundary conditions $F$-$F$ and minimum for boundary conditions $p$-$p$.

The above conclusions are more or less in the realm of expectation. However, specific numerical values have been determined for the case mentioned above. Similar diagrams for $k_0 = 100\,\text{t/m}^2$ and $n_h = 0$ (case b) and $n_h = 100\,\text{t/m}^3$ and $k_0 = 0$ (case c) have been reported elsewhere (Prakash, 1985). Results as above will become readily usable by field engineers when these are plotted in

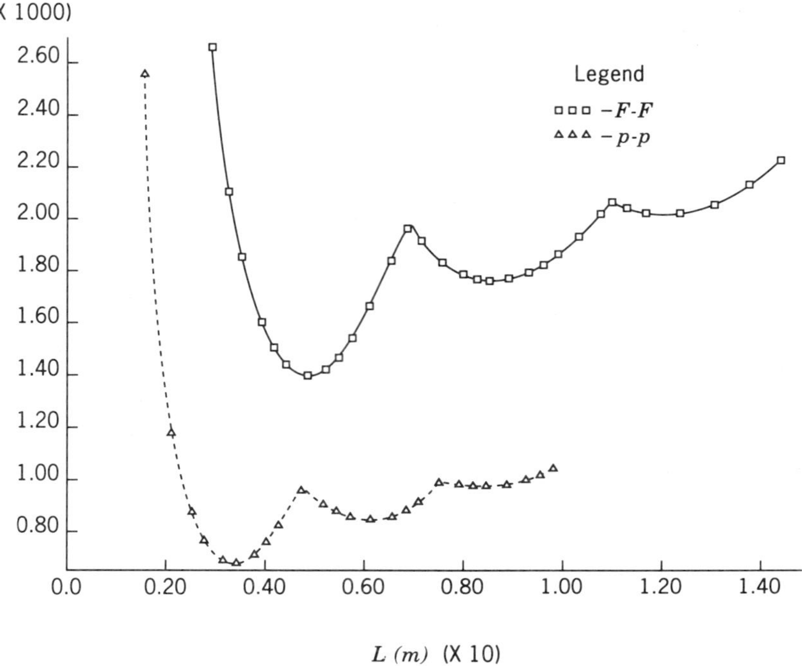


**Figure 10.7** Critical load of case d for a pile with different boundary conditions when $k_0 = 100$ and $n_h = 100\,\text{t/m}^3$ (Prakash, 1987).

non-dimensional forms as in Figures 10.2 and 10.3 for case b and c (Figure 10.4) respectively.

## 10.2 PARTIALLY EMBEDDED PILES

Column instability is usually a problem in the design of structures supported by piles that are partially free standing. Furthermore, for structures such as piers that are subjected to both vertical and lateral loads, a flexural analysis of the piles may control the design of the foundation. Generally, the analysis is highly indeterminate and unwieldy unless some simplifying conditions are imposed (Davisson and Robinson, 1965). In Figure 10.8, $L_u$ is the unsupported pile length above the ground level. The vertical load tends to magnify the deflection caused by $Q$ and $M$.

***Solutions for Constant k*** Davisson and Robinson (1965) have presented solutions for buckling loads of partially embedded piles. The axial load on the pile is constant and the pile is relatively long. In this analysis, it has been assumed that the actual pile in Figure 10.8a is equivalent to a pile of length $L_e$ fixed at the tip

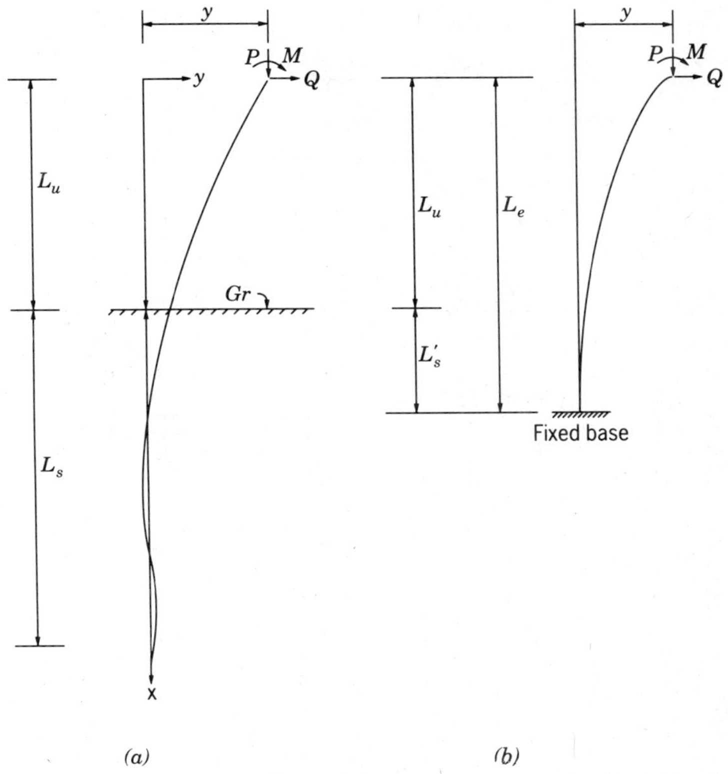


**Figure 10.8** Partially embedded pile (a) Actual Pile, (b) equivalent system (Davisson and Robinson, 1965).

(Figure 10.8b). The depth $L'_s$ may be viewed as one that will make the buckling loads of the actual system equal to the equivalent system.

By solving equation (10.1) for the freestanding length, the solution has been developed in nondimensional form with the help of the following functions:

$$S_R = \frac{L'_s}{R} \tag{10.13}$$

$$J_R = \frac{L_u}{R} \tag{10.14}$$

$L'_s$ = equivalent length of embedded portion of pile (Fig. 10.8)
$L_u$ = unsupported pile length

and $R$ is defined in equation 10.2 with $L_s$ = embedded length.

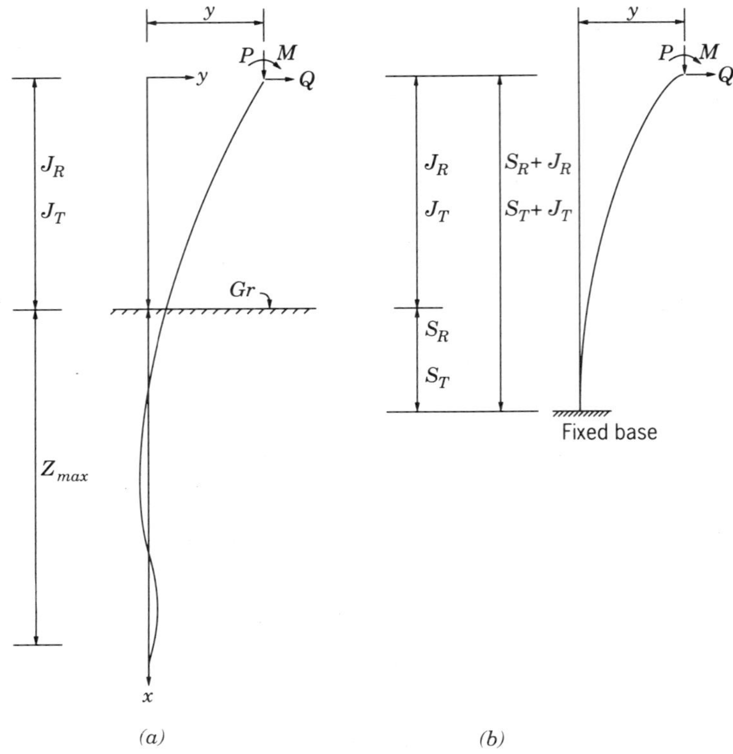


**Figure 10.9** Nondimensional representation of partially embedded pile (a) Actual pile, (b) equivalent system (Davisson and Robinson, 1965).

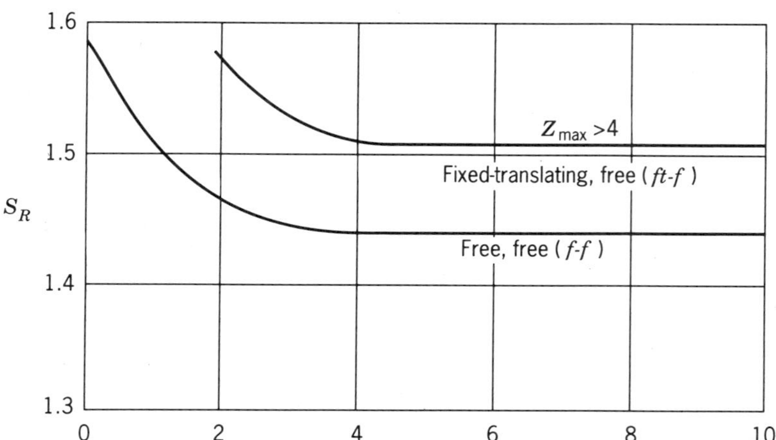


**Figure 10.10** Dimensionless depth of fixity for buckling $k=$ constant (Davisson and Robinson, 1965). (Reprinted by permission of University of Toronto, Canada.)

With the above substitutions, the dimensions in Figure 10.8 may be expressed in nondimensional parameters in Figure 10.9. The equivalent length of the freestanding length is now equal to $(S_R + J_R)$.

The relationship between $S_R$ and $J_R$ were developed by Davisson and Robinson (1965) for two cases of loading on the pile top, $Q$ shear only and $M$ moment only. It was found that $S_R$ varies within a narrow range of approximately 1.33 to 1.6. Therefore, for practical case, a value of 1.33 for $S_R$ has been recommended.

For the two boundary conditions in Figure 10.10, the critical buckling load $P_{cr}$ may then be computed from equation

$$P_{cr} = \frac{\pi^2 E_p I_p}{4(S_R + J_R)^2 R^2} \tag{10.15}$$

***Solutions for Linearly Varying k*** Solutions for a long pile ($Z_{\max} > 4$, with $L_s$ = embedded length) for the case $k = n_h \cdot x$ are shown in Figure 10.11. The equivalent length of embedded portion of pile has been defined as (see Figure 10.9).

$$S_T = \frac{L_s'}{T} \tag{10.16a}$$

$$J_T = \frac{L_u}{T} \tag{10.16b}$$

The buckling load is

$$P_{cr} = \frac{\pi^2 E_p I_p}{4(S_T + J_T)^2 T^2} \tag{10.17}$$

Lee (1968) carried out model tests on 1/4-in. to 1/2-in. diameter piles in dry sand. He found good agreement between the measured and computed buckling loads.

## 10.3 EFFECT OF AXIAL LOAD TRANSFER

In the solutions in the preceding sections, it has been assumed that the axial load is constant along the pile, that is, no load transfer occurs along the pile shaft. This condition is applicable for relatively short or stiff end-bearing piles. In floating piles and compressible end-bearing piles, considerable load transfer occurs along the shaft. The effect of axial load transfer on the critical buckling loads of fully and partially embedded piles has been investigated by Reddy and Valsangkar (1970). The following idealized axial load distributions has been assumed:

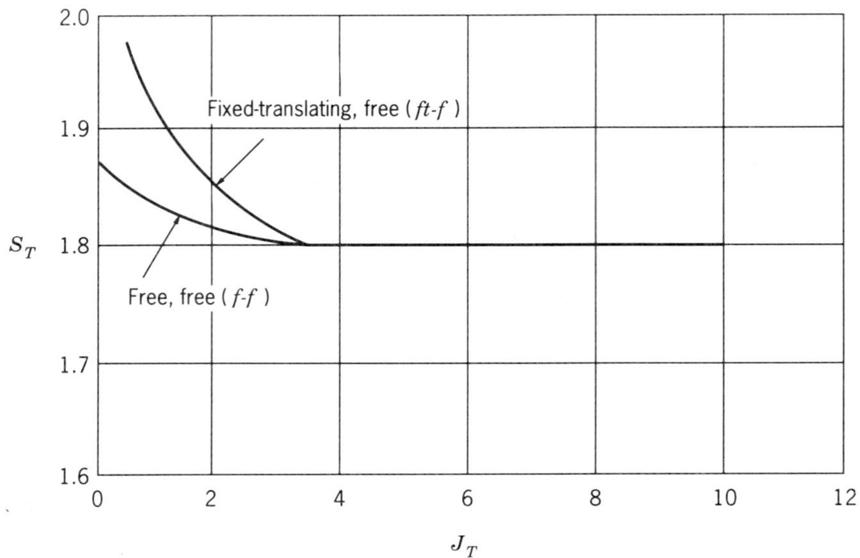


**Figure 10.11** Dimensionless depth of fixity for buckling. Linearly varying $k$ (Davisson and Robinson, 1965). (Reprinted by permission of University of Toronto Press, Canada.)

### 10.3.1 Fully Embedded Piles

$$P = P_0(1 - \psi x/L) \tag{10.18a}$$

and

$$P = P_0(1 - \psi(x^2/L^2) \tag{10.18b}$$

where

$P_0$ = load at pile head
$x$ = depth below surface
$L$ = pile length
$\psi$ = parameter $(0 \leqslant \psi \leqslant 1)$

For $\psi = 0$, the pile is an end-bearing pile and for $\psi = 1$, the pile is a friction pile.

### 10.3.2 Partially Embedded Piles

$$P = P_0\left[1 - \psi\left(\frac{x}{L_t} - n\right)\right] \tag{10.19}$$

where

$L_t$ = total length of pile $(L_s + L_u)$, Figure 10.8
$n$ = ratio of unsupported length to total length, $L_u/(L_s + L_u)$

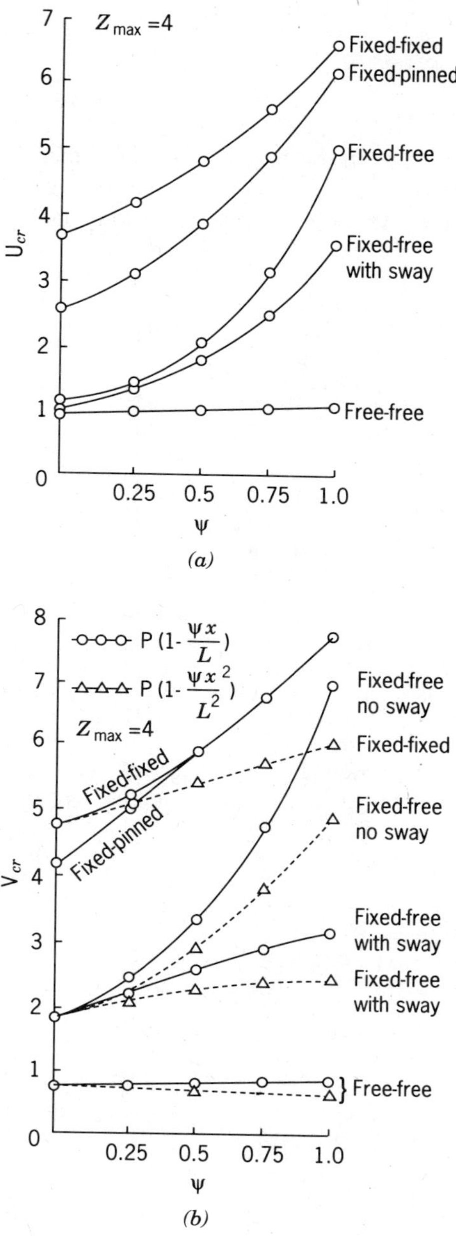


**Figure 10.12** Effect of skin friction on buckling loads of fully embedded piles for (a) Constant soil modulus, (b) linear soil modulus (Reddy and Valsangkar, 1970).

In this case, the dimensionless length is defined as

$$Z_{\max} = L_t/T \tag{10.20}$$

Also, $\psi$ can be greater than one.

For long piles ($Z_{\max} > 4$), the variation of the dimensionless buckling loads $U_{cr} = P_{cr}R^2/E_pI_p$ and $V_{cr} = P_{cr}T^2/E_pI_p$ with $\psi$ is shown in Figure 10.12 for fully embedded piles. For $\psi = 0$ and appropriate boundary conditions, the solutions in

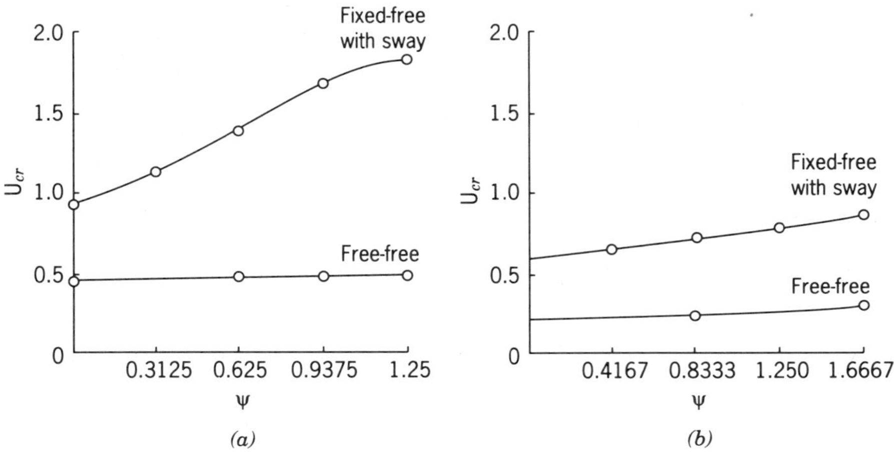


**Figure 10.13** Effect of skin friction on buckling loads for partially embedded long piles $Z_{\max} = 4$ for $k =$ constant (a) $n = 0.2$, (b) $n = 0.4$ (Reddy and Valsangkar, 1970).

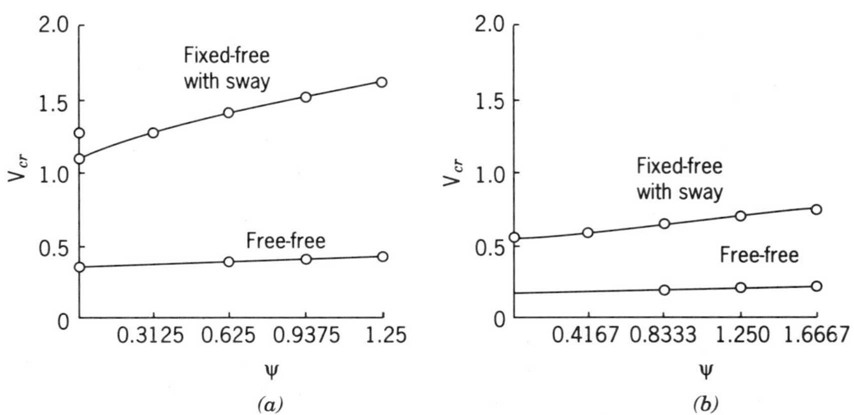


**Figure 10.14** Effect of skin friction on buckling loads for partially embedded long piles $Z_{\max} = 4$ for $k = n_h{\cdot}x$: (a) $n = 0.2$, (b) $n = 0.4$ (Reddy and Valsangkar, 1970).

Figures 10.2 and 10.12a agree ($k$ constant with depth). Similarly for $k$ increasing linearly with depth, the limiting solutions in Figures 10.3 and 10.12b agree. For $\psi$ greater than 0, considerable increase in the buckling load occurs because of load transfer. For fixed-translating top and free bottom ($ft$-$f$), the increase in buckling load is about three times for friction piles ($\psi = 1$) as compared to end bearing piles ($\psi = 0$) for $k =$ constant. Corresponding solutions for partially embedded long piles ($Z_{\max} = 4$) for constant $k$ and $n = 0.2$ and 0.4 are shown in Figure 10.13(a) and (b). Similar solutions for $k = n_h \cdot x$ and $n = 0.2$ and 0.4 have been plotted in Figure 10.14(a) and (b). The values of $\psi$ may be estimated by a suitable distribution of skin friction.

## 10.4 GROUP ACTION

Model tests by Toakley (1964) with groups of two and three strip piles in soft silt showed the critical load is reduced by group action. However, full scale tests by Hoadley et al. (1969) showed little interaction between closely spaced piles.

In practice, both vertical and horizontal loads will act on a group of piles. The change in the value of $k$ (soil modulus) due to group action was described in Chapter 6. It is recommended that the same value of soil modulus be used for computing the buckling loads of piles in a group as for computing lateral deflection. The presence of lateral load is equivalent to introduction of eccentricity in the vertical load, which reduces the critical buckling load.

## REFERENCES

Davisson, M. T., "Estimating Buckling Loads for Piles," *Proceedings Second Pan American Conference on Soil Mechanics and Foundation Engineering*, Sao Paulo, Vol. 1, (1963), pp. 351–371.

Davisson, M. T. and Gill, H. L., "Laterally Loaded Piles in a Layered Soil System," *J. Soil Mech. Found. Div.* Vol. 89, No. SM3, (1963), pp. 63–94.

Davisson, M. T. and Robinson, K. E., "Bending and Buckling of Partially Embedded Piles," *Proceedings 6th International Conference on Soil Mechanics and Foundation Engineering*, Montreal, Canada, Vol. 2, (1965), pp. 243–246.

Forsell, C., "Berakning av palar 1918" Stockholm.

Forsell, C., "Knacksakerhet nos Palar Och Palgrupper" Uppsal No. 10, Festskrift kungl. Vag-och Vattenbyggna-dskarem 1926, Stockholm.

Grandholm, H., "On Elastic Stability of Piles Surrounded by a Supporting Medium," *Ing. Vet. Akad.*, Hand. 89, (1929), Stockholm.

Hetenyi, M., *Beams on Elastic Foundations.* University of Michigan Press, Ann Arbor (1946).

Hoadley, P. J., Francis, A. J., and Stevens, L. J., "Load Testing of Slender Steel Piles in Soft Clay," *Proceedings 7th International Conference on Soil Mechanics and Foundation Engineering*, Mexico, Vol. 2, (1969), pp. 123–130.

Lee, K. L., "Buckling of Partially Embedded Piles in Sand," *J. Soil Mech. Found. Div.*, ASCE, Vol. 94, No. SM1, (1968), pp. 255–270.

Poulos, H. G. and Davis, E. H., *Pile Foundations.* Wiley, New York (1980).

Prakash, Sally, "Buckling Loads for Fully Embedded Piles," M. S. Thesis University of Missouri-Rolla (1985).

Prakash, Sally, "Buckling Loads of Fully Embedded Piles," *Int. J. Computer Geotech.* Vol. 4, (1987), pp. 61–83.

Reddy, A. S. and Valsangkar, A. J., "Buckling of Fully and Partially Embedded Piles," *J. Soil Mech. Found. Div.*, ASCE, Vol. 96, No. SM6, (1970), pp. 1951–1965.

Toakley, A. R., "The Behavior of Isolated and Group of Slender Point Bearing Piles in Soft Soil," M. S. Thesis, University of Melbourne, (1964), Australia.

# 11

# CASE HISTORIES

Pile foundations behavior has been studied for decades, but there are several gaps in the proper and quantitative understanding of the response of piles, both under static and dynamic loads. Field tests are the best method of study of their response, but these are expensive. Therefore, study of case histories is important.

The four main sections of this chapter provide actual case histories for piles that were designed to resist (1) axial compressive loads, (2) axial pullout loads, (3) lateral loads, and (4) dynamic loads. In the beginning of each section, we provide information on soil conditions, pile geometry and installation methods, predicted pile capacities, and measured pile loads. Finally, a comparison is made between the estimated and measured loads.

In Section 11.3, the predicted load deflection of a pile group under sustained and cyclic lateral load is compared with the measured data. A single pile load-deflection data was used as a basis for group predictions. In Section 11.4, the natural frequency of oscillations of piles is predicted. These values were compared with the measured frequencies.

There is limited information available on full sized pile tests. Also the method of interpretation used in this chapter may not be the only method used by researchers. In practice these interpretation methods are reasonable.

## 11.1 PILES SUBJECTED TO AXIAL COMPRESSION LOADS

This section provides case histories for (1) a site where cast-in-place belled and bored concrete piles were installed, (2) a site where expanded base compacted piles were installed, and (3) five sites where closed-ended steel pipe piles were driven. These cases present the ultimate pile capacities estimated from the methods presented in Chapter 5. These values have then been compared with the ultimate pile capacities obtained from full-scale pile load tests.

### 11.1.1 Cast-in-Place Belled and Bored Piles

At a petrochemical complex located in central Alberta, Canada, a total of about 1500 *cast-in-place belled and bored concrete* piles were installed during 1981 to 1983 to support heavy equipments and building loads. Sharma et al. (1984) present the details of soil conditions and pile design.

***Soil Conditions*** The generalized soil conditions consisted of about 30 ft (9.2 m) deep clay till having an average undrained strength of 1.36 ksf (65 kPa) underlain by weathered clay shale bedrock having an average undrained strength of 4.7 ksf (225 kPa) (Figure 11.1).

***Pile Geometry and Installation Method*** The piles were about 20 in. (500 mm) shaft diameter and about 48 in. (1200 mm) bell diameter. The piles were about 31 ft (9.5 m) long and bearing into the clay shale bedrock.

***Pile Load Test*** An axial compression pile load test was carried out as per ASTM D1143-81. A sono tube was installed outside the pile to a depth of 5.0 ft (1.5 m) below ground surface. This feature was installed in the test pile because in the

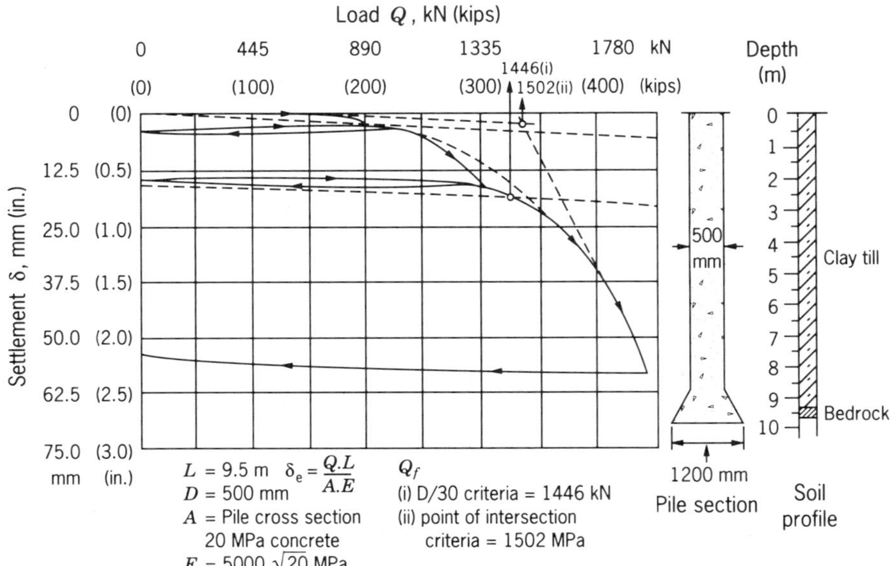


**Figure 11.1** Load-settlement curve for a cast-in-place bored and belled concrete pile (Sharma et al., 1984).

estimation of theoretical pile capacity, adhesion in the top 5.0 ft (1.5 m) of pile length was neglected in order to take into account the seasonal variations for this site. Figure 11.1 shows the load settlement curve and generalized soil conditions for this case.

### *Estimation of Pile Capacity*

(a) *Shaft Friction Capacity, $Q_f$*

$$Q_f = p \sum_{L=0}^{L=L} c_a \Delta L \tag{5.46}$$

$c_a/c_u = 0.5$ (from Table 4.7)

$c_a = 0.5 \times 1.36 = 0.68$ ksf

$L =$ pile length—(depth of seasonal variation $+ 2 \times$ pile shaft diameter)

from Table 5.9

$L = 31 - (5 + 2 \times 20/12) = 22.67$ ft.

$p = \pi(20/12) = 5.24$ ft where $B = 20/12 = 1.67$ ft

Substituting these values in equation (5.46) will yield the following:

$$Q_f = 5.24 \times 0.68 \times 22.67 = 80.8 \text{ kips}$$

(b) *End-bearing Capacity, $Q_p$*

$$Q_p = A_p c_u N_c \tag{5.45}$$

$$A_p = \frac{\pi}{4}(48/12)^2 = 12.57 \text{ ft}^2$$

$$c_u = 4.7 \text{ ksf}$$

$D_f/B = 31/1.67 = 18.6$, $N_c = 9$ from Table 5.7
From Table 5.8 for pile base diameter $> 3$, $N_c = 6$
The lower of the above two $N_c$ values is 6. Therefore $N_c = 6$.
Substituting these values in equation (5.45) will yield the following:

$$Q_p = 12.57 \times 4.7 \times 6 = 354.5 \text{ kips}$$

$$(Q_v)_{\text{ult}} = A_p c_u N_c + p \sum_{L=0}^{L=L} c_a \Delta L \tag{5.47}$$

Substituting the values calculated in (a) and (b) above will yield the following:

$$(Q_v)_{\mathrm{ult}} = 354.5 + 80.8 = 435.3 \text{ kips (1937 kN)}$$

***Pile Capacity from Pile Load Test*** The load settlement curve is shown in Figure 11.1. The failure load, $(Q_v)_{\mathrm{ult}}$, interpreted by Fuller and Hoy, Vander Veen and Brinch Hansen's 80 percent criterion are 425, 428, and 452 kips, respectively. These methods of interpretation have been presented in Chapter 9. The above indicates that $(Q_v)_{\mathrm{ult}} = 435$ kips (1936 kN) would be a reasonable average value from the pile load test result.

***Pile Capacity: Estimate Versus Load Test Result*** The above analysis indicates the following:

Estimated: $(Q_v)_{\mathrm{ult}} = 435.3$ kips (1937 kN)
Load Test: $(Q_v)_{\mathrm{ult}} = 435$ kips (1935.8 kN)

These values are plotted in Figure 11.4.

### 11.1.2 Expanded Base Compacted (Franki) Piles

At a refinery project site in Regina, Saskatchewan, Canada, a total of about 2000 expanded base compacted (Franki) piles were installed during 1986 to 1988 to support heavy structural loads. Sharma (1988) presents the details of soil conditions and pile design data for this case.

***Soil Conditions*** The subsoil conditions at this site consisted of about 25 ft (7.6 m) deep high-plasticity clay having an average undrained strength of 1.36 ksf (65 kPa) underlain by about 15 ft (4.6 m) silt having an undrained strength 0.8 ksf (38 kPa). Below this silt stratum existed a 20-ft (6 m) thick silty sand deposit with average Standard Penetration Test ($N$) values of 13.4. The generalized soil conditions are shown in the borehole log (Figure 11.2).

***Pile Geometry and Installation Method*** The piles were 20 in. (500 mm) shaft diameter. The installation method consisted of preboring through the high-plasticity clay and then driving the casing. The pile base formation was started at 45 ft (13.7 m) depth with 118,000 ft-lb (160 kN-m) energy. Three concrete buckets, each with 5 cu ft (0.14 cu m), concrete were used in the base. The last (the third) bucket of concrete required 30 blows of 118,000 ft-lb impact energy for concrete expulsion into the base.

***Pile Load Test*** An axial compression pile load test was carried as per ASTM D1143-81. Figure 11.2 shows the load settlement curve and generalized soil conditions for this case.

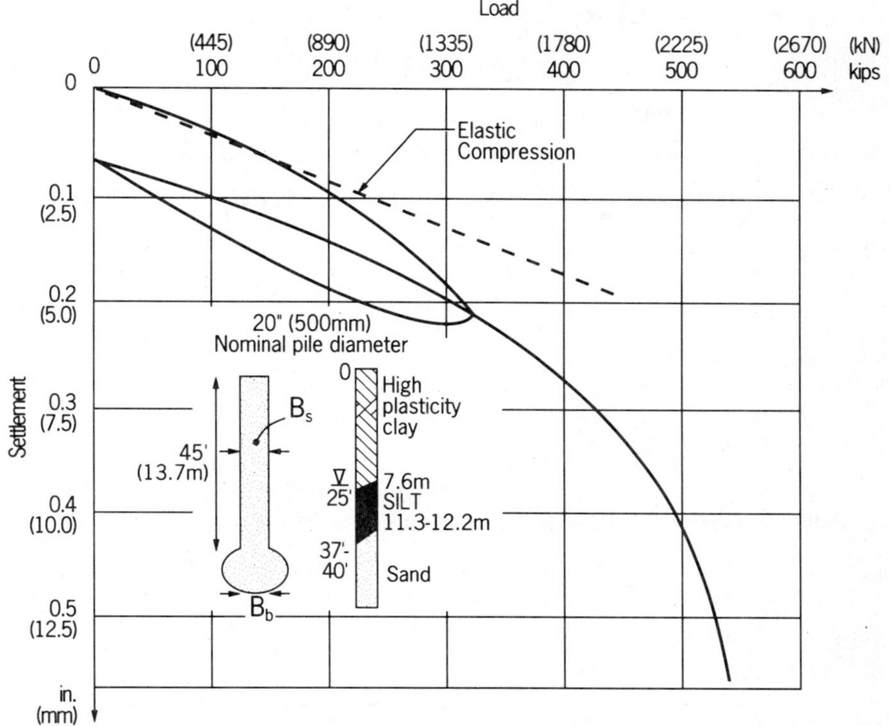


**Figure 11.2** Load-settlement curve for an expanded base compacted pile (Sharma, 1988).

### *Estimation of Pile Capacity*

(a) *Shaft Friction Capacity*

$$Q_f = p \sum_{L=0}^{L=L} c_a \Delta L \tag{5.46}$$

The values of $c_a/c_u$ for expanded base compacted piles are not available. However, a review of Figure 4.27 and Table 4.7 indicates that for clay $(c_a/c_u) = 0.5$ and for silt $(c_a/c_u) = 1$ will provide reasonable estimates for $c_a$ values.

FOR CLAY $\quad c_a = 0.5 \times 1.36 = 0.68\,\text{ksf}$

$$L_1 = 25\,\text{ft}$$

$$c_a L_1 = 17\,\text{k/ft}$$

FOR SILT $\quad c_a = 1 \times 0.8 = 0.80\,\text{ksf}$

$$L_2 = 15\,\text{ft}$$

$$c_a L_2 = 12\,\text{k/ft}$$

$$p = \pi B = \pi \times \frac{20}{12} = 5.24\,\text{ft}$$

Substituting these values in equation (5.46) will yield the following:

$$Q_f = 5.24\,(17 + 12) = 151\,\text{kips}\,(472\,\text{kN})$$

(b) *End-bearing Capacity*, $Q_p$

$$(Q_p)_{\text{all}} = W \times H \times N_b (V)^{2/3}/K \tag{5.68}$$

$$WH = 118{,}000\,\text{ft-lb} = 118\,\text{ft kips}$$

$$N_b = 30/5 = 6\ \text{blows/ft}^3$$

$V = 4 \times 5 = 20\,\text{ft}^3$ (This consisted of 1 bucket during driving and 3 buckets during base formation; each bucket has 5 cu ft of concrete. From Table 5.14, $K = 2.5N$ for prebored compacted shaft pile. For $N = 13.4$:

$$K = 13.4 \times 2.5 = 33.5$$

Substituting these values in equation (5.68), we get:

$$(Q_p)_{\text{all}} = 118 \times 6\,(20)^{2/3}/33.5 = 157\,\text{kips}\,(698\,\text{kN})$$

$$Q_p = 2.5 \times 157 = 392.5\,\text{kips}\,(1747\,\text{kN})$$

A factor of safety of 2.5 has been used to obtain $Q_p$. This was discussed in Section 5.1.15.

$$(Q_v)_{\text{ult}} = Q_p + Q_f$$

$$(Q_v)_{\text{ult}} = 392.5 + 151 = 543.5\,\text{kips}\,(2419\,\text{kN})$$

***Pile Capacity from Pile Load Test*** The load settlement curve is shown in Figure 11.2. The failure load, $(Q_v)_{\text{ult}}$, interpreted by the Butler and Hoy, Davisson, and Fuller and Hoy methods were 530 kips (2359 kN), 540 kips (2403 kN), and 540 kips (2403 kN), respectively. These methods were presented in Chapter 9. The above indicates that $(Q_v)_{\text{ult}} = 537$ kips (2390 kN) would be a reasonable value from the pile load test result.

***Pile Capacity: Estimated versus Load Test Result*** The above analysis indicates the following:

Estimated: $(Q_v)_{\text{ult}} = 543.5$ kips (2419 kN)
Load Test: $(Q_v)_{\text{ult}} = 537$ kips (2390 kN)

These values are fairly close to each other and have been plotted in Figure 11.4.

**TABLE 11.1 Soil Conditions and Pile Geometry for Case 3[a]**

| Site | Soil Conditions | | | Steel Pipe Pile Dimensions | |
|---|---|---|---|---|---|
| and Location | Average Depth (ft) | Soil Type | Av. $N$(App) | OD (in.) | Length (ft) |
| Site A, North York | 0–33 | Fill (silt, sand, clay, and organics) | — | 11.75 | 66 |
| | 33–79 | Wet sand | 20 | | |
| Site B, Mississauga | 0–33 | Fly ash | 30 | 9.6 | 30 |
| | 33–deeper | Shale bedrock | | | |
| Site C, Owen Sound | 0–15 | Sand fill | 5 | 9.6 | 154 and 110 |
| | 15–75 | Silt to clayey silt | 10 | | |
| | 75–108 | Sandy silt | 20 | | |
| | 108–144 | Silt and clay till, | — | | |
| | 144–deeper | very dense bouldery till | — | | |
| Site D, Lake Ontario Toronto Toronto | 0–26 | Hydraulic fill | 20 | 12.75 | 39 |
| | 26–38 | Fine to med. sand | 10 | | |
| | 38–deeper | Shale bedrock | — | | |
| Site E, Hamilton | 0–43 | Fill | — | 12.75 | 59 |
| | 43–59 | Hard to stiff silts and clays | — | | |
| | 59–deeper | Very dense silt till | | | |

[a]These data have been summarized from the information reported by Cheng and Ahman (1988).

### 11.1.3 Driven Closed-ended Steel Pipe Piles

This case presents case histories at five sites where the ultimate bearing capacity of driven closed-ended steel pipe piles was evaluated by both dynamic measurements (case method) and static load tests. These piles were driven to different depths into different soil types at various locations in Southern Ontario, Canada. Cheng and Ahman (1988) present the details of soil conditions, load test data, and pile design information for these piles.

***Soil Conditions*** The soil conditions at the five sites A, B, C, D, and E are summarized in Table 11.1. The Standard Penetration Test ($N$) values for cohesionless materials at these sites are available. However, undrained strength values for cohesive soils and rock core strengths are not available for these sites.

***Pile Geometry and Installation Methods*** Piles at all these sites consisted of driven closed-ended steel pipe piles. In driving these piles, the strain and acceleration of the piles by the pile driver were measured. From strain measurements, the force at pile top and from the acceleration measurement the velocity of the pile being driven can be obtained. This information was then used to estimate ultimate pile capacity by the Case Method as presented below.

***Pile Load Tests*** Axial compression pile load tests were carried out as per ASTM D1143-81. Figure 11.3 shows the load settlement curves for the pile load tests carried out at the five sites.

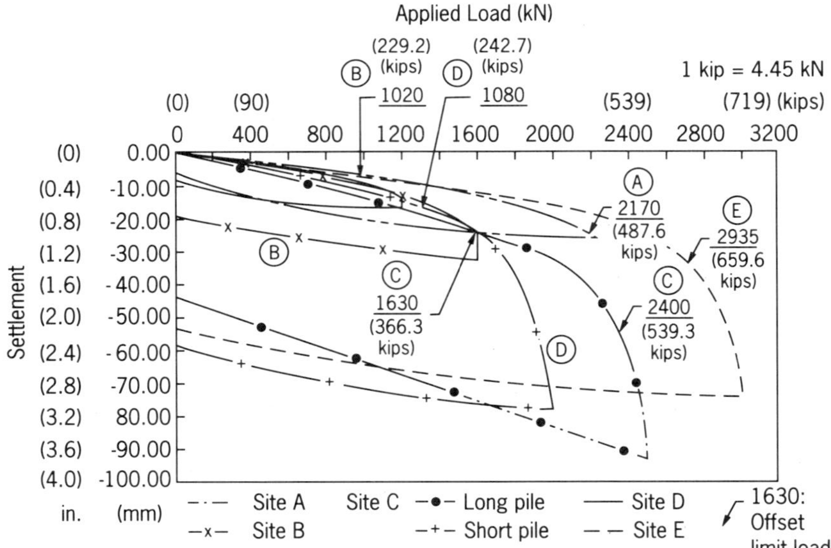


**Figure 11.3** Load-settlement curves for driven pipe piles cited in case 3 (Cheng and Ahman, 1988).

***Estimation of Pile Capacities*** As mentioned above, the force and velocity measurements were carried out for each pile during their installation by dynamic monitoring method. This method has been described in Chapter 5 (Section 5.1.2). These measurements were then used to estimate the static ultimate pile capacity by using the following relationship (called Case Method):

$$(Q_v)_{\text{ult}} = R_r(t_m) = 1/2(1 - j_c)\left[F(t_m) + \frac{MV_r}{L}v_T(t_m)\right] + 1/2(1 + j_c)\left[F\left(t_m + \frac{2L}{V_r}\right) - \frac{MV_r}{L}v_t\left(t_m + \frac{2L}{V_r}\right)\right] \tag{5.33}$$

Various terms of this equation have been explained in Chapter 5. The estimated $(Q_v)_{\text{ult}}$ values from this equation for each of the five sites (A to E) are listed in Table 11.2 under "Case method".

***Pile Capacity from Pile Load Test*** The load settlement curves are shown in Figure 11.3. The failure load $(Q_v)_{\text{ult}}$ as estimated by Davisson's method are presented in Table 11.2.

***Pile Capacity: Estimated versus Load Test Results*** The estimated (calculated) versus load test results are presented in Table 11.2 and have also been plotted in Figure 11.4. These values are close to each other.

Finno et al. (1989) report a comparison of measured capacity and 22-predictions made by different investigations on 2–50 ft long piles. Considerable variation in the measured and predicted values was observed. This highlights the importance of case studies in pile foundations.

**TABLE 11.2 Ultimate Load Capacities: Estimated versus Load Tested**

| | Ultimate Capacity, $(Q_v)_{\text{ult}}$ | |
|---|---|---|
| Site | Estimated (Case Method) (kips) | Load Test (kips) |
| A | 445.6 | 467.6 |
| B | 264.0 | 229.2 |
| C | 516.9 | 539.3 |
| | 387.7 | 366.3 |
| D | 222.5 | 242.7 |
| E | 629.2 | 659.6 |

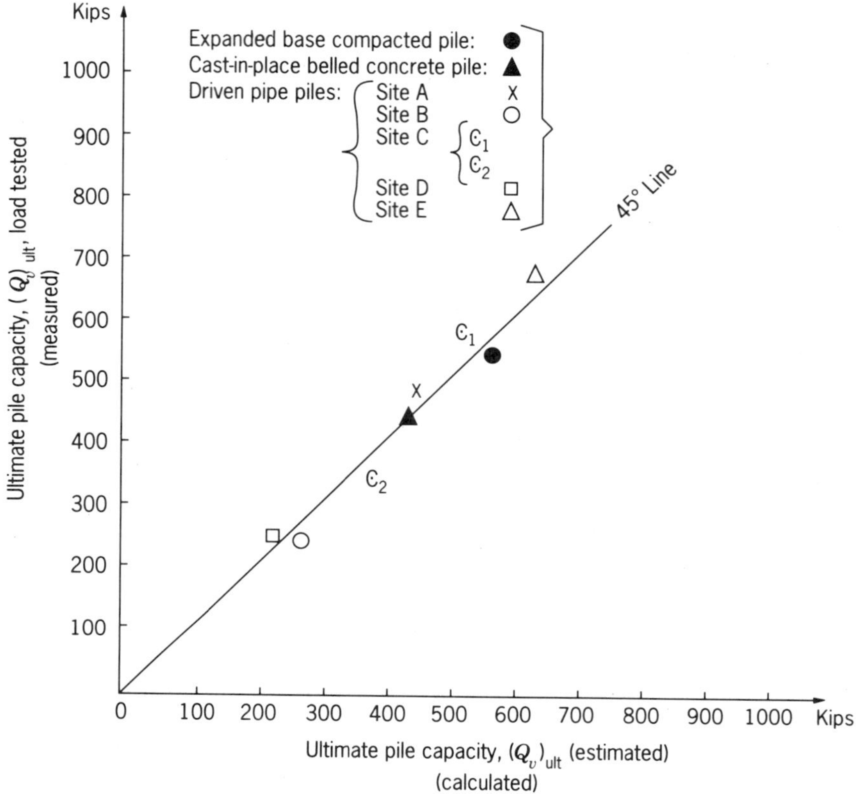


**Figure 11.4** A comparison of measured and calculated axial compression load capacities.

## 11.2 PILES SUBJECTED TO PULLOUT LOADS

This section presents four cases where driven steel HP piles were used to resist tension forces. The results of full-scale pile loads are compared with the predicted (estimated) ultimate pullout capacity based on available soil parameter. The detailed information on these cases are presented by Hegedus and Khosla (1984).

***Soil Conditions*** Figures 11.5 and 11.6 summarize the soil conditions for pile sites 1 to 4. At site 1 the soils were primarily silty clay with undrained strength of 4.1 kips/sq ft (196 kPa). Site 2 consisted of medium to dense silty sands and dense nonplastic silts. At site 3 the pile was installed through silty clay having undrained strength of 2.8 kips/sq ft (134 kPa). At site 4 the pile was installed through silty clay fill underlain by silty sand and sand with gravel. Table 11.3 provides a summary of soil parameters at these four sites.

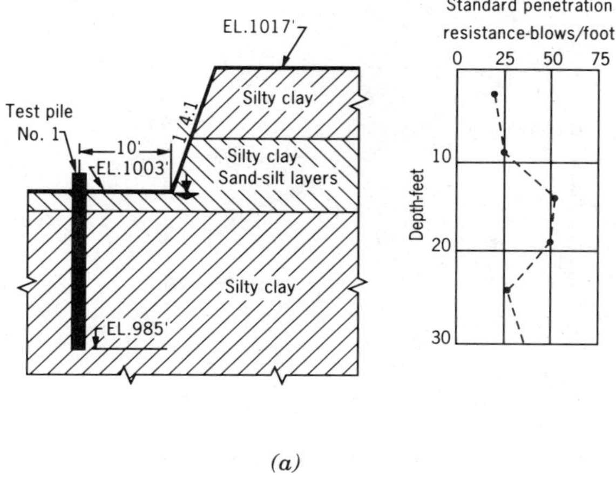


*(a)*

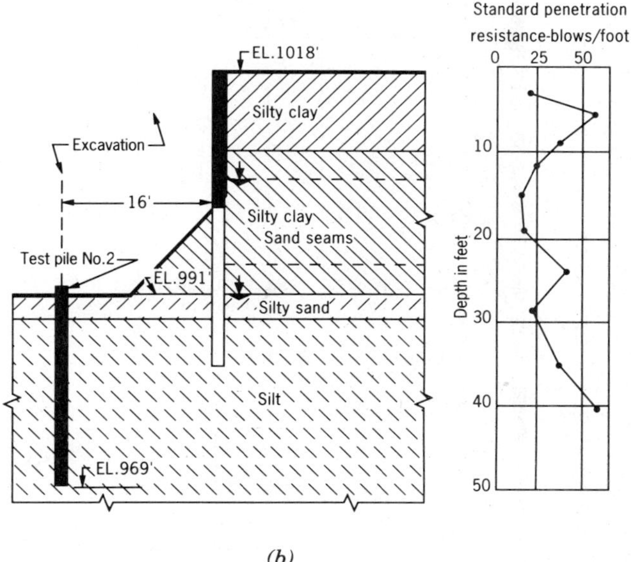


*(b)*

**Figure 11.5** Summary of soil conditions and pile details for test piles at (a) Site 1 and (b) site 2 (Hegedus and Khosla, 1984).

***Pile Geometry*** The piles installed at these four sites were driven HP sections. The nominal section for test piles at sites 1,2 and 3 was HP 10 × 42 (254 mm × 19.0 kg) while the section for the test pile at site 4 was HP 12 × 74 (304.8 mm × 33.6 kg). At location 3 the actual piles and the test pile was installed at 30° inclination from the horizontal. At other three sites the piles were vertical.

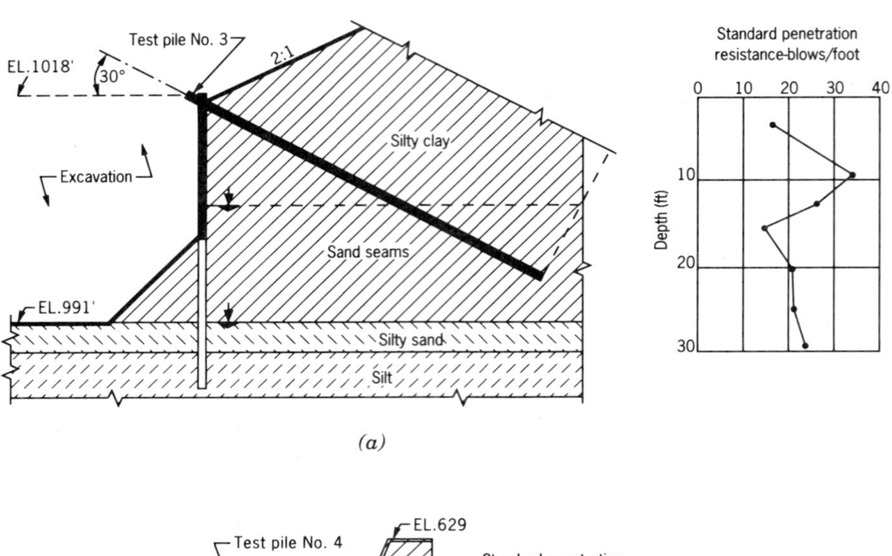


(a)

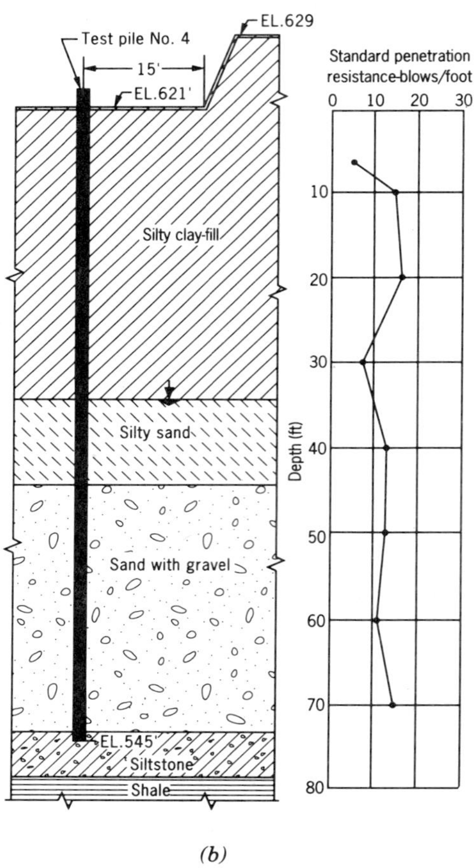


(b)

**Figure 11.6** Summary of soil conditions and pile details for test piles at (a) Site 3 and (b) site 4 (Hegedus and Khosla, 1984).

**TABLE 11.3 Summary of Soil Parameters for Four Sites**

| | Site 1 | Site 2 | Site 3 | Site 4 | |
|---|---|---|---|---|---|
| Soil parameter (1) | Silty Clays (2) | Silt or Silty Sand (3) | Silty Clay (4) | Silty Clays (5) | Sand with Gravel (6) |
| Moisture content, in percentage | 16 | 12 | 19 | 24 | 14 |
| Dry unit weight, in pounds per cubic foot | 116 | 125 | 113 | 88 | 106 |
| Liquid limit | 29 | N.A. | 32 | 35 | N.A. |
| Plasticity index | 12 | N.P.[a] | 14 | 19 | N.P. |
| Unconfined strength, in tons per square foot | 4.1 | — | 2.8 | — | — |
| Angle of shearing resistance—drained, in degrees | — | 34 | — | — | 33 |
| Angle of shearing resistance—remolded drained, in degrees | 25 | — | 25 | 27 | — |
| Angle of shearing resistance between pile and soil, in degrees | — | 20 | — | — | 20 |
| Overconsolidation ratio (OCR) | 2.6 | — | 1.5 | 1.0 | — |

[a]Nonplastic material.
Note: 1 lb/ft$^3$ = 0.0157 Mg/m$^3$; 1 ton/ft$^2$ = 95.8 kN/m$^2$.
Hegedus and Khosla, 1984.

***Pile Load Tests*** At each of these four sites one test pile was installed. These piles were then subjected to pullout test loads. The loads were applied to these piles by hydraulic jack and their butt movements were measured. The loading schedule for these piles is shown in Table 11.4. Typical load deformation curves for these piles are shown in Figures 11.7 through 11.10. The ultimate pullout load, $P_u$, for each of these test piles was interpreted by the following three methods:

1. Tangent method: Load corresponding to the intersection of lines drawn tangent to initial and final sections of load-deformation curve.
2. AASHTO method: Load corresponding to 0.25 in. (6.35 mm) net (gross minus rebound) settlement.
3. Davisson's method: This method has been described for axial compression pile load in Chapter 9.

**TABLE 11.4 Loading Schedule for the Four Test Piles**

| Test Pile Number (1) | Test Load (tons) (2) | Duration (min) (3) |
|---|---|---|
| Site 1 | 4 | 30 |
| | 8 | 30 |
| | 12 | 30 |
| | 16 | 70 |
| | 20 | 67 |
| | 24 | 240 |
| | 28 | 65 |
| | 32 | 1440 |
| Site 2 | 6 | 30 |
| | 11 | 30 |
| | 17 | 30 |
| | 23 | 60 |
| | 28 | 60 |
| | 34 | 60 |
| | 39 | 60 |
| | 45 | 1440 |
| Site 3 | 10 | 10 |
| | 20 | 10 |
| | 30 | 30 |
| | 40 | 60 |
| | 50 | 60 |
| | 60 | 120 |
| Site 4 | 47 | 30 |
| | 75 | 30 |
| | 102 | 60 |
| | 128 | 60 |
| | 155 | 30 |

Note: 1 ton = 907 kg.
Hegedus and Khosla, 1984.

Table 11.5 provides a summary of the interpreted ultimate pullout capacities $P_u$ for test piles at these four sites.

***Estimation of Pile Capacity*** As discussed in Chapter 5 (Sections 5.2.1 and 5.2.4), the ultimate pullout capacities can be estimated by using the following equations:

*Piles in Cohesionless Soils*

$$P_u = p(2/3)k_s \tan\delta \sum_{L=0}^{L=L} \sigma'_{v\ell} \Delta L + W_p \qquad (5.74)$$

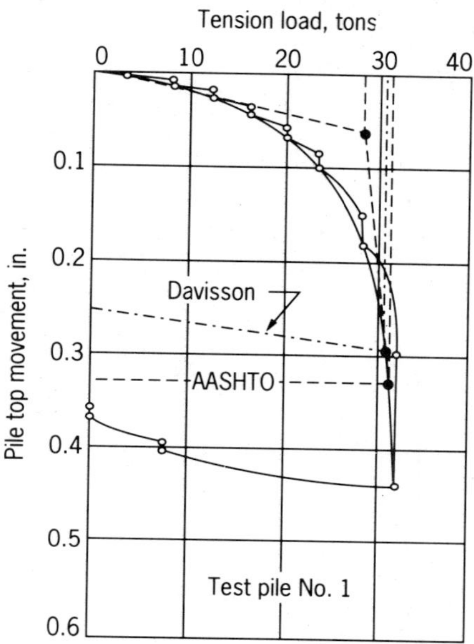


**Figure 11.7** Load-deformation curve for test pile at site 1 (Hegedus and Khosla, 1984).

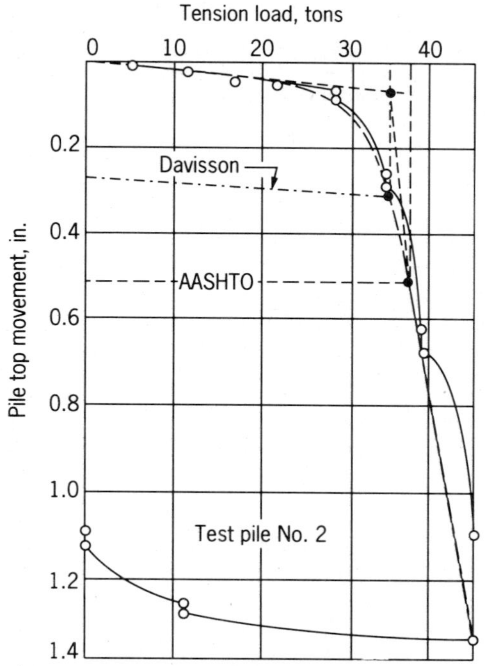


**Figure 11.8** Load-deformation curve for test pile at site 2 (Hegedus and Khosla, 1984).

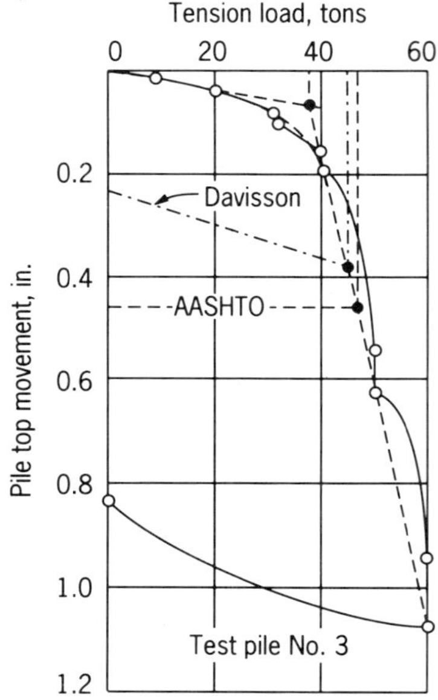


**Figure 11.9** Load-deformation curve for test pile at site 3 (Hegedus and Khosla, 1984).

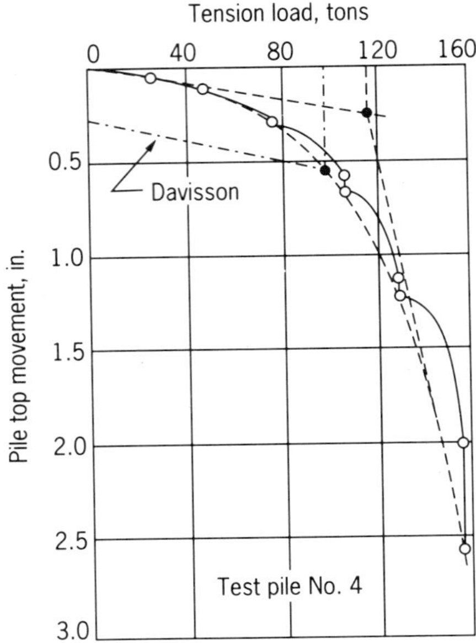


**Figure 11.10** Load-deformation curve for test pile at site 4 (Hegedus and Khosla, 1984).

**TABLE 11.5 Summary of Ultimate Pullout Capacities for Four Test Piles**

| Ultimate Pullout Capacity (tons) (1) | Test Pile at Site Number 1 (2) | 2 (3) | 3 (4) | 4 (5) |
|---|---|---|---|---|
| Tangent method | 28.5 | 34.5 | 37.5 | 115.0 |
| AASHTO method | 31.0 | 37.5 | 46.0 | — |
| Davisson's method | 30.5 | 34.0 | 44.0 | 95.0 |
| Average | 30.0 | 35.3 | 42.5 | 105.0 |

Note: 1 ton = 907 kg.
Hegedus and Khosla, 1984.

*Piles in Cohesive Soils*

$$P_u = p \sum_{L=0}^{L=L_e} c_a \Delta L + W_p \tag{5.77}$$

where pL = 2(a + b)L, (See Section 5.2.7) $\phi' = \delta$ in clay and $\tan\delta = \tan\phi$ in sand.

Other terms in these equations have been discussed in Sections 5.2.1 and 5.2.4.

**TABLE 11.6 Predicted Ultimate Pullout Capacities for Four Test Piles**

| Site | Soil Conditions | Method of Calculation | Pile Length ft | $P_u$ (tons) |
|---|---|---|---|---|
| 1 | Silty clay | Equation (5.77) | 18 | 31 |
| 2 | Silty sand and cohesionless silt | Equation (5.74) | 22 | 35.1[a]<br>35.7[b] |
| 3 | Silty clay | Equation (5.77) | 43 | 50.8 |
| 4 | Silty clay over cohesionless soils | Equation (5.74)[c] | 76 | 167.7 |

[a] $p$ in equation (5.77) is $[2(a+b)]$.
[b] $p$ in equation (5.77) is actual pile surface.
[c] For cohesive soils, in this case, $c_u$ value was not available. Therefore the mobilized friction $\phi'$ concept was used in the analysis.
Hegedus and Khosla, 1984.

Also as discussed in Chapter 5 (Section 5.2.7) for H or HP pile sections, "soil plug" develops between the pile flanges. The $P_u$ can then be estimated from equations (5.74) and (5.77) with the following assumptions:

1. For piles in cohesionless soils, the failure takes place along pile perimeter and applicable friction is between the pile surface and the surrounding soil.
2. For piles in cohesive soils, the failure takes place in soil where soil plug is formed between flanges. Soil–pile adhesion is used along the flange surfaces.
3. For piles in stratified (layered) deposits, as is for site 4, a combination of the preceding failure surfaces should be considered. This means in cohesive soils the soil will adhere to the pile surface, and soil-to-soil friction shall apply. In cohesionless deposits, soil-to-pile material friction takes place. The effective stress concept utilizing mobilized friction angle $\phi'$ has been used since undrained strength for clays are not available. Using the above concepts and the soil parameters given in Table 11.3, the ultimate pullout pile capacities $P_u$ were estimated for each of the four cases (Hedgedus and Khosla, 1984). These have been summarized in Table 11.6.

***Pile Capacities: Estimated versus Load Test Results*** The estimated (calculated) versus load test ultimate pullout capacities $P_u$ are plotted in Figure 11.11. The estimated and measured (load tested) values for piles in cohesive and cohesionless soils (sites 1, 2, and 3) are in good agreement. However for the pile in stratified soils (site 4) the estimated and load test values do not agree. Full-scale pile load tests are the only way to obtain a reasonable value of ultimate pullout capacities for such cases.

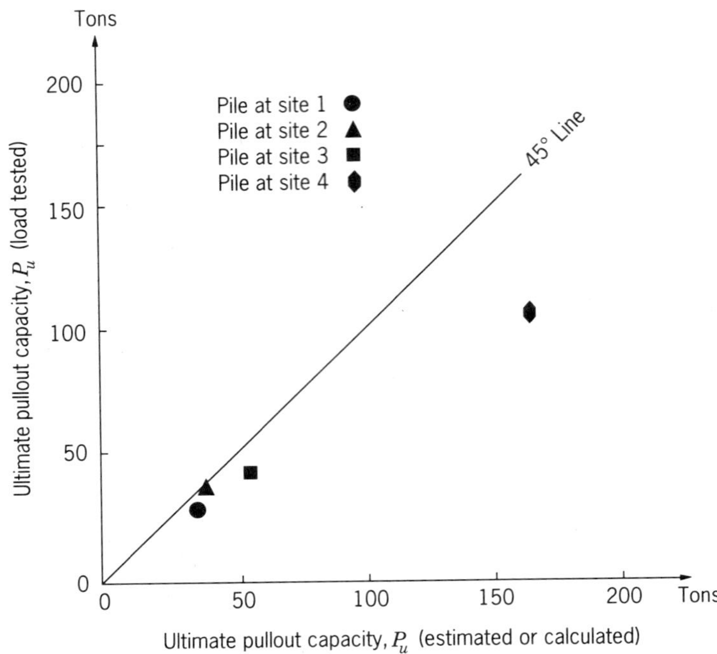


**Figure 11.11** A comparison of measured and calculated ultimate pullout capacities.

## 11.3 PILES UNDER LATERAL LOADS

Two pile groups load tests under static lateral loads have been analyzed in this section taking into effect: soil property determined from the single pile test and realistic group effects. A comprehensive test program on single piles and pile groups in over-consolidated clay has been conducted at the University of Texas (Brown, 1985). Static and cyclic lateral loading tests have been reported by Brown et al. (1987).

In these tests the pile group consisted of nine steel pipes, 10.75 in. (273 mm) in outside diameter, with wall thicknesses of 0.365 in. (9.27 mm). The piles were installed in October 1979, in a 3 × 3 arrangement with a nominal spacing of three-pile diameters on centers. The piles were driven closed ended into a layered system of overconsolidated clays to a depth of 43 ft. Prior to pile driving, a pilot hole 8 in. (203 mm) in diameter by 10 ft. (3.05 m) deep was excavated to facilitate vertical alignment of each pile.

Stiff, preconsolidated clays and silty clays of the Pleistocene-age Beaumont Clay formation extend to a depth of about 24 ft (7.3 m) below final grade, thus encompassing the zone of primary importance during lateral loading. Underlying the Beaumont is the Montgomery formation, a similar but older Pleistocene deposit. Both of these formations are deltaic terraces, deposited during interglacial periods and preconsolidated by desiccation during periods of glaciation (when the sea level was lowered).

The undrained shear strength increased mildy with depth with an initial value at the pile top. Both single pile and a 9-pile group were tested with *no fixity* at the pile head (in both cases). Both static and cyclic lateral load test data has been reported. The load had been applied 1 foot above the ground level.

***Prediction*** The single-pile test data (curves A and B, Figure 6 of Brown et al., 1987) has been analyzed to determine the soil property as:

1. The deflection $y$ at the load point in a fully embedded pile is given by:

$$y = A_{yc} Q_g R^3/EI + B_{yc} M_g R^2/EI \qquad (6.78)$$

and moment, $M = Q \times 1'$lb-ft

where

$A_{yc}$ and $B_{yc}$ = deflection coefficients

$Q_g$ = applied load at pile top

$M_g$ = applied moment at pile top

$R$ = relative stiffness factor

$$= 4\sqrt{EI/k} \qquad (6.80b)$$

$k$ = soil modulus assumed constant with depth

$EI$ = flexural stiffness of the pile

2. On the basis of pile tests on groups in sand, Prakash (1962, 1981) and Davisson (1970) had recommended as follows (see Chapter 6, Table 6.6): "If

**TABLE 11.7 Results of Pile Group Calculations (First cycle of loading)**

| Serial No. | Deflection ($y$) (in.) | Single Pile $Q$(lb) | Predicted Displacement, $y'$ |
|---|---|---|---|
| 1 | .10 | 4,200 | 0.27 |
| 2 | .20 | 7,200 | 0.54 |
| 3 | .24 | 8,300 | 0.65 |
| 4 | .30 | 9,400 | 0.81 |
| 5 | .40 | 11,600 | 1.08 |
| 6 | .50 | 13,200 | 1.36 |
| 7 | .60 | 14,600 | 1.63 |
| 8 | .70 | 15,800 | 1.90 |
| 9 | .75 | 16,300 | 2.04 |
| 10 | .80 | 16,800 | 2.18 |
| 11 | .90 | 17,800 | 2.45 |
| 12 | 1.00 | 18,600 | 2.72 |
| 13 | 1.10 | 19,600 | 3.00 |
| 14 | 1.20 | 20,400 | 3.27 |
| 15 | 1.25 | 20,800 | 3.41 |

(After Prakash et al., 1988).

the spacing of piles in the direction of load is 3d the effective value of $k(k_{\text{eff}})$ is 0.25 $k$ where $d$ is diameter of the pile." The pile spacing in this test series is $3d$.

3. For a spacing of $3d$ in the pile group, soil modulus

$$k_{\text{eff}} = 0.25k \tag{11.1}$$

Then substitutions of values from equation (11.1) in equation (6.78) we obtain the following relationship:

$$y' = A_{yc}\frac{QR_{\text{eff}}^3}{EI} + B_{yc}\frac{MR_{\text{eff}}^2}{EI} \tag{11.2}$$

Davisson and Gill (1963) calculated the $A$ and $B$ coefficients for clays as:

$$A_{yc} = 1.4$$
$$B_{yc} = 1.0$$

Equation (11.2) has been solved for $R$ for several values of $y$.

$$\text{Letting } K_{\text{eff}} = 0.25k$$
$$R_{\text{eff}}^3 = 2.827R^3$$
$$\text{and } R_{\text{eff}}^2 = 2R^2$$

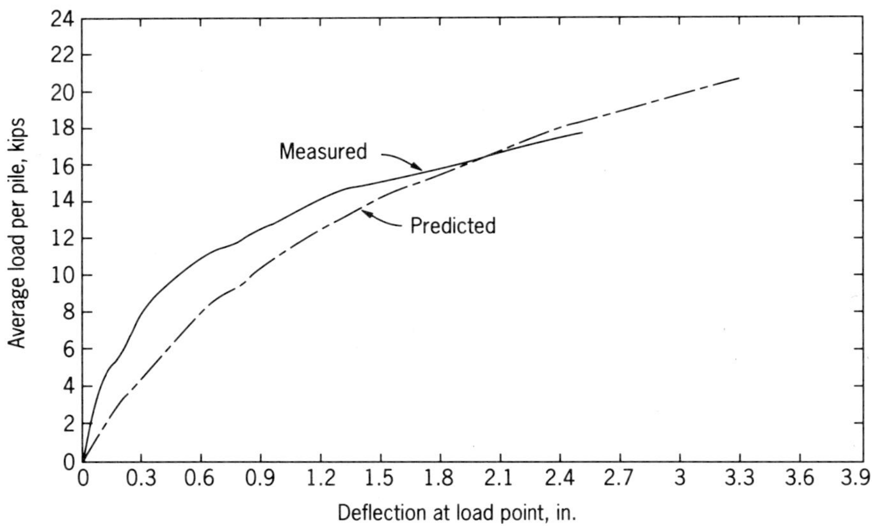


**Figure 11.12** Predicted lateral load-deflection response of a pile group in clay and measured response after 1-cycle of loading (Prakash et al., 1988).

Therefore,

$$y' = 2.827\,A_{yc}Q_gR^3/EI + 2B_{yc}M_gR^2/EI \tag{11.3}$$

where $y'$ = deflection of the pile group at the same load per pile as on single pile. The values of $y'$ so computed have been listed Table 11.7. The predicted and measured load deflection curves are plotted in Figure 11.12. (Prakash et al. 1988, 1989).

***Cyclic Load Tests*** The cyclic load test data of single pile had been analyzed in the same manner and the corresponding results are shown in Table 11.8. The predicted and observed load deflection curves are plotted in Figure 11.13.

***Discussion*** The full-scale pile tests and the model pile tests of Prakash (1962) differ in the following respect (Prakash et al., 1988):

1. The full-scale pile tests are in overconsolidated clay, while the model pile tests were in sand.
2. The full-scale pile tests were performed with two-directional loading, while the model pile tests were performed with one-directional loading.
3. The full-scale pile tests are performed with complete control of moment at the point of load application ($M = 0$) while the model tests were performed with indeterminate moment (or rotation condition). The rotation of the pile cap had been monitored, however.

**TABLE 11.8 Results of Pile Group Calculations (100 cycle of loading)**

| Serial No. | Deflection ($y$) (in.) | Single Pile $Q$(lb) | Predicted Displacement (in.) |
|---|---|---|---|
| 1 | 0.1 | 3,800 | 0.270 |
| 2 | 0.2 | 6,300 | 0.541 |
| 3 | 0.3 | 8,000 | 0.814 |
| 4 | 0.4 | 9,000 | 1.087 |
| 5 | 0.5 | 10,000 | 1.361 |
| 6 | 0.6 | 10,700 | 1.635 |
| 7 | 0.7 | 11,500 | 1.910 |
| 8 | 0.8 | 12,200 | 2.184 |
| 9 | 0.9 | 12,700 | 2.459 |
| 10 | 1.0 | 13,300 | 2.734 |
| 11 | 1.1 | 13,900 | 3.009 |
| 12 | 1.2 | 14,300 | 3.285 |
| 13 | 1.3 | 14,800 | 3.560 |

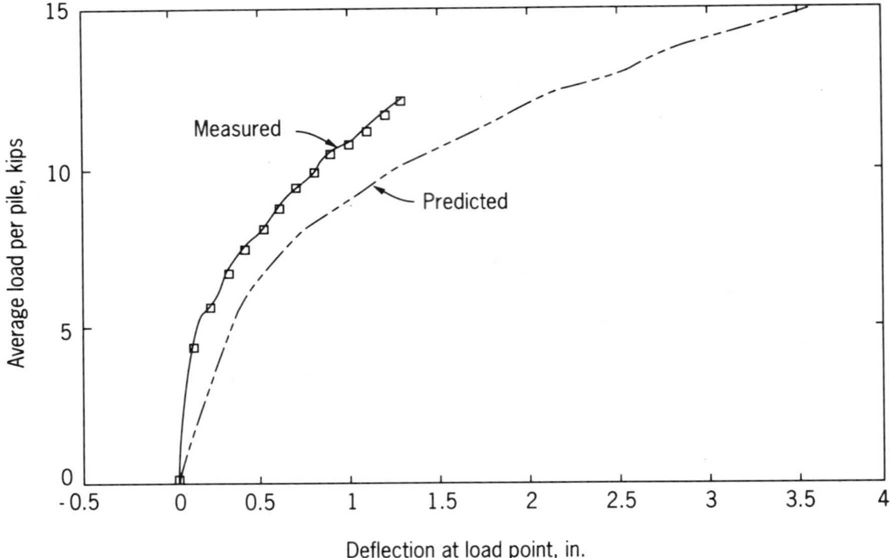


**Figure 11.13** Predicted lateral load response of pile group in clay and measured response after 100 cycles (Prakash et al., 1988).

Despite the above variations in the test conditions, it is interesting to note that:

1. The results of load deflection predicted for a full-scale pile group agree very well with the measured load deflection in clays.
2. The most significant conclusion that can be drawn on the basis of this comparison is that the analysis of single pile and pile groups according to theory of modulus of subgrade reaction predicts the behavior well, provided a reasonable value of soil modulus is estimated.
3. The interaction effects under lateral loads both in sands and clay are of the same order.

Prakash and Prakash (1989) analyzed the test data on 9-pile group reported by Brown et al. (1988). The soil conditions consisted of 9.5 ft of medium dense sand underlain by very stiff clay.

The predicted and measured load deflection curves for 1-cycle of loading and 100 cycles of loading tallied fairly closely.

In both these analyses, the single pile was the basis of predictions. Since the soil modulus is strain (deflection) dependent, there is a need to generate data on the dependence of subgrade modulus ($k$ or $n_h$) with deflections, both from pile tests and based on analytical studies.

**TABLE 11.9 Predicted and Measured Natural Frequencies of Single Pile**

| Serial No. | G (psi) | $(G/G_{max})^b$ | $\gamma_\theta$ | Amplitude (in.) × $10^{-3}$ | Frequency Measured | Frequency Predicted |
|---|---|---|---|---|---|---|
| 1 | 5772[a] | 0.88 | 0.0126 | 1.8 | 34 | Reference |
| 2 | 5705 | 0.87 | 0.014 | 2.0 | 33 | 31.4 |
| 3 | 3583 | 0.55 | 0.0455 | 6.5 | 31.5 | 26.6 |
| 4 | 3019 | 0.46 | 0.063 | 9.0 | 30.0 | 25.0 |
| 5 | 2625 | 0.40 | 0.0735 | 10.5 | 27.5 | 23.5 |

[a] Calculated value
[b] (See Prakash et al., 1988)

## 11.4 PILES UNDER DYNAMIC LOADS

Woods (1984), presented pile load tests in lateral vibrations in a soft clay at Belle, Michigan Figure 7.46a. The natural frequency decreases with the level of excitation indicating a nonlinear behavior of the soil–pile system. The pile was 14 in. outside diameter with 0.375 in. wall thickness and 157 ft long pipe.

The test data have been reworked using the results of the first test (Table 11.9) as reference (Prakash et al., 1988):

Prakash et al. (1988) have shown that the computed and measured amplitude-frequency relationships for loads higher than the reference case (see Table 11.9) are close to each other. Thus the single pile test forms the basis for any further predictions.

However, the behavior of pile groups under vibrations is difficult to predict. The group interaction factors are frequency dependent. As explained in chapter 7, several arbitrary corrections are applied to stiffness and damping of pile groups to match the predicted values with the measured values.

Correlations of pile group response with single pile response under earthquake type excitation is subject of a comprehensive study at University of Missouri-Rolla currently (1990).

## 11.5 OVERVIEW

Typical case histories of piles under vertical compressive and tensile loads and lateral static loads have been presented. There is very limited data on dynamic pile tests.

There have been several cases where the predictions may not match with the performance.

Davisson (1989) reports three case histories, widely separated geographically, where analyzer $(Q_v)_{ult}$-values were unconservatively different from static load test results. In Figure 11.14 the static load test failure load $(Q_v)_{ult}$ has been plotted on the vertical axis, and the PDA $(Q_v)_{ult}$-value on the horizontal axis.

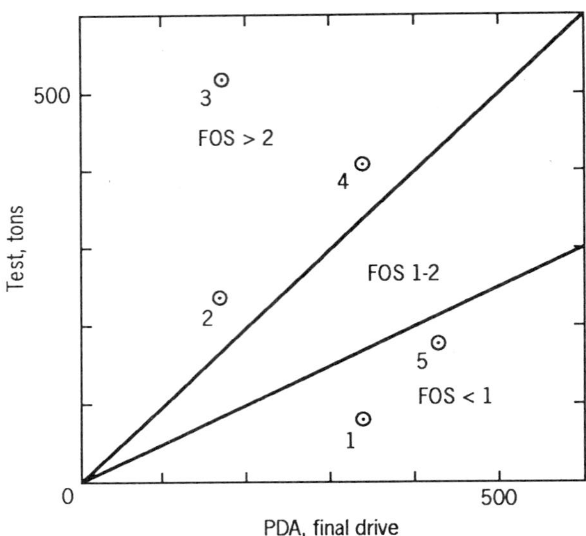


**Figure 11.14** Comparison of load, predicted by pile driver analysis (PDA) and measured from load test (Davisson, 1989).

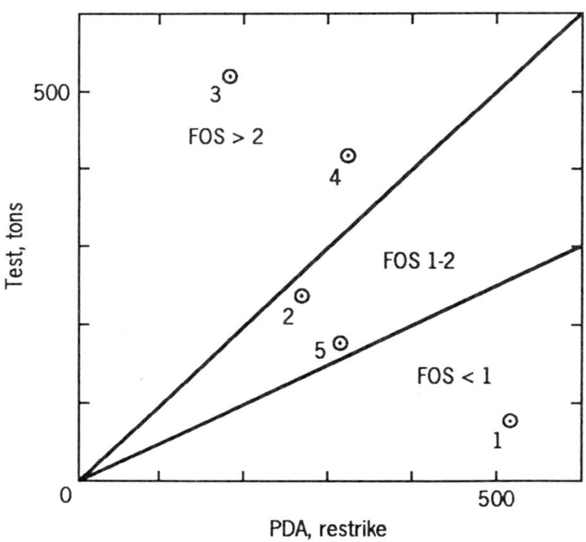


**Figure 11.15** Comparison of load, predicted by pile driving analyser (PDA) on re-strike and measured from load test (Davisson, 1989).

Five load test points are shown with an adjacent number representing the chronological order of testing. Test No. 1 was conducted after the PDA indicated the desired ultimate load (340 tons) at a depth less than the design depth, with a resulting failure. Figure 11.14 shows the results based on data taken at the time of driving. PDA predicted load capacities, if correct, would lie on the correlation line (45°). If the working load is taken as half of the PDA predicted failure load, then points on the line drawn 22.5 degrees from the horizontal would be on the verge of failure at the working load. Thus, the graph has been divided into three zones. The upper left half above the correlation line represents a zone where a working load taken as half the PDA $(Q_v)_{\text{ult}}$-value involves a FOS exceeding 2. Just below the correlation line is a zone where a similarly derived working load has a FOS between 1 and 2. The lower zone represents FOS values below 1. It may be seen that use of PDA $(Q_v)_{\text{ult}}$-values would have resulted in failures under service load for 2 of the 5 tests (Davission, 1989).

Figure 11.15 is similar to Figure 11.14 except that restrike results (redriven at up to several days after original driving) were used, so as to incorporate the effects of soil freeze. *On this basis one of the 5 tests would have resulted in a failure at the working load, with one other very close to that result.*

CAPWAP results based on the *original final* driving resistance have been shown in Figure 11.16. It is seen that one of the 5 tests would still result in a failure at the working load with one other would have a FOS below 1.5. CAPWAP results based on *restrike* data show similar results (Davisson, 1989). $(Q_v)_{\text{utl}}$-value would result in a failure at the working load with one other very

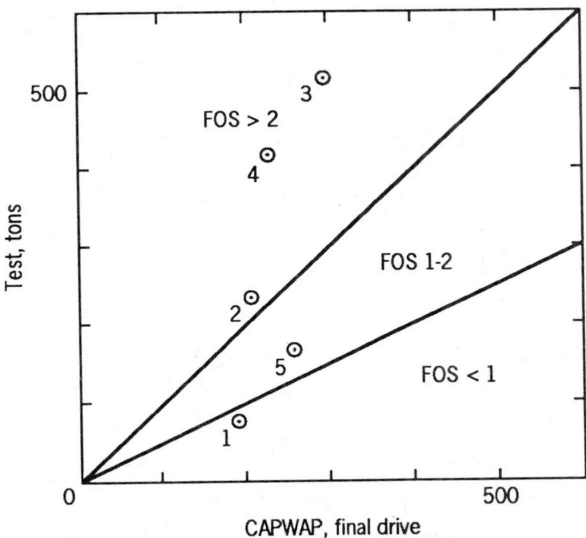


**Figure 11.16** Comparison of load, predicted by CAPWAP analysis and measured from load test (Davisson, 1989).

close to that result. The foregoing represents an unsatisfactory performance for the PDA. This is especially true in that each successive test should have allowed a better assessment to be made of the subsoil conditions.

Davisson (1989) concludes on the basis of the foregoing case history that soil deposits exist that will defeat all of the geotechnical engineer's tools except for the static load test. Therefore, *no major projects should be without static load tests.*

## REFERENCES


Brown, D. A. "Behavior of a Large-Scale Pile Group Subjected to Cyclic Lateral Loading," Ph.D. Thesis 1985, University of Texas, Austin, TX.

Brown, D. A., Morrison, C., and Reese, I. C., "Lateral Load Behavior of Pile Groups in Sand," *J. Geotch. Eng. Div.* ASCE, 1988, Vol. 114, No. 11, pp. 1261–1276.

Brown, D., Reese, L. C., and O'Neill, M. W., "Cyclic Lateral Loading of a Large Scale Pile Group," *J. Geotech. Eng. Div.*, ASCE, 1987, Vol. 113, No. 11, pp. 1326–1343.

Cheng, S. S. M. and Ahman, S. A., "Dynamic Testing versus Static Load Tests: Five Case Histories," *Proceedings Second International Conference on Case Histories in Geotechnical Engineering*, Vol. II, St. Louis, MO, 1988, pp. 1343–1348.

Davisson, M. T., "Lateral Load Capacity of Pile Groups," HRR, No. 333, 1970, pp. 104–112.

Davisson, M. T., "Foundations in Difficult Soils–State of the Practice Deep Foundations–Driven Piles," Seminar on Foundations in Difficult Soils, Metropolitan Section, ASCE, April 1989, New York.

Davisson, M. T. and Gill, H. L., "Laterally Loaded Piles in a Layered Soil System," *J. Soil Mech. Found. Div.*, ASCE, Vol. 89, No. SM3 1963, pp. 63–94.

Finno, R. J., Jacques, A., Hsin-Chih, C., et al., "Summary of Pile Capacity Predictions and Comparison with Observed Behaviour", Symp. Predicted and Observed Axial Behavior of Pile, *Pile Prediction Symp.* Evanston (IL) June 1989 pp. 356–386.

Gle, D. R., "The Dynamic Lateral Response of Deep Foundations," Ph.D. Thesis, University of Michigan, Ann Arbor, 1981, 278 pp.

Hegedus, E. and Khosla, V. K., "Pullout Resistance of H Piles," *J. Geotech. Eng.*, ASCE, Vol. 110, No. 9, 1984, pp. 1274–1290.

Prakash, S., "Behaviour of Pile Groups Subjected to Lateral Loads," Ph.D. Thesis, University of Illinois, Urbana, 1962.

Prakash, S., *Soil Dynamics*, McGraw-Hill Book Co., New York 1981.

Prakash, S. and Prakash, Sally, "Re-analysis of Piles Under Static and Dynamic Loads," Proceedings International Conf. on Piling and Deep Foundations, London May 1989 Volume 1 pp. 355–361.

Prakash, S., Sreerama, K., and Prakash, Sally. "Predictions and Performance of Typical Piles Under Static and Dynamic Loads," *Proceedings Second International Conference on Case Histories in Geotechnical Engineering*, St. Louis, MO 1988, Vol. 111, pp. 1757–1762.

Prakash, S., Sreerama, K., and Prakash, Sally, "Discussion on Cyclic Lateral Loading of a Large-Scale Pile Group," Dan A. Brown et al., paper no. 21927, 1988a, *J. Geotech. Div.*, ASCE, Vol. 115, No. 5 May, 1989, pp. 747–749.

Sharma, H. D., "Static Pile Capacity Based on Penetrometer Tests in Cohesionless Soils," *Proceedings First International Symposium on Penetration Testing*, Orlando FL., 1988, pp. 369–374.

Sharma, H. D., Sengupta, S., and Harron, G., "Cast-in-Place Bored Piles on Soft Rock Under Artesian Pressures," *Can. Geotech. J.*, Vol. 21, No. 4, 1984, pp. 684–698.

Woods, R. D., "Lateral Interaction between Soil and Pile," *Proceedings International Symposium Dynamic Soil Structure Interaction*, Minneapolis, MN, 1984, pp. 47–54.

# AUTHOR INDEX

# SUBJECT INDEX